NEUROBIOLOGY OF MONOTREMES

To Mark Joseph Rowe
1943–2011
Friend, colleague and mentor

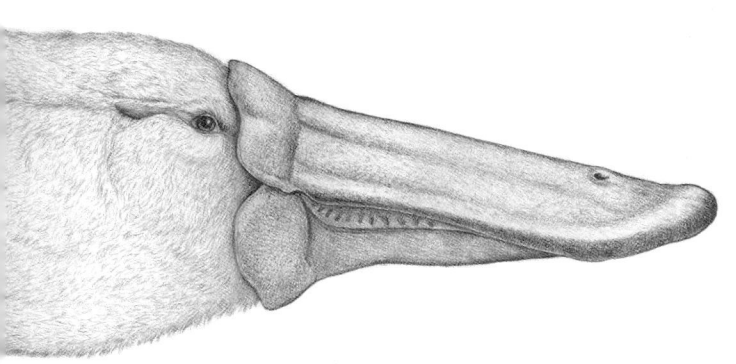

NEUROBIOLOGY OF MONOTREMES

Brain Evolution in Our Distant Mammalian Cousins

Editor: Ken Ashwell

CSIRO PUBLISHING

© Ken Ashwell 2013

All rights reserved. Except under the conditions described in the *Australian Copyright Act 1968* and subsequent amendments, no part of this publication may be reproduced, stored in a retrieval system or transmitted in any form or by any means, electronic, mechanical, photocopying, recording, duplicating or otherwise, without the prior permission of the copyright owner. Contact **CSIRO** PUBLISHING for all permission requests.

National Library of Australia Cataloguing-in-Publication entry

Ashwell, Ken W. S., editor.

Neurobiology of monotremes : brain evolution in our distant mammalian cousins / edited by Ken Ashwell.

9780643103115 (hardback)
9780643103153 (epdf)
9780643103160 (epub)

Includes bibliographical references and index.

Monotremes.
Neurobiology.
Neuroanatomy.
Neurophysiology.

Ashwell, Ken W. S., editor.

599.1

Published by

CSIRO PUBLISHING
150 Oxford Street (PO Box 1139)
Collingwood VIC 3066
Australia

Telephone: +61 3 9662 7666
Local call: 1300 788 000 (Australia only)
Fax: +61 3 9662 7555
Email: publishing.sales@csiro.au
Website: www.publish.csiro.au

Front cover and title page: illustrations by Anne Musser

Set in 10/13 Palatino
Edited by Joy Window
Cover and text design by James Kelly
Typeset by Thomson Digital
Printed in China by 1010 Printing International Ltd

CSIRO PUBLISHING publishes and distributes scientific, technical and health science books, magazines and journals from Australia to a worldwide audience and conducts these activities autonomously from the research activities of the Commonwealth Scientific and Industrial Research Organisation (CSIRO). The views expressed in this publication are those of the author(s) and do not necessarily represent those of, and should not be attributed to, the publisher or CSIRO. The copyright owner shall not be liable for technical or other errors or omissions contained herein. The reader/user accepts all risks and responsibility for losses, damages, costs and other consequences resulting directly or indirectly from using this information.

Original print edition:
The paper this book is printed on is in accordance with the rules of the Forest Stewardship Council®. The FSC® promotes environmentally responsible, socially beneficial and economically viable management of the world's forests.

CONTENTS

	List of contributors	vii
	Preface	ix
	Acknowledgements	xi
1	Classification and evolution of the monotremes A. M. Musser	1
2	Behaviour and ecology of monotremes Stewart C. Nicol	17
3	Embryology and post-hatching development of the monotremes Ken W. S. Ashwell	31
4	Overview of monotreme nervous system structure and evolution Ken W. S. Ashwell	47
5	Peripheral nervous system, spinal cord, brainstem and cerebellum Ken W. S. Ashwell	69
6	Diencephalon and deep telencephalic structures Ken W. S. Ashwell	107
7	Cerebral cortex and claustrum/endopiriform complex Ken W. S. Ashwell	131
8	Visual system Ken W. S. Ashwell	161
9	Somatosensory and electrosensory systems Ken W. S. Ashwell and Craig D. Hardman	179
10	Auditory and vestibular systems Ken W. S. Ashwell	219
11	Chemical senses: olfactory and gustatory systems Ken W. S. Ashwell	235

12	The hypothalamus, neuroendocrine interface and autonomic regulation Ken W. S. Ashwell	251
13	Monotremes and the evolution of sleep Ken W. S. Ashwell	275
14	Reflections: monotreme neurobiology in context Ken W. S. Ashwell	285
15	Atlas and tables of peripheral nervous system anatomy Ken W. S. Ashwell and Anne M. Musser	299
16	Atlas of the adult and developing brain and spinal cord of the short-beaked echidna (*Tachyglossus aculeatus aculeatus*) Ken W. S. Ashwell and Craig D. Hardman	315
17	Atlas of the adult and developing brain of the platypus (*Ornithorhynchus anatinus*) Ken W. S. Ashwell and Craig D. Hardman	387
	List of abbreviations used in brain and embryo atlas plates	437
	Index of brain and embryo atlas plates	447
	References	465
	Glossary	495
	Appendix tables and figures	505
	Index	515

LIST OF CONTRIBUTORS

Ken W. S. Ashwell
Department of Anatomy, School of Medical Sciences, University of New South Wales

Craig D. Hardman
Department of Anatomy, School of Medical Sciences, University of New South Wales

A. M. Musser
Australian Museum, Sydney

Stewart C. Nicol
School of Zoology, University of Tasmania

PREFACE

What is the place of the modern monotremes in mammalian brain evolution? For many years during the 19th and 20th century they were seen as quaint remnants of the ancestral mammalian condition, indicative of the original state of the mammalian body plan. The platypus was seen as a bizarre, almost chimeric, pastiche of features familiar from Northern Hemisphere mammals, while the short-beaked echidna was viewed as a shambling, animated pincushion. Adjectives such as 'prototypical' and 'primitive' were often used in the neuroscience literature in association with both the platypus and echidnas. It was felt that if we could understand the nervous systems of these supposedly 'archaic', 'archetypal', 'ancient', 'prehistoric' and 'primordial' mammals, then we would be able to glean clues not just as to how the first mammalian brains were built and worked, but how the brains of so-called advanced mammals (i.e. primates and of course humans) subsequently evolved.

Far from being primitive mammals, the modern monotremes are extraordinarily successful creatures with remarkable adaptations to their environment. Certainly the monotremes have many anatomical features in common with extinct primitive mammals, but then so do humans (look at your own primitive and generalised forelimb). Any modern creature is a mosaic of primitive and advanced characteristics. Primitive features are inherited unchanged from a distant ancestor, simply because those features served their purpose well enough in the competition for survival, whereas advanced features or specialisations are critically important for continued existence and have been relentlessly honed for their purpose by natural selection.

The remarkable specialisations of the living monotremes are nowhere more evident than in their nervous systems. The discovery of mammalian electroreception in, first, the platypus and then the short-beaked echidna highlighted the highly specialised nature of the monotreme nervous system. Although a welcome improvement over the chauvinistic bias of earlier studies, the pendulum of scientific opinion has perhaps swung too far, in that electroreception is now seen almost as the defining and characteristic sensory modality of the modern monotremes, at the expense of olfaction and general somato- or touch sensation in the head, trunk and limbs. Electroreception has taken on almost mythic stature in the popular scientific literature, so much so that the story of the clever platypus navigating its stream bed habitat using nothing more than an uncanny sense for electrical fields is familiar to every Australian school child. These cute and cosy wildlife stories ignore the fact that most of the prey of the platypus emit little or no electrical signals and that the evidence for field navigation with electroreception in the wild is non-existent. The evidence for use of electroreception by echidnas in the natural setting is even weaker.

The focus on electroreception has also strongly influenced conceptions of monotreme evolutionary history, with several authors maintaining that the earliest monotremes were electroreceptive and essentially platypus-like in both morphology and habits. As will be argued in this book, this contention flies in the face of the embryological evidence and is founded on a logical flaw, i.e. the absence of fossil evidence to the contrary. As such, the *platypus first* hypothesis is a misconception that arises from the patchy tachyglossid fossil record.

I hope that the reader will come to share something of the fascination that I feel for the remarkable monotremes and their extraordinary nervous systems and behaviour. In the absence of a complete fossil record, we can only indirectly infer much of their neural evolution. Nevertheless, the modern monotremes are critically important animals for understanding mammalian brain evolution, not because they are primitive living fossils, but because their story tells a tale of ancient origins and astonishing sensory adaptations to their environment.

Ken Ashwell

ACKNOWLEDGEMENTS

There are many colleagues and organisations without whose help this book would not have been possible. Some of these have given me access to the many collections of monotreme embryos, hatchlings, histological sections, brains, spinal cords, bills and skulls that go to tell the story of the monotreme nervous system. Others are colleagues and mentors who have provided helpful advice and guidance during the many years of work that have led to this book.

I am particularly indebted to Dr Peter Giere, who was a wonderfully helpful host and colleague during the 3-month stay that my wife and I made in Berlin while I worked my way through the monotreme collection at the Museum für Naturkunde. His cheerful disposition and help with negotiating local bureaucracy made our sojourn in the poor (but sexy) city a great pleasure. Professor Ulrich Zeller, also of the MfN, kindly allowed access to a collection of sections through the heads of platypus hatchlings, the original specimens of which came from the American Museum of Natural History. The embryological collection at the MfN in Berlin is an extraordinary but under-utilised resource, and is key to any future understanding of monotreme brain evolutionary developmental biology. It is a mystery to most of the Australian scientists I discuss this collection with that such a resource ended up on the other side of the planet from the monotreme homeland!

Of course, there are many monotreme collections in Australian museums, including skulls, skins, intact embryos and hatchlings and adult monotremes in various states of dissection. I am indebted to Drs Sandy Ingleby and Anja Divljan of the Australian Museum, Sydney who have helped me use the AM collections with many studies of monotreme and marsupial biology over the last decade. Perhaps the largest collection of intact monotreme eggs, embryos and hatchlings is held at the National Museum of Australia in Canberra, incorporating the bottled specimens of the former Museum of Anatomy of the same city. I am very grateful to Anthea Gunn, Anne Kelly and Sara Kelly for their help in accessing that collection. It is a great pity that this resource has not been sectioned and immunostained with modern techniques.

Monotreme fossil material is sparse, particularly that which pertains to the brain. I was very lucky to have had the kind help of the remarkable Don Squires at the Tasmanian Museum and Art Gallery and radiology staff at the Calvary Hospital in Hobart to examine and CT scan the *Megalibgwilia* specimen in the TMAG collection. The late Peter Blias of the South Australian Museum was very helpful in organising access to the South Australian fossil long-beaked echidnas for analysis and CT scanning. I am also grateful to Peter Murray for conversations on the subject of fossil echidnas while I was in Alice Springs studying dromornithids.

Many of my neuroscientist colleagues have been an inspiration over the many years that led to this book. Charles Watson and George Paxinos were not only my original partners in the grants that began this work, but have been a source of encouragement and insightful advice over the years. I am particularly thankful to them for the opportunity to bring their insights into neuromeric brain mapping into the context of monotreme neurobiology and brain evolution. The system of nomenclature and abbreviations used in this book is derived from their work, with some monotreme-specific modifications.

I am very grateful to the many staff who have helped me scan all of the images that have been used in this book, in particular Dr Maria Sarris and her staff at the Histology and Microscopy Unit at the School of Medical Sciences at UNSW, as well as Andy (Huazheng Liang) in the Paxinos Laboratory at Neuroscience Research Australia.

This work would not have been possible without the financial support of the Australian Research Council who provided the funds for the work on the chemoarchitectural mapping of the echidna brain, and I am particularly indebted to the Alexander von Humboldt Foundation for its financial support of my studies in Düsseldorf and Berlin. German science has always had a close interest in monotreme and marsupial neurobiology and the AvH Foundation has been a generous sponsor of the work that led to this book.

Finally, I would like to thank the very helpful staff at CSIRO Publishing (John Manger, Tracey Millen, Briana Melideo), who had the vision to support the publication of this work.

Ken Ashwell

The premammalian characters of the platypuses only identify the antiquity of their lineage as a separate branch of the mammalian tree ... in opposition to the myth of primitivity: [it really is] a superbly engineered creature for a particular, and unusual, mode of life. The platypus is an elegant solution for a mammalian life in streams – not a primitive relic of a bygone world.

S. J. Gould (1991)

Monotremes display a mosaic of primitive, advanced and unique features. This mixture should endear them to those who believe in a serendipitous, rather than an orthogenetic, evolutionary process.

L. R. Walter (1988)

1

Classification and evolution of the monotremes

A. M. Musser

Summary

The higher taxon Monotremata includes the extant platypuses and echidnas as well as their extinct relatives. Living monotremes are highly specialised animals occupying distinct ecological niches, and are known only from Australia and New Guinea. Platypuses are semi-aquatic insectivores/carnivores, and echidnas are spine-covered, terrestrial insectivores. There has been extensive interest in and debate over monotreme origins since their discovery by western science in the late 1800s. Monotremes lay soft-shelled eggs from which the embryonic young hatch, in contrast to marsupial and placental mammals that bear live young. The many 'primitive' (plesiomorphic) features of monotremes occur in concert with unique specialisations, a phenomenon known as 'mosaic evolution'. Monotreme taxonomy and classification have been confounded in part by this mosaic of features, and in part from a dearth of adequate comparative material. However, new discoveries of early mammals and near-mammals and development of cladistic methodologies for determining relationships have shed much-needed light on monotreme origins. Australosphenidans – recently discovered Mesozoic mammals from the Southern Hemisphere with tribosphenic-like teeth – may share a unique relationship with monotremes, and higher level classification schemes have followed based on this presumption, although the debate is far from resolved. Research into unique monotreme attributes – in particular, an electrosensory ability unknown in other mammals – will help science understand the success and longevity of this oldest of mammalian groups.

Introduction

Living monotremes include a single species of platypus, *Ornithorhynchus anatinus* (Ornithorhynchidae); and two genera of echidnas (Tachyglossidae) in four species: the long-beaked *Zaglossus* (three species) and the short-beaked *Tachyglossus* (a single species with several subspecies) (see below). *Ornithorhynchus* is a semi-aquatic insectivore/carnivore that feeds primarily on benthic invertebrates, restricted to the waterways of eastern Australia (including Tasmania but excluding northern Queensland). Echidnas are terrestrial insectivores, feeding on ants, termites and other soft-bodied invertebrates. The short-beaked echidna *Tachyglossus aculeatus* is found in all Australian states as well as in New Guinea, in areas with suitable habitat. Until very recently, long-beaked *Zaglossus* species were thought to have survived only in New Guinea, although mainland Australian *Zaglossus* fossil material is known from the Pleistocene and Holocene. A skin and skull of what appears to be an overlooked museum specimen of *Zaglossus bruijnii*, however, has been reported from a collection made in 1901 at Mount Anderson in the West Kimberley region of Western Australia; this remarkable find suggests that *Zaglossus* survived until at least the early 20th century in this isolated area of WA, and has raised hopes that Australian *Zaglossus* may still be extant (Helgen *et al.* 2012).

Monotremes are the only extant egg-laying mammals, pointing to origins deep within the mammalian tree. Features shared by monotremes and other living mammals include the presence of hair or fur, and the ability to produce milk to feed their young, suggesting that these attributes appeared early in mammalian evolution. Such features, however, may also have been present in the small, advanced cynodont reptiles that were the direct ancestors of mammals, and in the Late Triassic–Early Jurassic 'mammaliaforms' – early lineages of mammal-like taxa that lacked 'advanced' features of later therian mammals like marsupials and placentals.

Monotremes have their origins in the Mesozoic, the era comprised of the Triassic, Jurassic and Cretaceous periods. Mesozoic mammals or near-mammals, however, are almost always found as fragmentary fossils, often known only from small jaws and teeth (although this situation has dramatically improved over the past decade). Determining relationships from incomplete fossil taxa in order to understand monotreme phylogeny has been an ongoing exercise involving anatomists, palaeontologists and molecular biologists. To this writer, at least, it is a debate that has not yet been resolved.

Monotremes have traditionally been thought of as uniquely and quintessentially Australian (inclusive of New Guinea). However, a large ornithorhynchid species (*Monotrematum americanum*) is now known from the Paleocene of Patagonia in southern South America (Pascual *et al.* 1992a, b). This occurrence suggests an eastern Gondwanan distribution possibly from the late Mesozoic through earliest Cenozoic encompassing Australia, South America and, by inference, Antarctica (the land connection between Australia and Patagonia throughout much of this time period). Ornithorhynchids are therefore Gondwanan species with a relictual distribution in eastern Australia.

As we will see, the known evolutionary history of monotremes is lengthy, unique and in many ways surprising, extending back over 100 million years and encompassing at least three continents. Living monotremes are therefore relict species, and platypuses and echidnas may be but two long-separate branches of a larger, deeply rooted family tree.

Basic monotreme anatomy and physiology

The 'primitive' (plesiomorphic) features found in ornithorhynchids and tachyglossids have generated a great deal of debate over the place of monotremes in mammalian evolution, and whether monotremes are in fact fully mammalian as traditionally defined (see Musser 2006a, b for a review of relationship hypotheses). Egg-laying would almost certainly have been the plesiomorphic mammalian reproductive condition. Later therian mammals bore live young (e.g. altricial young in Marsupialia and well-developed young nourished by a placenta in Placentalia). The body temperature of monotremes is lower than that of therian mammals and may be variable, a physiological feature often cited as plesiomorphic and somewhat reptilian (e.g. Augee *et al.* 2006). Plesiomorphic osteological features include skull features such as paired pilae antoticae, which are ossifications in the skull floor along the sella turcica as in Sauropsida (among Mammalia, this is found only in monotremes);

retention of a large septomaxilla, a bone of the snout found almost exclusively in pre-mammalian taxa; and absence of a bony auditory bulla. Plesiomorphic postcranial features include the retention of additional bones of the shoulder girdle in both monotremes (lost in therian mammals), the morphology of the humerus in living monotremes, and the archaic form of the femur in ornithorhynchids, almost identical to that of the pre-mammalian cynodont *Tritylodon*. The bones of the mammalian middle ear (malleus, incus and stapes), developed from the accessory jaw bones of pre-mammalian ancestors, were apparently not incorporated into the middle ear in early monotremes, which instead had a 'pre-mammalian' ear/jaw apparatus with at least some comparatively unreduced accessory jaw bones (Rich *et al.* 2005).

An unusual plesiomorphic feature once thought to be a monotreme autapomorphy is the presence of a crural (extratarsal) spur on the inner side of the ankle in both living monotreme families, comprised of an os calcaris and cornu calcaris (generally functional and venomous only in male *Ornithorhynchus*). Extratarsal spurs are found in some basal Mesozoic mammals, including eutriconodonts, symmetrodonts and multituberculates but not, as far as known, in either pre-mammalian Cynodontia or Theria (Hurum *et al.* 2006). The presence of ankle spurs may therefore be a basal mammalian feature – possibly a defence against such threats as dinosaurian predators – retained in monotremes but lost or never possessed in crown therians (Hurum *et al.* 2006).

Certain anatomical and physiological features of monotremes are highly sophisticated, relating to the great degree of specialisation in both living families and their ancestors. Perhaps the most remarkable specialisation is a sensory system incorporating both electroreceptors and mechanoreceptors in the skin of the beak or bill (e.g. Scheich *et al.* 1986; Pettigrew 1999). These sensory receptors need a moist environment in which to function optimally; the platypus operates in an aquatic environment, keeping the bill wet while feeding, whereas the tip of the beak in echidnas is kept moistened by mucus.

The electroreceptive system of the platypus is by far the most highly developed, and is critical to the navigational skills of the platypus as it finds its way through dark or turbid waterways. It is tempting to think of the extent of development of the electrosensory organs in ornithorhynchids as an ancient adaptation to the prolonged winter darkness of the southern polar regions during the early evolution of the group. This electro/mechanoreception is present but much less well-developed in the terrestrial tachyglossids, and receptors are concentrated at the tip of the beak.

The development of electroreception/mechanoreception is correlated with enlargement of the trigeminal nerve (5n), which passes to the bill or beak through foramina on the rostrum and via the mandibular canal (e.g. Manger 1994). The presence of an enlarged or hypertrophied mandibular canal is a key monotreme synapomorphy that may identify fossil taxa as either belonging within Monotremata (e.g. the Early Cretaceous Victorian monotreme *Teinolophos trusleri*, which has an enlarged mandibular canal: Rich *et al.* 2005) or excluded from Monotremata (e.g. the Early Cretaceous *Kollikodon ritchiei* from Lightning Ridge, New South Wales, which lacks an enlarged mandibular canal: Musser 2003, 2006a, b) (Fig. 1.1).

Monotreme evolution and diversification

Among extant monotremes, platypuses appear to be the older and more archaic type despite their aquatic specialisations. Ornithorhynchids have retained a greater number of plesiomorphic characters (see Musser 2006a, b), and their fossil history extends back to the earliest Paleocene (61–63 million years ago (mya)) (Pascual *et al.* 1992a, b). In addition, platypus-like monotremes are known from the Early Cretaceous of Australia (115–108 mya) (Archer *et al.* 1985). In contrast, the oldest echidna fossils are currently thought to be Miocene in age (no older than 15 million years old), and may in fact be much younger (see Musser 2006a, b).

Aquatic adaptations in *Ornithorhynchus* include a dorsoventrally flattened body, waterproof fur, webbed feet and a large bill comprised of soft skin richly imbued with mechano- and electroreceptors (Fig. 1.2a). *Ornithorhynchus* spends most of its waking hours foraging along the bottom of rivers, streams and lakes feeding on benthic invertebrates. Its habitat is generally limited to waterways, surrounding banks and river bottoms (platypuses may disperse or otherwise travel overland on occasion). Waterways provide a relatively constant and predictable environment,

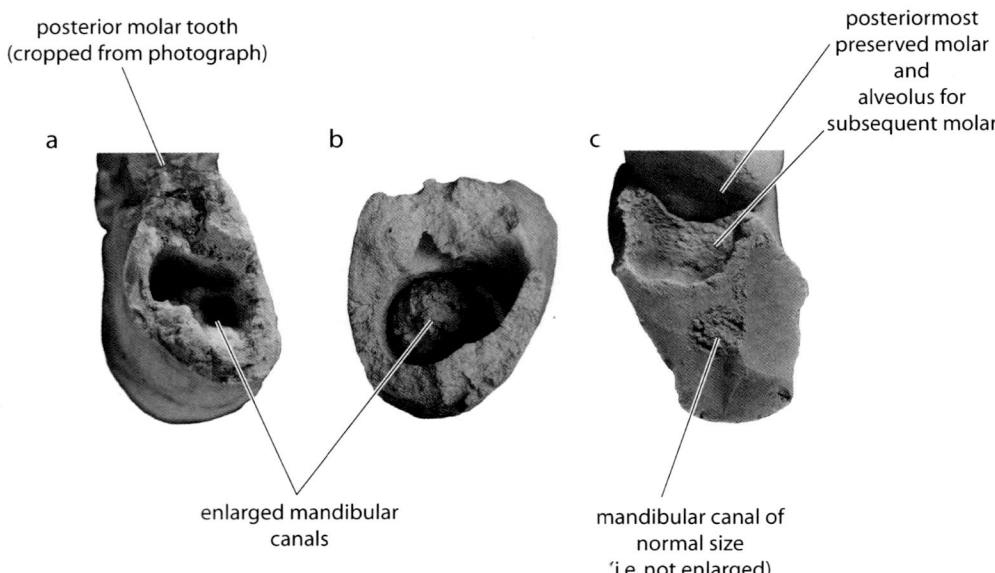

Figure 1.1: Comparison of the mandibular canals of *Steropodon galmani* (AMF66763), *Obdurodon dicksoni* (QMF18981), and *Kollikodon ritchiei* (AMF96602) (a–c respectively) illustrating development of the internal mandibular canal running through the dentary (not to scale). Enlarged mandibular canals, a synapomorphy of Monotremata, are shown in (a) *Steropodon* and (b) *Obdurodon*. The mandibular canal of (c) *Kollikodon* is not enlarged, and *Kollikodon* is excluded from Monotremata. All views are of the posterior ends of the dentaries, and all jaws are naturally broken, fragmentary specimens. The jaw fragments of *S. galmani* and *K. ritchiei* have preserved dentition, while the jaw fragment assigned to *Ob. dicksoni* is edentulous. AMF: Australian Museum. QMF: Queensland Museum.

and platypuses appear to have utilised this niche and adapted to it very early in their history. Water temperatures remain fairly consistent daily and yearly, and seasonal change is not as pronounced as it is in terrestrial environments. Platypuses generally sleep, rest or care for young in burrows dug into river banks where soil is well consolidated by tree roots (e.g. Grant 2007).

The feeding behaviour of *Ornithorhynchus* is fairly proscribed: it dives to capture small benthic invertebrates, fills its cheek pouches with its prey, rises to the surface to process its food, and then dives to begin the cycle again. *Ornithorhynchus* is a voracious feeder, and a hungry platypus can be quite single-minded if foraging in a predator-free environment (personal observation at Blue Lake, Jenolan Caves, NSW).

Interestingly, *Ornithorhynchus* may navigate subterranean cave streams in order to utilise underground stream banks for burrow construction (first reported by Hamilton-Smith in 1968). Caves offer near-complete protection from predators, and burrows may be used for many generations (personal observation, Buchan Caves, Victoria, where numerous platypus burrows and resting depressions were observed). Resting depressions seen at Buchan are of special note; platypuses do not normally rest on exposed ground, but several rounded depressions that were almost certainly constructed by platypuses were seen at Buchan, where platypuses could rest 'above ground' in the total darkness of the caves. As with cave-dwelling bats, platypuses must navigate the channels of cave streams in pitch-black darkness without the use of sight, suggesting that electroreception may play a part in this underground navigation.

Tachyglossids are well adapted for a terrestrial, insectivorous lifestyle. They have stout, hemispherical bodies, a thick body covering of keratin spines over the head, back and tail, a long rostrum prolonged into a 'beak', large forepaws armed with stout claws for digging, and hind feet that have been rotated rearwards (an adaptation for digging vertically into the ground). Both *Zaglossus* and *Tachyglossus* are known as fossils from Australia and New Guinea, although the fossil record is much longer for long-beaked species (?Miocene to Recent) than for *Tachyglossus* (Pleistocene to Recent). Several species of long-beaked

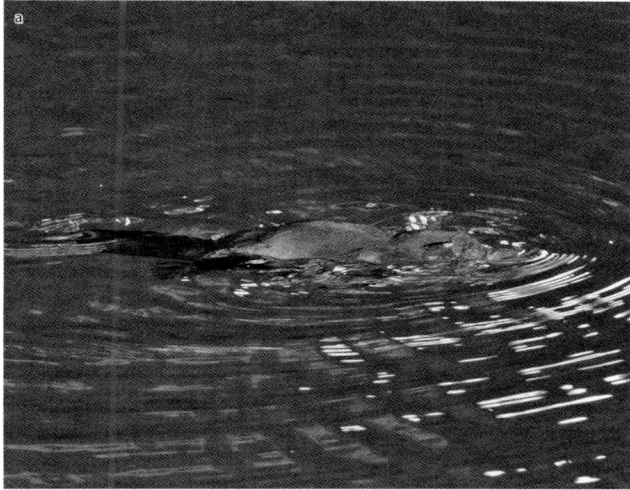

Figure 1.2: Body form in living monotremes (a, b, respectively). (a) shows the dorsoventrally flattened body form of the platypus, *Ornithorhynchus anatinus*. This animal is actively processing its food at the surface (note concentric ripples emanating from bill). The white patch is under the eye, and the pinna-less ear opening is posterior to the eye, in a groove housing both eye and ear. (b) illustrates the hemispherical (dome-shaped) body form of tachyglossids (here, the short-beaked echidna *Tachyglossus aculeatus aculeatus*). This individual from Jenolan, New South Wales shows the stout form, robust forelimbs, back-turned hind limb, vertical ear opening and small eyes typical of tachyglossids. The snout is moistened by mucous secretions. Photographs by A. M. Musser.

echidnas, including species of *Zaglossus*, were once common on mainland Australia (see below).

The short-beaked echidna, *Tachyglossus aculeatus*, is a successful and widespread species (Fig. 1.2b).

Tachyglossus has been divided into five subspecies (per Augee *et al.* 2006): *T. a. acanthion* (Northern Territory, northern Queensland, inland Australia and Western Australia); *T. a. aculeatus* (eastern New South Wales and Victoria; southern Queensland); *T. a. lawesii* (New Guinea lowlands); *T. a. multiaculeatus* (South Australia, especially Kangaroo Island); and *T. a. setosus* (the island state of Tasmania). *Tachyglossus* has benefitted from the spread of insect prey such as ants and termites that expanded in range with the gradual drying of the Australian continent over the past 15 million years, and is now Australia's most widely distributed native mammal. *Tachyglossus* may be exceptionally long-lived, with some individuals reaching over 50 years of age (Hulbert *et al.* 2008).

There are three extant species currently proposed for the genus *Zaglossus*, all of which are considered endangered: *Z. bruijnii* (western New Guinea and possibly Western Australia); *Z. bartoni* (central and eastern New Guinea); and the poorly known *Z. attenboroughi* (known only from the Cyclops Mountains along the north coast of New Guinea) (Flannery and Groves 1998). The single 20th century record of mainland *Zaglossus bruijnii* in the Kimberley region supports close past geographical ties with western New Guinea (Helgen *et al.* 2012). New Guinea *Zaglossus* species remain under threat from hunting and development pressure (Opiang 2009). In addition to an elongated, down-curved beak, *Zaglossus* species have much smaller spines than *Tachyglossus*.

Recent (and much-needed) fieldwork on New Guinea *Zaglossus* has confirmed a wild diet of earthworms and wood-boring grubs (Opiang 2009). Females are heavier than males, and some females retain spurs into adulthood. The suggestion that *Zaglossus* species are less adept at digging has been shown to be erroneous, since *Z. bartoni* is now known to construct large underground dens or burrows, most likely to avoid predation (Opiang 2009).

Echidnas search for their insect prey in a slow and seemingly deliberate way, in contrast to the much more active feeding mode of the platypus, and the electrosensory apparatus of the snout is much reduced (needing a moist environment in which to function; see Chapter 9). Echidnas operate in an arguably more complex environment than do platypuses: echidnas must navigate forest, woodland or desert floors (habitats subject to greater environmental and climatic change than the comparatively constant waterways) and echidnas may interact with large terrestrial vertebrate species such as wombats or kangaroos (and introduced animals like wild dogs, foxes and cattle). This may be one reason for differences in the architecture of the brain between *Ornithorhynchus* and tachyglossids, although few studies have been conducted on comparative monotreme intelligence and behaviour (see Chapters 2 and 7).

Phylogenetic relationships within Monotremata

Although the bauplans (body plans) of ornithorhynchids and tachyglossids differ in numerous ways, living monotremes share several unique features. Synapomorphies include reduced or absent dentition, long secondary palates and long vomers, early fusion of skull bones, extreme platybasy (flat skull bases), and the abovementioned sensory system incorporating electroreception and mechanoreception (e.g. Musser 2006a, b).

The many neurological differences between platypus and echidnas are in some instances quite profound (see Chapter 14), and confidently determining relationships between these two specialised end-branches of a very ancient group may need to wait until adequate fossil material of earlier monotreme species is discovered. A partial list of differences in addition to external appearance and respective specialisations includes the development of teeth in young platypuses (echidnas are completely edentate), retention in ornithorhynchids of a jugal bone and autapomorphic development of a dumb-bell shaped pre-vomer bone ('os paradoxum') (both bones absent in echidnas), the order and makeup of sex chromosomes (Grützner *et al.* 2004; Rens *et al.* 2007), and extensive olfactory development in echidnas (discussed in detail in Chapter 11). These differences appear to reflect a long period of independent evolution between the two surviving monotreme families, although this view has been challenged by recent molecular research (e.g. Phillips *et al.* 2009; see discussion below).

The monotreme fossil record

The monotreme fossil record is scant (*contra* Phillips *et al.* 2009) and site-specific, with substantial chronological gaps. Monotreme fossils are known from the Early Cretaceous and Cenozoic of Australia (Oligocene to Recent), the Pleistocene-Holocene of New Guinea, and the early Paleocene of Patagonia, Argentina. There are no monotreme fossils yet known from Antarctica, although their presence in Patagonia indicates an Antarctic distribution as well. Late Cretaceous monotremes and Australian Paleocene monotremes are unknown. Jurassic or Triassic monotremes are not yet known, although many morphological and molecular analyses suggest monotremes may have originated during the Jurassic if not earlier (e.g. Musser 2006a, b).

The earliest monotremes were neither platypus nor echidna, and the 'archetypal monotreme' remains unknown. Consensus on monotreme origins and membership within monotreme families has been difficult to reach. This is in part because early monotreme taxa are so fragmentary, and in part because such 'stable' features as mammalian dentitions and middle ear conformation, thought to be too complex to have evolved more than once, now appear to be more homoplastic than previously thought.

Mesozoic monotremes

Discoveries of Australian Mesozoic monotremes over the past few decades have greatly expanded our understanding of early monotreme evolution. The oldest known monotremes are Early Cretaceous in age (Aptian–Albian), from coastal Victoria and Lightning Ridge, New South Wales (Rich *et al.* 1999, 2001b, 2002, 2005; Archer *et al.* 1985). Mesozoic monotremes are known almost exclusively from isolated lower jaws, some with partial dentition. In addition to the lower jaws recovered, a tachyglossid-like humerus from Victoria (*Kryoryctes cadburyi*) has also been reported, although this may not be that of a monotreme (Pridmore *et al.* 2005).

The oldest known monotreme is *Teinolophos trusleri* from the Aptian of Victoria (115–108 mya), described from several isolated lower jaws and a few molar teeth (Rich *et al.* 1999, 2002, 2005) (Fig. 1.3a–c). The tiny *Teinolophos*, first described as an early therian eupantothere (Rich *et al.* 1999), has subsequently been described as a 'gopher-like' monotreme (Rich *et al.* 2001b) a less specialised basal monotreme, probably some sort of insectivore given its small size and triangulated teeth (Musser 2006a, b) or the oldest known platypus (Rowe *et al.* 2008; see below). Its double-rooted molars are similar to the teeth of the platypus-like *Steropodon galmani* from Lightning Ridge, and occlusion was substantially orthal rather than transverse. In contrast, the molar teeth of true platypuses (*Obdurodon* species see below) are multiple-rooted and occlusion involves a strong transverse component (Musser 2006a, b).

Teinolophos shares three synapomorphies with other monotremes: (1) an enlarged mandibular canal (2) a posterointernal process on the medial side of the jaw; and (3) the distinctive monotreme dental pattern (rectangular molar teeth with V-shaped lophs) (Rich *et al.* 2005). On present evidence, its identity as a monotreme appears secure. Notably, there are facets on the medial side of the dentary for accessory jaw bones, unlike living monotremes, indicating that *Teinolophos* had a highly plesiomorphic lower jaw comprised of the dentary and possibly angular, coronoid and splenial bones (Rich *et al.* 2005; contra Rowe *et al.* 2008). It had been thought that all therian mammals had lost these accessory jaw bones, suggesting that evolution of the archetypal mammalian middle ear occurred independently in monotremes and therian mammals (Rich *et al.* 2005; see discussion in Chapter 10) however, at least one adult Mesozoic therian mammal has now been reported to have retained an ossified Meckel's cartilage, although this may be paedomorphic (Ji *et al.* 2009).

Steropodon galmani from Lightning Ridge, New South Wales was the first Mesozoic monotreme to be discovered (Archer *et al.* 1985) (Fig. 1.3d–f). *Steropodon* is slightly younger than *Teinolophos*, between 108 and 103 million years of age (middle Albian). Once considered an early ornithorhynchid (Pascual *et al.* 1992b), *Steropodon* was put into its own family, Steropodontidae, on the basis of estimated divergence times between platypuses and echidnas that excluded *Steropodon* from Ornithorhynchidae (Flannery *et al.* 1995). Steropodontids have an ornithorhynchid-like dentition (Archer *et al.* 1985) and an anterior twist to the dentary correlated with the development of an expanded rostrum (perhaps an incipient bill) (personal observations). However, as with *Teinolophos*, the lower jaw of *Steropodon* is archaic, with recessed areas for accessory jaw bones (recognised independently by Luo *et al.* 2002 and Musser 2003), although *Steropodon* and *Teinolophos* differ over which accessory bones may have been retained (personal observations).

Steropodontids and ornithorhynchids may have shared a basic dental pattern and perhaps a similar lifestyle, but possession of accessory jaw bones in *Steropodon* suggests major differences in jaw mechanics and feeding behaviour. Steropodontids, with at least a partial complement of accessory jaw bones and deeply rooted molar teeth, would have had a more orthal masticatory stroke. Ornithorhynchids, which have lost almost all evidence of accessory jaw bones (a facet for a coronoid bone on the dentary has been reported for the Miocene ornithorhynchid *Obdurodon dicksoni*: Musser 2006b), have a more transverse jaw action as evidenced by extensive transverse wear on the shallow-rooted molars of *Obdurodon* species.

Steropodon, although probably platypus-like, is therefore clearly more plesiomorphic than the more derived ornithorhynchids. Based on the form of its jaw and teeth, *Steropodon* could be described as a 'proto-platypus', outside of Ornithorhynchidae but perhaps ancestral to later ornithorhynchids. It is included in the order Platypoda *sensu* McKenna and Bell (1997) as a platypus-like monotreme. It may have had an incipient bill, and may therefore have been at least partly aquatic.

Possession of a bill – suggesting a taxonomic/ecological link to Ornithorhynchidae – is central to the debate over membership of Platypoda, the platypus-like monotremes (McKenna and Bell 1997). In ornithorhynchids and steropodontids, the anterior rami of the dentaries twist laterally to splay out and form some sort of bill. Although Rowe *et al.* (2008) consider *Teinolophos* a basal platypus, the preserved anterior ends of the dentaries in *Teinolophos* appear to turn medially rather than twisting laterally, suggesting that *Teinolophos* did not have a bill, incipient or otherwise. Although *Teinolophos* has an enlarged mandibular canal, suggesting a highly sensitive snout, its general lower jaw morphology is more similar to that of some basal mammals rather than that of known monotremes – neither platypus-like, nor highly reduced as in echidnas. At present, therefore, *Teinolophos trusleri* should be considered a basal monotreme of uncertain affinities, pending discovery of more diagnostic fossil material.

Cenozoic monotremes: Ornithorhynchidae + Tachyglossidae

Ornithorhynchidae are the true platypuses, morphologically distinct from other monotremes

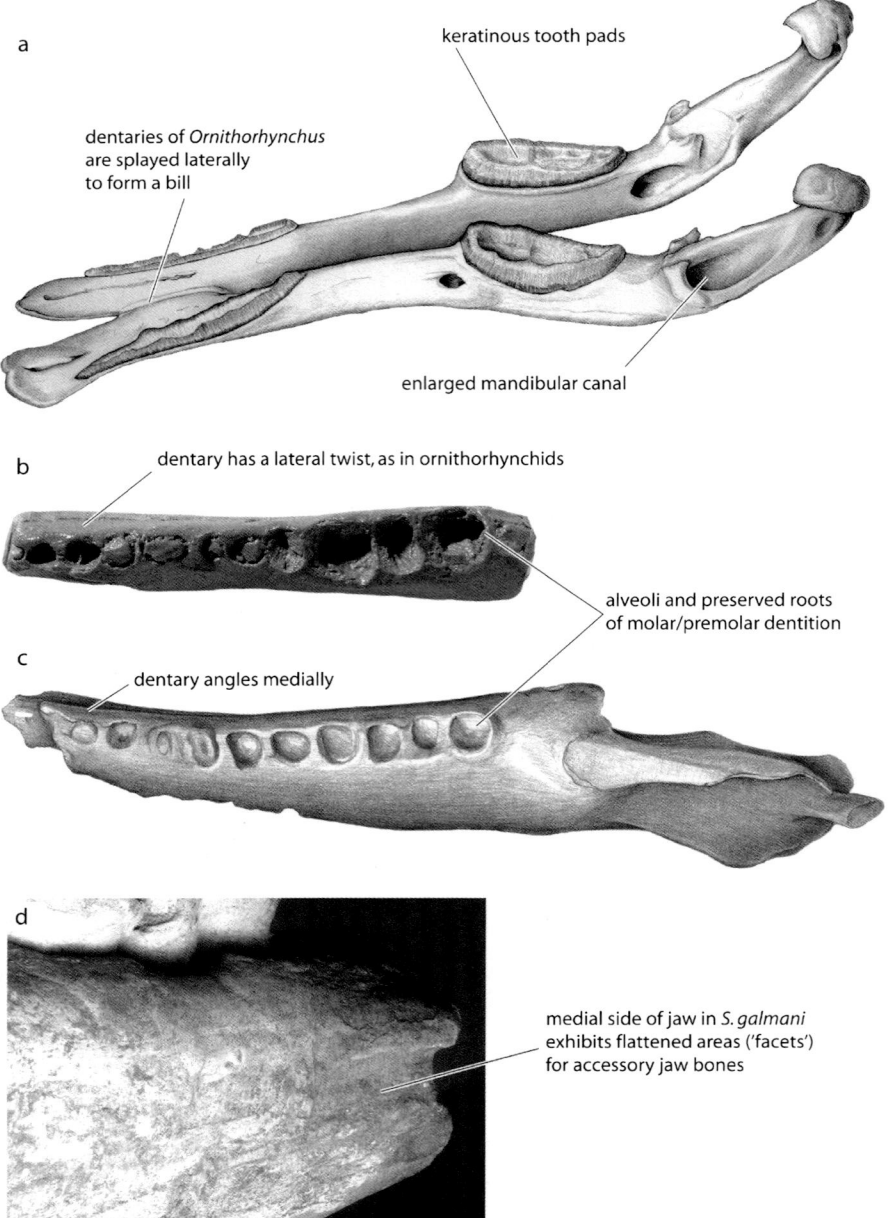

Figure 1.3: Comparison of the lower jaws of *Ornithorhynchus anatinus* (the living platypus), an undescribed steropodontid from Lightning Ridge, New South Wales (AMF97263), *Teinolophos trusleri* (NMVP212933) and *Steropodon galmani* (AMF66763) (a–d respectively) (not to scale). (a) (mandible of *Or. anatinus*) and (b) (occlusal view, left dentary, steropodontid) showing the lateral twist of the dentary in Platypoda, the platypus-like monotremes, suggesting the presence of a splayed bill. Image (c), left dentary (occlusal view) of *T. trusleri* does not appear to splay outward, suggesting that *T. trusleri*, at least on this evidence, did not possess a splayed bill as in Platypoda. Drawing (d) (medial side of the dentary of *S. galmani*) illustrates faceting for accessory jaw bones as in basal mammals or mammaliaforms; this undescribed feature was originally said to be absent in *S. galmani* (Archer *et al.* 1985). Illustration (a) was drawn by A. M. Musser (personal files). Illustration (c) was made by A. M. Musser in collaboration with J. A. Hopson and T. H. Rich (Rich *et al.* 2005) and is reproduced with permission of T. H. Rich. The s.e.m. (scanning electron micrograph) of AMF66763 (d) was photographed by S. Lindsay, Australian Museum. AMF: Australian Museum Fossil Collection. NMVP: National Museum of Victoria.

and occupying an ecological niche as semi-aquatic insectivores/carnivores. Early ornithorhynchids (*Monotrematum* and *Obdurodon* species) retained teeth as adults, unlike the extant *Ornithorhynchus*, which begins life with a transitory dentition that is lost when the young animal begins to feed on its own.

The early Paleocene *Monotrematum sudamericanum*, a monotypic species from Patagonia, southern Argentina, is the oldest known representative of the family. *Monotrematum* is the first monotreme discovered outside the Australasian region, and is the largest platypus known (Pascual *et al*. 1992a, b; Forasiepi and Martinelli 2003). *Monotrematum* has only been recovered from Banco Negro Inferior (Salamanca Formation) near Golfo de San Jorge, one of the oldest Cenozoic mammal-bearing sites in South America (Pascual *et al*. 1992a, b, 2002; Forasiepi and Martinelli 2003). It is represented by isolated upper and lower molars bearing the distinctive ornithorhynchid dental pattern (Pascual *et al*. 1992a, b, 2002), and by two fragmentary femora almost identical to the femur of *Ornithorhynchus* except for their markedly larger size (Forasiepi and Martinelli 2003). The discovery of *Monotrematum* confirms a past Gondwanan distribution for ornithorhynchids, with a past range encompassing Australia, Antarctica and Patagonia at the least.

Monotrematum shares several synapomorphies with ornithorhynchids to the exclusion of other monotremes, including reduction of upper molar number to two (Musser and Archer 1998) and development of extensive transverse wear on molar teeth. Morphological differences between *Monotrematum* and other ornithorhynchids are minor. *Monotrematum* differs from species of *Obdurodon* mainly in that the two principal lingual cusps of upper molars are more cone-like than pillar-like (Pascual *et al*. 1992b); lower molar roots may not be as fully divided as in *Obdurodon* species (Pascual *et al*. 2002); and most of the material recovered is much larger in size than the material recovered for *Obdurodon* species or *Ornithorhynchus* (Pascual *et al*. 1992b; Musser and Archer 1998; Pascual *et al*. 2002; Forasiepi and Martinelli 2003) (Fig. 1.4a, b). *Monotrematum* is otherwise almost indistinguishable from other ornithorhynchids, based on admittedly incomplete material, and should be considered a true ornithorhynchid monotreme. The overall similarity between the molars of *Monotrematum* and those of later *Obdurodon* species is so striking, in fact, that *Monotrematum* and *Obdurodon* might be considered congeneric (Musser and Archer 1998; Musser 2006a). *Monotrematum* is included here within Ornithorhynchidae as its most basal member.

The Oligocene to Miocene *Obdurodon* species were small to mid-sized platypuses, and are more plesiomorphic than *Ornithorhynchus* in most respects as far as can be determined (Woodburne and Tedford 1975; Archer *et al*. 1978; Archer *et al*. 1992; Musser and Archer 1998; Musser 2006a, b) (Fig. 1.4c to e). There are currently two named species, and at least one other species (*Ob*. sp. A personal observations). *Obdurodon insignis*, from the Oligocene of central Australia, is known from molar teeth, a partial lower jaw, and a partial pelvis (Woodburne and Tedford 1975; Archer *et al*. 1978). *Ob. insignis* appears to have been a gracile species, unlike the larger and more robust *Obdurodon dicksoni*, from the Miocene Riversleigh World Heritage Fossil Site in north-western Queensland. *Ob. dicksoni* is known from a complete upper skull and isolated teeth (Archer *et al*. 1992; Musser and Archer 1998).

The well-preserved skull of *Ob. dicksoni* differs from that of *Ornithorhynchus* in being more dorsoventrally flattened, and the bill is proportionately larger and wider (Musser and Archer 1998). On this evidence, *Ob. dicksoni* was almost certainly a semi-aquatic predator like *Ornithorhynchus* (Musser and Archer 1998). It is doubtful that *Ob. dicksoni* was the ancestor of later *Ornithorhynchus*, since its hypertrophied bill and dorsoventrally flattened skull appear to be autapomorphic specialisations rather than ancestral characteristics (Musser and Archer 1998). *Ob. dicksoni* is more plesiomorphic than *Ornithorhynchus* in having adult dentition, and is the only Cenozoic mammal known to have retained a coronoid (Musser 2006a, b).

Ornithorhynchus, edentate as an adult with a lightened skull and more streamlined body form, is the most highly specialised ornithorhynchid. The fossil record for the genus goes back to the Pliocene (Bow Local Fauna) although well-preserved skulls are known only from the Pleistocene to the present (e.g. Musser 2006a, b). There are no notable differences between Pleistocene and Recent *Ornithorhynchus* in the fossil material recovered to date.

In sum, Ornithorhynchidae comprises *Monotrematum sudamericanum*, *Obdurodon* species and *Ornithorhynchus anatinus*, all of which share synapomorphic features unique to platypuses. Ornithorhynchidae is an eastern Gondwanan clade whose distribution encompassed Australia, Antarctica and southern South America during the latest Mesozoic to earliest Cenozoic, whose origins are most likely mid- to Late Cretaceous (i.e. Mesozoic rather than Cenozoic). Inclusion of the 61–63 million-year-old *Monotrematum* as an ornithorhynchid suggests that a family-level split occurred before the end of the Cretaceous

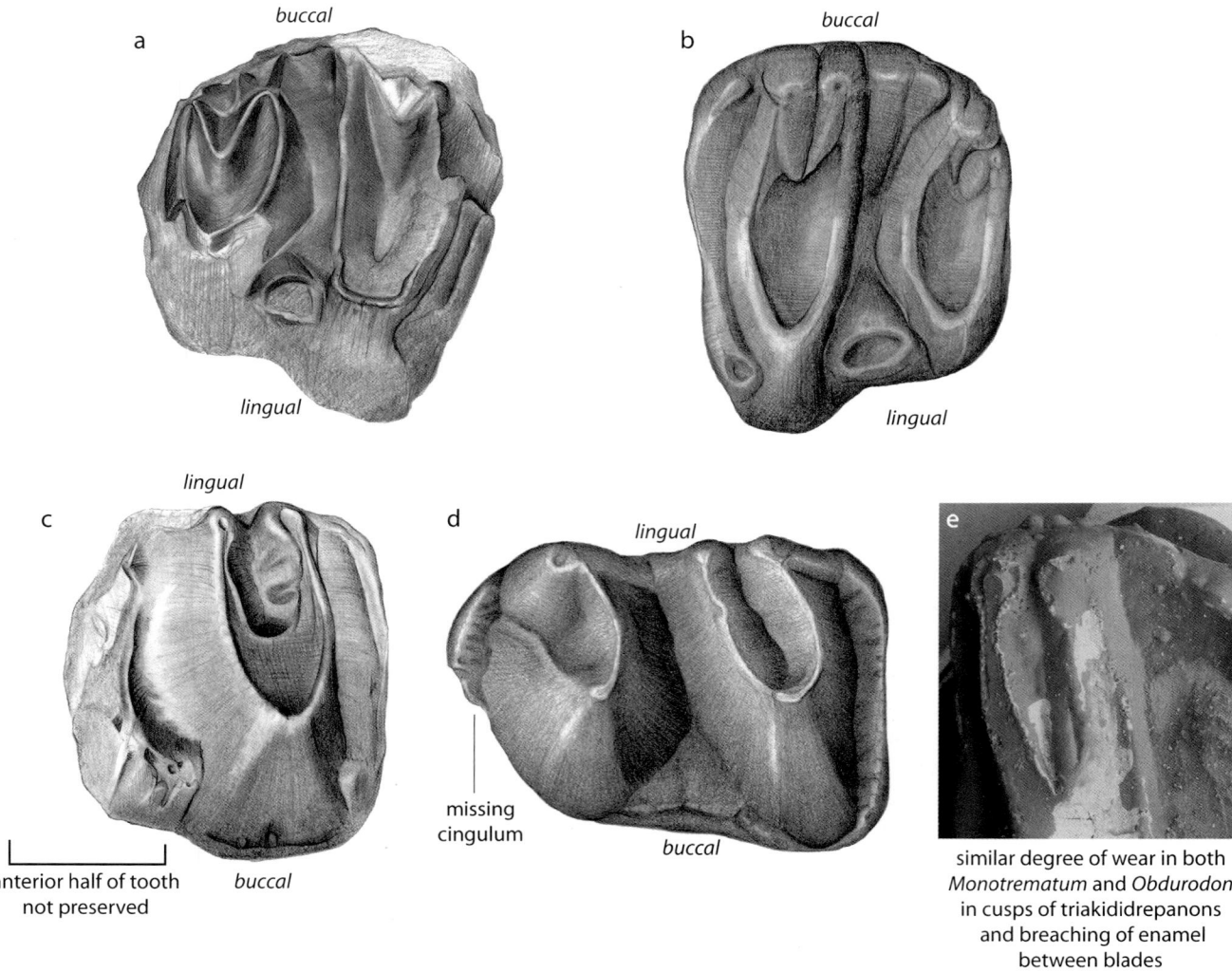

Figure 1.4: Comparisons between the molar teeth of *Monotrematum sudamericanum*, *Obdurodon* sp. A and *Obdurodon dicksoni* (not to scale). (a) Right upper second molar, *M. sudamericanum* (MPEF-PV1634) and (b) and left upper second molar, *Ob.* sp. A (SAM P30158). Although (a) is fragmentary, it shows strong similarities to (b), the m2 of *Ob.* sp. A from the Oligocene of South Australia. In addition to similarities in general form and shape of the triakididrepanon basins, there is almost identical wear in the cusp between triakididrepanons. (c) Left lower first molar of *M. sudamericanum* (MPEF-PV1635) and (d) left lower first molar of *Ob. dicksoni* (QMF18985). The buccal cingulum of m1, well formed in *S. galmani*, has been greatly reduced in *M. sudamericanum* as in species of *Obdurodon*. Although basin wear is not seen in (d), basin wear can be seen in (e), another lower molar of *Ob. dicksoni* (QMF18616) (both QMF18985 and QMF18616 are from Ringtail Site, Riversleigh World Heritage Area). (a)–(d) were drawn from the original specimens (Musser 2006a, b). The s.e.m. of QMF18616 (e) was photographed by S. Lindsay, Australian Museum. AMF: Australian Museum palaeontology collection. MPEF-PV: Museo Paleontológico Egidio Feruglio, Paleontología Vertebrados, Trelew, Argentina. QMF: Queensland Museum palaeontology collection. SAMP: South Australian Museum palaeontology collection.

65 mya, and the occurrence of *Monotrematum* in the late early Paleocene of Patagonia, not long after the end-Mesozoic extinction event, indicates trans-Antarctic dispersal of ornithorhynchids before the end of the Mesozoic. The presence of Late Cretaceous ornithorhynchids is consistent with the results of several molecular studies, although not with others (see below).

The distribution of early ornithorhynchids was 'Weddellian': a cool-temperate palaeo-province comprising Australia/Antarctica and southern South America, dominated by forests of southern beech (*Nothofagus*) (e.g. Musser 2006a, b). There is no evidence that ornithorhynchids spread north from Patagonia further into South America. The extinction of South American ornithorhynchids probably occurred with the arrival of therian mammals (marsupials and placentals) that replaced the existing community of archaic, pre-therian mammals, including ornithorhynchids, gondwanatheres, dryolestoids and possibly docodonts (e.g. Bonaparte 1990; Pascual *et al.* 1992a, b).

During the Cenozoic, after Australia had rifted from Antarctica, ornithorhynchids spread as far north as Riversleigh in north Queensland, and as far west as the Lake Eyre region of South Australia (Musser 2006a, b). During the Cenozoic, platypuses were part of a diverse mammalian fauna comprised almost entirely of marsupials (including many early representatives of endemic Australian groups) along with bats (Chiroptera) and later rodents. Their Australian distribution contracted severely due to the drying of the continent that began in the Miocene and which continues to the present day.

The survival of ornithorhynchid monotremes in Australia can be attributed to Australia's geographic isolation and to a lack of competition from other mammals occupying the niche utilised by ornithorhynchids (Musser 1998). The absence of ornithorhynchids in the north, including New Guinea as far as known, may in part be related to the warmer temperatures of northern or equatorial regions. Ornithorhynchids evolved in the cooler climes of the 'deep south', and thrive today in upland and/or more southern parts of eastern Australia. The present limited distribution of *Ornithorhynchus* along the eastern part of Australia is precarious and relictual, subject to continued contraction due to climate change and other environmental stresses (e.g. Musser and Temple-Smith 2008), and its status needs to be carefully monitored.

Echidnas have a much shorter fossil history than ornithorhynchids, certainly dating from the Pleistocene, but possibly dating from as far back as the Miocene. The shorter fossil record for tachyglossids is due in part to the lack of teeth in echidnas, since harder enamelised teeth fossilise more readily than other skeletal elements. It may also reflect the general supposition that echidnas are a more recent group than platypuses (but see below). The ancestor of the first tachyglossids is unknown, and there is no dental trail. The 'hypothetical' ancestral type, however, may have been a robustly built terrestrial insectivore, perhaps with vestigial dentition or a reduced dentition specialised for terrestrial invertebrate prey, as in numbats and other myrmecophages.

Long-beaked echidna fossils are more numerous than those of short-beaked echidnas, although there are few complete skulls or skeletons, and there may have been at least three genera of long-beaked echidna on the Australian mainland and in Tasmania. Currently recognised species include *Zaglossus bruijnii*, *Megalibgwilia owenii* (= *Megalibgwilia ramsayi*) and the huge '*Zaglossus*' *hacketti* from Western Australia (possibly not a species of *Zaglossus*: see below).

The supposed differences between *Megalibgwilia* and *Zaglossus* are based on presumed diet (Griffiths *et al.* 1991). However, these fossil echidnas are quite similar in most respects, and may not warrant generic distinction (personal observations). The precise diet of any extinct species is also notoriously hard to determine. '*Zaglossus*' *hacketti*, the immense echidna from the Pleistocene of Western Australia (Krefft 1868), appears to be genuinely different at the generic level. However, although its limb proportions are known, there is unfortunately no skull for '*Z*'. *hacketti*, and comparisons cannot be made with the contested skulls placed in either *Megalibgwilia* or *Zaglossus*.

Short-beaked echidnas are known from the Pleistocene to the present, and all material can be referred to *T. aculeatus* (Murray 1978b; Pledge 1980). However, Pleistocene *T. aculeatus* from Naracoorte, South Australia are up to 10% larger than the largest of living species, a phenomenon found in many Pleistocene taxa (Pledge 1980).

There is debate over the age of the Deep Lead mine shaft at Gulgong, New South Wales, where the oldest echidna fossils have been found (the long-beaked *'Zaglossus' robustus*: Dun 1895). This now-collapsed and inaccessible site may be either Pleistocene in age, as originally suggested (Dun 1895), or perhaps as old as Miocene (around 15 million years old) (Woodburne *et al.* 1985). Sediments and fossil material resemble those from Pleistocene sites such as Wellington Caves, and it is possible that the earlier Pleistocene determination is the correct one (Augee *et al.* 2006). Perhaps new methods of dating fossil material will help to clarify the date of this significant deposit.

Mammalian evolution during the Mesozoic

Monotremes are surviving Mesozoic mammals, as are marsupials and placentals, whose lineage can be traced back to the Early Cretaceous of China (e.g. Luo 2007). In order to understand the early evolution of monotremes, a brief summary of early mammalian evolution is presented below.

About two-thirds of Mesozoic mammaliaform genera (Mammalia + early mammal-like forms that technically had not reached full mammalian status) have been described over the past 25 years, greatly enhancing our understanding of these early mammals and their worlds (Luo 2007). Many Mesozoic mammals were not simply small terrestrial insectivores – the traditional view of Mesozoic Mammalia – but instead show surprising ecomorphological diversification. New discoveries are showing early mammals to be a varied and interesting group, with many occupying specialised niches much as platypuses and echidnas do. There were early gliders, arboreal mammals, fossorial mammals and even swimming/diving forms (e.g. Luo 2007). These taxa came from several unrelated, archaic groups (e.g. pre-mammalian cynodonts, docodonts, symmetrodonts and multituberculates). The platypus-like, swimming/burrowing docodont *Castorocauda* from the Middle Jurassic of China was, like ornithorhynchids, otters and beavers, specialised for a semi-aquatic lifestyle (Ji *et al.* 2006). Fossorial myrmecophages occupied the niches of ant and termite specialists (e.g. the somewhat echidna-like but toothed *Fruitafossor* from the Late Jurassic of Colorado, which may have been myrmecophagous: Luo and Wible 2005).

The Mesozoic worlds of these early mammals are being reconstructed through discoveries of beautifully preserved fossils that help piece together whole ecosystems, such as the Early Cretaceous fossil deposits found in the province of Liaoning in northeastern China. The rich Liaoning deposits have preserved plants, invertebrates and vertebrate species. The vertebrate fauna included many dinosaur groups, early birds that were almost as diverse as the dinosaurs, many species of reptiles and amphibians, and several key species of Mesozoic mammals (e.g. Hu *et al.* 2005). Mammal species from Liaoning include the very large, badger-sized eutriconodont *Repenomamus* (found with the remains of a young dinosaur, *Psittacosaurus*, in its abdominal region (Hu *et al.* 2005), the oldest known marsupials and placentals, *Sinodelphys* and *Eomaia* respectively) (Luo *et al.* 2003; Ji *et al.* 2002) and the early therian *Akidolestes*, a symmetrodont with some monotreme-like (but apparently convergent) postcranial features of the lumbar vertebrae, pelvis and hind limb (Li and Luo 2006). The Lightning Ridge biota of north-central New South Wales, where plants, invertebrates and vertebrate animals have been preserved in opal, may likewise prove to be a window into the world of early monotremes like *Steropodon*.

Monotremes as Mesozoic mammals

Several anatomical features of monotremes are similar to those of Late Triassic–Early Jurassic mammaliaforms, before a Middle Jurassic diversification of more advanced mammals (e.g. Luo 2007). This may point to an origin in the Late Triassic to Early Jurassic for Monotremata, or a later origin from archaic Late Triassic–Early Jurassic stock. The level of jaw development in Early Cretaceous monotremes is noteworthy because of its undeniably plesiomorphic character, as in Late Triassic–Early Jurassic mammals, mammaliaforms and advanced cynodonts but not later therian mammals (Rich *et al.* 2005; Musser 2006a, b). By the Early Cretaceous, most mammals had just a single paired bone, the dentary, comprising the lower jaw (e.g. Kielan-Jaworowska *et al.* 2004). It is also of interest that monotremes have a crural spur, a feature found in basal Mammalia, or Mammaliaformes, but not in Theria (Hurum *et al.* 2006).

Australian Mesozoic biotas are not yet as well known as those of such places as Liaoning, and

Australian Mesozoic mammals are still extremely rare (to date known only from the Albian–Aptian of Victoria and New South Wales). In addition to *Teinolophos* and *Steropodon*, there are just five other named Mesozoic mammalian or mammaliaform taxa: the ausktribosphenids *Ausktribosphenos nyktos* (Rich *et al.* 1997, 1999, 2001a) and *Bishops whitmorei* (Rich *et al.* 2001a); the Victorian multituberculate *Corriebaatar marywaltersae* known from a single tooth (Rich *et al.* 2009); the enigmatic *Kollikodon ritchiei*, also from Lightning Ridge, first described as a monotreme (Flannery *et al.* 1995) but subsequently revised as a non-monotreme stem mammal or near-mammal (Musser 2006a, b); and *Kryoryctes cadburyi*, named on the basis of an isolated, tachyglossid-like humerus (Pridmore *et al.* 2005). Aside from *Kryoryctes*, Australian Mesozoic mammals are represented only by jaws and teeth.

The ausktribosphenids *Kollikodon*, *Teinolophos* and *Steropodon* all have comparatively archaic dentaries, with evidence of facets for accessory jaw bones (Kielan-Jaworowska *et al.* 1987; Rich *et al.* 2005; Musser 2006a, b). Known Australian Cretaceous mammals were therefore relict taxa, more typical of Jurassic mammals or mammaliaforms than of most other Early Cretaceous mammals. Such relictualism is seen in other Early Cretaceous Australian taxa, including labyrinthodont amphibians from the Early Cretaceous of Victoria (Warren *et al.* 1991) and dicynodont therapsids (Thulborn and Turner 2003), a phenomenon most likely due to Australia's isolated geographic position at the far south-east of Gondwana from the late Jurassic onwards.

Competitors for early monotremes would have included other early mammals (some of which, like *Castorocauda* or *Fruitafossor*, may have occupied unique and monotreme-like ecological niches), perhaps small dinosaurs, other small reptiles and birds (several of which may have been insectivorous, like many birds today). Predators would have included dinosaurs, other large reptiles like crocodiles (known from the Early Cretaceous of both New South Wales and Queensland), and the massive, crocodile-like labyrinthodont amphibians. The environment itself would have been challenging. Australia during the Early Cretaceous was tethered to Antarctica close to the South Pole, with known monotreme distributions near to or within the Antarctic Circle (Musser 2006b). The gradual loss of geographic land connections between parts of eastern Gondwana, causing progressive isolation, and the vast extent of cold, dark southern forests in the region created a unique environment with a biota limited by these factors that continued to evolve in isolation over the course of most of the Cenozoic (see Chapter 14 for the biogeography of Australasia during the Cenozoic).

Debates over origins

Relationships between monotremes and other mammals as well as within Monotremata are the subject of considerable debate, and have been reviewed in several publications (e.g. Luo *et al.* 2001, 2002; Musser 2006a, b). In addition to disagreements over morphological interpretations, reasons for disagreement over relationships include alternative results from computer-assisted phylogenies, which can be subjective rather than objective, and molecular studies using various techniques that in some instances have produced widely differing results.

Mammals can be divided into three groups based on level of organisation: (1) Early Jurassic–Late Triassic 'near-mammals' and mammaliaforms such as symmetrodonts, docodonts and advanced, mammal-like cynodonts (2) mid- to Late Jurassic–Early Cretaceous mammals such as the Southern Hemisphere Australosphenida and (3) advanced Cretaceous mammals including Marsupialia and Placentalia. Although once thought to be closely related to therian mammals, particularly to Marsupialia, current consensus is that monotreme relationships lie with basal Mammalia, perhaps with the Southern Hemisphere Australosphenida.

Monotremata–Australosphenida

Australosphenida is a recently erected higher level taxon (currently an infraclass) comprised of several archaic Jurassic–Cretaceous southern mammals, and is envisaged as an endemic southern radiation of non-therian tribosphenic mammals (Luo *et al.* 2001, 2002). Monotremes are included in Australosphenida by Luo *et al.* (2001, 2002) although there has been disagreement about their inclusion, particularly from those working on Victorian Mesozoic monotremes (e.g. Rich *et al.* 2002 Woodburne *et al.* 2003). Monotremes do not possess true tribosphenic teeth although molar cusps are triangulated, the V-shaped structure of molar teeth differs from that of true tribosphenic molars, which are distinguished by possession of trigonids and talonids ('mortar and pestle' occlusion).

Australosphenida is characterised by possession of tribosphenic or near-tribosphenic teeth and jaws in taxa that have plesiomorphic jaw structure (i.e. retention of additional bones on the medial side of the lower jaw). Tribosphenic teeth are otherwise a hallmark of therian mammals ('Tribosphenida'), with more advanced jaw structure. Aside from questions over inclusion of monotremes, australosphenidans include the ausktribosphenids *Ausktribosphenos nyktos* and *Bishops whitmorei* from the Early Cretaceous of southern coastal Victoria (Rich *et al.* 1997; Rich *et al.* 1999) *Asfaltomylos patagonicus* from the Late Jurassic of Patagonia (Martin and Rauhut 2005) and *Ambondro mahabo* from the Middle Jurassic of Madagascar (Flynn *et al.* 1999). A dual origin of tribosphenic molars has therefore been proposed, with two infraclasses: Australosphenida (southern/Gondwanan) and Boreosphenida (a later northern radiation of more advanced, therian mammals) (Luo *et al.* 2001, 2002). Dental features said to be present in Australosphenida but not in Boreosphenida per Luo *et al.* (2001, 2002) include a well-developed mesial cingulid wrapping around and extending to the lingual side of the trigonid; a mesiodistally short, buccolingually broad talonid; and reduced height of the trigonid (but see Rich *et al.* 2002; Woodburne *et al.* 2003).

Monotremes may or may not share a close relationship with ausktribosphenids. Ausktribosphenids are not monotremes because they lack key monotreme synapomorphies (e.g. enlarged mandibular canals, posterointernal mandibular processes or the non-tribosphenic monotreme dental pattern). Certain dental features are similar in monotremes and ausktribosphenids (e.g. some elements of molar structure) however, differences are potentially significant, and the lack of material apart from teeth and lower jaws suggests a cautious approach in interpreting relationships (Musser 2006a, b). An alternative scenario is that monotremes may represent an independent lineage of archaic mammals that independently acquired triangulated molars (Pascual *et al.* 2002). Phylogenetic analyses have produced conflicting results in large part because some key characters have been given alternate character scores based on differing interpretations of researchers. Although the Australosphenida concept is intriguing, it may be premature to suggest major classification changes at the infraclass level based on such fragmentary material.

Composition of Ornithorhynchidae

Recent debate has arisen over the composition of Ornithorhynchidae and over divergence times between ornithorhynchids and tachyglossids. This discussion has arisen in part because of the discovery of Mesozoic monotremes (all of which are fragmentary and some of which are inadequately described and thus difficult to interpret), and in part because molecular studies investigating divergence times have produced widely conflicting dates for the phylogenetic split between ornithorhynchids and tachyglossids.

Two recent molecular studies present diametrically opposed views of the composition of Ornithorhynchidae, and consequently of the timing of the phylogenetic split between platypuses and echidnas. Rowe *et al.* (2008), using a relaxed molecular clock, include both the Early Cretaceous *Teinolophos* and *Steropodon* in Ornithorhynchidae on the basis of a single morphological feature, an enlarged mandibular canal, which they consider an ornithorhynchid synapomorphy. They therefore conclude that *Teinolophos* and *Steropodon* were platypuses both morphologically and ecologically, thus removing *Steropodon* from its own family, Steropodontidae (Flannery *et al.* 1995), and adding *Teinolophos* to Ornithorhynchidae as its most basal member. Under the proposal by Rowe *et al.* (2008), the origins of Ornithorhynchidae predate the Aptian, with echidnas and platypuses last sharing a common ancestor over 115 mya (Rowe *et al.* 2008). However, an enlarged mandibular canal is a monotreme synapomorphy that is not restricted to platypuses (Rich *et al.* 2005; Musser 2006a, b).

A subsequent study by Phillips *et al.* (2009), also using a relaxed molecular clock approach but including multigene evaluation, reaches a much different conclusion. Their divergence dates place the platypus–echidna divergence in the mid-Cenozoic (19–48 mya) rather than the Mesozoic, bringing the origins of Ornithorhynchidae to the Cenozoic and suggesting platypuses diverged from echidnas during either the Oligocene or Miocene. This excludes not only *Teinolophos* and *Steropodon* from Ornithorhynchidae but the

Paleocene *Monotrematum* as well (described as 'platypus-like' but not an ornithorhynchid).

By restricting Ornithorhynchidae to Oligocene–Miocene *Obdurodon* species and the extant *Ornithorhynchus*, Phillips *et al.* (2009) exclude *Monotrematum* from Ornithorhynchidae without justification on morphological grounds. Morphological differences between *Monotrematum* and *Obdurodon* species are in fact quite minor, perhaps even interspecific rather than intergeneric. Ornithorhynchid-like features in *Monotrematum* include teeth and femora almost indistinguishable from those of *Obdurodon* and *Ornithorhynchus* respectively. Based on present evidence, then, *Monotrematum* should be retained in Ornithorhynchidae as its most basal member, as discussed above. Suggestions by Phillips *et al.* (2009) that *Monotrematum* is demonstrably more plesiomorphic than other ornithorhynchids and that it lacks sufficient morphological data for inclusion within Ornithorhynchidae are not accepted here. Extending the temporal range of ornithorhynchids to at least the latest Mesozoic suggests that the ornithorhynchid–tachyglossid split occurred within the Mesozoic and not during the Cenozoic, as proposed by Phillips *et al.* (2009).

Camens (2010) cites a tachyglossid-like humerus from the Early Cretaceous of Australia, *Kryoryctes cadburyi*, as evidence that tachyglossids may have been present in Cretaceous Australia, challenging the idea of Cenozoic origins for extant monotreme families (e.g. Phillips *et al.* 2009, 2010). However, as noted by Phillips *et al.* (2010), Pridmore *et al.* (2005) discuss the possibility that *Kryoryctes* may not be monotreme, or alternatively suggest that it may belong to the similarly sized *Steropodon* (highlighting the difficulties inherent in assigning taxonomic names to such materials as isolated limb elements). Several plesiomorphic features of *Kryoryctes* are in fact similar to those of pre-mammalian cynodonts and some very basal mammals. Echidna-like monotremes may have been present in Australia since the Cretaceous, but *Kryoryctes* does not provide compelling evidence for this hypothesis.

Rowe *et al.* (2008) and Phillips *et al.* (2009, 2010) both challenge current views of Ornithorhynchidae, a family distinguished until now by well-defined dental, cranial and postcranial features: a wide bill formed by splayed upper and lower jaws, used to probe riverbeds for benthic invertebrates; a dorsoventrally flattened body (as in many aquatically adapted animals); and wide, multiple-rooted molar teeth in all toothed fossil species (Griffiths 1978; Musser 2006a; Grant 2007). The divergent views presented by Rowe *et al.* (2008), Phillips *et al.* (2009) and Camens (2010) beg the question of just what a platypus *is*, both anatomically and ecologically. I argue here that the membership of Ornithorhynchidae must be determined by possession of clearly platypus-like morphology, which therefore determines the palaeoecological niche of the species.

Echidnas as derived platypuses: an aquatic origin for echidnas?

Phillips *et al.* (2009) suggest that Tachyglossidae originated from within Ornithorhynchidae, making them 'derived platypuses' (an idea first proposed by Pascual *et al.* 1992a, b in describing *Monotrematum*). However, this is difficult to envision given the extreme specialisations in both ornithorhynchids and tachyglossids: a highly derived, terrestrial ant and termite specialist feeder from a semi-aquatic insectivore/carnivore feeding on benthic invertebrates (Musser 2006b and Chapter 14). Features said to be shared by both ornithorhynchids and tachyglossids presented to support the 'aquatic ancestry' hypothesis include a dorsoventrally flattened body shape and reversed hindfoot posture (said to have been derived from the partly reversed hindfoot posture of ornithorhynchids) (Phillips *et al.* 2009, 2010). Echidnas, however, are not dorsoventrally flattened, but have a dome-shaped body (i.e. there is no dorsoventral compression comparable to that of the platypus as claimed by both Phillips *et al.* 2010 and Camens 2010), and supporting comparative anatomical studies have not been made on hindfoot structure.

Could echidnas be as old or older than platypuses, and could echidnas have been present in Australia since the Early Cretaceous, as suggested by Camens (2010)? A Late Cretaceous origin for Ornithorhynchidae may suggest a Late Cretaceous origin for Tachyglossidae *if* echidnas were derived off the platypus lineage. However, it remains a strong possibility that echidnas were derived from an as-yet-unknown monotreme ancestor, particularly given the lengthy gaps in the Australian mammalian fossil record. The inclusion of echidnas within Ornithorhynchidae (as derived platypuses and, by extension, members of Platypoda) is not supported here on morphological grounds and on the numerous anatomical transformations needed for this to occur.

Taxonomy and classification

Revised taxonomic diagnoses

Revised, morphology-based diagnoses of Monotremata, Platypoda, Ornithorhynchidae, Tachyglossa and Tachyglossidae are given below. This proposed taxonomy is based on the most recent information available and has been modified from McKenna and Bell (1997). I emphasise that this is a preliminary or interim diagnosis, noting that large-scale projects examining monotreme relationships are in preparation, and that new discoveries will continue to surface. As our understanding progresses, monotreme taxonomy will undoubtedly change.

Monotremata: basal Mammalia with (1) an enlarged mandibular canal and (2) a dental pattern characterised by one to two V-shaped lophs (dentition secondarily lost in Tachyglossa).

Platypoda (Steropodontidae/Ornithorhynchidae): monotreme mammals with either an incipient or well-developed bill and other anatomical and physiological specialisations related to a semi-aquatic lifestyle. Early members of Platypoda retained a complement of accessory jaw bones (e.g. *Steropodon*), progressively lost in later members (reduced to just the coronoid in toothed ornithorhynchids and lost completely in *Ornithorhynchus*). *Steropodontidae* (Flannery *et al.* 1995): Platypoda that differ from ornithorhynchids in having relatively unreduced m1; long-rooted molars; deep mandibles with accessory jaw bones in addition to the coronoid; and no pseudoentoconids. *Ornithorhynchidae*: Platypoda with a wide, splayed bill; dorsoventrally flattened body form and other aquatic adaptations; distinctive dental pattern with shallow roots; and reduction in molar number (dentition is vestigial in *Ornithorhynchus*).

Tachyglossa: edentulous monotremes adapted for an insectivorous or specialised myrmecophagous diet and characterised by robust, dome-like, spine-covered bodies. Synapomorphies: brain transversely expanded and cranium dome-like ('bird-like'); large hemispherical braincase (larger than that of the platypus); feeble masticatory musculature; highly reduced (vestigial) lower jaw; absence of a sagittal crest (gyri of the brain mould the cranium); and auditory ossicles and middle ear area larger than in platypuses. *Tachyglossidae*: as for Tachyglossa.

Prototheria (Monotremata) **incertae sedis** *(basal Monotremata)*: *Teinolophos trusleri* (Rich *et al.* 2001a, b, 2005).

Note: the Early Cretaceous taxon *Kollikodon ritchiei* from Lightning Ridge, New South Wales has been described as an aberrant, uniquely specialised monotreme (Flannery *et al.* 1995). *Kollikodon* does have an abrupt disjunction between molar and premolar teeth similar to that in ornithorhynchids (Flannery *et al.* 1995) and an expanded, ornithorhynchid-like maxilla (Musser 2006b). However, in almost all other features, *Kollikodon* is not monotreme-like (Musser 2006a, b; *contra* Flannery *et al.* 1995). The lower jaw does not have an enlarged mandibular canal – a key monotreme synapomorphy – contrary to the description published by Flannery *et al.* (1995). In addition, the molar cusps clearly lack a pattern of reversed triangles; the remarkably bunodont cusps are much more similar to those in certain early mammals, mammaliaforms or pre-mammalian Cynodontia with bunodont molariform teeth (Musser 2006a, b). *Kollikodon* is therefore not considered a monotreme here.

2
Behaviour and ecology of monotremes

Stewart C. Nicol

Summary

The extant monotremes share a number of general features that must reflect their common ancestor and constrain their ecology and behaviour. All of the extant monotremes have diets which consist principally of invertebrates: platypuses feed mainly on benthic macroinvertebrates; long-beaked echidnas feed on invertebrates in the soil, leaf litter and decaying logs; and short-beaked echidnas, although principally feeding on ants and termites, also eat a range of other soil invertebrates. Platypuses and long-beaked echidnas are predominantly nocturnal, and although the short-beaked echidnas are often seen during the day, most of their activity occurs during the night. The long-beaked echidnas, apart from a possible recent population in the Kimberley region in northwest Australia, are restricted to New Guinea, and platypuses occur in freshwater environments in eastern Australia, but short-beaked echidnas occur throughout Australia and in New Guinea.

Monotremes are solitary and females have well-established home ranges, while males have larger home ranges and compete for matings. All of the monotremes have very large testes, indicating strong competition between males, and adult males have a crural or femoral gland connected by a duct to a spur on the ankle. The glands increase in size during the breeding season and are clearly important in breeding behaviour. The platypus and short-beaked echidnas are seasonal breeders, but the data for long-beaked echidnas are equivocal. Gestation is very short compared with lactation, and late in the gestation period a pregnant female echidna will enter a nursery burrow to lay her single egg, while the female platypus lays two eggs in her nest or burrow. The eggs are incubated in the female's pouch and hatch after 10 or 11 days. The lactation period for platypuses is about 19 weeks, but for echidnas there are large differences between populations from different parts of Australia: in eastern Australian echidnas the lactation period may be as little as 20 weeks, and in Western Australian and Kangaroo Island as much as 30 weeks. There are also differences in other aspects of maternal care between these echidna populations.

All the monotremes have low metabolic rates, but echidna metabolic rates are extremely low, and short-beaked echidnas in cooler areas show long periods of hibernation. This low energy way of life makes the possession of a large, energetically expensive brain hard to explain, and it is possible that their behaviour, particularly their social behaviour, is more complex than previously thought.

Introduction

For the neurobiologist, perhaps the most remarkable characteristics of the monotremes are their capacity for electroreception and their large, highly encephalised brains (Chapter 7). The high encephalisation of monotremes seems particularly incongruous given the plesiomorphic ('primitive') aspects of the monotreme skeleton and reproduction, which have lead to a popular depiction of them as 'living fossils', a term introduced by Darwin in a reference to the platypus:

> ... *some of the most anomalous forms now known in the world, as the Ornithorhynchus and Lepidosiren (South American lungfish), which, like fossils, connect to a certain extent orders now widely separated in the natural scale. These anomalous forms may almost be called living fossils; they have endured to the present day, from having inhabited a confined area, and from having thus been exposed to less severe competition.*
> (Darwin 1859)

Gould (1992) has written on the confusion and semantic contortions of scientists who have not been able to reconcile the apparent complexity of monotreme brains with their status as 'primitive' mammals. So why do monotremes have large brains? Do they show particularly complex behaviour or cognitive abilities? Because of the taxonomic significance of the monotremes there are several reports on the behaviour of zoo animals, e.g. echidnas at the Prague (Dobroruka 1960), Zurich (Hediger and Kummer 1961) and Kansas (Brannian and Cloak 1985) zoos, while Fleay (1980) recounts numerous observations on captive platypuses. There are four laboratory studies on learning in echidnas (Buchmann and Rhodes 1979; Burke *et al.* 2002; Saunders *et al.* 1971a, b). Significantly, most of these studies preceded any systematic studies of ecology and behaviour in the field, and so I will summarise this material before providing the ecological context.

Behavioural observations on captive animals

Dobroruka (1960) described interactions between one female and three male echidnas at Prague Zoo. As well as observing copulation, he also described one animal (sex not noted) everting its cloaca and wiping a strong smelling secretion on the ground, suggesting some form of chemical communication and territorial marking. Hediger and Kummer (1961) reviewed what was known about sensory biology, feeding, hibernation, territoriality, and social behaviour, but noted that captive echidnas often showed stereotypic behaviour, which was attributable to their captive conditions.

Experimental behavioural studies have only been carried out on short-beaked echidnas. Brattstrom (1973) made detailed observations of the behaviour of captive echidnas, mostly in small laboratory pens, with no shelters or retreats, and often with many animals caged together. Some observations were made in outdoor pens and in the wild. He described a range of behavioural postures and social behaviour patterns, and dominance hierarchies. 'Much of the dominance-subordinance relationships between echidnas is resolved by the subordinate individual recognising some dominance feature (apparently size) of another and avoiding any contact with the dominant animal' (Brattstrom 1973). He concluded that: 'Many complicated behavioural postures such as grooming, aggression, courtship, and maternal behaviour are missing in the echidna. The echidna thus seems to have a behaviour that is not only simpler than that found in most mammals, but perhaps simpler also than that of many lizards', but did note that 'many postures elicited in captivity may never occur in the wild state and vice versa'. In order to investigate possible territorial behaviour, and to re-examine social behaviour under more natural conditions, Augee *et al.* (1978) conducted studies of echidnas in a large enclosure at Taronga Zoo. Observations on paired encounters at feeding sites did not support a simple size-related dominance order as suggested by Brattstrom (1973), and the only aggressive behaviour was an encounter between two females, one of which was carrying a pouch young. Perhaps one of the most interesting features of such studies is that echidnas, which are solitary in the wild, tolerate being housed communally, and will group together, even when given access to multiple refuges.

Laboratory tests of monotreme cognition

In non-human primates at least, brain size is an excellent predictor of cognitive ability (Deaner *et al.* 2007), and there have been a small number of laboratory studies of cognition in echidnas. In tests involving remembering the position of a food reward in a T-maze, the acquisition rate of choice behaviour was almost identical to that of rats with similar levels of training (Saunders *et al.* 1971b), and in spatial habit reversal tasks echidnas showed a significant reduction in the number of errors committed on successive reversals (Saunders *et al.* 1971a), with the authors commenting 'the rapid reduction in errors resembled the mammal-like performance reported for many mammalian species'.

Although vision has been thought to be of little importance in echidnas, Gates (1978) demonstrated visual discrimination at least comparable to that of the rat. Using two choice doors with a food reward, he concluded that echidnas could discriminate between symbols of different brightness, differences in orientation, and of various complex shapes. He found (not surprisingly) that although echidnas lack a corpus callosum they showed interocular transfer of discrimination.

Buchmann and Rhodes (1979) extended the laboratory investigation of echidna learning by employing operant techniques: echidnas were trained to press a treadle for a food reward in response to visual, tactile and positional cues. After the animals achieved a satisfactory (criterion) level of performance, the cues were reversed. The reversal indices obtained were similar to those obtained on a range of placental and marsupial species, and while overall performance compared favourably with results from cats and rats, echidnas appear to be capable of achieving criterion performance in a smaller number of reversals than other mammals (Buchmann and Rhodes 1979).

Burke *et al.* (2002) attempted to study spatial memory performance of echidnas in terms of their foraging ecology by testing their learning response to the positioning of food in two-way and four-way mazes. Working from the premise that echidnas feed on ants and termites, and that these are concentrated in spatially isolated, temporally stable patches (nests), which contain too much food to be depleted, but which will be defended when disturbed, thus limiting feeding time, they made two predictions: when there was a short time between consecutive food rewards (the retention time) echidnas would prefer to forage at another location, but at a later time, when nest defences had returned to normal, echidnas would prefer to forage at the same location. When studied in controlled settings, returning to a previously rewarding location is referred to as adopting a win–stay strategy, and avoiding a previously rewarding location is referred to as a win–shift strategy (Burke *et al.* 2002). Echidnas were better able to learn to avoid a previously rewarding location (to 'win–shift') than to learn to return to a previously rewarding location (to 'win–stay'), at short retention intervals (2 minutes) but were unable to learn either of these strategies at retention intervals of 90 minutes.

Some comparative psychologists have claimed that we can test animals for general problem solving and associative-learning abilities with 'unnatural' tasks in 'arbitrary environments', i.e. using laboratory tests (Plowright *et al.* 1998), and Doré and Dumas (1987) have argued that using a Piagetian perspective on cognitive development is, despite its apparent anthropocentrism, truly comparative. However, laboratory tests carry the risk of posing problems in an 'unfair' manner, because of certain perceptual or cognitive predispositions of the animals being tested (Roth and Dicke 2005). This must be particularly so for monotremes, whose sensory modalities include electroreception and a strong reliance on olfaction.

Many recent papers have stressed the importance of field-based studies:

> ... there is much to be gained by doing research on cognitive processes as they are used in animals' natural habitats ... although there are many obvious advantages to the controlled environment of the laboratory, who knows what we lose when we measure cognitive processes under such restricted and unnatural (for the animal) conditions? Although field work is expensive and difficult to do, we should do more of it. (Hulse 2006)

> Animal cognition is missing an important stream of evidence for cognitive modelling if it treats field data as second rate or preliminary: carefully recorded observations from the field can and should be used to build and test theories, and are necessary to enable appropriate, ecologically valid experiments to be designed where experimentation is subsequently possible. (Byrne and Bates 2011)

Roth and Dicke (2005) argue that intelligence may be defined and measured by the speed and success of how animals, including humans, solve problems to survive in their natural and social environments. Thus in order to understand the evolution, structure and function of the monotreme nervous system, it is important to understand their ecology and behaviour in their normal environments.

General biology of the monotremes

The extant monotremes share several general features that must reflect their common ancestor (see Chapter 1) and constrain their ecology and behaviour. All monotremes lack teeth and their diets consist principally of invertebrates. Most power is supplied by the front limbs: the forefeet of short- and long-beaked echidnas are used for digging and breaking open logs, ant nests or termite mounds and platypuses use their front limbs for both swimming and digging. Monotremes also have normal body temperatures of 30–32°C and low metabolic rates, and are long-lived for mammals of their size. Metabolic and life history details of the monotremes are summarised in Table 2.1. Basal metabolic rate (BMR) is generally believed to be an indicator of metabolic capacity (White and Seymour 2005). Although there has been debate about how the metabolic constraints on life history variables relate to BMR (Johnson et al. 2001; Mueller and Diamond 2001), in placental mammals energy expenditure on reproduction is positively correlated with energy expended on maintenance, such that high maintenance species harvest more energy and expend more on reproduction than low-maintenance species (McNab 2002) while a low BMR optimises longevity.

The metabolic and thermal physiology of monotremes has been a long-standing area of investigation. The metabolic rates measured for short-beaked echidnas by Martin (1903) are very close to those found by a range of more recent investigators (reviewed by Nicol and Andersen 2007), ~30% of those predicted for placental mammals of the same mass (Dawson et al. 1979) – or 40% of the value predicted from the more representative placental dataset used by Weisbecker and Goswami (2010) – and the long-beaked echidna has an even lower metabolic rate (Dawson et al. 1979; McNab 1984).

The platypus metabolic rate, although low, is similar to that of many marsupials and placental mammals of the same mass, and platypus are continuously active for many hours in thermally challenging cold-water environments (Bethge et al. 2003). Short-beaked echidnas show much lower levels of activity and further reduce energy expenditure by employing torpor and hibernation, but there is no evidence that long-beaked echidnas or platypus employ torpor (Nicol et al. 2008). The low body temperature of monotremes, and consequent likelihood of overheating in warm environments, would be expected to be important in affecting monotreme distribution and behaviour. However, short-beaked echidnas, although they have no sweat glands (Augee 1976) and no means of increasing evaporative heat loss, are the most widely distributed mammal species in the Australasian ecozone. The other two monotreme species possess sweat glands (Grant and Dawson 1978; Griffiths 1978) but have a much more restricted distribution.

Distribution and diet

Platypuses occur in lotic (flowing rivers and streams) and lentic (still ponds and lakes) permanent

Table 2.1. Life history data for extant monotremes

Short-beaked echidna and platypus data from the review by Nicol and Andersen (2007), Grant (2004) and Williams et al. (2012), long-beaked echidna data from Dawson et al. (1979) and McNab (1984).

	Short-beaked echidna	Long-beaked echidna	Platypus
Adult mass (kg)	2–7	5–15	0.9–2.5
No. of young	1 (2*)	1	1–2 (3*)
Time to reach adult mass (years)	4–5		1
Age at sexual maturity (years)	4–6		2–3
Age at weaning (days)	145 (Tas) 210 (KI, WA)		100–120
Mass at weaning (kg)	0.7–2.1		0.9 (NSW)
Life span (years)	45–50?		Male 11 Female 21
BMR (kJ.kg^{-1}.d^{-1})**	84 (45%)	60 (39%)	178 (71%)
FMR (kJ.kg^{-1}.d^{-1})	240		1033

* very rarely; ** percentage of placental value as calculated from the regression for placental mammals (Weisbecker and Goswami 2010); BMR = basal metabolic rate; FMR = field metabolic rate.
Tas: Tasmania; KI: Kangaroo Island; WA: Western Australia; NSW: New South Wales

freshwater environments in the Australian east, from Cooktown in north Queensland to Tasmania, and are common in water bodies east of the Great Dividing Range. Mitochondrial DNA reveals that there are two major clades: one from mainland Australia and another from Tasmania and King Island (Gongora et al. 2012). Within the mainland there is a genetically divergent lineage north of the Burdekin River in north Queensland, and platypuses above this break are also much smaller (Gongora et al. 2012; Kolomyjec 2010) (see caption for Fig. 2.1). Platypus distribution is limited by the availability of permanent water and, in the northern part of its range, by maximum air temperature (Klamt et al. 2011; Kolomyjec 2010). It is possible that a limiting factor is dissipation of metabolic heat produced during active swimming, when heat production increases by four times over basal (Bethge et al. 2001), in which case water temperature during the warmest period would be the proximate limiting factor. However, higher rates of heat production have been measured in walking platypuses (Bethge et al. 2001) and ability to travel between refuge water bodies without overheating could be a significant constraint. Much smaller body size at the northernmost part of its range is consistent with temperature being an important factor in limiting platypus distribution. Dependence on permanent water, and intolerance of high temperatures mean that some platypus populations may be at considerable risk from a warming climate (Klamt et al. 2011; Kolomyjec 2010).

Platypuses prefer to inhabit areas with river or stream banks consolidated by roots of vegetation where they construct resting and nesting burrows (Grant and Temple-Smith 1998). The burrows provide shelter from predators and create a microhabitat with a constant temperature, protecting the platypuses from large ambient temperature variations and preventing heat loss and unnecessary energy expenditure (Bethge et al. 2004; Grant 1983). Burrows are often found in undercut stream and riverbanks with plenty of ground vegetation and riparian vegetation overhanging the bank and water (Serena 1994; Serena et al. 1998). The overhanging vegetation provides shelter from predators during foraging and when the platypus is entering or leaving the water. Vegetation also controls bank erosion and provides shading, food and habitat for in-stream macroinvertebrate

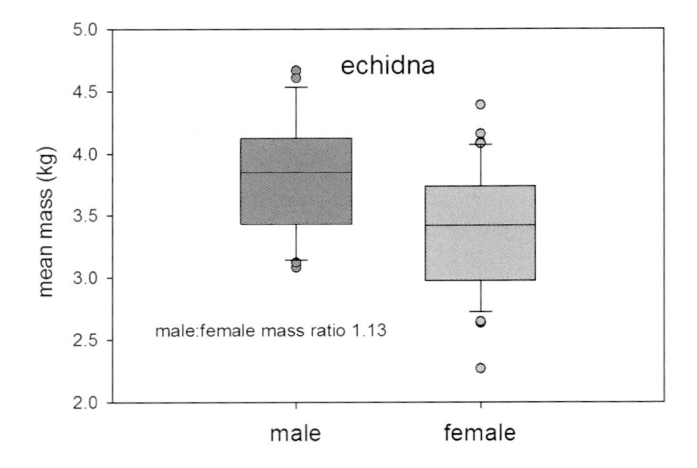

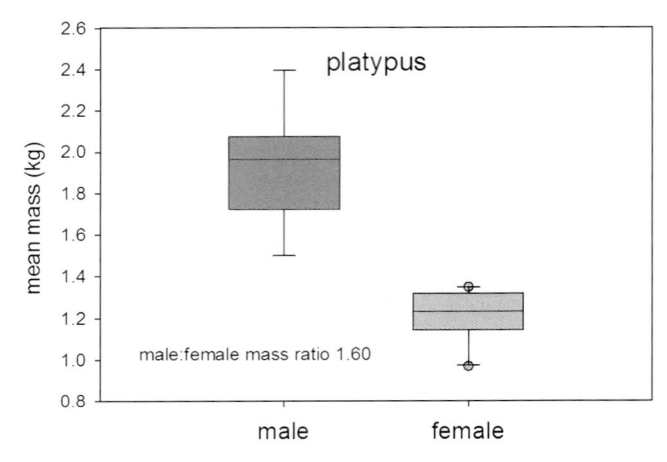

Figure 2.1: Sexual size dimorphism in adult monotremes. Both datasets are from Tasmanian studies; echidnas 25 M and 45 F (Nicol et al. 2011), platypus 11 M and 9 F (Olsson Herrin 2009). Echidna male to female mass ratio is 1.13:1, although in practice echidna sexual dimorphism is even less marked as both sexes show a large annual cycle of body mass (males ± 15% and females ± 20%) (Nicol and Morrow 2012) and the echidna data in this graph represent long-term mean masses for individuals over 4–15 years. For the platypuses, male to female mass ratio is 1:1.60. Serena and Williams (2013) found essentially the same mean masses and mass ratio in two Victorian populations, but for a northern Queensland population mean female mass was only 0.754 ± 0.084 kg and male mass 1.12 ± 0.198 kg, with a male to female mass ratio of 1.47:1 (Kolomyjec 2010). Box boundaries represent 25th and 75th percentiles, the line shows the median, errors bars indicate the 90th and 10th percentiles, and symbols show outliers.

organisms consumed by the platypus (Grant and Temple-Smith 1998; Serena et al. 2001). In some areas, predominantly in Tasmania, burrows have been found above ground in scrub and dense vegetation (Otley et al. 2000) and even in caves (Munks et al. 2004).

The platypus is an opportunistic feeder with a diet consisting mainly of benthic macroinvertebrates, especially insect larvae (Faragher et al. 1979; Grant and Temple-Smith 1998; McLachlan-Troup et al. 2010; Olsson Herrin 2009; Serena et al. 2001), although free-swimming species such as shrimps, crayfish, beetles, water bugs and tadpoles may also be taken. Captive platypuses have been observed to catch and consume small fish (Grant and Temple-Smith 1998) and platypuses from the Plenty River and Lake Lea respectively in Tasmania had fish scales and small fish otoliths in their cheek pouches, while a sample from one male included a small but intact trout jawbone (Olsson Herrin 2009). Most commonly, platypuses feed by sweeping their bill from side to side in the substrate, with their eyes, ears and nostrils closed, and locate dietary items with the electro- and mechanoreceptors in the bill (Fjällbrant et al. 1998; Manger and Pettigrew 1995; Pettigrew 1999; Proske and Gregory 2004). Food items are stored in the cheek pouches and then brought up to the water surface to be masticated (Grant 1989; Grant and Temple-Smith 1998).

The short-beaked echidna is the most widely distributed Australasian native mammal (Augee 2008) and occurs from New Guinea to Tasmania. The species *Tachyglossus aculeatus* has been divided into five subspecies on the basis of geographic distribution, the relative amounts of spines and hair, and the claws on the hind feet. Modified hairs form strong sharp spines, and in most subspecies hair is present between the spines. The hairiest of the echidnas (*T. a. setosus*) occurs in Tasmania and the Bass Strait islands, and the fur often hides most of the spines, while *T. a. acanthion*, which is found in the Northern Territory, northern Queensland, inland Australia and Western Australia, is nearly hairless on its dorsal surface, with a pelage consisting of spines and bristles (Augee et al. 2006; Griffiths 1989). The claws on the hind feet are used for grooming, and the lengths of the claws vary with the relative amounts of hair and spines. As well as these morphological differences there are also important differences in other aspects of their biology, such as the aspects of maternal care, lactation period and use of hibernation.

Although the echidna is usually described as being myrmecophagous, i.e. with a diet consisting principally of ants and termites, studies in New England (NSW) (Smith *et al.* 1989), the Strathbogie Ranges (Vic) (Harrison 1997) and at our study site in the Tasmanian midlands (Sprent 2011) found that underground larvae of cockchafer beetles and Lepidoptera ('pasture grubs') occurred in up to 50% of scats and formed a significant proportion of the diet. Dietary data from across Australia suggests that in areas where sufficient termites are available they form the majority of the diet, with a smaller proportion of ants (Griffiths 1978). Where termites are absent or in low availability ants are the most common dietary item, but these are supplemented with pasture grubs (Sprent and Nicol unpublished observations). Concentrated food sources, such as termite mounds or galleries, large ant nests, or ant nests in rotting logs, are vigorously attacked and broken open using the front claws and the beak, which is inserted into small gaps and used to prise them apart. When the food is more dispersed, echidnas will feed by probing the soil with their beaks, often leaving a trail of characteristic beak-shaped holes in the ground.

Although the genus *Zaglossus* has been divided into three species with several subspecies on the basis of morphology (Flannery and Groves 1998) there is so little information on their field biology that I will treat them here as a single species, and refer to it as *Zaglossus*, or the long-beaked echidna. Long-beaked echidnas are classified as Critically Endangered on the IUCN list of threatened species, and are listed in Appendix II of CITES (Leary *et al.* 2008). They are confined to altitudes between 600 and 4150 m in the New Guinea highlands, where they are found in a range of ecosystems from montane forests to alpine grasslands (Flannery and Groves 1998; Opiang unpublished observation; Wilson and Reeder 2005), although the discovery of a museum specimen of a long-beaked echidna collected in the West Kimberley in 1901 raises the possibility that they may still survive on mainland Australia (Helgen *et al.* 2012).

Long-beaked echidnas are largely nocturnal, and rarely active during the day (Opiang 2009). During the day they shelter in dens often more than 0.5 m below the ground and accessed by tunnels several metres long. They feed by probing their long beak into leaf litter and soil and by using their strong claws to tear open rotten logs. In some areas at least earthworms appear to be an important part of the diet: villagers in the Wharton Ranges claimed that long-beaked echidnas eat earthworms and scarab larvae, analysis of seven scats showed them to contain the chaetae of oligochaete worms, and the stomach of an animal from Mt Suckling also contained earthworms (Griffiths 1978). However, analysis of scats from long-beaked echidnas from the Crater Mountain Wildlife management area showed a range of soil- or log-dwelling arthropods, including mole-crickets, scolopendromorph centipedes, and scarab larvae (Opiang unpublished observation). In the only published study on the ecology of long-beaked echidnas based on long-term observation in the wild, Opiang (2009) found home-range areas of up to 168 ha; unfortunately there were insufficient observations in the study to compare home ranges of males and females.

Timing of reproduction

The platypus and short-beaked echidna and are both seasonal breeders. Platypuses breed in late winter and early spring but breeding times vary with latitude, occurring earliest in Queensland and progressively later in New South Wales, Victoria and Tasmania (Connolly and Obendorf 1998; Grant 1984; Grant 2004; Munks *et al.* 2004). In New South Wales, platypuses normally mate around August to October, lactating females are found from late September to early March, and the weaned young emerge from late January to March. The lactation period for platypuses has been estimated to be between 3 and 4 months for wild platypuses in New South Wales (Grant 2004), 114–127 days for platypuses in Taronga Zoo (Hawkins and Battaglia 2009), and at Healesville Sanctuary in Victoria 140–150 days (Fleay 1944) and 135–145 days (Holland and Jackson 2002). Grant *et al.* (2004) suggested that these differences may be due to nestlings in the wild being weaned more rapidly than those bred in captivity, but it is possible that there are real differences between populations from different geographic areas, as is the case with echidnas.

Short-beaked echidna breeding times also vary with latitude, but the pattern is the reverse of platypus:

breeding begins later in more northerly populations. Mating in Tasmania extends from early June until early September (Morrow et al. 2009), on Kangaroo Island between early June and early August (Rismiller and McKelvey 1996; Rismiller and McKelvey 2003), at Mount Kosciusko between late June and early August (Beard et al. 1992), and in south-east Queensland between mid-July and early September (Beard and Grigg 2000). A second mating in October following the loss of the first young has been observed in Queensland (Beard and Grigg 2000) and in Tasmania (Harris and Nicol unpublished observations). Lactation duration differs dramatically between geographic areas: in echidnas from Kangaroo Island and Western Australia, young are weaned at 200–210 days (Abensperg-Traun 1989; Rismiller and McKelvey 2003), in south-east Queensland 150–165 days (Beard and Grigg 2000), and in the Tasmanian midlands the lactation period is only 125–145 days, although young reach similar masses at weaning (Morrow et al. 2009, Morrow and Nicol 2013).

These differences in reproductive timing within and between the two species presumably reflect the necessity for maximum food requirement, which occurs in late lactation, to coincide with maximal food availability. By mating in mid-winter, Tasmanian echidnas ensure peak lactation occurs when ecosystem productivity is highest (Morrow and Nicol 2013). Even so, this is only made possible by a lactation period that is shorter than the other geographic subspecies. Platypuses have a higher metabolic rate than echidnas and Holland and Jackson (2002) estimated that daily food requirements in platypuses during late lactation were nearly 100% of body weight.

There is little information on long-beaked echidna reproduction, and it is not even clear whether they are seasonal or asynchronous breeders. Testicular samples from six long-beaked echidnas suggested that males become reproductively active in late June, reach a peak in July, and showed testicular regression in October (Griffiths 1978). Follicular development of two females collected in July and October also suggested seasonal breeding (Griffiths 1999), with the breeding season being similar to that of short-beaked echidnas. However, Opiang (2009 and unpublished observations) observed one female to be lactating in April 2002, April 2005, September 2011 and June 2012, while another was lactating in late October and early December 2010. Assuming that the lactation period is at least as long as Kangaroo Island short-beaked echidnas, this is difficult to reconcile with seasonal breeding and a mating period of June–September.

Mating system

All of the monotreme species are solitary and males have exceptionally large testes (Griffiths 1978; Rose et al. 1997), suggesting a promiscuous mating system with a high degree of sperm competition (Dixson and Anderson 2004; Preston et al. 2003). Monotreme mating systems are characterised by roving males (see Clutton-Brock 1989): male assistance is not required for rearing of young, and females are solitary. In such a mating system, both sexes are promiscuous and males range widely in search of oestrous females, defending them against other males (Clutton-Brock 1989). Consistent with this, one of the best known features of the echidna mating system is the occurrence of 'mating trains', groups of males following or clustered around a single female (Morrow et al. 2009; Rismiller and McKelvey 2000). On Kangaroo Island, up to 11 males have been observed in a train (Rismiller and McKelvey 2000), and at Dryandra in Western Australia seven males were found with a female (C. Cooper personal communication). Studies of echidna populations at Mount Kosciusko in the Australian Alps and at an elevated site in south-east Queensland never found more than one male with a female (Beard et al. 1992; Beard and Grigg 2000), leading Taggart et al. (1998) to suggest that there are fundamental differences in echidna courtship behaviour depending on climatic conditions, and that the absence of trains during courtship in eastern echidnas indicates that two different forms of mate selection are operating in echidna populations. However, multi-male mating groups are common at our Tasmanian study site (Morrow et al. 2009), and we have found groups with four males on several occasions. Most of these groups have been radio-tracked to shelters such as hollow logs, so are less obvious than groups or trains in the open, and in fact at our Tasmanian field site many matings occur in the female hibernaculum (Morrow and Nicol 2013). Although Rismiller and McKelvey (2000) claimed that Kangaroo Island females mate only once in a breeding season, and with the largest and lead male in the train, by taking swabs from the female reproductive tract to check for the presence of sperm we have shown that in our population females mate more than once, and with more than one male (Morrow et al. 2009; Morrow and Nicol 2009).

The roving male mating system in echidnas is further demonstrated by the fact that male echidnas have much larger home ranges than females. At our Tasmanian study site average home range (90% kernels) for males was 107 ± 48 ha, twice that of females (48 ± 28 ha) (Nicol et al. 2011), and while female home-range sizes were closely dependent on habitat quality, those of males were not (Sprent and Nicol 2012). Females appear to be distributed according to resource availability, and show only slight overlap of home ranges, while male home ranges cover those of many females, and substantially overlap each other.

Spatial organisation and communication between individual short-beaked echidnas appears to involve the use of latrines. Griffiths (1968) noted that caves in Western Australia 'contained pounds of echidna scats' and in the Northern Territory Griffiths et al. (1990) found collections of up to 35 individual scats. Large accumulations of echidna scats have been noted at several sites within Tasmania (Grove et al. 2006; Sprent et al. 2006). Latrines typically occupy an area of ~25 cm in diameter, cleared of vegetation and sticks under conspicuous features such as large logs, and are used continuously between August and February. One latrine cleared eight times over a 3-year period yielded more than 1 kg of faeces (Sprent et al. 2006).

Less information is available on the platypus mating system. Observations on captive platypuses suggest that mating is controlled by female receptivity (Hawkins and Battaglia 2009), with females having a single receptive period of 4–6 days, although Hawkins and Battaglia 2009 report that one female had two receptive periods 1 month apart. Platypuses commonly have two young (Hawkins and Battaglia 2009), and although there is one report of a female having young by different males in one season (Akiyama 1998) the evidence is equivocal (T. Grant personal communication). As with the echidna, home range use provides some insights into the mating system. Generally, home ranges of adult females overlap substantially, but in small streams adult male platypuses are intolerant of other adult males (Serena 1994). Platypuses feed in water bodies but use refuge burrows in the earth or vegetation on the banks of rivers and lakes when they are not foraging (Grant and Temple-Smith 1998; Otley et al. 2000); the amount of bank, and thus of potential burrow sites relative to foraging area, will vary with the type of water body, leading to variations in the degree of social tolerance between habitats. Gardner and Serena (1995) used radio-tracking to study the spatial organisation of platypuses in two southern Victorian streams, and found that while home ranges of only some males were mutually exclusive, where home ranges overlapped males avoided each other by spending most of their time in different parts of the shared area. All male home ranges overlapped the ranges of two or more adult females. In narrow lotic habitats, male platypuses engage in territorial defence, and population densities of only one or two individuals per kilometre of small waterway are the consequence (Gardner and Serena 1995). Where male home ranges overlap on river systems there appears to be temporal separation (Gust and Handasyde 1995). In broader habitats and weir systems, where prey availability is higher, population densities are much higher and territorial defence may not be a viable strategy (Grant and Carrick 1978). Consequently, platypuses tolerate more home-range overlap (Gust and Handasyde 1995; McLeod 1993). At Lake Lea, a shallow lake system in Tasmania, which provides a very large foraging area but a relatively small bank area, platypuses showed a high tolerance for sharing burrow areas. This seems to be partly accomplished through temporal separation, with some males being predominantly nocturnal, particularly during the breeding season (Bethge et al. 2009). Some of the platypuses at Lake Lea showed activity rhythms clearly linked to the lunar cycle (Bethge et al. 2009).

In male mammals, reproductive success is determined by both overt contest (pre-copulatory) competition and covert sperm competition (Preston et al. 2003). In short-beaked echidnas and platypuses, roving males compete for matings, but while platypus show significant sexual dimorphism, in echidnas sexual dimorphism is low (Fig. 2.1), suggesting that contest competition is more important in platypuses than echidnas. High sexual dimorphism is consistent with strong pre-copulatory competition between males (Weckerly 1998), and in the platypus the male crural system, which consists of a pair of venom glands connected by ducts to hollow spurs on each rear ankle, is used aggressively during the mating season and males are often found with punctures in their bodies, particularly in the tail region (Grant and Temple-Smith 1998). Although it has generally been believed to be vestigial (Augee et al. 2006) male short-beaked echidnas also have a functional crural system (Krause 2009; Morrow unpublished observations). Seasonal variation in the crural system appears to be controlled by testosterone and in both platypuses and short-beaked echidnas the crural glands reach their maximum size during the mating period (Grant and Temple-Smith 1998; Morrow unpublished observations). While platypus spurs are used in aggressive encounters between males the role in echidnas is less clear. To envenomate, the male platypus erects his spurs, wraps both legs around the victim, and then drives the spurs in, using repeated jabs to inject the venom (Temple-Smith 1973). The attachment of the echidna spur to the tarsus is much less rigid than that of the platypus (Hurum et al. 2006), and it is difficult to see how they could be used aggressively; while there are several records of humans being (extremely painfully) envenomated by platypus (Whittington and Belov 2007) in over 20 years of handling male echidnas during the mating period I have never seen any indication of the spurs being used aggressively, although they sometimes show traces of blood around the base (G. Morrow personal communication). Camera-trap recordings of echidnas in mating groups show that males spend a considerable amount of time grooming themselves using the grooming claws on their hind legs, and the location of the spur means that it is likely that during this process they also spread spur fluid on themselves, and possibly also on the female as well (Morrow unpublished observation).

Platypus venom contains a range of peptides including defensin-like peptides, c-type natriuretic peptides, and L-to-D-peptide isomerise (Koh et al. 2009), and while echidna venom contains fewer proteins and peptides it does contain a defensin-like peptide and L-to-D-peptide isomerise (Koh et al. 2011). The echidna crural gland transcriptome differs greatly from that of the platypus, with no correlation between the 50 most highly expressed genes, and the highly expressed genes in echidna crural glands appear to be mainly associated with steroidal and fatty acid production (Wong et al. 2012). While gas chromatography–mass spectroscopy of echidna venom showed no signs of steroids or fatty acids (R. Harris personal communication), the glands at the base of the spur and around the cloaca secrete a huge range of compounds that may be involved in chemical communication, including volatile carboxylic acids, aldehydes, ketones, fatty acids, methyl esters, ethyl esters, terpenes, nitrogen- and sulfur-containing compounds, alcohols, and aromatics. They also produce solid exudates containing long chain and very long chain monounsaturated fatty acids, sterols, and sterol esters. While there is a high degree of compositional overlap between compounds produced by male and female echidnas, there are significant variations and many of these compounds could be involved in mediating mating behaviour and other social interactions (Harris et al. 2012).

Mating behaviour

A detailed description of mating behaviour has been given for captive platypuses at Taronga Zoo (Hawkins and Battaglia 2009) which is in accord with more fragmentary observations in the wild and other captive observations (De-La-Warr and Serena 1999; Grant 1984). Mating has only been observed in the water, and begins with a period of pre-mating interaction, initiated by both male and female – nuzzling, swimming together, tail biting and spiralling as the female dives, pulling the male after her. The male forcibly curls his tail under the female's belly and clasps her tail between his hind feet. Once intromission is achieved the female continues to drag the male as she swims and they remain conjoined for up to 20 minutes (Hawkins and Battaglia 2009).

Detailed observations of echidna mating have been made on wild echidnas on Kangaroo Island:

> *When the female was receptive she responded to prodding by a male by lying flat on the ground with her spines in a relaxed position. If only one male was present, he would dig on one side of the female, stroking her spines with his forefoot and*

attempting to lift her tail with his hindfoot. When several males were with the female, all would dig beside her while attempting to push the other males aside. When one male remained, he continued digging and lifting until his tail was placed under the female's tail, cloaca on cloaca. Pairs remained coupled … for 30–180 minutes. (Rismiller and McKelvey 2000)

Similar behaviour has been noted from captive echidnas at Perth Zoo, which were housed as pairs (Ferguson and Turner 2012). In the weeks preceding mating the male would follow the female around, interacting with her for periods of less than 2 minutes. This was followed by a 6–9-day period of intense courtship behaviour culminating in mating, with the male repeatedly probing the female with his beak, pawing at her flank and tail, and attempting to dig under the female and lift her tail. Most matings at our Tasmanian field site occur in hollow logs or other retreats, and we have been able to record some using motion-sensitive infrared cameras, but late matings with active females are sometimes observed in the open. In late August 2012 a mating group with two males was found in an open area (R. Harris personal communication). One male had been found with the same female and at least two other males 3 days previously. This male remained motionless, lying on his belly, throughout the observation period. The second male was very active, and spent a few minutes sniffing the air and the female and walking around her before positioning his forelegs and head over her lower back and tail. He then used his hind legs to elevate the female's tail, and manoeuvred himself into a sitting position behind her, with his tail and hind legs under the female's tail, and one foreleg placed on her back. Duration of copulation was ~50 minutes.'

In Tasmania, mating behaviour is complicated by the fact that for females the hibernation and mating periods overlap (Morrow and Nicol 2009; Nicol and Morrow 2012). Reproductively active males arouse from hibernation in early June, and usually leave the hibernaculum almost immediately after becoming euthermic. After ~25 days of feeding they then seek out females, most of which are still hibernating – mean date of arousal from hibernation for females is July 23 (Nicol and Morrow 2012). Although the female may re-warm during mating, it is difficult to see how she may have any pre-copulatory choice. Males may guard the female for several days after mating, before looking for another mating opportunity, while if mating occurs before late July, the female is likely to re-enter hibernation, even if pregnant. Hibernation extends the gestation period by 1 day for every day of hibernation (Nicol and Morrow 2012), which means that although matings occur over a period of 13 weeks, egg-laying occurs over an 11-week period (Morrow et al. 2009).

Maternal care

During the breeding season the female platypus builds a complex nesting burrow lined with plant material in which she lays up to three eggs (Grant et al. 1983). In captive platypuses at Taronga Zoo, burrow building occurred over 4–6 days with females carrying nesting material in their tails to line the burrow (Hawkins and Battaglia 2009). During the first 20 days of burrow occupancy, which includes the 10 or 11-day period of egg incubation, the female may leave the burrow for very short periods to groom in the water, but not to feed (Hawkins and Battaglia 2009). From day 20 the mother starts to leave the burrow daily to feed, with the time spent feeding increasing until at 41 days she only visits the young every second day. Towards the end of lactation the mother is in the burrow for as little as 1.4 hours. When they are ~19 weeks of age and ~60% of adult mass, the young emerge from the burrow and begin feeding on the normal adult diet, although for some females there may be an overlap of ~10 days during which the mother also feeds the young (Hawkins and Battaglia 2009), and a juvenile platypus captured in the Shoalhaven River had milk in the stomach and invertebrates in the cheek pouches (Grant 2007). Juveniles caught in the upper Shoalhaven River between 1 and 2 months after emergence showed a significant sexual dimorphism in body mass (males 779 ± 127 g; females 588 ± 88 g) (Grant and Temple-Smith 1998).

Maternal care differs between the echidna subspecies (Morrow et al. 2009). One key difference in maternal care between the geographical subspecies occurs during the first weeks of lactation. Echidnas from Kangaroo Island (KI), Western Australia and New South Wales may forage while they have an egg in the pouch, and deposit the young in a nursery burrow when it is between 45 and 55 days of age (45–50 days on KI, 50–55 days in WA) (Abensperg-Traun 1989; Griffiths 1978; Rismiller and McKelvey 2009). In contrast, echidnas from south-east Queensland, Mt Kosciusko and Tasmania enter freshly built nursery burrows before egg-laying and remain there while incubating their eggs and during the first few weeks of lactation (Beard and Grigg 2000; Beard *et al*. 1992; Morrow *et al*. 2009). Echidna nursery burrows are simpler than those of platypus, and unlined (Morrow and Nicol 2013). Mothers plug the burrow with earth when they leave to feed and, as lactation progresses, they are absent for increasingly long periods, and spend less time with the young during feeding. In late lactation (100–150 days after the young has hatched), Tasmanian echidna mothers are typically absent for 6 days, and spend less than 3 hours with the young while feeding it (Morrow and Nicol 2013). After depositing the young in the burrow Kangaroo Island mothers returned at 5-day intervals and stayed with the young for ~2 hours (Rismiller and McKelvey 2009).

Although it has been believed that weaning of echidna young is abrupt, with the young beginning to forage on their own only after being abandoned by the mother (Augee *et al*. 2006) this does not seem to be the case. A Victorian echidna that successfully raised twin young fed them for 60 days after they first emerged from the burrow (Pierce *et al*. 2007). At our Tasmanian study site, young echidnas began foraging outside the burrow when they were between 125 days and 145 days old, and had a body mass exceeding 1 kg. Foraging trips lasted for 1–4 hours and the mother was never present. This behaviour lasted for a period of ~12 days with the mother typically returning to the nursery burrow twice to feed the young during this period. At the end of this period the young abandoned the burrow and did not have any further interaction with the mother (Morrow and Nicol 2013).

It is not surprising that weaning in echidnas and platypuses is not abrupt; the transition from milk to invertebrates, particularly ants, which have a very low nutritional value (Redford 1987), requires significant adjustments to the digestive system (see Chapter 12), and the young must take some time to learn the skills of feeding themselves efficiently. Even so, most echidna young at our field site lose body weight after weaning. Similarly, compared to their relatively high body fat reserves after weaning, juvenile platypuses in the upper Shoalhaven River in New South Wales had very low body fat following their first winter (Hulbert and Grant 1983). It is noteworthy that there is no evidence for a period of maternal instruction for either platypus or echidna.

Conclusions

One of the aims of this chapter was to provide some ecological and behavioural context to why the monotremes possess relatively large brains. Brain tissue is energetically expensive and requires nearly an order of magnitude more energy per unit weight than several other somatic tissues during rest (Mink *et al*. 1981). The energy usage of monotreme brains can be estimated from brain mass, using the equations from Hofman (1983), and then adjusting brain metabolic rates from the placental T_b of 38°C to the monotreme value of 32°C. These calculations estimate that the percentage of basal oxygen consumption used by the brain is 5.8% for the platypus, 8.5% for the short-beaked echidna, and 9.5% for the long-beaked echidna. Most mammals lie in the range from 2 to 8% (the mean value for 240 mammals is 4.6%) with only primates and cetaceans having values above 8% (Hofman 1983). However, these estimates depend on several assumptions about the scaling of brain metabolism, and a more direct analysis is provided by simply plotting brain mass as a function of BMR (Fig. 2.2). This graph again shows that the monotremes, because they have brain sizes similar to those of the relatively large-brained placentals but very low basal metabolic rates, have very large brains relative to their metabolic rate. One possible explanation for this is that additional cognitive capacity is required for processing input from electroreceptors, but this seems unlikely, as the platypus, with a lower brain:BMR ratio, has 40 000 electroreceptors, while the long-beaked echidna has 2000 and the short-beaked echidna 400 (Pettigrew 1999).

The expensive tissue hypothesis states that an increase in brain size must be accommodated by an increase in total metabolic rate or by a reduction of the demands of the other expensive organs, such as heart, liver, kidney and gastrointestinal tract (Aiello and Wheeler 1995). Thus it is argued that the relatively

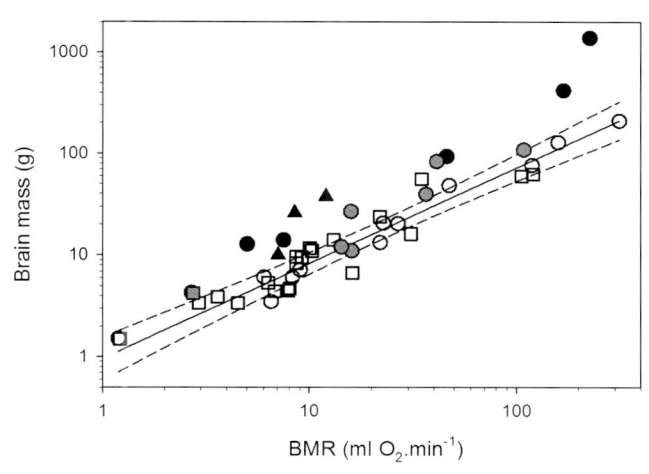

Figure 2.2: Brain mass as a function of basal metabolic rate in mammals. Both BMR and brain mass scale with body mass, and the relationship between brain mass and BMR illustrates the relative investment of energy in the brain. The regression line and 99% confidence limits have been fitted to the data for 12 non-primate, non-myrmecophagous terrestrial placental mammals from 11 families and nine orders. The two echidnas (*Tachyglossus aculeatus* and *Zaglossus* sp.) and the primates (*Perodicticus potto*, *Nycticebus coucang*, *Macaca mulatta*, *Pan troglodytes*, *Homo sapiens*) lie well above the upper confidence limit. The giant anteater (*Myrmecophaga tridactyla*), the hairy-nosed wombat (*Lasiorhinus latifrons*) and tamandua (*Tamandua tetradactyla*) also lie above the upper confidence limit. Circles: placental mammals, data from McNab and Eisenberg (1989); squares: marsupials, data from Ashwell (2008b); Weisbecker and Goswami (2010). Triangles: monotremes, data from McNab and Eisenberg (1989), Nicol and Andersen (2007). In order of increasing brain mass they are platypus, short-beaked echidna, long-beaked echidna. Solid shading: primates; grey shading: myrmecophages; open symbols: other terrestrial mammals. Body masses have been restricted to a range of 250 to 70 000 g to avoid any effects associated with very low or very high body masses. Metabolic rates were recalculated for the body mass used for brain mass measurement assuming a within-species exponent for metabolic rate of body mass of 0.70 (Kozlowski and Konarzewski 2005; Sieg *et al.* 2009; White and Seymour 2005). To minimise errors associated with this correction, datasets have been restricted to cases where body masses for BMR measurement and body masses used for brain weight were within 30% of each other. Under these circumstances, any errors in BMR correction are trivial.

large brain sizes of humans and other primates could not have been achieved without a shift to a high quality diet, allowing a reduction in gut size. It is doubly puzzling then that the short-beaked echidna has a brain of similar size to that of a similar sized placental carnivore but a metabolic rate only 45% of the placental prediction (Table 2.1), and has a diet of extremely low energy density and digestibility. In placentals there is a positive correlation between brain size and BMR, but this relationship does not hold for marsupials (Weisbecker and Goswami 2010). It may be that it is only in the large-brained primates that the brain requires a sufficiently large fraction of total body oxygen for this to be a constraint on natural selection (Weisbecker and Goswami 2011). However, the echidnas have brain size to BMR relationships similar to those of primates, suggesting that there must be very considerable fitness benefits for the echidnas to maintain such large brains, i.e. the cognitive benefits must outweigh the metabolic costs (Isler and van Schaik 2006).

The fitness benefits must be considerable for short-beaked echidnas, because the species seems to be specialised to minimise energy expenditure, and many aspects of their ecology and behaviour are correlated with small brain size in other mammals. Insectivorous placental mammals have smaller brains than carnivores and omnivores (Gittleman 1986), possibly because a larger brain may be necessary to handle a resource that requires more complex foraging strategies, and for a range of birds and non-primate mammals large relative brain size is associated explicitly with pair-bonded monogamy (Dunbar and Shultz 2007), while the monotremes are all solitary. However, there are two aspects of the life history of monotremes that are consistent with correlations seen with large brain size in other mammals: age at weaning (Weisbecker and Goswami 2010) and longevity (González-Lagos *et al.* 2010). However, longevity is also correlated with a low basal metabolic rate (Hofman 1983; White and Seymour 2004).

Dunbar and Shultz (2007) distinguish between constraints on brain evolution and selective pressures. Thus the positive correlation between brain size and age at weaning for both placentals and marsupials (Weisbecker and Goswami 2010) may reflect a metabolic constraint, and the long lactation period of monotremes may mean that the absence of this constraint allows other selective agents to operate. The selective agents are likely to be consistent with the cognitive buffer hypothesis – a large brain facilitates construction of behavioural responses to unusual, novel or complex socioecological challenges (González-Lagos *et al.* 2010; Sol 2009). It could be argued that its widespread distribution is consistent with the short-beaked echidna having flexible behaviour as well as dietary requirements that can be met in essentially all Australian terrestrial environments.

The large brain to BMR ratio in the monotremes, particularly the echidnas, suggests that their behaviour, particularly their social interactions, are more complex than previously thought, and this is supported by the echidna's use of latrines and complex chemical signals (Harris *et al.* 2012). Tantalising insights into the complexity of the behaviour of the short-beaked echidnas are now being provided by the use of GPS loggers and camera traps (Nicol unpublished observations). As well as providing us with details of foraging behaviour and spatial ecology, they also show details of male–female and male–male interactions. Although logistically and technically more difficult to use in the New Guinea highlands, these techniques are currently being trialled with long-beaked echidnas. Platypuses present a different set of problems, but more information on their behaviour in the wild is being slowly accumulated by a small number of dedicated field researchers.

Acknowledgements

I would like to thank the following people who have provided unpublished data and helpful discussion: Christine Cooper, Rachel Harris, Gemma Morrow, Muse Opiang, Melody Serena, Jenny Sprent and Phil Withers. Thanks to Rachel Harris, Tom Grant and Gemma Morrow for commenting on the manuscript.

3

Embryology and post-hatching development of the monotremes

Ken W. S. Ashwell

Summary

Although the embryology of the monotremes has been studied for more than a century, the difficulties of obtaining embryonic and hatchling specimens and the problems of correlating embryonic stages with the time since conception have hindered the establishment of precise timetables of monotreme development. Monotreme development is naturally divided into intrauterine, incubation and post-hatching phases, lasting between ~3 and 4 weeks, 10 days, and between 4 and 5 months, respectively. Like marsupials, monotremes have an extended intrauterine phase of development, and are hatched with immature forebrains, but also with a suite of special adaptations to facilitate survival in the nest or pouch. The newborn monotreme (puggle) has a brainstem sufficiently mature to handle the requirements of lung ventilation and the limited movements required to maintain position in the nest or pouch and locate the mothers nipple-less mammary areolae. Post-hatching life also involves a protracted period of dependency, during which the young monotreme relies on milk from maternal mammary glands, while gaining fat reserves, establishing thermoregulation and achieving central nervous system maturity. Segmental organisation of the brain appears to be similar in monotremes to therians, but gene expression patterns have not been analysed in the monotreme nervous system.

The history of monotreme embryology

Much less is known about monotreme reproduction than any other group of mammals (Griffiths 1999). In large part this arises from the difficulty of breeding monotremes in captivity. All three genera of monotremes have been maintained in captivity since the early 20th century and yet very few monotreme young had been bred and raised in zoos and wildlife parks until the late 20th and early 21st century. Although there are identifiable annual breeding seasons (see Chapter 2), individual females appear to breed unpredictably, perhaps due to the temporal and spatial patchiness of key resources. The result is that, even in the 21st century, we have only a vague knowledge of key features in monotreme development, in particular the timing of major developmental events (see Figs. 3.1 and 3.2 and Appendix Figs. 1, 2 and 3).

The major focus of scientific study of monotreme reproduction for the first century after European colonisation was the question of whether monotremes were oviparous, ovoviviparous or viviparous, and this was not resolved until Caldwell's work culminating in the famous 1884 telegram 'Monotremes oviparous, ovum meroblastic' sent to the British Association meeting in Canada (Temple-Smith and Grant 2001), followed shortly afterwards by his 1887 publication (Caldwell 1887). Coincidentally with Caldwell's study, Haacke also reported the discovery of an egg in the pouch of an echidna (Haacke 1885).

In the late 19th and early 20th century, several thousand specimens illustrating the development of echidna and platypus were collected and became part of important collections. The most important of these is the JP Hill collection (Carter 2008; Richardson and Narraway 1999), which was housed in the Hubrecht International Embryological Laboratory in Utrecht during the late 20th century, before moving to the Museum für Naturkunde in Berlin in 2004 (Giere and Zeller 2005). The magnitude of the harvest of embryological material and manner in which the tissue was collected would not be acceptable to the modern Australian public, so this sectioned embryological material remains an important scientific resource that is unlikely to be matched until captive breeding of monotremes becomes routine. Our knowledge of the development of the monotreme nervous system is largely based on analysis of these European-based embryological collections, with some supplementary information from collections of intact embryos and hatchlings in Australian museums (National Museum of Australia, Canberra; Australian Museum, Sydney).

Analysis of the European embryological collections has resulted in staging systems for monotreme embryos and hatchlings (Semon 1894; Flynn and Hill 1942; Werneburg and Sánchez-Villagra 2011), but these are naturally based on a small number of available specimens and therefore have limitations. All the specimens are wild-caught and are of unknown timing relative to the maternal reproductive cycle. As noted by Hughes and Hall (Hughes and Hall 1998), such archived material is of limited value in comparison with the detailed studies of timed embryological sequences for placental species, but remains the best available source of information on monotreme development.

General features of monotreme reproduction

Monotremes are distinguished from other mammals by oviparity and by the possession of a true cloaca. They have hair and mammary glands, but the latter do not present nipples. The female reproductive tract opens into the cloaca and there are left and right reproductive tracts, with each possessing an ovary, oviduct, uterus and cervix (Temple-Smith and Grant 2001). Only one side of the reproductive tract is functional in the platypus (the left), whereas both are functional in the short-beaked echidna. However, this does not limit the number of eggs produced by the female platypus, in that the platypus usually produces two ova (Burrell 1927), whereas the short-beaked echidna produces only one (Griffiths 1968, 1978). Insemination may be intrauterine (Temple-Smith and Grant 2001).

All the modern female monotremes build burrows during the breeding season, for the protection of the female and her young. Before egg-laying, the female short-beaked echidna develops a pouch (marsupium or incubatorium) into which the egg is laid directly from the protruded end of the cloaca (Temple-Smith and Grant 2001). The long-beaked echidna may also develop a pouch (Griffiths 1978; Temple-Smith and Grant 2001), but no pouch develops in the female platypus. The function of the pouch appears to be to maintain hydration of the leathery-shelled egg. In the platypus, this hydration appears to be achieved by the female bringing moist vegetation into the burrow.

Phases of monotreme development

Development of placental and marsupial young naturally divides into pre- and postnatal phases, reflected

Phase	Stage	Greatest length (mm)	Dorsal contour length (mm)	Head length (mm)	Age (days)	Key developmental events
Intrauterine phase	U	Not applicable	Not applicable	Not applicable	H-13	primitive streak formation
						3 to 4 somite pairs
					H-12	5 somite pairs neural groove formation
						10 somite pairs
					H-11	11 somite pairs
						12 somite pairs
Incubation phase	early pharyngeal arch In[a]	5.0	8.0		H-10	laying of egg, 18 to 20 somite pairs, trigeminal ganglion primordium is up to 2 mm diameter, anlagen of fore-, mid- and hindbrain are present
		6.0		2.0		otic vesicle formation
					H-8	optic vesicle, olfactory pit and forelimb bud formation
		7.0	10.0			five brain vesicles present; prominent midbrain flexure; five rhombomeres present
	late pharyngeal arch In[b]	8.0		2.5	H-6	
			12.0			primordia of os caruncle and egg tooth begin to appear
		9.0				hindbrain has a thin roof
						forelimbs have partially separated digits
		10.0	14.0	3.0	H-4	
prehatching	In[c]	11.0	16.0			
				3.5	H-2	manus is almost completely pronated
		12.0	18.0			prominent os caruncle and egg tooth
		13.0		4.0		
Post-hatching phase	PH	14.0	20		H-0/PH0	hatching
		16.0	30	6.0	PH2	forelimbs have five separated digits bearing blunt recurved epitricheal claws
		20	40	8.0	PH4	bill starts to elongate
		30	70	14.0	PH6	bill shields separate from skin of the head; webbing appears on forelimb
		40	90		PH10	webbing appears on hindlimb
		50	110	20	PH20	external ear reopens
		60		30		os caruncle is lost
		70	190		PH50	
		80		40		
		90	230			hairs on the back, head and forelimb
		100				forelimb is positioned below the body in preparation for knuckle-walking.
			290	50	PH100	
		150				hindlimb rotates towards adult position (complete by PH180)

Figure 3.1: Time line of major developmental events in the platypus. Dimensions of embryos and hatchlings are correlated with deduced developmental age and correlated with major developmental events based on Hughes and Hall (1998) and Manger et al. (1998a). The intrauterine phase is a period of relatively slow development, whereas the incubation phase involves not only the first appearance of the major organ primordia (pharyngeal arches, ear or otic vesicle, and eye or optic cup) but also their elaboration to a stage equivalent to the end of the embryonic phase in placentals (i.e. formation of major body parts, but before cellular differentiation).

Period	Phase	Stage	Greatest length (mm)	Dorsal contour length (mm)	Head length (mm)	Age (days)	Key developmental events
Intrauterine period	U	6	Not applicable	Not applicable	Not applicable		primitive streak formation
		7				H-13	3 to 4 somite pairs head region appears
		8					5 somite pairs neural groove formation mandibular arch bud forms
		9				H-12	10 somite pairs 2nd pharyngeal arch forms
		10					11 somite pairs
		11				H-11	12 somite pairs
		12					laying of egg 19 somite pairs
Incubation period	early pharyngeal arch In^a	13	5.0	8.0	2.5	H-10	21 somite pairs anterior neuropore closure
		14	6.0		3.0		otic vesicle formation forelimb forms as ridge
		15	7.0	10.0	3.5	H-3	optic vesicle, olfactory pit and forelimb bud formation
	late pharyngeal arch In^b	16					29 to 30 somite pairs and maxillary process reaches the anterior level of the lens placode
		17	8.0	12.0	4.0	H-6	forelimb is elongated and paddle-shaped
		18	9.0				39 somite pairs and lens vesicle is obvious
		19	10.0	14.0	5.0	H-4	digital plate forms at end of forelimb
	prehatching In^c	20					digital grooves appear in manus
		21	11.0	16.0			all pharyngeal clefts have closed
		22	12.0	18.0	6.0	H-2	os caruncle formation and the contour of the lens becomes obvious
		23	13.0				digits appear on manus
		24	14.0	20	7.0	H-0/PH0	hatching
Post-hatching period	PH	25	16.0	50	10.0	PH2	lower eyelid covers half the eye
		26	20			PH4	
		27	30	100	20	PH6	os caruncle declines
		28	40	120		PH10	
		29	50	140	30	PH20	poison sporn is visible on medial hindlimb marsupium dents on the belly
		30	60	160		PH50	snout develops a well-defined beak
		31	70	200	40		hairs on the back, head and forelimb
		32	80				first hairs on the belly
			90				
		33	100	270	60	PH100	first hairs on the hindlimb
			150				hairs cover the entire hindlimb

Figure 3.2: Time line of major developmental events in the short-beaked echidna. Developmental stages are correlated with greatest length, dorsal contour length and head length, deduced age and major developmental events based on Werneburg and Sánchez-Villagra (2011). The tempo of development in the three phases is broadly similar to the platypus, but specific differences within the nervous system will be discussed in later chapters.

in the usage of embryonic days (E) and postnatal days (P) to count days of development and compare developmental tempo in the scientific literature. In the monotremes, development may be divided into three distinct phases (intrauterine or gestational – U, incubation – In, and a post-hatching or lactational – PH) by the key events of the laying, and hatching, of the egg, respectively (Figs. 3.1, 3.2, 3.3 and 3.4). The latter two phases are more accessible to scientific enquiry than the former and more certainty exists concerning their duration. The so-called embryonic period of marsupials (actually pre-embryonic and embryonic phases) roughly corresponds to the intrauterine and incubation phases of the monotremes, whereas the embryonic period of placentals (actually pre-embryonic, embryonic and foetal in the case of most placentals) corresponds to both intrauterine and incubation phases of the monotremes, as well as the first week or two of the monotreme post-hatching phase.

Seasonal timing of monotreme reproduction in the wild

In both the platypus and echidnas, the timing of mating allows for the post-hatching growth of the young to occur during the time of maximal available resources (see also Chapter 2).

In the wild platypus, breeding times appear to occur earlier in the northern parts of its distribution and later in the southern. In New South Wales, platypuses mate in August or September (i.e. late winter and early spring) and young first appear in streams in January or February (i.e. mid to late summer) (Grant and Temple-Smith 1983; Grant *et al*. 1983; Grant and Temple-Smith 1998). This period of 20–23 weeks is probably made up of 3 or 4 weeks intrauterine phase, 10 days incubation and 15–18 weeks post-hatching or lactational phase.

In short-beaked echidnas, hibernation and reproduction may be closely intertwined and even overlap (Morrow and Nicol 2009; Nicol and Morrow 2012). This is particularly the case in cold climates and ensures that the maximum growth rate of the young corresponds to the time of greatest ecosystem productivity. In free-ranging Tasmanian echidnas (*Tachyglossus aculeatus setosus*) males enter hibernation in mid-February and females enter hibernation in mid-March. In reproductive years, males arouse from hibernation in early winter (June to July) and begin to seek females. Matings occur before females have completed hibernation (in July) and the females often return to hibernation both between matings and when already pregnant. Competition for females may be so intense that males will mate with torpid females (Morrow and Nicol 2009). Most egg-laying of Tasmanian echidnas occurs within 20–24 days of arousal from incubation, consistent with an intrauterine phase of 3–3.5 weeks duration, although some egg-laying occurs even earlier, consistent with females mating while torpid.

In the Australian Alps, short-beaked echidnas breed in July to August, a timeframe that is also coincident with the usual period of hibernation in this region (Beard *et al*. 1992). The difference from Tasmanian echidnas appears to be that mainland alpine echidnas do not re-enter hibernation immediately after mating: females will remain active after mating and only re-enter hibernation in the subsequent winter after the young are weaned. Egg-laying usually occurs at around the beginning of September, with hatching around the middle of September. Young echidnas remain in the burrow until about the middle of October and are dependent on their mother until late December.

Short-beaked echidnas of Kangaroo Island (*Tachyglossus aculeatus multiaculeatus*) also breed in the winter months (late June to late July). Egg-laying occurs 15–23 days after the males have dispersed from the area, with a suggested period between mating and egg-laying of 18–27 days (Rismiller 1992).

The question of whether the long-beaked echidna has a mating season remains open (see also Chapter 2). Griffiths (1978) concluded that *Zaglossus* do have a breeding season, probably in the months of June to July, but this conclusion is based on observations of reproductive organs from a very limited number of animals. More field studies are needed to clarify this (see Chapter 2).

Conception to egg-laying: the intrauterine phase

This is the least readily observed developmental phase and its duration is the subject of the greatest uncertainty. The intrauterine (or gestation) phase begins at insemination and ends at the laying of the egg. In the case of the platypus, best estimates of the

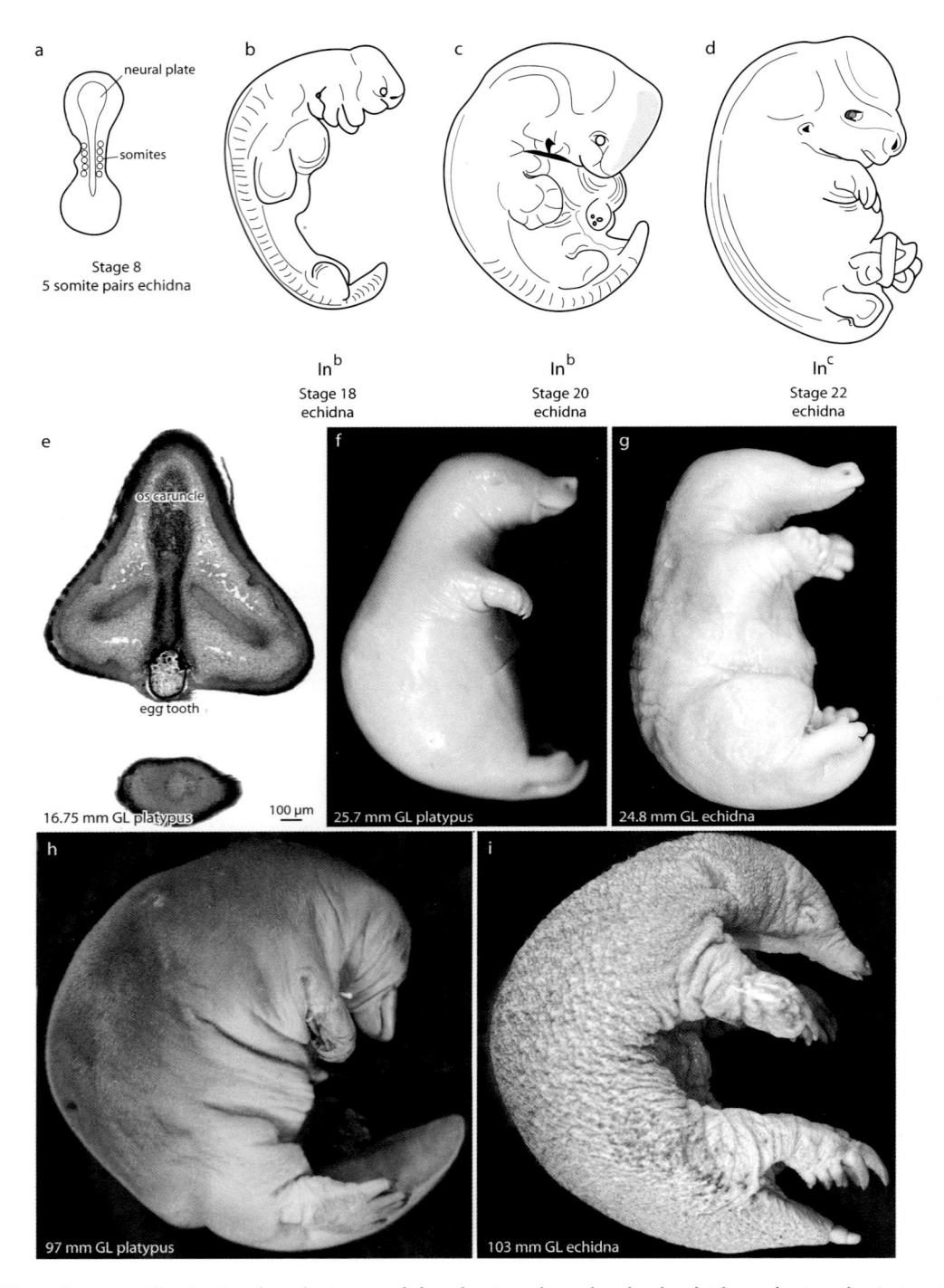

Figure 3.3: Line diagrams illustrating key features of developing short-beaked echidnas during the intrauterine (a), middle incubation (b, c), and late incubation (d) phases. Figure (e) shows the os caruncle and egg tooth in the snout of a recently hatched platypus, while (f) and (g) show spirit preserved hatchlings (~35 days after hatching, M5017 and M5014 of the Australian Museum, respectively). Note the pronation (rotation) of the forelimbs and the presence of epitricheal claws in both new hatchlings, contrasting with the poor development of the hindlimb. Figures (h) and (i) show preserved platypus (M2781, AustMus) and short-beaked echidna (M2165, AustMus) young of between ~10 and 11 weeks post-hatching. At this point the young echidna has left the pouch, but both platypus and echidna would still be confined to the nursery burrow.

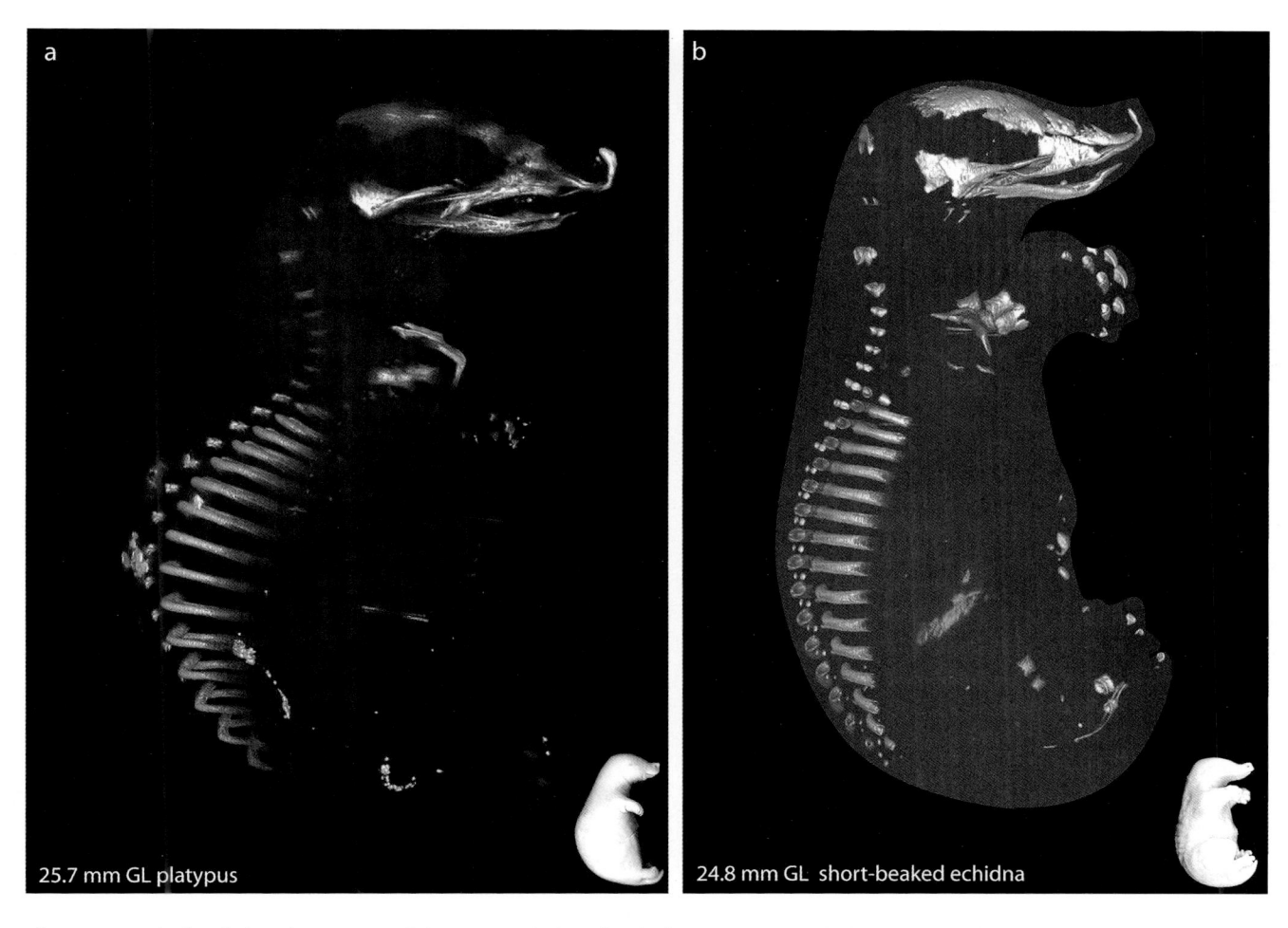

3 – EMBRYOLOGY AND POST-HATCHING DEVELOPMENT OF THE MONOTREMES 37

Figure 3.4: Skeletal development of the recently hatched platypus (a) and short-beaked echidna (b). These illustrations show the skeleton as revealed by computerised reconstruction following high resolution microCT of M5017 and M5014 of the Australian Museum collection. The insets in the lower right-hand corner show the matching lateral view photomicrographs of each hatchling. Note the os caruncle on the dorsal tip of the snout and the ossification of the phalanges of the distal forelimb.

duration of the uterine phase are based on captive breeding of platypuses at Healesville Sanctuary in regional Victoria (Holland and Jackson 2002) and Taronga Zoo in Sydney (Hawkins and Battaglia 2009), supplemented by deductions from the timing of changes in wild populations. The female platypus has a receptive period of ~46 days in the spring of each year and commences burrow preparations immediately after mating. The intrauterine period in the platypus has been estimated at 15–21 days (Holland and Jackson 2002; Hawkins and Battaglia 2009). In the short-beaked echidna, the intrauterine period is estimated as between 20 and 27 days (Rismiller and McKelvey 2000; Temple-Smith and Grant 2001).

Pre-embryonic and embryonic development during the intrauterine phase proceeds from cleavage of the monotreme zygote (stage 1; Werneburg and Sánchez-Villagra 2011) to 3 days after the formation of the primitive streak (Hughes and Hall 1998; stage 12 echidna, Werneburg and Sánchez-Villagra 2011). The pre-embryonic stage, by definition, proceeds from the zygote (stage 1 echidna, Werneburg and Sánchez-Villagra 2011) to the first appearance of organ primordia (stage 7 echidna, Werneburg and Sánchez-Villagra 2011). It begins with zygote cleavage to produce blastomeres, proceeds through the generation of endodermal and ectodermal layers, and concludes with the formation of the primitive streak and mesoderm to produce the three-layered embryo. The first definitive organ primordium to develop (at the end of the pre-embryonic stage) is the neural plate and folds and these are flanked by the first paraxial somites.

The final intrauterine stage of the platypus occupies the last 3 days before laying of the egg (Hughes and Hall 1998) and probably a similar period in the echidna (stages 8 to 12 Werneburg and Sánchez-Villagra 2011); but it should be noted that the 3-day figure is an estimation based on the relative paucity of post-primitive streak intrauterine embryos in European collections and the primordial state of organogenesis in monotreme embryos at the time of laying (Hughes and Hall 1998). The earliest embryo proper has a tadpole-shaped brain plate (4.7 mm wide and 4.0 mm long in the platypus) that is yet to develop obvious primary brain regions (Fig. 3.3a). The platypus embryo at this stage shows strikingly large trigeminal ganglionic primordia, lying alongside the prospective hindbrain (Hughes and Hall 1998), but these are not so prominent in the short-beaked echidna (Werneburg and Sánchez-Villagra 2011; see Chapter 9). Twelve somite pairs are reached by the end of the intrauterine period.

Development within the egg: the incubation phase

This period is estimated (with reasonable accuracy for both platypus and echidna) to last 10 or 11 days (Hawkins and Battaglia 2009, Renfree et al. 2009). The incubation phase can be conveniently divided into three subphases (Hughes and Hall 1998; Werneburg and Sánchez-Villagra 2011; Figs. 3.1, 3.2 and 3.3b to d; Appendix Figs 1 to 3): (1) an early pharyngeal arch subphase (stages 12 to 15; In^a) (2) a mid-incubation late pharyngeal arch subphase (stages 16 to 20; In^b); and (3) a late pre-hatching stage (stage 21 to 24; In^c). Each of these is probably between 3 and 4 days long, but this is based on an assumption of steady, daily progression through the stages.

The early pharyngeal arch subphase (In^a) is defined by the successive appearance of the first four pharyngeal arches, but also includes the initial morphogenesis of the auditory, visual and olfactory sensory apparatuses from the relevant placodes. Pharyngeal arches 1 to 3 make their first appearances successively between stages 12 and 13, with the fourth arch emerging at stage 14. The external bulging rudiment of arch 5 is a unique feature of the monotreme embryo (Hughes and Hall 1998), but the functional significance of this for the adult monotreme is unknown. The front opening of the neural tube (anterior neuropore) closes at stage 13 and the three primary brain vesicles (fore-, mid- and hindbrain) enlarge. The trigeminal ganglion is already prominent, particularly in the platypus, but other sensory apparatuses begin to emerge at this time. The otic pit develops at stage 13, becoming an otic (ear) vesicle at stage 14; whereas the optic (eye) vesicle grows out from the diencephalon at stage 15, coincidentally with the formation of an olfactory (nasal) placode, surrounded by medial and lateral nasal swellings (Werneburg and Sánchez-Villagra 2011). The forelimb and tail buds appear at the end of this subphase.

During the mid-incubation (late pharyngeal arch) subphase (In^b), all the pharyngeal arches are present and development of their neural, muscular and skeletal derivatives is underway. This subphase ends (at stages 20 and 21, Werneburg and Sánchez-Villagra 2011) when the pharyngeal arches become obscured by the disappearance of the intervening pharyngeal clefts, and the overgrowth of arches 3 and 4 by the second arch, so as to form a cervical sinus and give a smooth contour to the neck. In the brain, telencephalic vesicles expand from the forebrain and the cerebellar primordium, the rhombic lip, develops over the fourth ventricle. The lens vesicle appears at stage 18 (Werneburg and Sánchez-Villagra 2011). This subphase also encompasses the first development of the fore- and hindlimbs (Fig. 3.3b, c). The forelimb paddle develops partly separated digits by the end of this subphase; whereas the hindlimb foot-plate is less advanced, showing only the beginnings of digital ray modelling by stage 21. The primordia of the os caruncle and egg tooth also make their first appearances at the end of this subphase. These will be essential for breaking through the developmental membranes at the time of hatching.

The late pre-hatching stage (In^c) prepares the young monotreme for life in the nest or pouch (Fig. 3.3d). The manus (forefoot) plates of the forelimb become pronated (rotate to face towards the tail) and develop the primordia of epitricheal claws to facilitate grasping and clinging (digito-palmar prehension), whereas the hindlimb is still little more than a paddle.

The upturned snout develops a prominent os caruncle and the upper jaw grows a sharp, recurved egg tooth to break through the egg membranes (Fig. 3.3e). Fusion of the derivatives of the pharyngeal arches and frontonasal process is completed to form protective rims to the olfactory and auditory apparatuses and the oral cavity. In the central nervous system, some divergence between echidna and platypus emerges at this age. The platypus cerebral cortex remains relatively thin and unfolded until after hatching, whereas infolding and thickening of the iso- and allocortex of the echidna begins during this subphase (see Chapters 7 and 14 for more details), but these differences have no impact on the viability of the hatchling, because only hindbrain neural systems are essential for post-hatching survival. In all monotremes, the brainstem respiratory and cardiovascular centres must reach a sufficient state of maturation to control ventilation and cardiovascular function before hatching (see discussion in Chapter 5).

The largest dimension of the monotreme egg is ~18mm (see Appendix Table 1) so the newly hatched monotreme is believed to have a greatest length of between 14 mm and 15 mm (see Appendix Figs 1 to 3).

Challenges for the newly hatched

By the end of the incubation period, the young monotreme must be able to break through the egg membranes, but it does not face the challenge of migrating to the pouch against gravity that many young diprotodont marsupials do. It is therefore unlikely that newly hatched monotremes require a fully functional vestibular system, but actively mobile prehensile forelimbs would be necessary for maintaining position in the nest or pouch, or moving to the milk source. Like all mammalian young born in an immature state, the naked, newly hatched monotreme (Fig. 3.3f, g; Fig. 3.4a, b) is ill-equipped to deal with the problems of dehydration and thermoregulation, but these problems are obviated by the moist, warm environment provided by the mother in the nest (platypus) or her pouch (echidnas).

On the other hand, the peculiar structure of the monotreme mammary gland provides particular problems for the young monotreme. In newborn marsupials, the rostral snout and mandible are formed into an oral plate that allows firm attachment to the nipple. Unlike marsupials and placentals, monotreme mammary glands lack nipples, so the young monotreme is unable to form a long-standing bond with a nipple and must repeatedly locate the milk source. In any event, the presence of the egg-tooth in the oral cavity of the newly hatched monotreme would preclude attachment to a nipple. We know that young hatchlings grow rapidly; echidnas for example increase in body weight more than 30-fold in the first 10 days (Rismiller and McKelvey 2003), so feeding must be highly effective. On first principles, one would postulate that the newly hatched monotreme might be able to use a chemical sense such as olfaction, or a physical sense, such as trigeminal tactile or thermal sensation, to locate the milk source. The olfactory epithelium is present and may be connected with the forebrain primordium at birth, but none of the central structures necessary for processing olfactory information are present at hatching (see Chapter 11 for details). By contrast, the precursors of receptors in the skin of the bill or beak, trigeminal nerve pathways and nerve cells of brainstem trigeminal nuclei are present at hatching (see Chapter 9) and in close proximity with those brainstem gigantocellular neurons that provide reticulospinal pathways for controlling forelimb movements. This suggests that trigeminal sensation provides a more likely mechanism for sensing the milk source and mediating the areola-seeking behaviour.

Lick, sip or suck: how does the young monotreme access milk?

The young monotreme must survive on milk secreted by mammary glands that have ducts opening to the skin surface at two special areas of the skin surface (the mammary areolae). In both platypus and echidna the areolae lie on either side of the midline and in the echidna these are located within the pouch (Griffiths 1965). Growth rates for newly hatched monotremes are impressive (see below) and the milk ejection reflex is robust, such that young echidnas may take in 7–10% of their body weight in 30 minutes (Griffiths 1965). Griffiths estimated the intake for a captive pouch young echidna at 2 g of milk per minute for a 400 g puggle. Furthermore, the suckling is quite audible and reaches a frequency of 6 kHz, suggesting that this is a vigorous process.

Griffiths believed that the young echidna actively sucks milk, but the maternal areolae are nipple-less and the young have an oral fissure apparently unsuited to forming a suction seal (Fig. 3.3f, g). This should be contrasted with the oral opening of newborn marsupials, where the oral margin forms a tight seal around the base of the nipple and the tip of the nipple is held between palate and tongue for weeks of intense feeding. If licking movements were used, then a very different set of musculature would need to be developed to suckling. Therian young (marsupials and placentals) push the teat against the secondary palate with the front part of the tongue. It has been argued that the anterior or front parts of the pouch young marsupial tongue can move without attachments to the hyoid bone because of connective tissue attachments to the floor of the mouth (Smith 1994) and that the marsupial mode of suckling requires a more developed tongue musculature than in monotremes, but the differences may be more complex than that. Sticking out the tongue for licking requires an active genioglossus or circular intrinsic muscles. Similarly, tongue retraction to return milk to the oral cavity would require precocious function of the internal longitudinal, hyoglossus or styloglossus musculature, so muscles adjusting the position of the tongue would be expected to be more functionally mature in newly hatched monotremes who protrude the tongue than in newborn therians, in whom the tongue need only press the teat against the palate. This issue has not yet been investigated.

So does the young monotreme literally suck milk from the areolae, or does its nuzzling or licking of the areola stimulate a let-down reflex that produces milk to be sipped, sucked or licked from the areola surface? This question is unlikely to be answered for either the echidna or platypus without much closer and intrusive observation than is current practice in captive breeding programs.

Post-hatching life of the platypus

The young platypus develops in a nest, so its post-hatching life may be divided into a lactational period from hatching to weaning, and a juvenile period from weaning to sexual maturity. The lactational period has been estimated for captive platypuses at 135–145 post-hatching days (Holland and Jackson 2002), or 114–127 (Hawkins and Battaglia 2009). Young platypuses emerge from the burrow at ~130 days after laying of the egg (Holland and Jackson 2002 Hawkins and Battaglia 2009). In the wild, female platypuses reach adult size ~17 months after hatching, but male platypuses require a longer period (Grant and Temple-Smith 1983). The juvenile period sees an increase in body weight from ~1.0 to 2.0 kg.

At the time of hatching, the young platypus has a greatest length of less than 14 mm and closer to between 12 mm and 13 mm (Appendix Table 2). This is the maximum size that a pre-hatching monotreme could reach, because the monotreme egg is 15 by 17 mm (Hughes *et al.* 1975; Appendix Table 1). For the first 6 hours after hatching, the young platypus will have an attached yolk navel, as a remnant of the extra-embryonic yolk sac (Hughes and Hall 1998). At first, the external morphology of the newly hatched platypus is remarkably similar to that of newborn marsupials and shows little to suggest the aquatic life of the mature platypus, apart from the webbing ridge at the margins of the manus (forefoot) plate (Hughes and Hall 1998). The head is not dorsoventrally compressed as seen in the adult, but is more circular in outline as for marsupials. The other distinctive features of the newborn monotreme are the os caruncle and egg tooth (see Fig. 3.4). The egg tooth is shed within 2 days of hatching, probably because its continued presence interferes with feeding, but the os caruncle will remain as a protuberance on the snout until 11–14 weeks (Manger *et al.* 1998a). At birth, the forelimbs are pronated and the digits bear epitrichial, recurved claws, whereas the hindlimbs, although longer than the forelimbs, lack claws (Manger *et al.* 1998a).

The distinctive bill of the platypus develops by progressive flattening of the snout and jaw over the first post-hatching month and elongation of the bill occurs mainly in the second to sixth month (Fig. 3.3h; Appendix Table 2; Manger *et al.* 1998a). The epitricheal claws of the forelimb are replaced with true nails by PH2 (post-hatching day 2), and these elongate by PH4 (Manger *et al.* 1998a). True webbing first appears between the digits of the forelimb at PH7 and between the digits of the hindlimb at PH10 and reaches adult proportions by 6 weeks after hatching. The adult position of the forelimbs is attained at 6 months.

Post-hatching life of the short-beaked echidna

For the short-beaked echidna, post-hatching life can be further divided into three subphases (Rismiller and McKelvey 2003). The first of these is growth in the pouch (post-hatching stage a or PH^a extending from PH1 to PH55); the second is growth in the nursery burrow (PH^b extending from ~PH55 to weaning at PH205; Fig. 3.3i) the third is growth during subadulthood (PH^c extending from PH205 to sexual maturity at 3–5 years).

The newly hatched short-beaked echidna has a greatest length of 14 mm (dorsal contour length of 20–24 mm) and weighs ~0.3–0.4 g (Griffiths 1978; Rismiller and McKelvey 2003). Body weight increases 10-fold between PH1 and PH5 and 2.6 fold between PH5 and PH10 (Appendix Table 3). Daily weight increase is 0.8 g/day for the first 5 days and rises to 3.3 g/day by PH15. A body weight of 100 g is reached at ~PH30 and ~200 g is attained at the time of departure from the pouch (PH55) (Appendix Table 3). Much like the newly hatched platypus, the early echidna hatchling retains a prominent os caruncle (see Fig. 3.4b), but also has a distinct pouch on its abdomen. Structural changes during early pouch life include the loss of the os caruncle, development of the first true claws on the forelimb and emergence of the poison sporn on the hindlimb (Werneburg and Sánchez-Villagra 2011). Structural changes during later pouch life prepare the young echidna for life outside the pouch with the thermoregulatory demands that that entails. These include the accumulation of subcutaneous fat reserves, with the resulting extension of skin folds over the body and the emergence of hairs on the back, forelimb, neck, head, belly and hindlimb, in that order.

There is a wide variation in the body weight of young echidna at weaning (750 g to 2.2 kg), and this is strongly correlated with maternal weight (Rismiller and McKelvey 2003), highlighting the importance of maternal nutrition for the growth of young. Growth of juvenile echidnas is also variable, with some subadult echidnas losing weight between weaning and 1 year of age.

Comparisons with marsupials and placentals

Both monotremes and marsupials are born in an immature state and this naturally raises questions as to how similar the newborn of the two groups are (Griffiths *et al.* 1969). Monotremes and marsupials are similar in that both have the pre-embryonic stage of development occupying the major part of gestation (50–80%), with the embryonic stage occupying only 10–25% of gestation (Hughes and Hall 1998). Both exhibit precocious locomotor and respiratory adaptations to survival after birth, but some of these are quite different between the two groups. The newborn marsupial has an oral plate for attachment to the nipple, whereas the monotremes do not. Conversely, the peri-hatching monotreme has an os caruncle and an egg tooth (Fig. 3.3e) to enable it to tear through surrounding membranes at hatching, whereas these are absent from the marsupial head. Both monotremes and marsupials have strong forelimbs with epitricheal claws. In both groups, these facilitate emergence from membranes and allow the young of some marsupial families to climb maternal fur to the pouch. Both monotreme and marsupial young are also capable of digito-palmar prehension to enable them to grasp maternal fur to retain position in the pouch or nest.

Compared to human development, the state of development of the full-term intrauterine monotreme embryo exhibits: (1) retardation of formation of the neural folds; (2) retardation of definitive embryonic folding; and (3) delayed heart development (Hughes and Hall 1998). On the other hand, the intrauterine monotreme embryo shows acceleration in rostrocaudal growth relative to human embryos of similar development of organ systems (Hughes and Hall 1998). The rate of somitogenesis in monotremes also appears to be faster in relation to morphogenesis of the central nervous system than in human embryos. Smith (2001) reported that monotremes and marsupials share an accelerated development of the pharyngeal arches relative to the neural tube, but Werneburg and Sánchez-Villagra (2011) were not able to detect any acceleration of the mandibular arch development relative to the neural tube.

At the point of hatching, the monotreme lungs have established a functional respiratory membrane at the early terminal air sac stage of lung development (Ferner *et al.* 2009) and the new hatchling relies on a mesonephros-based urinary excretion system, rather than a mature kidney. Out-of-phase myogenesis also enables the generation of striated muscle myotubules for the breach of embryonic membranes and the demands of early post-hatching mobility (Hughes and Hall 1998).

Early development of the monotreme nervous system

Very few studies have examined the development of the monotreme nervous system in any detail and the difficulties of obtaining specimens mean that no studies of gene expression have been undertaken in these embryos. Nevertheless, the timecourse of development of the key components of the nervous system can be determined from the available preserved material and correlated with the developmental schema outlined above.

The earliest events in morphogenesis of the nervous system are the formation of the neural plate and the formation and migration of the neural crest. The neural plate gives rise to the brain and spinal cord by the process of neurulation, whereas the neural crest cells are generated from the paired lateral ridges of the folding neural plate. The neural crest cells give rise to the neurons of the peripheral nervous system (sensory ganglion cells of the spinal and cranial nerves; autonomic ganglia; enteric neurons), Schwann cells for peripheral nerve myelination, and adrenal medullary neurons, among other non-neural elements. The neural plate in monotremes first appears shortly after formation of the primitive streak, at the beginning of the late intrauterine phase (stages 7 and 8 of Werneburg and Sánchez-Villagra 2011; Fig. 3.3a) and all the derivatives of the neural crest are distributed throughout the embryo by the time of laying of the egg, which is presumed to be a few days later.

During the last few days of the intrauterine phase, the neural plate folds to form a tube. This process begins at the future cervical region and proceeds rostrally and caudally. The precise time or stage at which the closure of the rostral and caudal neuropores occurs in monotremes is not known, but the illustrations of Werneburg and Sánchez-Villagra (2011) suggest that both neuropores are still open at stage 13 in the short-beaked echidna, i.e. at the beginning of the first incubation subphase (In[a]). There are no data available at present to be able to determine whether the rostral or caudal neuropore closes first. Formation of the fore-, mid- and hindbrain vesicles coincides with elaboration of the pharyngeal arches during In[a].

The lateral wall of the neural tube is divided into basal and alar primary longitudinal zones (basal and alar plates), separated by a groove called the sulcus limitans (Fig. 3.5a, b). The tube also has roof and floor plates dorsally and ventrally and the latter is an important organiser of the developing nervous system. The basal plate gives rise to motor elements of the future nervous system (e.g. cranial nerve motor nuclei, ventral horn motor neurons), whereas the alar plate gives rise to sensory neurons (e.g. trigeminal, vestibular and cochlear nuclei of the brainstem and the dorsal horn neurons of the spinal cord) (Fig. 3.5c, d). At the front or rostral end of the neural tube, the precise trajectory of the sulcus limitans becomes difficult to define and the alar and basal plates converge at the midline in the space between the sites of the developing neurohypophysis (posterior pituitary) and future anterior commissure (Puelles et al. 2004).

Neuromeric organisation of the developing monotreme brain

All amniote embryos exhibit segmental organisation during neural tube development (Fig. 3.6). These segments give rise to discrete parts of the adult brain and underlie the modular organisation of the brain. The brain segments consist of rhombomeres and the isthmic segment in the hindbrain; the mesencephalic segments in the midbrain and prosomeres and secondary prosencephalon in the forebrain. The boundaries between these are defined in murine and chick embryos on the basis of differential expression of key genes (*Pax-6*, *Pax-3* & *Pax-7*, *Nkx-6.1*, *Nkx-2.1*, *Gbx-2*; Puelles et al. 2000, 2004), but these have not been analysed for monotreme embryos.

The rhombomeres give rise to constituent neurons of the medulla and pons (e.g. pontine nuclei, trigeminal sensory and motor nuclei, facial, abducens and ambiguus nuclei, pontine and medullary reticular formation). The rhombencephalon also develops a distinct rhombic lip over the sides and roof of the fourth ventricle during the middle of the incubation phase in monotremes. This structure is known to produce neurons for the cerebellum and precerebellar nuclei (i.e. pontine, inferior olivary and lateral reticular nuclei among others). The isthmic region

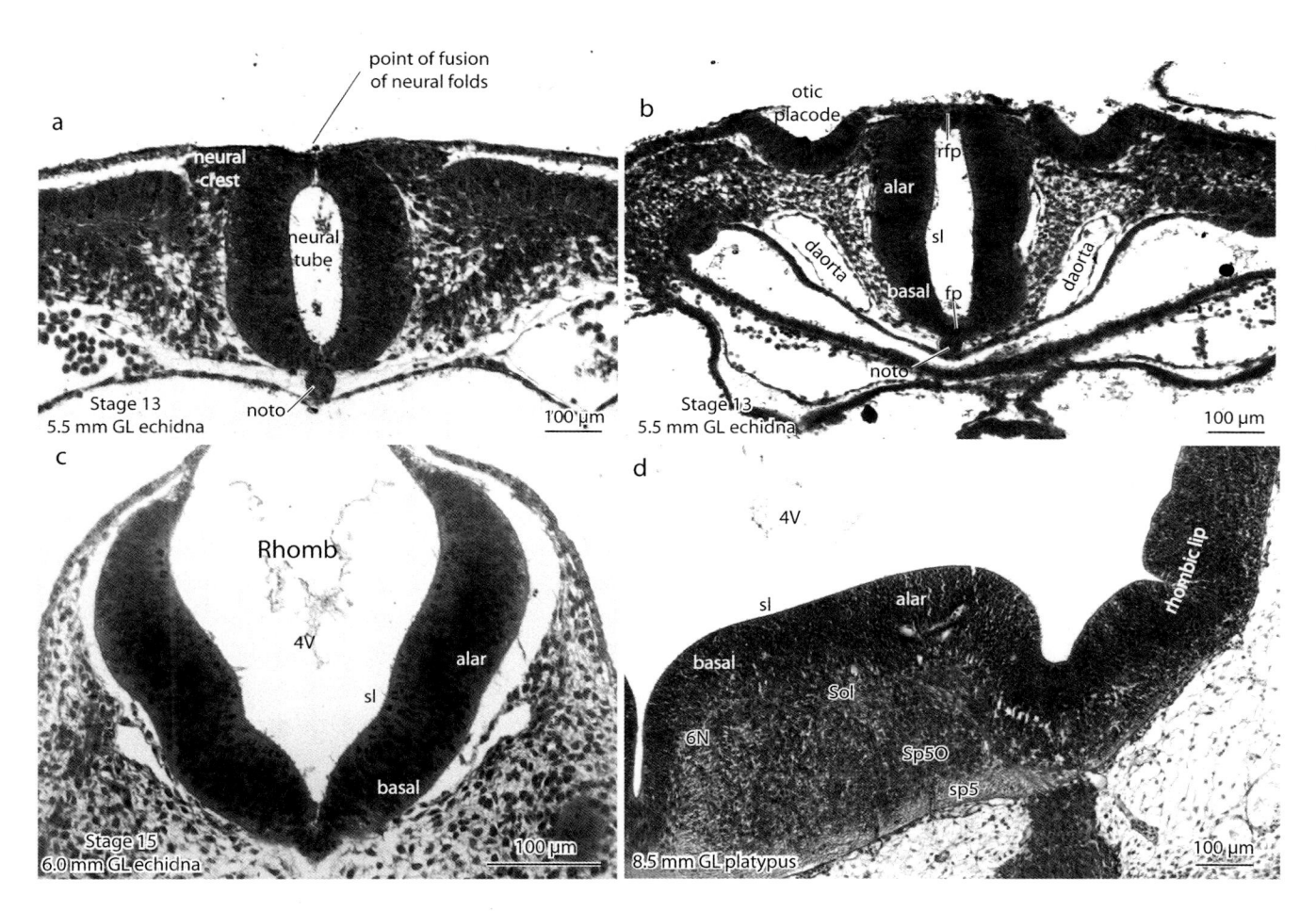

Figure 3.5: Transverse sections through the spinal cord and hindbrain of early to mid incubation monotremes. Shortly after laying of the egg (a, b), neural tube closure has just finished and some neural crest cells remain alongside the developing spinal cord (a). These cells will probably give rise to dorsal root (spinal sensory) ganglion cells. The sulcus limitans (sl) demarcates dorsal (alar) and ventral (basal) compartments of the lateral neural tube wall in the caudal hindbrain (b). Distinct roof (rfp) and floor (fp) plates are visible and the notochord (noto) lies beneath the floor plate. The alar and basal compartments continue into the open medulla (c) and will give rise to sensory (e.g. solitary viscerosensory and trigeminal sensory nuclei) and effector nerve cells (e.g. abducens and facial motor nucleus) of the brainstem nuclei (d). The midline is to the left of frame in image (d). Note the rhombic lip (a derivative of the alar plate) at the lateral margin of the brainstem. All specimens are from the MfN collection. 4V – fourth ventricle; 6N – abducens nucleus; daorta – dorsal aorta; Sol – nucleus of the solitary tract; sp5 – spinal trigeminal tract; Sp5O – oral part of spinal trigeminal sensory nucleus.

gives rise to the trochlear nucleus; whereas the isthmic and rhombomere 1 segments together give rise to the constituent neurons of the cerebellum (Puelles *et al.* 2004).

The mesencephalic segments gives rise to the oculomotor nucleus, caudal midbrain tegmentum and the tectum. In the forebrain, prosomere 1 gives rise to the pretectum and subcommissural organ, prosomere 2 gives rise to the dorsal thalamus and epithalamus (habenular nuclei) and prosomere 3 gives rise to the ventral thalamus (prethalamus) and fields of Forel. The medial parts of the mesencephalic segment and prosomeres 1 and 2 together give rise to the dopaminergic neurons of the substantia nigra and ventral tegmental area.

The secondary prosencephalon at the very front end of the neural tube (Puelles *et al.* 2004) gives rise to the hypothalamus (including mammillary, posterior, chiasmatic and preoptic regions), a flattened sheet called the pallium (i.e. cortical regions such as the isocortex, hippocampus and olfactory bulb) and

44 NEUROBIOLOGY OF MONOTREMES

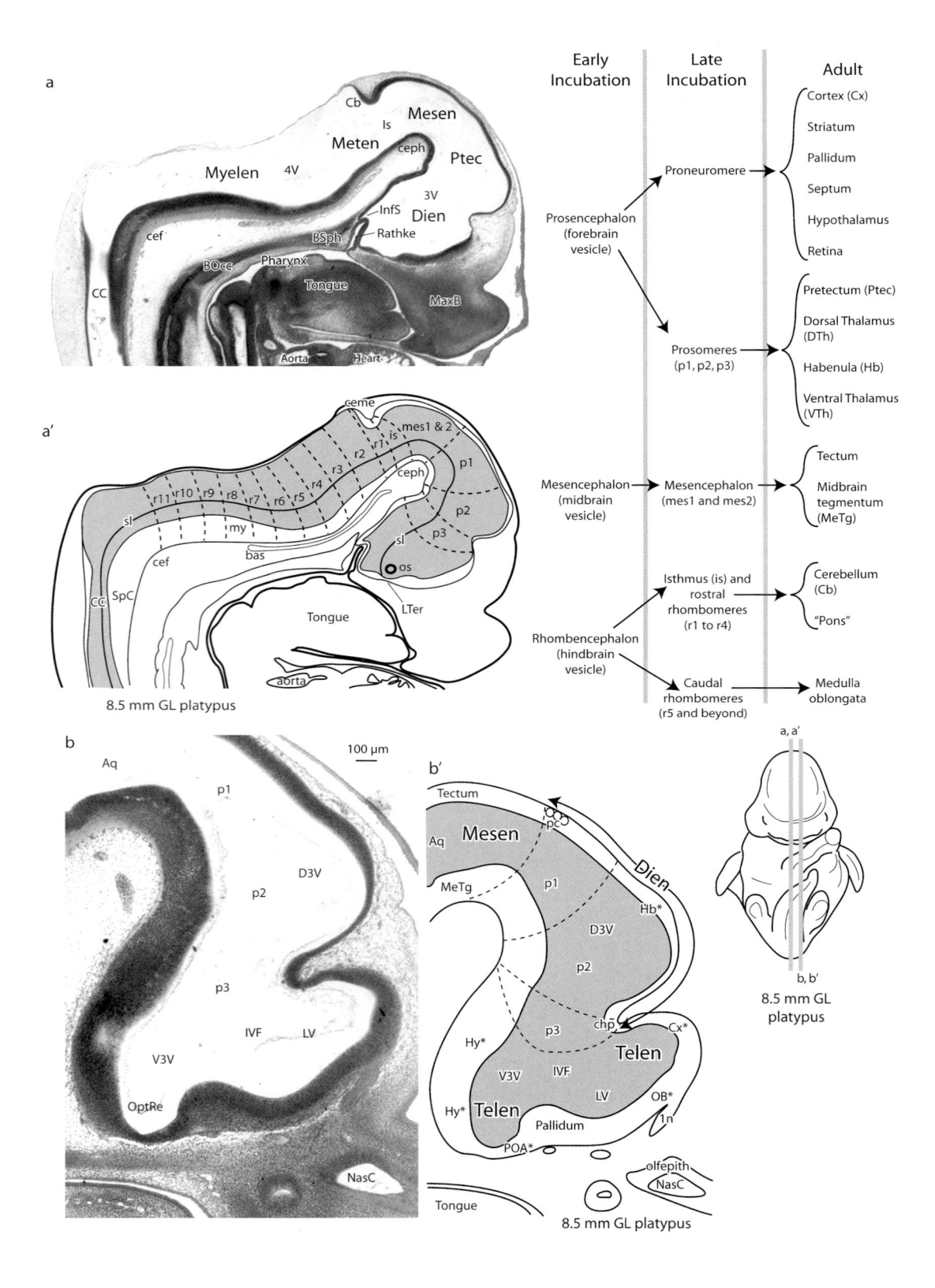

deeper ganglionic or subpallial (i.e. striatal, pallidal and septal) regions. The point of origin of the eye cup, which grows out from the side of the forebrain proneuromere in the early incubation subphase, is marked by the position of the optic stalk (os in Fig. 3.6a). This will connect the eye cup and forebrain and will provide a guide for growth of optic nerve axons during post-hatching life.

Overview of post-hatching nervous system development

The details of development of the nervous system components will be discussed in the relevant chapter, but a précis of key events and their relative timing will be provided here in the context of the development of external features and behaviour.

At the time of hatching, the hindbrain regions of both platypus and short-beaked echidna are structurally similar, as dictated by the evolutionarily conserved nature of the hindbrain and the similar demands of post-hatching life. By contrast, the forebrain has already begun to show differences that will be further elaborated to produce the rather different forebrains of the two species (see Chapter 7). The key developmental events in the early post-hatching period involve the differentiation of the forebrain and the cerebellum.

By the peri-hatching period, the forebrain has begun differentiation into pallial, ganglionic and midline septal primordia. Pallial differentiation to produce the isocortex and hippocampus appears to be more advanced in the short-beaked echidna than the platypus, perhaps reflecting a precocious progression towards the highly folded cortex of the mature echidna. Nevertheless, in both newly hatched monotremes most of the pallial wall is made up of a poorly developed neuroepithelium, albeit with a slightly thicker preplate zone external to the neuroepithelium in the short-beaked echidna. Like other mammals, proliferation of stem cells to generate the neurons and glia of the cortex occurs in both the ventricular germinal zone of the vesicle wall and the subventricular zone outside it, with the latter making its first appearance a few days after hatching. The cortical plate of young neurons emerges at the end of the first week after birth and the first fibre bundles in the pallium and striatum also appear at this time. The second post-hatching week is a period for rapid elaboration of the input (afferent) and output (efferent) connections of the cortex and the cortical plate develops an external compact cell zone overlying a loose-packed zone, much as is seen in 2-to-3-week postnatal diprotodont marsupials. In the platypus a distinctive subventricular/subcortical zone underlies the future electroreceptive part of the

Figure 3.6: Regional organisation in the developing brain of the mid incubation phase monotreme as illustrated by midline (a, a') and parasagittal (b, b') sections through an 8.5 mm GL (greatest length) platypus. The small diagram to the lower right shows the approximate position of the planes of section on a line drawing of an 8.5 mm GL platypus embryo (MfN collection). The line diagrams a' and b' illustrate neuromeric organisation of the brainstem (into rhombomeres 1 to 11 – r1 to r11; isthmic – is; and mesencephalic – mes segments) and the diencephalon (into prosomeres p1 to p3; for the pretectum, dorsal thalamus and ventral thalamus, respectively). The tabular illustration to the right summarises the transformation of the primary brain vesicles (fore-, mid- and hindbrain), through neuromeric regions to adult structures. Embryonic neuroepithelial precursors of the cortex (Cx), hippocampus (Hi), hypothalamus (Hy), olfactory bulb (OB), preoptic area (POA) etc. are denoted by the abbreviation with an asterisk. 1n – olfactory nerve fibres; 3V – third ventricle; 4V – fourth ventricle; Aq – cerebral aqueduct; bas – basilar artery; BOcc – basal occipital bone; BSph – basal sphenoid bone; CC – central canal of the spinal cord; cef – cervical flexure; ceme – cerebellomesencephalic fissure; ceph – cephalic flexure; chp – choroid plexus; Dien – diencephalon; DTh – dorsal thalamus; Hb – habenular nuclei; Hi – hippocampus; Hy – hypothalamus; InfS – infundibular stalk; is – isthmic segment; Is – isthmus; IVF – interventricular foramen; LH – lateral hypothalamus; LTer – lamina terminalis; LV – lateral ventricle; MaxB – maxillary bone; mes – mesencephalic segment; Mesen – mesencephalon; Meten – metencephalon; MeTg – mesencephalic tegmentum; mge – medial ganglionic eminence; Myelen – myelencephalon; NasC – nasal cavity; olfepith – olfactory epithelium; OptRe – optic recess of third ventricle; os – optic stalk; p1 to p3 – prosomeres 1 to 3; pc – posterior commissure; Ptec – pretectum; r1 to r11 – rhombomeres; Rathke – Rathke's pouch; SpC – spinal cord; sl – sulcus limitans; VTh – ventral thalamus.

developing primary somatosensory cortex for up to 4 months after hatching (see Chapter 7 for a detailed discussion).

The diencephalon of the peri-hatching period exhibits the segmental organisation into the three diencephalic neuromeres (p1, p2 and p3; Fig. 3.6b) described above, but postmitotic neuronal populations are small in number at the time of hatching. The first 2 weeks after hatching are a period of intense production of thalamic neuronal populations accompanied by the establishment of thalamocortical and corticothalamic pathways. The generation of huge numbers of granular cells for the ventral posterior medial (VPM) thalamic nucleus is a particularly striking feature of the development of the platypus thalamus during the first 4 weeks after hatching (see Chapter 6).

The cerebellum and precerebellar nuclei also have the bulk of their development in the post-hatching period. Although the macroneurons of the cerebellum (Purkinje cells and deep cerebellar nuclei) are generated in the late incubation period, small neurons of the cerebellar cortex are not produced until the first few weeks after hatching. The first 14 days of the post-hatching period is also the period when precerebellar neurons migrate from the rhombic lip to their final settling sites in the pons and medulla oblongata.

Questions for the future

Young monotremes must access copious amounts of milk to achieve the observed growth rates, but precisely how the newly hatched monotreme stimulates the areola to commence let-down and how milk is drawn into the oral cavity are questions that remain unexplored. The oral fissures of the newly hatched monotremes appear poorly equipped for forming a suction seal against the areola, so licking seems a more likely mechanism.

The triphasic nature of monotreme development is unique among vertebrates, but the details of how developmental stages of monotremes correlate with maternal reproductive cycles, seasonal changes and maternal behaviour is only poorly understood. The factors that determine reproductive success of captive monotremes is of critical importance for zoos and wildlife parks who hold these animals and may have long-reaching importance for the survival of endangered monotremes, like the long-beaked echidna.

A related question concerns the mechanisms by which seasonal change stimulates the onset of mating. Since mating appears to occur shortly after the winter solstice, it is possible that mating behaviour is triggered by changes in day length as seen in small marsupial carnivores.

4

Overview of monotreme nervous system structure and evolution

Ken W. S. Ashwell

Summary

Although the brains of the modern platypus and echidnas have very distinctive features, they can be readily subdivided according to the schema applied to therian brains. The mammalian brain is derived developmentally from a segmentally organised neural tube, with distinct segments for the hindbrain (rhombomeres 1 to 11, and an isthmic segment), midbrain (two mesomeric segments) and forebrain (three prosomeres and a proneuromere region). Although the gene expression patterns that underlie these subdivisions have not been studied in monotremes, the anatomical features of internal ridges and depressions that reflect the segmentation can be identified at some stage in the developing monotreme brain.

All the modern monotremes (and also extinct members of the group) have larger brains than many modern and extinct marsupials and some placentals, although this large brain size is achieved by rather different patterns of developmental growth in the platypus and echidnas (extensive growth in cortical thickness in the former and an emphasis on cortical folding in the latter). Analysis of mammalian phylogeny on the basis of neural characters emphasises the abundance of plesiomorphic or primitive features in the monotreme central nervous system, but this analysis does not give due consideration to the sensory specialisations of the group (see Chapter 14).

Deducing the timing of monotreme brain evolution and the sequence of emergence of the distinctive neurological features of the platypus and echidnas is a frustratingly difficult task, because of the paucity of the fossil record. Nevertheless, many of the distinctive neurological features that characterise the modern platypus and reflect trigeminal specialisation are evident by the middle Miocene. Similarly, by the early Pleistocene echidnas have the olfactory and telencephalic adaptations that characterise the modern group.

Overview of nervous system structure and function

The vertebrate nervous system is divided into the central nervous system, including the brain and spinal cord, and the peripheral nervous system, including the nerves and neuron clusters (ganglia) outside the central nervous system. The traditional schema for dividing the central nervous system was based on the secondary brain vesicles seen during late embryonic life (refer back to Fig. 3.6), dividing the brain into the myelencephalon (or medulla oblongata), the metencephalon (or pons and cerebellum), the mesencephalon (or midbrain), the diencephalon and the telencephalon (further divisible into pallial and subpallial parts). Although modern conceptions of brain development are redrafting the classical brain divisions, the traditional divisions are still commonly used in standard neuroanatomical textbooks and will be applied occasionally in subsequent chapters on monotreme brain structure. Where the modern, molecular-based conceptions are relevant to properly interpreting monotreme brain structure and development (e.g. in the diencephalon and hypothalamus), they will be discussed in detail.

Although modern molecular markers have yet to be applied to monotreme embryos, it is possible to deduce major developmental regions from available archived material and apply modern terminology to the developing and adult central nervous system. In the following overview, the broad structure and function of the longitudinal subdivisions of the central nervous system will be discussed in turn from caudal to rostral, using modern conceptions of the developmental segmentation of the central nervous system.

Spinal cord

Functions of the spinal cord include the initial processing of somatosensory information. This includes pain; temperature; simple and discriminative touch; vibration from the skin surface, muscle beds and bones; and proprioceptive (joint position or muscle tension) information from joints and muscles. Motor neurons of the spinal cord (lower motor neurons) project to muscle fibres in the neck, trunk, tail or limbs and are under control of descending upper motor neuron pathways from the brainstem (tectospinal, reticulospinal, rubrospinal or vestibulospinal tracts) or cortex (corticospinal tracts). The spinal cord also provides circuitry for reflexes that optimise muscle tension or length during contraction; or withdraw body parts from noxious and damaging agents (i.e. sharp, abrasive or excessively hot or cold objects).

The spinal cord is derived from the caudal end of the neural tube and is the most conserved part of the central nervous system during evolution. It is organised in a segmental fashion, with spinal nerves emerging between adjacent vertebrae to provide somatosensation and motor control to the axial body and limbs. The cervical region of the spinal cord has eight spinal nerves in all mammals and these emerge either between the base of the skull and the first cervical vertebra (spinal nerve C1) or below the seven cervical vertebrae (spinal nerves C2 to C8). More caudally, there are usually 12 thoracic spinal nerves, three to six lumbar spinal nerves, four or five sacral spinal nerves and a variable number of coccygeal spinal nerves, although the actual number of thoracic and lumbar vertebrae, and hence spinal segments, can vary between echidna subspecies (see Chapter 5; Griffiths et al. 1991). Enlargements are present at segmental levels serving the forelimb (approximately segments C5 to T1) and the hindlimb (segments L3 to S2). The length of the spinal cord varies substantially between mammals and is inversely correlated with the degree of flexibility of the vertebral column. Where the spinal cord is very short, the distance between its caudal end and the point of emergence of spinal nerves from between vertebral segments is made up of loosely bound dorsal and ventral roots known as the cauda equina (horse's tail). The short-beaked echidna has a particularly short spinal cord (see Chapter 5), whereas the platypus spinal cord extends for almost the entire length of the vertebral column. The transverse width of the platypus spinal cord at lower cervical levels is only 4.2 mm (only one specimen; Appendix Table 4) compared to 6.9 ± 0.9 mm for the short-beaked echidna (Appendix Table 5).

The internal structure of the spinal cord consists of a central 'H' shaped region of grey matter, housing neuronal cell bodies, axon terminals and the proximal parts of neuronal dendritic trees; and a surrounding white matter zone containing axonal conduits (columns, funiculi or tracts) for transmitting information up and down the spinal cord.

The dorsal part of the spinal cord grey matter (dorsal horn) is concerned with processing sensory information from receptors in the skin, joints, muscles and bones (somatosensation). The incoming fibres carrying this information are the central or proximal processes of dorsal root (spinal ganglion) cells and may either terminate in the dorsal horn or ascend in the dorsal column white matter tracts to the brainstem. These inputs are collectively called general somatic afferents because they come from the limbs, skin surface or wall of the body (soma) and are concerned with general sensation, i.e. touch, pain, temperature, vibration. At middle levels (thoracic to upper lumbar) and far caudal levels (mid-sacral segments), the spinal cord also receives information from internal organs (viscera) of the thoracic and abdominopelvic cavities (general visceral afferents). These fibres convey information about the state of filling or emptying of internal organ cavities as well as pain sensation, traction and vibration on the peritoneal folds supporting or enclosing the organs.

The ventral part of the spinal cord (ventral horn) contains motor neurons that send their axons directly to somatic (skeletal) muscle. These nerve cells (lower motor neurons) and their outgoing axons are known collectively as general somatic efferents. Alpha motor neurons control extrafusal muscle fibres that produce the bulk of voluntary contraction, whereas gamma motor neurons control intrafusal muscle fibres that dynamically adjust the sensitivity of muscle stretch receptors.

The dorsal and ventral horns are separated by a region called the intermediate grey zone, containing neuronal circuitry concerned with processing sensory information from internal organs (intermediomedial cell column) and (in the thoracic and upper lumbar segments) an intermediolateral cell column for sympathetic autonomic control of sweat glands, blood vessels and internal organs of the thoracic and abdominopelvic cavities. This latter group of cells forms a horn of grey matter (lateral horn) projecting into the lateral white matter funiculus at those segments. A further group of effector neurons of the parasympathetic division of the autonomic nervous system is located in the mid-sacral segments. These effector nerve cells for pelvic and caudal abdominal viscera are collectively known as general visceral efferent neurons. Rostral thoracoabdominal viscera receive parasympathetic control from the vagus nerve.

Brainstem and cerebellum

The brainstem is traditionally divided into medulla, pons and midbrain on the basis of the classic subdivisions visible in the human brain, but modern gene-mapping studies of vertebrate brain development (see review in Watson 2012a) leads to an alternative schema of subdivision, one that removes the pons as a distinct region and emphasises the segmental nature of the embryonic midbrain, isthmus and hindbrain (see Fig. 4.1a and b). The cerebellum is usually considered separate from the brainstem, but both the cerebellum and precerebellar nuclei are in fact derivatives of the alar plates of the embryonic hindbrain.

Hindbrain

The hindbrain serves four functions: (1) it provides a conduit for nerve pathways going up and down the central nervous system; (2) it acts as an integrative centre, processing sensory information and making commands to internal organs to control vital body functions, often without conscious awareness; (3) it provides initial processing of sensory information from the cranial nerves; and (4) it controls the muscles and glands of the head and neck.

The hindbrain is derived from the embryonic hindbrain vesicle. In traditional human neuroanatomy the hindbrain has been divided into a rostral pons and a caudal medulla oblongata, with the pons getting its name because of the (bridge-like) large bundles of axons from the pontine nuclei crossing to the opposite middle cerebellar peduncle to enter the cerebellum. Studies of gene expression during developmental have revealed that the hindbrain part of the brainstem is more properly considered to be divided into 12 rostrocaudal segments (one isthmic segment, plus rhombomeres 1 to 11) (Watson 2012a). Pontine nuclei usually lie alongside rhombomeres 3

50 NEUROBIOLOGY OF MONOTREMES

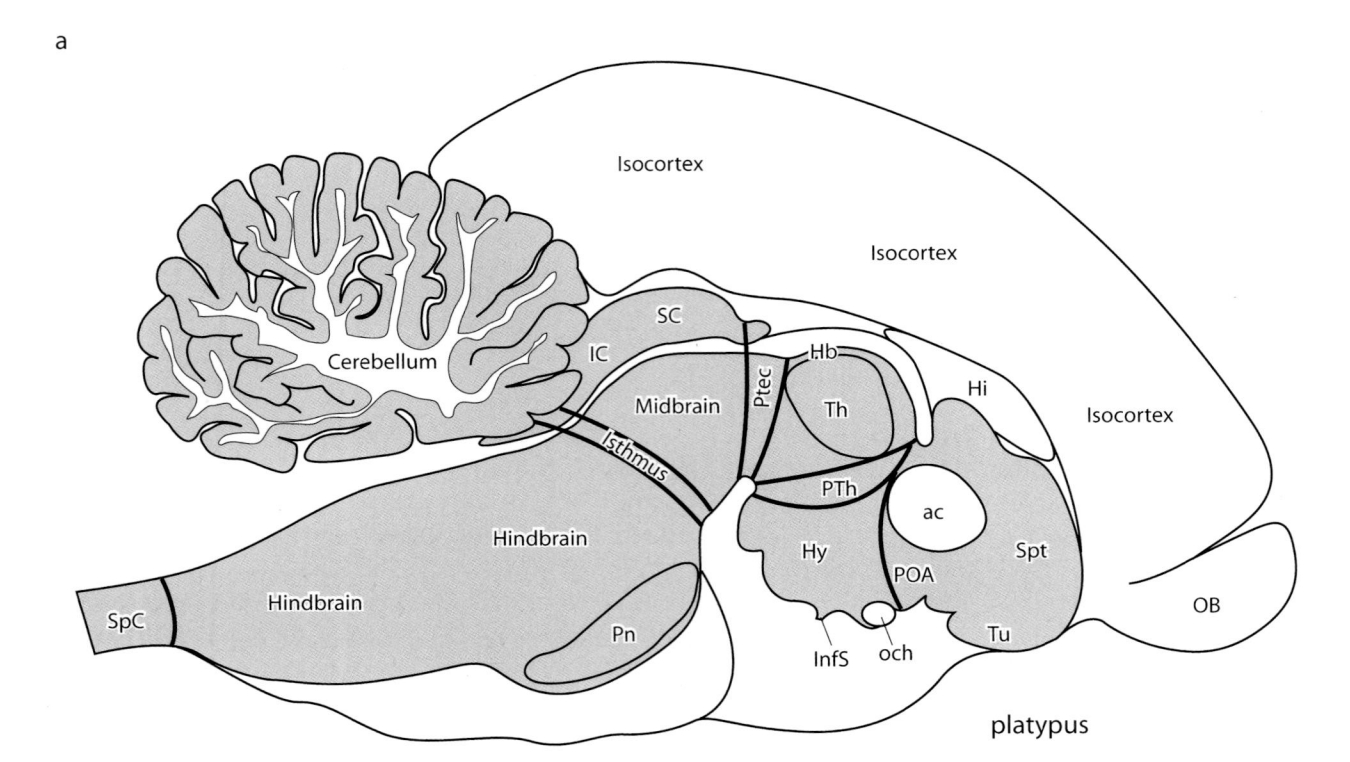

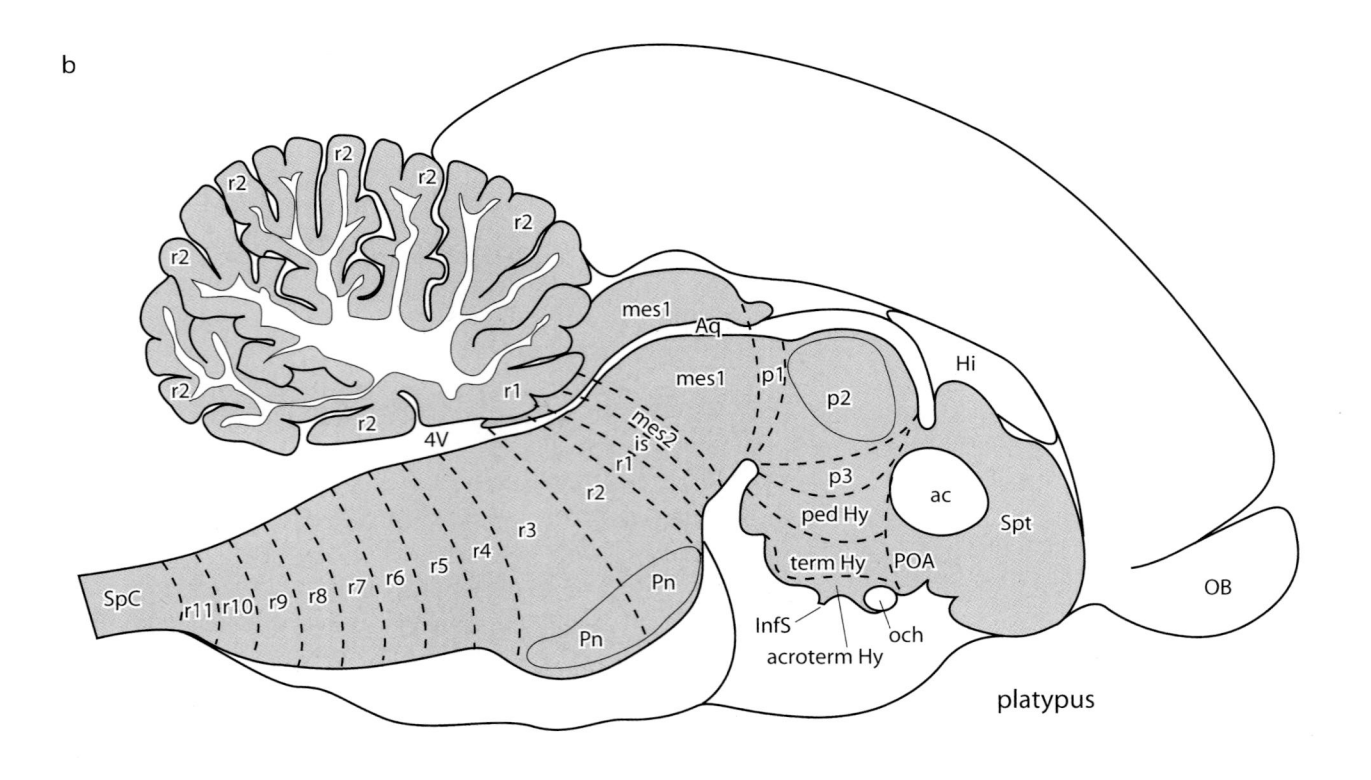

Table 4.1. Derivatives of hindbrain embryonic segments[1]

Segment	Alar plate derivatives (sensory nuclei, cerebellum and precerebellar nuclei)	Basal plate derivatives	Associated adult fibre tract/nerve
Isthmus	Rostral part of cerebellar vermis, parabigeminal nucleus	Trochlear nucleus, dorsal raphe nucleus	–
Rhombomere 1	Caudal cerebellar vermis, cerebellar hemispheres	Locus coeruleus, interpeduncular nucleus, paramedian raphe nucleus	Crossing of superior cerebellar peduncle
Rhombomere 2	Rostral pontine nuclei	Main part of motor trigeminal nucleus, paramedian raphe nucleus	Entry of trigeminal nerve
Rhombomere 3	Caudal pontine nuclei, principal trigeminal nucleus	Paramedian raphe nucleus	–
Rhombomere 4	Oralis part of spinal trigeminal nucleus	–	Exit of facial nerve
Rhombomere 5	Oralis part of spinal trigeminal nucleus, superior olivary complex, nucleus of trapezoid body	Abducens nucleus, raphe magnus nucleus	Genu of facial nerve
Rhombomere 6	Oralis part of spinal trigeminal nucleus	Facial nucleus, raphe magnus nucleus	–
Rhombomere 7	Interpolaris part of spinal trigeminal nucleus	Compact part of nucleus ambiguus, raphe pallidus nucleus	Exit of glossopharyngeal nerve
Rhombomere 8	Interpolaris part of spinal trigeminal nucleus, rostral part of inferior olivary complex	Subcompact part of nucleus ambiguus, raphe pallidus nucleus	Exit of vagus nerve
Rhombomere 9	Interpolaris part of spinal trigeminal nucleus, middle part of inferior olivary complex	Loose part of the nucleus ambiguus, raphe obscurus nucleus	Exit of spinal accessory nerve
Rhombomere 10	Caudalis part of spinal trigeminal nucleus, caudal part of the inferior olivary complex	–	–
Rhombomere 11	Caudalis part of spinal trigeminal nucleus	–	Crossing of pyramidal tract

1 Based on rodent rhombomeres as described in Watson (2012a).

Figure 4.1: Schematic longitudinal section through the brain of a platypus showing the main anatomical subdivisions (a) and the putative division of the brain according to neuromeric origin (b). Midline grey matter structures are represented by grey shading. Note that rhombomeric boundaries are often crossed by migration during development so segmental boundaries often become somewhat blurred. The front of the brain is to the right in both illustrations. Note that the cerebellum has derivatives from rhomberes 1 and 2. ac – anterior commissure; acroterm Hy – acroterminal part of the hypothalamus; Aq – cerebral aqueduct; Hb – habenula; Hi – hippocampus; Hy – hypothalamus; IC – inferior colliculus; InfS – infundibular stalk; is – isthmic segment; mes1, mes2 – mesencephalic neuromeres 1 and 2; OB – olfactory bulb; och – optic chiasm; ped Hy – peduncular hypothalamus; Pn – pontine nuclei; POA – preoptic area; Ptec – pretectum; PTh – prethalamus; r1 to r11 – rhombomeres 1 to 11; SC – superior colliculus; SpC – spinal cord; Spt – septum; term Hy – terminal hypothalamus; Th – (dorsal) thalamus; Tu – olfactory tubercle.

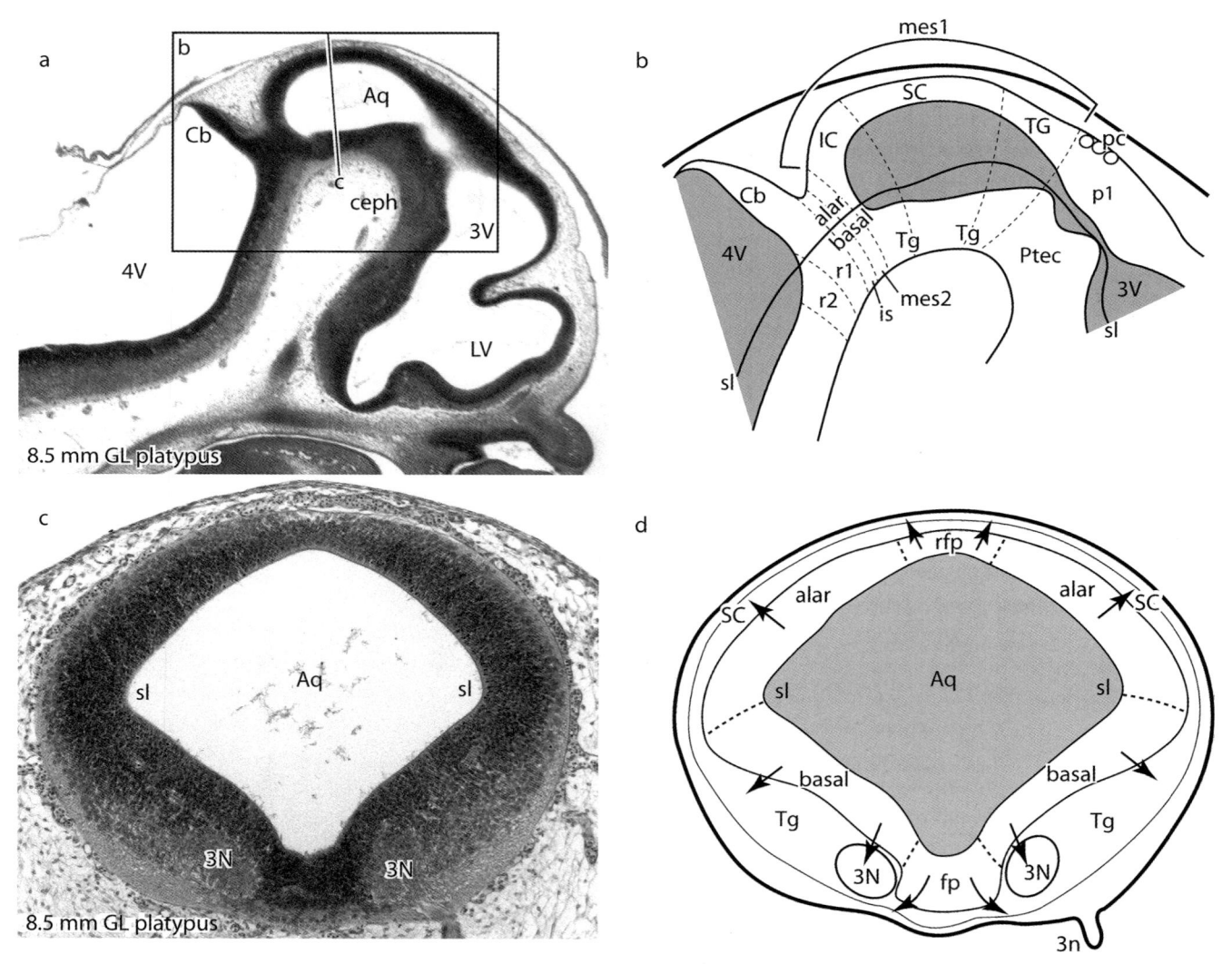

Figure 4.2: Developmental origins of the subdivisions of the adult midbrain using the platypus as an example. Figure (a) shows a sagittal section through the developing brain of an 8.5 mm greatest length (GL) platypus embryo. The inset in (a) covers the region from the caudal diencephalon (pretectum – Ptec) to the rostral hindbrain (including cerebellum) and is enlarged as (b) showing the neuromeric subdivisions of the midbrain vesicle. The midbrain is composed of two developmental regions (mesencephalic neuromeres). The more rostral of these is the larger mes1 (mesomere 1). Its dorsal alar plate derivatives include (from rostral to caudal) the tectal grey (TG), superior colliculus (SG) and inferior coliculus (IC). The basal plate of mes1 is much smaller than the alar plate region and gives rise to nerve cells of the midbrain tegmentum (Tg). The smaller mes2 (mesomere 2, or the preisthmus) is a thin zone sandwiched between the caudal mes1 and the isthmus (see Table 4.2 for derivatives). The line marked (c) in figure (a) is the plane of section for photomicrograph (c) of a transverse section through the embryonic midbrain and the matching line diagram (d). Figure (d) shows the roof plate (rfp), alar plate (alar), basal plate (basal) and floor plate (fp) regions of the embryonic midbrain and the direction of migration of their neuronal derivatives (arrows) in the developing monotreme midbrain. 3N – oculomotor nucleus; 3n – oculomotor nerve; 3V – third ventricle; 4V – fourth ventricle; Aq – cerebral aqueduct; Cb – cerebellar primordium; ceph – cephalic flexure or bend; LV – lateral ventricle; p1 – prosomere 1; pc – posterior commissure; sl – sulcus limitans.

Table 4.2. Components of the midbrain

Mesomere	Rostrocaudal embryonic segments	Radial embryonic divisions	Component adult structures
Mesomere 1	Tectal grey	Roof plate	Tectal grey commissure
		Alar plate	Tectal grey, tectal grey commissural nucleus, DMPAG, DLPAG, VLPAG
		Basal plate	Red nucleus magnocellular part, substantia nigra, VMPAG, mesencephalic reticular formation, Edinger Westphal nucleus
		Floor plate	Ventral tegmental area, rostral linear nucleus
	Superior colliculus	Roof plate	Tectal commissure
		Alar plate	Superior colliculus, bed nucleus of the brachium of the superior colliculus, DMPAG, DLPAG, VLPAG
		Basal plate	Red nucleus magnocellular part, substantia nigra, pararubral nucleus, oculomotor nucleus
		Floor plate	Ventral tegmental area, rostral linear nucleus
	Inferior colliculus	Roof plate	Intercollicular commissure
		Alar plate	Inferior colliculus, bed nucleus of the brachium of the inferior colliculus, subbrachial nucleus, DMPAG, DLPAG, VLPAG
		Basal plate	Red nucleus magnocellular part, substantia nigra, rostral linear nucleus
		Floor plate	Ventral tegmental area
Mesomere 2	Preisthmus	Roof plate	–
		Alar plate	Sagulum nucleus, cuneiform grey, subcuneiform nucleus
		Basal plate	Ventral periaqueductal grey, deep mesencephalic nucleus, substantia nigra, ventral tegmental area, rostral linear nucleus
		Floor plate	Interfascicular nucleus

Based on mesomere derivatives as described for the mouse by Puelles et al. (2012c).

and 4, but these can vary so much in size between species that their position is not a consistent or suitable topographic subdivision. The relationship between the major nuclei of the brainstem and embryonic hindbrain segments is summarised in Table 4.1.

Cerebellum

The cerebellum is mainly concerned with coordination of movement and the planning of motor activity. The cerebellum and the precerebellar neurons that project to it are derived from the rhombic lip, the edge of the alar plate of the embryonic hindbrain, so the cerebellar and precerebellar systems can properly be considered part of the adult hindbrain. The two segments that contribute most to the cerebellar systems are the alar plate of the isthmus, which produces the rostral part of the cerebellar vermis (midline), and the alar plate of rhombomere 1, which gives rise to the caudal cerebellar vermis and the cerebellar hemispheres. The more caudal precerebellar neurons are derived from the alar plates of the caudal rhombomeres. The adult cerebellum is attached to the brainstem by large bundles of fibres (superior, middle and inferior cerebellar peduncles) that carry information into and out of its interior.

Midbrain

The formation of the midbrain during development is under the influence of an organiser (isthmic organiser) at the junction between the midbrain and hindbrain (Puelles *et al.* 2012c). The developing mammalian midbrain (or mesencephalic) part of the embryonic neural tube can be divided into four rostrocaudal segments, mainly on the basis of the larger alar plate component (Fig. 4.2). The three rostral segments are most readily named according to the alar plate derivatives (tectal grey, superior colliculus, inferior colliculus) and are grouped together as mesomere 1. The most caudal part of the true midbrain is the derivatives of the much smaller mesomere 2 or preisthmus. Within each mesomere there are four radial subdivisions (roof plate, alar plate, basal plate, floor plate) with derivatives that may overlap due to developmental movements (Table 4.2).

The alar part of the midbrain is primarily a visual reflex centre (superior colliculus) and auditory relay centre (inferior colliculus). Basal parts of the midbrain are concerned with eye movement (oculomotor nucleus), motor control of the body as a whole (substantia nigra), and reward pathways to the forebrain striatum and cortex (from the ventral tegmental area).

Diencephalon

The diencephalon is one of the three parts of the forebrain, the other two being the hypothalamus and the derivatives of the telencephalic vesicle. The diencephalon includes derivatives of the three prosomeres of the embryonic brain that can be mapped by gene-expression patterns (Fig. 4.1a, b; Table 4.3; Puelles *et al.* 2012b). Each of the three prosomeres has alar and basal components, but the alar components are much bigger and give rise to the bulk of the adult diencephalon (p1 – pretectum and associated nuclei; p2 – dorsal thalamus and epithalamus; p3 – ventral or prethalamus). The basal components of each prosomere give rise to a small tegmental region (previously attributed to the midbrain) that underlies the larger alar derivatives.

The traditional view has been that the hypothalamus is a derivative of the diencephalic brain vesicle, but gene-mapping studies suggest that the hypothalamus should be considered a derivative of the proneuromere rostral to the p3 and therefore is a discrete part of the forebrain, quite separate from the diencephalon (Puelles *et al.* 2012a; see below).

Derivatives of prosomere 1

The bulk of prosomere 1 gives rise to the pretectum, a large region mainly concerned with visual reflexes. A host of additional nearby nuclei that have traditionally been considered part of the midbrain are also considered to be p1 derivatives (Puelles *et al.* 2012b; see Table 4.3). Some of these are also concerned with the control of eye movements and intraocular muscles, but many are part of systems that regulate activity in the motor and behavioural striatum through dopaminergic pathways. Some of the nerve cell groups at the midbrain/p1 junction (e.g. Edinger-Westphal, substantia nigra, ventral tegmental area)

Table 4.3. Derivatives of the three prosomeres of the diencephalon

Prosomere 1	Prosomere 2	Prosomere 3
• Pretectal nuclei • Subcommissural organ • Nuclei of the posterior commissure (PCom, MCPC) • Nucleus of Darkschewitsch, interstitial nucleus of Cajal • Rostral periaqueductal grey • Prosomere 1 parts of the Edinger-Westphal, substantia nigra, ventral tegmental area • Parvicellular part of the red nucleus	• Thalamus (divided into 7 groups – anterior, medial, lateral, ventral, posterior, intralaminar, and midline groups) • Epithalamus (as traditionally described: habenular nuclei, paraventricular nuclei, pineal gland, stria medullaris thalami tract)	• Reticular nucleus • Zona incerta • Pregeniculate nucleus • Nucleus of the H field of Forel • H1 and H2 fields of the zona incerta • Subgeniculate nucleus

Based on prosomere derivatives as described for the mouse by Puelles *et al.* (2012b).

appear to be derived from both midbrain and p1 regions of the embryonic brain. Derivatives of p1 are demarcated by the posterior commissure at their dorsocaudal boundary and the fibre bundle called the fasciculus retroflexus at their rostral border.

Derivatives of prosomere 2

The greater part of prosomere 2 gives rise to the thalamus (Puelles *et al.* 2012b) formerly or alternatively called the dorsal thalamus (Jones 2007). In both monotremes, the two halves of the thalamus are partially fused across the third ventricle (the massa intermedia). The thalamus serves as a gateway for most sensory information (visual, auditory and somatosensory input, but not olfaction), channelling that information to the appropriate part of the cerebral cortex. The thalamus is also part of looped circuits that run from the cerebral cortex through the striatum and pallidum or through the pontine nuclei and cerebellum. These looped circuits allow the cerebral cortex to request and retrieve motor programs or routines from the striatum and cerebellum.

In traditional divisions of the diencephalon (Jones 2007), the epithalamus (habenular nuclei, paraventricular nuclei, stria medullaris tract and pineal gland) has been considered as a separate part of the diencephalon from the thalamus, but Puelles and colleagues (Puelles *et al.* 2012b) regard the habenula as a specialised part of prosomere 2 and have called for the term 'epithalamus' to be discarded. Nevertheless, debate on this point is ongoing, so the term has been retained in this book. The pineal gland is integral to circuits that regulate sleep–wake cycles through its secretion of melatonin, whereas the habenular nuclei are part of a pathway from the septum, basal forebrain and hypothalamus to the interpeduncular nucleus, ventral tegmental area and raphe nuclei of the brainstem. This latter pathway probably contributes to the physiological expression of emotions.

Derivatives of prosomere 3

The embryonic prosomere 3 region gives rise to the nuclei of the prethalamus, some of which were formerly called the ventral thalamus. Since the prethalamus develops *rostral* to the thalamus in the neuraxis, the adjective *ventral* is inappropriate. The components of the prethalamus form a shell around the thalamus, with the laterally placed reticular nucleus component being in continuity with the more medially placed zona incerta. The reticular nucleus receives collateral branches from thalamocortical axons and is populated by GABAergic neurons that have a profound inhibitory effect on the relay neurons of the thalamus (Jones 2007). It plays a central role in producing the slow rhythmic activity of the cerebral cortex in synchronised sleep. The zona incerta is also populated by GABAergic neurons, but has very different connections from the reticular nucleus and receives input from diverse areas of the cerebral cortex and spinal cord. The pregeniculate nucleus was formerly called the ventral lateral geniculate nucleus, because of its

Table 4.4. Components of the embryonic hypothalamus and its adult derivatives

Rostrocaudal Divisions	Alar plate derivatives	Basal plate derivatives
Acroterminal zone	• Retinal outgrowth • Vascular organ of the lamina terminalis • Suprachiasmatic nucleus	• Median eminence • Arcuate nucleus • Median mammillary nucleus
Terminal hypothalamus	• Rostral paraventricular nucleus • Rostral subparaventricular nucleus	• Tuberal nucleus • Mammillary area • Ventromedial hypothalamus • Dorsomedial hypothalamus
Peduncular hypothalamus	• Caudal paraventricular nucleus • Caudal subparaventricular nucleus	• Retrotuberal area • Peri-retromammillary area • Retromammillary area • Subthalamic nucleus

Based on derivatives of the primordial hypothalamus as described for the mouse by Puelles *et al.* (2012a).

a

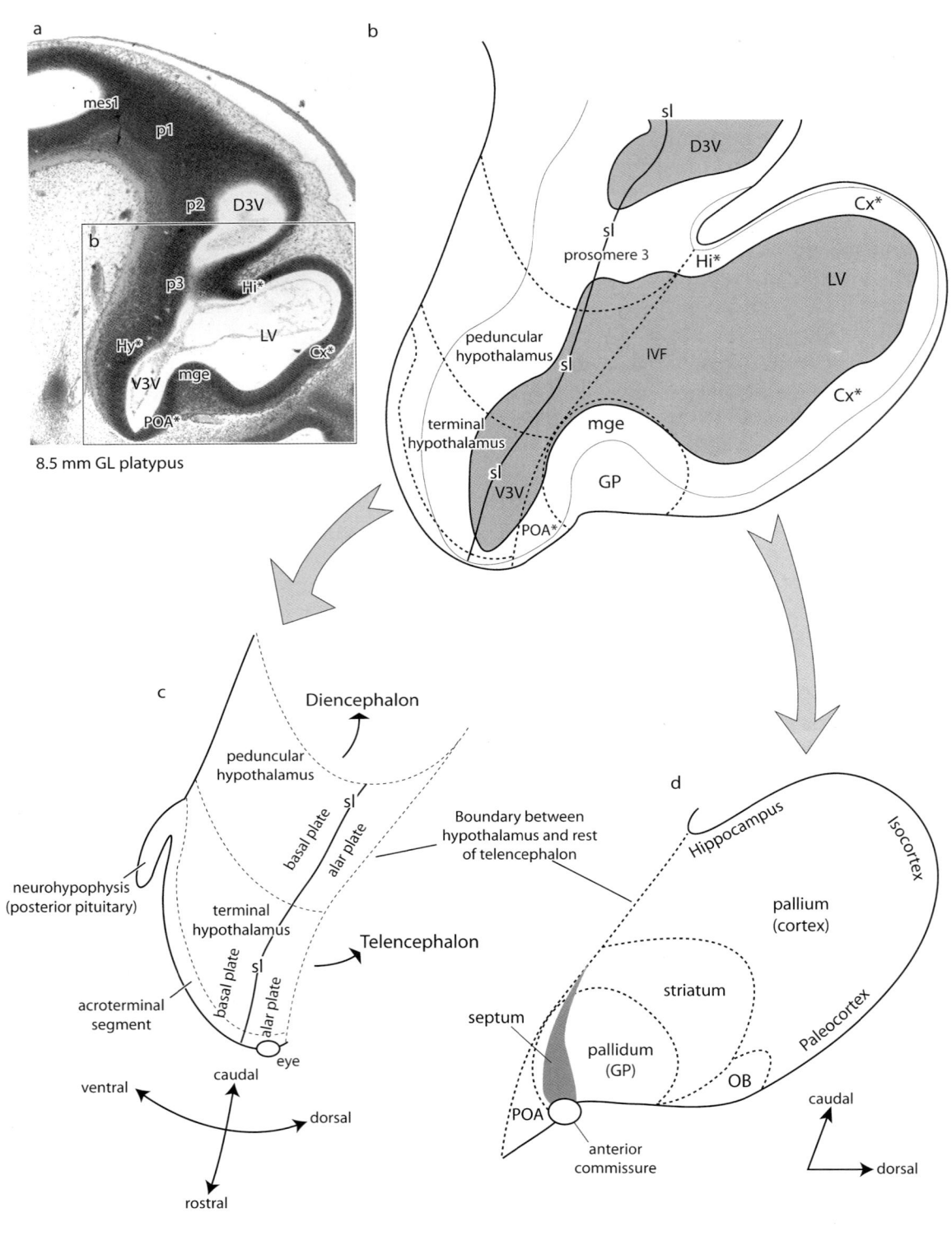

mes1
p1
p2 D3V
p3 Hi*
Hy* LV Cx*
V3V mge
POA*

8.5 mm GL platypus

b

sl
D3V
sl
prosomere 3
Cx*
peduncular hypothalamus
Hi*
sl
LV
IVF
terminal hypothalamus
mge
Cx*
sl
V3V GP
POA*

c

Diencephalon
peduncular hypothalamus
basal plate sl alar plate
Boundary between hypothalamus and rest of telencephalon
neurohypophysis (posterior pituitary)
terminal hypothalamus
Telencephalon
basal plate sl
acroterminal segment
alar plate
eye

ventral ↔ dorsal
caudal / rostral

Components of hypothalamus

d

Hippocampus
Isocortex
pallium (cortex)
striatum
septum
Paleocortex
pallidum (GP)
OB
POA
anterior commissure

caudal / dorsal

Components of rest of telencephalon

input from the retina, but has very different projections from the lateral geniculate nucleus (a part of the dorsal thalamus). Like the reticular nucleus and zona incerta, the pregeniculate nucleus is full of GABAergic neurons, as part of intrinsic inhibitory circuitry (Puelles *et al.* 2012b).

Hypothalamus

The hypothalamus (Figs. 4.1 and 4.3; Table 4.4) plays a central role in controlling the automatic functions of the nervous system, the blood pressure and heart rate changes that accompany emotional responses and the glands of the endocrine system, as well as a range of complex behaviours that maintain the body and the species (nutrient balance, water intake, thermoregulation and mating). The hypothalamus is closely influenced by the limbic system and controls the endocrine system, allowing it to influence the internal environment of the body in response to emotional states.

As noted above, the hypothalamus was previously regarded as a component of the diencephalon, but modern gene-mapping studies of embryonic development have shown that it is a distinct region of the forebrain, quite separate from the diencephalon. The hypothalamus has both alar and basal plate derived components (Puelles *et al.* 2012a) and three rostrocaudal segments (acroterminal, terminal and peduncular; see Figs. 4.1 and 4.3). Flexion of the embryonic forebrain results in the alar-derived components of the hypothalamus lying closer to the snout than the basal derivatives. The retina is also an embryonic outgrowth of the alar plate component of the acroterminal hypothalamus.

Telencephalon

The telencephalon as an adult brain region (Figs. 4.1 and 4.3) has traditionally been defined as those mature derivatives of the embryonic telencephalic brain vesicle, but recent gene-mapping studies have challenged the long-held distinction between telencephalic and traditional diencephalic components (Puelles *et al.* 2012a). Nevertheless, for practical purposes of considering adult brain structure it is still useful to bear in mind that the telencephalon consists of a dorsal, sheet-like, laminated pallium (the cortex of mammals), ventral striatal and pallidal masses, and a rostromedial septal region, all surrounding the lateral ventricle (Fig. 4.1). In modern conceptions, the preoptic area (formerly regarded as part of the hypothalamus) is also included in the telencephalon (Puelles *et al.* 2012a).

The pallium can be divided into isocortex, meaning that this part of the pallium has six layers at some stage during development, and allocortex (hippocampus or archicortex, and olfactory or paleocortex) that have between three and five layers. The isocortex contains discrete regions for major senses (hearing, vision and somatosensation), motor regions, and

Figure 4.3: The embryonic hypothalamus and telencephalic vesicle and their derivatives. Figure (a) shows a sagittal section through an 8.5 mm greatest length platypus embryo, with the neuromeres labelled. The box in (a) shows the position of the region illustrated in (b). Figure (b) shows a line diagram delineating the boundaries between regions of the front of the forebrain. Note the course of the sulcus limitans (sl) through the dorsal and ventral third ventricle (D3V and V3V), respectively and its termination behind the primordium of the preoptic area (POA*). Figures (c) and (d) show the zones of the embryonic hypothalamus and telencephalic vesicle, respectively. The hypothalamus is divided into alar and basal plate regions by the sulcus limitans. Note that, because of the curved course of the sl, the alar plate (and hence the dorsal direction) is facing towards the front of the head. Conversely, the basal plate (and hence the ventral direction) is facing towards the brainstem. At right angles to the sl, the hypothalamus is divided into three rostrocaudal segments: peduncular, terminal and the thin acroterminal, proceeding from caudal to rostral. The rostrocaudal and dorsoventral patterning divides the hypothalamus into six zones (see Table 4.4 for derivatives of each). The eye (strictly speaking, the optic stalk) is a derivative of the alar part of the acroterminal segment while the neurohypophysis (posterior pituitary) is a derivative of the basal part of the acroterminal segment (c). Figure (d) shows the derivatives of the telencephalic vesicle. Note that not all of these will appear in any one sagittal section (e.g. a and b), because the striatum and pallidum arise in distinctly different ganglionic eminences. The laminated sheet-like part of the telencephalon (the pallium) is further divided into the isocortex and two allocortical regions (hippocampus or archicortex, and paleo- or olfactory cortex). The olfactory bulb (OB) is also a layered structure derived from the telencephalic vesicle. Cx* – cortical primordium; GP – globus pallidus; Hi* – hippocampal primordium; Hy*–hypothalamic primordium; IVF – interventricular foramen; LV – lateral ventricle; mge – medial ganglionic eminence; mes1 – mesomere 1; p1, p2, p3 – prosomeres 1, 2 and 3.

association cortex for higher order processing of sensory information or executive functions directing complex behaviour. The hippocampus is concerned with laying down new memories, while the paleocortex is concerned with processing of olfactory information.

The striatum and pallidum are components of what were traditionally called the basal ganglia. They develop from bulges (lateral and medial ganglionic eminences) of the telencephalic vesicle into the interior of the embryonic lateral ventricle. The striatum is divided into a dorsal component (the caudatoputamen) and a ventral component formed by the nucleus accumbens, olfactory tubercle and granule cell clusters. The pallidum includes a dorsal component, the globus pallidus that can be divided into medial and lateral subdivisions; and a ventral pallidum. Modern functional conceptions of the striatum and pallidum group the dorsal striatum and pallidum together as a dorsal striatopallidal system (see Chapter 6), mainly concerned with cognition and motor behaviour; and the ventral striatum and pallidum together as a ventral striatopallidal system, mainly concerned with motivation and reward.

The septum or septal area consists of four groups of nuclei (lateral, medial, posterior and ventral septal nuclei) arranged in a wall (or septum) medial to the rostral end of the lateral ventricle. The septum is concerned with regulating complex behaviours to do with social behaviour, learning and maintaining a constant internal environment (homeostasis). The outflow of information from the septum includes a descending projection to the hypothalamus and brainstem involved in the control of neuroendocrine and autonomic function, and an ascending projection to the cerebral cortex and striatum concerned with learning and memory (see review in Medina and Abellán 2012).

The preoptic area was traditionally regarded as part of the hypothalamus, but is more correctly regarded as a separate element of the telencephalon. The preoptic area is involved in autonomic regulation and homeostasis. It is particularly notable because the area contains sexually dimorphic regions in therians, i.e. nuclei that show clear differences between genders. At present sexual dimorphism of the region has not been reported in monotremes.

The basal telencephalon also includes diverse groups of neurons that project their axons to the cortex. These include neurons that use acetylcholine (the basal nucleus of Meynert) as well as other neurons using the excitatory neurotransmitter glutamate or the inhibitory neurotransmitter gamma aminobutyric acid (GABA) in their projections to the cortex.

Brain chemistry of monotremes

Studies of brain chemistry in the monotremes are limited to molecular phylogenetic analysis of brain gangliosides and neurotrophins.

Brain ganglioside composition is important in thermoregulation and has been shown to correlate with thermal adaptation. Rahmann *et al.* (1984) have noted four trends in brain gangliosides with progression to (so-called) more complex nervous systems: (1) an increase in concentration; (2) a decrease in the number of single fractions; (3) a change in polarity of the molecules; and (4) an alteration of the preponderance of one of three biosynthetic pathways. Relative to therians, the monotremes have more individual ganglioside fractions, and monotremes, marsupials and so-called 'lower' placentals have a lower percentage of alkali-labile gangliosides than more encephalised therians (Rahmann *et al.* 1986).

Neurotrophins are a family of proteins that serve as survival factors in the development of the peripheral nervous system. Types of neurotrophins include nerve growth factor (NGF), brain-derived neurotrophic factor (BDNF), neurotrophn-3 (NT-3) and neurotrophin-4 (NT-4). In the peripheral nervous system, the neurotrophins promote the survival and maintenance of sensory ganglion cells and autonomic ganglion cells. In the central nervous system, they probably promote neuronal differentiation and plasticity.

Neurotrophins are highly conserved molecules in vertebrate evolution, reflecting their fundamental and critical role in neural development in all vertebrates. Many mutations in neurotrophins are lethal because loss of neurotrophic function leads to catastrophic effects in the developing nervous system. Nevertheless, there are some significant differences in neurotrophin structure between therians and monotremes (Kullander *et al.* 1997). Kullander and colleagues concluded that the sequences of monotreme neurotrophins are more closely associated with birds than therian mammals, but that overall the functions of the neurotrophins are very similar across all vertebrates. NGF is probably more divergent than BDNF and NT-3, perhaps because evolutionary constraints on the amino acid sequence of NGF have been lower than for other neurotrophins.

Cellular populations of the monotreme brain

Structure of the monotreme cortical neurons, in particular those that have very distinctive architecture with potential functional implications, will be discussed in Chapter 7. At this point it is appropriate to consider the morphology of the non-neuronal (i.e. glial) and vascular support cells of the monotreme brain and the possible evolutionary significance of their features.

In therians, glial cells include astrocytes and oligodendrocytes (grouped together as macroglia, derived locally from neuroectoderm during embryonic development); and microglia (derived from the mesoderm and invading the central nervous system during development). Other key support cells in the brain include the ependymal cells and choroid plexus that form the interface between the brain tissue and cerebrospinal fluid spaces. Nothing is currently known about the fine structure or function of the ependyma or choroid plexus in monotremes, but there is an intriguing report that suggests significant differences between the monotremes and therians in macroglial structure and function.

In therians, astrocytes form a glial limiting membrane around central nervous system blood vessels and provide the structural matrix for the brain and spinal cord. Therian astrocytes are further divided into protoplasmic astrocytes that mainly reside in the grey matter and fibrous astrocytes that are mainly found in the white matter. Protoplasmic astrocytes have shorter processes with many short branches, whereas fibrous astrocytes have long thin processes with few branches. Astrocytic processes terminate in expansions called end-feet that surround brain capillaries and the inner surface of the pia mater, contributing to the blood–brain and blood–cerebrospinal fluid barriers.

In therians, oligodendrocytes are smaller than astrocytes and have more irregularly shaped and darker staining nuclei. Oligodendrocytes play a key role in the myelination of axons. Microglia are smaller still and have a primary function of phagocytosis and immune protection of the brain and spinal cord.

Although limited in scope, the electronmicroscopic study by Lambeth and Blunt (1975) suggests that there may be significant differences between monotremes and therians in the structure and function of macroglia. In the platypus, glial cells can be readily distinguished from neurons because they are smaller than neurons, have denser cytoplasmic and nucleoplasmic matrices, and lack Nissl substances and synaptic contacts. In the grey matter, glia are found surrounding neurons and around blood vessels. They contain microtubules, but these are not circumferentially arranged (as in therian oligodendrocytes), nor are they associated with filaments (as in immature therian astrocytes). The platypus glia have neither the filaments or glycogen of mature therian astrocytes, nor the cytoplasmic and nucleoplasmic densities of therian oligodendrocytes.

Significantly, Lambeth and Blunt found only a single type of macroglia in the platypus, but two types were found in the brain of an immature echidna. In the echidna they studied, lighter glial cells had round, pear-shaped or occasionally indented nuclei. A darker type of glial cell was also found in the echidna brain, but was less common than the lighter type. Dark glial cells of the echidna brain were less regular in appearance than the more lightly stained type and had a more homogeneously dispersed chromatin in the nucleus. Microtubules and mitochondria were also more numerous in the darker type. Although two types of glia were found in the immature echidna, neither type closely resembled the astrocytes or oligodendrocytes of therians. The light cells contained neither the filaments nor glycogen of therian astrocytes. Conversely, the darker cells had microtubules, but none of the rounded shape or circumferential arrangements of microtubules that therian oliogodendrocytes do.

Lambeth and Blunt reached the conclusion that the monotremes have only a single macroglial cell type, albeit with light and dark variants in the immature brain. If this were indeed the case, then monotremes would share this unusual feature with cyclostomes.

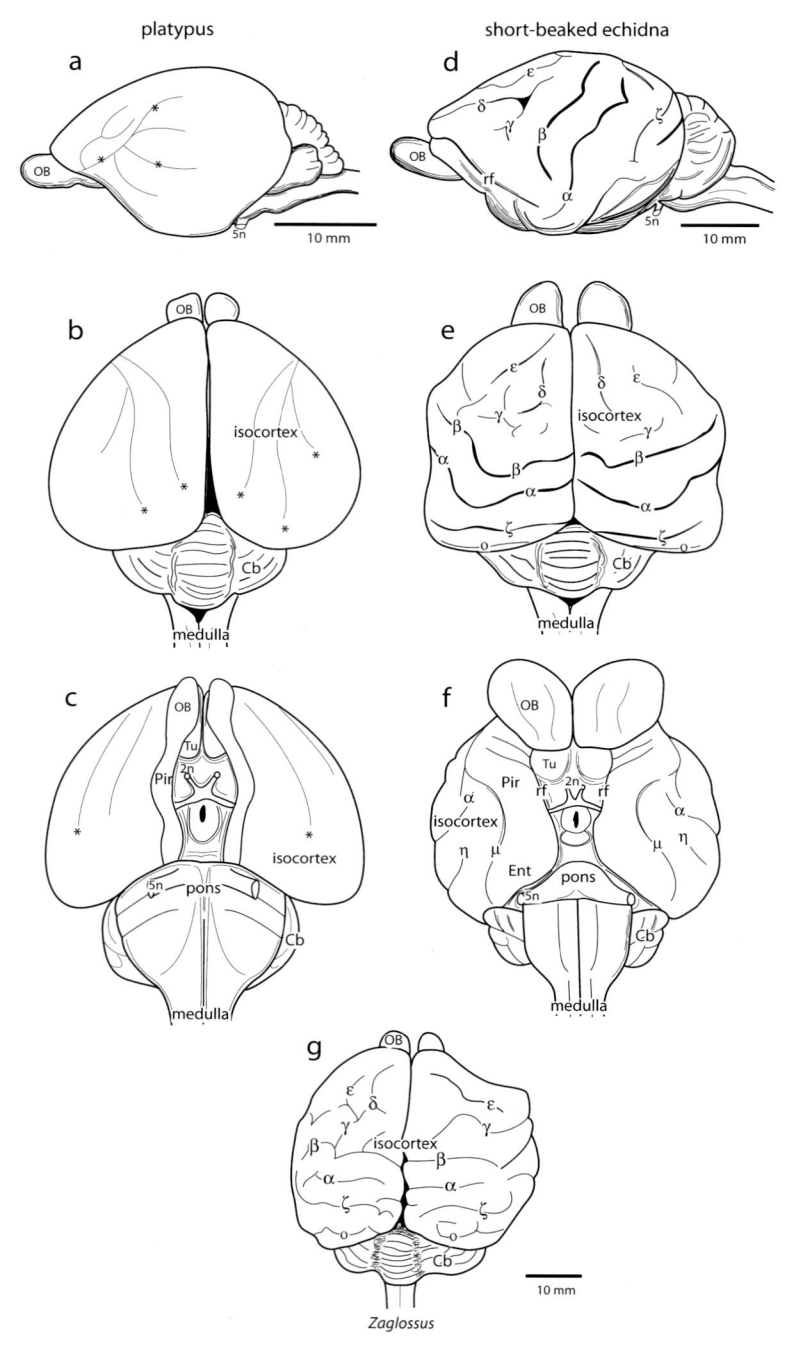

Figure 4.4: External features of the brain of the platypus in left lateral (a), dorsal (b) and ventral (c) views. The line diagrams are based on photographs of platypus brains held in the National Museum of Australia. Asterisks indicate the vascular grooves on the platypus cortical mantle. Line diagrams (d), (e) and (f) show left lateral, dorsal and ventral views, respectively, of the brain of the short-beaked echidna, also based on brains held in the National Museum of Australia. The line diagram labelled (g) is based on illustrations of the dorsal view of the brain of a long-beaked echidna (genus *Zaglossus*) from Kolmer (1925) (reproduced in Griffiths 1978). Greek letters indicate the major sulci of the cortical surface for both the short and long-beaked echidna. Note the considerable variation in the sulci between the two sides of the *Zaglossus* brain. 5n – trigeminal nerve; Cb – cerebellum; Ent – entorhinal cortex; OB – olfactory bulb; Pir – piriform cortex; rf – rhinal fissure.

Furthermore, the single glia would need to serve not just the myelination function of oligodendrocytes, but also the neuronal support and homeostatic roles of astrocytes. In this respect, the single type of monotreme macroglia may resemble the Schwann cells of the peripheral nervous system, which serve the dual functions of both myelination and neuronal support. Given that many other non-mammalian vertebrates have distinct astrocytes and oligodendrocytes, these findings have profound implications for the phylogeny of monotreme glia, but it would be unwise to make broad-reaching conclusions based on a single study. Further studies of glia in the monotremes, using modern molecular markers of glial types, are necessary to confirm whether the conclusions that Lambert and Blunt have reached are justified.

Vasculature of the monotreme brain

The vasculature of the monoteme central nervous system is actually more like that of placental mammals than marsupials. Marsupials have a distinctive pattern of central nervous system vascularisation whereby capillary loops arise from branching paired vessels (an artery and its accompanying vein). The paired artery and vein branch at contiguous points, and the terminal capillaries link the ends of the branches (Wislocki and Campbell 1937). Among the mammals, this feature is peculiar to marsupials (both Australian and American): the capillary networks of the platypus and echidna central nervous system share the meshwork pattern of placental central vasculature (Sunderland 1941).

External morphology of the monotreme brain

Figure 4.4 shows the main external features of the brains of the platypus, and short- and long-beaked echidnas.

Although several studies of the platypus brain were published in the 19th century (Elliot Smith 1896b, 1899), Hines (1929) provided the first detailed description of the brain of the platypus at both the macroscopic and sectional level. The isocortex of the platypus is large (4100 mm^3 and 48% of brain volume, Pirlot and Nelson 1978) and dominates the external appearance (average length ± s.d. of 25.7 ± 3.2 mm and average width of 30.6 ± 1.7 mm for adults; Appendix Table 4), but it is relatively smooth or lissencephalic. Although the platypus brain does not have the well-defined sulci (grooves) of the echidnas, there are consistent vascular grooves that radiate from the frontal pole across the surface of the forebrain and limit electrophysiological access to the cortex of the frontal pole (Bohringer and Rowe 1977). Hines also mentioned a cruciate sulcus between the frontal and parietal regions, but this is not consistently observed and has not been included in Fig. 4.4. On the other hand, the olfactory cortex (including the olfactory bulb, anterior olfactory area, piriform cortex and entorhinal cortex) is separated from the isocortex by a prominent rhinal fissure. Monotremes do not have the corpus callosum of placentals, so the main commissure for transfer of information between the hemispheres is the anterior commissure (average area ± s.d. of 5.4 ± 1.7 mm^2 in adults; Appendix Table 4), which can be seen on the midline cut surface of the brain (Flower 1865; Elliott Smith 1910).

The other distinctive external feature of the platypus nervous system is the large size of both the trigeminal ganglion and trigeminal nerve. The pronounced development of the trigeminal sensory nuclear column (see Chapter 9) also produces a pair of prominent bulges on the lateral edges of the rostral brainstem. The external appearance of the platypus cerebellum is broadly similar to that of other mammals, but the cerebellum is wider in the transverse dimension due to the presence of lateral extensions of the nodulus and pyramis (see Chapter 5).

Ziehen (1897, 1901) who focused on external morphology and Abbie (1934), who made an extensive study of the brainstem, made the earliest detailed studies of the brain of the short-beaked echidna. The outstanding feature of the telencephalon of both short- and long-beaked echidnas is the large size and complexly folded topography of the cerebral hemispheres (volume of 11 400 mm^3 or 43% of brain volume, Pirlot and Nelson 1978; mainly isocortex, but also the allocortical regions devoted to olfaction and forming a large ventrally projecting piriform lobe). The cerebral hemispheres of the adult short-beaked echidna are an average ± s.d. of 34.0 ± 2.8 mm in length and 40.9 ± 4.1 mm in width (Appendix Table 5).

Terminology for the (approximately) 10 sulci of the echidna cerebral hemispheres is based around a system of Greek letters assigned in descending order of prominence of the sulci (Rowe 1990; Griffiths 1978). Nine of these sulci are visible on the dorsal (α, β, γ, δ, ε, ζ, o) or ventral (μ, η) surfaces, while sulcus ψ is best seen on the medial surface of each hemisphere. Sulci α and β are of the greatest functional significance because they define the boundaries between motor and somatosensory, and motor and (pre)frontal cortex, respectively. Although there is some variation in the pattern of cortical sulci between subspecies of short-beaked echidna and even between the two sides of the one animal, the pattern of cortical sulci α, β and ζ is broadly similar both within *T. aculeatus* and between short- and long-beaked echidnas (Figs. 4.4 and 4.6). The cross-sectional area of the major commissural connection in the short-beaked echidna, the anterior commissure (Elliott Smith 1910), is 8.6 ± 2.9 mm^2, almost 60% larger than that in the platypus (see Appendix Tables 4 and 5). The trigeminal nerve and ganglion is large in the echidnas, but not as striking as in the platypus.

Brain characters in mammalian evolution: how do the monotremes compare?

The monotremes have traditionally been seen as primitive mammals, although this perception is deeply flawed and ignores their many sophisticated specialisations in both body structure and neural systems (see Chapter 1). In truth the monotremes (like any other vertebrate) present a mosaic of primitive and derived features. The details of how this mosaicism is reflected in the nervous system of modern monotremes will be discussed in detail in Chapter 14, but neural characters have also been used to construct mammalian phylogenies and to assess the position of the monotremes in mammalian evolution (Johnson *et al*. 1982a, b; 1994) and these analyses are relevant to the present discussion of monotreme origins and evolution.

Johnson and colleagues identified nine derived neural states among mammals that are characteristic of certain restricted groups of mammals (Johnson *et al*. 1994), but none of these is diagnostic or specific for the monotremes. In fact, only one neural character of the 24 analysed by Johnson and colleagues in their study was considered to be derived in the monotremes. This is the divided (bifurcate) course of the optic tract as it enters the diencephalon (Table 4.5, discussed in more detail in Chapter 8). In monotremes, the pronounced separation of the visual components of the dorsal thalamus and prethalamus (i.e. the lateral geniculate nucleus and pregeniculate nuclei, respectively) leads to an equally pronounced bifurcation of the optic tract to distribute axon terminals from retinal ganglion cells to these nuclei. In all other mammals, the lateral geniculate nucleus and pregeniculate nucleus (also called the ventral lateral geniculate nucleus in some species) are close together in the adult brain. The separation of the lateral geniculate and pregeniculate nuclei is a consequence of the enlargement of the somatosensory components of the dorsal thalamus (ventral posterior medial nucleus), particularly in the ornithorhynchid lineage where the evolution of electrosensory abilities has recruited ever more neurons to this part of the dorsal thalamus (see Chapters 5 and 9). Nevertheless, perhaps influenced by the preconception of monotremes as 'primitive', Johnson and colleagues had 'some trepidation' in scoring the monotreme condition as derived and that of other mammals as primitive (Johnson *et al*. 1982b).

These observations are consistent with the long-standing perception that the monotremes have retained many plesiomorphic neural traits, but ignore the presence of electroreception in the group or the clear specialisation of the echidnas for olfaction. Sample trees obtained by Johnson and colleagues maintain the segregation of the monotremes and marsupials, and support the closer affiliation of marsupials with placentals rather than with the monotremes.

Encephalisation of the monotremes

There are several quantitative measures of brain size that have been used in making comparisons between mammals. *Absolute brain size* is the mass or volume of the brain considered regardless of body mass. It is a measure that is often ignored in comparative neuroscience, but Deaner and colleagues (Deaner *et al*. 2007) have noted that overall brain size is a superior predictive factor of cognitive ability among primates. *Relative brain size* is brain mass relative to that expected of the 'average' animal of the same type and body size (Striedter 2005). This latter is measured by plots of

Table 4.5. Characters of monotreme v. marsupial and placental brains[1]

Character order	Arteriovenous loops in brain vessels	Mitral cells of olfactory bulb form a monolayer	Course of olfactory tract past accessory olfactory formation	Presence of jugal cones with oil droplets	Bifurcate optic tract in thalamus	Hemispheres joined by corpus callosum	Presence of fasciculus aberrans	Somatosensory cortex with barrels	Emergence of 7n over or beneath 5n	Medial or lateral ventral nuclei of inferior olive
Monotremata	0	0	0	0	1	0	0	0	0	0
Didelphimorphia	1	1	0	0	0	0	0	0	0	0
Dasyuromorphia	1	1	0 to 1	1	0	0	0	0 to 1?	0	0
Diprotodontia	1	1	0 to 1	1	0	0	1	0 to 1	0	0
Rodentia	0	1	3	0	0	1	0	0 to 1	3	3
Cetacea	0	1	5	?	0	1	0	0	3	3
Artiodactyla	0	1	0 to 2	0	0	1	0	0	3	3
Primates	0	1	3 to 4	0	0	1	0	0	3	3
Carnivora	0	1	0 to 2	0	0	1	0	0	3	3
Chiroptera	0	1	3 to 4	0	0	1	0	0	3	1 to 3

1 Adapted from Johnson et al. (1982b).

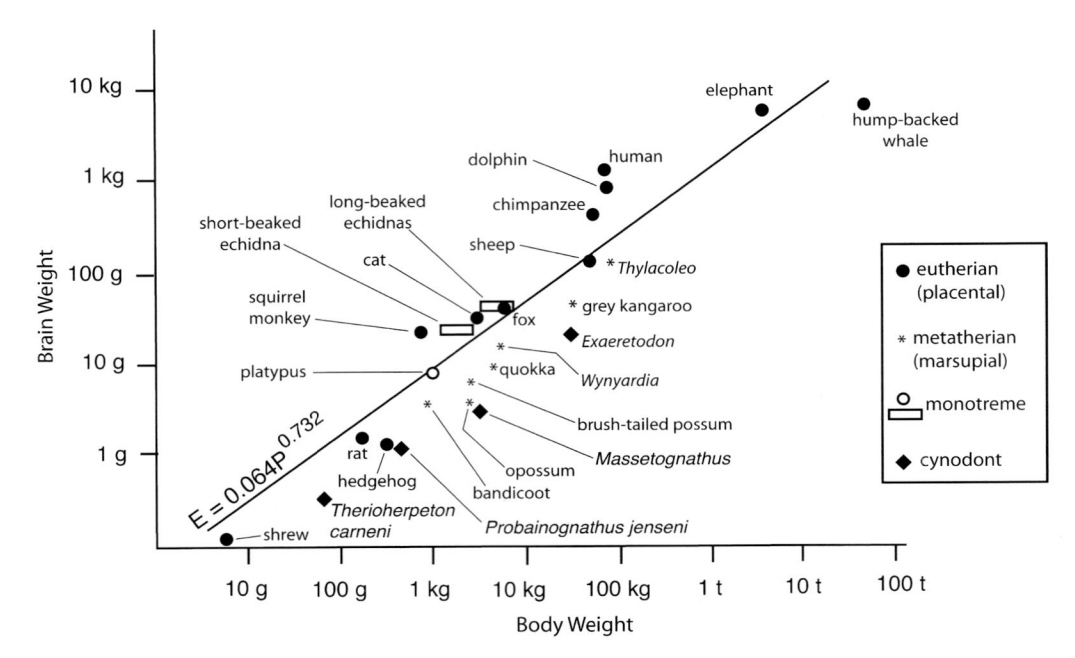

Figure 4.5: Brain size plotted against body weight for the monotremes and selected placentals, marsupials and probable mammalian ancestors (cynodonts). Data are derived from Hofman (1982) for placental brains and the opossum (*Didelphis virginiana*), Haight and Murray (1981) and Meyer (1981) for living and extinct Australian marsupial brains, Pirlot and Nelson (1978) for platypus and the author's own data on endocranial volume for the short- and long-beaked echidnas. Values for cynodonts are derived from cranial endocast volumes determined by Quiroga (1979, 1980a, 1980b, 1984). Note that both the platypus and the echidnas have brain sizes larger than many modern and extinct marsupials and comparable to many less encephalised placentals. All modern and recent mammals have brain sizes several orders of magnitude greater than cynodonts.

mammalian brain and body mass collected from a vast array of animal groups and scaled logarithmically to accommodate the large range in masses, where the line of best fit indicates the 'expected' brain mass for a particular body mass (Striedter 2005). The underlying assumptions of these comparisons are that a given volume of brain tissue has essentially the same metabolic activity and information processing capacity in all mammals and that a larger brain implies greater behavioural complexity and/or flexibility. Assumptions of uniform metabolic activity among mammals may well be flawed, particularly in the case of the monotremes, given that their usual core temperature is well below that of either marsupials or placentals. Nevertheless, large brains represent a significant investment in structural protein and lipid during development, not to mention blood flow and nutrients throughout adult life (Aiello and Wheeler 1995; Isler and van Schaik 2006, 2009a, b), and, under the expensive organ hypothesis, this would detract from reproductive potential and Darwinian fitness unless the large brain conferred some other advantage.

Brain weights of the monotremes have been plotted against body weight to compare the relationship between these two parameters in a wide variety of mammals and their putative ancestors (Hassiotis *et al.* 2003; Fig. 4.5). Values for the short-beaked echidna lie within a region close to those of New World primates such as the squirrel monkey, and lie significantly above both the regression line describing therian mammals and the cluster of points representing living and extinct marsupials (Hassiotis *et al.* 2003). Similarly, the long-beaked echidnas have brain sizes comparable to, or greater than, many carnivores (e.g. domestic cat and European fox). On the other hand, the platypus lies on the regression line describing therians.

Figure 4.5 also incorporates brain size values for some representative cynodonts (Hassiotis *et al.* 2003), which represent putative ancestors for all living mammals (Musser 2003). Both the short- and long-beaked echidnas in particular have greatly expanded brains compared to the common ancestors of therian and prototherian mammals, mainly due to expansion of the highly gyrified (but not particularly thick) cortex. The brain of the platypus has a more modest enlargement, probably due to thickening of the unfolded (lissencephalic) cortex. In the case of the echidnas, this represents at least a 10-fold expansion of brain weight compared to cynodonts of similar body weight, whereas the platypus brain is only about three or four times larger than a cynodont of similar body weight. By contrast, all the living and extinct marsupials (even carnivores that would be expected to be more highly encephalised) have brain weights only two or three times greater than the cynodonts.

Encephalisation indices have also been calculated (Hassiotis et al. 2003) according to the method outlined by Hofman (1982). The short-beaked echidna has values of the encephalisation measure C (where $C = $ brain weight/body mass$^{0.732}$) that range from 0.050 to 0.094 and the platypus has values of C around 0.060. Values for brain mass of the long-beaked echidnas are not readily available, but can be deduced from the endocranial volumes for modern long-beaked echidnas (44–60 mL, unpublished observations by the author) and best estimates of body weight (averaging around 9 kg; see Chapter 2). For the long-beaked echidnas, values of C lie within the range of 0.056 to 0.076. Both long and short-beaked echidnas and the platypus have values of C that place them in a similar range to placental ungulates, carnivores and prosimians, three or four times higher than didelphid marsupials and about twice as high as lipotyphlid insectivores (Hassiotis et al. 2003).

Paleoneurology of the monotremes

Paleoneurology is the study of the evolution of the nervous system as revealed by analysis of fossils. Of necessity, it depends on deducing soft tissue structure (i.e. the shape and size of the brain, spinal cord and peripheral nerves) from remnant hard tissues (the skull and vertebrae). The fossil record for the monotremes is scanty, even more so when one is considering evidence for evolution of the brain. There is only one ornithorhynchid fossil of any antiquity that throws light on monotreme brain evolution and all the available tachyglossid fossils are too recent to give any insights into early brain evolution in this group.

Evolution of the brain in ornithorhynchids

Most cranial fossils of ornithorhynchids, particularly those from the Cretaceous and early Tertiary, are fragmentary and confined to the jaw and teeth (Musser 1998, 2003; Rowe et al. 2008). The notable exception is the beautifully preserved skull of *Obdurodon dicksoni* from Riversleigh in Queensland (Macrini et al. 2006; Musser 1998, 2003 and this book; Musser and Archer 1998) that allows deductions about brain size, shape and trigeminal specialisation.

The specimen in question consists of the skull, dentary and dentition of a middle Miocene platypus, which is extraordinarily like the modern platypus in many features (hypertrophied bill, overall brain shape and trigeminal specialisation), but does have some features (e.g. retention of functional teeth, and the robust flattened skull and dentary) that suggest that *Obdurodon dicksoni* had some differences in diet and lifestyle from modern platypuses (Musser and Archer 1998). Musser and Archer suggested that *Obdurodon dicksoni* may have foraged at the water surface or high in the water column, rather than in the river substrate where the modern platypus searches for most of its food.

A digital cranial endocast of *Obdurodon dicksoni* was extracted from high resolution computerised tomography of the skull by Macrini and colleagues (Macrini et al. 2006). They noted several neurological similarities between *Obdurodon dicksoni* and modern platypuses. These include well-developed casts of the paraflocular parts of the cerebellum; an ossified falx cerebri (the membrane between the two cerebral hemispheres); three endoturbinals (elevations) in the nasal cavity; a reduction of the size of the olfactory bulb; and large impressions on the skull due to the trigeminal nuclei, trigeminal ganglion and trigeminal nerve divisions (particularly maxillary and mandibular). Interestingly, the endocranial volume of *Obdurodon dicksoni* is ~50% larger than for an average modern platypus (15.4 mL compared to 9.7 mL), although the encephalisation of *Obdurodon dicksoni* is probably comparable to both modern platypus and therians, because the larger endocranial volume is probably

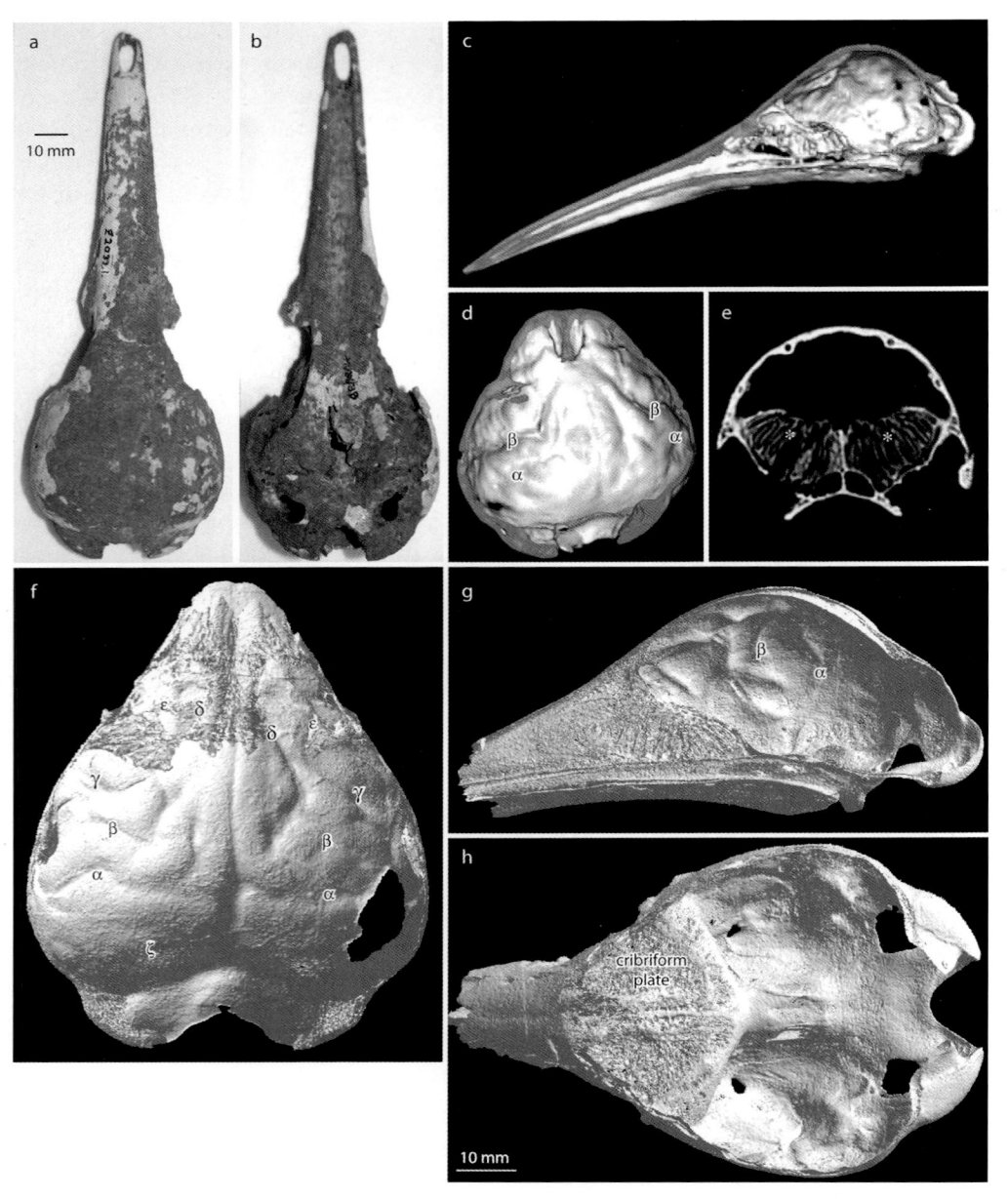

Figure 4.6: Photographs and CT scans of extinct long-beaked echidnas from the Australian Pleistocene. Images (a) and (b) are dorsal and ventral views of Z2031.1, a long-beaked echidna from Montagu in Tasmania (Murray 1978a, b). This specimen was assigned *Zaglossus ramsayi* by Murray (1978a, b), but has subsequently been named *Megalibgwilia ramsayi* by Griffiths *et al.* (1991). Image (c) shows the interior of the right half of the digitally reconstructed skull of Z2031.1 following computerised tomography of the specimen. Image (d) shows an interior view of the dorsum of the skull indicating the ridge formed by the α sulcus that separates motor and primary somatosensory cortex and the β sulcus that marks the rostral boundary of the motor cortex. Image (e) shows a transverse section through the nasal cavity of Z2031.1, showing the vertically oriented plates that carry the extensive olfactory epithelium of this long-beaked echidna. Images (f), (g) and (h) show reconstructions of the skull of *Megalibgwilia ramsayi* (P23144) from the Henschke fossil cave, Naracoorte, South Australia (Griffiths *et al.* 1991) following microCT. Note the impression of cortical sulci on the interior of the dorsal skull (f) and the interior of the right half of the skull (g). This specimen also has an extremely large cribriform plate area (h), indicating that the Pleistocene long-beaked echidnas shared the olfactory specialisation of modern short-beaked echidnas.

due to the larger body mass of *Obdurodon* (over 2 kg compared to 1.4 kg for the modern platypus).

Some distinctive neurological features of *Obdurodon* include the arrangement of the sphenorbital fissure, the depth of the trigeminal nuclear impression and the size of the hypophyseal (pituitary) fossa. The sphenorbital fissure (also called foramen pseudoopticum – see Table 5.1 – by other authors; de Beer and Fell 1936) transmits the optic, oculomotor, trochlear, ophthalmic division of trigeminal and abducens nerves in *Ornithorhynchus* and has no internal subdivision in the modern platypus, but in *Obdurodon* has a clearly visible bony division. Macrini and colleagues also felt that the trigeminal nuclei impressions on the endocast were flatter in *Obdurodon* than in *Ornithorhynchus* and this may suggest that trigeminal specialisation was less advanced in *Obdurodon* than in the modern platypus. Finally, the hypophyseal fossa for the pituitary gland occupies a larger percentage of the endocranial volume in *Obdurodon* (0.17%) than in mature and juvenile *Ornithorhynchus* (0.10% and 0.08%, respectively).

Evolution of the brain in tachyglossids

The earliest definitive tachyglossid is a specimen (*Zaglossus robustus*) dated to the middle Miocene (13 to 14 mya). It consists of only a partial skull that does not allow any conclusions regarding brain size or shape, although the shape of the beak suggests that this animal resembled the Pleistocene long-beaked echidnas in both structure and habits (Musser 2003; see also Chapter 1). During the Pliocene and Pleistocene, three distinct types of echidnas emerged: robust forms with an upright stance like *Zaglossus hacketti* from Western Australia; medium-sized to large-sized long-beaked echidnas like the modern New Guinean forms; and the smaller, more specialised short-beaked echidnas (Musser 2003; see Chapter 1).

Most of our paleoneurological consideration must be confined to the Pleistocene tachyglossids (*Megalibgwilia ramsayi*) from South Australia (P20488, P23144, P22811) and Tasmania (Z2031) (Griffiths *et al*. 1991; Murray 1978a, b). These specimens range in age from 13 000 to 120 000 years before present (Griffiths *et al*. 1991). Preliminary unpublished investigations of these by the author suggest that the neurology of these echidnas is very similar to that of the long-beaked echidnas and also rather like that of modern short-beaked echidnas. The endocranial volume of these specimens ranges from 48.5 to 60.5 mL, well within the range for modern long-beaked echidnas (39.5 to 63.5 mL). Computerised tomography of the skulls allows the interior of the skull to be studied in three-dimensional reconstructions. The pronounced gyrification of the tachyglossid brain leaves distinctive impressions on the inside of the skull (Fig. 4.6); these allow the allocation of sulci to the missing brain and in turn permit the deduction of the position of major functional areas on the isocortex. The remarkable similarities to modern tachyglossids in endocranial volume, shape and conformation of the cribriform plate and cortical gyrification suggest that the Pleistocene echidnas had very similar neurological organisation and behaviour to modern long-beaked echidnas.

Questions for the future

Although the results of traditional embryological analysis of archived monotreme material are consistent with the segmental organisational plans that have been applied to therian embryos in recent years, there are significant questions remaining, particularly with respect to the development of the rostral hindbrain. The rostral rhombomeres (r1 and r2) give rise to the rostral trigeminal nuclei and the cerebellum in therians. In the case of the platypus, the rostral trigeminal sensory nuclei are particularly large and one would expect enlargement of the developmental regions that contain those precursor populations. Therefore analysis of gene expression and rhombomeric segmentation in the developing platypus brain and the distribution of adult neural derivatives from those segments are likely to reveal unique features of the rostral hindbrain in the platypus (see also Chapters 5 and 9).

The intriguing findings of Lambeth and Blunt (1975) regarding macroglial ultrastructure need to be repeated and extended because they have major significance for the structure and function of the monotreme nervous system at the cellular level. If their interpretations are correct, then the neurons of monotremes may have some unique glial support.

Our understanding of brain evolution in the monotremes is greatly limited by the paucity of intact cranial fossils. The presence of a well-preserved ornithorhynchid fossil in the form of the Miocene *Obdurodon dicksoni* has contributed to a biased interpretation of monotreme evolution, namely that the ancestral monotreme was rather *platypus-like* and that modern echidnas are derived from a semi-aquatic ancestor, despite embryological evidence to the contrary (see Chapter 14). Better quality cranial fossils of Cretaceous and early Tertiary monotremes are likely to dispel this notion.

5

Peripheral nervous system, spinal cord, brainstem and cerebellum

Ken W. S. Ashwell

Summary

The monotremes have a similar number and disposition of cranial nerves to other mammals, but the openings for the exit of those nerves from the skull and musculature supplied by the cranial nerves are somewhat different (see also Chapter 15). Development of the peripheral nervous system begins during the early incubation phase of development (~5.0–5.5 mm greatest length or GL) when the neural crest cells begin migration through the trunk. Peripheral nerves grow into the limb buds from ~6.0 mm GL and rapidly establish the main nerve branches. Adult spinal cord structure in the monotremes is similar to that of therians and the grey matter at the centre of the spinal cord can be subdivided into Rexed's laminae according to similar criteria. The short-beaked echidna has remarkably sophisticated dorsal column sensory pathways (for proprioception) that are comparable in size to primates and placental carnivores. Spinal cord development in monotremes follows a similar developmental trajectory to marsupials, with the motor centres in cervical level spinal cord reaching functional maturity before other systems.

The monotreme brainstem is organised along very similar lines to other vertebrates, but in the platypus there are some specific features that reflect the enlargement of the trigeminal sensory system due to electroreception. Monotremes have some anatomical differences in the distribution of catecholaminergic, serotonergic and cholinergic nerve cell groups compared to therians, but these have not yet been convincingly linked to functional diversity.

The monotreme cerebellum has sometimes been described as reptilian or even (and this makes no evolutionary sense!) avian, but it is clearly mammalian in structure with similar cellular structure and corticonuclear subdivisions to therians. Very little is currently known about the sensory input to the various parts of the monotreme cerebellum, and the formation of the cerebellum from the embryonic rhombic lip of the brainstem is almost completely unexplored, despite the clear benefit of analysing this important developmental system in monotremes. As in marsupials, cerebellar development is largely a post-hatching event in the monotremes.

Peripheral nervous system

Structure, function and development of trigeminal system receptors

The ultrastructure and function of the peripheral sensory receptors in the bill and beak will be discussed in Chapter 9 as part of a consideration of the trigeminal somatosensory system. The very special features of the development of the snout mechano- and electroreceptors will also be considered there and in Chapter 14 in the context of the evolution of electroreception.

Cranial nerves

Both platypus and echidnas have 12 cranial nerves. Neither monotreme has a terminal nerve (cranial nerve 0) in adulthood, although this nerve is present during incubation (Ashwell 2012b) and probably provides a migration path for gonadotrophin-releasing-hormone cells to the hypothalamus, as in therians. The details of the structure and function of cranial nerves 1 (olfactory), 2 (optic), 5 (trigeminal), and 8 (vestibulocochlear) will be discussed in Chapters 11, 8, 9 and 10, respectively. Only a brief overview of cranial nerve anatomy will be presented here and the details of peripheral course and muscular supply by the cranial nerves will be considered in Chapter 15 (see Fig. 15.1 and Table 15.1). The key details of the cranial nerves for the short-beaked echidna are summarised in Table 5.1.

There are no reliable estimates of the number of olfactory axons in the echidna, but it is likely to be large given the complex nasal topography and gyrified olfactory bulb of the echidna (see Chapter 11). There is some disparity in the estimated number of myelinated axons in the optic nerve, but the older estimate of Schuster (1910) is actually more consistent with the estimated retinal ganglion cell population (Stone 1983; see Chapter 8 for details). The nerves that supply the intrinsic and extrinsic musculature of the eye (oculomotor, trochlear and abducens) are noticeably small in the echidna, consistent with the limited eye movement and doubtful accommodative capacity of the ciliary muscle. On the other hand, the trigeminal nerve of the echidna is large (43 410 myelinated axons), as would be expected given the role of trigeminal sensation from the beak in assessing the environment, but this pales beside estimates of the number of axons in the platypus trigeminal nerve, which reach an astonishing 1.4 million (Manger and Pettigrew 1996). The facial nerve not only supplies the muscles of facial expression, but also contains parasympathetic axons to the sublingual glands. It is quite large in the echidna, reflecting the development of superficial facial muscles as the primitive platysma of the neck has moved across the face and snout (Griffiths 1978).

The vestibulocochlear nerve will be discussed in detail in Chapter 10. Very little is known about the glossopharyngeal, vagal and accessory nerves in the echidna. Griffiths (1968) states that the accessory nerve supplies muscles of the pharynx, but most likely it supplies muscles of the neck and rostral shoulder (see Chapter 15). The hypoglossal nerve is substantial in size in the short-beaked echidna because it supplies a sophisticated mechanism for extruding and retracting the tongue, and grinding invertebrates against the palate. The details of these will be covered in more detail in Chapter 15.

Peripheral nerves of the trunk and limbs

The physiology of the peripheral nerves will be discussed in Chapter 9 and the course of peripheral nerves in the monotremes will be illustrated in Chapter 15. Few details of the microscopic structure of the peripheral nerves are available for monotremes, but the axonal diameters of dorsal and ventral roots in the short-beaked echidna are quite similar to those of therians (Ashwell and Zhang 1997), suggesting comparable motor function.

In the echidna, myelinated fibres in the ventral (motor) root have two peaks in size distribution. One is at ~1–4 μm diameter (presumably the axons of γ motor neurons) and makes up about a third of all

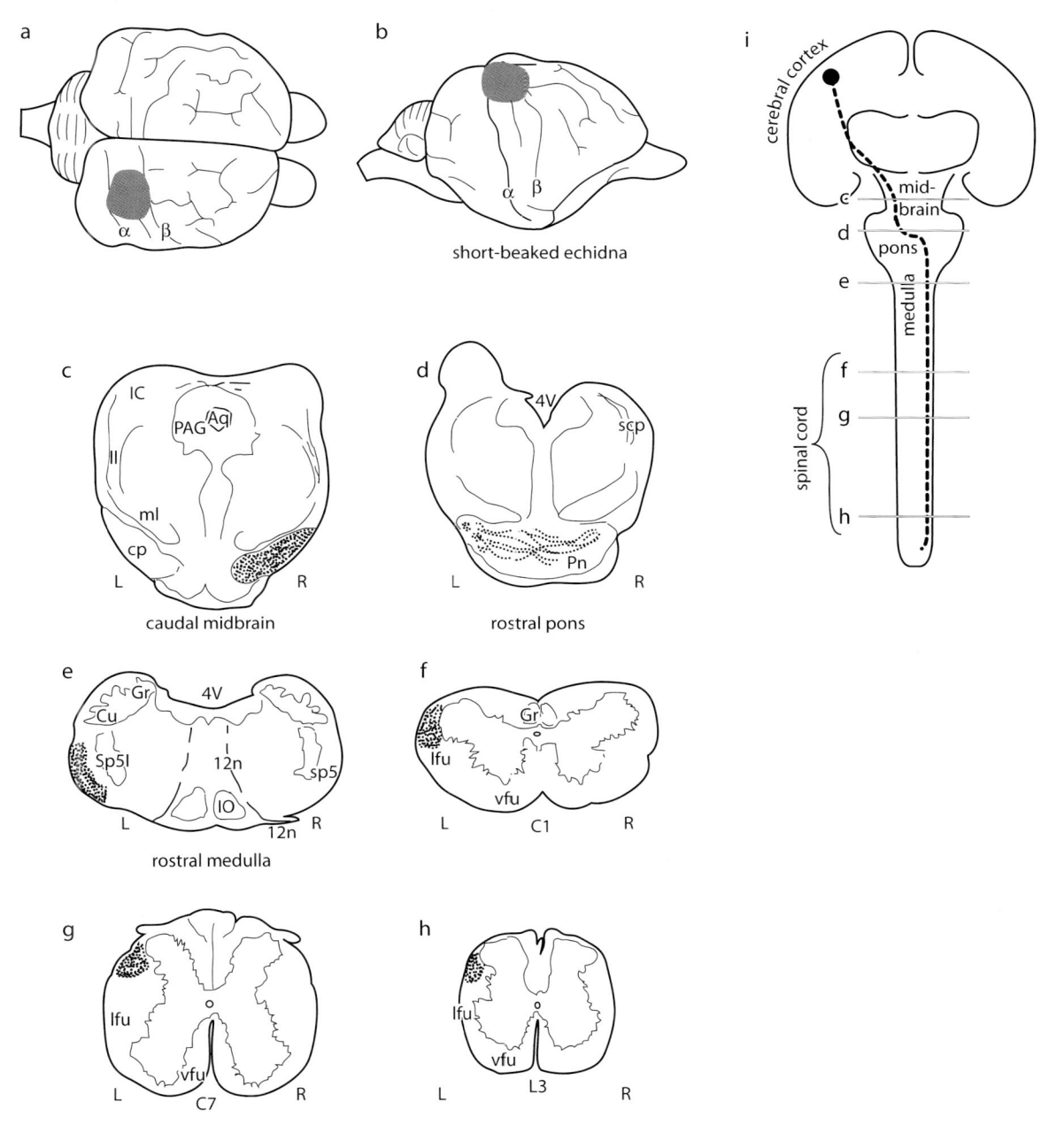

Figure 5.1: The course of the corticospinal tract in the short-beaked echidna illustrated in line diagrams based on data in Goldby (1939). Goldby removed part of the motor cortex that corresponded to the hindlimb and tail representation (a, b). Degenerating axons were traced through the internal capsule, cerebral peduncle (c), a decussation in the rostral half of the pons (d), down the ventrolateral part of the caudal pons and medulla (e) external to the trigeminal spinal tract (sp5), and down the entire length of the spinal cord in the dorsolateral part of the lateral funiculus (f to h). Diagram (i) shows a schematic representation of the course of the corticospinal tract through the neuraxis of the echidna with representative section levels labelled (c to h). Note that the echidna does not have a pyramid-shaped collection of corticospinal axons as therians do. L and R indicate left and right sides of the brainstem and spinal cord. 4V – fourth ventricle; 12n – hypoglossal nerve; Aq – cerebral aqueduct; cp – cerebral peduncle; Cu – cuneate nucleus; Gr – gracile nucleus; IC – inferior colliculus; IO – inferior olivary nuclear complex; lfu – lateral funiculus; ll – lateral lemniscus; ml – medial lemniscus; PAG – periaqueductal grey; Pn – pontine nuclei; scp – superior cerebellar peduncle; Sp5I – nucleus of the trigeminal spinal tract, interpolar part; vfu – ventral funiculus.

myelinated fibres. The other is at ~7–14 μm diameter (presumably α motor neuron axons) and makes up ~58% of all myelinated fibres. The α motor neurons control the bulk of the skeletal muscle fibres, whereas the γ motor neurons control the muscle fibres within muscle spindles that adjust the length sensitivity of those organs. This bimodal distribution is quite similar to those in the domestic cat and suggests the presence of similar proportions of α and γ motor neurons in monotremes and therians (Ashwell and Zhang 1997).

In the echidna dorsal (sensory) roots, the myelinated fibres also have a bimodal distribution of diameter, with distinct groups at 1–5 μm diameter (45% of population) and 6–12 μm diameter (55% of population). This bimodal distribution is similar to that seen in rodents, carnivores and primates (Ashwell and Zhang 1997). Although limited in scope, these findings suggest that the echidna has similar populations of sensory myelinated axons in different fibre classes to those in placentals.

Development of peripheral nerves

There are two areas of peripheral nerve development that are of particular significance in monotreme neurobiology. The first of these is the trigeminal system, because of its role in electroreception and somato (touch) sensation from the bill and beak; the other is the spinal cord, because of its role in early motor capabilities of the newly hatched monotreme. The development of trigeminal receptors will be discussed in Chapter 9 and the evolutionary significance of the tempo of snout receptor development will be considered in Chapter 14.

Development of the postcranial peripheral nervous system

Peripheral nerves are made up of sensory nerves (peripheral processes of the dorsal root, or spinal, ganglia), skeletal motor nerves from the ventral horn motor neurons, autonomic nerves from the ganglia of the pre- and para-aortic sympathetic ganglia, and autonomic nerves to the parasympathetic ganglia adjacent to visceral organs. The dorsal root and autonomic ganglia of the peripheral nervous system are derived from the neural crest cells, which leave the rim of the folding neural plate immediately before neural tube closure. In the monotremes, neural tube closure occurs during the early incubation phase (at ~5.0–5.5 mm GL for cervical and thoracic segments; Ashwell 2012d). At this early stage, neural crest cells are confined to the angle between the dorsal neural tube and the epidermis.

The first axons from the spinal cord enter the base of the forelimb bud at ~6 mm GL (Ashwell 2012d; see also Appendix Fig. 3 for a summary diagram). These pioneer axons appear to be exclusively motor in function at first, because the dorsal root ganglia have no associated axons at this time. Mesenchymal condensations of the developing limb bones emerge rapidly around the end of the first third of incubation (~8.0 mm GL) and these form detailed cartilaginous models very quickly (probably within a day or ~1 mm growth in GL). The process of limb bone formation may be slightly more advanced in the echidna forelimb than in the platypus (by about a day of development), but the hindlimbs of both species are both very poorly differentiated at this time. Although the differentiation of the long bones and musculature of the hindlimb is delayed (by between ~2 and 3 mm of GL growth or between 2 and 3 days) relative to the forelimb in both species, invasion of the respective limb buds by nerves of the brachial and lumbosacral plexuses is almost contemporaneous and quite rapid. The rate of progression of nerves down the hindlimb is only ~1 day of development behind that of the forelimb. Growth of the nerves down the whole length of the limbs occurs in only 2 days and is complete well before hatching in both the fore- and hindlimb, i.e. by ~11 or 12 mm GL.

Spinal cord

Spinal cord structure and function

Gross anatomy of the spinal cord is rather different between the platypus and echidna (see Appendix Tables 4 and 5). In the platypus, the spinal cord runs almost the entire length of the vertebral column, reaching sacral levels (Ariëns Kappers et al. 1960). By contrast, the short-beaked echidna has one of the shortest spinal cords, relative to the vertebral column, of any mammal. The echidna spinal cord is only ~10 cm long and terminates at the level of the seventh thoracic vertebra, but the dural sac that surrounds the

Table 5.1. Cranial nerves of the short-beaked echidna

Nerve	Number of myelinated axons[1]	Foramen of passage from skull interior	Function
1. Olfactory	–	Foramen olfactorium advehens (platypus)[5] Cribriform plate of lamina infracribrosa (echidna)	Olfaction
2. Optic	14 950; 14 709[1] 28 585; 46 566[4]	Foramen opticum/pseudoopticum[6,7]	Vision
3. Oculomotor	1978; 1849[4]	Foramen pseudoopticum (platypus)[6] Foramen pseudosphenoorbitale (echidna)[7]	Supplies inferior oblique and all recti muscles (except superior); parasympathetic axons to sphincter pupillae and ciliary muscle
4. Trochlear	404; 223[4]	Foramen pseudoopticum (platypus)[6] Foramen pseudosphenoorbitale (echidna)[7]	Supplies superior oblique muscle to produce intorsion and depression of eye
5. Trigeminal	– ; 43 410[4]	5oph – foramen pseudoopticum (platypus)[6] Foramen pseudosphenoorbitale (echidna)[7] 5max – foramen rotundum (platypus), foramen pseudosphenoorbitale (echidna)[7] 5man – foramen pseudoovale[7]/ovale[8]	Mechano- and electrosensation from the beak and head; proprioception from masticatory apparatus; motor control of muscles of mastication (derived from the embryonic first pharyngeal arch)
6. Abducens	671; 474[4]	Foramen pseudoopticum (platypus)[6] Foramen pseudosphenoorbitale (echidna)[7]	Supplies external rectus muscle to abduct (turn out) the eye
7. Facial	4281; 3535[4]	Internal acoustic meatus	Supplies muscles of facial expression (large platysma, buccinator, frontalis platysmae, sphincter colli) derived from the embryonic second pharyngeal arch; parasympathetic supply of sublingual gland
8. Vestibulocochlear	24 565; 23 523[4]	Internal acoustic meatus	Auditory and vestibular function; vestibular axons are grouped into ascending and descending roots
9. Glossopharyngeal	2740; 2170[4]	Jugular foramen	Supplies muscles derived from the embryonic third pharyngeal arch; parasympathetic axons to the parotid gland
10. Vagus	3617; 3042[4]	Jugular foramen	Innervates laryngeal and pharyngeal musculature; parasympathetic axons to thoracic and upper abdominal viscera
11. Accessory	3227; 2180[4]	Jugular foramen	Innervates muscles of rostral shoulder and neck[2]
12. Hypoglossal	5214; 2865[4]	Hypoglossal canal	Muscles of the tongue (sternoglossus superior and inferior, laryngoglossus, genioglossus, styloglossus, myloglossus, annulus inferior, intrinsic longitudinal and circular muscles)[3]

1. Gates (1973); 2. Griffiths (1968); 3. Griffiths (1978); 4. Schuster (1910); 5. Zeller (1988); 6. de Beer and Fell (1936); Kuhn (1971); 7. Zeller (1989); 8. Murray (1978a, b); Griffiths *et al.* (1991).

spinal cord ends at the level of sacral vertebra three, giving it perhaps the longest lumbar subarachnoid cistern (~25 cm) of any mammal (Ashwell and Zhang 1997). Hassiotis and colleagues (Hassiotis et al. 2004a) have suggested that the short spinal cord of the short-beaked echidna is an adaptation to enable it to better adopt the characteristic curled defensive posture of the short-beaked echidna. The spinal cord lies posterior or dorsal to the vertebral column, so extreme flexion of the vertebral column during the defensive posture tends to stretch the cord and roots. This may amount to a 15% increase in length or ~6 cm increase in length in a large adult. The cauda equina that contain the dorsal and ventral roots of caudal spinal cord segments are collectively as thick as the spinal cord in the short-beaked echidna, but contain discrete nerve bundles that are loosely attached and relatively free to move relative to each other. This is in contrast to the spinal cord where axons are tightly bound together with a delicately interlaced glial and capillary network. Hassiotis and colleagues have suggested that the long cauda equina of the short-beaked echidna are better able to tolerate stretching during defensive curling than an equivalent length of spinal cord would be. In other words, the echidna has shifted the length of motor and sensory pathways from spinal cord white matter into the cauda equina to minimise the risk of central axon trauma and avulsion during defensive curling. One would hypothesise that this shift is more extreme in the short-beaked than the long-beaked echidna, but no data are currently available concerning cord length in *Zaglossus*.

During normal locomotion, the cervical spinal cord of the short-beaked echidna also has a pronounced dorsally concave flexure in the region of the first thoracic segment, to allow the snout to be directed forwards (Ashwell and Zhang 1997).

Cyto- and myeloarchitecture of the short-beaked echidna spinal cord

Like all other mammals, the spinal cord of the short-beaked echidna has eight segments in the neck or cervical region (i.e. C1 to C8). The number of vertebrae and spinal segments in the rest of the trunk may vary between echidnas, reflecting the anatomical variation across the geographical range of the species. The thoracic (T), lumbar (L), sacral (S) and caudal (Cau) regions of the cord may have respectively 14–17, 2–4, 3 or 4, and 9–12 segments, depending on the number of vertebrae in each region. The cord has enlargements at the segmental level of the brachial plexus (cervical segment five to thoracic segment two) and the lumbosacral plexus (lumbar segment one to sacral segment one).

The same criteria for dividing the grey matter of the therian spinal cord into 10 Rexed's laminae are applicable to the short-beaked echidna (Ashwell and Zhang 1997). These features are summarised in Table 5.2 and illustrated in atlas plates Ec-Ad31 to 33. In addition to the grey matter layers, two nuclei are present throughout the rostrocaudal extent of the cord: the lateral spinal nucleus in the lateral funiculus and the intermediomedial nucleus (presumptive viscerosensory nucleus) at the medial end of lamina R7. Ashwell and Zhang identified an internal basilar nucleus in cervical segments down to C5 and a lateral cervical nucleus in cervical segments down to C3. Clarke's column (also known as nucleus dorsalis or nucleus thoracicus) is the site of the neurons that give rise to the dorsal spinocerebellar tract for conscious and non-conscious proprioception. It is found at T2 to L1 segments in the medial part of lamina R5 (see plate Ec-Ad32). Ashwell and Zhang (1997) did not find any central cervical nucleus in the spinal cord of the echidna.

Termination of small dorsal root ganglion cell axons

The B4 isolectin derived from *Griffonia simplicifolia* is a plant compound that labels small dorsal root ganglion cells that project unmyelinated axons (C fibre afferents) into the dorsal horn and serve the sort of pain sensation that indicates peripheral tissue damage. In the spinal cord of the echidna, lectin-labelled axons terminate in Lissauer's zone and laminae R1 and R2 of the dorsal horn, much as these axons do in placentals. However, the lectin label in the echidna spinal cord has several unusual features (Ashwell and Zhang 1997) that may have functional implications. In the short-beaked echidna, bundles of lectin-labelled axons also run around the lateral rim of the dorsal horn to end in discrete patches in the lateral parts of laminae R3, R4 and R5 throughout the entire rostrocaudal extent of the spinal cord. This deeper pattern of lectin labelling has not been reported for placentals and may therefore represent a

Table 5.2. Features of Rexed's laminae in the spinal cord of the short-beaked echidna[1]

Rexed's lamina	Function	Features in short-beaked echidna
R1	Also known as the marginal zone, it may participate in transmission of cold, itch and chemically induced pain.	Large fusiform and occasional multipolar neurons (10 to 12 µm by 15 to 25 µm) elongated tangential to the surface of the dorsal horn.
R2	Also known as substantia gelatinosa, it contains nociceptive (pain) axons responding to high intensity stimulation (outer zone) or low-threshold mechanosensation (inner zone).	Small round neurons (6 to 8 µm by 9 to 15 µm). Divided into a densely packed outer zone and a less compact inner zone.
R3	Neurons here respond to weak mechanical stimulation. Contains spinospinal, spinocervical and spinothalamic neurons.	Contains longitudinally running myelinated axons.
R4	Forms the base of the head of the dorsal horn. Neurons respond to proprioceptive and cutaneous stimuli. Neurons project to spinal cord, lateral cervical nucleus, dorsal column nuclei and thalamus.	Curves ventrally around the medial border of the dorsal horn. Contains similar cells to layer R3, but these are more diffusely spread.
R5	Forms the neck of the dorsal horn. Noxious and light touch sensation. Neurons project to lateral cervical nucleus, dorsal column nuclei, reticular formation of brainstem, cerebellum and thalamus.	Heterogeneous neuropil with large multipolar neurons (up to 13 by 30 µm diameter). Contains neurons of Clarke's column (neurons of origin for dorsal spinocerebellar tract) from T2 to L1.
R6	Forms the base of the dorsal horn. Neurons respond to cutaneous and proprioceptive inputs from limbs. Cells project to cerebellum.	Only present in levels C5 to T1 and L3 to S1. More compact appearance with heavily stained cells than adjacent layers.
R7	Intermediate zone of grey matter, contains viscerosensory and visceromotor neurons and interneurons.	Contains intermediolateral (sympathetic preganglionic) nucleus in T2 to L3 levels. An intermediomedial (viscerosensory) nucleus is present at all spinal cord levels.
R8	Contains commissural (crossing-over) neurons. Some cells project to reticular formation and thalamus.	Has a more heterogeneous appearance than R7.
R9	Cluster of motor neurons embedded within lamina R8.	Contains large clusters of large multipolar neurons (35 to 50 µm by 60 to 80 µm), distributed into medial and lateral motor columns for control of axial and limb muscles, respectively.
R10	Surrounds central canal. Neurons may respond to noxious stimuli.	Large multipolar neurons (up to 40 µm by 60 µm) in the ventral part of cervical, thoracic and lumbar segments.

1 Ashwell and Zhang (1997).

real functional difference in this monotreme. Lectin-labelled axons terminate in a region of the dorsal horn called the internal basilar nucleus. They also run ventrally to reach and perhaps cross to the other side of the spinal cord in the dorsal white commissure. The deeper terminating lectin-labelled fibres seen in the echidna may serve visceral nociception or indicate unusually deeply terminating cutaneous C fibres.

Motor pathways and locomotion in monotremes

Descending pathways from the cerebral cortex (corticospinal tract), vestibular nuclei (medial and lateral vestibulospinal tracts), red nucleus (rubrospinal tract), superior colliculus (tectospinal tract) and reticular formation (pontine and medullary reticulospinal tracts) all influence motor activity in the spinal cord in different and specific ways. Data on motor pathways in the monotremes are very limited, but available information does indicate some significant and functionally relevant features.

The corticospinal tract

The corticospinal tract allows the primary motor cortex (and adjacent regions) to directly influence α motor neurons in the spinal cord. The only data on the course and termination of the corticospinal tract in the monotremes come from a silver degeneration study from the early 20th century (Goldby 1939). Goldby removed part of the cerebral cortex high on the dorsolateral surface of the right hemisphere between the α and β sulci (Fig. 5.1a, b). This position is likely to correspond to the hindlimb or tail regions of the primary motor cortex.

Degenerating fibres from this cortical lesion were seen to pass through the internal capsule and the cerebral peduncle, before beginning to decussate (cross to the other side of the brainstem) among the pontine nuclei, just rostral to the level of the trigeminal nerve (Fig. 5.1c, d; see also Addens and Kurotsu 1936). The decussation appears to be complete, in that all fibres cross to the other side; no fibres were seen to continue on the same side of the brainstem (Goldby 1939). The degenerated axons pass through the ventrolateral white matter of the medulla, where most fibres lie external to the spinal trigeminal tract (Fig. 5.1e; see also Addens and Kurotsu 1936), rather than form a pyramidal tract close to the midline as in therians. Some corticospinal tract fibres even intermingle with the axons of the spinal root of the trigeminal nerve. As fibres descend into the spinal cord they lie in the dorsal part of the lateral funiculus, alongside the dorsal horn (Fig. 5.1f, g, h). Goldby traced these degenerating corticospinal axons as far as the lumbosacral enlargement of the spinal cord (Fig. 5.1i).

The above description highlights several unusual features of the corticospinal tract in the echidna relative to therians. These include: (1) the relatively high level of decussation (at the pontine level); (2) the absence of a pyramidal tract close to the midline in the medulla; (3) the absence of an ipsilateral (uncrossed) component; and (4) the penetration of the corticospinal tract to the lumbosacral level of the spinal cord.

The corticospinal tract decussation in the echidna is higher than in any therian, and the extreme lateral position of the tract is unlike any other mammal (Goldby 1939; Hassiotis et al. 2004a). Goldby notes that a high decussation of the corticospinal tract is characteristic of a small number of highly specialised mammals, which probably developed these corticospinal specialisations at a very early period in mammalian evolution (Goldby 1939). The corticospinal tracts of some bats and edentates have a decussation just caudal to the pons and there is a tendency in some of these mammals for fibres from this high decussation to take up a lateral position in the medulla, e.g. in an armadillo, *Lysiurus* (*Cabassous*) *unicinctus*, and the pangolin, *Manis tricuspis* (Goldby 1939). Like the short spinal cord and long cauda equina of the short-beaked echidna, the high corticospinal decussation may be advantageous for mammals that use the highly flexed defensive posture because both the armadillo and pangolin are capable of pronounced vertebral flexure, as is the echidna (Hassiotis et al. 2004a). Nevertheless, the precise nature of the advantage that this may confer is not clear at present. The penetration of the corticospinal tract to the most caudal levels of the spinal cord in the short-beaked echidna is also quite unusual, because the tract penetrates to only cervical or rostral thoracic levels in many marsupials (Ashwell 2010a). Extension of the corticospinal tract throughout the entire cord length is usually associated with a very direct effect of cortical motor areas on the motor neurons concerned with hind limb function, as is seen in primates.

Other descending motor pathways

A tectospinal tract has been identified in the monotremes, arising from the sixth layer (deep grey layer) of the superior colliculus in the platypus (Hines 1929). Once this tract emerges from the superior colliculus, it swings ventrally around the central grey matter and crosses to the opposite side. The position of the tract in the spinal cord white matter has not been identified. In therians, the size of the superior colliculus and the number of neurons projecting in the tectospinal tract are positively correlated with predatory behaviour (Barton and Dean 1993), because the tract allows rapid transmission of information from the superior colliculus to the spinal cord in visually guided prey capture, but this flies in the face of current conceptions of the platypus as an electrically guided predator that relies on sophisticated cortical processing of information (see Chapter 9).

A rubrospinal tract is also present in the platypus (Hines 1929) and the short-beaked echidna (Abbie 1934). In therians, this pathway is involved in locomotor and skilled reaching movements. In the platypus, the rubrospinal tract arises from nerve cells in the ventromedial part of the red nucleus, decussates immediately dorsal to the interpeduncular nuclei and can be followed into the caudal medulla where it lies dorsal to the sensory trigeminal nuclear column. No data are available concerning its position, extent or function in the spinal cord of either monotreme.

No data are available concerning the reticulospinal pathways in either monotreme group, but they are probably important in the regulation of swimming and walking movements.

Posture and energetics of monotreme locomotion

While on the topic of the control of movement, it is worthwhile considering monotreme locomotion and its energy costs. The platypus is semi-aquatic, whereas the short-beaked echidna is entirely terrestrial, so the locomotor demands in the two species differ significantly. Both species also engage in digging, particularly so in the case of the echidna, so the forelimbs must be equally effective as powerful digging tools. While both monotremes have laterally extended humeri, their limbs do not sprawl like lizards or have a reptilian posture, because the manus is held under the glenoid cavity of the shoulder, just like other mammals.

In the platypus, swimming movements are by alternate rowing motions of the forelimbs, while the hindlimbs are held against the body and serve no propulsive purpose (Burrell 1927). During swimming, the forefeet are stretched laterally and the rowing movements depend on the lateral extension of the humerus. The forefeet have webbing that extends well beyond the distal ends of the digits and claws.

The locomotor mechanisms underlying these forelimb movements in the platypus and echidna are currently unknown. Pattern generators in the cervical and upper thoracic levels of the spinal cord probably directly control the alternate rowing movements of the platypus forelimb, but reticulospinal pathways would be crucial to modulating the movements to change direction.

The metabolic cost of terrestrial locomotion for the platypus is ~2.1 times higher than for swimming (Fish et al. 2001). Both the platypus and echidna walk with a splayed gait, but there are differences between them. The platypus must knuckle walk, with the digits of the manus distally flexed where they contact the ground (Burrell 1927), because the extensive webbing would interfere with the locomotor cycle if the digits were extended. This means that the locomotor cycle on land must include the contraction of deep antebrachial digital flexors throughout the reaching and adduction phases. In both the platypus and echidnas (long- and short-beaked), there is substantial long-axis rotation of the humerus over the positioned manus during locomotion. When the platypus is moving slowly, the ventral surface of the body rubs against the ground, but when the platypus is moving at higher speeds, the body is carried slightly above the ground. The platypus never trots, meaning that it never demonstrates a gait where only two feet are on the ground over a complete stride cycle (Fish et al. 2001). This is possibly less demanding of locomotor control than the echidna's gait.

In the short-beaked echidna, the humerus is also held to the side, at a right angle to the sagittal plane, but the femur is directed anterolaterally at angles from 35° to 50° (Jenkins 1970). The degree of the humeral movement (a rotation around the proximo-distal humeral axis) is small during locomotion, amounting to only 40–50°. The humeral rotation during the stance phase progressively shifts the antebrachial (forearm) axis from a position in front of the humerus, to parallel to the humerus, and then behind the humerus.

The main locomotory movement of the femur in the echidna is also a rotation around its long axis, along with some elevation and depression of its distal end (Jenkins 1970). Femoral rotation is ~45°, but the femoral long axis also sweeps posteriorly by ~10–15° during the stance phase of locomotion. Unlike the platypus, the echidna always holds its body above the ground, even when stationary or walking slowly (Jenkins 1970).

The adaptation of the platypus for aquatic locomotion has considerable energetic costs compared to the echidna, in that the slope of the increase in metabolic costs with increasing speed of locomotion for the platypus is higher than for similarly sized mammals, including the echidna. These energetic costs are related to the sprawled stance, meaning that the platypus is unable to take advantage of pendular limb movements that shift energy between kinetic and potential forms, and energy storage in elastic tissue of the joints and muscles. Nevertheless, both the platypus and echidna have lower energy consumption for a given locomotion speed than therians of similar body weight (Edmeades and Baudinette 1975; Fish et al. 2001), largely due to their low resting metabolic rate.

Contractile properties of skeletal muscle fibres in monotremes

If locomotion energy costs are lower for monotremes than other mammals, how do the individual muscles of monotremes compare functionally with therians? The fast-twitch skeletal muscle fibres of placentals share similar Ca^{2+} activation characteristics at the level of the contractile filaments of the muscle. Bakker and colleagues have examined the contractile properties of muscle fibres from the extensor digitorum longus (EDL) of the short-beaked echidna and established that these fibres have quite similar contractile properties to those found in rodents (Head et al. 2000; Bakker et al. 2005). The echidna muscle fibres develop a similar mean maximal force (21 N/cm^2) to rodent EDL and also display similar contractile sensitivity to Ca^{2+}. One distinct difference they did notice with echidna EDL fibres is the presence of myofibrillar force oscillations, which have only been seen in slow and intermediate twitch fibres in placentals, but which are present in fast-twitch fibres in the echidna. The functional significance of this is unknown, but the overall similarity of the fast-twitch responses in echidna EDL to both marsupials and placentals suggests that the unique features of mammalian muscle emerged before the divergence of the monotreme and therian lineage.

Sensory pathways in monotremes

In placentals, the dorsal column pathways carry somatosensory information about discriminative aspects of touch, vibration, proprioception from joints and muscle spindles (Voss 1963), and visceral nociceptors.

The dorsal column is divided into the medially placed fasciculus gracilis serving sensation from the caudal segments and the lateral wedge-shaped fasciculus cuneatus for the rostral segments of the spinal cord. In the short-beaked echidna, the dorsal column is entirely sensory in function (Goldby 1939), unlike in rodents where some descending motor axons are located there. Ashwell and Zhang (1997) showed that the dorsal columns of the short-beaked echidna are comparable in cross-sectional area (2.17 mm^2) to the dorsal columns of therians of a similar body weight (e.g. the domestic cat – 2.46 mm^2 and macaque – 2.57 mm^2). This is a particularly significant finding given that the domestic cat and macaque are considered to have quite sophisticated discriminative touch for exploring their environment.

The fibre calibre of dorsal column axons in the short-beaked echidna is also broadly like in therians, but Ashwell and Zhang (1997) were unable to find any regional variation in axonal calibre as has been reported for the dorsal columns of the domestic cat. Almost two-thirds of fasciculus gracilis axons in the echidna were in the δ calibre range (2–5 μm diameter) and ~6% were in the larger β calibre range (6–12 μm diameter). The proportion of fibres in the β calibre range of the echidna is slightly lower than in the domestic cat (12–21% depending on region) and may suggest that the overall speed of conduction of impulses in the dorsal column pathway of the short-beaked echidna is slower than in the cat. On the other hand, the axonal density in the fasciculus gracilis of the echidna is slightly higher than that in the cat (8.7 compared to 5.4 axons per 100 $μm^2$), suggesting that overall there may be more axons serving discriminative touch in the short-beaked echidna.

The functional significance of all of the above is that the short-beaked echidna is well equipped by extensive sensory pathways in the spinal cord to explore its environment through touch. One would anticipate that it has a sophisticated ability to differentiate between tactile textures in the ground or termite mounds that it touches (see Chapter 9). Proprioception is probably also critical in the control of activity in the panniculus carnosus, the muscular apron that controls the elevation and movement of the spines during defence and locomotion through confined spaces.

Spinal cord development

The spinal cord is derived from the caudal end of the neural tube and its constituent neurons are generated from the walls of this tube after closure of the caudal neuropore. As noted in Chapter 3, the developing spinal cord is divided into alar and basal plates by a groove known as the sulcus limitans, which runs the entire length of the developing cord. The alar plate gives rise to sensory or afferent components of the adult spinal cord (dorsal horn), whereas the basal plate gives rise to motor or effector components (ventral horn). The other key neural components of the spinal sensorimotor system are the spinal or dorsal root ganglia, which are derived from the neural crest.

Pre-hatching development of the monotreme spinal cord

Closure of the back opening of the neural tube, the caudal or posterior neuropore, occurs at the beginning of the incubation phase of monotreme development and the major goal for development of the spinal cord and its associated nerves during the next 10 days of incubation is to prepare for the demands of early hatchling life (Ashwell 2012d; see also Appendix Fig. 3 for summary). These motor challenges of the peri-hatching period include: (1) rupturing or tearing of the membranes of the egg by the os caruncle, egg tooth and claws, presumably by rhythmic flexion and extension cycles of the head and neck in concert with alternating slashing and reaching movements of the pronated forelimbs, although this has never been observed; (2) achieving sufficient motor control in the forelimb for digito-palmar prehension of the pronated distal forelimb and flexion/adduction of the proximal forelimb, so that the new hatchling can move to the areolae of the mammary glands and maintain its position within the nest or pouch by grasping maternal hair or vegetation; and (3) control of ventilation of the lungs, mainly (it is presumed) by activation of the diaphragm. Other possible functions of the hatchling spinal cord might be control of the enteric nervous system, although this is most likely regulated predominantly by parasympathetic pathways through the vagus nerve (but see discussion of the development of the vagal sensorimotor complex below). As noted in Chapter 3, structural and functional development of the hind limb lags behind that of the forelimb by as much as a week during incubation, so the bulk of spinal cord development during the incubation phase involves production of key neurons and the establishment of the essential pathways in the cervical and upper thoracic levels only.

Returning to the specific events of spinal cord development during incubation, the key events are (1) neural tube closure; (2) migration of the neural crest; (3) generation of ventral and dorsal horn neurons; and (4) formation of local circuit and intersegmental connections. The following description applies to both platypus and echidna, unless otherwise specified.

Neural tube closure occurs first at the cervical level of the spinal cord (by 5.5 mm GL) and at this time neural crest cells at the cervicothoracic junction level are confined to the angle between the dorsal neural tube and epidermis. Production (neurogenesis) of the ventral horn motor neurons at the caudal cervical level proceeds rapidly, so that postmitotic neurons begin to settle outside the neuroepithelium from 6.0 mm GL (Fig. 5.2a), which is probably equivalent to only 1 day or less after neural tube closure. Differentiation of these first motor neurons must occur rapidly, because the first axons approach the base of the forelimb bud at the same time (Ashwell 2012d) and before the dorsal root ganglion cells have produced peripheral sensory processes. The development of the dorsal horn of the spinal cord appears to be delayed relative to the ventral horn, because a distinct motor neuron column emerges at cervical segmental levels before postmitotic cells have even begun to accumulate in the dorsal or sensory part of the mantle zone. The cervical level is also much more advanced developmentally than the lumbosacral level.

By 8.0 mm GL, sensory components of the spinal cord are more advanced developmentally. The dorsal root ganglia have migrated into a more ventral position and expanded greatly in size. Their central processes have penetrated the dorsolateral edge of the spinal cord to form the superficial intersegmental axon zone known as the dorsolateral tract or zone of Lissauer. It also appears that ascending sensory tracts from neurons in the grey matter itself have started to develop at this stage, because the ventral white commissure (presumably representing the proximal

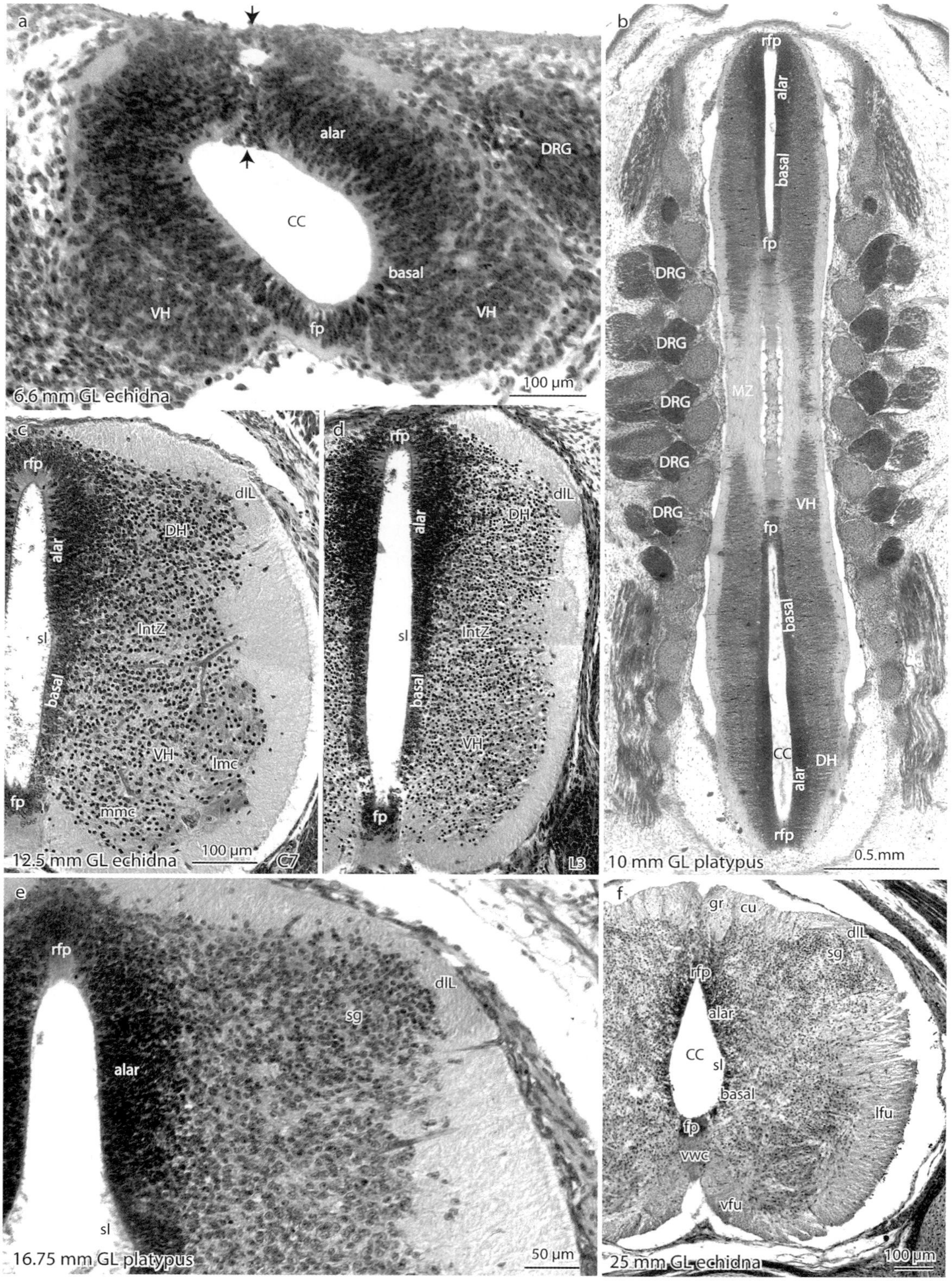
a
alar
DRG
CC
basal
VH
fp
VH
100 µm
6.6 mm GL echidna
b
rfp
alar
basal
fp
DRG
DRG
DRG
DRG
DRG
MZ
VH
fp
basal
CC
alar
DH
rfp
10 mm GL platypus
0.5 mm
c
rfp
dlL
DH
alar
IntZ
sl
basal
VH
lmc
mmc
fp
12.5 mm GL echidna
100 µm
C7
d
rfp
dlL
DH
alar
sl
IntZ
basal
VH
fp
L3
e
rfp
dlL
sg
alar
sl
16.75 mm GL platypus
50 µm
f
gr
cu
dlL
sg
rfp
alar
CC
sl
basal
lfu
fp
vwc
vfu
25 mm GL echidna
100 µm

segments of the spinothalamic and spinoreticular tracts) also emerges at this time.

During the middle third of incubation (~8.0–11.0 mm GL; Fig. 5.2b), the main structural changes in the spinal cord are an increase in the area of marginal zone white matter, particularly the ventral and lateral white matter columns or funiculi; the development of motor neuron cytoplasm; and the dispersion of the ventral horn neurons by an increase in axons and dendrites between the cell bodies. The dorsal horn remains undifferentiated during this period, but some distinction between medial and lateral motor columns begins to emerge at around 9–10 mm GL.

Even during the last third of incubation (12.0–15.0 mm GL; Fig. 5.2c and d), the spinal cord remains poorly differentiated (Ashwell 2012d). The basal plate neuroepithelial layer for the ventral horn has begun to thin and become discontinuous, suggesting that neurogenesis of motor neurons is coming to a close. By contrast, the alar plate neuroepithelial layer is still quite thick, consistent with the delayed neurogenesis and differentiation of the dorsal horn neurons. No laminae of Rexed can be distinguished in the dorsal horn during the last third of incubation. By contrast, motor neurons of the future Rexed's lamina R9 of the ventral horn have developed cytoplasm and can be divided into medial (for axial musculature) and lateral (for forelimb musculature) motor columns at the cervical level (Fig. 5.2c). Ashwell (2012d) also identified a condensation of neurons in the region of the future Clarke's column (the site of origin of the dorsal spinocerebellar tract) at thoracic segmental levels, but no lateral horn (intermediolateral nucleus for sympathetic outflow) could be found. Overall, the state of differentiation of the monotreme spinal cord at hatching is roughly equivalent to early foetal rodents – a mouse spinal cord at 13 days postconception (pc) and a rat spinal cord at 15.5 pc (Ashwell 2012d).

Post-hatching development of the monotreme spinal cord

The most significant change in the structure of the grey matter of the spinal cord during the first post-hatching week is the differentiation of layered structure in the dorsal horn (Ashwell 2012d). This is best illustrated by the emergence at 16.75 mm GL (only a few days after hatching; Fig. 5.2e) of the substantia gelatinosa (a small cell zone in the dorsal horn that receives small unmyelinated afferents from the skin) at all segmental levels of the platypus spinal cord. Over the first 8 weeks after hatching, the neuroepithelial zones become exhausted, the dorsal horn layers and identified neuronal populations (e.g. the lateral spinal nucleus) become progressively differentiated, and the white matter progressively increases in cross-sectional area as ascending sensory and descending motor pathways are added. Demarcation of distinct gracile and cuneate fasciculi in the dorsal columns by an interfascicular sulcus is initially visible during the first few post-hatching weeks (Fig. 5.2f), but becomes more defined during the second post-hatching month.

Figure 5.2: Development of the monotreme spinal cord as shown in transverse sections through the cord at incubation and early post-hatching ages. Shortly after closure of the neural tube (a) the proliferative zones of the alar and basal plates (alar, basal) are thick. Motor neurons have already begun to settle in the ventral horn (VH), but there is no dorsal horn and the neural crest cells for the dorsal root ganglia (DRG) are alongside the spinal cord. The floor plate (fp) is present, but the roof plate is incompletely formed. Arrows indicate the point of closure of the neural tube. Fibres have begun to accumulate in a marginal zone (MZ), the future white matter, by the end of the middle third of incubation (b), but the dorsal horn is small and less differentiated than the ventral horn. Immediately before hatching in the echidna (c, d), distinct medial (mmc) and lateral (lmc) motor columns (for the supply of the trunk and limb muscles, respectively) appear in the ventral horn of the upper cord (c), but not the lower (d). Shortly after hatching (e), discrete layers within the dorsal horn (DH), including the substantia gelatinosa (sg), begin to emerge. Accumulation of white matter is relatively slow, such that the white matter is still of small cross-sectional area by the end of the first post-hatching week (f). Nevertheless, key parts of the white matter are visible, such as the dorsolateral tract of Lissauer (dlL) that contains incoming axons as they ascend or descend before termination, and the ventral white commissure (vwc) where spinothalamic axons will cross. CC – central canal; cu – cuneate fasciculus; gr – gracile fasciculus; IntZ – intermediate zone of grey matter; lfu – lateral funiculus of white matter; rfp – roof plate; sl – sulcus limitans; vfu – ventral funiculus of white matter.

82 NEUROBIOLOGY OF MONOTREMES

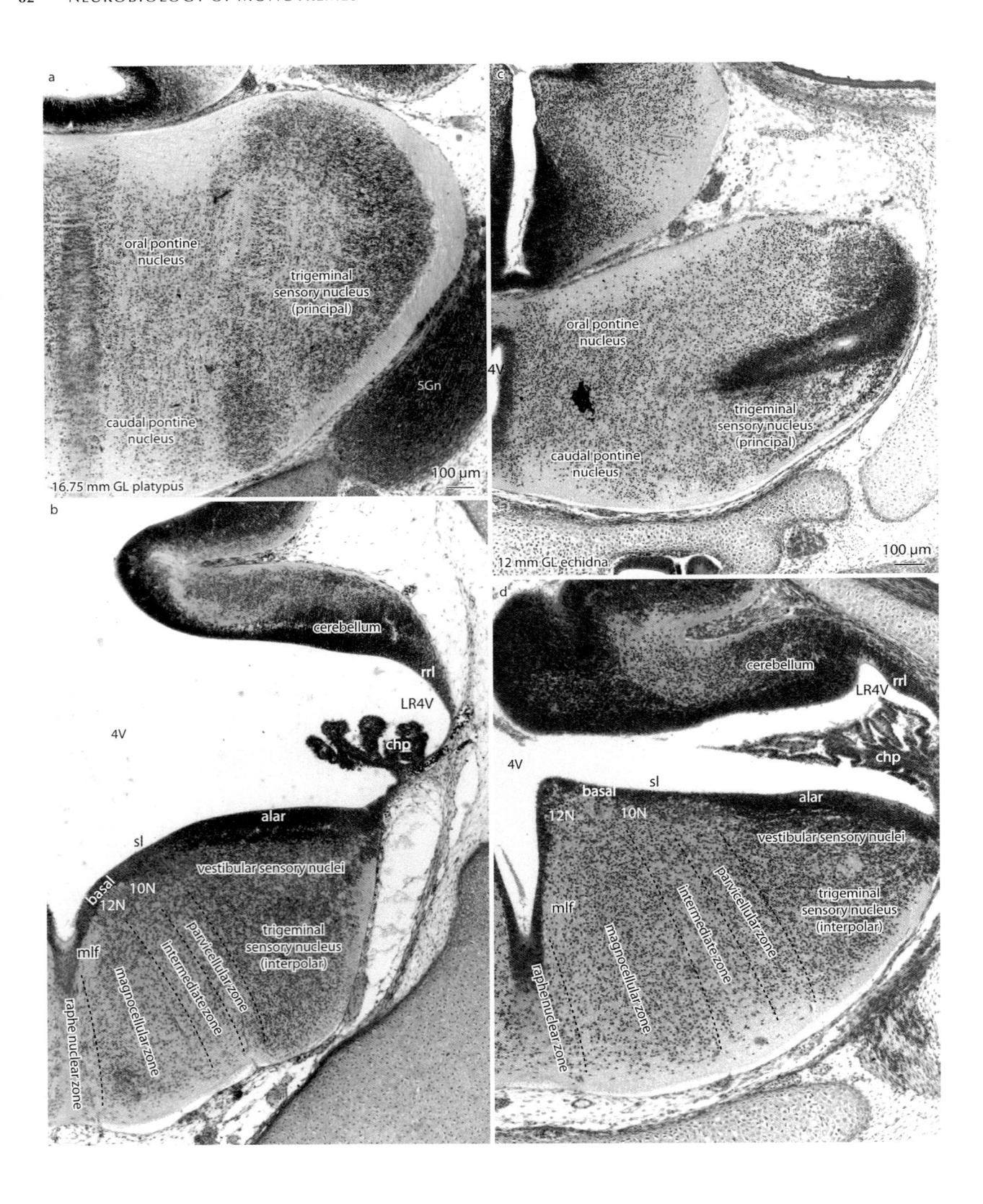

Functional capacity of the developing monotreme spinal cord

To summarise the state of spinal cord development leading up to hatching, the main feature appears to be the advanced maturation of the ventral horn (both in absolute terms and relative to the dorsal horn) at brachial levels, such that medial and lateral motor columns are clearly distinguishable by the time of hatching (Ashwell 2012d). This advanced maturation of the ventral horn is presumably associated with the demands of the peri-hatching period. By contrast, dorsal horn development and expansion of the white matter tracts are poorly advanced by hatching. This is a very similar pattern of spinal cord differentiation to that seen in newborn diprotodont marsupials, which are capable of climbing from the urogenital sinus to the pouch and attaching to the maternal nipple. Although behavioural observations of new monotreme hatchlings are not available, it seems reasonable to propose (on the basis of the structural evidence) that the brachial spinal cord of monotreme hatchlings contains all the necessary motor systems required for breaking through the reproductive membranes, maintaining the position of the newly hatched monotreme in the pouch and stimulating the maternal areolae to elicit milk release.

In the absence of any physiological studies, it is difficult to reach firm conclusions about spinal cord functional development and the nature of locomotor regulatory centres in hatchling monotremes. Nevertheless, based on morphological criteria and by extrapolation from marsupials (Comans et al. 1988; Harrison and Porter 1992; Ho 1997, 1998; Ho and Stirling 1998) one would expect that the spinal cord of the newly hatched monotreme would have central pattern generators in either the caudal brainstem and/or cervical spinal cord to regulate forelimb movements. Spinal cord centres are probably controlled by reticulospinal projections from the magnocellular reticular nuclei of the medulla to the brachial cord to allow trigeminal and perhaps vestibular stimuli to influence patterned motor behaviour. Propriospinal pathways would also be required to coordinate activity in the different segmental levels of the spinal cord, and some form of proprioceptive feedback from the forelimb musculature would be necessary to adequately control power and duration of contraction of the limb muscles.

Brainstem

Overview of brainstem structure and function

As considered in Chapters 3 and 4, the developing brainstem is organised by dorsoventral and rostrocaudal compartmentation. The dorsoventral compartmentation comprises alar and basal plate derivatives. In the case of the rostral hindbrain, where the cavity of the fourth ventricle system opens by lateral movement of the alar plates, sensory components become displaced to the side whereas basal plate derivatives (motor and autonomic effector nuclei) retain a position close to the midline. The sensory and motor components of the developing brainstem are best seen around the time of hatching in the monotremes (Fig. 5.3). The rostrocaudal compartments of the brainstem are the rhombomeric segments (see Chapter 4). These are visible only transiently, at the very beginning of incubation, and have been obscured by the time of hatching.

Figure 5.3: Structure of the peri-hatching monotreme brainstem as seen in frontal sections through the pontine flexure (a, c) and rostral medulla (b, d) of the brains of a 16.75 mm GL platypus (a, b) and 12.5 mm GL short-beaked echidna (c, d). At the time of hatching the brainstem is still very immature in both monotremes: proliferative zones of the alar and basal plates (alar and basal, respectively) are still actively producing neurons, and fibre tracts are incompletely formed. The division of the brainstem into wedge-shaped functional regions radiating from the floor of the fourth ventricle (4V) is particularly evident in the medulla (b, d). The midline is occupied by the primordium of the raphe nuclei (including serotonergic neurons). To the side of the raphe zone is the reticular formation, divided into a medial zone occupied by magnocellular reticular nuclei (the major source of brainstem pathways to the spinal cord), an intermediate zone (future autonomic and visceromotor nuclei) and a lateral zone of parvicellular reticular nuclei. Sensory nuclei (e.g. trigeminal and vestibular) are located in the most lateral wedge of the brainstem. 5Gn – trigeminal ganglion; 10N – vagus nerve nucleus; 12N – hypoglossal; chp – choroid plexus; LR4V – lateral recess of fourth ventricle; mlf – medial longitudinal fasciculus; rrl – rostral rhombic lip; sl – sulcus limitans.

There have been several studies of physiological phenomena mediated by the monotreme brainstem (e.g. lung ventilation and response to blood gas changes; Frappell et al. 1994) but very little is known about the functional activity in the brainstem that mediates that physiology.

Sensory nuclei of the monotreme brainstem

Sensory nuclei of the brainstem are concerned with processing information arriving in the brainstem by sensory nerves and may be divided into general somatic afferents, concerned with inputs from receptors in the surface of the head (trigeminal sensory nuclear complex); special somatic afferents, concerned with input from the auditory and vestibular sensory apparatus (cochlear and vestibular nuclei); general visceral afferents, concerned with input from the thoracic and abdominal viscera (caudal nucleus of the solitary tract); and special visceral afferents, concerned with input from the taste receptors on the tongue and adjacent oral surfaces (rostral nucleus of the solitary tract).

The vestibular nuclei will be considered along with the cerebellar and precerebellar systems later in this chapter, the special features of the trigeminal sensory nuclear complex in monotremes will be discussed in Chapter 9, and the cochlear nuclei will be discussed along with the rest of the auditory system in Chapter 10.

The nucleus of the solitary tract is the viscerosensory nucleus of the medulla. It has a large bundle of myelinated fibres at its core (the solitary tract). On the basis of the relationship of the nucleus to the embedded solitary tract, Hines (1929) divided the nucleus of the solitary tract of the platypus into a dorsolateral zone caudally, where cells are of mixed size (large and small), and a ventromedial region more rostrally, where neurons are all small. Abbie arrived at a similar subdivision for the nucleus of the solitary tract in the echidna (Abbie 1934).

Motor nuclei of the mammalian brainstem

In the broadest meaning of the term 'motor' (i.e. neurons concerned with producing an effect – effector neurons), motor nuclei of the brainstem may be divided into a somatic efferent column (oculomotor, trochlear, abducens and hypoglossal nuclei) lying close to the midline, a branchiomotor efferent column (motor trigeminal, facial motor and spinal accessory nuclei) concerned with supply of muscles derived from the pharyngeal arches and a visceromotor efferent column (superior and inferior salivatory nuclei, lacrimal nucleus and dorsal motor nucleus of the vagus) concerned with supply of salivary and lacrimal glands and viscera of the trunk.

Motor nuclei of the platypus brainstem

The somatic efferent column consists of the eye muscle and hypoglossal nuclei. The oculomotor nucleus of the platypus is made up of four components at the level of the superior colliculus of the midbrain: two lateral subnuclei, in rostral and caudal positions; and two medial subnuclei, in dorsal and ventral positions (Hines 1929). These individual subnuclei probably provide discrete topographic supply of the extraocular muscles (medial rectus, superior rectus, inferior oblique and inferior rectus), but this has never been traced in the monotremes. The trochlear nucleus (which supplies the superior oblique extraocular muscle) lies on either side of the midline at the rostrocaudal level of the inferior colliculus. Hines (1929) described the trochlear nucleus of the platypus as having two groups of neurons: one is more spherical, contains large triangular neurons, and is more lateral, rostral and ventral than the other, which is made up of smaller neurons. As in other vertebrates, the axons of the trochlear nerve swing dorsally to decussate in the superior medullary velum of the fourth ventricle and emerge lateral and caudal to the inferior colliculus. The abducens nucleus (for supply of the lateral rectus muscle) lies at the level of the facial motor nucleus in the platypus and echidna and its axons emerge caudal to the pontine nuclei (as in other mammals).

The hypoglossal nucleus of the platypus is an elongated group of large neurons that lie immediately lateral to the dorsomedial tegmental bundles of the medulla (Hines 1929). The hypoglossal column of the platypus extends for ~2.5 mm in its rostrocaudal extent and its constituent motor neurons are larger than all other neurons of the brainstem except the magnocellular reticular nucleus. Nevertheless, Hines noted that the motor neurons of the hypoglossal nucleus of the platypus were rather small compared to those in humans. As in other mammals, the monotreme hypoglossal nucleus has an admixture of smaller neurons that probably help mediate reflexes involving the tongue musculature.

During development, the branchiomotor group of nuclei migrate substantial distances from their origin in the floor of the fourth ventricle and are found within the ventral brainstem tegmentum. In the platypus, the trigeminal motor nucleus (which supplies the muscles of mastication and some others) is a large spherical mass that lies dorsal to the decussation of the trigeminothalamic axons and medial to the axons of the ascending auditory axons of the lateral lemniscus (Hines 1929; see atlas plate Pl-Ad21). As in rodents, the trigeminal nucleus of the monotremes has discrete parts that probably serve control of different masticatory muscles. In both platypus and echidnas, the facial motor nucleus (which supplies muscles of facial expression, plus some deeper muscles) is divided into dorsal and ventral parts (Hines 1929; Abbie 1934; see atlas plate Ec-Ad25). The precise role of the two components in the monotremes is not known with certainty, but (based on tracing studies in therians) it is likely that the dorsal subnucleus is concerned with the supply of deep facial muscles (stapedius, styloideus), while the ventral subnucleus supplies more superficial facial muscles. Like in marsupials, the efferent axons of the facial motor nucleus emerge as the facial nerve dorsal to the trigeminal sensory nuclear column, rather than ventral to the trigeminal sensory complex as in placentals. This key difference in the path of emergent facial nerve axons is probably due to the early ventral migration and maturation of the trigeminal sensory nuclei in the monotremes and marsupials (because of the sensorimotor demands of early postnatal life) so that these sensory nuclei have moved out of the dorsal tegmentum before the facial axons have grown out. The nucleus ambiguus (which supplies the muscles of the pharynx, soft palate, larynx and upper oesophagus) is difficult to identify in the platypus and often merges with the adjacent magnocellular reticular nucleus (Hines 1929). Hines suggested that it is an insignificant structure compared to other mammals.

The visceromotor column contains the parasympathetic preganglionic neurons that drive the secretion of exocrine (salivary and lacrimal) glands and movement of smooth muscle in the head, neck, thorax and upper abdomen. Hines identified the largest visceromotor nucleus, the dorsal motor nucleus of the vagus (or simply the vagus nerve nucleus), as a complex group of neurons dorsal to the nucleus of the solitary tract (Hines 1929). Hines also described a ventrolateral part which forms a separate nucleus that was labelled as the motor nucleus of the ninth (glossopharyngeal) nerve, but the function of this subdivision is unknown. The platypus also has an Edinger-Westphal nucleus (which supplies the sphincter pupillae and ciliary muscle of the eye) situated at the level of the oculomotor nucleus (Hines 1929), even though the accommodative ability of the monotreme eye remains controversial (see Chapter 8).

Motor nuclei of the short-beaked echidna brainstem

The hypoglossal nucleus of the short-beaked echidna is large (see atlas plate Ec-Ad28), as one would expect given the importance of the tongue for feeding (Abbie 1934). Abbie thought that the lateral part of the hypoglossal nucleus gave rise to axons that joined the vagal nerve, consistent with a suggestion that the hypoglossal nucleus in this species contains undescended motor neurons that migrate to the nucleus ambiguus in other mammals. This could also explain the relatively small size of the nucleus ambiguus in the echidna.

Abbie noted that the dorsal nucleus of the vagus has dorsal small-celled and ventral large-celled components (Abbie 1934), but the functional significance of this division is unknown at present. As in the platypus, the facial nucleus of the echidna is usually described as having distinct dorsal and ventral parts, although Abbie (1934) disagreed with the usual division and described the nucleus as 'V'-shaped with the point of the 'V' lying caudally and each limb extending rostrally as dorsal and ventral components. The trigeminal motor nucleus of the echidna is remarkably large given the virtually atrophic state of the jaw musculature (Abbie 1934).

The abducens, trochlear and oculomotor nuclei are all rather small in volume and contain small motor neurons (Abbie 1934), consistent with the poor development of extra-ocular muscles in the echidna. The Edinger-Westphal nucleus, for control of the ciliary and sphincter pupillae smooth muscle of the eye to effect accommodation and pupil constriction, is also tiny (see Chapter 8 for consideration of visual accommodation in the echidna).

86 NEUROBIOLOGY OF MONOTREMES

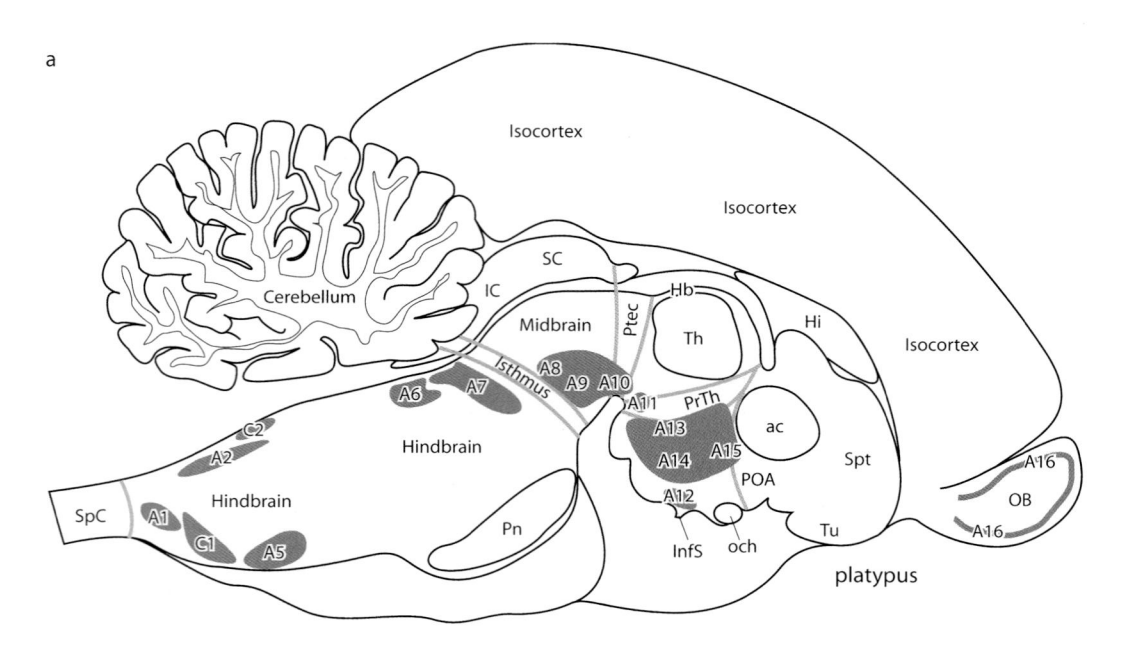

a

Caudal rhombencephalic group

A1	Caudal ventrolateral tegmental group
A2	Caudal dorsomedial group
C1	Rostral ventrolateral tegmental group
C2	Dorsal rostromedial group

Rostral rhombencephalic group

A5	Fifth arcuate nucleus
A6	Locus coeruleus
A7	Subcoeruleus

Midbrain group

A8	Retrorubral area
A9	Substantia nigra
A10	Ventral tegmental area

Forebrain group

A11	Caudal diencephalic group
A12	Tuberal group
A13	Zona incerta
A14	Rostral periventricular group
A15	Postcommissural group
A16	Olfactory bulb

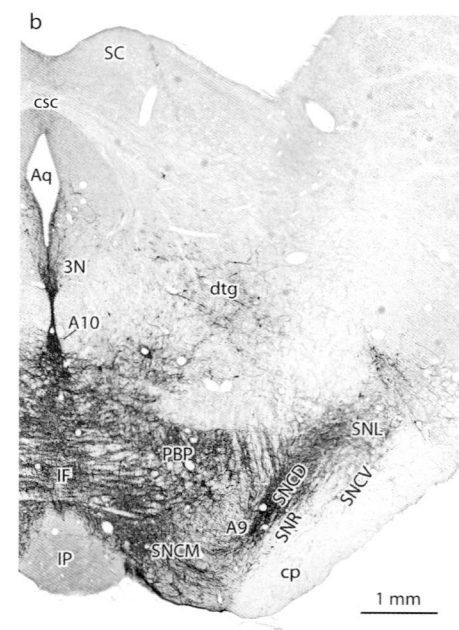

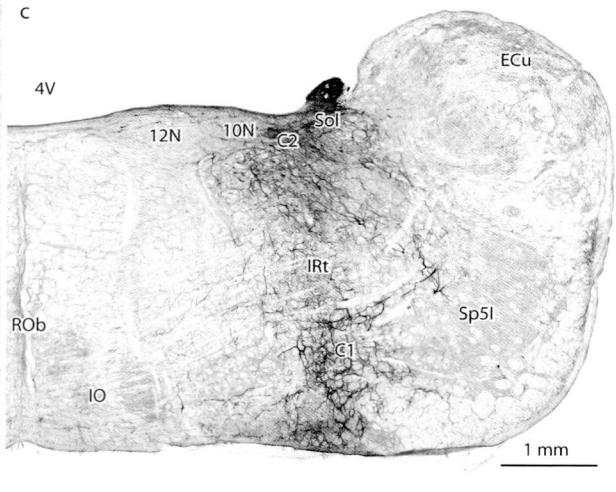

Catecholaminergic, serotonergic and cholinergic cell groups in the brainstem

The monoaminergic cell groups of the brain are critically important for neural plasticity and learning, regulation of the autonomic nervous system in general and the cardiovascular regulatory systems in particular, and control of sensory attention and sleep. The cholinergic neurons of the brainstem include motor neurons and preganglionic parasympathetic neurons that use acetylcholine as a neurotransmitter (branchiomotor and visceromotor nuclei, respectively) as well as tegmental cholinergic neurons (PPTg and LDTg) that project to diverse areas of the brainstem, diencephalon, hypothalamus and basal forebrain. The arrangement and morphology of catecholaminergic, serotonergic and cholinergic neurons have been studied in both the platypus and short-beaked echidna (Manger et al. 2002a–c).

Catecholaminergic cell groups

Catecholaminergic neurons of the mammalian brainstem are involved in diverse functions including motor regulation, behavioural reinforcement and reward, sleep and cardiovascular control. Manger et al. (2002b) used the immunohistochemistry of the biosynthetic enzyme tyrosine hydroxylase to map the distribution of catecholaminergic neuronal groups in the monotreme brainstem. Tyrosine hydroxylase is a key enzyme in the pathway that produces noradrenaline, adrenaline and dopamine, so all three chemical groups of neurons are identified by this method. Catecholaminergic neuronal groups are classified according to an alphanumerical system (A1 to A17; C1 to C3) and can also be described on the basis of their (sometimes distinctive) nuclei (Table 5.3, Fig. 5.4).

Broadly speaking, the catecholaminergic cell groups of the monotreme brainstem are similar to those in therians, but several groups named in therians are not found in the monotremes. These are two noradrenaline (A3, A4) groups and one adrenaline (C3) group, but note that these are also not delineated in humans (Paxinos et al. 2012). In fact, the A3 group has only been identified by induced fluorescence in rats (Dahlström and Fuxe 1964a, b) and has not been identified by immunohistochemical techniques, so it may not represent a real group. The A4 group is also not present in all therians and the C3 group has been identified only in rodents, so the apparent absence of these groups in monotremes may have no great functional significance (Manger et al. 2002b).

The A6 group or locus coeruleus (from the Latin for 'blue spot') is of particular functional significance, because it is a relatively small nucleus in the rostral hindbrain that engages in noradrenergic projections to virtually every part of the central nervous system and plays a key role in diverse functions from sleep to pain, attention and cognitive processing (Aston-Jones 2004). Morphologically, this nucleus and its constituent neurons are very similar in the monotremes to those in other mammals. A more important question is how the functional activity of the locus coeruleus in monotremes like the echidna is similar or different to that in therians. This question is particularly pertinent because of the potential role of locus coeruleus activity during sleep/wake cycles in the echidna (see Chapter 13; Siegel et al. 1996; Manger et al. 2002b).

Figure 5.4: Distribution of catecholaminergic neurons in the brainstem of the platypus, mapped onto a longitudinal representation (a) and based on the data in Manger et al. (2002b). Illustrations (b) and (c) show catecholaminergic neurons (as revealed by immunohistochemistry for the synthetic enzyme tyrosine hydroxylase) in the midbrain (b) and caudal medulla (c) of the short-beaked echidna. Note the concentration of catecholaminergic neurons in the wedge-shaped intermediate reticular zone (IRt) of the medulla for regulation of autonomic and visceromotor function (c). 3N – oculomotor nucleus; 4V – fourth ventricle; 10N – vagus nerve nucleus; 12N – hypoglossal nucleus; ac – anterior commissure; Aq – cerebral aqueduct; cp – cerebral peduncle: csc – commissure of the superior colliculus; dtg – dorsal tegmental tract; ECu – external cuneate nucleus; Hb – habenular nuclei; Hi – hippocampus; IC – inferior colliculus; IF – interfascicular nucleus; InfS – infundibular stalk; IO – inferior olivary nuclear complex; IP – interpeduncular nuclei; OB – olfactory bulb; och – optic chiasm; PBP – parabrachial pigmented nucleus; POA – preoptic area; Pn – pontine nuclei; Ptec – pretectum; PrTh – prethalamus; ROb – raphe obscurus nucleus; SC – superior colliculus; SNCD – substantia nigra, compact part, dorsal tier; SNCM – substantia nigra, dorsal tier, medial part; SNCV – substantia nigra, dorsal tier, ventral part; SNL – substantia nigra, lateral part; SNR – substantia nigra, reticular part; Sol – nucleus of the solitary tract; Sp5I – nucleus of trigeminal spinal tract, interpolar part; SpC – spinal cord; Spt – septum; Th – (dorsal) thalamus; Tu – olfactory tubercle.

Catecholaminergic neurons of the hypothalamus and olfactory bulb will be discussed in Chapters 6 and 11, respectively.

Serotonergic cell groups

The raphe nuclei (from the Latin for 'seam') of the brainstem collectively form a narrow, sagittally oriented plate in the midline of the midbrain and hindbrain. The raphe nuclei contain mainly serotonin as their neurotransmitter. In placentals they have been shown to receive afferents from the cerebral cortex, hypothalamus and other parts of the reticular formation. Efferents from the raphe nuclei differ substantially depending on the particular nucleus.

Table 5.3. Catecholaminergic cell groups of the mammalian brainstem

Region	Alphanumeric group	Nucleus or area	Comments on features in the monotremes[1]
Caudal rhombencephalon	A1	Caudal ventrolateral tegmental group	Caudal continuation of the C1 group (see below).
	A2	Caudal dorsomedial group	Divided into three subdivisions: (1) within the area postrema; (2) within the vagal sensorimotor complex (10N and Sol); and (3) caudal to the vagal sensorimotor complex
	A3		Not found in platypus and short-beaked echidna, but may not be a true group in other mammals
	C1	Rostral ventrolateral tegmental group	Spread along the ventral floor of the medulla at the level of the inferior olive
	C2	Dorsal rostromedial group	Ventral and medial to the vagal sensorimotor complex. Morphologically similar to C1 group
	C3		Not found in platypus and short-beaked echidna, but not present in all therians
Rostral rhombencephalon	A4		Not found in platypus and short-beaked echidna, but not present in all therians
	A5	Fifth arcuate nucleus	Found in ventromedial medulla from the level of the rostral tip of the inferior olive to the caudal limit of the motor trigeminal nucleus
	A6	Locus coeruleus	Located in the caudal part of the ventrolateral quadrant of the periaqueductal grey matter
	A7	Subcoeruleus	Associated with the axons of the brachium conjunctivum (superior cerebellar peduncle). It forms a ventral and lateral extension of the locus coeruleus.
Mesencephalic (midbrain)	A8	Retrorubral area	Scattered neurons caudal to the red nucleus. In the short-beaked echidna, the group is larger in volume, but of lower spatial density than in the platypus.
	A9	Substantia nigra	The main component of A9 is within the pars compacta of the substantia nigra. Scattered catecholaminergic neurons are also found within the pars reticulata of the substantia nigra (A9v) and within the pars lateralis of the substantia nigra (A9l).
	A10	Ventral tegmental area	The main group of A10 is located between the red nucleus and the interpeduncular nuclei. A caudal group of A10 (A10c) lies close to the midline mixed with neurons of the median raphe. A dorsocaudal group (A10dc) is within the midline of the periaqueductal grey matter, dorsal to the oculomotor nucleus.

1 Manger et al. (2002b).

Broadly speaking, caudal raphe nuclei project to the spinal cord, intermediate nuclei project to the cerebellum, and rostral raphe nuclei project to the locus coeruleus, periaqueductal grey, thalamus, striatum, hypothalamus, hippocampus, amygdala and cerebral cortex. In therians, the serotonergic neurons have been allocated into groups according to an alphanumeric system (Table 5.4; Fig. 5.5).

Manger and colleagues (Manger et al. 2002c) identified two broad groups of serotonergic neurons in the brainstem of the platypus and short-beaked echidna. The most rostral group consists of two groups of raphe nuclei (the four divisions of the dorsal raphe nucleus – lateral, medial, ventromedial and interfascicular divisions; and the median raphe nucleus), as well as serotonergic cells in the region dorsal to the medial lemniscus (supralemniscal region). The caudal group consists of three divisions, largely defined by the raphe obscurus nucleus, raphe pallidus nucleus, and the raphe magnus nucleus, but with ancillary nuclei in the paramedian brainstem.

Overall, the topography of serotonergic neurons in the midbrain and hindbrain of the monotremes is similar to that in therians (but see discussion of hypothalamic serotonergic neurons in Chapter 12) and each of the serotonergic nuclei identified by Manger and colleagues have direct homologues in other mammals. Nevertheless there are some significant differences. First, there do not appear to be serotonergic neurons in the caudal linear nucleus in the monotremes, although the nucleus itself can be identified (see atlas plates Ec-Ad19 and 20). In placentals, the caudal linear nucleus is a small midline group of nerve cells in the isthmus that lies rostral and dorsal to the decussation of the superior cerebellar peduncle. In placentals, serotonergic neurons in this nucleus project to the striatum to modulate motor activity. The absence of serotonergic neurons of the caudal linear nucleus in monotremes suggests that this group only emerged in the therian lineage after divergence from monotremes. Alternatively, its functions may be served by other serotonergic neurons, e.g. the midbrain median raphe nucleus.

The second significant point concerns the lateralisation (degree of dispersal to the side) of serotonergic neurons in the monotreme hindbrain. In rodents, serotonergic neurons are more lateralised in the more caudal raphe nuclei relative to the rostral nuclei. By contrast, in monotremes the serotonergic neurons in the rostral dorsal raphe nuclei are more lateralised than the caudal clusters of serotonergic neurons. The functional significance of this (if any) remains unknown. Most importantly, there is no evidence to support the contention (Bowker and Abbott 1990; Bjarkam et al. 1997) that so-called 'neurologically advanced' mammals (usually taken to mean primates) have more lateralised groups of serotonergic neurons than 'neurologically primitive' mammals (usually and inappropriately meaning rodents, the monotremes and marsupials). Lateralisation of serotonergic neurons has been reported in both monotremes (Manger et al. 2002c) and marsupials (Ferguson et al. 1999) to such an extent that one can conclude that there is no consistent evolutionary trend in lateralisation of serotonergic neuron groups.

Cholinergic cell groups of the brainstem

The cholinergic system of the brain is a diverse collection of neurons involved in learning, memory and consciousness, as well as somatic and visceral motor control. They can be identified by immunohistochemistry for choline-O-acetyl-transferase, the biosynthetic enzyme for the neurotransmitter acetylcholine, or by immunohistochemistry and *in situ* hybridisation for the vesicular acetylcholine transporter that is concentrated in neurons using acetylcholine. Cholinergic neurons of the monotreme forebrain will be discussed in Chapters 6 and 12, but those in the brainstem will be discussed below.

Brainstem cholinergic neurons mainly include the motor nuclei (motor trigeminal, facial, oculomotor, trochlear, abducens, nucleus ambiguus, dorsal nucleus of the vagus nerve, and hypoglossal nucleus), but also the parabigeminal nucleus, pedunculopontine tegmental nucleus and laterodorsal tegmental nucleus (Table 5.5; Fig. 5.6). Broadly speaking, the cholinergic neurons of the monotreme brainstem are much like those in therians, but Manger and colleagues did not find a parabigeminal cholinergic group and not all branchiomotor and visceromotor groups were identified as strongly cholinergic (e.g. spinal accessory nucleus, salivatory and lacrimal nuclei). Some cholinergic nuclei (e.g. the medial habenular nucleus) have projections to other brain regions (e.g. by the tract called the fasciculus retroflexus to the interpeduncular nuclei) so their target structures have high levels of acetylcholinesterase (AChE), the degradative enzyme for acetylcholine (see Fig. 5.6b).

Brainstem development in monotremes

The broad features of development of the brainstem are the same in monotremes as in all other mammals, but there are two specific topics that are worth considering in detail: (1) how does the development of the brainstem during the incubation phase prepare the hatchling for the demands of the immediate post-hatching period; and (2) how are the special features of the trigeminal pathway developed during incubation and early lactation phase life?

Development of the monotreme brainstem during incubation

Closure of the anterior neuropore occurs at the beginning of the early pharyngeal subphase of incubation life (i.e. at ~5.0–5.5 mm GL; see Appendix Figs. 1 to 3). In the first subphase of incubation, the brainstem is in the form of a neural tube with alar and basal plate regions separated by a sulcus limitans. Dorsal spreading and thinning of the roof of the fourth ventricle begins at ~6.0 mm GL, within about a day of anterior neuropore closure, so that the dorsal sensory parts of the rostral rhombencephalon move to the side of the brainstem. As in other vertebrates, neurogenesis and differentiation of motor nuclei from the basal plate occurs earlier than for sensory nuclei from the alar plate. Migration of the motor basal plate derivatives to the mantle layer begins at ~6.5 mm GL, whereas very few neurons have migrated to the sensory regions by 7.3 mm GL.

Neurons of the brainstem are derived either from the ventricular germinal zone of the floor of the fourth ventricle (usually for the larger neurons) or from the rim of the fourth ventricle, a special proliferative region known as the rhombic lip (usually for the production of smaller neurons). In the case of the monotremes, generation of neurons from the ventricular germinal zone (VGZ) begins at ~6.5 mm GL and extends to about the end of the first week of post-hatching life, although this assessment is based on the rather indirect evidence of changes in the thickness of VGZ. Clearly it would be preferable to confirm this timing using modern techniques for tagging newly generated neurons, but this is not ethically possible for monotremes at present.

The rhombic lip is the region at the lip or rim of the fourth ventricle, and thickens as the hindbrain begins to unfold to form the rhombic fossa floor (Ashwell and Hardman 2012b). Its temporospatial continuity with the region that produces the neural crest before neural tube closure has been recognised (Nichols and Bruce 2006), highlighting the possible link between the cell migrations that produce the sensory and autonomic neurons of the peripheral nervous system and those that give rise to the small neurons of the hindbrain. Analysis of rhombic lip migrations in the mouse hindbrain has revealed four bands of migrating neurons (Nichols and Bruce 2006), accenting the important contribution that the rhombic lip makes to the component circuitry of the hindbrain. These migrations initially move in a radial direction, before turning rostrally or ventrally to reach their target nuclei. Some of the migration occurs along the surface of the hindbrain, where the migrating neurons can be clearly seen beneath the pial membrane, whereas other streams move within the mantle zone of the hindbrain. The latter are very difficult to see in the developing hindbrain without advanced neuron-tagging techniques, so their presence in monotreme hindbrain can only be inferred indirectly from the accumulation of neurons at the final resting place.

The four bands of rhombic lip migrations are (1) dorsal (giving rise to small neurons of the gracile, cuneate, cochlear, vestibular nuclei as well as cerebellar granule cells); (2) dorsal intermediate (giving rise to small neurons of the trigeminal sensory, parvicellular reticular, and deep cerebellar nuclei); (3) ventral intermediate (giving rise to small neurons of the lateral and intermediate reticular nuclei); and (4) ventral (giving rise to the small neurons of the raphe and pontine nuclei) (Nichols and Bruce 2006). Study of archived sections of monotreme embryos and hatchlings suggests that the rhombic lip migration process for all but the dorsal band is much like that in all other mammals, but the large size of the rostral rhombic lip and its contribution to the sensory nuclei of the rostral hindbrain in the platypus through the dorsal band is strikingly unusual (see below and Chapter 9).

In both the platypus and short-beaked echidna, the rhombic lip begins to thicken and expand dorsolaterally at the end of the middle third of incubation and this process extends into the last 4 days before hatching (Ashwell and Hardman 2012b, Appendix Fig. 3). Rostral and (caudo)lateral subregions can be distinguished within the rhombic lip in both species. The rostral part of the rhombic lip is associated with rhombomeres 1 to 3 and extends from the isthmus to the pontine angle. It includes the region at the lateral angle of the fourth ventricle that was formerly known as the germinal trigone (corresponding to rhombic lip of rhombomeres 2 and 3, Nichols and Bruce 2006); whereas the (caudo)lateral rhombic lip lies along the caudolateral rim of the caudal fourth ventricle and is associated with rhombomere 4 and beyond.

Table 5.4. Serotonergic neurons of the monotreme brainstem

Major group	Alphanumeric group	Nucleus	Comments on features in the monotremes[1]
Caudal group	B1	Raphe pallidus nucleus Caudal ventrolateral nucleus	The raphe pallidus is located medial and ventral to the inferior olivary nuclear complex. The raphe pallidus neurons are vertically oriented bipolar, whereas caudal ventrolateral neurons are bipolar or tripolar and oriented along the ventral surface of the medulla.
	B2	Raphe obscurus nucleus	Located along the midline between the rostral part of the cervical spinal cord and the mid-level of the nucleus ambiguus. Contains two types of cells: bipolar cells in a midline vertical column and triangular cells in a lateral group.
	B3	Raphe magnus nucleus Rostral ventrolateral medulla Lateral paragigantocellular reticular nucleus	These neurons are found at the level of the facial nucleus. The raphe magnus is a column of bipolar cells located along the midline. The rostral ventrolateral nucleus is in continuity with the B1 group of the lateral medulla. The lateral paragigantocellular reticular group is located to the side of the midline.
Rostral group	B4	Central grey of the medulla oblongata	Not mentioned and not illustrated by Manger et al. (2002c)
	B5	Pontine median raphe nucleus	Continuous with the midbrain median raphe nucleus
	B6	Pontine dorsal raphe nucleus	Scattered neurons along the pontine midline. It lies in continuity with the caudal end of the midbrain dorsal raphe nucleus.
	B7	Midbrain dorsal raphe nucleus	Located within the periaqueductal grey matter. It extends from the most rostral tip of the trigeminal motor nucleus to the level of the oculomotor nucleus and is divided into lateral, median, ventromedial and interfascicular divisions.
	B8	Midbrain median raphe nucleus Caudal linear nucleus	Located in the midline of the brainstem from the trigeminal motor nucleus to the decussation of the superior cerebellar peduncle.
	B9	Medial lemniscus/supralemniscal region	Located lateral to the midline of the rostral brainstem. The supralemniscal component lies between the medial lemniscus and the brachium conjunctivum (superior cerebellar peduncle).

1 Manger et al. (2002c).

Development of the trigeminal sensory nuclei complex

In both platypus and short-beaked echidna, large neurons of the trigeminal sensory nuclei begin to accumulate in the trigeminal sensory nuclear column at the beginning of the middle third of incubation (~8.0 mm GL). This appears to coincide with the emergence of a spinal trigeminal tract alongside the column as incoming axons from the trigeminal ganglion grow into the brainstem. Even at this early stage, all components of the complex can be seen, including the mesencephalic trigeminal nucleus (actually displaced sensory ganglion cells that have migrated into the rostral rhombencephalon, isthmus and midbrain), the principal nucleus (adjacent to the trigeminal nerve entry zone) and the oral, interpolar and caudal parts of the nucleus of the trigeminal spinal tract (Ashwell and Hardman 2012b).

The appearance of the trigeminal sensory nuclear complex in the platypus and echidna begins to diverge at the end of the middle subphase of incubation (i.e. ~10 mm GL), due to differences in the contribution of the rhombic lip to the sensory nuclei complex (Ashwell and Hardman 2012b). The rostral part of the rhombic lip is much thicker and wider in the platypus (being several hundred μm wide in both its dorsal and lateral extent, compared to less than 100 μm in the short-beaked echidna), whereas the (caudo)lateral rhombic lip is similar in thickness in both species (less than 100 μm thick). Neurons migrate from both rostral and caudolateral regions into the surrounding ventral hindbrain and cerebellum. Given the large size of the rostral rhombic lip in the platypus, it is not surprising that the most rostral component of these migratory streams (i.e. from the rhombic lip of putative rhombomeres 2 and 3 into the rostral hindbrain) is also much larger in the platypus than in the short-beaked echidna. In fact, the rostral migratory stream appears to be much larger in the platypus than in any other mammal described to date.

By the time of hatching in the platypus, the rostral migratory stream has deposited a huge mass of granular cells into two groups of sensory nuclei of the rostral hindbrain. One of these is a large cluster within the Pr5 at the most rostral end of the hindbrain, where the migrating neurons initially form a dense clump of granular cells in the lateral part of the nucleus. This is never seen at any stage in the short-beaked echidna. Given the large size of the oral and interpolar parts of the nucleus of the trigeminal spinal tract in the mature platypus, there may also be significant rhombic lip migrations to those nuclei, but this has not been specifically traced in the platypus. The second group of rhombic lip derivatives is a migrating stream to the developing cochlear nuclei, which will form dorsal and a little caudal to the Pr5 in the platypus. These are also much larger in the platypus than in the short-beaked echidna.

Development of the nucleus of the solitary tract in monotremes

The viscerosensory nucleus of the solitary tract acts in concert with the dorsal motor nucleus of the vagus to mediate visceral reflexes (e.g. lung, stomach and intestinal) and the combined visceral nuclear complex is often referred to as the vagal sensorimotor complex. Precocious development of the vagal sensorimotor complex in monotreme hatchlings would be a prerequisite for control of visceral function, but the vagal complex in both platypus and short-beaked echidna at the time of hatching appears to be quite poorly differentiated (Fig. 5.3b, d). There is certainly no clear separation of the motor or sensory components of the complex at hatching, suggesting that the differentiation of the constituent neurons at this time is very limited. Separation of the neurons in the vagal sensorimotor complex by neuropil is much more advanced by the end of the first post-hatching week, suggesting that the maturation of the constituent neurons is mainly a post-hatching event.

Development of motor nuclei in the monotreme brainstem

The motor nuclei of the monotreme brainstem include the eye, trigeminal and facial motor nuclei, for the control of the extraocular, masticatory and facial muscles, respectively; the hypoglossal nucleus for control of the tongue musculature; and the nucleus ambiguus for control of the muscles of the soft palate, pharynx and larynx.

The motorneurons of the somatic efferent column (oculomotor, trochlear, abducens and hypoglossal nuclei) develop early in incubation phase development (~8.0–9.0 mm GL) and the motor neurons settle quickly into the component nuclei (by 10.0 mm GL).

Table 5.5. Cholinergic neurons of the monotreme brainstem

Nucleus or area	Comments on features in monotremes[1]
Parabigeminal nucleus	Cholinergic neurons not identified in platypus and short-beaked echidna.
Pedunculopontine tegmental nucleus	These small cholinergic neurons (mean neuronal area of 272 μm^2 in platypus and 222 μm^2 in echidna) are associated with the brachium conjunctivum (axons of superior cerebellar peduncle). The cholinergic cells of this group are formed into a distinct arc in the echidna, but are more loosely clustered in the platypus.
Laterodorsal tegmental nucleus	These moderately size cholinergic neurons (mean neuronal area of 583 μm^2 in platypus and 347 μm^2 in echidna) are closely associated with the locus coeruleus.
Oculomotor nucleus and Edinger-Westphal nucleus	Manger *et al.* (2002a) did not differentiate these two distinct nuclei in their description of cholinergic neurons. Cholinergic motor neurons in the oculomotor nucleus control the extraocular muscles except the superior oblique and external rectus. The cholinergic neurons in the Edinger-Westphal nucleus are parasympathetic preganglionic neurons that control the ciliary muscle and sphincter pupillae of the eye.
Trochlear nucleus	Contains moderately large motor neurons (mean neuronal body area of 647 μm^2 in platypus and 527 μm^2 in echidna). Constituent neurons control the superior oblique muscle that intorts and depresses the eye.
Motor trigeminal nucleus	Contains very large motor neurons (mean neuronal body area of 920 μm^2 in platypus and 907 μm^2 in echidna). It supplies the muscles of mastication.
Facial (motor) nucleus	Divided into dorsal and ventral components. Each contains very large motor neurons (neuronal body area of 906 and 753 μm^2 in platypus, and 1172 and 1019 μm^2 in echidna, respectively). Constituent neurons supply the muscles of facial expression.
Abducens nucleus	Contains small motor neurons (mean neuronal body area of 487 μm^2 in platypus and 484 μm^2 in echidna). Constituent neurons supply the external (lateral) rectus muscle for abduction of the eye.
Salivatory and lacrimal nuclei	Not identified by Manger and colleagues, but may lie within the region delineated as dorsal facial nucleus by them. These nuclei contain preganglionic parasympathetic neurons for control of salivary and lacrimal glands.
Vagal sensorimotor complex	Not differentiated into sensory and motor components by Manger *et al.* (2002a). The motor component contains preganglionic neurons that control the thoracic and abdominal viscera.
Nucleus ambiguus	Large motor neurons (mean neuronal body area of 870 μm^2 in platypus and 956 μm^2 in echidna). It supplies the muscles of the larynx, pharynx and soft palate.
Medullary tegmental nucleus	A loosely associated group of neurons located in the middle of the medullary tegmentum at a rostrocaudal level close to the middle of the hypoglossal nucleus. The neurons are morphologically similar to the nucleus ambiguus and lie in continuity with the nucleus ambiguus and the ventral horn motor neurons.
Hypoglossal nucleus	Contains large motor neurons (mean neuronal body area of 704 μm^2 in platypus and 862 μm^2 in echidna). It supplies the intrinsic and extrinsic muscles of the tongue.
Spinal accessory nucleus	Not specifically identified by Manger and colleagues. These neurons supply musculature of the neck and rostral shoulder and probably lie in the groups identified as ventral horn by Manger and colleagues.

1 Manger *et al.* (2002a).

The medial longitudinal fasciculus (a pathway that connects the eye muscle nuclei with each other and the cervical spinal cord for coordination of eye and head movements) is also visible well before hatching (by ~10.0 mm GL) and may be active in the control of the neck movements of the hatchling.

Neurogenesis of the trigeminal and facial motor nuclei begins during the middle of incubation and motor neurons have begun to settle in the tegmentum by ~10 mm GL. Nevertheless, the facial motor neurons have yet to complete settling in the ventral tegmentum by the time of hatching, suggesting that they may not be functional at the time of emergence from the egg.

The nucleus ambiguus is difficult to differentiate from the surrounding reticular formation until late in lactational phase life.

Development of the reticular formation in monotremes

The reticular formation may play a critically important role in the developing monotreme by providing a processing centre for mediating the effect of somatosensory or vestibular input on motor function of the neck and forelimbs. This would be central to allowing the newly hatched monotreme to regulate its forelimb and neck movements to move around the nest or pouch and bring the snout in contact with the maternal mammary areolae.

As in therians, the reticular formation of the monotreme brainstem is divided into a series of radial wedges best seen during development (Fig. 5.3b, d): a median or paramedian raphe nuclei zone; a medial magnocellular ('big cell') compartment that gives rise to the reticulospinal pathways; an intermediate zone with autonomic function; and a lateral parvicellular ('small cell') zone that processes sensory information from the sensory (afferent) nuclei column along the side of the brainstem. This radial division of the brainstem is clearly present by the time of hatching, but the main division into medial magnocellular and lateral parvicellular zones is visible as early as the middle of incubation. The constituent neurons of the circuitry that would be required for mediating simple sensorimotor reflex loops at the brainstem level are certainly present around the time of hatching, but it is impossible at present to be sure whether the relevant circuitry is fully connected.

The state of the monotreme brainstem at hatching: meeting the demands of post-hatching life

How does the process of brainstem development prepare the newly hatched monotreme for survival in the external world? How does the brainstem of the newly hatched monotreme compare to that of the newborn marsupial? Both monotremes and marsupials emerge into the external environment in a relatively immature state and both must break through developmental membranes at the time of hatching or birth. Given the prominent egg tooth and os caruncle of monotreme hatchlings (Griffiths 1978), one might deduce that monotreme hatching is a more physically demanding process than birth from the urogenital sinus is for marsupials. On the other hand, the mother monotreme has already placed the egg in the pouch (short-beaked echidna) or vegetation nest (platypus) before hatching, so the newly hatched monotreme does not face quite the same challenge of climbing to the pouch that newborn diprotodont marsupials (e.g. wallabies, kangaroos, possums) do. There are also other challenges peculiar to monotreme hatchlings. The teat-less structure of the monotreme mammary gland means that the newly hatched monotreme cannot permanently attach to a nipple, and is forced to repeatedly re-locate the milk source to access an ongoing supply of nutrition. Despite this, juvenile monotremes are very effective at sucking, even from the time of hatching, as evidenced by the milk-filled gastrointestinal tract of newly hatched echidnas (Griffiths 1968, 1978; see discussion in Chapter 3).

The extent of maturation of the monotreme brainstem at the time of hatching can be judged indirectly from the anatomical organisation of the brainstem and the degree of differentiation of sensory and motor nuclei (see Fig. 5.3 and the Ec-In and Pl-Ha atlas series). The process of hatching probably requires central pattern generators in the brainstem to produce the necessary rhythmic movements of the head, neck and forelimbs to break through the developmental membranes. The critical question for life after hatching is whether the brainstem provides the required sensory networks for analysing information about the environment of the newly hatched monotreme, neural machinery for ensuring adequate motor coordination under the direction of that sensory input, and the central pattern generators for the control of head, neck and forelimb movements to allow movement to the milk source and stimulate let-down. It is evident from the structure of the brainstem around hatching that there are already substantial populations of neurons settled in the medial magnocellular and lateral parvicellular components of the reticular formation. The former occupy a larger cross-sectional area of the brainstem and are particularly important because they provide the descending projection to the spinal cord for the control of cervical enlargement motor neurons and forelimb movement (through reticulospinal pathways), but the latter must also be important for interpreting sensory input through the trigeminal nuclei and regulating the magnocellular component. Significantly, the cerebellum and precerebellar systems are very immature in the peri-hatching period, and there are no descending pathways from the forebrain to either drive or coordinate motor activity by the hatchling.

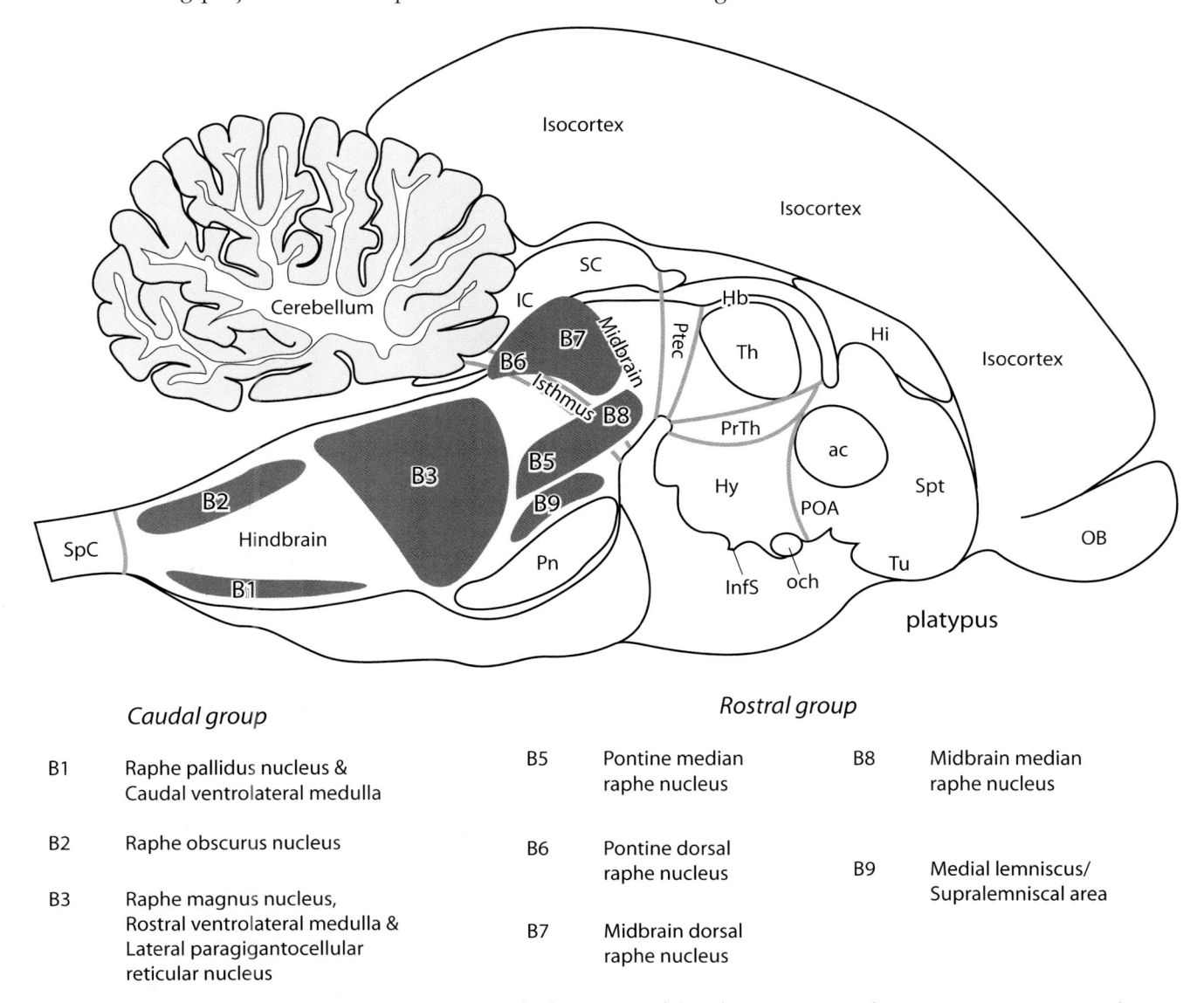

Caudal group

B1	Raphe pallidus nucleus & Caudal ventrolateral medulla
B2	Raphe obscurus nucleus
B3	Raphe magnus nucleus, Rostral ventrolateral medulla & Lateral paragigantocellular reticular nucleus

Rostral group

B5	Pontine median raphe nucleus
B6	Pontine dorsal raphe nucleus
B7	Midbrain dorsal raphe nucleus
B8	Midbrain median raphe nucleus
B9	Medial lemniscus/ Supralemniscal area

Figure 5.5: Distribution of serotonergic neurons in the brainstem of the platypus, mapped onto a representative midsagittal section and based on the data in Manger *et al*. (2002c). ac – anterior commissure; Hb – habenular nuclei; Hi – hippocampus; IC – inferior colliculus; InfS – infundibular stalk; OB – olfactory bulb; och – optic chiasm; POA – preoptic area; Pn – pontine nuclei; Ptec – pretectum; PrTh – prethalamus; SC – superior colliculus; SpC – spinal cord; Spt – septum; Th – (dorsal) thalamus; Tu – olfactory tubercle.

Cerebellum and precerebellar nuclei

Cerebellum structure and function in mammals

The cerebellum consists of a highly folded layered cortical structure with associated deep cerebellar nuclei (medial, interposed and lateral). The folded cerebellar cortex consists of midline vermal and paravermal zones, flanked by cerebellar hemispheres. Three functional subdivisions are usually recognised in the mammalian cerebellum, although it should be noted that the boundaries between these are not always clear-cut. The *vestibulocerebellum*, comprising the flocculus and nodule (10Cb) of the cortex (flocculonodular lobe) and the medial or fastigial deep cerebellar nucleus, receives and processes information concerned with vestibular input and control of balance and eye movement. The *spinocerebellum*, made up of the anterior lobe, vermal and paravermal zones of the cerebellar cortex and the interposed deep cerebellar nuclei, is concerned with processing information from the spinal cord for the coordination of axial or proximal limb musculature. The *pontocerebellum* or *cerebrocerebellum* is composed of the cerebellar hemispheres and lateral deep cerebellar nucleus, and is concerned with the coordination of distal musculature for relatively fine movements. It receives descending control from the contralateral cerebral cortex (via the pontine nuclei and pontocerebellar tract) and feeds back to the contralateral motor cortex through a loop involving the motor thalamic nuclei. Each of these functional zones in the cerebellum therefore comprises cerebellar cortex and one of more associated deep nuclei (i.e. corticonuclear zones) that may be recognised on chemoarchitectural, hodological (i.e. connectional) and electrophysiological grounds.

External features of the monotreme cerebellum

Figure 5.7 shows line diagrams summarising the major fissures and parts of the cerebellum of the platypus and echidna. Despite the larger body size of the short-beaked echidna compared to the platypus, the size of the cerebellum in the two species is broadly similar (compare cerebellar length, width and height in Appendix Tables 4 and 5).

As in other vertebrates, 10 vermal lobules (numbered according to the schema of Larsell 1970) may be identified in both species, although early studies of the monotreme cerebellum were characterised by robust argument over the homology of lobules with those in other vertebrates (Larsell 1970). The area of the midline (vermal) cortex is only slightly larger in the short-beaked echidna (mean ± s.d. of 93.5 ± 13.3 mm^2, Appendix Table 5) compared to the platypus (mean ± s.d. of 86.6 ± 20.0 mm^2, Appendix Table 4).

Hemispheric regions are also identifiable, with ansiform lobules, paramedian lobules and lobulus simplex (Larsell 1970; Holst 1986; Fig. 5.7) and an extensive multifoliated paraflocculus. In an early description of the monotreme cerebellum, Dillon identified a ventral cerebellar lobe as a peculiar feature of the monotremes (Dillon 1962), but this has not been supported by later studies (Holst 1986; Ashwell *et al.* 2007a).

In the platypus, the flocculus (of the vestibulocerebellum) is a small lobule situated behind the laterally projecting paraflocculus (Hines 1929), which in this respect resembles the relations of the avian flocculus and paraflocculus (Larsell 1970). The nodule of the platypus is directly continuous with the flocculus, with little or no constriction between them, also like the flocculonodular lobe of birds (Larsell 1970). The primary and secondary fissures of the cerebellar vermis are of equal depth, but there is no doubt as to the identity of each (Fig. 5.7c). The primary fissure is important for separating the anterior lobe (vermal lobules 1Cb to 5Cb and cortical tissue to the side of those) from the posterior lobe.

Holst (1986) noted that, in contrast to some previous claims, the cerebellum of the echidna is fully

Figure 5.6: Distribution of cholinergic neurons (darkly shaded regions) in the brain of the platypus, mapped onto a longitudinal representation (a) and based on the data in Manger *et al.* (2002a). Illustrations (b) and (c) show photomicrographs of sections through the midbrain (b) and medulla (c) of the short-beaked echidna. These have been reacted for acetylcholinesterase (AChE), the degradative enzyme for acetylcholine. 3N – oculomotor nucleus; 3V – third ventricle; 4N – trochlear nucleus; 5N – motor trigeminal nucleus; 6N – abducens nucleus; 7N – facial (motor) nucleus; 10N – vagus nerve nucleus; 12n – hypoglossal nerve; 12N – hypoglossal nucleus; ac – anterior commissure; Acb – nucleus accumbens; Amb – nucleus ambiguus; CPu – caudatoputamen; Hb – habenular nuclei; Hi – hippocampal formation; Hy – hypothalamus; IC – inferior colliculus; IP – interpeduncular nucleus; LDTg – laterodorsal tegmental nucleus; mlf – medial longitudinal fasciculus; OB – olfactory bulb; och – optic chiasm; pc – posterior commissure; Pi – pineal gland; Pn – pontine nucleus; Ptec – pretectum; PTg – pedunculotegmental nucleus; PrTh – prethalamus; SC – superior colliculus; Sp5I – nucleus of the trigeminal spinal tract, interpolar part; SpC – spinal cord; Th – (dorsal) thalamus.

mammalian in organisation: it shows a large suite of mammalian features – well-developed hemispheres, 10 vermal lobules, ansiform and paramedian lobules, multi-foliated paraflocculus, deep pre-pyramidal fissure and a parafloccular fissure restricted to the hemisphere (Holst 1986; Ashwell et al. 2007a). The medullary (central white matter) ray pattern of the echidna cerebellum is also clearly similar to that of other mammals, in that branching of medullary rays in the vermis occurs to more than one folium (Holst 1986). The superior vermis of the echidna cerebellum is highly peaked due to being sandwiched between the caudal extensions of the occipital lobe of isocortex. The flocculus is represented by only a very small region of cerebellar cortex adjacent to the pons in both the platypus and echidna.

Corticonuclear zones in the monotreme cerebellum

Chemically based architecture and tracing of input and output pathways of the cerebellar cortex reveal a longitudinally running zonal organisation (Fig. 5.8) linking the cerebellar cortex with the deep cerebellar nuclei. This zonation appears to be equally applicable to the monotreme cortex, although the full range of chemical markers and tracing studies have not yet been undertaken in the monotremes. In contrast to early reports (Hines 1929; Abbie 1934) both the platypus and short-beaked echidna (like other mammals) have three sets of deep cerebellar nuclei (medial, interposed and lateral nuclei; Ashwell et al. 2007a; see atlas plates Ec-Ad25, Ec-Ad26, Pl-Ad25).

In general, the vertebrate cerebellar cortex can be divided into medial (vermal), intermediate and lateral zones, projecting to medial, interposed and lateral deep cerebellar nuclei, respectively. In acetylcholinesterase reacted and parvalbumin or calbindin immunoreacted sections, parasagittal bands may be identified in the white matter of the anterior vermis and paravermis of the echidna cerebellum (Ashwell et al. 2007a). Putative homologies with the corticonuclear zones described in placental cerebellum may be assigned on the basis of similarities to banding in primate cerebellum (Voogd 2004). These include a midline band of moderate AChE activity and low parvalbumin and calbindin immunoreactivity, which may be traced through several vermal lobules in the rostral and middle cerebellum (putative m compartment). Lateral to this is a relatively broad region (~2 mm wide) of lower AChE reactivity, but stronger parvalbumin and calbindin immunoreactivity, in line with the medial deep cerebellar nucleus. This may be the homologue of compartments A, X and B of placental cerebellum, but it does not have any internal subdivision in the monotremes. This is flanked in turn by a region of moderate AChE reactivity, moderate parvalbumin immunoreactivity, but poor calbindin immunroeactivity, which may be traced to the medial part of the interposed nucleus and may be homologous to compartments C1 and C2 in placentals (Voogd 2004). Further lateral is another zone of relatively low AChE reactivity, but moderate parvalbumin and calbindin immunoreactivity (putative C3 homologue), which may be traced to the main part of the interposed nucleus. This in turn is flanked by a parasagittal region of moderate AChE reactivity, which is bordered laterally by the granule cell layer of the ansiform lobule and may be traced into the lateral deep cerebellar nucleus (putative D compartments).

Cellular structure of the cerebellar cortex in monotremes

The vertebrate cerebellar cortex has three layers; molecular, Purkinje cell and granule cell. The outermost molecular layer mainly contains the parallel fibres from the granule cells as well as terminal climbing fibre axons and a small population of neurons (basket and stellate cells). The Purkinje cell layer contains the cell bodies of the Purkinje cells and the interposed cell bodies of candelabrum cells and Bergmann glia. The granule cell layer contains the abundant granule cell bodies, mossy fibre terminations and smaller numbers of local circuit neurons (unipolar brush cells, Golgi cells and Lugaro cells).

The overall appearance of the cerebellar cortex in sections stained for Nissl substance is much like in therians (see atlas plates Ec-Ad25 and Pl-Ad25), but detailed cytological data are lacking. The cellular constituents of the platypus cerebellum have not been studied in any detail, but the chemoarchitecture of the cerebellum of the short-beaked echidna (Ashwell et al. 2007a) has received some attention. In short, the cytology of the echidna cerebellum shows several similarities to therians (see below), but also some significant differences, suggesting that the echidna cerebellum may have functional differences in local

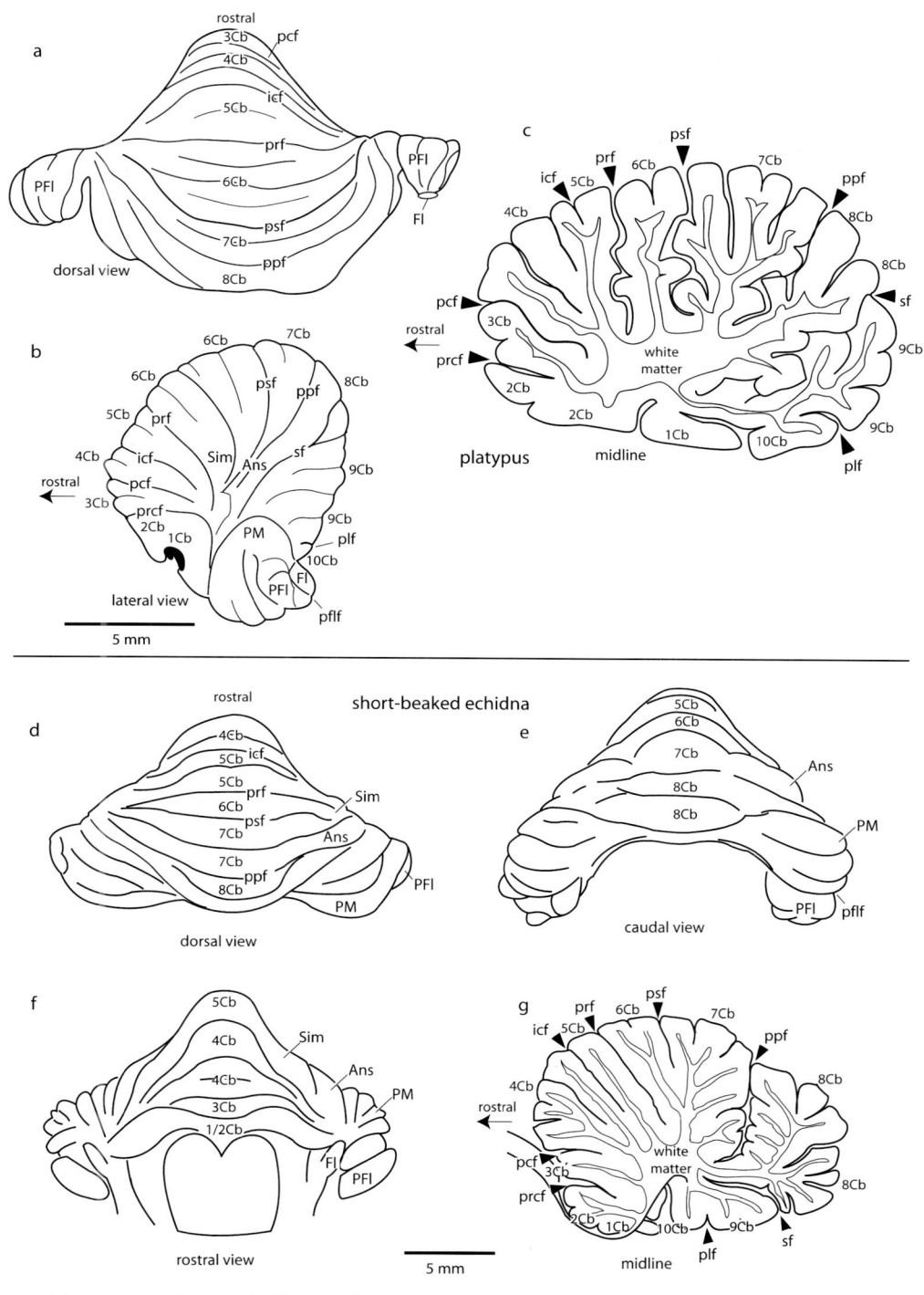

Figure 5.7: External features of the cerebellum in the platypus (a to c) and short-beaked echidna (d to g). The line illustration of the platypus cerebellum is based on Hines (1929), with nomenclature for lobes and fissures as applied by Larsell (1970). The line illustration of the echidna cerebellum is based on Holst (1986) and Ashwell *et al.* (2007a) and has been reproduced from Ashwell *et al.* (2007a) with kind permission from S. Karger AG, Basel. Ans – ansiform lobule; Fl – flocculus; icf – intraculminate fissure; pcf – preculmen fissure; PFl – paraflocculus; pflf – paraflocccular fissure; plf – posterolateral fissure; PM – paramedian lobule; ppf – prepyramidal fissure; prf – primary fissure; prcf – precentral fissure; psf – posterior superior fissure; sf – secondary fissure; Sim – simplex lobule.

circuit neurons compared to the therian cerebellar cortex.

Immunoreactivity for the SMI-32 antibody (raised against non-phosphorylated neurofilament protein) labels the proximal dendrites, somata and proximal axons of Purkinje cells in the short-beaked echidna, but the numerical density of SMI-32 immunoreactive Purkinje cells is much lower in the echidna cortex compared to corresponding regions of the laboratory rat (Ashwell *et al.* 2007a). Ashwell and colleagues also noted that the flocculus and nodule of the echidna have only weakly immunoreactive Purkinje cells, whereas these regions show strongly SMI-32 immunoreactive Purkinje cells in the laboratory rat. This may reflect differences in axonal calibre and length between Purkinje cells in functionally divergent regions of cerebellar cortex, but overall the dendritic trees of the neurofilament containing Purkinje cells in the echidna are very similar to those in the Purkinje cells of other mammals.

Table 5.6. Nuclei projecting to the cerebellum of the short-beaked echidna[1]

Nucleus	Ipsilateral projection	Contralateral projection
Inferior olivary nuclear complex	+	++++
Pontine nuclei	++	++++
Reticulotegmental nucleus	+	+
Lateral reticular nucleus	++	+++
Principal trigeminal nucleus	+	+
Spinal trigeminal nucleus	+	+/−
Gracile nucleus	+	+
Cuneate nucleus	+	++
External cuneate nucleus	++++	+++
Vestibular nuclei • Lateral (LVe) • Medial (MVe) • Spinal (SpVe) • Superior (SuVe)	+ + + +	+ + + +
Cochlear nuclei	+	+
Superior olivary nucleus	+/−	+/−
Nucleus of solitary tract	+/−	+/−
Vagal nerve nucleus	+/−	+/−
Paratrigeminal nucleus	−	+
Arcuate nucleus	+	+
Parabrachial nuclei	++	−
Intercalated perihypoglossal nucleus	−	+
Nucleus of Roller	−	+
Nucleus prepositus hypoglossi	+	++
Raphe nuclei	++	++
Reticular formation • Oral reticular nucleus of pons • Caudal reticular nucleus of pons • Parvicellular reticular nucleus • Gigantocellular reticular nucleus • Medullary reticular nucleus • Linear nucleus of the medulla	+ + + ++ + −	+ + ++ + + ++

1 Based on injections into B and C1, C2 corticonuclear zones of the cerebellar hemisphere; see Holst (1986) and Fig. 5.8. Number of + indicates relative number of neurons projecting to the cerebellum.

In broad terms, the pattern of laminar distribution of tyrosine hydroxylase immunoreactivity (the biosynthetic enzyme for production of catecholamine neurotransmitters) in the echidna cerebellum is also similar to that in therian mammals. Tyrosine hydroxylase immunoreactive axons are mainly distributed to the Purkinje cell layer and upper granule cell layers in therians and the echidna (Nelson et al. 1997; Ashwell et al. 2007a), but there are significant topographic differences. In the echidna for example, most tyrosine hydroxylase immunoreactive axons are distributed to the lateral parts of the cerebellum (flocculus, paraflocculus, paramedian and ansiform lobules; Ashwell et al. 2007a). By contrast, most tyrosine hydroxylase axons are found within the midline or vermis (lobules 5Cb and 6Cb) in the opossum and cat (Nelson et al. 1997). The echidna cerebellum also has large numbers of tyrosine hydroxylase immunoreactive Purkinje cell somata, particularly in the paraflocular and paravermal regions.

Immunoreactivity for the calcium-binding proteins parvalbumin and calbindin is found in many Purkinje cells of the echidna, in a similar pattern to some avian species and therian mammals (Celio 1990), where calbindin appears to play a complementary role to parvalbumin in Ca^{2+} buffering in Purkinje cells during periods of intense activity (see discussion in Ashwell et al. 2007a). On the other hand, no calbindin immunoreactivity has been found in the bodies of any non-Purkinje cells in the echidna cerebellar cortex, although strong calbindin immunoreactivity was found in terminals on scattered Lugaro cells in the granule cell layer, much like in therian cerebellum (Ashwell et al. 2007a).

Probably the most striking difference in calcium-binding protein immunoreactivity between the echidna and therians concerns calretinin. In primate cerebellar cortex, calretinin has been found in virtually all neurons and fibre systems associated with the cerebellar granular layer (monodendritic cells, granule cells, parallel fibres, Golgi and Lugaro cells and mossy fibres; Fortin et al. 1998), as well as some Purkinje cells during development (Yew et al. 1997). In cetacean and elephant cerebellum, calretinin is also present in climbing fibres and different groups of granular layer cells (Maseko et al. 2013; Kalinichenko and Pushchin 2008). Immunoreactivity for caletinin is also seen in a wide variety of cells (stellate cells, basket cells and the avian equivalent of Lugaro cells) in chick cerebellum (Rogers 1989). By contrast, immunoreactivity for calretinin was never seen in any Purkinje cells of the echidna cerebellar cortex and only occasionally in scattered cell bodies in the molecular and granule cell layers of limited regions of the cerebellar cortex (Ashwell et al. 2007a).

Connections of the monotreme cerebellum

Primary vestibular axons (i.e. directly from the vestibular ganglion) pass to the nodulus, uvula, flocculus and lingula of the platypus cerebellum (Hines 1929; Larsell 1970), bringing information about balance and acceleration directly into the cerebellum for processing. Secondary vestibular axons from the vestibular nuclei are also distributed to the cerebellum and the medial (fastigial) deep cerebellar nucleus.

Given the large size of the trigeminal sensory complex in the platypus, it is not surprising that there is a large trigeminal input (carrying touch and presumably electrosensory information from the bill) to the cerebellum in the platypus (Larsell 1970). These axons appear to be branches of the ascending trigeminal nerve axons that will terminate in the rostral trigeminal sensory nuclear complex. The trigeminocerebellar axons have not been traced by modern techniques in either group of monotremes, but Larsell has linked the large size of lobule 4Cb in the platypus with the large trigeminal input.

The precise course of spinocerebellar axons (carrying touch and proprioceptive information from the spinal cord) in the monotremes is still a matter of uncertainty (Larsell 1970), but the dorsal spinocerebellar tract is likely to enter through the inferior peduncle and the ventral spinocerebellar tract probably enters through the white matter of the superior medullary velum (superior cerebellar peduncle). Olivocerebellar and cuneocerebellar axons have been included in the restiform body of the inferior cerebellar peduncle by both Hines (1929) and Abbie (1934). Abbie also identified reticulocerebellar axons in the inferior cerebellar peduncle. Hines and Abbie both identified tectocerebellar axons that enter the cerebellum through the superior cerebellar peduncle. These last would carry integrated visual and (probably) somatosensory information about the position of objects in space around the organism into the cerebellum for use in motor coordination.

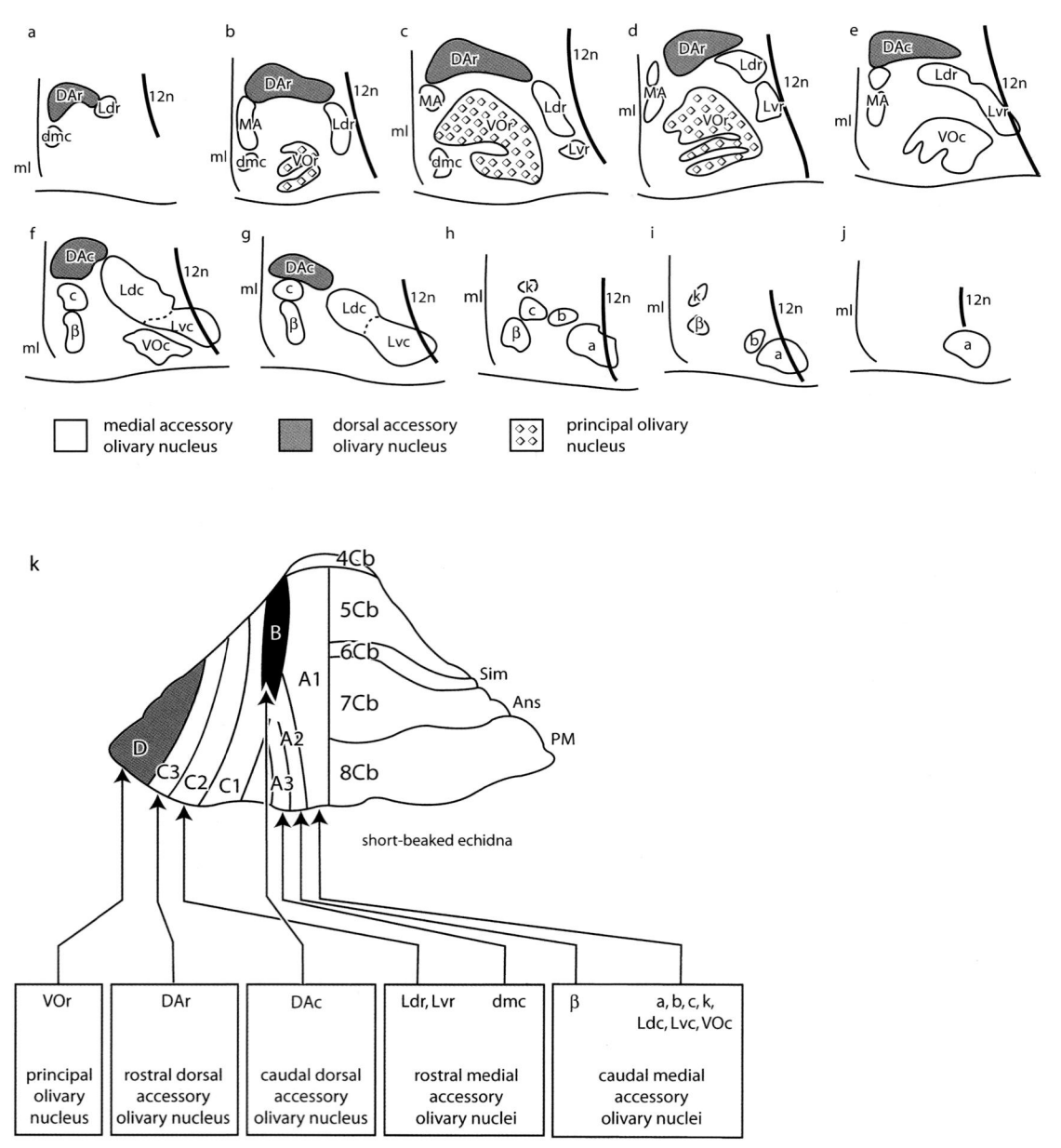

Figure 5.8: A diagrammatic representation of the inferior olivary nuclear complex and the climbing fibre projection from the inferior olivary nuclear complex to the cerebellum in the short-beaked echidna (adapted from Holst 1986). The illustrations at the top (a to j) show a series of transverse section through the inferior olivary complex in sequence from rostral to caudal. The monotreme inferior olivary complex includes (1) a centrally placed VOr (the principal nucleus homologue); surrounded by (2) DA (the dorsal olivary nuclei homologue) with rostral and caudal components (DAr, DAc); and (3) a loose archipelago of nuclei including the VOc, a, b, c, k, β, dmc, Ld, Lv, and MA, together making up the putative ventral olivary nuclei homologue. The dorsal and ventral lateral subnuclei can be further subdivided in rostral and caudal components (Ldr, Lvr, Ldr, Lvc) based on projections to the cerebellum (see diagram k). The line diagram (k) shows the projection of the components of the inferior olivary nuclear complex to the corticonuclear zones of the cerebellum (olivocerebellar projection), as mapped onto a dorsal view of the cerebellar cortex. The olivocerebellar projection is based on data in Holst (1986). 12n – hypoglossal nerve; 4Cb to 8Cb – lobules of cerebellar cortex; Ans – ansiform lobule; ml – medial lemniscus; PM – paramedian lobule; Sim – simplex lobule.

Precerebellar nuclei

The precerebellar nuclei are the groups of neurons that project axons to the cerebellum (mossy and climbing fibre afferents). They are developmental derivatives of the rhombic lip, the same embryonic precursor structure which gives rise to the neurons of the cerebellum itself. Precerebellar nuclei have been identified for the short-beaked echidna with the aid of retrograde tracing of horseradish peroxidase injected into the cerebellar cortex (Holst 1986). These include what might be called the major precerebellar nuclei (inferior olivary nuclear complex, pontine nuclei, reticulotegmental and lateral reticular nuclei) because of their extensive projections to the cerebellum, as well as minor precerebellar nuclei which have a small projection to the cerebellum in addition to their major projection to other targets. These are summarised in Table 5.6.

The inferior olivary nuclear complex contributes climbing fibres to the cerebellar cortex with collaterals to the deep cerebellar nuclei. The overlying structure of the inferior olivary complex (IO) of the echidna is an arch with a ventral neuronal group between the two arch pillars (Fig. 5.8a to j; Kooy 1916; Abbie 1934; Holst and Watson 1973; Holst 1986; Ashwell et al. 2007b). The arch can be divided into medial, dorsal and lateral components, with the lateral component further subdivided into dorsal (Ldr/Ldc) and ventral subnuclei (Lvr/Lvc). At the rostral end of the complex, a dorsomedial column can be identified medial to the medial arch component. At the caudal end of the complex, the medial arch separates into subnuclei k, c and β. The ventral IO can be divided into caudal and rostral components (VOc and VOr of Fig. 5.8). Holst (1986) described the caudal ventral olivary nucleus as being associated with the ventral part of the lateral arch, but Ashwell and colleagues (Ashwell et al. 2007b) reported that the caudal ventral olive was usually discrete from that part of the arch in their material. The division of the IO into discrete subnuclei is consistent with a fully mammalian organisation and stands in contrast to the poorly differentiated IO of amphibians and reptiles (Bangma and ten Donkelaar 1982; see review in Holst 1986).

Consistent with the subnuclear division of the IO, Holst (1986) found a topographic olivocerebellar projection between the IO components and parasagittal zones in the cerebellar cortex (Fig. 5.8k). The caudal portions of the dorsal and lateral components of the lateral arch of the echidna IO (i.e. LAd and LAv), subnuclei a, b, c and the dorsal cap of Kooy (k), the most caudal part of the VO and a small part of the medial arch (MA) all appear to project to the vermis (putative zone A) and seem to be homologous to the caudal medial accessory olivary nucleus of the therian IO (Holst 1986). The caudal DA projects to the B zone of the rostral vermis, while zones A2 and A3 receive projections from the β subnucleus and dorsomedial cell column. The caudal DA of the echidna therefore appears to be homologous to the caudal dorsal accessory olivary nucleus of placentals, while the β subnucleus and dorsomedial cell column may be homologous to the ventromedial part of the caudal MAO (medial accessory olivary nucleus) of placentals. The rostral parts of the echidna LA and rostral MA appear to be homologous to the rostral MAO of placentals in that they project to the C2 zone. The rostral DA of the echidna IO projects to the putative C3 zone of the cerebellar hemisphere and can be considered homologous to the rostral part of the dorsal accessory nucleus of the placental IO (Holst 1986). The VOr appears to be homologous to therian principal olivary nucleus and projects to the paramedian lobule and crus I and II of the ansiform lobule and lobulus simplex parts of the cerebellar hemisphere (putative zones D1 and 2) (Holst 1986).

Vestibular nuclei of monotremes

Both the platypus and echidna have four vestibular nuclei (Hines 1929; Ashwell et al. 2007b) like therians do. Hines identified medial, spinal, lateral and superior vestibular nuclei. The lateral vestibular nucleus has large neurons (as in therians) and gives rise to the lateral vestibulospinal tract for controlling muscles along the body axis to maintain balance.

The vestibular nuclear complex of the echidna brainstem extends for between ~5 and 6 mm in the rostrocaudal direction (Ashwell et al. 2007b: atlas plates Ec-Ad24, Ec-Ad25, Ec-Ad26). Those authors identified a superior vestibular nucleus (SuVe) at the level of the medial parabrachial nucleus, medial and lateral vestibular nuclei (MVe, LVe) at mid-rostrocaudal levels and a spinal vestibular nucleus (SpVe) extending from middle levels to the caudal extremity of the complex. Large neurons are found mainly in the lateral and spinal nuclei.

Development of the cerebellum and precerebellar nuclei

The vertebrate cerebellum develops from two proliferative zones in the hindbrain. The first of these is the ventricular germinal zone of the cerebellar primordium, which generates the GABAergic neurons of the cerebellum (Purkinje cells, nucleo-olivary neurons of the deep cerebellar nuclei, interneurons of the deep cerebellar nuclei, and Golgi, Lugaro, basket and stellate cells) (Carletti and Rossi 2008).

The second is the rhombic lip, a very active germinative zone along the lateral margin of the fourth ventricle of the hindbrain. The rostral rhombic lip generates the glutamatergic neurons of the cerebellum (Carletti and Rossi 2008), including all the cerebellar granule cells, which amount to more than 80% of all the neurons in the brain (Herculano-Houzel and Lent 2005) so it is a region with an astounding capacity for neuron generation. The cerebellar granule cells are produced from the external granular layer, which migrates over the developing cerebellum from the rostral rhombic lip (rhombomeres r0 and r1). The rostral rhombic lip also gives rise to other glutamatergic neurons, such as the unipolar brush cells and projection neurons of the deep cerebellar nuclei (Carletti and Rossi 2008).

The rhombic lip also generates the cells of the major precerebellar nuclei (the cell groups in the hindbrain that send input to the cerebellar cortex; Altman and Bayer 1987a–d; Ambrosiani *et al.* 1996; Harkmark 1954; Rodriguez and Dymecki 2000). Distinct migratory streams have been identified in rodents, transferring young neurons from the proliferative zones around the margin of the fourth ventricle to their final resting sites throughout the hindbrain and isthmus. The anterior extramural migratory stream transfers neurons to the basal pontine grey and reticulotegmental nucleus of the pons (Altman and Bayer 1987d); the posterior extramural migratory stream carries neurons to the lateral reticular and external cuneate nuclei (Altman and Bayer 1987c); and the intramural migratory stream transfers neurons to the inferior olivary nuclear complex (Altman and Bayer 1987b).

The developing monotreme cerebellum shows several key features in common with the developing cerebellum of marsupials and placentals (Altman and Bayer 1978a, b; 1985a, b; 1987a–d; Ashwell 2012f; see Appendix Fig. 3). These include: (1) the same primary and secondary proliferative zones (i.e. cerebellar ventricular germinal zone, rhombic lip or germinal trigone, and the external granular layer); (2) the same transitory cell populations (i.e. nuclear and cortical transitory zones); and the same transient fibre zones (i.e. intermediate fibres between the nuclear and cortical transitory zones, hook bundle of Russell, and marginal zone external to the nuclear transitory zone). These similar features strongly suggest that the monotreme cerebellum is subject to the same developmental processes observed in other mammals.

Any differences between monotreme and placental cerebellar development are more a matter of timing rather than process. Expansion of the rhombic lip and generation of the nuclear and cortical transitory zones of the developing monotreme cerebellum occur before hatching, whereas formation and migration of the external granular layer over the cerebellum, generation of the granule cells of the cerebellar cortex, and folding of the cerebellum to form fissures and folia occur after hatching. The external granular layer (for production of small cerebellar neurons of the granular layer) persists for up to 4 or 5 months of life after hatching. In many respects, the timing of the development of the monotreme cerebellum is much more like that in poly- and diprotodont marsupials (i.e. opossums and wallabies) than in placentals. Marsupials also have the greater part of cerebellar development taking place after birth, whereas generation of granule cells is usually the only part of cerebellar histogenesis that occurs postnatally among placentals, and that only in altricial species like rodents and humans. On the other hand, the monotreme cerebellum is slightly advanced at hatching compared to that in newborn marsupials, in that the nuclear and cortical transitory zones are already evident by the last third of incubation, whereas these transitory populations emerge postnatally in all the marsupials studied to date (Ashwell 2012f).

The first extramural (marginal) migratory streams of precerebellar neurons become visible streaming around the curve of the brainstem at the end of the first post-hatching week in the monotremes (Ashwell

2012f). As in placentals (Nichols and Bruce 2006), these consist of an anterior extramural migratory stream destined for the pontine and reticulotegmental nuclei and a posterior extramural migratory stream from the more caudal rhombic lip, destined for the external cuneate and lateral reticular nuclei. Intramural (mantle II) and perigerminal zone (mantle I) migrations that pass through the interior of the brainstem cannot be clearly seen, but this may be due to the limitations of the thickly sectioned material in the Hill collection.

Questions for the future

The structural similarity of the monotreme brainstem and spinal cord at hatching to that in marsupials naturally raises the question of how the hatchling nervous system regulates movements during early life. Does the monotreme nervous system rely on central pattern generators in the spinal cord and brainstem to produce rhythmic movements to break through developmental membranes and stimulate milk let-down? How does sensory input at the brainstem level influence hatchling behaviour? Is sensory input confined to trigeminal stimulation or do other sensory modalities contribute?

Although the overall structure of the monotreme cerebellum is very similar to that in therians, the significant differences in cellular chemoarchitecture point to potential functional differences in cerebellar microcircuitry. Immunoreactivity for zebrinII/aldolaseC has yet to be analysed in the monotremes and would provide important clues concerning corticonuclear compartmentation. Anterograde tracing of mossy and climbing fibre inputs to the cerebellum would also provide new perspectives on the phylogeny of cerebellar input in mammals. The unusual rhombic lip morphogenesis in the platypus is suggestive of unique developmental mechanisms for the microneuron populations of the brainstem and cerebellum. Recent advances in the molecular analysis of mammalian cerebellar and precerebellar development would benefit from a molecular study of monotreme rhombic lip morphogenesis.

6

Diencephalon and deep telencephalic structures

Ken W. S. Ashwell

Summary

The diencephalon includes the pretectum, dorsal thalamus, prethalamus and epithalamus. The monotreme pretectum and prethalamus are morphologically very similar to those in therian mammals, but the dorsal thalamus has some distinctive anatomical features that separate the monotremes both from the therians and from each other. In particular, the expansion of the trigeminal somatosensory thalamus in the platypus has distorted the topography of the caudal parts of the dorsal thalamus and shifted the positions of the visual nuclei of the diencephalon. This process is visible during post-hatching development of the diencephalon. Deep structures of the telencephalon include the striatopallidal complexes (dorsal and ventral striatum and pallidum), the amygdala and septal nuclei, as well as scattered populations of cholinergic neurons (the basal nucleus of Meynert). Initial information suggests that the striatum of the short-beaked echidna has similar functional subdivisions (into association, sensorimotor and limbic components) to those seen in therians. The striatopallidal components of the forebrain develop from the lateral and medial ganglionic eminences of the subpallium. The ganglionic eminences appear around the middle of incubation in the monotremes, but the generation of neurons for the striatum and pallidum and the formation of connections between them is a prolonged process, taking until the second month of post-hatching life.

Approach to the diencephalon and telencephalon

The diencephalon is one of the three main parts of the forebrain, the other two being the telencephalon and the hypothalamus (Puelles *et al.* 2012a, b). Although the preoptic area and hypothalamus have traditionally been included in the diencephalon, modern molecular analysis of neural development suggests that they are more properly considered as derivatives of the forebrain *rostral* to the true diencephalon (Puelles *et al*. 2012a). The diencephalon consists of three divisions derived from the respective prosomeres (Puelles *et al*. 2012b). Prosomere 1 gives rise to the pretectum, prosomere 2 gives rise to the (dorsal) thalamus, and prosomere 3 gives rise to the prethalamus (formerly known as the ventral thalamus).

The telencephalon is composed of the layered pallium (cortex) divided into medial, dorsal, lateral and ventral pallium; and all other telencephalic regions, collectively called the subpallium and including the striatopallidal complexes, amygdala and the septum. The pallium or cortex will be discussed in Chapter 7 for the isocortex, transitional cortex and hippocampal formation, and Chapter 11 for the olfactory cortex. The preoptic area and hypothalamus will be considered in Chapter 12 in the context of the interface between the nervous and endocrine systems and control of the autonomic nervous system.

In this chapter, the available information on the monotreme diencephalon will be reviewed in order of the prosomeres. The structure, function and development of the basal parts of the telencephalon (i.e. subpallium: striatum, pallidum, septum, amygdala) will also be reviewed, along with the cholinergic neurons of the basal forebrain.

Pretectum

The pretectum is the general term for the part of the neuraxis that is derived from prosomere 1. This region consists of groups of nerve cells loosely clustered around the posterior commissure, and situated between the midbrain and the thalamus. The main nuclear components of prosomere 1 are the pretectal nuclei, but many other smaller nuclei (subcommissural organ, nuclei of the posterior commissure, rostral periaqueductal grey, nucleus of Darkschewitsch, interstitial nucleus of Cajal) are also derived from this region of the developing brain (Puelles *et al*. 2012b).

Fibre bundles define the boundaries of the pretectal area in the mature brain. The rostral boundary of the pretectum with the thalamus is defined by the caudal limits of the habenulointerpeduncular tract (fasciculus retroflexus) and the parafascicular nucleus of the thalamus that surrounds it. The posterior commissure marks the caudal boundary between the pretectum and the midbrain.

Conceptions of how many nuclei make up the pretectal area have changed in recent years. Developmental analysis now suggests that the pretectal area of tetrapods is derived from three developmental regions arranged in a rostrocaudal sequence related to the posterior comissure: precommissural, juxtacommissural and commissural (Puelles *et al*. 2012b). The anterior pretectal nucleus is the largest nucleus of the adult pretectal region and is the main derivative of the most rostral part of the pretectum, the precommissural developmental region. It is heavily reactive for acetylcholinesterase in therians, contains both glutamatergic and GABAergic neurons and projects to the thalamus and the zona incerta (Puelles *et al*. 2012b). The smaller pretectal nuclei (medial pretectal nucleus, olivary pretectal nucleus and the nucleus of the optic tract) are interposed between the larger anterior pretectal nucleus and the tectal grey of the midbrain (which was formerly included in the pretectal area as the posterior pretectal nucleus). They are probably derived from the juxta- or commissural developmental areas. Puelles and colleagues have recently included two newly discovered nuclei (principal pretectal and subpretectal) within the pretectal area (Puelles *et al*. 2012b).

Organisation of pretectal nuclei in the monotremes

The pretectum of the living monotremes have been considered to be poorly differentiated compared to placentals and living diapsids (Butler and Hodos 2005). It has also been suggested that the region was better developed in the ancestral mammals and has been secondarily reduced in size in the monotremes, but cyto- and chemoarchitectonic analysis of the pretectum in the platypus and echidna indicates that the region is as differentiated in the monotremes as in most modern therians (Ashwell and Paxinos 2007). In fact Ashwell and Paxinos were able to identify four of the major pretectal nuclei found in therians.

The position of the monotreme pretectum was originally identified by the distribution of degenerating retinal ganglon cell axons, labelled by silver impregnation following removal of one eye (Campbell and Hayhow 1971, 1972). In the echidna, Campbell and Hayhow (1971) identified degenerating fibres coursing along the dorsum of the diencephalon/midbrain junction. Campbell and Hayhow labelled this region as the rostral superior colliculus (their Fig. 4), even though this region shows none of the cyto- or chemoarchitectonic features of the superior colliculus in enzyme or immunochemically stained sections. Ashwell and Paxinos noted that this region appears to be a misidentified pretectal area, because it is coextensive with an area identified as including the dorsal part of the anterior pretectal nucleus (APTD), medial pretectal nucleus (MPT), and olivary pretectal nucleus (OPT). The region of degenerating terminals that Campbell and Hayhow identified as either the dorsal terminal nucleus or nucleus of the optic tract (their Fig. 5) appears to correspond to the nucleus of the optic tract (OT) (Ashwell and Paxinos 2007).

In a subsequent study of the distribution of retinodiencephalic axons in the platypus, Campbell and Hayhow (1972) identified degenerating axons running along the dorsal surface of the midbrain/diencephalon junction, but considered only a small region of this terminal field as the pretectum. Nevertheless, it is clear from their diagrams that the retinofugal fibres are distributed to an area that encompasses the APTD, MPT, OPT and OT (Ashwell and Paxinos 2007).

Chemoarchitecture of the monotreme pretectum

Ashwell and Paxinos (2007) used topography and chemoarchitecture to identify all four pretectal nuclei usually found in placentals. The APT of the monotreme brain has strong acetylcholinesterase activity, much like the same region in therians. In both the platypus and echidna, the APT has neurons immunoreactive for parvalbumin, also much like in rodents. Ashwell and Paxinos did note a difference between the platypus and echidna in the alignment of the APT: in the platypus the nucleus is largely horizontal, with division into medial and lateral subnuclei (Fig. 6.1a); in the echidna the APT is more vertically oriented, with division into dorsal and ventral subnuclei, as in rodents..

The MPT and OPT in the platypus and echidna also have broadly similar chemoarchitecture to the rodent pretectal nuclei of the same name. On the other hand, in the nucleus of the optic tract (OT) there is a significant difference between the platypus and short-beaked echidna. Like the laboratory rat, the platypus OT has a distinct core and shell structure, based on differential staining for parvalbumin (strong in the core, but weak in the shell), whereas this is not seen in the echidna (Ashwell and Paxinos 2007).

Size of the pretectal region in the two monotremes relative to placentals

The pretectal regions of the platypus and echidna have a similar rostrocaudal extent (~2.5 mm). The transverse extent and thickness of the pretectum at its maximum width is also similar in the two mammals (between 3 and 4 mm transversely and 1.5 mm dorsoventrally, in both species), even though the anterior pretectum has a slightly different orientation in the two. Surprisingly the dimensions of the pretectum are similar in both platypus and echidna, even though the body weight of echidnas (2–3.5 kg) is much greater than the platypus (500 g to 1 kg). By contrast, the pretectum of the laboratory rat (body weight range of 270–310 g) extends for almost 2 mm in the rostrocaudal direction, has a mediolateral extent of ~2 mm, and has a dorsoventral thickness of 1.5–2 mm. The pretectum of the platypus (Fig. 6.1a) is at least similar in size to (if not larger than) the pretectum of domesticated rats of similar body weight, and the pretectum of the short-beaked echidna is comparable in size to that of domesticated placental mammals of similar body weight (e.g. cat and rabbit, Ashwell and Paxinos 2007). There is, therefore, no basis for arguing that the pretectum of either monotreme is smaller or less differentiated than the homologous region in the placental brain (Ashwell and Paxinos 2007).

Thalamus

The thalamus (also known as dorsal thalamus) is a group of nuclei that relays ascending sensory information (mainly visual, auditory and somatosensory, but also gustatory and viscerosensory), as well as feedback from the striatopallidal complexes, and other information, to the cerebral cortex (Table 6.1). Developmentally, the dorsal thalamus is derived from prosomere 2 of the embryonic brain (Puelles *et al.*

2012b). The nuclei of the thalamus project to the cerebral cortex in a highly organised fashion, such that each sensory recipient nucleus projects to a specific area of cortex. There are also many thalamic nuclei that do not receive specific sensory input, but are instead part of higher order associative corticothalamic circuitry (Puelles *et al.* 2012b). In marsupial and placental mammals, the nuclei of the (dorsal) thalamus are divided into six functional/anatomical sets: specific sensory relay, motor relay, limbic relay, association, and midline and intralaminar/parafascicular groups.

The initial impression from viewing Nissl-stained sections through the monotreme dorsal thalamus is that the structure is monotonously homogeneous. This impression is reinforced by the absence of an internal fibre system (the internal medullary lamina) that is found in all therians. The fibre tracts that form discrete shells or laminae in the therian thalamus (e.g. the mammillothalamic tract to the anterior nucleus of the thalamus) appear to be more diffusely spread through the thalamus in the monotremes (Mikula *et al.* 2008), so the defining landmarks that separate nuclei in therians are at best indistinct. Nevertheless, the impression of uniformity is dispelled by closer examination of sections immunostained for functionally significant chemical markers like neurofilament and calcium-binding proteins.

The platypus and echidna also have quite different internal thalamic topography due to expansion of the trigeminal somatosensory thalamus in the platypus. The absence of an internal medullary lamina means that identification of the thalamic nuclei on the basis of their relationship to the internal fibre system is not possible, so establishing homologies must depend on hodological and chemoarchitectural data.

Specific sensory relay thalamic nuclei

As their name implies, the specific sensory relay group of thalamic nuclei receive information relayed from particular sense organs and, in turn, relay this information to specific sensory areas of the isocortex. The nuclei in this group include the ventral posterior nucleus for somatosensory information, the (dorsal) lateral geniculate (LG) nucleus for visual information, and the medial geniculate (MG) nucleus for auditory information.

The ventral posterior nucleus differs markedly between platypus and echidna. In the platypus, this nucleus is an enormous, laterally extended, pear-shaped structure (Mikula *et al.* 2008; Fig. 6.1b; atlas plate Pl-Ad17). In all likelihood, most of this ventral posterior is probably homologous to the VPM of placental thalamus (the medial component of the ventral posterior nucleus) because most of it is probably concerned with information from the bill and head. Its extreme enlargement makes the dorsal thalamus protrude into the cerebral hemispheres far more caudally than is common in other mammals. Mikula and colleagues parcellated the ventral posterior thalamic nucleus of the platypus into three incompletely separated divisions: a medial region of small densely packed neurons, a central region of less densely packed cells and a dorsal region of larger more dispersed neurons. They suggested that the small-celled medial part may be homologous to the taste and viscerosensation nucleus of therians (VPPC) and the remaining part is homologous to the ventral posterior nucleus of therians, although they did not consider that the ventral posterior nucleus could be subdivided into 'head' recipient VPM and 'limb and trunk' recipient VPL subnuclei. There are many parvalbumin immunoreactive small neurons embedded in the parvalbumin immunoreactive neuropil of this nucleus. Running across the ventral posterior nucleus from dorsolateral to ventromedial are four bands of enhanced parvalbumin immunoreactivity (Fig. 6.1b, atlas plate Pl-Ad17) that also correspond to increased staining for myelin, cytochrome oxidase, calbindin and GABA. The functional significance of these bands is currently unknown, but the hypothesis that naturally springs to mind is that the parvalbumin immunoreactive bands in the ventral posterior thalamic nucleus are functionally linked with the bimodal (i.e. both mechano- and electroreceptive) zones of the S1 cortex (see Chapter 9).

In the short-beaked echidna, Ashwell and Paxinos (2005) identified a ventral posterior nucleus of the thalamus (Fig. 6.1c, d; atlas plates Ec-Ad16 and 18) corresponding to the nuc. thal. vent. (lat.) of Abbie (1934), the VB of Welker and Lende (1980), the V of Ulinski (1984) and the V-VP of Regidor and Divac (1987). This is a lens-shaped region that has high levels of activity for succinate dehydrogenase and cytochrome oxidase (Wong-Riley 1989), indicating dense terminations of

Table 6.1. Dorsal thalamic nuclei in mammals

Nucleus group	Nucleus	Input[1]	Output[1]	Features in monotremes
Sensory relay	VP	Medial and spinal lemniscus (body and neck), trigeminothalamic tract (head)	Primary somatosensory cortex (S1)	Strongly labelled for CO, parvalbumin and neurofilament protein. Can be divided into VPM (for head) and VPL (for body) in the echidna. Difficult to subdivide in the platypus (it may be mainly VPM).
	LG	Retinal ganglion cell axons	Primary visual cortex	Known as LGNb by Campbell and Hayhow (1971, 1972). Contains aggregations of parvalbumin, calbindin and neurofilament immunoreactive neurons. It projects to the visual cortex in the echidna.
	MG	Inferior colliculus	Primary auditory cortex	Difficult to identify in both monotremes. Corresponds to a cluster of parvalbumin immunoreactive neurons in the echidna. Its connection to the cortex has not been specifically tested in the echidna.
Motor relay	VA	Mainly from globus pallidus	Motor cortex	Not identified by Mikula et al. (2008) in the platypus. It may be the anterior pole of the VL in both monotremes. Projects to the motor cortex of echidna.
	VL	Mainly from cerebellum	Motor cortex	Large in the platypus. It has scattered neurofilament immunoreactive neurons in the echidna. Its anterior pole may include the VA. Projects to the motor cortex in the echidna.
	VM	Mainly from substantia nigra reticular part	Motor cortex	Identified mainly on the basis of position in both platypus and echidna. Its projection to the cortex has not been analysed in either monotreme.
Limbic relay	Anterior	Mammillothalamic tract from hypothalamus	Cingulate cortex	Scattered neurons among the terminating axons of the mammillothalamic tract. Unlike therians it cannot be easily subdivided. Projects to the medial frontal cortex and probably to the cingulate cortex (not tested as yet).
	LD	Pretectal nuclei?	Cingulate cortex and peristriate cortex	Contains dispersed calbindin immunoreactive neurons in both platypus and echidna. It projects to the caudal parts of the frontal cortex in the echidna. It may also project to the cingulate cortex, but this has never been tested.
Association relay	MD	Prefrontal cortex, orbitofrontal cortex, piriform cortex, basal forebrain nuclei, amygdala	Prefrontal cortex	It contains neurons immunoreactive for parvalbumin, calbindin and calretinin. Its neurons project to the extensive frontal cortex of the echidna.

(Continued)

Table 6.1. (Continued)

Nucleus group	Nucleus	Input[1]	Output[1]	Features in monotremes
	LP	Superficial layers of superior colliculus	Visual cortex	Contains large numbers of neurofilament and parvalbumin immunoreactive neurons in the echidna. Projects to visual cortex and occipital pole (putative sensory association cortex) of the cortex.
	Po	Spinal cord and trigeminal sensory nuclei	Cortical sensory areas	Contains large numbers of neurofilament and parvalbumin immunoreactive neurons in the echidna. Projects to the visual cortex and occipital pole (putative sensory association cortex) of the cortex.
Midline	Re	Hypothalamus, basal forebrain, amygdala and hindbrain cholinergic and monoaminergic neuronal groups	Medial frontal and entorhinal cortex, nucleus accumbens	Part of massa intermedia. No data available on connections in monotremes.
	IMD			Part of massa intermedia. No data available on connections in monotremes.
Intralaminar/ parafascicular	PVA/PVP			May be considered part of the epithalamus (Jones 2007). Contains populations of calbindin and calretinin immunoreactive neurons in the echidna. No data available on connections in monotremes.
	CL	Brainstem reticular formation, raphe nuclei, cholinergic nuclei of the brainstem, spinal trigeminal nuclei, locus coeruleus	Prominent projection to the striatum and medial frontal, entorhinal and olfactory cortex	Strongly reactive for acetylcholinesterase. Contains populations of calbindin immunoreactive neurons. No data available on connections in monotremes.
	CM			Strongly reactive for acetylcholinesterase. Contains populations of calbindin immunoreactive neurons. No data available on connections in monotremes.
	PF			Distinguished by its close relationship to the fasciculus retroflexus. No data available on connections in monotremes.

1 Based on findings in therians (reviewed in Puelles et al. 2012b; Mai and Forutan 2012).

mitochondria-rich sensory axons (Fig. 6.1c; Regidor and Divac 1987; Ashwell and Paxinos 2005). Neurons of the ventral posterior nucleus also degenerate following ablation of the S1 cortex (Welker and Lende 1980). Ashwell and Paxinos divided the ventral posterior nucleus into a medial subnucleus (VPM, presumptive 'head' area) where cell density is higher and a more cell sparse lateral subnucleus (VPL, presumptive 'body/limbs' area). The densest plexus of myelinated fibres is within VPM, whereas the VPL is paler with most myelinated fibres appearing to simply traverse this nucleus on the way to the VPM. Consistent with the density of myelinated axons, the VPM is particularly strongly reactive for cytochrome oxidase, in contrast to the neighbouring medial lemniscus. VPL is less reactive for cytochrome oxidase than VPM, but nevertheless more reactive than most thalamic nuclei. Neurons in the VPL and VPM engage in projections to the somatosensory cortex and have high concentrations of neurofilament protein (Fig. 6.1d).

The LG ((lateral geniculate nucleus) of both monotremes corresponds to LGNb, identified as a retinorecipient nucleus by Campbell and Hayhow (1971, 1972). Note that the retinorecipient LGNa of Campbell and Hayhow is homologous to the pregeniculate nucleus (PrG) of the ventral or prethalamus (see below). The LG can be identified in the platypus brain on the dorsal aspect of the thalamus (see atlas plates Pl-Ad17 and 18), but it is very small and only recognisable by following the retinofugal axons of the optic tract to it (Campbell and Hayhow 1972; Mikula et al. 2008). In the short-beaked echidna, the LG nucleus lies on the dorsolateral margin of the caudal dorsal thalamus (see atlas plates Ec-Ad19 and 20) and is the only one of the two visual thalamic nuclei (LGNa and LGNb of Campbell and Hayhow 1971) to undergo degeneration following ablation of the primary visual cortex (Welker and Lende 1980). It is much larger than the LG of the platypus and its constituent neurons form lobular masses that merge with the adjacent LP (lateral posterior) nucleus. The LG contains aggregations of parvalbumin, calbindin and neurofilament protein immunoreactive neurons in proportions much like those seen in the VP complex, and consistent with its role as the visual thalamocortical relay nucleus (Ashwell and Paxinos 2005).

Identification of the auditory relay nucleus of the thalamus (medial geniculate – MG) in the monotremes is controversial. Hines (1929) originally identified the MG as a small olive-shaped body in the ventrolateral region of the caudal limit of the diencephalon, but Mikula and colleagues were unable to identify MG in either platypus or echidna (Mikula et al. 2008). Nevertheless, Ashwell and Paxinos (2005) later identified a nuclear mass in the caudal dorsal thalamus of both species (see atlas plates Pl-Ad19 and 20; and Ec-Ad19 and 20) that has similar chemoarchitectural features (clustered neurofilament, parvalbumin and calbindin immunoreactive neurons) to the other sensory relay nuclei of the dorsal thalamus (VP and LG). This issue will be discussed further in Chapter 10.

Motor relay thalamic nuclei

Therian motor relay nuclei include the ventral anterior (VA) nucleus, the ventral lateral (VL) nucleus, the ventromedial (VM) nucleus, and the submedius nucleus, but the distinction between the VA and VL is not clear in some therians (e.g. rodents, Puelles et al. 2012b). The major inputs to the motor thalamic nuclei are from the pallidum (globus pallidus), the cerebellum, the red nucleus, and the substantia nigra, with each of those inputs predominating in a particular part of the motor thalamus: the VA mainly receives input from the globus pallidus; the VL is the main target for cerebellar input; and the substantia nigra reticular part projects to the VM. The motor thalamic nuclei project to the motor parts of the cerebral cortex (primary motor, premotor and supplementary motor areas) in a topographically ordered fashion.

Neither Mikula and colleagues (Mikula et al. 2008) nor Ashwell and Paxinos (2005) identified VA in the monotreme dorsal thalamus, but it is possible that the rostral pole of the large monotreme VL (see Fig. 6.2d, atlas plates Pl-Ad13 to 16 and Ec-Ad15 to 17) is homologous to the therian VA. The VL is not particularly distinguished chemoarchitecturally and has only scattered neurofilament immunoreactive neurons (Ashwell and Paxinos 2005). This is surprising given that it must engage in robust connections as part of looped circuits through the striatopallidal complexes (see below) and the large calibre axons of those circuits should have a high content of neurofilament proteins. The VM nucleus has been identified in both platypus and echidna on the basis of position and cytoarchitecture, but not the submedius nucleus. None of the chemical markers used to date in the monotremes particularly distinguish these motor nuclei.

Limbic relay thalamic nuclei

In the therian dorsal thalamus, the limbic relay nuclei include the anterior and possible lateral dorsal (LD) nuclei. The anterior group of nuclei includes the anterodorsal (AD), anteroventral (AV) and anteromedial (AM) nuclei. The group receives the

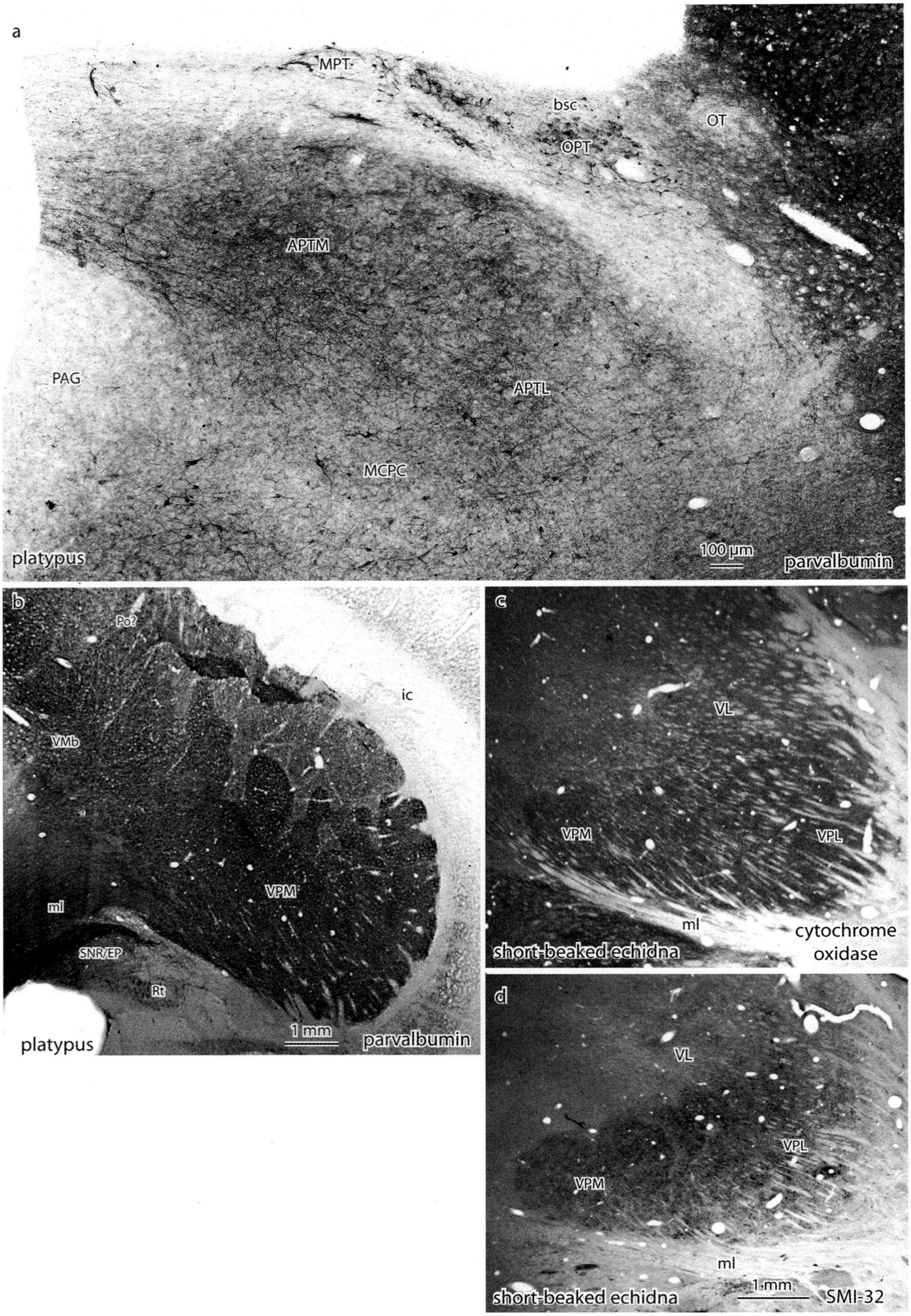
a
MPT
bsc
OPT
OT
APTM
PAG
APTL
MCPC
platypus
100 µm
parvalbumin
b
Po?
ic
VMb
VPM
ml
SNR/EP
Rt
platypus
1 mm
parvalbumin
c
VL
VPM
VPL
ml
short-beaked echidna
cytochrome oxidase
d
VL
VPL
VPM
ml
short-beaked echidna
1 mm
SMI-32

mammillothalamic tract from the hypothalamus and projects to the so-called limbic cortex close to the midline of the brain (cingulate gyrus and retrosplenial cortex). The anterior group of nuclei stain prominently for Nissl substance in most therians, in particular the anterodorsal nucleus, which in rodents contains large neurons (Puelles *et al.* 2012b). The anterior nuclei of rodents as a group stain moderately strongly for acetylcholinesterase, in particular the lateral part of the anteroventral nucleus. The LD may also be considered as a member of the limbic relay group (Jones 2007) because it is connected to many of the same cortical fields that the anterior nuclei contact (Puelles *et al.* 2012b).

There is no nucleus in the dorsal thalamus of either the platypus or echidna that looks quite like the anterior nuclei of the therian thalamus, leading to some disagreement as to exactly where the homologous nucleus lies in the monotremes and where its boundaries lie. Mammillothalamic tract fibres can certainly be identified entering and ascending through the dorsal thalamus in the monotremes, but these axons do not form the forked pattern of white matter that embraces the anterior nucleus in primates. Instead, the mammillothalamic tract breaks up into small bundles that radiate through the dorsal thalamus (Ashwell and Paxinos 2005; Mikula *et al.* 2008). Mikula and colleagues identified scattered neurons among the axons of the mammillothalamic tract of the platypus thalamus that they considered to resemble the anteroventral nucleus of the therian thalamus. They did not identify any subdivision into AD, AV and AM nuclei as in therians. In the echidna, Ashwell and Paxinos identified a putative anterodorsal nucleus, but the anterior nuclei of the echidna (see atlas plate Ec-Ad13) do not appear to share the strong AChE activity of rodent anterior nuclei. The boundary between the anterior nucleus of the echidna and the adjacent paratenial nucleus is also difficult to define (Ashwell and Paxinos 2005; Mikula *et al.* 2008).

The LD nucleus has been identified in both platypus and echidna. Mikula and colleagues identified an LD in the platypus on the basis of the presence of relatively dispersed neurons that are all calbindin immunoreactive (Mikula *et al.* 2008). This is also the case in the echidna (Ashwell and Paxinos 2005).

Association relay thalamic nuclei

This group includes the mediodorsal (MD) nucleus, lateral posterior (LP) nucleus, and pulvinar or posterior (Po) thalamic nucleus. The group is characterised in primates by reciprocal connections with the prefrontal and multimodal sensory association cortex. The main cortical target of the MD nucleus in therians is the frontal, orbitofrontal and olfactory cortical areas. The LP nucleus projects to the primary and secondary visual cortical areas, while the Po nucleus connects with the somatosensory cortex (Puelles *et al.* 2012b).

Mikula and colleagues identified an MD nucleus in the platypus (labelled M in their paper), but did not delineate an LP or Po and did not provide any details of the chemoarchitecture of the MD (Mikula *et al.* 2008). In the dorsal thalamus of the short-beaked echidna, MD, LP and Po have all been identified (Ashwell and Paxinos 2005; Fig. 6.2e–h; see atlas plates Ec-Ad14 to 20). In fact, Regidor and Divac (1987) identified a very large MD in the echidna, but their delineation of MD is too generous and appears to include regions more properly considered parts of the CM, CL, Re and VM nuclei (see below). The MD contains neurons immunoreactive for all three calcium-binding proteins (parvalbumin, calbindin and calretinin), but no neurofilament immunoreactive cell bodies.

Figure 6.1: Photomicrograph of the pretectum as seen in a parvalbumin immunoreacted frontal section through the platypus brain (a). Photomicrographs (b) to (d) show the VP thalamus in the platypus (b) and short-beaked echidna (c, d). Photomicrograph (b) is of a parvalbumin immunoreacted section and shows the variegated pattern within the large ventral posterior thalamic nucleus (VP) of the platypus. Photomicrographs (c) and (d) show cytochrome oxidase and neurofilament protein immunoreacted (SMI-32 antibody) sections, respectively. The midline is to the left in all images. The scale bar in (d) also applies to (c). APTL – anterior pretectal nucleus, lateral part; APTM – anterior pretectal nucleus, medial part; bsc – brachium of the superior colliculus; ic – internal capsule; MCPC – magnocellular nucleus of the posterior commissure; ml – medial lemniscus; MPT – medial pretectal nucleus; OPT – olivary pretectal nucleus; OT – nucleus of the optic tract; PAG – periaqueductal grey matter; Po? – probable posterior thalamic nucleus; Rt – reticular nucleus; SNR/EP – substantia nigra, reticular part/entopeduncular nucleus; VL – ventral lateral thalamic nucleus: VMb – basal ventromedial nucleus; VPL – lateral segment of VP; VPM – medial segment of VP.

116 NEUROBIOLOGY OF MONOTREMES

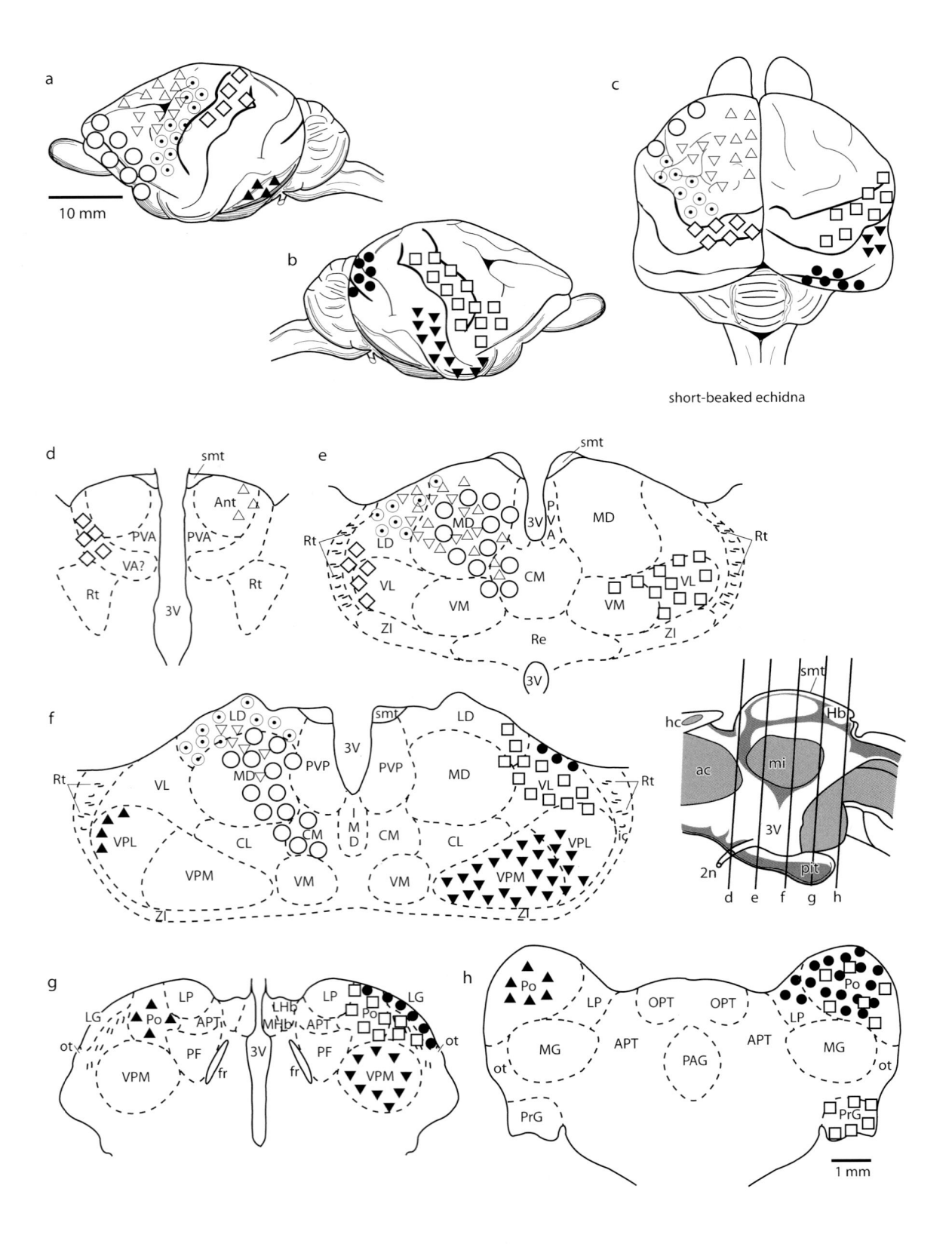

short-beaked echidna

Neuronal density is quite low in the MD relative to the VP and reticular nucleus and calcium-binding protein immunoreactive neurons make up only a small proportion of the neurons present. Neurons in the MD are large, with parvalbumin and calbindin immunoreactive neurons being larger than the average for all neurons in the nucleus. The MD projects to the large frontal cortex of the echidna (Divac et al. 1987a; Regidor and Divac 1987; Welker and Lende 1980).

The LP and Po of the echidna are characterised by large numbers of neurofilament immunoreactive neurons (Ashwell and Paxinos 2005), consistent with the presence of large axon projections from these nuclei to the cerebral cortex. This is also supported by findings from retrograde degeneration studies (Welker and Lende 1980; see below). Both the LP and Po of the echidna also have large populations of parvalbumin and calbindin immunoreactive neurons. Although these studies are limited in scope, there is good reason from the data to believe that the LP and Po of the monotremes serve similar roles to the homologous nuclei in therians. They probably engage in a reciprocal flow of information between the thalamus and multimodal sensory areas of the isocortex.

Midline nuclei

Midline thalamic nuclei are paired structures that are often connected across the midline. This group includes the xiphoid nucleus (Xi), reuniens and ventral reuniens nucleus (Re and VRe), and rhomboid nucleus (Rh). Some authors also include the paraventricular nuclei (anterior and posterior; PVA, PVP) in this group (Puelles et al. 2012b), but others consider the paraventricular nuclei to be part of the epithalamus (Jones 2007). In therians, the paraventricular nuclei are known to receive inputs from many areas of the brain (hypothalamus, basal forebrain, hindbrain cholinergic and monoaminergic nuclei) and connect to medial frontal and entorhinal cortex and the ventral striatum (Puelles et al. 2012b), but their role remains obscure.

Both the platypus and echidna have a large massa intermedia (the region of the thalamus that is fused across the midline) so there is a potentially large region occupied by these midline nuclei (Hines 1929; Abbie 1934). The main question is: how reliable are the subdivisions that have been applied? Mikula and colleagues identified a diffuse region of paraventricular nuclei in the platypus and Ashwell and colleagues identified anterior and posterior paraventricular nuclei in the short-beaked echidna. Mikula and colleagues also identified a medioventral nucleus in the platypus (Mikula et al. 2008), which they suggested is possibly homologous to the therian Re. Ashwell and Paxinos also identified Re and a midline intermediodorsal nucleus (IMD) in the short-beaked echidna, but the data on midline nuclei in monotremes are extremely limited. The PVA and PVP of the short-beaked echidna contain populations of calbindin and calretinin immunoreactive neurons, but few parvalbumin immunoreactive cells and no neurofilament immunoreactive cells.

Intralaminar/parafascicular nuclei

In therians, the intralaminar nuclei are those groups of neurons associated with the internal medullary

Figure 6.2: Projections from the dorsal thalamus to the cerebral cortex in the short-beaked echidna, based on the degeneration study by Welker and Lende (1980). The line diagrams (a), (b), and (c) show the areas removed by Welker and Lende as depicted on left hemisphere, right hemisphere and dorsal views, respectively. The series of line diagrams (d) to (h) show transverse sections through the rostrocaudal extent of the thalamus with symbols indicating the position of degenerating neurons. The inset to the right shows the position of the line diagrams (d) to (h) on a view of the midline of the diencephalon of the short-beaked echidna. 2n – optic nerve; 3V – third ventricle; ac – anterior commissure; Ant – anterior nucleus (unspecified); APT – anterior pretectal nucleus; CL – central lateral nucleus; CM – central medial nucleus; Hb – habenula; hc – hippocampal commissure; ic – internal capsule; IMD – intermediodorsal nucleus; LD – lateral dorsal nucleus; LHb – lateral habenular nucleus; LP – lateral posterior nucleus; MD – mediodorsal nucleus; MG – medial geniculate nucleus; MHb – medial habenular nucleus; mi – massa intermedia; OPT – olivary pretectal nucleus; ot – optic tract; PAG – periaqueductal grey; pit – pituitary; PF – parafascicular nucleus; Po – posterior thalamic nucleus; PVA – paraventricular nucleus (anterior); PVP – paraventricular nucleus (posterior); Re – reuniens nucleus; Rt – reticular nucleus; smt – stria medullaris thalami; VA? – possible ventral anterior nucleus; VL – ventral lateral nucleus; VM – ventral medial nucleus; VPL – ventral posterior lateral nucleus; VPM – ventral posterior medial nucleus; ZI – zona incerta.

lamina. Although there is no internal medullary lamina in the monotremes, this group of nuclei is still present, although poorly defined. This group includes the central lateral (CL), paracentral (PC), and central medial (CM) nuclei, all of which are marked by high activity for acetylcholinesterase in therians. The parafascicular nucleus is a caudal extension of this group and surrounds the fasciculus retroflexus (the pathway from the habenular nuclei to the interpeduncular nuclei). The intralaminar/parafascicular group are distinguished by strong projections to the striatum as well as to the cortex (Jones 2007; Puelles et al. 2012b).

The absence of an internal medullary lamina in the monotreme thalamus naturally makes delineation of the intralaminar/parafascicular group difficult, but possible homologues have been identified in the monotremes. Ashwell and Paxinos identified CM, CL and PF (parafascicular) nuclei in the short-beaked echidna (Ashwell and Paxinos 2005). The CM and CL of the short-beaked echidna are strongly reactive for AChE, as would be expected if they are homologous to the therian nuclei of the same name. CM and CL of the echidna also have populations of calbindin immunoreactive neurons.

Connections of the monotreme dorsal thalamus

Anatomical data on the connections of the dorsal thalamic nuclei with the cerebral cortex in monotremes are confined to the short-beaked echidna (Divac et al. 1987a; Ulinski 1984; Welker and Lende 1980), although connections in the platypus can be inferred from physiological studies (Bohringer and Rowe 1977; Krubitzer et al. 1995).

From the point of view of cortical regions covered, the degeneration study of Welker and Lende (1980) gives the most comprehensive insight into thalamocortical relationships (Fig. 6.2). Overall, the pattern of thalamocortical connections in the short-beaked echidna matches the deduced nuclear homologies based on chemoarchitectural data. In particular, the pattern of degeneration that Welker and Lende found following ablation of the primary somatosensory cortex confirms the position of the VP nuclear complex and provides an identification of the VP thalamus that is more consistent with the chemoarchitectural data from neurofilament and cytochrome oxidase activity than the retrograde tracing study of Ulinski (1984). The pattern of projection from the MD nucleus to the frontal cortex is also entirely consistent both with the homology of the large frontal cortex with placental prefrontal cortex and the homology of the MD with the therian MD.

The putative association nuclei of the echidna thalamus (LP, Po) appear to have diverse projections to the sensory cortex, consistent with a role for these nuclei in reciprocal connections with several cortical sensory areas (Fig. 6.2 g, h).

The most significant drawback of these studies to date is that they provide no insight into the thalamocortical projections of the midline and intralaminar nuclei and only limited information on the relationships of the motor thalamic nuclei. Of the midline group, only data on connections of the CM nucleus can be derived from Welker and Lende's study (Fig. 6.2e, f).

Ventral or prethalamus

The ventral or prethalamus is derived from prosomere 3 of the embryonic brain. Although developing rostral to the dorsal thalamus during embryonic life, the overgrowth of the dorsal thalamus shifts the ventral thalamus derivatives into positions ventral and lateral to the dorsal thalamus in the adult brain. Nuclei that make up the prethalamus in the adult brain include the reticular nucleus (formerly incorrectly called the reticular nucleus of the thalamus), the pregeniculate nucleus, subgeniculate nucleus, nucleus of the H field of Forel and the zona incerta.

Reticular nucleus

The reticular nucleus is the most prominent component of the therian prethalamus and surrounds the dorsal thalamus in the form of a shell, separated from the lateral surface of the dorsal thalamus by a fibre layer (the external medullary lamina) and itself surrounded by the internal capsule. The reticular nucleus receives collateral branches of thalamocortical axons and contains GABAergic neurons that have a strong inhibitory effect on the glutamatergic projection neurons of the dorsal thalamus. The inhibitory effect of the reticular nucleus appears to be particularly important in regulating thalamic activity during awake and sleep states (Jones 2007; Puelles et al. 2012b). The reticular nucleus is thought to act as a pacemaker of the low frequency oscillations (synchronous activity) that are seen in the EEG during the drowsy or sleep state of therians (see discussion in Puelles *et al.* 2012b and Chapter 13).

In both the platypus and echidna, the reticular nucleus (Rt) is relatively small and concentrated at rostral levels of the diencephalon (Ashwell and Paxinos 2005; Jones 2007; Mikula *et al.* 2008). In the platypus, the Rt ends at about halfway through the rostrocaudal extent of the dorsal thalamus, and at rostral levels (see atlas plate Pl-Ad13) has the appearance of the transient perireticular nucleus that is seen in developing therians (Mikula *et al.* 2008). In the short-beaked echidna, the reticular nucleus consists of only loose aggregations of neurons scattered in the lateral margins of the thalamus and the surrounding internal capsule (see atlas plate Ec-Ad13). There is no clear external medullary lamina and the reticular nucleus does not show any of the intense acetylcholinesterase activity that is seen in therian Rt. The monotreme Rt does have parvalbumin immunoreactive neurons like in therians (see atlas plate Pl-Ad13), but the arrangement of parvalbumin immunoreactive neurons is quite different from that in therian Rt. In the echidna, parvalbumin immunoreactive neurons of the Rt are arranged into loose aggregations aligned with the circumference of the thalamus, with dendritic processes forming a loose network, but in therians like the rat, parvalbumin immunoreactive neurons are arranged along the bands of thalamocortical and corticothalamic fibres at the lateral rim of the thalamus with a weakly immunoreactive external medullary lamina intervening between the thalamic nuclei and the Rt. The functional significance of this anatomical difference remains unknown.

Given the controversy surrounding sleep physiology in monotremes, the function of the reticular nucleus is of particular interest, but nothing is currently known of its physiology in the monotremes.

Pregeniculate nucleus

This nucleus was formerly called the ventral lateral geniculate nucleus (Hines 1929), reflecting its role as a site of termination of retinal ganglion cell axons and its proximity to the (dorsal) lateral geniculate nucleus, the main retinorecipient nucleus of the dorsal thalamus. In many therians, the pregeniculate nucleus is divided into medial parvicellular and lateral (retinorecipient) magnocellular components.

The pregeniculate nucleus (PrG) occupies very different positions in the platypus and echidna (Ashwell and Paxinos 2005; Jones 2007; Mikula *et al.* 2008). In the platypus, the pregeniculate nucleus (called LGNa by Campbell and Hayhow 1972) is relatively large and displaced dorsomedially towards the superior colliculus (see atlas plate Pl-Ad19 and 20) by the expansion of the VPM nucleus. It is partially layered, with alternating small and large cell regions (Mikula *et al.* 2008).

In the short-beaked echidna, the putative pregeniculate is suspended below the caudal extremity of the thalamus (called LGNa by Campbell and Hayhow 1971; and LGa by Ashwell and Paxinos 2005). It consists of dense clusters of small neurons with the highest numerical densities closest to the ventral rim of the nucleus (see atlas plate Ec-Ad19). Chemoarchitecturally, the pregeniculate nucleus has patches of strong acetylcholinesterase activity embedded within a less reactive matrix, a pattern that can also be seen in cytochrome oxidase reacted material. Particularly strong cytochrome oxidase activity is found around the ventral rim of the nucleus, where parvalbumin immunoreactive neurons are also concentrated. Ashwell and Paxinos have argued for the homology of the LGa with the therian pregeniculate nucleus on the basis of its retinorecipient status and chemical profile: a high proportion of parvalbumin and calretinin immunoreactive neurons, and a low proportion of calbindin immunoreactive neurons. This chemical signature is similar to the pattern seen in the magnocellular (retinorecipient) part of the rodent pregeniculate nucleus (Paxinos *et al.* 2009).

Subgeniculate nucleus

This is a nucleus that receives a visuotopic input from the retina and is interposed between the pregeniculate nucleus and the hypothalamus. It has a strong acetylcholinesterase signal in rodents and projects to the superior colliculus (Puelles *et al.* 2012b). The subgeniculate nucleus has not been identified in either the platypus or short-beaked echidna.

Nucleus of the H field of Forel and the zona incerta

In therians, the H field of Forel (also known as the prerubral field) is a poorly defined region between the medial dorsal thalamus and the hypothalamus (Puelles *et al.* 2012b). It is occupied mainly by axons of the superior cerebellar peduncle carrying information from the cerebellum back to the motor thalamic nuclei. The zona incerta is situated lateral to the H field and is traversed by fibres of the pallidothalamic and dentatothalamic (cerebellothalamic) pathways destined for the VL nucleus of the dorsal thalamus. The H field of Forel and zona incerta have been identified in both platypus (see atlas plate Pl-Ad15) and echidna (see atlas plate Ec-Ad17), but nothing is known about their connections or function in the monotremes.

Epithalamus

The epithalamus is the part of the diencephalon that lies dorsal to the dorsal thalamus. There is some disagreement concerning its prosomeric origin and adult affiliation. Puelles et al. (2012b) consider that the subdivision 'epithalamus' is no longer useful and group the habenular and paraventricular nuclei (traditionally regarded as part of the epithalamus) with dorsal thalamus derivatives of prosomere 2. Nevertheless, the term 'epithalamus' is still in common usage (Jones 2007) and the traditional description will be applied below.

In the traditional delineation, the neural elements of the epithalamus are the medial and lateral habenular nuclei, and the paraventricular nucleus (Jones 2007). The epithalamus also includes the glandular elements of the pineal body or gland and subcommissural organ. The habenular nuclei receive input from the stria medullaris thalami and give rise to the habenulo-interpeduncular tract (fasciculus retroflexus) from the ventral surface of the habenular nuclei (Jones 2007).

The habenula in therians is divided into a medial nucleus that stains densely in Nissl preparations and has intense acetylcholinesterase activity due to the concentration of cholinergic neurons. The medial nucleus forms the wall of the dorsal recess of the third ventricle. The lateral nucleus is positioned ventral and lateral to the medial nucleus and lies ventral to the stria medullaris thalami. The medial nucleus receives most of its input from the septum, whereas the lateral habenula receives input from the basal forebrain, preoptic area, lateral hypothalamus and entopeduncular nucleus (medial segment of the globus pallidus). The medial habenula projects to the interpeduncular nuclei (by cholinergic fibres in the fasciculus retroflexus or habenulointerpeduncular tract), whereas the lateral habenular outflow continues to the ventral tegmental area, substantia nigra and the raphe nuclei.

Hines (1929) described the lateral habenular nucleus of the platypus as a long slender arciform nucleus that curves around the medio-dorsal boundary of the medial nucleus and is confined to the most caudal part of the epithalamus. The habenular nuclei of the monotremes are very similar cytologically and chemically to the therian habenula. The medial habenula is characterised by strong acetylcholinesterase activity and dense aggregations of calbindin immunoreactive neurons, whereas the lateral habenula has clusters of calretinin immunoreactive neurons and occasional neurofilament immunoreactive cells (Ashwell and Paxinos 2005).

Hines thought that the posterior commissure of the platypus arose exclusively from the lateral habenular nucleus and that the projections from the cortex to the habenula were confined to the medial nucleus, but these conclusions have never been tested with modern tract-tracing techniques. Hines also identified a tectohabenular tract entering the caudal end of the lateral nucleus, and a contribution from the lateral olfactory pathway joining the stria medullaris thalamis to enter the medial nucleus. The major efferent bundle from the habenular complex, the fasciculus retroflexus, leaves the ventral surface of the habenula from the main part of the lateral nucleus (Hines 1929). The habenular commissure interconnects the lateral habenular nuclei.

The structure of the pineal gland and its recess of the third ventricle have been studied in both the platypus and echidna (Kenny and Scheelings 1979). In both the platypus and echidna, the pineal gland is well vascularised and sits between the habenula and posterior commissures. Bundles of axons (many of them myelinated) enter the gland from both commissures. The pineal gland plays a critical role in circadian rhythms and seasonal regulation of reproduction in therians. As discussed in Chapters 3 and 13, regulation of behaviour by seasonal rhythm is also critically important for the monotremes, but nothing is currently known about pineal gland physiology in the monotremes (see also Chapter 12).

The subcommissural organ (SCO) is one of the periventricular organs that surround the fluid-filled cavity of the ventricular system. It is found on the ventral surface of the posterior commissure, running the entire rostrocaudal extent of the fibre bundle. The function of the SCO is poorly understood, but there is a curious structural difference between the SCO of monotremes and marsupials (Kenny and Scheelings 1979). Large neurons are found within the SCO of all marsupials studied to date, but are absent from the SCO of both the platypus and echidna. Whether this difference has functional implications or is merely an incidental association remains to be determined.

Development of the diencephalon

Development of the diencephalon in all mammals follows a broadly similar sequence of developmental events, but the tempo of the progression through that sequence can differ significantly between the three groups of modern mammals. In placentals, the major morphogenetic events for the diencephalon all occur during the late embryonic and early foetal period, but in the marsupials and monotremes only the earliest stages of diencephalic morphogenesis (formation of the prosomere domains and some early fibre tracts) occur before birth or hatching. In both marsupials and monotremes, most diencephalic neurogenesis, all neuronal differentiation of the diencephalic nuclei, and the formation of connections with the cortex and striatum all occur while the young are nutritionally supported by maternal lactation (Ashwell 2012c). This means that the formation of connections between the relay neurons of the thalamus and the cerebral cortex is entirely postnatal, so the behavioural repertoire of the newly born or hatched is confined to whatever can be achieved at the brainstem or lateral hypothalamic level.

Pre-hatching development of the monotreme diencephalon

Development of the prosencephalon (forebrain) and diencephalon is morphologically similar in the platypus and short-beaked echidna during the initial third of incubation. The earliest stage at which a prosencephalic primordium can be recognised is at ~5.5 mm GL, a stage at which the anterior opening or neuropore of the neural tube is just about to close (Ashwell 2012c; see Appendix Fig. 2). Neuromeric segmentation that will define the developmental domains that give rise to the three major elements of the diencephalon is poorly developed at this stage, but the optic stalks that will form the retina have already grown from the side of the prosencephalon. The production of postmitotic neurons has apparently not begun at this early stage, because there is only a very narrow (10 μm wide) marginal layer outside the neuroepithelium of the prosencephalic wall with very few young neurons.

Neuromeric segmentation is more obvious by ~6.0 to 7.5 mm GL in both species (see atlas plate Ec-InA1, Ashwell 2012c), at which point the prosencephalon develops into the three prosomeres (for the diencephalon) and a more rostral secondary prosencephalon (for the hypothalamus, subpallium and pallium). Cell division (mitosis) is particularly active along the diencephalic ventricular wall at this time, but mitosis is also underway in a periventricular position ~10 to 20 μm away from the ventricular surface. Some mitoses on the ventricular wall at this stage of development have cleavage planes that are as much as 40° from a normal to the ventricular surface, suggesting that at least some asymmetric cleavage has begun to produce the proliferative population of the diencephalic subventricular zone.

Neurogenesis for the pretectal, dorsal thalamic and prethalamic nuclei begins in both platypus and echidna at ~8.0 mm GL, at which time a distinct subventricular zone of abventricular mitosis has emerged outside the ventricular germinal zone (Ashwell 2012c, Appendix Fig. 2). Neurogenesis of pretectal, dorsal thalamic and prethalamic nuclei has also proceeded to the extent that the first neurons of the presumptive posterior nuclear group and thalamic reticular nucleus have settled external to the dorsal and ventral thalamic neuroepithelium, respectively. This gives rise to a mantle layer of postmitotic neurons ~70–130 μm thick. Although specific tracts cannot be identified at this time, the marginal zone of the diencephalon has reached a thickness of 30–40 μm, suggesting that the number of axons in forebrain tracts has substantially increased. Development of the epithalamic primordium must be even slower, because no postmitotic cells are visible in the habenular region until several days later.

It is during the late embryo stage that the tempos of diencephalic development in the platypus and echidna begin to diverge. The progression of thalamic neurogenesis seems to be slower in the platypus, in that settling of thalamic nuclear groups in the 10.0 mm GL platypus is not greatly advanced from 8 mm GL embryos. In support of this impression, the progression of thalamic neurogenesis (as judged by the width of the mantle layer of postmitotic neurons) in the 10.0 mm GL platypus is no further advanced than in the 8.0 mm GL echidna (Ashwell 2012c). Around this late embryonic stage, diencephalic neurogenesis of the echidna thalamus may be ahead of that in the platypus by a day or two (equivalent to between ~1 and 2 mm of body length).

Diencephalic development in the peri-hatching period

The period around hatching is a time of intense neurogenesis in the monotreme diencephalon. Immediately before hatching, the thalamic neuroepithelium is still very thick in all prosomeric regions and postmitotic neuronal populations are confined to only a few aggregations in the mantle zone, but by only a few days after hatching, large populations of both dorsal and ventral thalamic neurons have settled in the mantle and distinct nuclear groups have begun to emerge (see summary in Appendix Fig. 2). The available data do not allow precise mapping of temporospatial gradients in monotreme thalamic nucleogenesis, but the sequential settling of diencephalic neurons suggests that there is a lateral-to-medial gradient in the way that thalamic neurons settle into their respective nuclei, as described in marsupial and placental mammals (Ashwell 2010b; Altman and Bayer 1979). This means that lateral nuclei in both the dorsal and ventral (pre)thalamus in the monotremes can be found in position at least a week before constituent neurons of the medial and midline thalamus have reached their final location.

The dorsal thalamus is essentially embryonic at the time of hatching and is unlikely to contribute to the behaviour of the hatchling monotreme. In particular, the dorsal thalamic nuclei for the ascending sensory pathways (MG, LG, ventral posterior) have only a small fraction of their adult complement at hatching. They are also yet to receive ascending afferent axons, and have not extended thalamocortical axons to the isocortex. The state of differentiation of dorsal thalamic nuclei of newly hatched monotremes is therefore reminiscent of that in newborn marsupials. Although behavioural studies of newly hatched monotremes have never been made, it is most likely that the behavioural repertoire of the newly hatched monotreme (e.g. breaking out of the egg, finding the milk source and perhaps stimulating milk ejection) is achieved predominantly with circuitry in the brainstem and spinal cord (but perhaps with some help from the lateral hypothalamus; see Chapter 12) and requires no processing of information by the higher neuraxis.

Diencephalic development during lactational phase development

All development of the definitive connections between the diencephalon and other parts of the neuraxis occurs during the lactational phase. Thalamocortical fibres and the internal capsule emerge and, particularly in the platypus, begin to expand rapidly at the end of the first week of post-hatching life. The collective thalamocortical fibres and internal capsule are much larger in the platypus than in the short-beaked echidna at this age, consistent with the greatly expanded ventral posterior thalamus complex in the platypus. During the first post-hatching week, the ventral posterior nucleus of the echidna thalamus shows the first signs of differentiation into VPM and VPL, but the VP(M) of the platypus remains a homogeneously dense cluster of granular cells. During the second week after hatching the dorsal thalamic ventricular germinal zone is almost completely exhausted and the last of the thalamic neurons (medial and midline dorsal thalamic nuclei) settle into position.

Unfortunately it is not possible to follow the development of the diencephalon of the echidna beyond the 25 mm stage, because of a lack of specimens, but the nuclei of the dorsal and ventral thalamus of the platypus have achieved an essentially adult structure by the middle of the second month after hatching. The components of the prethalamus (Rt and ZI) are clearly demarcated from the dorsal thalamus and an external medullary lamina has developed between the Rt and the ventral tier of the dorsal thalamus.

Septum

The septum is a region of the forebrain that forms the medial wall of the rostral lateral ventricle. It is actually a complex region that is divided into medial, lateral, posterior and ventral subregions (Medina and Abellán 2012). The medial subregion includes the medial septal nucleus and the diagonal band nuclei, the lateral subregion includes the lateral septal nucleus, the posterior subregion includes the triangular septal nucleus and the bed nucleus of the anterior commissure, and the ventral subregion includes the bed nucleus of the stria terminalis (see the discussion of the extended amygdala below).

Projections from the septal nuclei reach neuroendocrine and autonomic hypothalamic centres, playing important roles in the control of homeostasis and energy metabolism (see the review in Medina and Abellán 2012). The medial septal and diagonal band nuclei provide important ascending projections to the hippocampus for spatial learning and memory. They also control the hippocampal theta rhythm and associated learning and memory functions that accompany exploratory behaviour and rapid eye movement sleep (Medina and Abellán 2012).

The monotreme septum

Hines (1929) identified two septal nuclei in the platypus (medial and lateral). The medial septal nucleus lies in the ventromedial part of the septum and is separated from the nucleus accumbens by a small bundle of axons. The medial septal nucleus has smaller neurons than the lateral septal nucleus, which lies in the dorsolateral part of the septum. Hines traced many axons of the medial forebrain bundle from the lateral nucleus. Hines also noted a nucleus of the diagonal brand of Broca made up of scattered polymorphous pyramidal neurons lying lateral to the axons of the medial forebrain bundle.

In the short-beaked echidna, medial and lateral septal nuclei are also easily distinguished on the basis of both cytoarchitecture and chemoarchitecture (Fig. 6.3a, b) as are the diagonal band nuclei. The lateral nucleus can (as in rodents) be further subdivided into dorsal, intermediate and ventral subregions, but nothing is currently known about the connections or functional activity of these regions in monotremes.

Striatum and pallidum

The striatum and pallidum are deeply seated grey matter structures of the forebrain that engage in long recurrent circuits involving the cerebral cortex and thalamus. Although they have traditionally been regarded as motor nuclei, their function is now known to be much more extensive, including key roles in cognition, motivation and emotions (Gerfen 2004; Medina and Abellán 2012).

The striatum and pallidum are functionally grouped together under the term 'striatopallidal complex' (Gerfen 2004; Butler and Hodos 2005). The dorsal striatopallidal complex consists of the caudate nucleus and putamen (which are together known as the dorsal striatum) and the globus pallidus (dorsal pallidum). The dorsal pallidum is often divided into external and internal segments. In many mammals the internal or medial segment of the globus pallidus is known as the entopeduncular nucleus. The ventral striatopallidal complex is composed of the nucleus accumbens and olfactory tubercle (together known as the ventral striatum) and the ventral pallidum (scattered cells in the basal forebrain).

Both the degradative enzyme acetylcholinesterase and the synthetic enzyme choline acetyltransferase can serve as markers of dorsal striatal territory (Fig. 6.3a), because they indicate the presence of intrinsic cholinergic neurons. The dorsal striatum is also characterised by dense innervation with dopaminergic and serotonergic axons from cell groups based in the brainstem. It has intrinsic neurons that use neuropeptide Y and somatostatin, and contain NADPH diaphorase (Fig. 6.3b; Butler and Hodos 2005). The dorsal pallidum receives innervation from substance P and enkephalinergic axons.

The striatopallidal complexes are involved in looped circuits (Gerfen 2004) that connect the cortex with the striatum, pallidum and thalamus, and return to the cerebral cortex (Fig. 6.3c). Broadly speaking, there are three parallel loops through the striatopallidal complexes, although the loops are not entirely separate. One loop, concerned primarily with motor function, begins with the motor and somatosensory cortex, passes through the putamen and globus pallidus to the VA and VL nuclei of the dorsal thalamus and projects back to the motor and somatosensory cortex. Another loop concerned with reinforcement of rewarding behaviours runs from the limbic cortex, through the ventral striatum (nucleus accumbens) and ventral pallidum, and through the dorsal thalamus to return to limbic cortex. The third

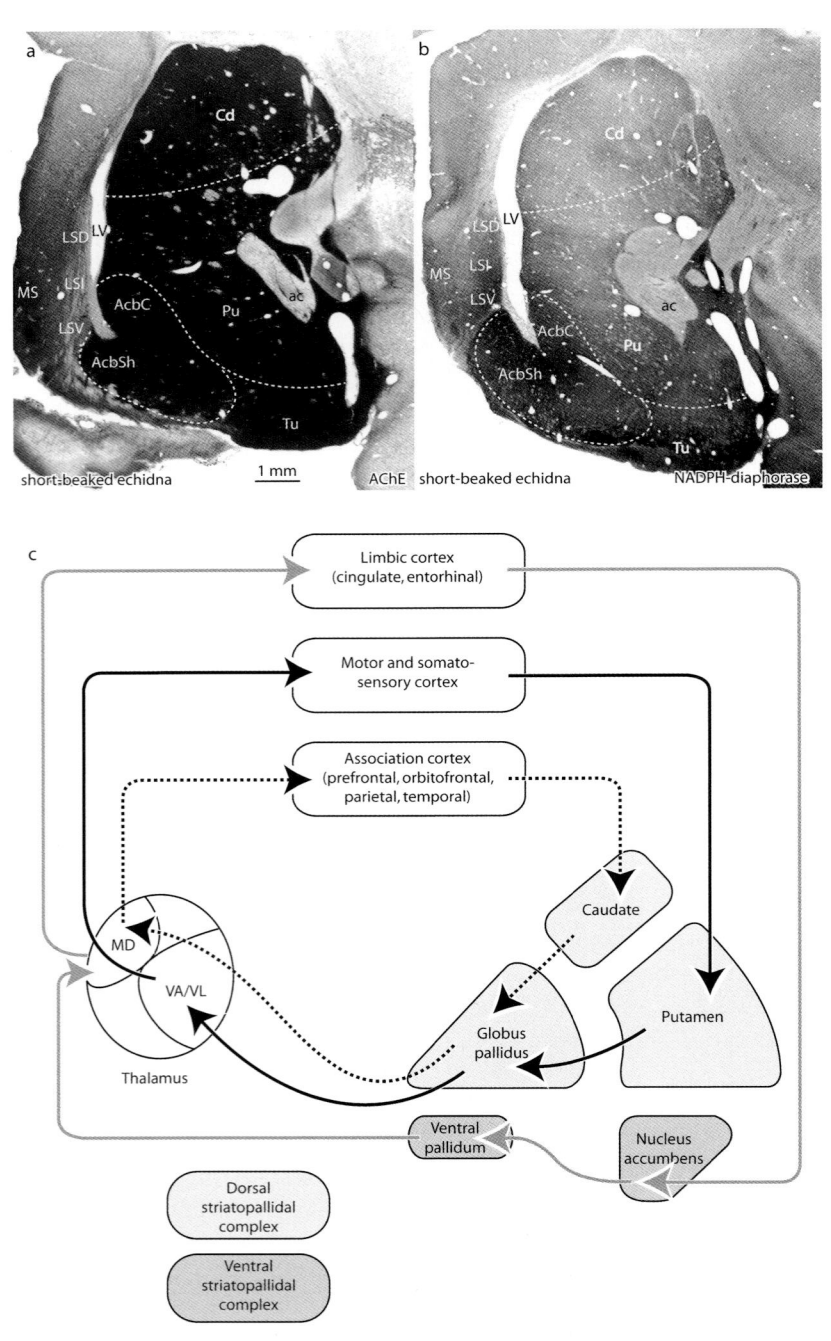

Figure 6.3: Photomicrographs showing the dorsal and ventral striatum of the short-beaked echidna in acetylcholinesterase (a) and NADPH (reduced nicotinamide adenine dinucleotide phosphate) diaphorase (b) reacted sections through the rostral forebrain. The dorsal striatum consists of caudate and putamen segments. The ventral striatum includes the nucleus accumbens (divided into core (AcbC) and shell (AcbSh) regions) and the olfactory tubercle (Tu). The sections also illustrate the distinction of medial and lateral septal nuclei. Line diagrams in (c) illustrate the idea of three parallel looped circuits from three broad regions of the cortex (limbic, sensorimotor and association) through the ventral and dorsal striatum, ventral and dorsal pallidum, thalamus, and returning to the cerebral cortex. ac – anterior commissure; Cd – caudate; l.s.d., LSI and LSV – dorsal, intermediate and ventral components of the lateral septal nuclear group; LV – lateral ventricle; MD – mediodorsal nucleus of the thalamus; MS – medial septal nucleus; Pu – putamen; VA/VL – ventral anterior and ventral lateral nuclei of the thalamus.

loop concerned with cognition (and language in humans) runs from the association cortex through the caudate nucleus, the globus pallidus, the mediodorsal nucleus of the dorsal thalamus and back to the association cortex.

Striatopallidal complexes in the monotremes

All the striatopallidal components found in placentals can be identified in the forebrains of both the platypus and echidna (Fig. 6.3a, b). Furthermore, at least in the case of the short-beaked echidna, there are broad chemoarchitectural similarities to those regions in placental brains.

Just as in therians, there is a dorsal striatum in the monotreme brain, with division into a dorsomedial caudate and a ventrolateral putamen (Hines 1929; Ashwell 2008a). The chemoarchitecture of the dorsal striatum has not been studied in the platypus, but has been analysed extensively in the short-beaked echidna (Ashwell 2008a). The dorsal striatum of the echidna can be divided into a rostrodorsomedial segment, with very little parvalbumin immunoreactivity in the neuropil and only a few parvalbumin immunoreactive neurons relative to calbindin neurons (a ratio of the number of parvalbumin to calbindin neurons of 0.09). This may serve an associative function, i.e. be homologous to the caudate of primates, and engage in looped circuitry starting in the large frontal cortex of the echidna. By contrast, the caudoventrolateral striatum has more parvalbumin immunoreactive neurons relative to calbindin immunoreactive neurons (a ratio of number of parvalbumin to calbindin neurons of 0.47), and may be the sensorimotor region that engages in a looped circuit starting in the sensorimotor cortex (Ashwell 2008a). This hypothesis is supported by the pattern of corticostriatal projections identified by Divac and colleagues in the short-beaked echidna (Divac et al. 1987b). Injection of labelled amino acids into the large frontal cortex of the echidna resulted in labelling of axons in the rostrodorsomedial dorsal striatum (putative caudate homologue), consistent with that part of the striatum being involved in association functions (see Chapter 7). Clearly more detailed study of cortico-striatal-pallido-thalamic relations in the echidna is needed to properly test the hypothesis.

Another important feature of the dorsal striatum of many therians (rodents, carnivores and primates) is the division into striosome (patch) matrix architecture (Gerfen 2004; Butler and Hodos 2005). Neurons within the patches or striosomes receive input from cortical areas and have reciprocal connections with the ventral part of the A9 (substantia nigra, compact part) dopaminergic cells in the brainstem. Neurons within the matrix also receive projections from the cortex, as well as cells in the intermediate (rather than ventral) part of the A9 group, and they project to the GABAergic neurons in the pars reticulata of the substantia nigra rather than the A9 region. In rodents, patch and matrix components of the caudatoputamen can be distinguished on the basis of differential staining for AChE and calcium-binding proteins like parvalbumin, calbindin and calretinin (Waldvogel and Faull 1993; Hiroi 1995; Gerfen 2004; Künzle 2005). Ashwell (2008a) was able to distinguish patch and matrix compartments in the dorsal striatum of the echidna using calbindin and AChE reacted sections, but the distinction is more difficult to make than in therians. Discrete regions of poor calbindin immunoreactivity (putative patch or striosome compartments) occupied a relatively small proportion of the cross-sectional area of the dorsal striatum compared to rodents (Ashwell 2008a). They were concentrated in the caudal and ventral parts of the dorsal striatum and were usually less than 200 μm in diameter. Differential staining of patch and matrix compartments for tyrosine hydroxylase was not visible in the echidna, although differential immunoreactivity was visible in ventral striatal versus pallidal areas (Ashwell 2008a).

Ancillary regions of the dorsal striatum such as the lateral striatal stripe (LSS) and the interstitial nucleus of the posterior limb of the anterior commissure (IPAC) can also be identified in the brain of the short-beaked echidna on the basis of differences in the strength of tyrosine hydroxylase immunoreactivity and the arrangement of myelin fascicles (Ashwell 2008a). However, the core and shell components of the accumbens (AcbC, AcbSh) are much more difficult to distinguish in the echidna than in rodents, where complementary neuropil immunoreactivity for calcium-binding proteins clearly demarcate these regions from each other (Gerfen 2004; Ashwell 2008a; Paxinos et al. 2009).

The globus pallidus of the short-beaked echidna also has similar topography and chemoarchitectural features to homologous regions of the rodent brain (Ashwell 2008a; Paxinos *et al.* 2009). Parvalbumin immunroeactivity is strong in the globus pallidus of the echidna, particularly in the neuropil of the dorsolateral segment. The low spatial density of calbindin immunoreactive neurons in the echidna globus pallidus and entopeduncular nucleus is very similar to the pattern observed in therians (Ashwell 2008a).

The broad similarities in the topography and chemoarchitecture of the echidna striatopallidal complexes to those in therian brains suggest that the organisation and chemoarchitectural features of the ventral and dorsal stiratopallidum emerged before the divergence of therian and prototherian lineages, perhaps as much as 240 mya (Ashwell 2008a).

Amygdala

The amygdala is involved in the modulation of neuroendocrine functions, visceral effector mechanisms, and complex patterns of integrated behaviours, such as defence, ingestion, aggression, reproduction, memory and learning (de Olmos *et al.* 2004). In the behavioural sphere, the amygdala is concerned with assigning emotional significance to experiences, and plays an important role in the expression of anger and fear. This role is exerted through a vast network of amygdalofugal connections (by the stria terminalis, fornix, and ventral amygdalofugal pathways among others) with many brain regions.

The amygdala is traditionally divided into an aggregated or clumped part in the temporal lobe (temporal amygdala) and an extended stream of neurons along the stria terminalis (extended amygdala) (de Olmos *et al.* 2004). Developmentally, the temporal amygdala includes both pallial or superficial, and subpallial or deeper structures (Puelles *et al.* 2000). The lateral pallial components include the basolateral nuclei (ventral, anterior and posterior) and the cortex/amygdala transition zone. Ventral pallial components include the basomedial nuclei (both anterior and posterior), nuclei of the olfactory tracts (nucleus of the lateral olfactory tract and bed nucleus of the accessory olfactory tract) and the cortical amygdala (anterior and posterior). The subpallial components include the central and medial nuclei and intra-amygdalar parts of the bed nuclei of the stria terminalis. The amygdala receives a strong input from both the main and accessory olfactory pathways. In rodents, the olfactory cortical amygdala includes the anterior cortical amygdala, cortex/amygdala transition zone, posterolateral cortical amygdala and amygdalopiriform region. The vomeronasal input (accessory olfactory pathway) is distributed to the posteromedial cortical amygdala (Martínez-García *et al.* 2012).

Apart from the amygdaloid nuclei in the temporal lobe (temporal amygdala), there is an arc of nuclei that makes up the extended amygdala, running between the temporal amygdala and septal nuclei. This arc includes two bands. One is continuous with the central amygdala and includes the nucleus accumbens shell, the interstitial nucleus of the posterior limb of the anterior commissure (IPAC), the lateral bed nuclei of the stria terminalis, and the sublenticular extended amygdala. The other is continuous with the medial amygdala and includes the medial parts of the bed nuclei of the stria terminalis and the medial sublenticular extended amygdala (de Olmos *et al.* 2004). The extended amygdala nuclei associated with the central nucleus of the temporal amygdala has been considered to be related to fear responses, anxiety and ingestion, whereas the medial extended amygdala is related to defensive or aggressive behaviour and reproduction (Martínez-García *et al.* 2012; Medina and Abellán 2012).

The monotreme amygdala: temporal and extended

The temporal and extended amygdala of the platypus has been described by Hines using cell and fibre stains (Hines 1929). The temporal amygdala is separated from the piriform cortex at rostral levels by the lateral olfactory tract, and more caudally by a distinct amygdalar fissure. The temporal amygdala of the platypus has clear divisions into medial, central, cortical, lateral and basal components (see atlas plate Pl-Ad14) as in therians. The medial nucleus is made up of polymorphous and triangular neurons and has input from the lateral olfactory tract. Hines did not report any subdivision of the central nucleus, but divided the cortical amygdala into anterior and posterior divisions. The anterior cortical amygdala has small irregular cell-laminations, mainly made up of pyramidal neurons, whereas the posterior cortical amygdala has large pyramidal neurons and some polymorphous neurons arranged in regular rows.

The lateral amygdala of the platypus has large neurons, whereas the basal nucleus has even larger 'enormous' trapezoidal neurons (Hines 1929). Hines also noted the relationship of the bed nucleus of the stria terminalis with the medial group of the amygdaloid complex, but could not find any continuity of the bed nucleus of the stria terminalis with the septal region.

Ashwell and colleagues have studied the internal structure and chemoarchitecture of the temporal and extended amygdala of the short-beaked echidna in detail (Ashwell *et al.* 2005). Although the broad subdivisions of the placental temporal amygdala are present in the echidna (see atlas plate Ec-Ad12), there are some noticeable structural and chemoarchitectural differences from therian amygdala. In both absolute volume and as a percentage of the whole brain volume, the amygdalae of the platypus and platypus are substantially smaller than that for prosimians or anthropoidea of similar brain weight (Ashwell *et al.* 2005). The paired amygdalae of the echidna and platypus occupy only 0.7% and 0.53% of the total brain volume (respectively) compared to 1.4% for prosimians and 1.0% for anthropoidea (Stephan and Andy 1977). The small size of the amygdala in monotremes raises functional implications: (1) are there particular parts of the monotreme amygdala that are relatively small, or are both corticomedial and basolateral components equally small; and (2) how does this small size affect the ability of the monotremes to assign emotional tags to experiences? The expansion of the amygdala in primates (mainly basal and lateral components) probably occurred in response to the increasing complexity of social interaction in primate societies through the late Tertiary. Certainly, behavioural studies suggest that monotremes live mainly solitary lives with minimal social interaction except during the mating season, so the relatively small size of the amygdala may reflect the solitary life that they lead.

The medial nucleus of the amygdala is also remarkably small in the short-beaked echidna (Ashwell *et al.* 2005), a surprising finding given the pronounced development of the olfactory system that feeds sensory information into the corticomedial amygdala. In placentals, the medial nucleus has extensive connections with the hypothalamus and is concerned with the control and modulation of visceral functions (Price 2003), but it is not possible to deduce the functional implications of its small size in the echidna. On the other hand, it is possible that the expansion of cortical, basal and lateral components of the amygdala *relative* to both the central and medial components may parallel the expansion of frontal cortical tissue in echidnas (Ashwell *et al.* 2005). Expansion of the lateral and basal nuclei is thought to be associated with enlargement of those sensory association cortical areas (the inferior temporal cortex for visual recognition of objects, anterior superior temporal cortex for auditory association function and insular cortex for somatosensory association function) that provide the major input to the basal and lateral nuclei of the amygdala (Price 2003). So the large volume of isocortex rostral to the motor cortex in the echidna may provide some explanation for the relatively large proportion of the echidna amygdala occupied by the (presumably cortically connecting) basal and lateral nuclei. This hypothesis is consistent with the extensive projections from the medial and orbital prefrontal cortex to the basolateral amygdalar nuclei in rodents (Martínez-García *et al.* 2012).

Ashwell and colleagues were also unable to find the nucleus of the lateral olfactory tract (LOT) in the short-beaked echidna (Ashwell *et al.* 2005). In therians like rodents, where the LOT is prominent, the nucleus is characterised by a strikingly layered appearance in Nissl stains, an intense AChE reactivity of layer 2 and strong parvalbumin immunoreactivity of its deepest layer, but none of these features is seen in the echidna in any region of the rostral medial amygdala. The absence of the LOT is surprising given the large size of the olfactory system in the echidnas, but the LOT can also be small or absent in many placentals (e.g. anthropoid primates; Stephan and Andy 1977).

There are some unusual features of the chemoarchitecture of the amygdala in the short-beaked echidna. Immunoreactivity for parvalbumin and calretinin in both neuronal somata and neuropil has been extensively reported in the temporal amygdala of therians, but Ashwell and colleagues were unable to find any immunoreactivity for parvalbumin in either neurons or neuropil in any of the nuclei of the temporal amygdala (Ashwell *et al.* 2005).

Most of the components of the extended amygdala can be identified in the monotreme brain (Ashwell *et al.* 2005). These include the IPAC (as discussed above under the striatopallidal complexes) and the

128 NEUROBIOLOGY OF MONOTREMES

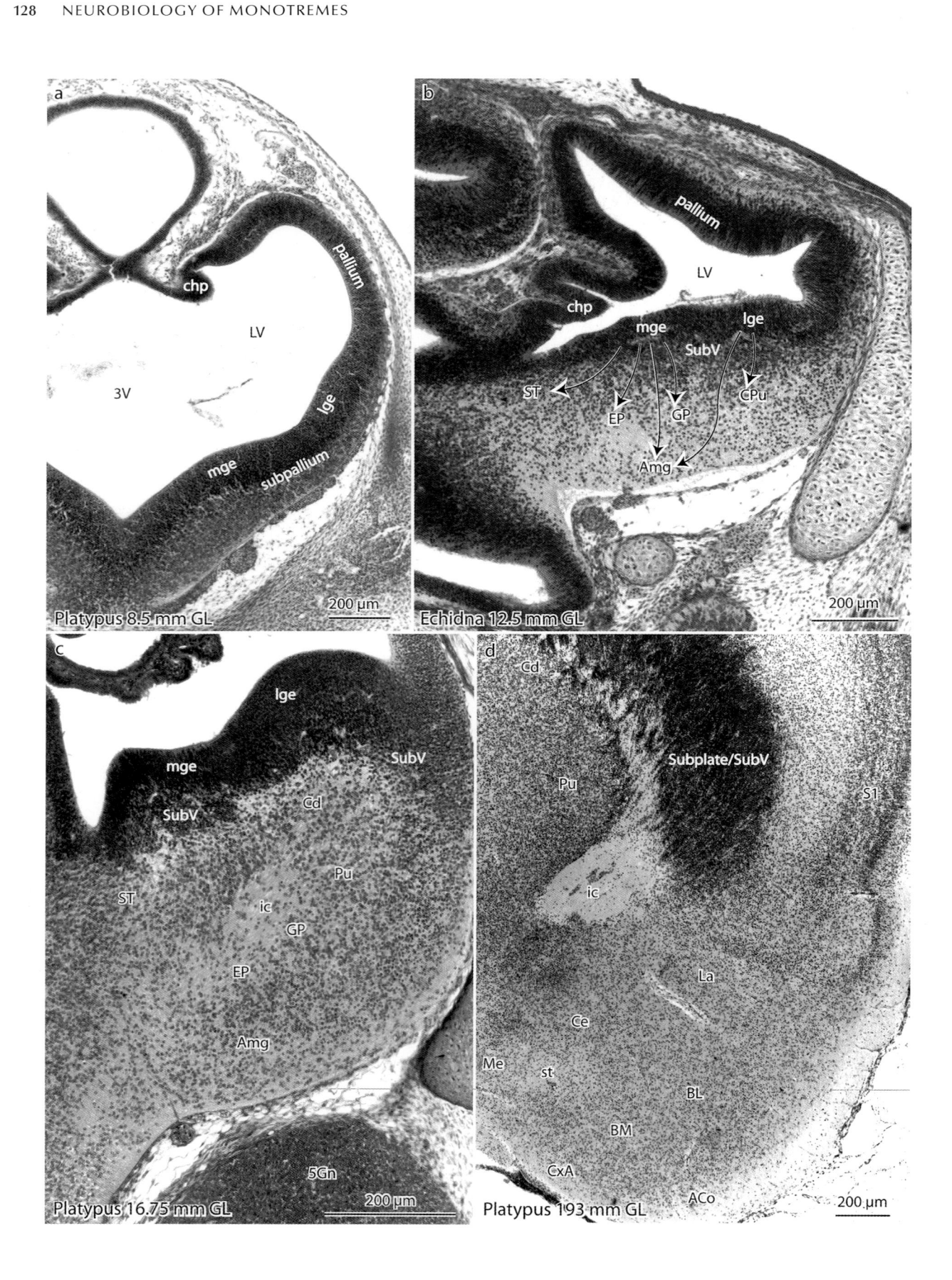

bed nucleus of the stria terminalis (ST), with the last divisible into medial and lateral components. The large anterior commissure of the echidna divides both medial and lateral components of the ST into dorsal and ventral limbs (see atlas plates Ec-Ad9 and 10).

Development of striatopallidal complexes and amygdala

The embryonic telencephalon in all mammals is divided into two major divisions: a pallium that will give rise to the cerebral cortex and other layered structures (e.g. olfactory bulb, claustrum, pallial parts of the amygdala and some of the septum); and a subpallium that will give rise to the striatopallidal complexes, extended and subpallial amygdala, most of the septum and the preoptic region. Proliferation and cell differentiation of the neurons of the subpallium lead to the accumulation of cells in the medial and lateral ganglionic eminences of the embryonic telencephalon (Medina and Abellán 2012). The lateral ganglionic eminence gives rise to the principal and projection neurons of the dorsal and ventral striatum, most of the neurons of the central and intercalated nuclei of the amygdala, and the striatal part of the olfactory tubercle. The medial ganglionic eminence gives rise to the globus pallidus, the ventral pallidum, the medial part of the extended amygdala, the ventral and medial parts of the temporal amygdala, and the sublenticular part of the extended amygdala.

Division of the embryonic monotreme telencephalon into pallial and subpallial segments is in place by the middle of incubation (8.5 mm GL platypus) and the subpallial ventricular zone can be divided into medial and lateral ganglionic eminences at about the same time (Fig. 6.4a; see the summary in Appendix Table 1). Generation of the striatal and pallidal neurons from the medial and lateral ganglionic eminences (Fig. 6.4b) appears to be a slow process in the monotremes, because the generative zones of the ganglionic eminences remain quite thick well into the first week of post-hatching life (Fig. 6.4c). The components of the striatal and pallidal region are also poorly differentiated for much of the first 2 weeks after hatching. It is not until the end of the first post-hatching month that internal detail of the amygdala begins to emerge (Fig. 6.4d).

Cholinergic cell groups in the basal forebrain

The cortically projecting cell groups of the basal forebrain include cholinergic, GABAergic and glutamatergic neurons distributed in a band from the preoptic region (sometimes included in the hypothalamus) forward to the pallidum and septum. The cholinergic cell groups are particularly important because of their projections to the cerebral cortex, and these neurons may play a role in different aspects of cognition, sensory attention, learning and memory (Medina and Abellán 2012). Cholinergic neurons use acetylcholine as their neurotransmitter and are identified by the presence of choline acetyltransferase, the synthetic enzyme for acetylcholine. The distribution and functional activity of cholinergic neurons in the monotremes is of particular significance because of the controversy over sleep physiology (see Chapter 13). Lesions of the cholinergic habenulointerpeduncular tract (fasciculus retroflexus) in therians decrease

Figure 6.4: Photomicrographs showing the main events in the development of the striatum, pallidum and amygdala of the monotreme forebrain. The division of the telencephalic wall into pallium and subpallium is evident by the middle of incubation (a) and the generation of neurons in the ganglionic eminences and the underlying subventricular zone (SubV) begins during the last third of incubation (b). Migration of postmitotic neurons (arrows in b) occurs coincidentally with the generation of neurons. Production of neurons and differentiation of the striatopallidal nuclei is a slow process, proceeding through the first week of post-hatching life (c) and is still underway at 5 weeks after hatching (d). The midline is to the left in all images. 3V – third ventricle; 5Gn – trigeminal ganglion; ACo – anterior cortical amygdala; Amg – early undifferentiated amygdala; BL – basolateral nuclei of amygdala; BM – basomedial nuclei of amygdala; Cd – caudate; Ce – central nuclei of amygdala; chp – choroid plexus; CPu – combined caudate and putamen before division; CxA – cortex/amygdala transition zone; EP – entopeduncular nucleus (medial division of globus pallidus); GP – globus pallidus; ic – internal capsule; La – lateral nuclei of the amygdala; lge – lateral ganglionic eminence; LV – lateral ventricle; Me – medial nucleus of amygdala; mge – medial ganglionic eminence; Pu – putamen; S1 – primary somatosensory cortex; st – stria terminalis; ST – bed nuclei of the stria terminalis; SubV – subventricular zone of pallium and ganglionic eminences.

the time spent in hippocampal theta and rapid eye movement sleep (Butcher and Woolf 2004).

Manger and colleagues have grouped cholinergic neurons in the monotreme forebrain into three groups: striatal, basal forebrain and diencephalic (Manger *et al.* 2002a; refer also to Fig. 5.6). Striatal cholinergic neurons of both the platypus and echidna are found in the islands of Calleja (granule cell clusters in the medial forebrain) and surrounding olfactory tubercle, the nucleus accumbens and the caudate and putamen (see below), all in similar positions to those in therians.

Basal forebrain cholinergic neurons include cell groups in the medial septal nucleus, nuclei of the vertical and horizontal limb of the diagonal band of Broca, magnocellular preoptic nuclei, substantia innominata/ventral pallidum, basal nucleus of Meynert, and nucleus of the ansa lenticularis. Manger and colleagues could only identify basal nucleus and medial septal nuclei groups with certainty in the monotremes. The basal nucleus of Meynert consists of small closely packed cholinergic neurons along the ventral border of the globus pallidus and putamen in both species. The medial septal nucleus contains cholinergic neurons throughout its rostrocaudal extent and is strongly reactive for the degradative enzyme acetylcholinesterase (see Fig. 6.3a).

Cholinergic neurons in the therian diencephalon are also located in the medial habenula and hypothalamus. The medial habenula contributes many cholinergic axons to the habenulointerpeduncular tract (fasciculus retroflexus), many of which terminate in the interpeduncular nucleus. The habenular cholinergic neurons are much the same morphologically in the monotremes as they are in therians, but Manger and colleagues noted the apparent absence of cholinergic neurons from the hypothalamus. Nevertheless, the operative significance of this absence is not clear and the difference may be functionally trivial. It should be noted that, in therians, hypothalamic cholinergic neurons are (1) not a major component of the cholinergic system; and (2) are found adjacent to the habenulointerpeduncular tract and contribute axons to that pathway (Butcher and Woolf 2004). Given that the cholinergic neurons of the medial habenula are robustly present in the monotremes and provide a strong input to the habenulointerpeduncular tract, hypothalamic cholinergic neurons may be superfluous for the monotremes.

Questions for the future

The structure and function of the dorsal thalamus of the monotremes remains a relatively unexplored area of comparative neuroscience and a potential testground for hypotheses of the molecular control of mammalian diencephalic development. In the foregoing discussion the patterns of diencephalic development in rodents and molecularly defined developmental regions have been used as a conceptual framework for a consideration of monotreme diencephalic development. Structural features of the monotreme diencephalon during late incubation and early lactation phases appear sufficiently similar to those in embryonic and foetal rodents to warrant hypotheses of similar molecular development in all mammals, but this deserves closer scrutiny. In particular, the molecular regulation of the morphogenesis of the large VP(M) of the platypus dorsal thalamus and the formation of its large thalamocortical projection may have unique features.

The anatomy of the striatopallidal complexes in the short-beaked echidna is similar to that of therians, raising the possibility that all mammals have similar sets of looped circuitry serving executive, motor and behavioural reinforcement functions. Nevertheless, several questions remain. Does the echidna have distinct cortico-striato-pallido-thalamic loops similar to those seen in therians? If so, what is the function of the association loop that would begin in the large frontal cortex? Is this loop involved in executive functions to do with exploiting invertebrate resources in the echidna's territory? How does this compare with loops arising in therian prefrontal cortex?

A further avenue for future study involves the monotreme amygdala. Preliminary studies suggest that certain components of the temporal amygdala (i.e. medial and central) are relatively small in echidnas, whereas the others are relatively large (i.e. lateral and basal). This difference highlights potential interactions between the cortex and amygdala in regulation of social behaviour. The monotremes provide an outgroup for comparisons of both molecular ontogeny and phylogeny of the mammalian amygdala.

7

Cerebral cortex and claustrum/endopiriform complex

Ken W. S. Ashwell

Summary

Both the platypus and echidnas have a six-layered isocortex like all therian mammals, but the two groups of living monotremes have very different cortical morphology: lissencephalic (smooth) and thick in the platypus, and gyrencephalic (folded) and thinner in the echidnas. The two major groups also have a very different distribution of sensory fields across the cortex: an extraordinary enlargement of the bill representation in the platypus primary somatosensory cortex, and a shift of the sensory areas for vision, touch and hearing towards the caudal half of the cortex in the echidnas. Anatomical indicators of cerebral metabolism suggest that the echidna cortex has a similar pattern of vascularisation and mitochondrial spatial density to that in marsupial and placental mammals. Synaptic morphology and distribution in the echidna somatosensory cortex are also much like the same region in therians, but the pyramidal neurons of the echidna cortex appear to include more atypical forms than in therians. Development of the isocortex in the monotremes depends on similar proliferative populations in the embryonic pallium to those found in therians, but with an unusual subcortical zone that may combine the properties of the subventricular and subplate regions of the developing cortex of therians. Much of the cortical development in monotremes occurs after hatching, in a protracted tempo similar to that seen in marsupials.

Cognitive abilities of the monotremes

Apart from its role in processing sensory information and controlling movement, the cerebral cortex is a major centre in cognitive function and decision-making, so it is useful to consider the behavioural evidence for monotreme cognition. This is confined to the short-beaked echidna, where studies of T-maze learning and operant techniques have been applied (Saunders et al. 1971a, b; Othmar et al. 1978). The latter used reversal learning tasks, where the subject must adapt to a changing association between a stimulus and a rewarding reinforcement.

Saunders and colleagues demonstrated that echidnas are capable of forming a position habit in T-mazes and show consistently improving performances on successive testing, much like therian mammals. Othmar and colleagues showed that echidnas are capable of performing successfully in instrumental reversal learning tasks, particularly when those tasks are based on spatial discrimination, but performed less well on visual discrimination tasks. Given the poor vision of echidnas, the latter is not surprising and is probably an unfair test to apply to them. Nevertheless, individual performances of echidnas on instrumental reversal learning tasks are comparable to those of nocturnal marsupials and even compare well with placental carnivores (Othmar et al. 1978).

Overview of the mammalian cortex

The cerebral cortex (pallium) is the layer of grey matter that covers the surface of most of the telencephalic part of the forebrain. In mammals, the cerebral cortex is divided into (1) iso- (or neo-) cortex; (2) allocortex; and (3) transitional regions between the isocortex and allocortex. The distinctive feature of the mammalian isocortex is that all parts of it possess six tangential layers at some stage during development, whereas the sauropsid (reptile and bird) dorsal cortex has only three layers (Aboitiz and Montiel 2007). The term 'isocortex' is preferable to 'neocortex', because the name 'isocortex' makes no assumptions concerning the phylogeny of the parts of the cerebral cortex, whereas 'neocortex' implies (incorrectly) that this type of cortex is a recent evolutionary development. 'Allocortex' is a term used for those parts of the cerebral cortex that have either three layers (archicortex or hippocampal formation) or five layers (paleocortex or primary olfactory cortex). A series of transitional regions between isocortex and allocortex is also recognised on the basis of progressive transformation between six-layered and three-to-five-layered structure. In rodents, these include the orbitofrontal cortex, agranular insular cortex, perirhinal cortex, cingulate cortex and retrosplenial cortex (Palomero-Gallagher and Zilles 2004). Another derivative of the pallium is the claustrum/endopiriform complex that lies deep to the insular and piriform cortex, but the precise derivation of this complex is subject to much debate (see below).

The standard mammalian isocortex has five neuronal cell body layers (2Cx to 6Cx) capped by the (relatively) neuron-free and axon-rich layer 1Cx (a molecular layer). The commonest type of neuron in the mammalian cortex is the pyramidal neuron, which receives its name because of its pyramid-shaped cell body or soma. Pyramidal neurons have a prominent apical dendrite running towards the pial surface of the cortex and a basal skirt of dendrites directed towards the underlying grey or white matter. Pyramidal neurons extend an axon out of the cortex to contact other neurons, but often this axon has a locally ramifying collateral branch, contacting local circuit neurons and other pyramidal neurons. Other neuronal types in the isocortex include the bipolar and bitufted neurons that have a vertically oriented dendritic tree giving them two distinct poles, and the multipolar stellate (star-shaped neurons) that are the major recipients of incoming information through afferent or corticopetal axons from the thalamus or brainstem. Some of the layers of the cortex are more obvious than others depending on the function of the particular cortical region being examined. Output layers predominate in motor regions, concerned with control of body movement, and input layers predominate in sensory regions to process incoming information about the external world or interior of the body.

The deepest layer of the isocortex is the polymorphic or multiform layer (6Cx) that contains fusiform or spindle-shaped modified pyramidal neurons. Above this is the internal or deep pyramidal layer (5Cx) that contains large pyramidal neurons engaged in long distance corticofugal connections (e.g. to the thalamus, striatum, brainstem or spinal cord). Layer 5Cx will often be thicker in motor cortex to accommodate the large number of large-bodied pyramidal neurons that provide the corticofugal motor pathways. Layers 5Cx and 6Cx are sometimes collectively called the infragranular layers.

Outside (or superficial to) the internal pyramidal layer is the (internal) granular layer (4Cx) that contains a high proportion of stellate (star-shaped) neurons and receives most of the incoming corticopetal (afferent) axons. Many afferent axons are carrying sensory information, so layer 4Cx will be thicker in sensory cortex. Highly specialised sensory cortex is often called granular cortex (or koniocortex from the Greek *konia* for dust) to signify the high spatial density of small-bodied (dust-like) stellate neurons and small pyramidal neurons. Conversely layer 4Cx may be very small or absent from motor and other cortices, leading to the name 'agranular cortex' for those regions.

External to 4Cx lies the external pyramidal layer (3Cx), populated by small pyramidal neurons that engage in short-range corticocortical (association) connections, as well as intrinsic (local circuit) neurons like chandelier cells, bipolar cells and double bouquet cells. Between 3Cx and the molecular layer is the external granular layer (2Cx), a layer of small pyramidal cells with extensive local intracortical ('within cortex') connections and non-pyramidal local-circuit neurons like basket cells that form basket-like axonal arbours around the cell bodies (somata) of small pyramidal neurons.

Cortical folding (gyrification) in mammals

The cerebral cortex is organised into functional columns only a few hundred µm wide, with neurons in a given column having similar physiological properties (see below and Rowe 1990). Since processing power is linked with the number of functional columns, one way to acquire more information-processing ability is to increase the surface area of the cerebral cortex. Gyrification, or the folding of the cerebral cortex, is therefore seen as a feature of advanced mammalian brains, but we should remember that it might not be the only approach to the demands of neural processing. For example, the platypus has quite a large encephalisation index (see Chapter 4) and yet its cortex is lissencephalic (i.e. smooth); its large brain is the result of thickening of the cortex, reaching as much as 4 mm thickness in some parts of the cortex of larger platypuses.

By contrast, all the echidnas have highly folded cerebral cortex of more moderate thickness (2 to 3 mm; Hassiotis *et al.* 2003). Short-beaked echidnas have a gyrification index (the ratio of the actual surface area of cortex, including that in the depths of sulci, to the externally visible cortical surface) comparable to that in many large-brained placentals. Measured values of gyrification index for three short-beaked echidnas range between 1.316 and 1.426 (Hassiotis *et al.* 2003). This is comparable to that in many highly encephalised placentals; for example, the gyrification index in the domestic cat is ~1.530. In fact, the gyrification index of the short-beaked echidna is comparable to that of many prosimians of similar brain weights (Hassiotis *et al.* 2003; Zilles *et al.* 1989). No quantitative estimates have been made of gyrification in long-beaked echidnas, but the external appearance of their brains suggests that their gyrification indices reach similar values.

Historical perspectives on cortical parcellation in monotremes

The most significant of the early systems for parcellating the cerebral cortex of the monotremes was that described by Abbie (1940). He described a series of obliquely running cortical regions in the monotreme cortex (see Fig. 7.1a–d and 7.2a–d), comprising two groups – one related to the hippocampus and known as parahippocampal cortex (PH), the other related to the piriform cortex and designated parapiriform cortex (PPy). Abbie described a distinct transition from one subdivision (parapiriform or parahippocampal) to the next, and he believed that this sequence indicated the successive degrees of modification associated with ensuing stages of differentiation during cortical evolution. Under Abbie's schema, the monotreme isocortex was made up of four parahippocampal regions (PH1 to PH4) and three parapiriform regions (PPy1 to PPy3). The boundary between PH4 (parahippocampal 4) and PPy3 (parapiriform 3) was along an oblique line that runs from the occipital pole of the cortex down to the rostral limit of the piriform cortex (Fig. 7.1a). In the short-beaked echidna this boundary lies in the depths of the α sulcus (Fig. 7.2a).

The schema used by Abbie is a reflection of the thinking about cortical evolution that was current in the early 20th century. The prevailing view at that time was that the neocortex (or neopallium) was formed partly from the parahippocampal pallium (Elliott Smith 1903, 1910, 1919) and partly from another cortical primordium related to the piriform cortex.

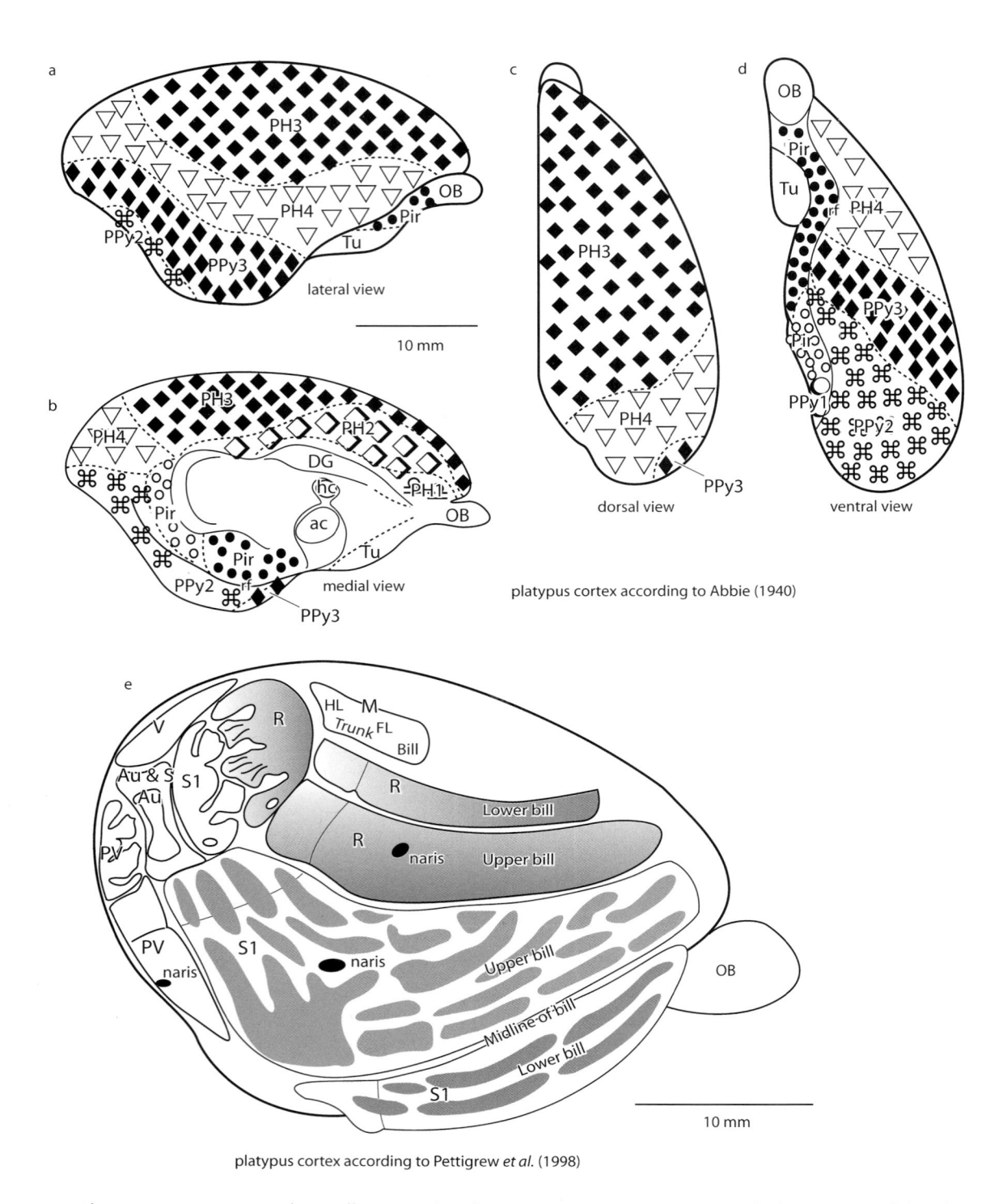

Figure 7.1: Changing conceptions of parcellation in the platypus cortex. Diagrams (a) to (d) show views of the platypus cortex with different symbols indicating the regions (PH1 to 4, PPy1 to 3) identified by Abbie (1940). Note that Abbie distinguished rostral and caudal olfactory areas (Pir) with different symbols. Illustration (e) shows the distribution of sensory areas (M – manipulation somatosensory; R – rostral somatosensory; S1 – primary somatosensory; PV – parietoventral somatosensory; Au – auditory; V – visual) on the isocortex of the platypus. These maps are based on findings reported in Krubitzer et al. (1995) and summarised in Pettigrew et al. (1998). Note the representation of discrete body areas (i.e. upper and lower bill, body and limbs) in M, R, S1 and PV. DG – dentate gyrus; naris – nostril; OB – olfactory bulb; Pir – piriform cortex; Tu – olfactory tubercle.

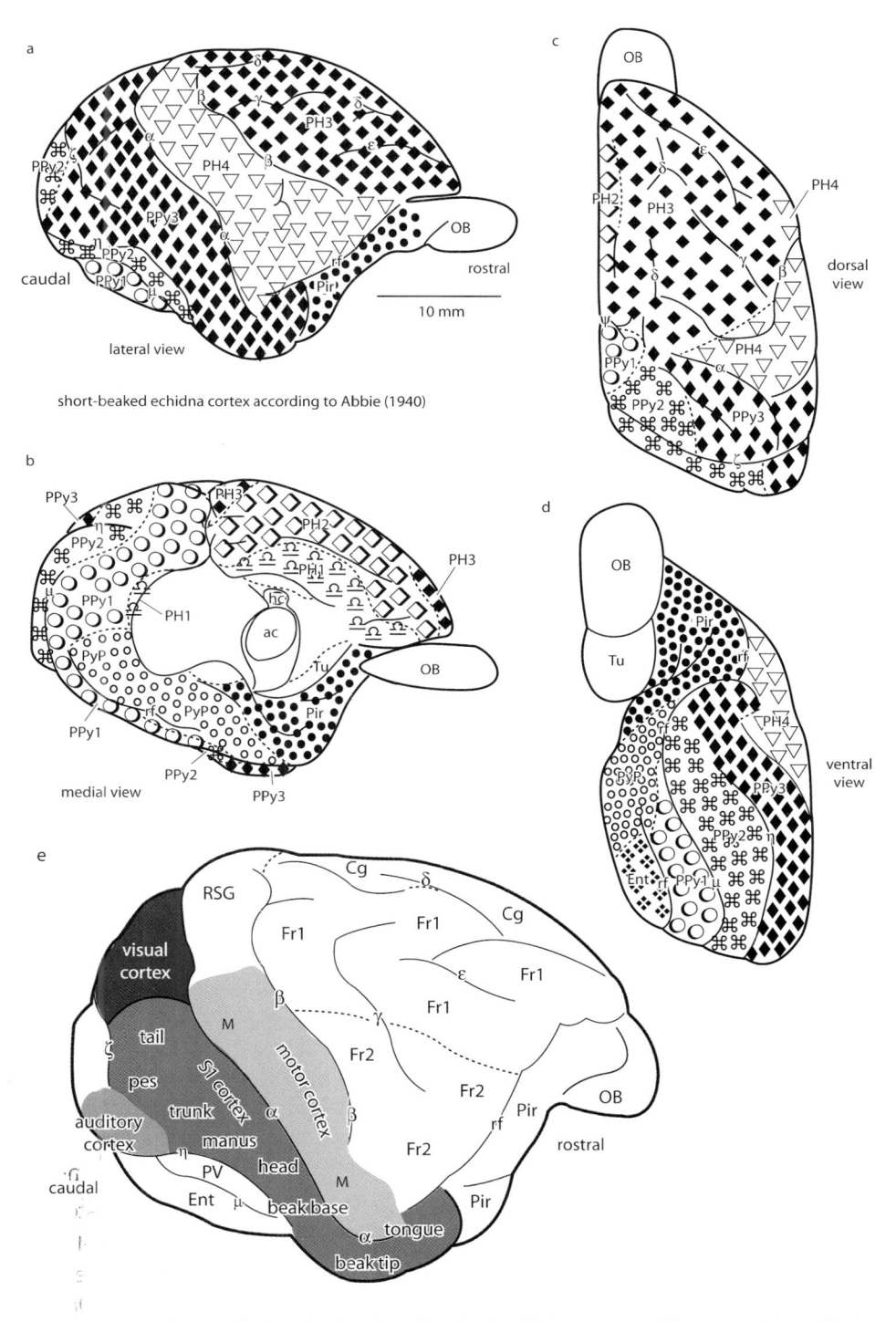

Figure 7.2: Changing concepts of parcellation in the short-beaked echidna cortex. Diagrams (a) to (d) show views of the platypus cortex with different symbols indicating the regions (PH1 to 4, PPy1 to 3) identified by Abbie (1940), while (e) shows a summary diagram of modern data concerning sensory and motor areas in the cortex (based on data from Lende 1964; Krubitzer et al. 1995). Note that the sensory regions R and M/R transition are on the wall of sulcus α, and are not visible on the external surface. Greek letters denote major sulci on the cortical surface. ac – anterior commissure; Cg – cingulate cortex; Ent – entorhinal cortex; Fr1, Fr2 – frontal areas 1 and 2; hc – hippocampal commissure; OB – olfactory bulb; Pir – piriform cortex; PV – parietoventral somatosensory area; rf – rhinal fissure; RSG – retrosplenial gyrus; Tu – olfactory tubercle.

Although the cytoarchitectural boundaries identified by Abbie are largely accurate, modern thinking on cortical evolution no longer holds that the isocortex is derived by successive modification of allocortex: isocortex may well have been present in the earliest ancestral mammals. There are also several limitations associated with the use of Abbie's isocortical subdivisions: (1) the nomenclature implies an unsupported hypothesis concerning cortical evolution; (2) it gives no indication of functional subdivisions within the isocortex; and, most importantly, (3) the defining cytoarchitectural features of particular regions are sometimes inconsistent between the two monotreme orders. To illustrate the last point, PH3 in the short-beaked echidna is essentially agranular, with a very prominent layer 5Cx filled with large pyramidal neurons. In the platypus, PH3 has a very well-developed granular layer that takes up 40% of the thickness of the cortex, but layer 5Cx has only moderately large pyramidal neurons and is relatively thin. Even without physiological data for this rostral part of the cortex in the echidna, such a cytoarchitectural disparity would suggest significant functional differences between the two species.

Physiological and chemoarchitectural data that have become available since the mid-20th century (summarised in Figs. 7.1e and 7.2e) indicate that Abbie's schema gives a reasonably accurate parcellation of the short-beaked echidna cortex, but has some significant limitations for the platypus. In the echidna, PH4 of Abbie's schema roughly corresponds to motor or manipulation cortex (Lende 1964; Krubitzer et al. 1995) and PPy3 includes somatosensory, visual and auditory cortex; but in the platypus both PPy3 and PH4 (and even some of PH3) would lie within the entire S1 (Bohringer and Rowe 1977; Krubitzer et al. 1995). In the following consideration of monotreme isocortex (and in atlases accompanying this book), the cortex will be parcellated primarily according to functional criteria where possible, with application of cyto- and chemoarchitectural criteria where functional data are not available (e.g. cingulate and far frontal regions).

Somatosensory cortex

The somatosensory cortex of the monotremes has received intense attention, mainly due to its potential involvement in electroreception and the importance of the monotremes for theories of cortical evolution. The functional details of somatotopy (body representation) and physiology of the somatosensory cortex will be discussed in Chapter 9. In this section the focus will be on cellular structure and chemical architecture and how this compares with therian somatosensory cortex.

Bohringer and Rowe (1977) found only a single somatosensory field in the isocortex of the platypus, but Krubitzer and colleagues (Krubitzer et al. 1995) identified three separate somatosensory fields (S1, PV and R); as well as a manipulation field (M), where neurons respond to deep stimulation, coextensive with motor cortex (see Fig. 7.1e). Cutaneous somatosensation covers parts of the platypus cortex that Abbie delineated as PPy2, PPy3 and PH4, whereas deep somatosensation extends onto areas that fall within PH3 of Abbie's parcellation. These are quite cytoarchitecturally diverse areas and the precise correlation of cytoarchitecture with physiology is often not fully addressed (Krubitzer 1998). The primary somatosensory cortex in the platypus has a dense input of myelinated axons and strong cytochrome oxidase activity, whereas R and PV are only lightly myelinated and moderately CO (cytochrome oxidase) reactive. In the accompanying atlas, the S1 cortex is identified on the basis of the thick layer 4Cx much like granular somatosensory cortex in therians. Rostral somatosensory cortex (R) lies more medially and rostrally, whereas PV cortex lies more caudally and laterally.

Lende (1964), Ulinski (1984) and Krubitzer and colleagues (Krubitzer et al. 1995) have successively studied somatosensory cortex in the short-beaked echidna. Lende's studies demonstrated the unusual position of the somatosensory cortex in the echidna, lying much further caudally than one would expect in a therian. The somatosensory cortex is positioned between the α and ζ sulci in the caudal half of the cortical mantle. The lateral border of the somatosensory cortex lies at the rhinal fissure (the boundary with the olfactory cortex) and the medial border abuts against the visual cortex. Somatotopy or body representation is arranged in a lateral-to-medial sequence with the snout and tongue laterally and the tail medially (Fig. 7.2e). Ulinski (1984) identified two cytoarchitectonic fields within somatosensory cortex: a caudal field (c), present throughout the post-α gyrus and extending onto the floor of the α sulcus, which has a well-developed granular layer (4Cx); and a layer 5Cx with only a few small pyramidal neurons. In Ulinski's description, the rostral field (r) runs from the depths of the α sulcus onto the caudal bank of the pre-α gyrus. It is distinguished from the more caudal somatosensory field because the rostral field has a thinner 4Cx and a higher proportion of medium-sized pyramidal neurons in 5Cx.

Krubitzer and colleagues (Krubitzer et al. 1995; Krubitzer 1998) identified S1, a rostral somatosensory field (R) and an additional PV area of somatosensory cortex in echidnas using microelectrode-recording techniques (see atlas plates Ec-Ad12 to 18). The R cortex contains neurons that respond to stimulation of deep rather than cutaneous receptors and receptive fields are larger than in S1. The PV cortex (Fig. 7.2e) contains a complete representation of cutaneous receptors, but the neurons have larger receptive fields than in either S1 or R.

While agreeing with the broad cytoarchitectural regions found by Ulinski and Krubitzer, Hassiotis and colleagues (Hassiotis et al. 2004b) noted the presence of a transition zone (M/R transition) between the M and R regions (see atlas plate Ec-Ad18). On the basis of chemoarchitectural features they suggested that the true boundary between M and the more caudal somatosensory fields (M/R transition and R) is located somewhere between the landmarks set by Ulinski and Krubitzer. Parvalbumin and neurofilament immunohistochemistry both suggest that the boundary between R and M lies on the rostral bank of the α sulcus. This boundary is marked by the transition from an absence of neurofilament immunoreactive neurons and poor parvalbumin immunoreactivity in M to the appearance of a distinct band of neurofilament immunoreactive 5Cx pyramidal neurons and intense parvalbumin immunreactive axons in R. The transition between these two areas is not clear-cut, hence the M/R transition zone that Hassiotis and colleagues describe. Histochemistry for acetylcholinesterase (AChE) lends further support to this proposition in that there is a boundary from an intense AChE activity in granular and infragranular layers in R and the M/R transition cortex to relatively poor activity in the M cortex about halfway along the mediolateral extent of the rostral bank of the α sulcus. The M/R demarcation shifts as one progresses from rostral to caudal cortex, being more lateral in rostral/inferior regions and more medial in caudal/superior parts of the α sulcus.

Visual cortex

In both the platypus and short-beaked echidna, the visual cortex is located at the far caudal pole of the cortical hemisphere (Fig. 7.1e and 7.2e; Bohringer and Rowe 1977; Rowe 1990; Krubitzer et al. 1995; Krubitzer 1998). Krubitzer identified two visuotopic (visual world) representations in the platypus cortex (Krubitzer et al. 1995). It is not clear whether multiple visual fields are present in the echidna cortex: Krubitzer mentioned two visual fields, but did not provide data to justify both (Krubitzer 1998). The details of the physiology of the visual cortex will be dealt with in Chapter 8, but the discussion here will focus on the cellular, fibre and chemical architecture of the region.

The visual regions in both the platypus and echidna are marked by strong myelination and high cytochrome oxidase (CO) activity of incoming axons from the LG (Krubitzer et al. 1995). In the platypus, the visual cortex is marked by a myelin-stained patch ~10 mm long and 4 mm wide as seen in tangential sections. The visual area of echidna cortex is also thickly myelinated and has a CO dense rectangle (~10 mm by 7 mm in diameter), indicating the high impulse transmission speed and high metabolic rate of the axons carrying the visual information.

Hassiotis and colleagues (Hassiotis et al. 2004b) identified possible V1 and V2 (primary and secondary visual) regions in the cortex of the echidna on the basis of myelination and neurofilament protein chemoarchitecture. Area V1 has the stronger input of myelinated axons of the two visual areas and is the likely homologue of therian V1. The possible V2 region lies immediately dorsal to V1 at the most caudal tip of the cortex. Area V1 also has a population of neurofilament immunoreactive neurons that may represent the cell bodies of long-distance corticofugal neurons for the transfer of visual information to other sensory areas.

Auditory cortex

The auditory cortex has been mapped in both the platypus (Bohringer and Rowe 1977; Krubitzer et al. 1995) and the short-beaked echidna (Lende 1964) using the standard free-field click stimulus. In the platypus, the auditory area is in the posterior rim of the cerebral cortex (Fig. 7.1e). Compared to therians and the short-beaked echidna, the auditory cortex of the platypus appears to be shifted posteromedially, perhaps due to the expansion of the bill representation of S1 (Rowe 1990). In the echidna, the auditory area is on the posterior pole of the cortex and is separated from both the visual and somatosensory cortex by the ζ sulcus.

Krubitzer has described a tonotopic (pitch-based) organisation of the auditory cortex in the platypus and echidna (Krubitzer et al. 1995; Krubitzer 1998; see Chapter 10). Krubitzer and colleagues also reported the existence of an area that is responsive to both auditory and somatosensory stimuli in addition to and surrounding the primary auditory cortex (Krubitzer et al. 1995). The details of tonotopy and possible multiple representations are discussed in Chapter 10.

The auditory cortex can be distinguished on cellular, fibre and chemical architectural grounds in both platypus and echidna. In sections stained for myelin and CO, the axons entering the auditory cortex are thickly myelinated and stain darkly for CO throughout all cortical layers (Krubitzer et al. 1995; Hassiotis et al. 2004b), indicating rapid impulse transmission and high metabolic activity of the axons carrying auditory information to the cortex. The auditory cortex of the echidna has a population of neurofilament immunoreactive pyramidal neurons, much like those seen in S1 and R, but not with the precise layer segregation seen in the somatosensory regions. The auditory cortex also has strong immunoreactivity for parvalbumin in both neuronal cell bodies and axons, suggesting high activity (Hassiotis et al. 2004b).

Motor cortex

Martin (1898) and Abbie (1938) made the earliest studies of motor cortex in the platypus. Martin (1898) identified a rather large excitable area in the front two-thirds of the platypus cortex, while Abbie (1938) later mapped the motor cortex with bipolar stimulation under ether anaesthesia, with results that largely agreed with Martin. Interestingly, many of the movements that both Martin and Abbie were able to induce by cortical stimulation involved the panniculus carnosus rather than the limbs, perhaps reflecting the extensive segmental supply of that muscle. A later study by Bohringer and Rowe (1977) found an excitable cortex that overlapped partially with the somatosensory cortex, but extended further anteriorly than the putative S1 (Fig. 7.3a). The forelimb, shoulder and bill have a large representation on the motor cortex (Fig. 7.3b), with the area for bill movement situated most laterally. They observed that the large representations of the bill and forelimb are consistent with the active exploratory bill movements and front-wheel-drive locomotion of the platypus. Bohringer and Rowe could not find an area for hindlimb or tail movement, an observation in harmony with the limited rudder-like role of the hindlimbs and tail when the platypus swims. The motor cortex of the platypus corresponds partly (but by no means congruently) to the manipulation field of Krubitzer and colleagues (Krubitzer et al. 1995).

As for the platypus, Goldby (1939) and Abbie (1938) made early maps of the excitable cortex of the short-beaked echidna on the basis of stimulation with bipolar and unipolar electrodes under ether anaesthesia. They identified a motor area in the cortex of the short-beaked echidna lying between the α and β sulci, and running lateromedially from the rhinal fissure to a point two-thirds of the way to the dorsomedial border of the hemisphere. The motor representation was almost exclusively contralateral and with somatotopy: the cranial end of the body was represented laterally and the tail medially. The region of cortex from which Abbie could elicit movements was six-layered and had a granular layer, unlike the agranular and large pyramidal neuron-dominated motor cortex of some therians. The more detailed study by Lende (1964) was largely in agreement with Abbie on the body representation in motor cortex (Fig. 7.2e), but also found that motor responses could be elicited from the somatosensory cortex (i.e. posterior to the α sulcus). Lende used this finding as evidence for a sensorimotor amalgam in the short-beaked echidna. However, Rowe (1990) has cautioned against the interpretation of these results as indicating distinct motor areas both in front and behind the α sulcus, because Lende did not demonstrate a discontinuity in the body representation in the depths of the α sulcus. An important additional point for comparison with therians is that no supplementary motor area has been identified in the echidna cortex by either Lende

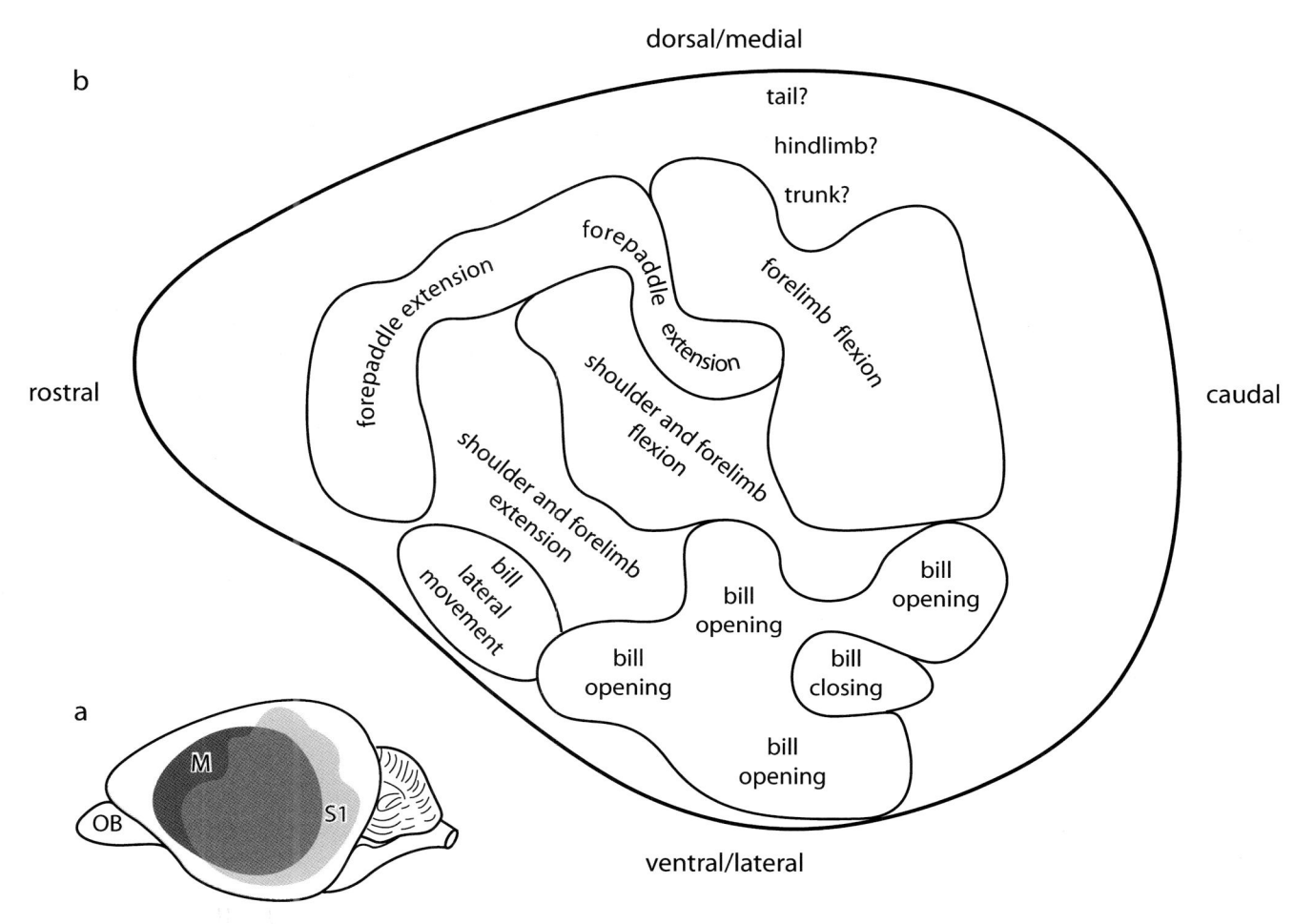

Figure 7.3: Distribution of motor cortex across the isocortex of the platypus, based on findings in Bohringer and Rowe (1977). Note the extensive overlap between motor (M) and somatosensory cortex (S1) depicted in (a). Although extensive areas of platypus cortex are devoted to control of the head, neck and forelimb (b), Bohringer and Rowe were unable to identify a motor area for the trunk, hindlimb or tail, but these are probably located in a small area of cortex close to the midline. OB – olfactory bulb.

(1964) or Krubitzer *et al.* (1995). In highly encephalised placentals (e.g. primates) the supplementary motor cortex is actually a complex of smaller areas concerned with movements on both sides of the body.

Cellular and chemical architecture of the monotreme motor cortex have been examined in several studies (Schuster 1910; Abbie 1940; Ulinski 1984; Krubitzer *et al.* 1995; Hassiotis *et al.* 2004b, 2005). The motor cortices of both the platypus and echidna have a prominent layer 5Cx with large pyramid-like neurons (with cell body diameter over 20 μm), a reduced layer 4Cx and relatively poor CO staining and myelination. Strangely, the motor cortex of the echidna does not have any neurofilament immunoreactive neurons (Hassiotis *et al.* 2004b, 2005). In fact there is a distinct boundary between the M cortex and R cortex, which can be defined by the appearance of neurofilament protein immunoreactivity in the sensory region R (see above), but not in M. The absence of neurofilament immunoreactive neurons in echidna motor cortex is a surprising finding given that abundant neurofilament protein is associated with large neurons with wide diameter axons, such as would be expected in a motor cortex that gives rise to large axon calibre pathways to control body movement. For example, giant Betz cells in the motor cortex of rodents and new and old world primates are intensely immunoreactive for

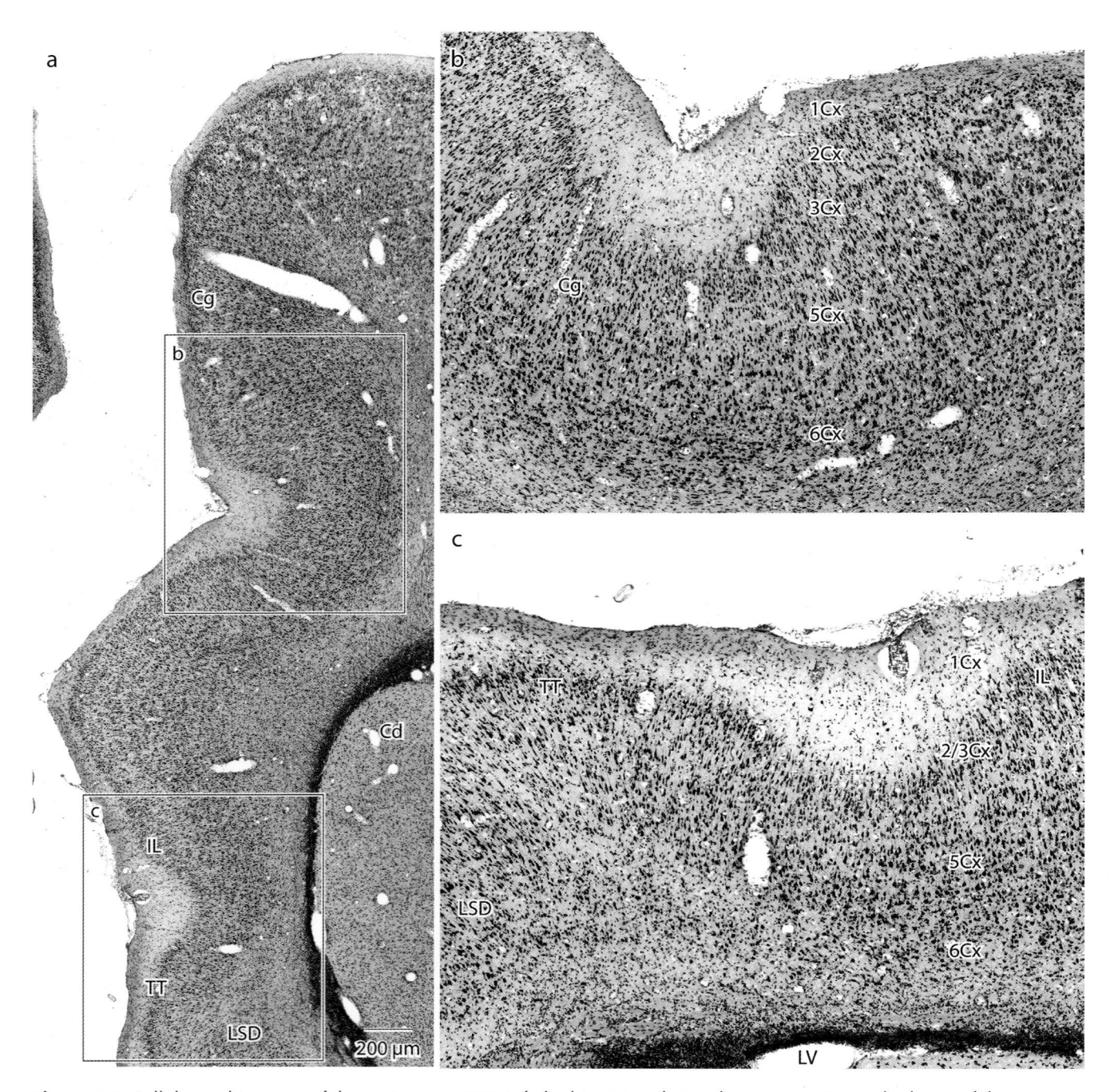

Figure 7.4: Cellular architecture of the tenia tecta (TT), infralimbic (IL) and cingulate cortex (Cg) in the brain of the short-beaked echidna. Photomicrograph (a) shows a low-power view of a Nissl-stained frontal section, with the positions of micrographs (b) and (c) indicated. Note the absence of a granular layer 4Cx in cingulate cortex. 1Cx to 6Cx – layers of cortex; Cd – caudate nucleus; l.s.d. –lateral septal nucleus, dorsal part; LV – lateral ventricle.

neurofilament protein (Kaneko *et al.* 1994; Tsang *et al.* 2000). Perhaps some of the motor control from the cortex in the short-beaked echidna depends on a series of shorter pathways through intermediate sites.

Is there association cortex in monotremes?

Association cortex in therians is that part of the cerebral cortex that does not directly receive sensory input from the relay nuclei of the dorsal thalamus and does not have a distinct motor role. Association cortex in placentals is divided into: (1) unimodal association cortex (i.e. cortex that engages in higher processing of a particular sensory modality); (2) multimodal association cortex (i.e. cortex that integrates sensory information from several modalities to form a model of the body and external world); (3) prefrontal cortex (which mediates the executive functions of planning, insight and foresight); and (4) limbic association cortex, which is concerned with emotions and memory. Unimodal and multimodal association cortex would be expected to be located around and between primary sensory areas, respectively; prefrontal cortex should be located in front of the motor cortex; and limbic association cortex should be located adjacent to allocortex.

Motor and sensory mapping studies in the platypus (Bohringer and Rowe 1977; Rowe 1990; Krubitzer et al. 1995) suggest that there is only a relatively small region of the cortex anterior of the motor cortex that might be homologous to therian frontal association cortex. On the other hand, the short-beaked (and probably the long-beaked) echidna has an extraordinarily large region of frontal cortex anterior to motor cortex. The connections, cellular architecture and possible function of this remarkable region will be considered separately below. The rostral frontal cortical field may be homologous to some or all of the orbital, medial and dorsolateral prefrontal cortex of placentals and this question will be considered in detail.

The monotreme cortex certainly contains limbic type association cortex (e.g. cingulate, retrosplenial and entorhinal cortex; Fig. 7.4, atlas plates Pl-Ad10, Ec-Ad15 and Ec-Ad21), but the precise homology of these regions to placental limbic cortex is unresolved. The cellular and chemical architectural features of these will be considered below.

The question of whether the monotreme has a unimodal or multimodal association cortex is best addressed on physiological grounds. Krubitzer and colleagues (Krubitzer et al. 1995) identified a region responsive to both auditory and somatosensory stimuli that surrounded the primary auditory cortex in both platypus (Fig. 7.1e) and echidna, as well as other regions of cortex that responded to visual plus auditory and visual plus somatosensory stimuli. These regions of bimodal responsiveness were architectonically distinct from the nearby primary sensory fields, leading Krubitzer and colleagues to conclude that these mutimodal regions were at least analogous (but not necessarily homologous) to placental multimodal association cortex.

The remarkable but inexplicable frontal cortex of the echidnas

The large and curious frontal cortex of the echidna, with no apparent motor or sensory role, has become almost as iconic a feature of the monotremes as platypus electroreception. Electrophysiological analysis has found that more than 50% of the rostral cortex has no attributable primary motor or sensory function and this region has often been considered as an expanded prefrontal cortex (Welker and Lende 1980; Divac et al. 1987a, b). If this interpretation is correct, then the proportion of isocortex occupied by the prefrontal area in the short-beaked echidna (around 50%) is larger than that in humans (a mere 29%) (Divac et al. 1987a, b). In this section, the evidence for prefrontal cortex homology is critically examined.

The first point to make is that the frontal cortex of the echidna is not structurally or chemically homogeneous. Hassiotis and colleagues (Hassiotis et al. 2004b) identified three subdivisions in the frontal cortex of the short-beaked echidna (Fr1, Fr2, Fr3) on the basis of cyto- and chemoarchitecture (Fig. 7.2e). The most dorsal region (Fr1) roughly corresponds to PH3 of Abbie (1940) and has a prominent patch of neurofilament immunoreactive neurons within it. The medial boundary between Fr1 and the neighbouring cingulate cortex (Cg1) is distinguished by the contrast in parvalbumin immunoreactivity (strong in Fr1, but weak or discontinuous in Cg1). The lateral border between Fr1 and Fr2 is marked chemically by the abundance of neurofilament immunoreactive pyramidal neurons in Fr1 and the relative scarcity of those neurons in Fr2. Fr1 also shows stronger myelin staining in layers 4Cx to 6Cx than is seen in Fr2. Hassiotis and colleagues identified another frontal region (Fr3; see atlas plate Ec-Ad05) on the underside of the frontal cortical fold. Abbie's study (1940) had focused on superficial cortex, so Fr3 does not have a corresponding region in his schema, but Fr3 is a large region of more than a hundred square millimetres area in each hemisphere. The Fr3 region is much thinner than Fr1 or Fr2, and has only a few neurofilament immunoreactive neurons.

If the large frontal cortex of the echidnas were homologous to the prefrontal cortex of therians, then one would expect that it would receive a projection from the mediodorsal thalamus or its homologue, because that is the thalamocortical relation for the prefrontal cortex in placentals. A cortical ablation study by Welker and Lende (1980) suggested that the anteromediodorsal thalamus is the major source of input to the frontal cortex (see Chapter 6 for a detailed consideration of thalamocortical relations). Divac and colleagues (Divac et al. 1987a) later used retrograde transport of injected horseradish peroxidase or fluorescent tracers to assess the non-thalamic sources of input to the frontal cortex in the short-beaked echidna. Several of the injection sites used by Divac and colleagues lay at the border of Fr1 and Cg1 (e.g. their animal Ta4) or penetrated to the subcortical white matter (and even as deep as the head of the caudate and olfactory bulb), so the results may not be clear-cut because of extension into several fields or involvement of axons-of-passage. Nevertheless, Divac and colleagues identified retrograde labelling of both stellate and fusiform neurons in the anteromediodorsal part of the dorsal thalamus. There seems to be a topographic correspondence to the thalamocortical connections such that the medial cortical injections (i.e. along the Fr1/Cg1 border or in Cg1) resulted in labelling of the medial part of the thalamus, whereas cortical injections in the lateral part of Fr1 resulted in labelling of the more lateral part of the anterior thalamus. Interhemispheric connections of the large frontal region are broadly symmetrical, such that retrogradely labelled neurons are found in the opposite hemisphere in the matching cortical region. Allocortical projections of the frontal field include the medial and lateral entorhinal cortex. Subcortical sources of input to the frontal cortex were also identified from the basal forebrain (the putative basal nucleus of Meynert), cortical and basal amygdala, ventral tegmental area, dorsal and laterodorsal tegmental nuclei of the brainstem, pedunculopontine tegmental nucleus, locus coeruleus, subcoeruleus, and the ventral and dorsal parabrachial nuclei of the pons.

In an accompanying publication, Divac and colleagues examined the efferent (outgoing) projections of the frontal cortex with injections of radioactively labelled amino acids (Divac et al. 1987b). Their injection in animal Ta6 appears to be mainly centred in the rostral parts of Fr1 but with some extension of the injection site into adjacent Cg1. This injection produced transport of label to a corresponding site in the contralateral Fr1 through the dorsal and rostral parts of the anterior commissure. Subcortical projections from the injection in this animal were to the medial division of the head of the mainly ipsilateral (but also contralateral) caudate nucleus and to a discrete region of the anteromediodorsal part of the thalamus. The projection to the medial caudate was patchy in a fashion reminiscent of the patch/matrix subdivisions of the therian striatum (see Chapter 6). Their injection site in another echidna (Ta5) was not illustrated, but was described as being more lateral than that for Ta6 and extending into the lateral orbital cortex. This may correspond to Fr2 (and perhaps Fr3) and label from this injection was also seen in symmetrically placed termination sites of contralateral frontal cortex. Subcortical projection from this injection site was also to the medial caudate and to a part of the anteromediodorsal thalamus slightly lateral to that labelled in Ta6. Contralateral cortical label resulting from both these injections was arranged in strips (~600 μm in width) consistent with a columnar organisation of the cortex like in therians.

The projection from the frontal cortex of the echidna to the dorsomedial caudate suggests the involvement of this part of the cortex in basal ganglia circuits through the dorsomedial or associative striatum (see also Chapter 6). This is rather similar to the topographic patterning observed in rodent and primate corticostriatal projection (Gerfen 2004; Haber et al. 2012), where projections to the mediodorsal striatum arise mainly from medial frontal (including prefrontal) and Cg1 regions of the cortex. The pattern of reciprocal thalamocortical and corticostriatal connections is therefore consistent with the Fr1 field of the echidna cortex being homologous to the prefrontal cortex of placentals. Such a large prefrontal cortex would presumably serve in decision-making and planning, but how this is used for cognition in the natural environment of the echidna remains unknown and untested. Hassiotis and colleagues (Hassiotis et al. 2004b) have also noted that consideration should be given to the possibility that Fr1 is homologous to anterior cingulate, orbital, prelimbic, premotor and/or rostral agranular insular cortex of placental mammals, because these regions also receive input from the mediodorsal nucleus of the dorsal thalamus in therians (Leonard 1969; Krettek and Price 1977; Groenewegen 1988; Ray and Price 1992).

Significant questions also remain concerning the possible homology of Fr2 and Fr3 (see also Chapter 11). In particular, the connections of Fr3 situated deep in a cortical groove have never been studied. From its position on the underside of the frontal fold, it is possible that Fr3 may be homologous to the orbital frontal cortex of rodents (LO, MO, VO) and primates (areas 11, 13 and 14). This is a tempting hypothesis because it would explain the expansion of the echidna frontal cortex to act as an integrative centre of chemosensory information and a limbic centre for regulation of behaviour. If this hypothesis were correct, then one would expect that Fr3 of the echidna receives both olfactory and gustatory input (perhaps in conjunction with input from other modalities, e.g. visual and somatosensory, for the purpose of sensory integration) and is involved in projections to the ventral striatum (nucleus accumbens). This is discussed in detail in Chapter 11 on olfaction and gustation.

In summary, the frontal cortex of the echidna rostral to the motor cortex has several key features: (1) it can be divided into at least three cyto- and chemoarchitecturally differentiated regions; (2) it engages in reciprocal connections with the anteromediodorsal part of the dorsal thalamus in a topographically ordered fashion; and (3) it projects to the dorsomedial striatum. Although the frontal fields of the echidna may be homologous to the dorsolateral, medial and orbital prefrontal cortices, it is also possible that the frontal cortex of the echidna reflects a specialisation unique to this group, with no clear homologue to therian prefrontal cortex. This is particularly salient given that the echidna cortex has expanded during evolution entirely independently of the cortex of therians.

Cingulate and retrosplenial cortex

In marsupials and placentals, the cingulate and retrosplenial cortex are part of the limbic cortex, but their affiliations in the monotremes are largely unexplored. In the short-beaked echidna, the cingulate cortex (Fig. 7.4) extends around the medial and dorsal circumference of the cortex almost as far as the most medial branch of the δ sulcus (Hassiotis *et al.* 2004b; Schuster 1910). The most medial part of the cingulate has an indistinct layer 2Cx and poorly differentiated layers 3Cx to 6Cx (Fig. 7.4b, c). As in placental anterior cingulate cortex, Hassiotis and colleagues could not distinguish a clear layer 4Cx, making the region agranular. The cingulate cortex gives way to the retrosplenial cortex, which is chemically similar to the cingulate cortex, at the rostrocaudal level of the ψ sulcus.

Insular cortex

Abbie (1940) was reluctant to delineate an insular cortex in the monotremes, given that he could not definitively identify a claustrum, but Hassiotis and colleagues were able to identify a discrete insular region of distinctive cyto- and chemoarchitecture in the depths of the rhinal fissure of the short-beaked echidna (Hassiotis *et al.* 2004b; Fig. 7.5a, b). The putative insular cortex (Ins) in the echidna is characterised by a thick layer 1Cx, indistinct layers 2Cx and 3Cx and prominent layers 5Cx and 6Cx. The insular cortex is also distinguished by having substantially less immunoreactivity for the calcium-binding protein parvalbumin than either the adjacent Fr3 or piriform cortex. Although tentatively identifying it as insular cortex, Hassiotis and colleagues could not exclude the possibility that their Ins cortex might be homologous to at least part of the orbital cortex of placentals, since therian orbital cortex occupies a similar position with respect to the placental rhinal fissure and piriform cortex.

Claustrum/endopiriform complex

The claustrum/endopiriform complex is a pallial structure that lies below the cerebral cortex in all therian mammals (Johnson *et al.* 1994). In therians, the claustrum/endopiriform complex is often divided into two regions: a dorsal claustrum, lying immediately deep to the insular cortex; and a ventral part (the endopiriform nucleus) that lies deep to the dorsal part of the piriform cortex. The dorsal or true claustrum has reciprocal connections with quite diverse regions of the isocortex, suggesting that it plays some

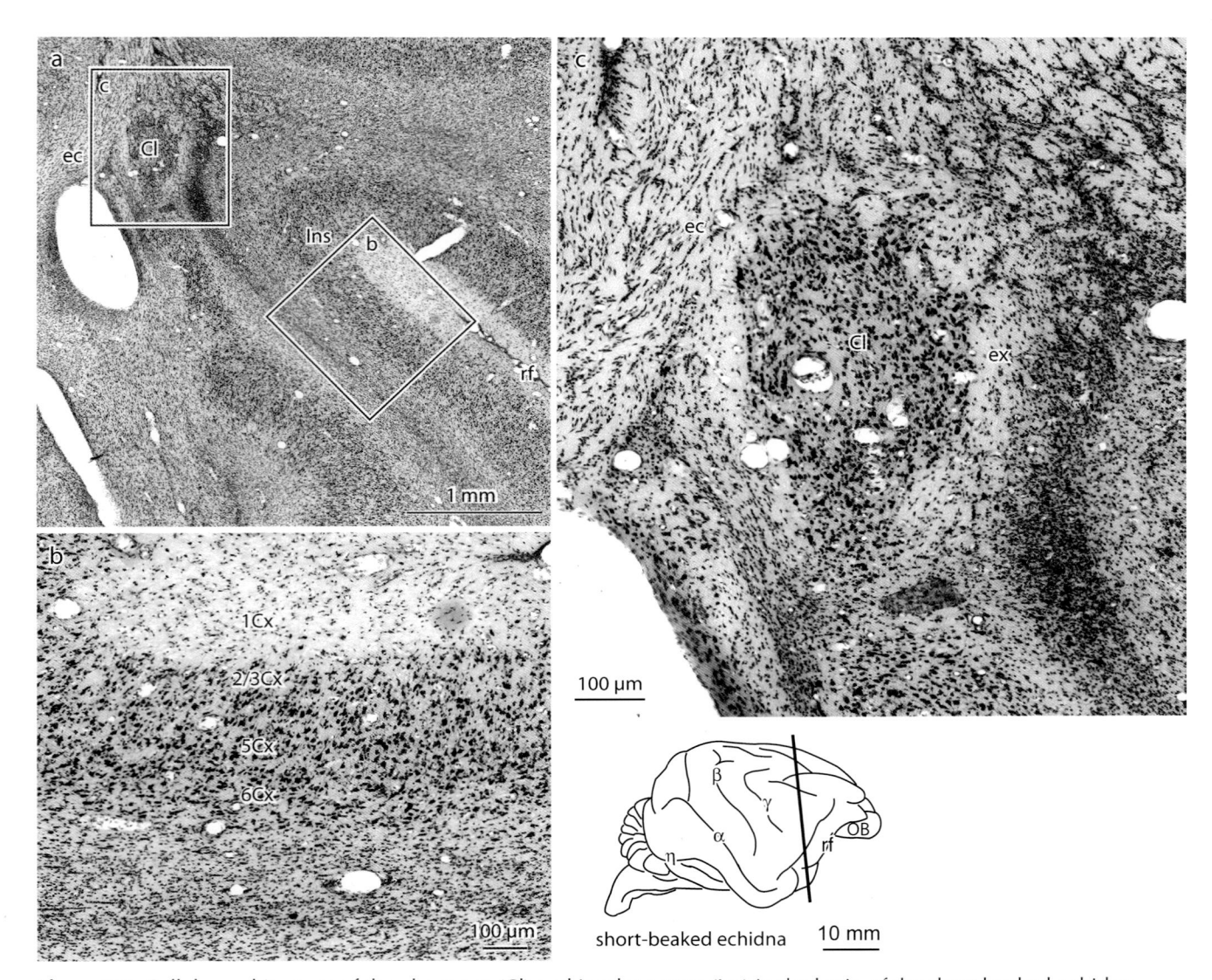

Figure 7.5: Cellular architecture of the claustrum (Cl) and insular cortex (Ins) in the brain of the short-beaked echidna. Photomicrograph (a) shows a low power view of a frontal Nissl-stained section through the front half of the brain (see the line diagram at lower right for the position of the section plane) and indicates the position of images (b) and (c). Photomicrographs (b) and (c) show higher power views of the cellular architecture of the insula and claustrum, respectively. Note the position of the claustrum between the external (ec) and extreme (ex) fibre capsules. 1Cx to 6Cx – layers of the insular cortex; rf – rhinal fissure.

integrative function in the processing of sensory, motivational and mnemonic information, but its precise role remains unclear.

On the basis of developmental expression of transcription factors, Puelles and colleagues (Puelles *et al.* 2000) classified the dorsal or insular claustrum as a derivative of the lateral pallium (represented by the rostral or anterior parts of the dorsal ventricular ridge of the reptilian telencephalon), whereas they thought that the endopiriform nucleus had a different origin, from the ventral pallium, making it homologous to the anterior or rostral parts of the dorsal ventricular ridge of the sauropsid telencephalon. The issue of whether monotremes have a true claustrum will be considered in detail below, but it is generally agreed that both platypus and echidna have an endopiriform nucleus. In this section we will discuss the findings concerning the endopiriform nucleus.

In the short-beaked echidna, the endopiriform nucleus lies within the interior of the extensive piriform lobe (Ashwell *et al.* 2004), but Ashwell and colleagues also identified a short dorsal extension of the endopiriform nucleus that intervenes between the dorsal claustrum and the insular cortex. In the endopiriform nucleus of the echidna there are very few parvalbumin immunoreactive neurons (in contrast to the rich population in the claustrum), but many calbindin immunoreactive neurons distributed throughout its dorsoventral extent. Calretinin immunoreactive neurons, which are rare in the monotreme isocortex and dorsal claustrum, are concentrated in the dorsal part of the endopiriform nucleus.

Ashwell and colleagues also noted the presence of neurons within the interior of the small piriform lobe of the platypus that were immunoreactive for calbindin and might constitute an endopiriform nucleus, but only a few calretinin immunoreactive neurons were present. Ashwell and colleagues (Ashwell *et al.* 2004) therefore left open the issue of whether the platypus has a true endopiriform nucleus.

The question of the monotreme claustrum

There has been a long-standing debate concerning whether one or more of the living monotremes has an insular claustrum. Abbie (1940) could not identify a claustrum in either the platypus or short-beaked echidna and hence did not describe any overlying insular cortex (but see above). Divac noted the apparent absence of the claustrum in the echidna as part of an experimental study of frontal cortex connections (Divac *et al.* 1987a, b). Johnson and colleagues (Johnson *et al.* 1982a, b, 1994) have used the apparent absence of the insular claustrum as one of the primitive characters of the monotremes (see Chapter 14 for discussion of these). Finally, Butler and colleagues have recently discussed the evolutionary significance of the apparent absence of a claustrum in the ornithorhynchid and tachyglossid monotremes (Butler *et al.* 2002).

Why should the apparent absence of such a tiny brain region be of such significance, and is it really missing? Butler and colleagues state that the 'presence of only a small claustrum – or the definitive lack of one – in monotremes is of pivotal importance for illuminating pallial evolution in amniotes' (Butler *et al.* 2002). The significance of the structure revolves around the possible homology of the mammalian claustrum to a large pallial structure (the anterior part of the dorsal ventricular ridge or ADVR) in turtles and diapsid reptiles, or to the neostriatum in birds (Butler *et al.* 2002). Gene expression patterns in mammalian and sauropsid embryos have led to the suggestion that the ADVR/neostriatum is a derivative of the same embryonic field that gives rise to both the basolateral amygdala and the claustrum formation of mammals (Puelles *et al.* 2000). If this were the case, then one would expect that monotremes have a claustrum at some stage during development.

Ashwell and colleagues challenged the prevailing view that the claustrum is missing from the monotreme brain. They reported that the short-beaked echidna (at least) has a definitive claustrum based on a combination of cellular, fibre and chemical architectural criteria, but the claustrum was less clear in the platypus brain (Ashwell *et al.* 2004).

In the short-beaked echidna, Ashwell and colleagues identified the putative dorsal or insular claustrum as a small cluster of neurons (or a group of clusters), positioned at the stem of the piriform lobe (Fig. 7.5c). This group lies below and to the side of the putamen and to the side of the interstitial nucleus of the posterior limb of the anterior commissure (IPAC). This putative dorsal claustrum is embedded in axons emerging from the anterior commissure and is related laterally to cortical tissue (putative insular cortex) in the depths of the rhinal fissure (Fig. 7.5a). They were able to exclude a striatal identity for these neurons because the claustral neurons were not tyrosine hydroxylase immunoreactive (like the striatum) and had a pattern of acetylcholinesterase activity that was similar to the adjacent cortical tissue rather that the putamen or IPAC. The dorsal claustrum of the echidna has populations of parvalbumin and calbindin immunoreactive neurons much like therian claustrum. These occupy complementary positions: parvalbumin neurons cluster in the equatorial region of the claustrum and calbindin neurons cluster dorsally and ventrally at the poles of the structure.

In most therians the claustrum is a thin sheet alongside the external surface of the putamen, so why does it adopt the clustered morphology in the short-beaked echidna? Ashwell and colleagues suggest that the unusual shape of the claustrum in the echidna is due to the unusual white matter topography of the monotreme forebrain. The expansion of the commissural fibre bundles approaching the large anterior commissure appears to have forced the dorsal claustrum to adopt a globular profile, or even to split into a chain of discrete neuronal clusters in some animals.

Is there any evidence that the putative claustrum of the echidna engages in reciprocal connections with isocortex? This appears to be the case for frontal cortex, based on the studies by Divac and colleagues (Divac et al. 1987a, b). Divac showed retrograde labelling of neurons (see Fig. 2E in Divac et al. 1987a) at the base of the piriform lobe in a region that corresponds to the caudal component of the claustrum as described by Ashwell and coleagues (Ashwell et al. 2004). This region also receives a projection from the frontal cortex (see Fig. 1d and e in Divac et al. 1987b). So at least for the frontal cortical regions, the putative echidna claustrum is engaging in reciprocal connections with other parts of the pallium.

In the platypus, the question of the presence or absence of a claustrum is much more difficult to answer. Ashwell and colleagues tentatively identified a dorsal or insular claustrum beneath layer 6Cx of the frontal cortical field, separated from the overlying isocortex by only a poorly defined extreme capsule. This is reminiscent of the rostral parts of the claustrum in rodents and may be a similar claustral topography to the situation in some placentals (e.g. lagomorphs, macroscelidea, chiroptera, sirenia) where the insular claustrum is embedded within the cortex and there is no intervening capsule (Johnson et al. 1994). In the platypus atlas (see Chapter 17), we have adopted a position that the platypus claustrum is present in the depths of the cortex, albeit poorly demarcated from the overlying isocortex.

To summarise the above discussion, the dorsal claustrum is certainly present in one of the living monotremes, refuting the contention that the claustrum is missing from all living monotremes (Ashwell et al. 2004; Butler et al. 2002). The presence of a claustrum in the short-beaked echidna strongly suggests that the claustrum was present in the brains of ancestral mammals, although the possibility cannot be excluded that the claustral/endopiriform complex of the echidna has developed by convergence.

Tenia tecta and infralimbic cortex

Hassiotis and colleagues also identified a tenia tecta in the limbic region of the echidna cortex (see atlas plate Ec-Ad04; Hassiotis et al. 2004b). The tenia tecta is a transitional region between the cingulate cortex above, the orbital cortex below and the hippocampal allocortex further caudally. The dense Nissl-stained cell band seen in this region of the echidna cerebral cortex is not as prominent as in rodents, but the chemical features of the region (strong labelling for acetylcholinesterase preparations and strong immunoreactivity for parvalbumin) are similar to those reported in rodents (Paxinos and Watson 2007; Paxinos et al. 2009). Dorsal to the tenia tecta, Hassiotis and colleagues delineated an infralimbic cortex. As in rodents (Palomero-Gallagher and Zilles 2004), this region of the echidna limbic cortex has only poor lamination and relatively poor immunoreactivity to parvalbumin in its neuropil, in contrast to the adjacent tenia tecta and prelimbic regions (Paxinos et al. 2009).

Neuronal morphology in the cerebral cortex of monotremes: Golgi and dye studies

Neuronal morphology in the cerebral cortex can be assessed by Golgi (silver) impregnation, dye injection of cells themselves, immunohistochemistry for specific markers that are concentrated in particular neuronal types (e.g. neurofilament protein in large pyramidal neurons, calcium-binding proteins in particular types of non-pyramidal neurons), and carbocyanine dye diffusion in postmortem tissue. Dye injection has been used in platypus S1 cortex (Elston et al. 1999) and Golgi impregnation and carbocyanine dye tracing have been applied to several cortical regions of the short-beaked echidna (Dann and Buhl 1995; Hassiotis and Ashwell 2003). Immunohistochemical findings will be considered in the subsequent section.

In the platypus, pyramidal neurons in four cortical regions have been injected with Lucifer Yellow and viewed in tangential sections in relation to cytochrome oxidase activity (Elston et al. 1999). The main focus of this study was to compare bias in the orientation of the basal dendritic skirt in bimodal (i.e. mechano- plus electrosensory) versus unimodal (solely mechanosensory) bands within S1 (PH4 of Abbie 1940), but the study also provided some useful information on the morphology of the basal dendrities of pyramidal neurons in the platypus S1 cortex. Pyramidal neurons in S1 have basal dendritic skirts that spread into the surrounding cortex, with differences in morphology depending on whether the neurons are within a cytochrome oxidase (CO) rich band, or not. Pyramidal neurons in CO-poor bands, where mechano- and electrosensory information is brought together, have basal dendritic skirts that are larger than those for pyramidal neurons in CO-rich bands, but note that this was only seen in one of two animals – the other showed no difference. Elston and colleagues speculated that the large basal fields serve averaging of stimuli to either improve signal-to-noise ratio or to enhance spatial discrimination (see discussion of integration of mechano- and electrosensory information in Chapter 9). They also noted some dendritic bias of pyramidal neurons, such that the basal fields of neurons at the borders of CO-rich bands tend to stay confined to those bands, but the bias was also seen in only one of two animals. These experiments need to be repeated to determine whether a real anatomical difference exists between CO-rich and -poor bands and the functional significance of this pattern.

In the short-beaked echidna there is much more information available on isocortical neuronal morphology. A Golgi study of neuronal morphology in four regions of the isocortex of the short-beaked echidna (frontal cortex – Fr1, motor cortex, S1, visual cortex) found eight classes of neurons (pyramidal, spinous bipolar, aspinous bipolar, spinous bitufted, aspinous bitufted, spinous multipolar, aspinous multipolar and neurogliaform) (Hassiotis and Ashwell 2003; see Table 7.1 for a summary). The key finding of this study was the high proportion of atypical pyramidal neurons (30–42% depending on the cortical region; Fig. 7.6). Atypical pyramidal neurons have inverted cell bodies (Fig. 7.6b, short or branching apical dendrites (Fig. 7.6c) and/or few basal dendrites (Fig. 7.6d). These are similar to the atypical pyramidal neurons of layer 2Cx of the cetacean visual cortex (Garey et al. 1985; Glezer and Morgane 1990), but are present throughout a much greater proportion of the cortical thickness (Hassiotis and Ashwell 2003).

Another important finding was the relatively low dendritic spine density on apical and basal dendrites of the isocortical pyramidal neurons of the echidna. Dendritic spines provide modifiable sites of excitable synaptic input and account for 70–95% of the synaptic input to placental pyramidal neurons (Horner 1993; Moutin et al. 2012), so the low spine density in the echidna isocortex may indicate functional differences in the synaptic input and electrotonic properties of the dendrites of echidna pyramidal neurons compared to therians (Hassiotis and Ashwell 2003).

Hassiotis and Ashwell (2003) also found that pyramidal neurons made up a lower proportion of the labelled cortical neurons (41% in frontal cortex, 49% in motor cortex, 35% in S1, and 34% in visual cortex). These values are both lower than reported frequencies for therians and the results obtained by Hassiotis and Ashwell using the same Golgi methods in the isocortex of the laboratory rat (75% in motor cortex and 78% in S1 barrel field).

The detailed morphology of pyramidal neurons obtained from Golgi impregnation is supported by carbocyanine dye tracing studies and immunohistochemistry. Dann and Buhl (1995) obtained retrograde labelling of pyramidal and pyramid-like neurons following postmortem insertion of the carbocyanine dye known as DiI in the visual and auditory cortex of the short-beaked echidna. The pyramidal neurons labelled in layers 5Cx and 6Cx look very similar to the atypical pyramidal neurons identified by Hassiotis and Ashwell (2003), in that the apical dendrites of carbocyanine-labelled pyramidal neurons are very short and often bifid.

Most data on isocortical neuronal morphology in monotremes are confined to the short-beaked echidna (Table 7.1), and this shortcoming needs to be addressed by morphological studies in the platypus. Although there are some significant morphological differences between the monotremes and therians in cortical neuronal morphology (e.g. the high proportion of atypical pyramidal neurons and the low dendritic spine density of apical and basal dendrites, discussed above), all the non-pyramidal neurons in the echidna isocortex are morphologically similar to their counterparts in therian cortices (Table 7.1; Hassiotis and Ashwell 2003). These broad similarities suggest that

Table 7.1. Types of neurons in the isocortex of the short-beaked echidna[1]

Neuron	Cell body position	Cell body shape and size	Dendrite and spine morphology; axonal ramification
Pyramidal	• Layers 2Cx to 6Cx	• Many are atypical in shape (tilted cell bodies) • Cell body size is similar across layers 2Cx to 5Cx in Fr and visual cortex	• Many (30 to 42%) are atypical (bifurcated apical dendrites) • Apical dendrites are narrow in M and S1 cortex • Dendritic spine length ranges from 1.5 to 3.5 µm • Spine density on apical and basal dendrites is low (8 to 16 per 10 µm length) • Spine morphology ranges from spikes to elongated clubs to mushrooms
Bitufted	• Mainly in 2Cx/3Cx (70%) • Some in 6Cx (20%) and 5Cx (10%)	• Large fusiform cell bodies (50 to 70 µm height and 20 µm width)	• Two primary dendrites, one at each pole of the cell body • Primary dendrites are short (only 20 µm) before branching into secondary dendrites • Dendritic trees extend for up to 300 µm in vertical span • Spinous and non-spinous forms found in all regions studied • Spinous forms have spines (1.3 to 2.6 µm length) only on secondary dendrites • Spine density of spinous forms is less than 15 per 10 µm dendrite length
Bipolar	• Mainly in 2Cx/3Cx (52%) • Some in 4Cx (19%), 5Cx (24%) and 6Cx (5%)	• Somata are usually spherical or ellipsoidal 20 µm vertically and 10 µm wide • No horizontal bipolar neurons found in echidna cortex	• Single primary dendrite (2 to 3 µm thick and 50 µm long) emerging from each pole of the cell body • Dendritic fields reach 350 µm in vertical span but have a narrow horizontal span (40 µm) • Spinous forms have spines (2.5 to 4.0 µm length) only on secondary dendrites • Spine densities of spinous forms are less than 15 per 10 µm dendrite length
Multipolar	• Layers 2Cx to 6 Cx	• Shape ranges from fusiform to irregular • Large (40 by 20 µm) and small (20 by 10 µm) types	• Up to four primary dendrites • Spherical dendritic trees • Vertical extent of dendritic trees reaches 270 µm for large spinous multipolar neurons and 550 µm for large non-spinous multipolar neurons of visual cortex • Spinous forms have spines only on secondary and tertiary dendrites • Spine morphology ranges from elongated necks to short and stubby stalks • Spine density of spinous forms is less than 15 per 10 µm length • Some axons have collateral branches in the same or infra-adjacent layer
Neurogliaform	• Mainly lower layers (5Cx, 6Cx)	• Small, round, elongated and angular somata (8 to 10 µm diameter)	• Very small dendritic tree (less than 30 µm from soma) • Up to eight primary processes • No spinous processes, but beaded primary processes • Axon branches only within the dendritic field

[1] Based on Hassiotis and Ashwell (2003).

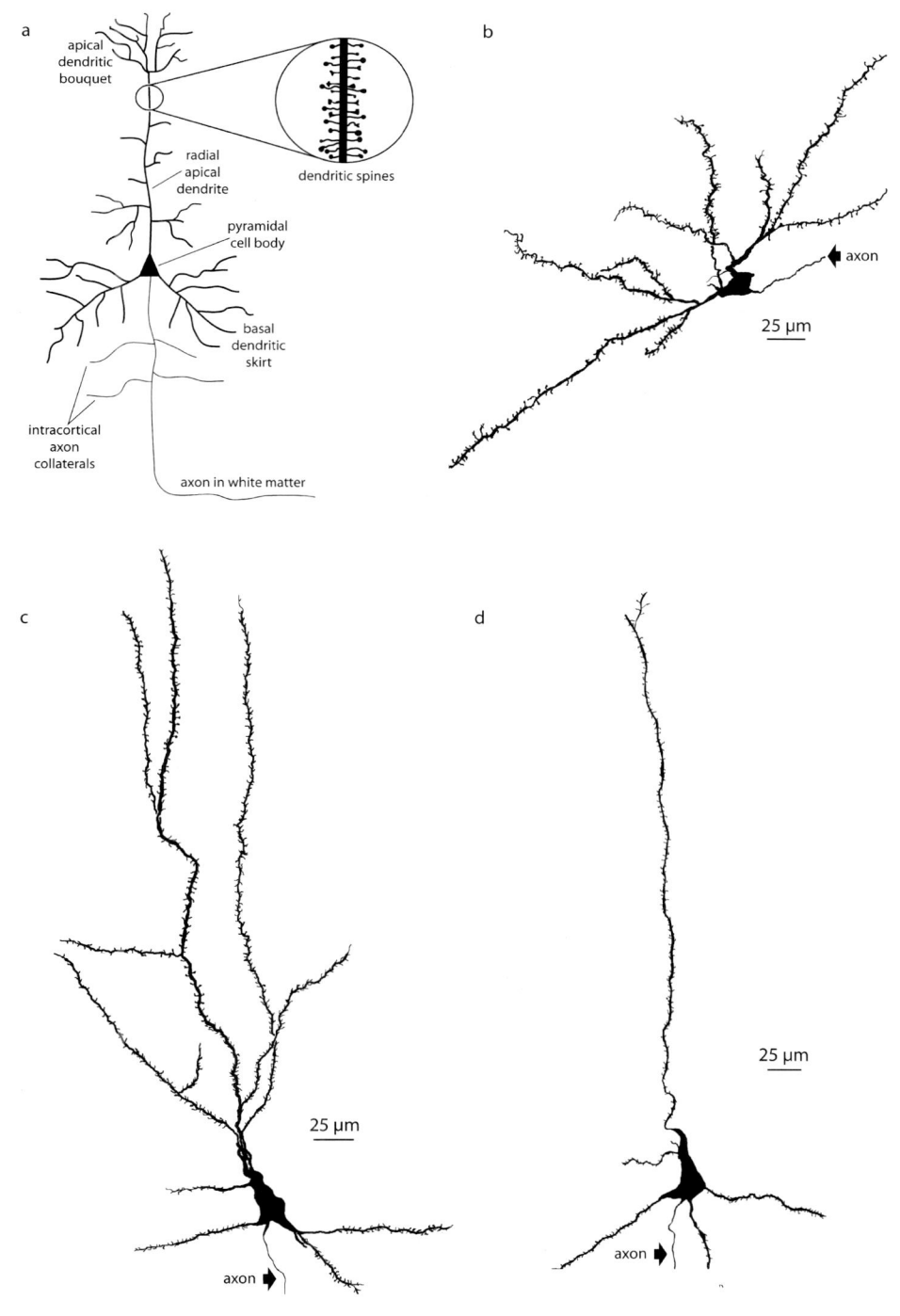

Figure 7.6: Illustrations of a *typical* pyramidal neuron in therian cortex (a) and examples of atypical pyramidal neurons encountered in the isocortex of the short-beaked echidna (b to d). The schematic illustration (a) shows the key features of a *typical* pyramidal neuron: pyramidal cell body, the presence of dendritic spines, a radially oriented apical dendrite, a terminal bouquet of apical dendritic branches in layer 1Cx, a skirt of basal dendrites, an axon descending to subcortical white matter, and intracortical axon collaterals. Atypical pyramidal neurons in echidna isocortex include inverted pyramidal neurons (b), pyramidal neurons with short, branched apical dendrites (c) and pyramidal neurons with limited basal skirts (d). Figures (b) to (d) are reproduced from Hassiotis and Ashwell (2003) with kind permission from S. Karger AG, Basel.

both primitive pyramidal and non-pyramidal neurons probably emerged as discrete morphological entities very early in mammalian cortical evolution, either at the time of divergence of the therian and prototherian lineages, or even earlier (Hassiotis and Ashwell 2003).

The high proportion of atypical pyramidal neurons in the echidna cortex could be explained by two hypotheses considered by Hassiotis and Ashwell (2003). The first is that the complete suite of 'typical' pyramidal neuron features as described for therians (i.e. a long, single, radially oriented apical dendrite; a terminal bouquet of apical dendritic branches in layer 1Cx; a skirt of basal dendrites; the presence of dendritic spines; an axon extending into the subcortical white matter; and intracortical axon collaterals; Nieuwenhuys 1994) may have arisen first among the earliest therians and *after* the divergence of the therian and prototherian lineages. The alternative hypothesis is that the 'typical' pyramidal neuron may have been present in large numbers in the cortex of the mid-Triassic cynodonts who gave rise to therian mammals and monotremes. In this second evolutionary scenario, pyramidal neurons in the monotreme lineage would have undergone secondary modification (e.g. the emergence of a bifurcated apical dendrite, reduction of the basal dendritic skirt, loss of the terminal dendritic bouquet in layer 1Cx). Hassiotis and Ashwell (2003) favoured the first hypothesis, because the similarity of non-pyramidal neurons in the echidna isocortex to those in eutheria strongly suggests that non-pyramidal neurons have changed little in morphology since the divergence of the prototherian and therian lineages.

Neuronal morphology in the cerebral cortex of monotremes: immunochemical studies

Neuronal types in monotreme isocortex have been identified using immuno- and enzyme-histochemical techniques (Hof *et al.* 1999; Hassiotis *et al.* 2004b, 2005), but the data are limited. Most of the available data concern the distribution and morphology of neurons immunoreactive for three calcium-binding proteins (parvalbumin, calbindin and calretinin) and neurofilament protein (SMI-32 antibody). Calcium-binding proteins are intracellular calcium-accepting molecules that act as buffers to protect neurons from surges in calcium levels inside the cell. In the cerebral cortex, these molecules are useful as markers of specific types of neurons that serve particular functions. In particular each of the three calcium-binding buffer proteins co-localises with the inhibitory neurotransmitter GABA in distinct subpopulations of non-pyramidal local circuit neurons (see discussion in Hof *et al.* 1999). Neurofilament proteins make up the structural scaffolding of neurons and are particularly concentrated in large neurons that maintain high diameter axons serving long-distance communication, so the presence of high concentrations of neurofilament protein in groups of neurons suggests that they are important for transferring bulk information between cortical regions or driving lower parts of the central nervous system.

Hof and colleagues (Hof *et al.* 1999) noted that the primary sensory and motor areas of the platypus cerebral cortex contain more parvalbumin and calbindin immunoreactivity than other cortical areas, with the primary somatosensory cortex having a particularly high concentration of parvalbumin immunoreactive neurons and axons. Hof and colleagues reported the presence of parvalbumin in large multipolar neurons in layers 3Cx to 6Cx that resemble basket cells from placental cortex, as well as small round interneurons in later 2Cx and 3Cx. Basket cells get their name from the basket-like terminals that the axons of these cells form around the cell bodies of pyramidal neurons in upper layers of the cortex. Calbindin was found in a large population of small and lightly stained round interneurons in layers 3Cx and 5Cx, as well as in some larger mulitpolar neurons in deeper cortical layers. By contrast, calretinin immunoreactivity is relatively rare in the platypus isocortex, but is found in significant numbers in small bipolar and round neurons in the piriform cortex. No data are available concerning neurofilament protein distribution in the platypus.

Hassiotis and colleagues found a rather different distribution of calcium-binding proteins in the echidna cerebral cortex from that reported by Hof and colleagues. Hof identified parvalbumin, calbindin and calretinin immunoreactive pyramidal neurons in the echidna cortex, but Hassiotis did not see these. Parvalbumin immunoreactive cell types in the echidna cerebral cortex are similar to the wide arbour neurons, bitufted and horizontally oriented cells found in placental cortex, and include some multipolar neurons (Hassiotis et al. 2004b). The morphology of calbindin immunoreactive neurons varies greatly by layer: calbindin immunoreactive neurons in the echidna are mainly multipolar types of neurons in layers 1Cx and 2Cx; mostly bipolar or bitufted types in layers 3Cx and 4Cx; and mainly multipolar types in layers 5Cx and 6Cx. Calretinin immunoreactive neurons are quite rare in the echidna cortex, never making up more than 1% of all neurons present even in the regions and layers where they are most prevalent. Hassiotis and colleagues identified horizontally oriented calretinin immunoreactive bipolar cells, tentatively identified as Cajal-Retzius cells, in layer 1Cx of most regions, similar to therian cortex.

In the short-beaked echidna, immunoreactivity to non-phosphorylated neurofilament protein (SMI-32 antibody) has been used to parcellate the isocortex and identify pyramidal neurons that engage in long-distance connections (Hassiotis et al. 2004b, 2005; see Table 7.2). Non-phosphorylated neurofilament proteins are important in the stabilisation of the protein skeleton of axons (Morris and Lasek 1982) and the presence of strong immunoreactivity to neurofilament protein has been associated with axons of large internal diameter and high impulse conduction velocity (Hoffman et al. 1987). Very few SMI-32 immunoreactive pyramidal neurons have been found in the echidna cortex and this may indicate that the echidna corticofugal pathways (e.g. descending motor pathways) have relatively few large calibre axons (Hassiotis et al. 2005). Nevertheless, significant groups of neurofilament immunoreactive pyramidal neurons are found in frontal regions (Fr1, Fr3) and primary sensory regions (S1, PV, R and M/R transition; visual and auditory), although (strangely) not in motor cortex (Hassiotis et al. 2004b). The last is surprising given that motor cortex is usually the site of origin of major long-distance descending motor pathways (i.e. corticospinal, corticostriatal, corticopontine tracts). Most neurofilament immunoreactive neurons are multipolar with occasional bipolar neurons. Typical pyramidal neurons are most common in Fr1 and R, but in other regions irregular multipolar, extraverted and inverted pyramidal neurons are seen. Labelled neurons in the echidna isocortex are mainly confined to layer 5Cx, with only a few neurons seen in other layers in somatosensory regions devoted to the snout and hindlimb and visual cortex. This stands in contrast to placentals where neurofilament immunoreactive neurons are positioned in both layers 3Cx and 5Cx, and are mostly of typical pyramidal shape (long single apical dendrite, wide basal dendritic skirt and pyramid-shaped cell body).

In cat and monkey cortex (Hendry et al. 1984) neuropeptide Y (NPY) neurons are known to be GABAergic inhibitory interneurons and many of the larger NPY neurons in rodent cortex also express nitric oxide synthase (Smiley et al. 2000). Broadly speaking, NPY immunoreactive neurons in the echidna cortex are very similar in shape and position to those in therian cortex. In echidna somatosensory cortex, most NPY immunoreactive neurons are in the deeper cortical layers, particularly 6Cx; but in motor cortex there are two distinct bands – one in layer 2Cx and upper 3Cx, the other in lower layer 5C and 6Cx. NPY neurons in upper layers are usually of bipolar shape with smooth dendrites, but multipolar neurons predominate in deeper layers. NPY neurons have no dendritic spines, although neurons in the lower layers tended to have beaded primary dendrites.

The enzyme reduced nicotinamide adenine dinucleotide phosphate diaphorase (NADPH-d), a synthetic enzyme for nitric oxide, has been used extensively to identify nitrergic (nitric oxide using) neuron types in therian cortex. Nitric oxide serves several roles in the central nervous system in general and cerebral cortex in particular, including vasodilation, presynaptic plasticity, sleep and cortical development (see Kirkcaldie 2012 for review). In rodent cortex, nitrergic neurons are mainly multipolar neurons of medium size, although a few smaller bipolar and monopolar cells in the deepest layer are also labelled by NADPH-d histochemistry. Many nitrergic neurons also use GABA, somatostatin and/or neuropeptide Y as neurotransmitters. Hassiotis and colleagues (Hassiotis et al. 2004b; 2005) observed a range of morphologies of NADPH-d reactive neurons in the echidna isocortex, varying from horizontally oriented bipolar cells, through pyramidal to multipolar neurons with up to four primary dendrites. Nevertheless, these cells are all similar to those seen in therian cortex (Hassiotis et al. 2005).

Table 7.2. Summary of chemically identified neurons in short-beaked echidna isocortex[1]

Chemical marker	Type of neuron	Distribution of labelled cells in cortical regions	Distribution in layers of cortex
Neurofilament protein (SMI-32 antibody)	• Mainly pyramidal, but some bipolar • Labels typical pyramidal neurons only in Fr1 and R • Rotated and inverted pyramidal neurons are labelled in other areas	• Fr1, Fr3, primary sensory regions (S1, R, PV, auditory, visual). Not present in Cg or entorhinal cortex	• Mainly layer 5Cx in Fr1, S1 forelimb region, R, PV, auditory and visual • Throughout layers 3Cx to 6Cx in S1 snout region • Throughout layers 3Cx to 5Cx in S1 hindlimb region
Parvalbumin	• Multipolar, but occasional bipolar and bitufted cells • Bitufted cells in motor cortex • Non-spinous stellate cells in Cg, visual and S1 • Some bipolar cells in visual cortex	• Most cortical areas, with high density in primary sensory (S1, visual, auditory, olfactory) and frontal areas	• Most cell bodies are located between 3Cx and upper 6Cx • In S1, highest spatial density in upper layers (3Cx to 4Cx)
Calbindin	• Multipolar cells in 1Cx and 2Cx • Bipolar cells in lower 3Cx and upper 4Cx • Multipolar cells in lower 5Cx and 6Cx	• Most cortical areas, but with regional differences in morphology • Bipolar cells most prominent in S1 and visual cortex	• All cellular layers (2Cx to 6Cx), but mainly in lower layers in S1 • Bipolar cells are most common in middle layers
Calretinin	• Very rare, mainly multipolar but some bipolar • Possible Cajal-Retzius cells in 1Cx of Cg and S1	• Rare, but present in low spatial density in most cortical areas	• Layers 3Cx and 5Cx in frontal cortex • Layers 3Cx and 4Cx in motor and visual cortex
Neuropeptide Y	• Small non-spinous bipolar in upper layers • Small non-spinous multipolar in deeper layers • Some fusiform horizontal spinous cells in subcortical white matter	• Most regions of cortex, but mainly in motor and primary sensory cortex	• Two bands in motor cortex: one band in 2Cx/upper 3Cx; the other in lower 5Cx/all of 6Cx • Deeper layers (6Cx) in S1
NADPH diaphorase	• Mainly multipolar, some are spinous • Some neurons in white matter	• Many regions, but mainly in motor and primary sensory areas	• Mainly in lower 3Cx and below

1 Based on Hof *et al.* (1999); Hassiotis *et al.* (2004b, 2005).

Columnar organisation and dendritic bundling in monotreme cortex

Electrophysiological studies of monotreme cortex have shown the same sort of columnar organisation that has been described for placental cerebral cortex. Columnar organisation of the cortex allows neural data concerned with similar aspects of a sensory stimulus (e.g. a particular orientation of a visual edge) to be processed by functionally linked neurons. Fine grain analysis of S1 somatosensory cortex in the platypus showed that neurons responsive to discrete mechanical stimulation of the bill surface were distributed most commonly at depths of 750–1750 μm from the pial surface, but confined to columns only a few hundred μm in width (Borhinger and Rowe 1977). Within each functional column, neurons shared similar receptive field sizes and locations on the body surface.

The columnar physiological organisation in placental cerebral cortex is associated with the bundling of the apical dendrites of pyramidal neurons. Dendritic clustering is best seen in tangential sections through lower layer 3Cx in agranular (i.e. motor) cortex or layer 4Cx in granular (e.g. somatosensory) cortex. Although the precise functional significance of dendritic clustering is uncertain, it is believed that clustering of apical dendrites permits exchange of information between the neurons grouped into a given functional columns (Feldman 1984). Although Bohringer and Rowe (1977) found electrophysiological evidence of columnar organisation in the platypus cortex and tracing studies suggest that corticocortical connections are organised in a columnar pattern, the anatomical correlate of dendritic clustering is only poorly developed in echidna cerebral cortex. Hassiotis and colleagues (Hassiotis et al. 2003) used nearest-neighbour analysis to compare apical dendritic bundling of pyramidal neurons in the motor cortex of the short-beaked echidna with the laboratory rat. This analysis involves the statistical comparison of the interdendritic distance with what would be expected if the dendrites were randomly distributed. For the rodent motor cortex, mean interdendritic distance (±s.d.) is 3.69 µm (±1.68), significantly lower than that obtained if dendrites are randomly distributed through the tangential section (4.44 ± 2.01 µm). By contrast, analysis of the echidna motor cortex found a mean interdendritic distance of 3.92 µm (±1.85), only slightly lower than for a random distribution of dendrites (4.14 ± 1.81 µm) in the same available section area. Although there was some suggestion of dendritic clustering in the agranular cortex of the echidna and peak interdendritic distance (at between 3 and 4 µm) was similar in both mammals, the deviation from random was much less pronounced in the echidna than in the laboratory rat. Similar findings were also obtained for analysis of dendritic clustering in two types of sensory cortex (visual and S1) in the echidna and laboratory rat. These findings suggest that the pyramidal neurons in the echidna may not participate in the sort of functional columnar organisation that pyramidal neurons in placental (or at least rodent) cortex do. Putting this another way, if there is functional organisation into cortical columns in echidna cortex, controlled cross-talk between members of a column is unlikely to be as effective as it is in therians (Hassiotis et al. 2003).

Carbocyanine dye tracing has also provided some evidence for columnar organisation in monotreme cortex. Dann and Buhl (1995) used insertion of the carbocyanine dye known as DiI (1,1'-dioctadecyl-3,3,3',3'-tetramethylindocarbocyanine perchlorate) to trace intracortical connections in visual and auditory cortices of the short-beaked echidna. Dye tracing in both the anterograde and retrograde directions resulted in discrete patches 500 µm to 1 mm in diameter (based on their illustrations). Anterograde label was distributed throughout all six layers of the target cortical column. Retrograde label was found in pyramidal neurons, mainly in the infragranular layers (layers 5Cx and 6Cx). While the results of Dann and Buhl do support the existence of columnar organisation in monotreme cortex, their illustrations suggest that the columns may be quite broad relative to those in therians, perhaps reflecting poor dendritic bundling.

Anatomical indicators of cerebral metabolic activity

The presence of a large and folded cerebral cortex in the short-beaked echidna naturally raises the question of how metabolically active this large cortical mantle is compared to placentals. The low body temperature of the echidna (around 32°C when physically active) implies that the metabolic demands of its large brain are relatively small compared to those of a therian of similar brain weight, but biochemical indicators of cerebral metabolism have never been tested in this species.

Nevertheless, Hassiotis and colleagues (Hassiotis et al. 2005) analysed several microanatomical features that reflect metabolic activity of the cerebral cortex (the percentage of cortical volume occupied by capillaries – the capillary volume fraction; the numerical density of mitochondria; and the percentage of cortical volume occupied by mitochondria – mitochondrial volume fraction). The capillary volume fraction for echidna cerebral cortex ranges from 1.18% in visual cortex to 1.34% in S1 cortex, similar to values obtained in therian cortex, suggesting that the echidna cortex is as well vascularised as in therians. Values for mitochondrial volume fraction in S1 of the echidna are slightly lower than those found for rodent visual cortex, but are higher than for the cetacean visual and auditory cortices (see discussion in Hassiotis et al. 2005). Mitochondrial numerical density in S1 of the echidna is comparable to cetacean visual and auditory cortices. Mitochondrial numerical density in the echidna sensory cortex is therefore below levels in rodent cortex, but nevertheless falls within the therian range as represented by data for rodent and cetacean cortex. This is in agreement with findings from a study of other short-beaked echidna tissue (liver, skeletal muscle, cardiac muscle, kidney) that showed similar total tissue mitochondrial surface membrane area to therians (Else and Hulbert 1985).

In summary, the anatomical indicators suggest that S1 cortex of the short-beaked echidna is as well vascularised as therian cortex of a similar nature and is as well supplied with mitochondria. Of course, the oxidative activity of the respiratory chain in the mitochondria would vary with body temperature (this being somewhat lower in the echidna under standard conditions), but clearly the echidna cerebral cortex has a similar metabolic potential to that in reference placentals.

Synaptic morphology in monotreme cortex

This is a critically important area for understanding the functional capacity of the monotreme cortex because synapses are the point of functional interaction between neurons, but the available data are confined to only one cortical region (snout representation of S1) of the short-beaked echidna (Hassiotis *et al.* 2004a, 2005). Both asymmetrical Gray type I and symmetrical Gray type II synapses are present in this region. The type I synapses make up ~72% of all profiles encountered in a traverse through the cortical thickness and have round vesicles ~25–35 nm in diameter and with up to 40 vesicles per synapse. Type II synapses accounted for ~28% of profiles and have elliptical or pleomorphic vesicles ~15 nm by 30 nm in diameter with up to 30 vesicles per synapse. Numerical density of synapses peaks in the upper part of layer 4Cx (presumably representing thalamocortical inputs) and in layer 1Cx, lower layer 2Cx and upper layer 3Cx. Synaptic density is lowest in deep layer 3Cx and declines rapidly as one progresses into the deeper part of 4Cx.

Synaptic morphology, spatial density and laminar distribution of synapses in the S1 region of the echidna are broadly similar to those in sensory regions of therian cortex (see Hassiotis *et al.* 2005 for review), but questions still remain concerning how the synaptic density on the spines of pyramidal and stellate neurons compares with placentals.

Hippocampal formation

The hippocampal formation plays a critically important role in the consolidation of new declarative memories (i.e. memories of facts and events) and is usually considered part of the limbic system.

The hippocampal formation (Witter and Amaral 2004) is a specialised region of allocortex divided into three subdivisions: (1) the dentate gyrus, which receives input from the entorhinal cortex and projects to pyramidal neurons of the hippocampus proper; (2) the hippocampus proper (or cornu Ammonis) consisting of pyramidal neurons and which receives input from the dentate gyrus and septal nuclei and projects to the subiculum; and (3) the subiculum, a transitional area between the hippocampus proper and the entorhinal cortex, which is the target of output from the hippocampus proper and in turn gives rise to the major output from the hippocampal formation. The hippocampus proper is usually divided into three regions (CA1 to CA3, for cornu Ammonis 1 to 3).

Morphologically the hippocampal formation of the monotremes (see atlas plate Ec-Ad17) is very similar to that in marsupials and placentals, but with a few differences. Hassiotis and colleagues noted that the pyramidal cell layer in CA1 of the short-beaked echidna is less tightly clustered than in placentals and the cytoarchitectural transition from CA1 to CA2 is also not as clear as in rodent hippocampus (Hassiotis *et al.* 2004b), although the CA2 region is better defined in acetylcholinesterase preparations of the echidna cortex. The most striking chemoarchitectural difference between the therian and echidna hippocampal formation is the poor level of immunoreactivity for parvalbumin in both the hippocampus and dentate gyrus of the echidna. In therian mammals both pyramidal neurons and dentate gyrus neurons are intensely immunoreactive for parvalbumin (Paxinos *et al.* 2009) and yet only isolated cells in the echidna hippocampus show immunoreactivity for this calcium-binding protein.

Some data are available on neuronal morphology in the hippocampus of the short-beaked echidna (Hassiotis *et al.* 2005). Non-phosphorylated neurofilament protein is present in the apical and basal dendrites of pyramidal neurons of cornu Ammonis, but not in the cell bodies. Immunoreactivity for calbindin is mainly found in neuropil of CA1 and labelling of cell bodies is strongest in CA2 and the granule cell layer of the dentate gyrus, with only weak labelling in the pyramidal cells of CA1 and CA3. Like the isocortex of this species, only a few calretinin immunoreactive cells are present in the echidna hippocampus. These are spine-free bipolar or multipolar neurons with only three primary dendrites and with the long axis of their dendritic fields oriented perpendicular to the pyramidal cell layer. Reactivity for NADPH-d is strong in the bodies of pyramidal cells of CA and in occasional large neurons of the polymorphic layer of the dentate gyrus. Strong reactivity is also present in the neuropil and cell bodies of the dentate gyrus.

No data are available on the connections of the hippocampal formation in the monotremes.

Piriform and entorhinal cortex

The unique features of these regions, particularly in the echidnas, will be considered in the discussion of the olfactory system in Chapter 11. At this point it is sufficient to note that the piriform and entorhinal cortex are present in the brains of both platypus and short-beaked echidna. In the platypus, these regions are relatively similar in structure to those in therians, whereas in the short-beaked echidna the piriform cortex forms an extensive olfactory lobe with complex internal differentiation. Limited data are available concerning connections of the regions. The entorhinal cortex of the short-beaked echidna projects to the Cg1 and/or Fr1 regions of the cortex (Divac *et al.* 1987a), perhaps providing olfactory input to the large frontal cortex and limbic cortex, but no study has ever applied tracer injections in the entorhinal or piriform cortex in either monotreme.

Development of isocortex in monotremes

Before reviewing current knowledge of how the monotreme cortex develops, we need to summarise the main points of the cellular and molecular development of therian isocortex.

The neurons of the adult mammalian isocortex arise from several proliferative cell populations in the embryonic pallium and subpallium that contribute both to growth of the brain and the generation of postmitotic neurons. There are three recognised sources of isocortical neurons: (1) the local (pallial) ventricular zone; (2) the local (pallial) subventricular zone (Cheung *et al.* 2007; Kriegstein *et al.* 2006); and (3) the subventricular zone of the lateral and medial ganglionic eminences of the ventral or subpallial telencephalon (Molnár 2011) (refer to Figs. 6.4 and 7.7). Experimental studies suggest that the larger glutamatergic projection neurons of the therian isocortex are derived from the pallial wall and migrate radially into position within the cortical plate, whereas most of the smaller inhibitory GABAergic interneurons of the isocortex are derived from the ganglionic eminences and migrate tangentially over long distances to reach the cortical plate (Carney *et al.* 2007; Nadarajah *et al.* 2003). The precise proportion of isocortical neurons that arise from subpallial sources (i.e. the ganglionic eminences) may differ between species.

The mammalian isocortex develops initially as a preplate region (Fig. 7.7b), which is later split into an overlying marginal zone (future layer 1Cx) and an underlying subplate by the insertion of the cortical plate neurons that will make up the definitive mature cortex (i.e. layers 2Cx to 6Cx). A process of inside-out neurogenesis forms the cortical plate, i.e. neurons destined for the deeper layer (e.g. 5Cx and 6Cx) are the first to be produced from the proliferative ventricular zone, whereas neurons destined for the more superficial layers (i.e. layers 4Cx to 2Cx) are generated later in development from the ventricular and subventricular proliferative zones. Some authors see the emergence of a pallial subventricular zone as a particularly important event in the evolution of a complex gyrencephalic isocortex (Charvet *et al.* 2009; Charvet and Striedter 2011; Cheung *et al.* 2010; Kriegstein *et al.* 2006).

An important element of therian isocortical development is the subplate (Montiel *et al.* 2011; Wang *et al.* 2010, 2011). This is a transient developmental region beneath the cortical plate that contains some of the earliest generated neurons and provides for transient targeting by developing corticopetal axons (e.g. thalamocortical axons from the VP or LG). Subplate neurons also pioneer the corticofugal projection and

156 NEUROBIOLOGY OF MONOTREMES

those early outgoing axons may provide guidance for thalamocortical axons.

The developing monotreme isocortex shows many of the features identified for therian cortical development (Ashwell and Hardman 2012a). The developing monotreme telencephalic vesicle differentiates into distinct pallial, ganglionic and septal components in both the platypus and echidna by between ~8.0 and 9.0 mm GL (Fig. 6.4a, Fig. 7.7a, Appendix Fig. 1). The only pallial proliferative zone at this time is the ventricular germinal zone (VGZ), which can be identified on the basis of mitotic figures along the ventricular surface. At this stage, cortical differentiation is most advanced around the ventral and lateral aspects of the pallium, where the VGZ has reached a thickness of up to 140 μm and a distinct preplate region ~60 μm thick and containing a few scattered cells (putative early generated Cajal-Retzius cells) has developed outside the VGZ.

Some differences in cortical development between the platypus and echidna begin to emerge towards the end of pre-hatching life (10–12.5 mm GL). The cortex of the platypus remains smooth at this stage (Fig. 7.7a), whereas the pallium of the echidna begins to fold (Fig. 7.7b). Both the platypus and echidna lack a cortical plate in the time leading up to hatching: the cortical mantle outside the VGZ consists of only scattered postmitotic cells forming a diffuse preplate zone at the margin of the pallium (Fig. 7.7b), with no apparent layering into cellular or fibre layers.

Cortical development after hatching (Fig. 7.7c, d; Appendix Fig. 1) is largely similar in the platypus and echidna, but the development of the isocortex in the echidna is a few days in advance of that in the platypus. In the first few days after hatching, two significant changes occur in both species. The first is the emergence of the cortical plate and the second is the first appearance of the pallial and subpallial subventricular proliferative zones (SubV). The SubV regions can be identified on the basis of mitotic figures that begin to appear away from the ventricular surface (Fig. 7.7d–f). At first, most of the SubV is (like that region in developing therian forebrain) external to the VGZ of the medial and lateral ganglionic eminences (subpallial SubV), or external to the cortical VGZ (pallial SubV). An unusual feature of the monotremes is that a special region of the SubV extends ventrolaterally from the palliostriatal angle (the boundary between the pallium and subpallium) as a sheet that extends between the postmitotic neurons of the developing putamen and the primitive lateral cortex (S1 and piriform cortex). Ashwell and Hardman (2012a) thought that this SubV extension is, given its site of origin, most probably derived from the lateral rim of the extensive lateral ganglionic eminence, but they could not exclude the possibility that some of the region is derived from pallial SubV or migration from the medial ganglionic eminence.

During the second week after hatching, stratification begins to emerge in the SubV and the region is invaded by afferent axons, suggesting that this subcortical zone is now serving purposes other than the simple generation of neurons and glia. On the basis of the large number of penetrating (presumably thalamocortical) axons, as well as the presence of abundant large cells during its early stages, Ashwell and Hardman (2012a) suggested that (at least at this stage) the subcortical zone may also be performing similar functions to the subplate of developing therian cortex. Nevertheless, further study is necessary to determine whether the putative subplate of the monotreme cortex has the same molecular markers of the therian subplate and performs the same

Figure 7.7: A montage of images showing key events in the development of the monotreme cerebral cortex. Photomicrographs (a) and (b) show the cortex during incubation in the platypus and echidna, respectively. Photomicrograph (c) shows the layers of the developing cortex in a newly hatched platypus. Line diagram (d) shows the distribution of mitoses (filled circles) in the ventricular germinal and subcortical/subventricular zone in an early post-hatching platypus. Photomicrographs (e) and (f) show examples of cell division (arrows) in the subventricular zone beneath the developing S1 cortex of a hatchling platypus. Illustration (d) is reproduced from Ashwell and Hardman (2012a) with kind permission from S. Karger AG, Basel. ccz – compact cortical zone of developing cortical plate; CPu – combined caudate and putamen; Cx* – isocortical primordium; chp – choroid plexus; Hi* – hippocampal primordium; IVF – interventricular foramen; lge – lateral ganglionic eminence; LV – lateral ventricle; mge – medial ganglionic eminence; PrePl – preplate of developing cortex; SubV – subventricular zone; VGZ – ventricular germinal zone.

shepherding functions for axons passing to and from the developing cortex.

Towards the end of the second post-hatching week, the developing S1 cortex of the platypus has a complex lamination. Beginning from the pial surface, there is a cell-sparse marginal layer ~40–50 μm thick overlying a cortical plate that is divided into a compact cortical zone ~30–35 μm thick and made up of densely packed small cells, and a loosely packed zone ~90 μm thick. This division of the cortical plate in dense and loosely packed layers is reminiscent of the structure of the developing marsupial cortex (Ashwell et al. 1996a, b; Marotte et al. 1997; Marotte and Sheng 2000; Leamey et al. 2007). Beneath the cortical plate is a fibrous 90 μm thick intermediate zone crossed by longitudinally and mediolaterally coursing fibres, indicating that connection formation is in process. The modified subventricular/subplate zone lies beneath the intermediate layer.

The subventricular/subplate zone persists for many months after hatching, but at ~3–4.5 months into the lactational phase, it starts to thin and retreat to the part of the cortical wall adjacent to the lateral ventricle. Axon bundles no longer run radially through the subventricular/subplate zone, but the overlying intermediate zone has now thickened considerably. Cortical layering approaches the adult pattern of lamination at ~4 months after hatching, with the compact cortical zone differentiated into layers 2Cx to 4Cx, and the loosely packed zone in the process of transforming into the subgranular layers (5Cx and 6Cx).

Ashwell and Hardman (2012a) were unable to comment on the development of the palliostriatal angle and claustrum in echidnas, because they could not study the region during the critical period from 25 mm body length to adulthood. Nevertheless, the palliostriatal angle, where the therian claustrum is generated, does appear to be an important region in the developing platypus, because it is the origin of the sheet of subventricular/subplate zone cells that extends beneath the trigeminal region of S1.

In summary, the developing cortex of both monotremes is very immature at the time of hatching, much like that seen in marsupials, and both species have a subventricular zone within both the ganglionic eminences and pallium during the post-hatching development. Ashwell and Hardman (2012a) were unable to establish whether the subventricular/subplate zone in the platypus has similar molecular features and functions to the therian subplate (Wang et al. 2011), because their study was performed on archived sectioned material. Nevertheless, the position and association of this region with prominent fibre bundles connecting to the behaviourally important bill representation of S1 suggests a significant role in monotreme isocortical development.

Development of the hippocampal formation

The hippocampal formation develops by a process of invagination of the dorsomedial pallial wall of the telencephalon (see atlas plate Ec-In10), followed by neurogenesis in two sites: the hippocampal VGZ for the macroneurons of cornu Ammonis; and the subgranular zone for the microneurons of the dentate gyrus.

The curvature of the hippocampal primordium can be distinguished in the developing monotreme brain at between ~8.0 and 9.0 mm GL, but there is no cortical plate in the hippocampus until after hatching. The cortical plate emerges around the end of the second post-hatching week. By ~6 months after hatching, the components of the hippocampus (CA and dentate gyrus) have largely completed differentiation. It is not known whether neural stem cells persist in the subgranular zone into adult life, as is described for therians.

Evolutionary considerations

The monotremes and therians have followed a prolonged period of separate evolution, perhaps as long as 150 million years. During this period there has been independent expansion of the isocortex in the Tachyglossidae and in highly encephalised placentals like cetaceans, carnivores and primates. Yet in many respects the cellular, chemical and ultrastructural features of the monotreme cerebral cortex are remarkably similar to those in placental isocortex. These striking similarities between monotreme and therian cortical architecture and neuronal types suggest that these were present in the cortex of the common ancestor for both lineages, since the detailed features are so alike that convergent evolution seems unlikely. On the other hand, some of the dissimilarities, like the puzzlingly large proportion of atypical pyramidal neurons and the poor development of apical dendritic bundling, would suggest that some very distinctive features of cytological organisation have emerged in the isocortex during the long period of separate evolution of the prototherian and therian lineages.

Questions for the future

The large frontal cortex of the echidnas has been an enduring source of fascination for comparative neuroscientists. There are several clues that suggest that its components are homologous with several components of the prefrontal and orbital cortex of placentals, but these correlations are hampered by a lack of detailed connectional studies in echidnas. Preliminary information suggests that the frontal cortex of echidnas has a similar engagement in striatopallidal circuits to the prefrontal cortex of placentals, but further studies of the functional role of the region are needed.

The question of the monotreme claustrum is an ongoing controversy in comparative neuroanatomy. In this discussion, the position has been taken that the short-beaked echidna is a real entity, but the presence of a claustrum in the platypus remains an open issue. As noted above, this question is critically important for understanding mammalian pallial evolution. For the purposes of the atlases in Chapters 16 and 17, the claustrum has been delineated in both species.

The presence of a pallial (subcortical) subventricular zone during development has been considered the hallmark of an advanced cerebral cortex and the monotremes clearly have a sophisticated mechanism for controlling the development of the somatosensory cortex and its topographic input. Nevertheless, nothing is currently known about the molecular control of monotreme cortical development and whether the subcortical zone of the monotreme cortex can be considered homologous with the both the subplate and pallial subventricular zones of placentals. The question of how these developmental mechanisms are related to those at play during placental cortical development is critically important for understanding mammalian cortical evolution.

8
Visual system

Ken W. S. Ashwell

Summary

None of the living monotremes place great reliance on vision, but the structure and genetics of their visual system is of some significance for the evolution of mammalian visual pathways. The visual pathways of both platypus and short-beaked echidnas show evidence of adaptation to their crepuscular and nocturnal lifestyles, with rod-dominated retinas. The suite of distinctive features of the monotreme eye includes the presence of a scleral cartilage, SWS2 cones, double cone photoreceptors and oil droplets in the cone photoreceptors (at least in the platypus) (Zeiss et al. 2011). Retinal topography is relatively unspecialised for all monotremes, with a low centre to periphery gradient of photoreceptor and retinal ganglion cell density, consistent with the relatively low visual acuity shown for the echidna and suspected for the platypus. The central visual pathways in both platypus and echidna are much like those in therians, with the exception that the pregeniculate and lateral geniculate nuclei have adopted distinctive positions in the two extant groups thanks to differential expansion of the caudal thalamus. The pretectal nuclei of the caudal diencephalon are just as differentiated as in therians, but questions remain concerning the capacity of the monotreme eye for accommodation. Two visual areas have been identified in the cerebral cortex, both with visuotopic (i.e. visual field based) organisation, one of which appears to be homologous to the primary visual cortex of therians.

Overview of the visual system

The mammalian visual system not only serves the gathering and processing of visual information from the external environment, but also allows visual information to influence the function of the hypothalamus and endocrine system. Anatomically it consists of the retina; visual pathways to the brain (optic nerve, optic chiasm and optic tracts); retinorecipient neuron groups in the thalamus, hypothalamus, pretectum and midbrain; and primary and association visual cortex for the detailed processing of visual information. Simple observation of monotreme behaviour indicates that vision is not a major sense for either the platypus or the echidnas, but the key position of Prototheria in mammalian evolution means that the structure and function of the monotreme visual system hold important clues for understanding how the mammalian visual system evolved.

Visual capabilities of the monotremes

Detailed testing of visual acuity in behaving platypus has never been undertaken, but a set of thorough observations is available for the short-beaked echidna. Behavioural studies of the short-beaked echidna by Gates (1973, 1978) showed that it is able to differentiate between vertically and horizontally oriented stripes and between triangles and circles, although Gates was uncertain as to whether shape discrimination was based on the entire shape of the object or some specific component of the shape (Gates 1973, 1978). The angle subtended by the line width used in the line discrimination studies at the distance where the subjects were able to discriminate vertical from horizontal lines was 0.499°, suggesting that visual discrimination in the short-beaked echidna is comparable to that in rodents. Gates also showed that interhemispheric transfer of visual information is effective in the short-beaked echidna (Gates 1973) despite the absence of a corpus callosum. Transfer was shown experimentally for a black–white brightness discrimination task, for vertical–horizontal line discrimination and for oblique line discrimination. On the other hand, Gates was unable to find any clear evidence for an optomotor response to moving vertical lines (i.e. head turning to follow vertical lines on a rotating drum) in the echidna (Gates 1973).

The role of vision for both the platypus and echidna in the wild is probably primarily for predator detection and avoidance, but direct scientific observations of this behaviour are limited. The flat cornea and protruding, laterally placed eye of the short-beaked echidna give it a panoramic view of the world unobstructed by any facial anatomy (Gates 1973).

Anyone who has tried to sneak up on an echidna in the wild knows that the animal is quick to visually detect even slight movement in its peripheral field and rapidly adopt defensive burrowing and body curling.

The monotreme eye and extraocular muscles

The eye of the platypus is spherical and very small (around 6 mm in diameter; see Table 8.1). The corneal curvature is relatively flat, whereas the lens curvature is rather different on the anterior and posterior surfaces. The anterior surface is flat, but the posterior surface is steeply curved, a feature that is reminiscent of the eyes of aquatic therians (Pettigrew *et al.* 1998). The lens is relatively thin (ratio of diameter to thickness of 1.38–1.50) and is suspended by zonular fibres arising in the ciliary body and a ciliary web (Walls 1942). The sclera surrounding the eye consists of a cartilaginous cup that is 400 μm thick behind the visual axis, but tapers to 25 μm thick towards its margin (Gunn 1884; Griffiths 1968). Extrinsic muscles of similar morphology to other mammals are inserted into the sclera and have the same skeletal attachments as in therians. The superior oblique has the same pulley arrangement characteristic of all other mammals, but the loop of the pulley system may be chondroid (i.e. at least partly cartilaginous) in the platypus, rather than fibrous as in other mammals, including the echidna. Lateral and medial palpebral muscles retract the upper eyelid, whereas the inferior palpebral muscle moves the lower lid (Saban 1969). The nictitating membrane is retracted by its own retractor membranae nictitantis. The highly vascular, pigmented choroid is 50 μm thick and is the sole source of nutrients to the overlying retina. The ciliary muscle is said to be absent and there is no dilator pupillae muscle, but a constrictor pupillae has been identified (Walls 1942). Taken together these observations suggest that the platypus is incapable of accommodation, but this has not been adequately tested by experiment.

The eye of the short-beaked echidna is 9 mm in diameter after fixation. As for the platypus, there is a complete cartilaginous cup between the collagenous sclera and the choroid layer (Gresser and Noback 1935; O'Day 1952; Fig. 8.1a), but this cartilage is much

Table 8.1. Structure of the eye and retina in platypus and short-beaked echidna

	Platypus	Short-beaked echidna
Eye diameter	6 to 8 mm[15]	8 to 9.5 mm[2-5]
Pupil size	1 mm[6]	? 2 mm
Anterior segment	Smooth muscle in pupillary sphincter[2,3] No dilator pupillae or ciliary muscle[2-5]	Smooth muscle in pupillary sphincter[2,3] No dilator pupillae or ciliary muscle[2-5]
Cartilage in sclera	Yes[4-6]	Yes[4,5]
Retinal thickness	175 µm[6]	70 to 80 µm[1,12,14]
Tapetum	Absent[6]	Absent[11]
Retinal topography	Area centralis and weak visual streak[10]	No fovea or area centralis, but a suggestion of a visual streak[12]
Photoreceptors	Rod dominated[9] LWS and SWS cones[15] Some double cones[9] Cone oil droplets[9]	Rod dominated[13] Some double cones[13] No oil droplets?[7,11,13]
Vascularity	Confined to optic nerve and choriocapillary layer[15]	Confined to optic nerve and choriocapillary layer[8,11]

1 Dreher et al. (1992); 2 Gates (1973) for review; 3 Gates (1978) for review; 4 Griffiths (1968) for review; 5 Griffiths (1978) for review; 6 Gunn (1884); 7 Locket (1985); 8 McMenamin (2007); 9 O'Day (1938); 10 Pettigrew et al. (1998); 11 Schwab and McMenamin (2005); 12 Stone (1983); 13 Young and Pettigrew (1991); 14 Young and Vaney (1990); 15 Zeiss et al. (2011).

thinner than in the platypus (only 27 µm behind the visual axis and 14 µm near the anterior lip). Also unlike the platypus, there is no nictitating membrane and only the lower lid has a tarsal plate for reinforcement. The cornea of the echidna is also sclerified, a feature that may be a protective adaptation to the formic acid defences of the ants that make up the echidna's diet (Griffiths 1978). The lens is very flat (ratio of diameter to thickness of 2.75; Walls 1942). Early reports claimed that the echidna does not have the necessary intraocular muscles required for accommodation (O'Day 1952), but Gates (1973, 1978) found that the echidna eye was indeed capable of accommodation and can change the refractive power of its eye by as much as 3.5–4.0 dioptres during recovery from ophthalamoscopic examination. Certainly the oculomotor and Edinger-Westphal nuclei in the midbrain that control the extraocular and intraocular muscles, respectively, are substantial in both the echidna and platypus, suggesting the necessary neural control is present. The extraocular muscles of the echidna include four rectus muscles, superior and inferior obliques and a retractor bulbi muscle (see Gates 1973 for review).

Retinal structure and function

The retina is the neural component of the vertebrate eye, responsible for the transduction of information contained in the impact of photons on photoreceptor outer segments into action potentials in the axons of the retinal ganglion cells. Figures 8.1b and c illustrate the cross-sectional structure of the mammalian retina, using peripheral primate retina as an example. The retina consists of a thin outer non-sensory pigmented epithelium and an inner sensory retina. Photons enter the retina from the vitreous (inner) surface and induce photochemical changes in the outer segments of the photoreceptors; these changes are converted to electrical signals at the axon terminals of the photoreceptors in the outer plexiform layer. This information is processed and transmitted by the bipolar and retinal ganglion cells before being transferred out of the retina by retinal ganglion cell axons that collect at the optic disk to form the optic nerve. The nuclei of photoreceptors are collected together in the outer nuclear layer, whereas nuclei of the bipolar cells and other significant retinal neurons (horizontal and amacrine cells) are aggregated into the inner nuclear layer. Axons of photoreceptors project into the outer plexiform layer and

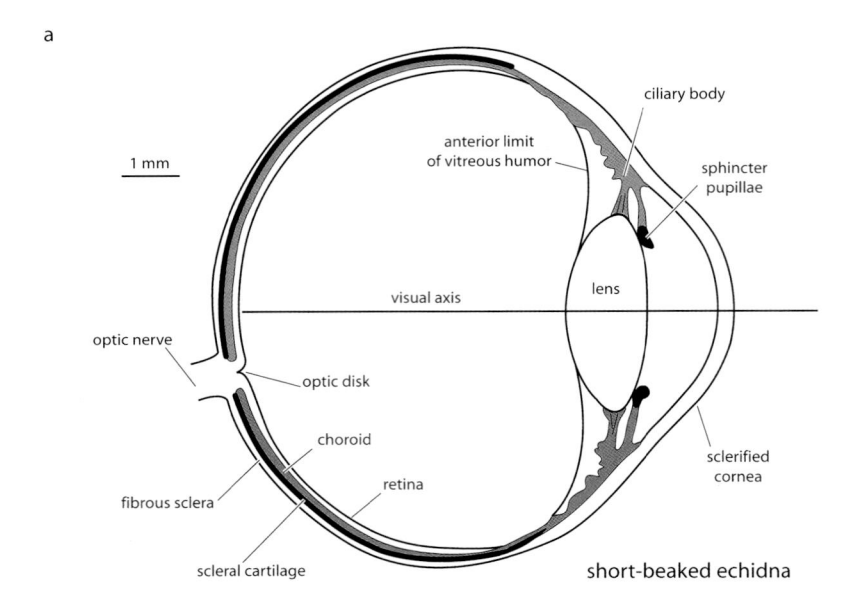

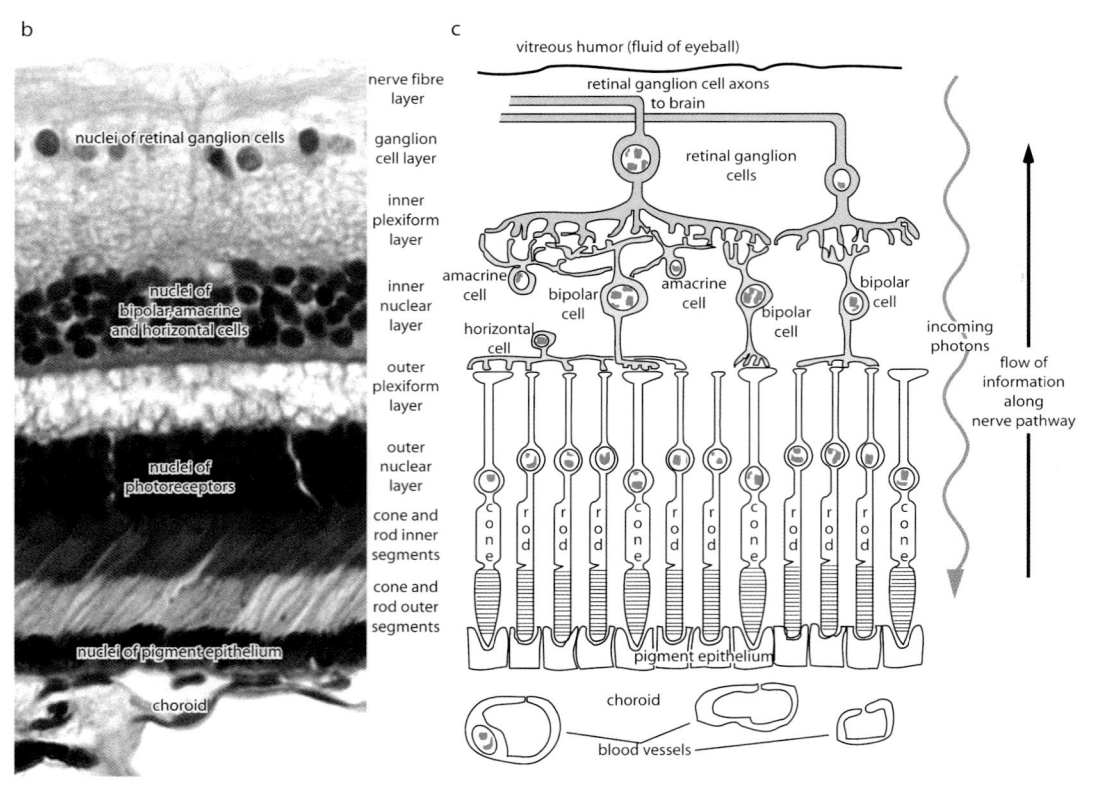

Figure 8.1: Structure of the monotreme eye as exemplified by a cross-section through the eye of the short-beaked echidna (a), based on an illustration by Walls (1942). Note the presence of the scleral cartilage, sclerified cornea and the relatively flattened lens. The layered structure of a typical therian retina is shown in (b) and (c) by a transverse section through the periphery of a primate retina. Photons pass through the retinal layers from above before striking the outer segments of the photoreceptors. The information in the photons is transduced to electrical signals by the photoreceptors and passed with processing through the sequence of retinal neurons (bipolar neurons to retinal ganglion cells) with assistance from horizontal and amacrine cells, before being carried out of the retina by retinal ganglion cell axons in the optic nerve.

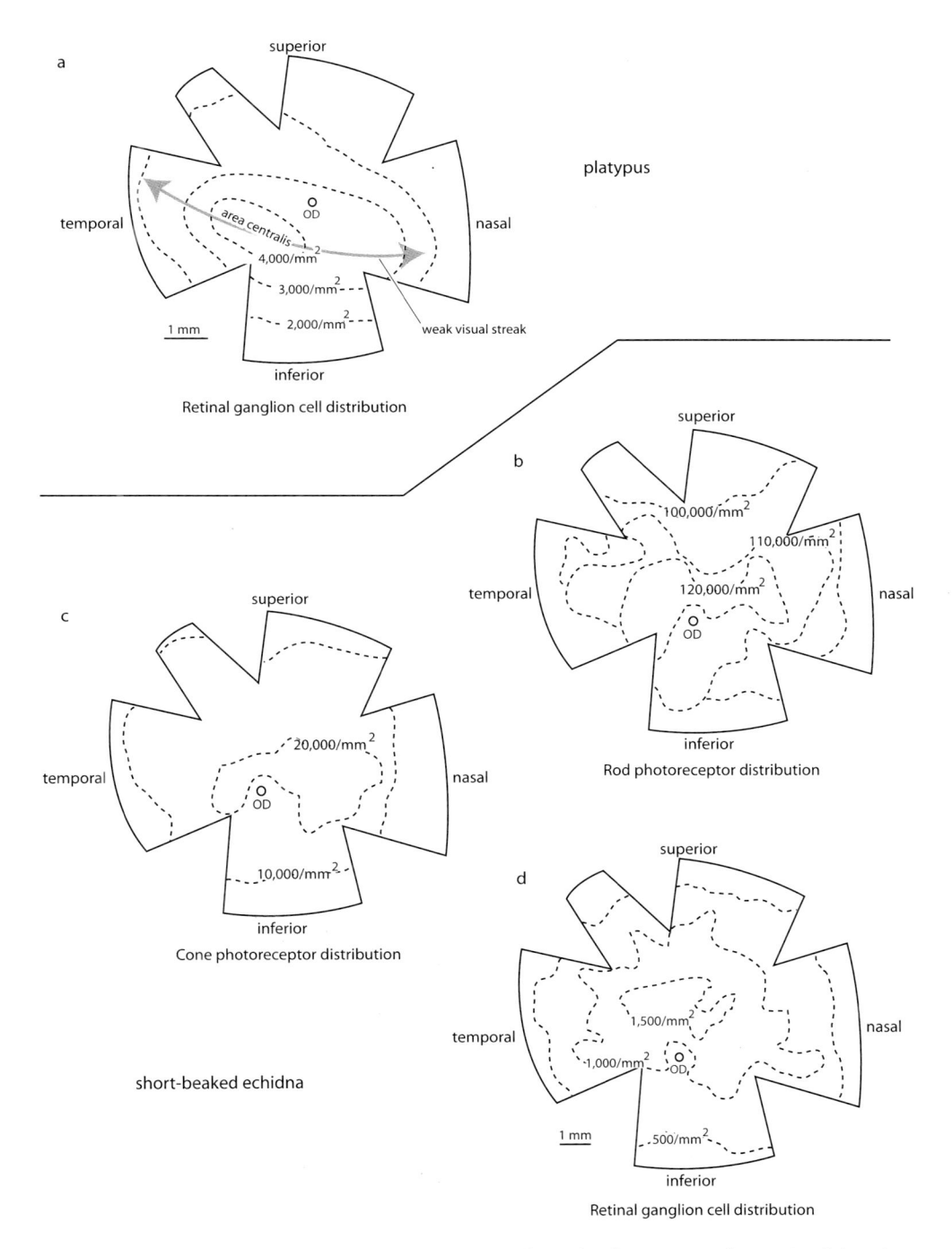

Figure 8.2: Distribution of photoreceptors and retinal ganglion cells in the flat-mounted retinas of the platypus (a) and short-beaked echidna (b and c). Map (a) is a schematic representation of retinal ganglion cell density based on an illustration in Pettigrew et al. (1998). Note the slight concentration of retinal ganglion cells in an area centralis and a weak visual streak inferior and temporal to the optic disk. Maps (b) and (c) show the distribution of rods and cone photoreceptors in representative flattened retinas of the short-beaked echidna, based on data in Young and Pettigrew (1991). Map (d) shows the distribution of retinal ganglion cells in the retina of the short-beaked echidna based on maps in Stone (1983). Note that retinal ganglion density is lower overall in the retina of the echidna than in the platypus. There is a relatively low centre to periphery gradient in the echidna retina of ~3:1 with the highest concentration immediately above the optic disk.

synapse with dendrites of the bipolar cells, whereas the axons of bipolar cells synapse with dendrites of the retinal ganglion cells in the inner plexiform layer. In many therian taxa (e.g. primates and carnivores), the retina has specialised regions of high ganglion cell density (area centralis, fovea and/or visual streak) and minimal convergence in retinal circuitry from photoreceptor to ganglion cell as an adaptation for high visual acuity. Some degree of topographic specialisation is also present in the retinas of the platypus and short-beaked echidna (Fig. 8.2).

The retina of the platypus has an internal area of ~40 mm^2 based on flattened maps (Pettigrew et al. 1998; Fig. 8.2a) and the retina of the short-beaked echidna has an internal area of 70 mm^2 (Stone 1983; Fig. 8.2b–d). When examined by ophthalmoscope, the retina of the echidna has a uniform lavender colour with no visible blood vessels (Gates 1973). Nutrient supply to the avascular monotreme retina is exclusively by diffusion from the choroid layer, a region that lies external to the pigment epithelium. Unlike the primate retina, there is no vascular network either on the vitreal surface of the retina or within the monotreme retina itself (McMenamin 2007). Interestingly, McMenamin (2007) noted several large calibre vessels with irregular cross-section in the choroid of the echidna that may be lymphatics, a type of vessel not previously described in the mammalian eye. The role of these remains unknown.

The monotreme retina has a complement of radial glia (the Müller cells) like all other mammals. Morphologically and immunochemically these are so similar to those in other mammals with avascular retinas as to be unremarkable (Dreher et al. 1992). Müller cells of the short-beaked echidna span the entire thickness of the retina (~70 μm) and have a diameter at the Müller cell body of 4–7 μm, slightly smaller than the Müller cells of those placentals that have been studied (e.g. guinea-pig – 5–10 μm, rabbit – 7–10 μm, domestic cat – 7–12 μm; Dreher et al. 1992). Diameter of the Müller cell trunk and endfeet also tends to be at the lower end of the range for mammals that have been studied (Dreher et al. 1992), but not so low as to invite speculation on functional differences.

The retinas of the platypus and echidna have the same pattern of nuclear layers described above (O'Day 1938; Locket 1985; Stone 1983; McMenamin 2007), but there are some important differences in neuronal populations of the outer and inner nuclear layers. The number of rows of nuclei in the outer nuclear, inner nuclear and retinal ganglion cell layers of the platypus retina are 3, 4 and 1, respectively (Griffiths 1978; Zeiss et al. 2011). Notably, the outer nuclear layer of the short-beaked echidna contains many fewer nuclei of photoreceptors than would be seen in a comparable region of a primate or carnivore retina (Stone 1983), suggesting a much lower spatial density of photoreceptors. The platypus retina has a rod and cone retina, with many cones being paired, with oil droplets present only in the chief cone (O'Day 1938, 1952; Zeiss et al. 2011). The short-beaked echidna was originally reported to have an all-rod retina (O'Day 1952), but Locket (1985) and Young and Pettigrew (1991) identified a cone-like photoreceptor that makes up 10–15% of photoreceptors (Fig. 8.2b, c). The cones identified by Young and Pettigrew in the echidna retina do not have oil droplets (unlike cones in the platypus retina), are often grouped as pairs, and are distributed at densities of 9000/mm^2 in the superior periphery to 22 000/mm^2 in the central retina. Note that Schwab and McMenamin (2005) claimed that oil droplets are present in cones of the echidna retina, but this statement is not backed by evidence. The density of rods in the retina of the short-beaked echidna ranges from 85 000/mm^2 in the superior nasal retina to 130 000/mm^2 in central retina. Nearest-neighbour analysis of cone distribution in the retina of the short-beaked echidna reveals a regular array of single and paired cones separated by an average distance of ~6.5 μm (Young and Pettigrew 1991).

Several cellular features of the bipolar and amacrine populations of the monotreme retina are unusual; in fact in several respects the inner nuclear layer neurons of the monotreme retina are more like those of non-mammalian tetrapods than those of theria. Therian retinas have a clear segregation of rod and cone pathways, with the rods serving scotopic (low light levels) and the cones serving photopic (high light levels) vision. In therians, rod photoreceptors synapse onto a single type of rod bipolar cell whose axons terminate in the inner plexiform layer, whereas cone photoreceptors synapse onto multiple types of cone bipolars, whose axons branch within the outer or middle strata of the inner plexiform layer.

In the retina of the short-beaked echidna (Young and Vaney 1990) bipolar cells exhibiting protein kinase C-like immunoreactivity (a marker for rod bipolar cells) have a very different morphology from those in marsupials and placentals. PKC-immunoreactive cells in the echidna are distributed at all levels of the inner nuclear layer, but mostly in the outermost part. Each of these cells has a single stout process that passes externally to give rise to several fine dendritic processes in the outer plexiform layer, with each of these processes ending in a club-like structure (Landolt's club) forming a knob 2.5 μm in diameter amongst the rod inner segments (Locket 1985; Young and Vaney 1990). Landolt's clubs have been identified in a variety of vertebrates (Selachii, Chondrostei, Dipnoi, Anura, Urodela, Squamata, Chelonia and Neognathae), but not in marsupial or placental mammals, nor in teleost fish. Young and Vaney also noted that the multistratified axonal pattern of PKC-immunoreactive bipolar cells in echidna retina is very different from the tightly stratified pattern seen in other mammalian retina, but similar to that of the domestic chicken retina, suggesting that the neuronal pathway that serves scotopic vision in the monotremes follows a sauropsid pattern rather than the recognised therian organisation. This in turn suggests that the segregated rod and cone pathways seen in Metatheria and Eutheria evolved after the divergence of the Prototherian line from the Therian line.

Another unusual feature of the monotreme retina is the presence of endogenous serotonin-like immunoreactivity in presumptive amacrine cells of the platypus retina (Young and Vaney 1990). These cells with serotonin-like immunoreactivity are 8–10 μm in diameter and are situated in the amacrine (inner) sublayer of the inner nuclear layer. Although serotonergic axons have been identified in the inner and outer plexiform layers of the retina of a microchiropteran bat, serotonergic cell bodies have never been found in the retinas of placentals or marsupials.

The retinal ganglion cell layer of the retina contains the cell bodies of the neurons that project to the central visual centres through the optic nerve. Pettigrew and colleagues (Pettigrew et al. 1998) identified a region of possible specialisation for acuity vision (elevated spatial density of retinal ganglion cells in the form of a roughly circular area centralis) in the temporal (lateral) side of the inferior retina of the platypus, with a suggestion of a weakly accentuated horizontal retinal or visual streak of slightly higher retinal ganglion cell spatial density passing inferior to the optic disk (Fig. 8.2a). Given the reversal of the visual field by the lens, the area centralis looks forward into binocular visual space at the tip of the bill, consistent with a possible role of this region in visually guided prey capture in the ancestral platypus. The area centralis has a concentration of retinal ganglion cells reaching 4000 per mm^2, not a high value compared to even the poorly visual mouse retina, where retinal ganglion cell spatial density may reach 8000 per mm^2 (Drager and Olsen 1981). Retinal ganglion cell density at the periphery of the platypus retina (i.e. superior and nasal retina) is a little over 1000 per mm^2, giving a centre to periphery gradient of ~3.5:1. Pettigrew and colleagues estimate that visual acuity with this peak spatial density of retinal ganglion cells would be 2 cycles per degree, comparable to the laboratory rat and sufficient to detect crustacean prey at a distance of 1 m in adequate illumination.

The retinal ganglion cell layer of the short-beaked echidna contains ~60 000 ganglion cells (Stone 1983; Fig. 8.2d). Unlike therians, the cell bodies of retinal ganglion cells in the short-beaked echidna do not exhibit size groupings; there is only a weak gradient in the spatial density of retinal ganglion cells between centre and periphery of the retina (~3 to 1), and there is no strong central concentration of retinal ganglion cells. Nevertheless, Stone (1983) reported weak evidence of a visual streak region of higher retinal ganglion cell density, between 7 and 8 mm wide, running across the retina 1 to 2 mm above the optic disk. The pattern of axon bundles in the innermost fibre layer of the echidna retina is not radially symmetrical, but elongated in the superotemporal to inferonasal axis (Stone 1983), consistent with the presence of a weak visual streak. There is no area specialised for high acuity vision (i.e. an area centralis) in the retina of the short-beaked echidna (Gates 1973, 1978; Stone 1983) but the weak visual streak may be of behavioural significance in detecting motion of potential predators along the visual horizon.

Molecular biology and genetics of photoreceptors

Colour vision in vertebrates relies on the differential expression of photoreceptor pigments in different classes of cone cells. All vertebrate photoreceptor pigments are based on an opsin protein that is covalently linked via a Schiff base to the chromophore molecule, retinal. Visual pigments are subdivided into a rod or Rh1 class that is generally restricted to rod photoreceptors and four different cone classes distinguished on the basis of the spectral sensitivity (λ_{max}) and the amino acid sequence of their respective opsins. Cone opsins include longwave-sensitive (LWS) with λ_{max} of 500–570 nm, midwave-sensitive (MWS or Rh2) with λ_{max} of 480–530 nm, and two shortwave-sensitive classes, SWS2 with λ_{max} of 400–470 nm and SWS1 with λ_{max} of 355–445 nm (Beazley et al. 2010). In placentals, there is a locus control region located upstream of the LWS pigment gene on the X chromosome that controls the expression of different pigments (Wakefield et al. 2008).

Wakefield and colleagues (Wakefield et al. 2008) consider that the structure and expression of monotreme cone opsin genes bridge the structural and functional gap between reptiles and therians. The LWS opsins of monotremes are possibly homologous to the LWS and MWS opsins of therians, but monotremes have an SWS2 opsin gene, similar to that of sauropsids and fish, but quite unlike the SWS1 opsin gene of therians. They suggest that most of the SWS1 gene has been lost in monotremes (Davies et al. 2007), with the exception of a pseudogene made of exon 5, and they argue that the continued presence of exon 5 favours the conclusion that this loss is a relatively recent event. However, given that there are no conclusive data on the rate of genetic change in monotremes, this deduction is open to question.

The platypus and echidna are dichromats, meaning that they rely on comparing the input from two types of cone photoreceptors to determine colours. The spectral peaks of the corresponding SWS2 and LWS pigments of the platypus are at 451 nm and 550 nm, respectively. The LWS pigment peak at 550 nm is similar to that found for LWS pigments in many marsupial and placental mammals and the SWS2 peak at 452 nm is at only a marginally longer wavelength than the violet-sensitive SWS1 pigments of some placental mammals (Hunt et al. 2007).

If we accept that the SWS1 opsin gene has been deleted during the course of monotreme evolution, then the common mammalian ancestor probably had an opsin complement that consisted of tandemly arranged SWS2 and M/LWS genes with a central LCR gene and an SWS1 gene (Wakefield et al. 2008). A duplication event of an ancestral cone visual pigment, followed by sequence divergence and selection might have given rise to the LWS and SWS2 visual pigments in modern monotremes, flanking the LCR (Wakefield et al. 2008).

By contrast, placentals have lost the SWS2 and midwave-sensitive (Rh2) classes of opsin genes and retain only the longwave-sensitive (LWS) and shortwave-sensitive (SWS1) genes. Most placentals are dichromats, with red–green colour blindness, with the exception of primates who recovered full trichromacy by a duplication of the LWS gene (Davies et al. 2007).

Optic nerve

The optic nerve of the platypus has been estimated to contain around 32 000 axons (31 300 and 32 700 in two animals; Bruesch and Arey 1942). This is somewhat less than the estimate of the number of retinal ganglion cells (40 000) obtained by Pettigrew and colleagues (Pettigrew et al. 1998), perhaps due to technical issues with the identification of smaller and/or unmyelinated axons by Bruesch and Arey (1942).

The optic nerve of the short-beaked echidna is much like that in other vertebrates in that the nerve is divided into multiple fascicles of closely packed myelinated fibres separated by connective tissue septa (Gates 1973). The cross-sectional area of dehydrated optic nerves is around 0.27 mm^2 and Gates estimated that there were 14 709 and 14 950 myelinated axons in the optic nerve in two animals (but note that Schuster 1910 obtained much higher estimates for two animals: 28 585 and 46 566). The peak optic fibre diameter is in the 1.8–3.3 µm range (comprising ~50% of all myelinated axons). Gates (1978) estimated that only ~150 axons of the optic nerve are distributed to the ipsilateral side of the brain, so the echidna has only a negligible capacity for binocular vision.

Why the echidna should have so few optic nerve axons compared to the platypus is unclear. Technical differences in the preparation of nerves and analysis may explain some of the reported difference. The higher estimates obtained by Schuster (1910) sit more comfortably with expectations based on the retinal ganglion cell population. Certainly, the discrepancy between the estimated number of retinal ganglion cells and optic nerve axons suggests that the nerve counts for both species are an underestimation. The important point is that both monotremes have only modest numbers of myelinated optic nerve axons, even fewer than many strains of the much smaller domestic mouse (32 000–87 000 depending on strain; see Watson 2012b for review). On the other hand, the optic nerve axons in the echidna are of much larger diameter than in the mouse, which has a peak fibre diameter around 0.8 µm.

Central visual pathways

Information from the retina is carried in retinal ganglion cell axons along the optic nerve to the optic chiasm, where a proportion of the axons will remain on the same side (ipsilateral projection) while the remainder will cross to the opposite side of the body (contralateral projection). After the optic chiasm, retinal ganglion cell axons continue in the two optic tracts. Our knowledge of which nuclei in the monotreme brain receive retinal input is confined to the findings of silver degeneration studies (Campbell and Hayhow 1971, 1972). In the case of the monotremes, who have little binocular vision, the proportion of uncrossed axons is very low (~1% or less) and is confined to fibres terminating in the ipsilateral lateral geniculate and pregeniculate nuclei of the dorsal and ventral thalamus, respectively (LGNb and LGNa of Campbell and Hayhow 1971, 1972). All other retinorecipient nuclei receive exclusively contralateral projections. Both the platypus and the short-beaked echidna have primary optic pathways projecting to the lateral geniculate nucleus of the dorsal thalamus, the pregeniculate nucleus of the ventral or prethalamus, the superior colliculus and the pretectal area, but there are some significant differences between the two monotremes in the pattern of termination of retinal projections. In the short-beaked echidna, there is a contralateral projection to a medial terminal nucleus at the boundary between the midbrain and diencephalon, four cell groups within the optic tract itself, and a nucleus within the brachium of the superior colliculus. This latter nucleus may be homologous to the placental dorsal terminal nucleus or be a displaced nucleus of the optic tract.

Although the platypus has a retinal projection to the medial terminal nucleus, no retinorecipient nuclei were found within the optic tract of the platypus and there was no putative dorsal terminal nucleus (Campbell and Hayhow 1972).

The arrangement of visual nuclei in the diencephalon of the platypus and echidna is both distinct from therians and rather different from each other. In the rostral diencephalon of the platypus, retinal axons terminate in nuclei that Campbell and Hayhow (1972) designated as lateral geniculate nuclei 'a' and 'b' (LGNa, LGNb). The nucleus they called LGNa is located on the caudal and dorsal surface of what they thought was the dorsal thalamus, but probably represents an extension of the prethalamus around the caudal extremity of the thalamus. This nucleus is adjacent to the superior colliculus and pretectal nuclei (see below) and is homologous to the pregeniculate nucleus (PrG) of therians (Mikula et al. 2008). Campbell and Hayhow (1972) divided this nucleus into medial and lateral subnuclei on the basis of an intervening axon layer, with larger neurons in the medial division. These subdivisions may correspond to the magno- and parvicellular divisions of the PrG in some therians. The true lateral geniculate nucleus of the dorsal thalamus is a small retinorecipient nucleus over the rostral part of the dorsal thalamus and was designated LGNb by Campbell and Hayhow (1972). Retinal projections to LGNb (more properly called the LG) appear to be from both eyes, but with the contralateral projection being by far the largest.

In the echidna diencephalon, retinal ganglion cell axons also terminate in two nuclei that have been identified by authors in the past as 'lateral geniculate' nuclei. The LGNa and LGNb of Campbell and Hayhow correspond to the LGa and LGb, respectively, of Ashwell and Paxinos (2005) and the PrG and LG of the present work. The PrG receives the strongest termination of optic axons and corresponds to the gen. lat. vent. nuc. of Abbie (1934). The LG proper nucleus lies on the dorsolateral margin of the caudal dorsal thalamus and is the only one of these two supposedly visual thalamic nuclei to undergo degeneration following ablation of the primary visual cortex (Welker and Lende 1980; see Chapter 6 and below). It is therefore the likely homologue of the therian dorsal lateral geniculate nucleus of the dorsal (or true) thalamus (Ashwell and Paxinos 2005; Mikula et al. 2008).

The LG of the echidna is substantially larger than the homologous nucleus in the platypus, reflecting the slightly larger visual cortex of the echidna, but contrast this with the supposedly fewer optic nerve axons in the echidna (see above). The chemical profile of the PrG (high proportion of parvalbumin and calretinin immunoreactive neurons and low proportion of calbindin immunoreactive neurons) is similar to the pattern seen in another retinorecipient ventral thalamic nucleus in rodents which does not project to visual cortex, i.e. the magnocellular part of the ventral lateral geniculate nucleus, now known as the pregeniculate nucleus (Paxinos *et al.* 2009), raising the possibility that the PrG of the echidna is homologous to the magnocellular part of the PrG of placentals (Ashwell and Paxinos 2005). The PrG of the echidna has a complex internal structure with dense clusters of small neurons and a complex variegated pattern of strongly acetylcholinesterase reactive blobs against a moderately reactive background. Cytochrome oxidase activity is also irregular, with strongest reactivity towards the ventral surface of the nucleus, where clusters of parvalbumin immunoreactive cell bodies are concentrated (Ashwell and Paxinos 2005). Mikula and colleagues (Mikula *et al.* 2008) noted that the LG and PrG are in continuity around the caudal rim of the thalamus in the echidna, but quite separate in the platypus. They suggested that the separation of these two retinorecipient nuclei in the platypus is due to the extraordinary expansion of the ventral posterior thalamic nucleus.

The pretectal nuclei are a group of retinorecipient nuclei in that region of the neuraxis derived from prosomere 1 (see Chapter 6). They are situated between the caudal thalamus and the midbrain, receive input predominantly from the retina and optic tectum, and are important for modifying motor behaviour in response to visual information (Watson 2012b). In placental mammals, four or five pretectal nuclei are generally recognised: the nucleus of the optic tract; the olivary pretectal nucleus; and the anterior pretectal, posterior pretectal nucleus (now grouped outside the pretectum and called the tectal grey; see Chapter 6) and medial pretectal nuclei (Butler and Hodos 2005). Earlier reports suggested that the pretectal region in monotremes was relatively undifferentiated (Hines 1929; Abbie 1934; Butler and Hodos 2005) and may even have been reduced in size during evolution, but subsequent chemoarchitectural studies have provided a clearer picture of the region's organisation in monotremes. Ashwell and Paxinos (2007) identified distinct anterior, medial, posterior (tectal grey) and olivary pretectal nuclei as well as a nucleus of the optic tract, all with largely similar topographical and chemoarchitectonic features to the homologous regions in therian mammals. The positions of these pretectal nuclei correspond to the distributions of retinofugal terminals identified by Campbell and Hayhow (1971, 1972). The overall size of the pretectum in both monotremes is at least comparable in size, if not larger than, the pretectum of representative therian mammals of similar brain and body size and is more than just an undifferentiated area pretectalis (Ashwell and Paxinos 2007). The presence of a differentiated pretectum with similar chemoarchitecture to therians in both living monotremes lends support to the idea that the stem mammal for both prototherian and therian lineages also had a functionally complex, structurally differentiated pretectum.

The superior colliculus of the midbrain is a more significant target for retinal ganglion cell axons in monotremes than the dorsal thalamus or prethalamus (Campbell and Hayhow 1971, 1972). The superior colliculus of the platypus has a large plateau-like dorsal surface (see atlas plates Pl-Ad19 and 20; Hines 1929). Hines identified eight layers in the superior colliculus of the platypus (in order from superficial to deep: zonal, superficial grey, superficial medullary or white, intermediate grey, intermediate medullary or white, deep grey, deep medullary or white, and periependymal grey). Axons of the brachium of the superior colliculus branch off the optic tract and run medially across the rostral superior colliculus before turning caudally to form a superficial fibrous layer (stratum zonale) over the superior colliculus (Campbell and Hayhow 1972). Beneath the zonal layer, retinal axons penetrate and terminate in the underlying superficial grey layer (stratum griseum superficiale or superficial grey layer of Hines 1929; SuG). This pattern of passage of retinal axons across the dorsal surface of the superior colliculus in the zonal layer stands in contrast to many therian mammals, where retinal axons traverse the superior colliculus in a deeper stratum opticum before terminating in the *overlying* superficial grey layer (Campbell and Hayhow 1972;

Johnson *et al.* 1994). Hines (1929) also identified a tectospinal tract in the platypus, arising from the sixth layer of the superior colliculus before decussating ventral to the oculomotor nucleus, and dorsal and ventral tectocerebellar pathways from the third layer. The role of the tectospinal and tectocerebellar tracts in either the platypus or echidna remains unknown, but they are recognised as important pathways for coordination of visually guided prey capture in therians (see Chapter 5).

In the short-beaked echidna (Campbell and Hayhow 1971), retinal axons in the brachium of the superior colliculus form a distinct bundle that courses across the surface of the superior colliculus in a stratum zonale as described for the platypus. The layers of the echidna superior colliculus correspond to those in the platypus. Axons progressively leave the superficial layer to terminate in the *underlying* superficial grey layer. As noted for the platypus, this pattern of passage and termination is unlike that in the therian superior colliculus.

Visual cortex

The visual cortex in both platypus and echidna occupies a relatively small proportion of cortical area, but is slightly larger in the echidna. The visual cortex of the platypus is very small and situated at the dorsomedial edge of the caudal cortex, medial to the auditory cortex (Bohringer and Rowe 1977; Krubitzer 1998). In the short-beaked echidna, the visual cortex is sandwiched between the α and ζ sulci and is somewhat larger than the auditory cortex.

Krubitzer (1998) identified at least two representations of the visual field in the cortex of the platypus. One representation located medial to the primary somatosensory cortex was denoted Vc (caudal visual cortex), the other situated rostral and medial to primary somatosensory cortex was denoted Vr (rostral visual cortex). The line between the two regions corresponds to the central area of the visual field, but Krubitzer was unable to obtain a precise topographic map for either region due to discontinuities in the visual field representation. Vc was reported to contain neurons that respond vigorously to visual stimuli and have receptive fields ranging in size from 15° to 20°. Neurons in Vr tend to have larger receptive fields and respond less vigorously to visual stimuli than those in Vc. Krubitzer has argued that Vc is homologous to the primary visual cortex of therians, based on the receptive field size, location and vigour of neuronal responses to visual stimuli.

One puzzle noted by Pettigrew and colleagues concerns the visual magnification factor of the cortical visual field representation in the platypus (Pettigrew *et al.* 1998). Visual magnification in the visual cortex is expressed as the number of millimetres of cortical distance subtended by a given angular dimension in the visual field. The smaller the value is, the smaller is the representation of the visual field on the cortex. For the platypus this is a mere 0.03 mm per degree, comparable to the laboratory mouse and hedgehog and much smaller than the tammar wallaby (0.4 mm per degree) or small primates like the marmoset (2.0 mm per degree). The explanation may lie in the relative importance of midbrain (i.e. retinotectal or -collicular) versus cortical (i.e. retinogeniculocortical) processing of visual information in the platypus. Therians with predominantly retinotectal visual processing (e.g. rodents and lagomorphs) have lower cortical magnification factors than those with predominantly cortical visual processing (e.g. primates and carnivores). Nevertheless, the cortical magnification factor in the platypus is so low that it is at the lower limit for all mammals studied and no more than in the tiny mouse brain! This discrepancy between cortical and potential retinal acuity might be explained if the peripheral components of the visual pathway, retinotectal pathway and visual cortex were adapted in the ancestral platypus for visual prey capture, but have now been superseded by trigeminal sensory systems (Pettigrew *et al.* 1998). If the cortical visual areas were competitively downgraded during the expansion of the cortical trigeminal representation, while the subcortical visual structures in the retina and midbrain were retained, then we would arrive at the structure and functional capabilities of the modern platypus visual pathway. This suggests that the ancestral platypus may have been a diurnally active hunter that used visual cues in prey capture. This would also explain the persistence of a significant tectospinal tract (see Chapter 5).

The visual cortex of the platypus is distinguished structurally by a heavily myelinated geniculocortical input and strong cytochrome oxidase activity in

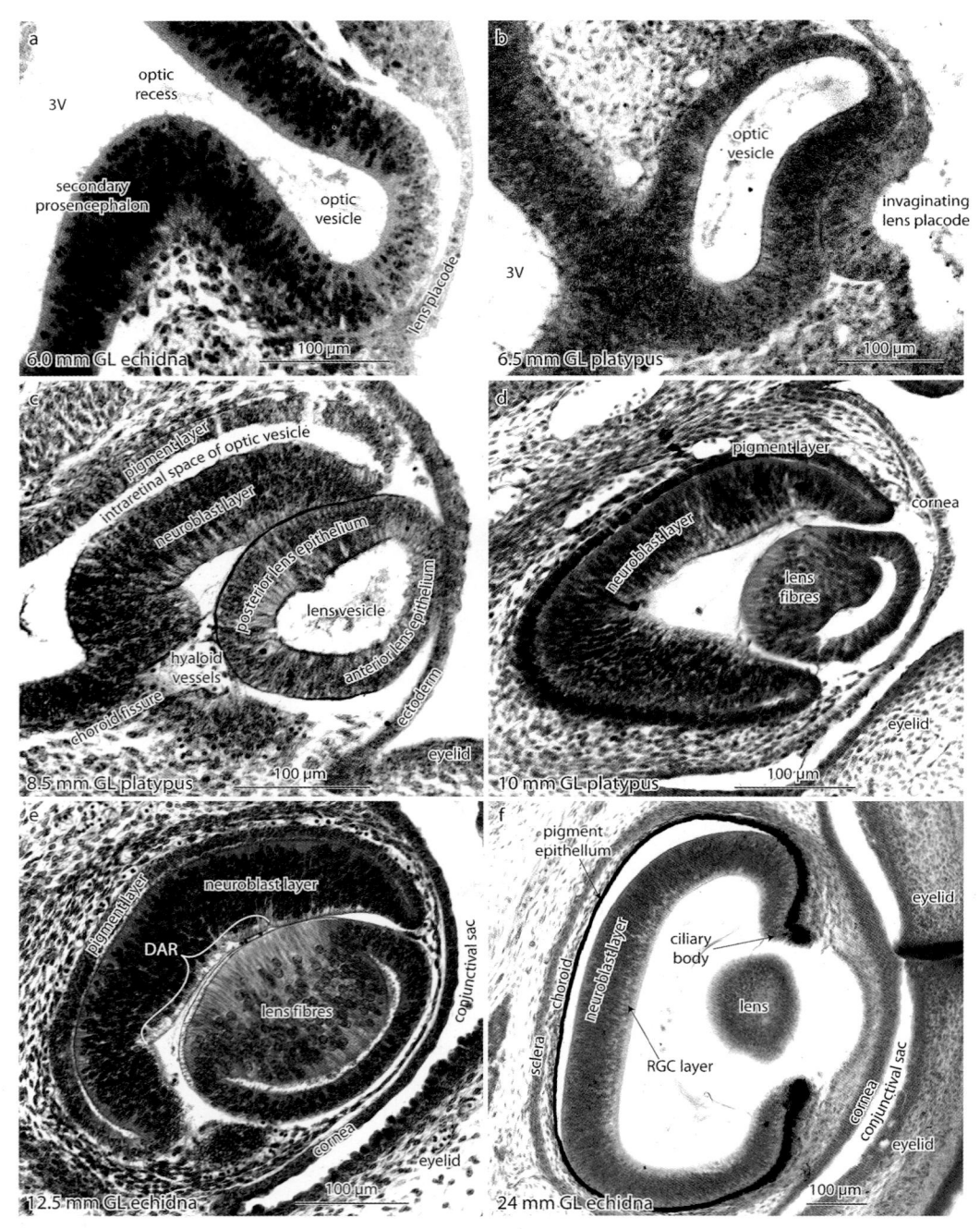

Figure 8.3: Development of the monotreme eye during incubation and the first week of the lactational phase. The eye develops as a lateral outgrowth of the secondary prosencephalon even before the anterior neuropore has closed (a). The optic vesicle outgrowth of the brain invaginates to form an optic cup (b). A lens placode (a) is induced to form from the ectoderm, before invaginating to form a lens vesicle (b, c). The inner layer of the optic cup will form the neural retina (neuroblast layer), whereas the outer layer will form the pigment layer (pigment epithelium of mature retina). Eyelids grow over the developing cornea in the last half of incubation (d, e) and fuse with each other to protect the eye during the early lactational phase (f). A developmentally advanced region (DAR) appears in the retina immediately before hatching (e). This is a site where neurogenesis is slightly advanced relative to the peripheral retina and will produce the poorly defined visual streak of the adult. Pigmentation of the pigment epithelium appears in the first post-hatching week (f). Superior is above in all photomicrographs. The midline is to the left in all photomicrographs.

afferent axons penetrating the granular layers. The caudal visual region was reported to be slightly more heavily myelinated than the rostral region (Krubitzer 1998), but the functional significance of this is unknown.

Krubitzer also found that there were two visual field representations in the short-beaked echidna (Krubitzer 1998), but the physiological evidence for this was not presented. Nevertheless, Hassiotis and colleagues (Hassiotis *et al.* 2004b) identified putative V1 and V2 regions in the cortex of the echidna on the basis of chemoarchitecture (see Chapter 7).

Connections of monotreme visual cortex

Information on the connections of visual cortex in monotremes is confined to studies in the short-beaked echidna. The cortical ablation study of Welker and Lende (1980) showed that removal of the putative visual cortex resulted in degeneration of the posterior pole region of the dorsal thalamus in a region, which both Ashwell and Paxinos (2005) and Jones (2007) interpreted as the LGb. On the basis of this, Ashwell and Paxinos, and Jones (2007) have argued that LGb is homologous to the therian dorsal lateral geniculate nucleus (hence the use of the term 'LG' in this work).

Columnar connections of visual cortex with neighbouring cortex have been found with carbocyanine dye tracing in the short-beaked echidna (Dann and Buhl 1995), much like that seen in therians, but the techniques used by Dann and Buhl do not permit conclusions to be reached about long-distance connections of the visual cortex.

Eye and optic nerve development

The vertebrate eye develops from two sequences during embryonic life: (1) formation of the retina, pigment epithelium and uveal tract; and (2) formation of the lens and cornea. The first of these begins with the evagination of the wall of the secondary prosencephalon to form an optic recess and optic process. In the platypus and echidna this begins at ~5.5 mm GL (Fig. 8.3a), even before closure of the anterior neuropore. The optic process engages in interactions with the surface ectoderm to form the lens (see below) and becomes invaginated itself to form an optic cup (Fig. 8.3b, c) with a choroidal fissure that transmits hyaloid vessels to the interior of the optic cup. The ventricular lumen of the optic cup is progressively reduced in size (Fig. 8.3c) until it is effectively obliterated by 10.0 mm GL (Fig. 8.3d). The inner surface of the optic cup (neural layer) thickens during the incubation phase to become the neuroblast layer that will produce the retina, whereas the outer surface of the cup (pigment layer) will give rise to the pigment epithelium. Shortly before the time of hatching (Fig. 8.3e), the retina has formed a thick neuroepithelium (neuroblast or cytoblast layer), but the pigment epithelium is quite thin. Pigmentation does not appear in the pigment epithelium until the end of the first post-hatching week (Fig. 8.3f).

Development of the lens, cornea and eyelids proceeds in parallel with that of the retina and pigment epithelium. As the optic stalk approaches the surface of the embryo at ~6.0 mm GL, the overlying ectoderm is induced to form a lens placode, which subsequently invaginates to form a lens vesicle (Fig. 8.3b, c). Cells of the posterior segment of the lens vesicle progressively elongate to obliterate the cavity of the vesicle (Fig. 8.3d, e) and form lens fibres. The cornea develops by movement of the ectoderm over the lens vesicle to form a continuous layer. During the first half of incubation the lens and cornea are uncovered, but eyelids form from 8.5 mm GL and completely cover the cornea by the time of hatching (Fig. 8.3f). These will remain sealed together throughout the early part of the lactational phase, not opening until about 6 weeks after hatching (Burrell 1927).

Development of retinal neurons is confined to the post-hatching period (Fig. 8.4, Appendix Fig. 2). Retinal neurogenesis in monotremes appears to follow a very similar sequence of events to those described for placentals (Braekevelt and Hollenberg 1970; Morest 1970; Rapaport *et al.* 1985; Robinson 1987; Robinson *et al.* 1985; Weidman and Kuwabara 1969). Just as in placentals, the late incubation and early lactational phase monotreme retina shows a developmentally advanced region (DAR) at the posterior eyecup (Figs. 8.3e, 8.4a). This region is the site where retinal neurogenesis is more advanced than in the rest of the retina and in some placentals ultimately gives rise to a specialised area in central retina for high acuity vision (e.g. the fovea of domestic cat and primates). In contrast to visually specialised placentals (e.g. carnivores and primates) in whom the DAR is clearly demarcated

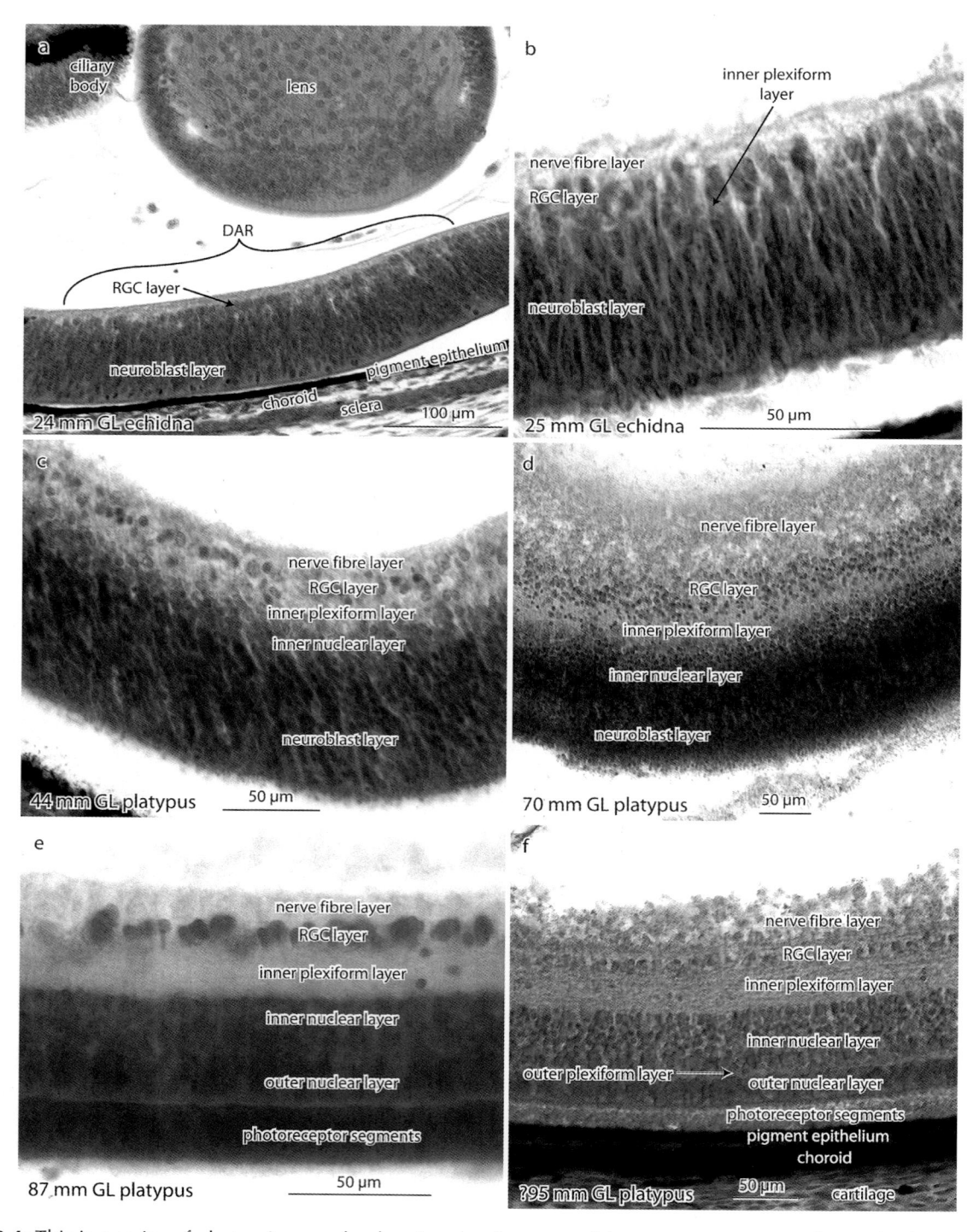

Figure 8.4: This is a series of photomicrographs showing development of the monotreme retina during the lactational phase of development. The area of retina adjacent to the optic disc has a developmentally advanced region (DAR), which is broader than in visually specialised placentals. Retinal neurons are produced in an inside-to-out sequence (as in other mammals), with a retinal ganglion cell layer apparent at the end of the first post-hatching week (a, b) and inner (c, d) and outer (e, f) nuclear layers emerging during the first and second post-hatching months, respectively. Note that photomicrograph (d) is a tangential section through the retina and the layers appear artefactually enlarged. The vitreal surface is uppermost in all images.

during development (Robinson 1987), the monotreme DAR is relatively broad during development (up to 300 μm across at the end of the first post-hatching week, Fig. 8.4a) and not well demarcated, in keeping with the rather poorly differentiated visual streak of the mature monotreme retina (Stone 1983; Robinson 1987). Proliferation of cells in the neuroblast layer progressively produces the layers of the retina in an inside-to-out sequence, much like that seen in placentals during foetal life (carnivores and primates) or foetal to early postnatal life (rodents). The retinal ganglion cell layer is the first to emerge, appearing at the end of the first post-hatching week (Figs. 8.3f, 8.4a, b). This layer becomes progressively more clearly defined as the cell-free space of the inner plexiform layer develops (Fig. 8.4b, c). The inner nuclear layer emerges around the end of the first post-hatching month (Fig. 8.4c, d), followed by the outer nuclear layer towards the end of the second post-hatching month (Fig. 8.4e, f). The cartilaginous cup around the eye develops during the second post-hatching month (Fig. 8.4f).

The optic nerve develops by growth of retinal ganglion cell axons along the optic stalk, but no axons are evident during the incubation phase (Fig. 8.5a). As noted above, retinal ganglion cells are not generated until after hatching, so the first pioneer axons to form the optic nerve are not clearly visible until about the end of the first post-hatching week (Fig. 8.5b). The optic chiasm can be identified at about the same time (Fig. 8.5c), indicating that retinal axon growth across the midline of the forebrain is quite rapid once the retinal ganglion cell axons reach the pial surface of the hypothalamus.

Development of central visual pathways and nuclei

Retinal ganglion cell axons reach the hypothalamus towards the end of the first post-hatching week. Suprachiasmatic nucleus neurons appear to be generated from around the time of hatching (Ashwell 2012e, Appendix Fig. 2), so the settling of these neurons in the anterior hypothalamus is essentially coincidental with the arrival of retinal axons. Nevertheless, it is unlikely that retinal input contributes significantly to the regulation of hypothalamic function at this stage of development, because the eye is still covered by fused eyelids.

The central visual pathways beyond the hypothalamus can be divided into the midbrain collicular and geniculocortical pathways. The midbrain collicular (or tectal) pathway is the major visual projection in monotremes and its development will be considered first. Neurons of the superior colliculus are generated from the neuroepithelium of the rostral mesencephalic neuromere (mes1; see Chapters 3 and 4). In the platypus and echidna, this process begins during the middle of the incubation phase (Fig. 8.5d), when the first postmitotic neurons begin to settle in the mantle layer around the mesencephalic neuroepithelium. Neurogenesis of superior colliculus neurons appears to last ~7–10 days, because the mesencephalic neuroepithelium remains quite thick and mitotically active through the first few post-hatching days (Fig. 8.5e, f), but distinct layers of the superior colliculus have emerged by ~PH5 to PH7 (Fig. 8.5g). Judging by the layers that appear first, the sequence of generation of the layers of the superior colliculus appears to be outside-to-inside (i.e. superficial layers generated before deeper layers), but this is difficult to say with certainty without labelling of postmitotic neurons.

The lateral geniculate nucleus (LGb or just LG) is part of the lateral group of nuclei of the dorsal thalamus. Constituent neurons of the LG are generated in a period from the last third of the incubation phase to the first post-hatching week in both the platypus and echidna. The LG therefore has only a small proportion of its neurons at the time of hatching and does not receive axons of the optic tract until early in the second post-hatching week. The LG is not a large or complex nucleus, even in the adult monotreme brain, so internal structure is not distinguishable until well into the lactational phase of development.

The development of the visual cortex is just as protracted as for the somatosensory cortex and occupies most of the lactational phase of monotreme development. The subventricular zone persists under caudal cortex until PH140, suggesting that production of small neurons for the visual cortex may well continue into the fourth or fifth month after hatching. No data are available concerning the timing of arrival of geniculocortical axons. Nevertheless, the protracted duration of visual cortex development is consistent with the late age of eye opening (~6 post-hatching weeks; Burrell 1927).

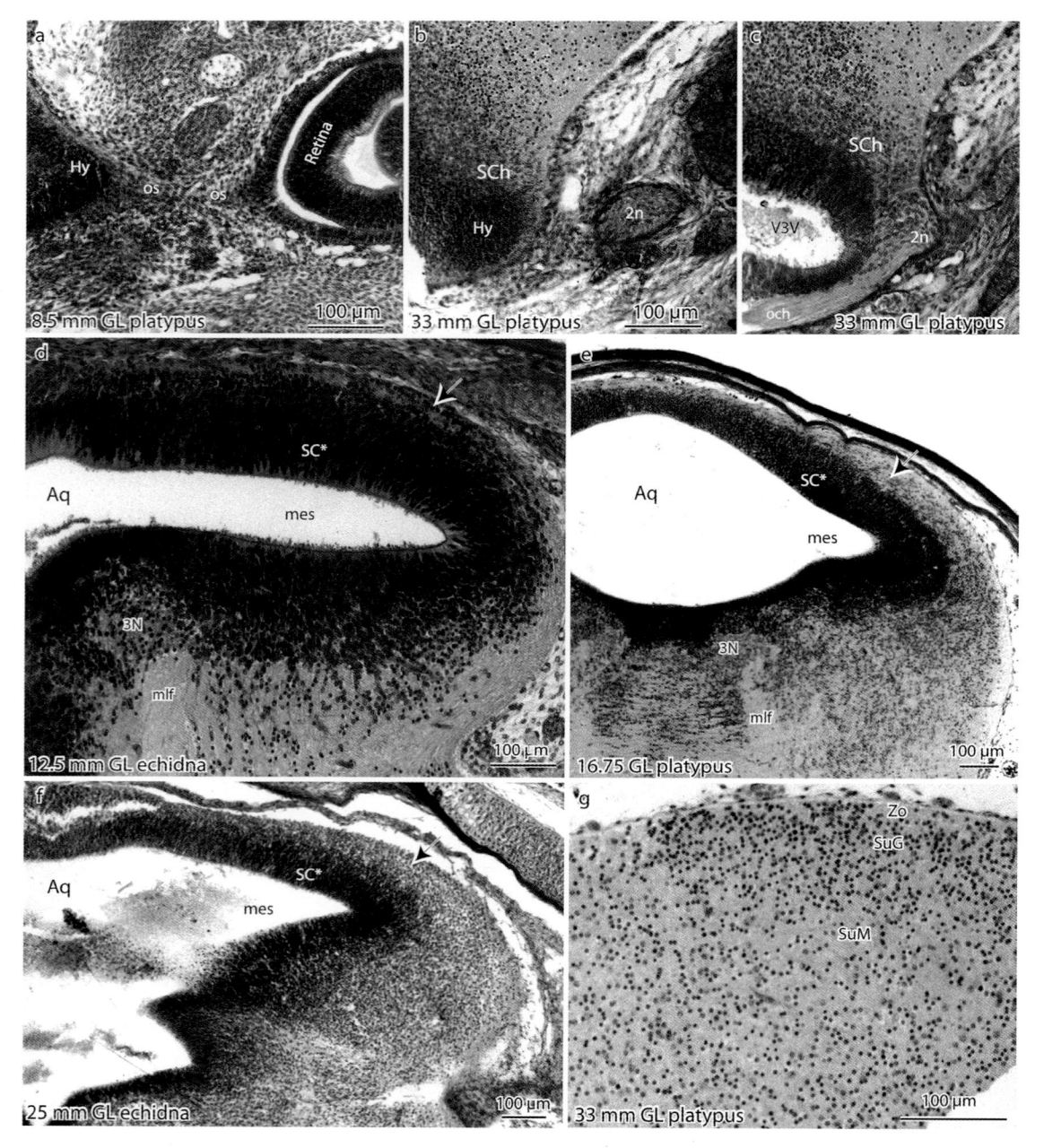

Figure 8.5: Development of the optic nerve, optic chiasm and superior colliculus in monotremes. The optic nerve is absent during the incubation phase (a), but appears soon after hatching (b) and axons quickly cross to form an optic chiasm (c). Photomicrographs d) to g) show generation of neurons of the superior colliculus from the mesencephalic neuroepithelium. Neurogenesis begins shortly before hatching (d), but continues to the beginning of the second week of post-hatching life (g). Arrows in (d) to (f) show the slow accumulation of neurons in the mantle layer of the superior colliculus. Stratification appears in the superior colliculus of the platypus at 33 mm GL, with emergence of zonal, superficial grey and superficial medullary layers. The superficial medullary layer corresponds to the stratum opticum of therians, but is much thinner because it does not carry the retinocollicular axons. 2n – optic nerve; 3N – oculomotor nucleus; Aq – cerebral aqueduct; Hy – hypothalamus; mes – mesencephalic neuromere; mlf – medial longitudinal fasciculus; och – optic chiasm; os – optic stalk; SC* – superior colliculus neuroepithelium; SCh – suprachiasmatic nucleus of the hypothalamus; SuG – superficial grey layer of superior colliculus; SuM – superficial medullary (white matter) layer of developing superior colliculus; V3V – ventral third ventricle; Zo – zonal layer of superior colliculus.

Evolution of monotreme vision

What do observations of the visual system of modern monotremes tell us about the evolution of vision in this group? The genetics of opsins in monotremes indicates that the monotremes and therians have followed very distinct paths in the evolution of colour vision. Although the monotremes are dichromatic, like most of the therians, they have achieved this by the evolution of a blue cone opsin that is encoded by the *SWS2* gene, rather than *SWS1* as in therians. The monotreme *SWS2* gene is situated close to the *LWS* gene, as in sauropsids and fish (Davies *et al.* 2007; Wakefield *et al.* 2008). These observations on cone opsin genes suggest that the common therapsid ancestor had a *SWS1* gene on one chromosome and the *SWS2* and *LWS* genes on another. The loss of the *SWS2* gene and the retention of the *LWS* gene on a sex chromosome appears to be unique to therians. Monotremes have retained other distinctive morphological features in their eyes (scleral cartilage, double cones and oil droplets) that were present in early tetrapods and have been lost in the therians (Zeiss *et al.* 2011).

Despite the unique features of the cone photoreceptor opsins and eye morphology in monotremes, the central visual pathways appear to be remarkably similar to those in therians. Most of the key elements of the retinorecipient regions are just as recognisable in monotremes as they are in therians, with the minor exception of the unusual topography of the pregeniculate nucleus. This suggests that the common ancestor of all mammals had a similar network of retinorecipient nuclei, which has changed relatively little in mammalian evolution, despite some extraordinary changes in visual proficiency and parallel processing among carnivores and primates. Both the platypus and echidnas have a similar predominance of the midbrain tectal (i.e. collicular) over cortical visual pathways to that seen in rodents.

Pettigrew and colleagues (Pettigrew *et al.* 1998) were surprised by the evidence for aquatic adaptation of the visual system in the platypus (i.e. lens curvature), given that platypuses close their eyes for most of the duration of their dives (Burrell 1927). They have suggested that an aquatic adaptation is a remnant of the ancestral monotreme that may have hunted for aquatic prey using vision. Given the predominance of midbrain collicular visual pathways in all extant monotremes, one would expect that visually directed hunting would be mediated by a well-developed tectospinal tract (as discussed above), but our knowledge of this pathway in monotremes is confined to tracing it in myelin stained sections (Hines 1929) and nothing is known of its functional capacity in the platypus.

Questions for the future

The monotremes occupy a unique place in mammalian evolution so there are many questions about the phylogeny of the mammalian visual system that would benefit from an analysis of their visual pathways. At present we have some relatively detailed information on the genetics and structure of photoreceptors in monotremes, but nothing is currently known about the structure and function of horizontal, non-serotonergic amacrine or retinal ganglion cells. Central visual centres are also very poorly understood in the monotremes apart from the basic topography and broad features of the distribution of visual information. Nothing is currently known about the processing of visual information in either diencephalic or midbrain visual centres. What parallel pathways are there for processing visual information in these species and how does it compare with the structure and function of magno- *v.* parvicellular pathways in therians? Do the differences in cone photoreceptors between monotremes and therians have matching peculiarities in the processing of colour information at the diencephalic or cortical level? What is the role of the pretectum and midbrain nuclei in accommodation and other visual reflexes? Does the tectospinal pathway play any significant role in monotremes?

9

Somatosensory and electrosensory systems

Ken W. S. Ashwell and Craig D. Hardman

Summary

In both the platypus and echidna, the trigeminal somatosensory pathways are the most behaviourally important sensory system. This is strikingly so in the platypus, where the trigeminal nerve carries electrosensory information from an extensive array of sensory glands on the bill. The platypus bill is a sophisticated electrical and mechanical sensory organ for determining the direction and range of actively motile prey. Although some have argued that the rudimentary electrosensory capability of modern echidnas is due to evolutionary regression, observations of embryonic and post-hatching development of the trigeminal pathways in the modern monotremes is most consistent with each of the modern groups having evolved from an ancestor with limited trigeminal specialisation.

Postcranial somatosensory pathways in both platypus and echidna are structurally and functionally similar to those in therians. The cerebral cortex of both species is reported to include as many as four somatosensory areas (S1, M, R and PV) each with a topographic body map, but this has been questioned (Rowe et al. 2004). In the primary somatosensory cortex of the platypus, bimodal neurons responsive to both mechanical and electrical stimuli are believed to underlie the ability of the platypus to calculate the distance to free-swimming prey.

Overview of the monotreme somatosensory system

The 'sixth' sense of monotremes

It has long been recognised that the platypus has a remarkable ability to locate invertebrate prey while submerged. Burrell (1927) speculated on the presence of a sixth sense that the platypus used underwater while its nostrils, eyes and ears are sealed shut. This sense must be so effective as to allow the platypus to catch as much as half its body weight in live prey each night (Burrell 1927). The recognition that the sensory glands of the platypus bill showed morphological similarities to ampullary electroreceptors of fish and amphibians has led to the discovery of an electrosensory capability first in the platypus and then in the short-beaked echidna. The value of an electrosensory organ for an aquatic mammal is self-evident provided that the field strength of emissions from the prey is matched by the sensitivity of the receptors (Taylor et al. 1992; Patullo and Macmillan 2004). In the case of the echidnas (both short- and long-beaked) the electrosensory organ would be used as a thrust probe in moist leaf litter, so its potential benefits in arid or dry sclerophyll environments are less clear.

Similarities and differences between monotreme and fish or amphibian electroreception

Electroreception has evolved independently several times among vertebrates. It has emerged perhaps once in agnathids, twice in teleosts, once in non-teleosts and either once or twice in amphibians (Butler and Hodos 2005). The key similarities and differences between the putative electroreceptors in monotremes and those in fish and amphibians are summarised in Table 9.1.

Electroreceptors fall into two broad categories: ampullary and tuberous receptors (New 1997; Butler and Hodos 2005). Ampullary electroreceptors (like those in the platypus) are open to the epidermal surface by a duct and are mainly responsive to low-frequency electrical rhythms that are produced by regular muscle contraction or tail flicks (0.1–50 Hz). Tuberous electroreceptors are more responsive to high frequency electrical discharge (50–200 Hz). Electroreception in monotremes is unusual in that it involves trigeminal pathways and significant isocortical processing, whereas electroreception in

Table 9.1. Similarities and differences in electroreception between monotremes and non-amniotes

	Monotremes[1]	Fish and amphibians[1,2]
Type of electroreceptor	Ampullary	Ampullary or tuberous
Associated nerve	Trigeminal (all divisions, but mainly maxillary and mandibular)	Octavolateralis system
Site of central processing	Cerebral cortex	Hindbrain
Distribution on body	Confined to bill or beak	Head and body (extent of the lateral line system)
Functionally significant distribution	Present in strips on bill (platypus) or hot spot on beak (echidna)	Distributed on body and/or head according to environment
Sensory transducer cells	Not present	Present
Cathodal or anodal sensitivity	Cathodal	Cathodal
Ratio of nerve terminal surface area/sensory epithelium area	Low	Low
Conductive secretion in gland duct	Present	Present
Relationship between duct length and salinity	Short for freshwater environment (platypus)	Short for freshwater environment Long for saltwater environment
Electroreceptor unit threshold	2 mV/cm for platypus	1 µV/cm to 1 mV/cm
Association with mechanoreception	Present	Absent
Frequency responsiveness	Maximal sensitivity below 100 Hz	Differs between species, but is low
Directionality detection	Direct detection of dipole position by head saccades and central integration of decay across bill	Approach source along field lines

1 Manger et al. (1996), 2 Butler and Hodos (2005).

non-amniotes involves the acoustico-lateralis system and exclusively hindbrain processing.

Somatosensory pathways

In the monotremes (as in therians), the somatosensory pathways convey electroreception, touch, vibration, proprioception, pain and temperature from the body surface, muscles and joints into the spinal cord and brain. These pathways can be divided into the trigeminal system (for the head/face) and spinal cord pathways (for the limbs and trunk). The trigeminal system is the only pathway for electroreception in the monotremes.

First order neurons for processing and conveying somatosensory (and electrosensory) information are located in the trigeminal ganglion for most of the head (Fig. 9.1), and the dorsal root or spinal ganglia for the limbs and trunk (Fig. 9.2). Other cranial sensory ganglia (i.e. glossopharyngeal, vagal) also probably convey somatosensory information from some limited parts of the head, pharynx and larynx, but these have not been studied in the monotremes and need not be discussed. Some first order neurons have central processes that form ascending pathways in the spinal cord (dorsal columns, Fig. 9.2a). First order neurons concerned with proprioception of the masticatory apparatus have cell bodies inside the rostral hindbrain and midbrain (mesencephalic nucleus of the trigeminal nerve, Fig. 9.1a).

Second order neurons are located in the trigeminal sensory nuclei of the pons and medulla for the trigeminal pathways from the head (Fig. 9.1a). In therians, second order neurons of the pain, temperature and simple touch pathways (spinothalamic and spinoreticulothalamic) are located in the spinal cord grey matter. The discriminative touch pathways (fasciculus gracilis and cuneatus) have second order neurons in the nucleus gracilis and cuneatus of the medulla, respectively (Fig. 9.2a). The spinothalamic pathway (not shown in Fig. 9.2a) is known as the spinal lemniscus in the brainstem and the output from the gracile and cuneate nuclei is known as the medial lemniscus (Fig. 9.2a).

Third order neurons of the spinothalamic/spinal lemniscus, dorsal column/medial lemniscal and trigeminal lemniscal pathways are located in the ventral posterior or ventrobasal thalamic nucleus group (lateral subdivision VPL for the trunk and limbs; medial subdivision VPM for the head/face) and project to the primary and secondary somatosensory cortex. Latencies of impulses along the electrosensory projection to the platypus cortex are low enough (only 9 ms) to indicate that the electrosensory pathway is likely to have the same three-neuron chain of all mammalian somatosensory pathways (Proske et al. 1992).

The bill receptors and trigeminal nerve of the adult platypus

The electrosensory apparatus of the platypus bill

The first reports of electroreception in monotremes came from Scheich et al. (1986) who reported that the platypus could detect weak electric fields with its bill at thresholds as low as 50 µV/cm, although the most consistent responses were evident at field strengths much higher than this (i.e. a 60% chance of eliciting a search response at a field strength of 1 mV/cm). The pathways for this electrosensory ability involve the trigeminal system, principally the maxillary and mandibular divisions of the trigeminal nerve. This is in contrast to electrosensory organs in mormyriform and gymnotiform fish and some amphibians, where electrosensation is associated with the acoustico-lateralis system and the cerebellum (Andres and von Düring 1988; Butler and Hodos 2005).

Gregory and colleagues (Gregory et al. 1987b; Gregory et al. 1988; Iggo et al. 1988) made recordings directly from isolated axons of the infraorbital nerve (a branch of the maxillary division of the trigeminal nerve) and reported that they could clearly distinguish between electro- and mechanoreceptor discharges.

Electroreceptor axons gave measurable responses to electrical stimuli in the millivolt range and were insensitive to non-damaging probing with a fine glass stylus (i.e. a gentle mechanical stimulus). Electroreceptor axons had resting discharge rates of 30–50 impulses per second, which increased up to a maximum in excess of 300 Hz when a strong cathodal electrical stimulus was applied. Maximal sensitivity for each electroreceptor axon corresponded to the position of the pore of an individual innervated gland duct. The threshold for step function stimulation of electroreceptors was 20 mV when the cathode was

182 NEUROBIOLOGY OF MONOTREMES

directly placed over the most sensitive spot. Platypus electroreceptors are noticeably responsive to rapidly changing voltages, such as would be encountered during the tail flick of freshwater shrimp or when the platypus sweeps its head from side to side. Gregory and colleagues (Gregory et al. 1989a) found a 1:1 response between the peak of the stimulating electric field waveform and electroreceptor discharge over frequencies from 12–300 Hz, with the lowest threshold at 50–100 Hz. It is notable that Gregory and colleagues felt that the thresholds they obtained for individual electroreceptors and their axons were substantially higher than the thresholds reported by Scheich and colleagues in their behavioural experiments.

By contrast, mechanoreceptor axons were unresponsive to electrical fields unless the voltages were three orders of magnitude greater than that required for the electroreceptor axons (Gregory et al. 1987b). Mechanoreceptors were also silent unless a mechanical stimulus was applied.

Sensory and non-sensory glands in the platypus bill

The sensory receptors in the platypus bill have long been recognised for their unusual structure and high spatial density (Bohringer 1976, 1981). The platypus bill has three types of cutaneous glands (Andres and von Düring 1988). These include: (1) a mucous gland without sensory innervation; (2) a mucous gland with sensory innervation; and (3) a serous gland with sensory innervation (Fig. 9.3). Each of these three types of gland is divided into four regions or segments: secretory tubules with a main and accessory segment; a subdermal component with a coiled and a narrow isthmic segment; a dermal component with a straight and subpapillary part; and an epidermal component with papillary, coiled sinus and pore regions. Andres and von Düring noted the similarity of these glands and their component segments to mammalian eccrine sweat glands, with the exception that the platypus glands have striking specialisations of the intraepidermal duct.

Mucous glands without sensory duct innervation open to the surface through a 50 to 80 µm wide pore complex that looks like a rose blossom when viewed from outside the skin. This appearance is due to the arrangement of superficial epidermal cells curled like the petals of a rose around the pore opening. The coiled secretory tubules of these glands have abundant mucous cells, cells with dark intracellular granules and cells with abundant mitochondria, the latter implying energy costly secretion (Andres and von Düring 1988).

Mucous glands with sensory innervation (putative electrosensory organs) have the largest surface pores (70 to 100 µm diameter for the whole surface complex) with a peony blossom-like appearance due to the presence of many more epithelial 'flakes' than for the non-sensory mucous gland (Andres and von Düring 1988). Scheich and colleagues suggested that the mucous sensory glands are the electrosensory organs

Figure 9.1: Somatosensory pathways for the mammalian head. Diagram (a) shows a schematic representation of the main trigeminal pathways in a representative placental mammal. Sensory information is passed from the skin surface through the trigeminal ganglion to the principal trigeminal nucleus (Pr5) or down the trigeminal spinal tract (sp5) to the nuclei of the trigeminal spinal tract (Sp5O, Sp5I, Sp5C). Information about tension or force of muscles is conveyed through the mesencephalic nucleus of the trigeminal nerve (Me5). The sequence of trigeminal nuclei is similar in monotremes, but with some profound specialisations of particular groups in the platypus (b to g). Neurons in somatosensory pathways have high levels of parvalbumin, a calcium-binding protein that stabilises physiologically active neurons. Immunoreactivity for parvalbumin is particularly strong in some of the trigeminal sensory nuclei of the platypus brainstem (b to f). Note the large size and overlapping of nuclei in the trigeminal sensory column as one proceeds from rostral to caudal (b to f). Somatosensory information is conveyed from the trigeminal sensory nuclei to the medial part of the ventral posterior nucleus of the thalamus (VPM in g) and from there to the primary somatosensory (S1) cortex (h). The second somatosensory area (S2) has not been definitively identified in monotremes. 5N – motor trigeminal nucleus; 8cn – cochlear division of vestibulocochlear nerve; Cu – cuneate nucleus; DC – dorsal cochlear nucleus; ECu – external cuneate nucleus; Gust? – possible gustatory (taste) area of cortex; LD – lateral dorsal nucleus of thalamus; mcp – middle cerebellar peduncle; Pr5mc – magnocellular part of Pr5; Pr5pc – parvicellular part of Pr5; s5 – sensory root of trigeminal nerve; VL – ventral lateral nucleus of thalamus; VLL – ventral nucleus of the lateral lemniscus; VMb – basal ventral medial nucleus of thalamus; VPPC? – possible parvicellular part of the ventral posterior thalamic nucleus.

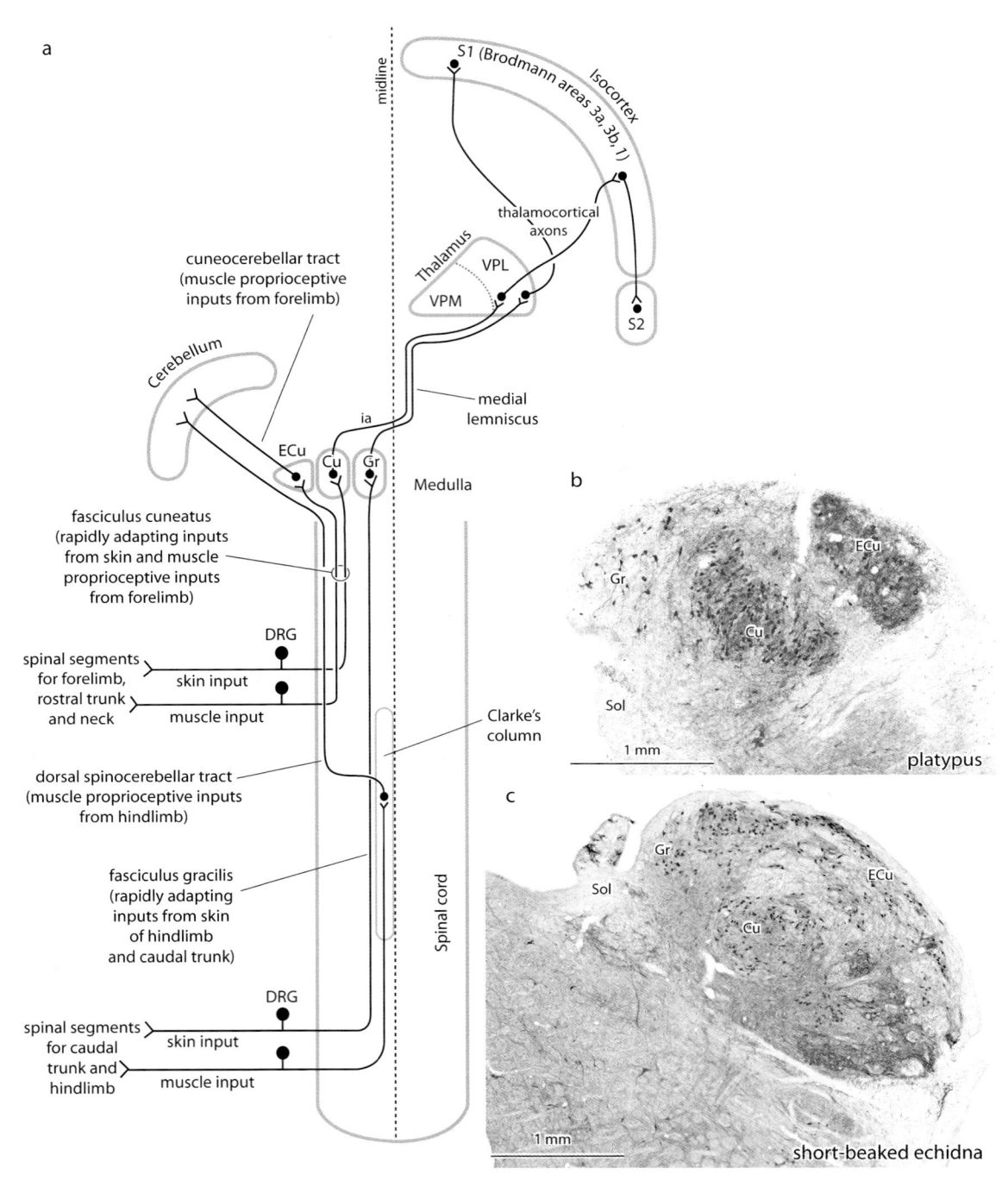

Figure 9.2: Pathways for discriminative touch and proprioception in mammals. The schematic diagram (a) shows the main pathways for these modalities in placental mammals (dorsal column pathways fasciculus gracilis and cuneatus for discriminative or fine touch and conscious proprioception; dorsal spinocerebellar and cuneocerebellar tracts for non-conscious proprioception). Conscious tactile information is channelled and sequentially processed in a series of relays from dorsal root ganglion cells (DRG), up the spinal cord, through medullary relay nuclei (gracile – Gr, cuneate – Cu), across the midline in the internal arcuate fibres (ia) to the lateral part of the ventral posterior nucleus of the thalamus (VPL) and then on through the thalamocortical axons to the S1 cortex. Photomicrographs (b) and (c) show the gracile, cuneate and external cuneate (ECu) nuclei in parvalbumin immunoreacted sections through the medulla of the platypus and echidna. Sol – nucleus of the solitary tract. Other abbreviations are the same as in the previous figure.

of the platypus bill on the basis of a structural similarity to the ampullary electroreceptors in freshwater fish (Scheich *et al.* 1986). The duct of the mucous gland is believed to provide a low resistance path for electrical fields to pass from the surface of the skin to the sensory nerve endings at the papillary part of the gland (Iggo *et al.* 1988). The secretory tubules of the sensory mucous glands have the same types of sensory cells as the non-sensory mucous glands, but with more translucent mucous cells. The papillary portion of the intraepidermal duct is also much longer (150 µm) in the sensory mucous gland than for non-sensory mucous glands (50 µm), thanks to the much longer epidermal papilla of the sensory type. This long epidermal papilla is surrounded by up to 30 myelinated nerve fibres forming a cuff around the base of the papilla. The axons lose their myelin as they penetrate the basal epidermal layer and terminate as thin naked sensory nerve endings. Andres and von Düring (1988) identified two types of axons: type 1 (80–90%), with prominent nodal enlargements of the axonal membrane due to invaginations of Schwann cell processes; and type 2, with narrower initial node segments. Both types have terminal bulbar swellings 5 µm in diameter and slender axonal processes extending into the papillary portion of the gland.

In contrast to electroreceptors in fish, the electrosensory terminals of the mucous sensory glands of the platypus are not associated with a sensory cell (Manger and Pettigrew 1996). This may account for the short latency of electroreceptor activation in the platypus (0.8 ms) compared to 10–50 ms for ampullary electroreceptors in fish (see discussion in Gregory *et al.* 1989a). Excitatory amino acids are used in synaptic transmission between receptor cells and nerve terminals in the electrosensory organs of fish and amphibians (Akoev 1995), but the mechanism of nerve terminal activation in the platypus remains unknown. Gregory and colleagues (Gregory *et al.* 1989a) have noted that the latency of 0.8 ms is rather long if the stimulus were acting directly on the axonal membrane, because utilisation time for impulse initiation in most mammalian peripheral nerve is only 0.1 ms. They speculated that the initiation of the impulse may require more than simple depolarisation of the axonal membrane, but the mechanism remains unknown.

The serous glands have a small pore (40 µm diameter) surrounded by a small rim or epithelial cells. Andres and von Düring (1988) described the gland cells of the secretory coil as being much like typical mammalian sweat glands. A plexus of up to 22 sensory myelinated axons surrounds the basal part of the papillary portion of the epidermal duct to form a cuff ~20 µm thick. Andres and von Düring identified four different types of axons supplying the serous gland duct.

The mechanosensory apparatus of the platypus bill

There are three types of mechanosensory structures in the platypus bill: vesicle chains (central and peripheral types), Merkel cells, and Paciniform (or lamellated) corpuscles (Manger and Pettigrew 1996). These are grouped together into specialised mechanoreceptor complexes (Fig. 9.3). The most distinctive of these is the push-rod mechanoreceptor (Poulton 1885; Quilliam 1979), a highly specialised organ distributed across the bill, but with concentrations at the lateral and rostral rim of the bill. The push-rod mechanoreceptor is found only in the monotremes, but has some resemblance to the Eimer organ of placental moles (Quilliam 1979).

The central structure of the push-rod mechanoreceptor is an epidermal rod ~400 µm long and 70 µm wide, made of stratified keratinocytes. Three types of mechanosensory structures (central vesicle chain, peripheral vesicle chain and a cluster of Merkel cells) are located within the rod, while three to six Paciniform corpuscles are located in the dermis immediately beneath each rod.

There is also a group of loosely attached connective tissue cells around the upper part of the rod that isolate the distal rod from the surrounding epithelium and facilitate its free movement relative to the rest of the skin (Manger and Pettigrew 1996). Each of the four mechanoreceptors associated with each epidermal rod is supplied with axons from a plexus in the nearby dermis. An average of nine axonal branches are distributed to the central vesicle chain, where an unmyelinated branch runs within the very centre of the rod to within a few cells' thickness of the epidermal surface. An average of 40 myelinated axons are distributed to the peripheral vesicle chain, where axons also run along the length of the rod, but around its external margin rather than the rod centre. Three

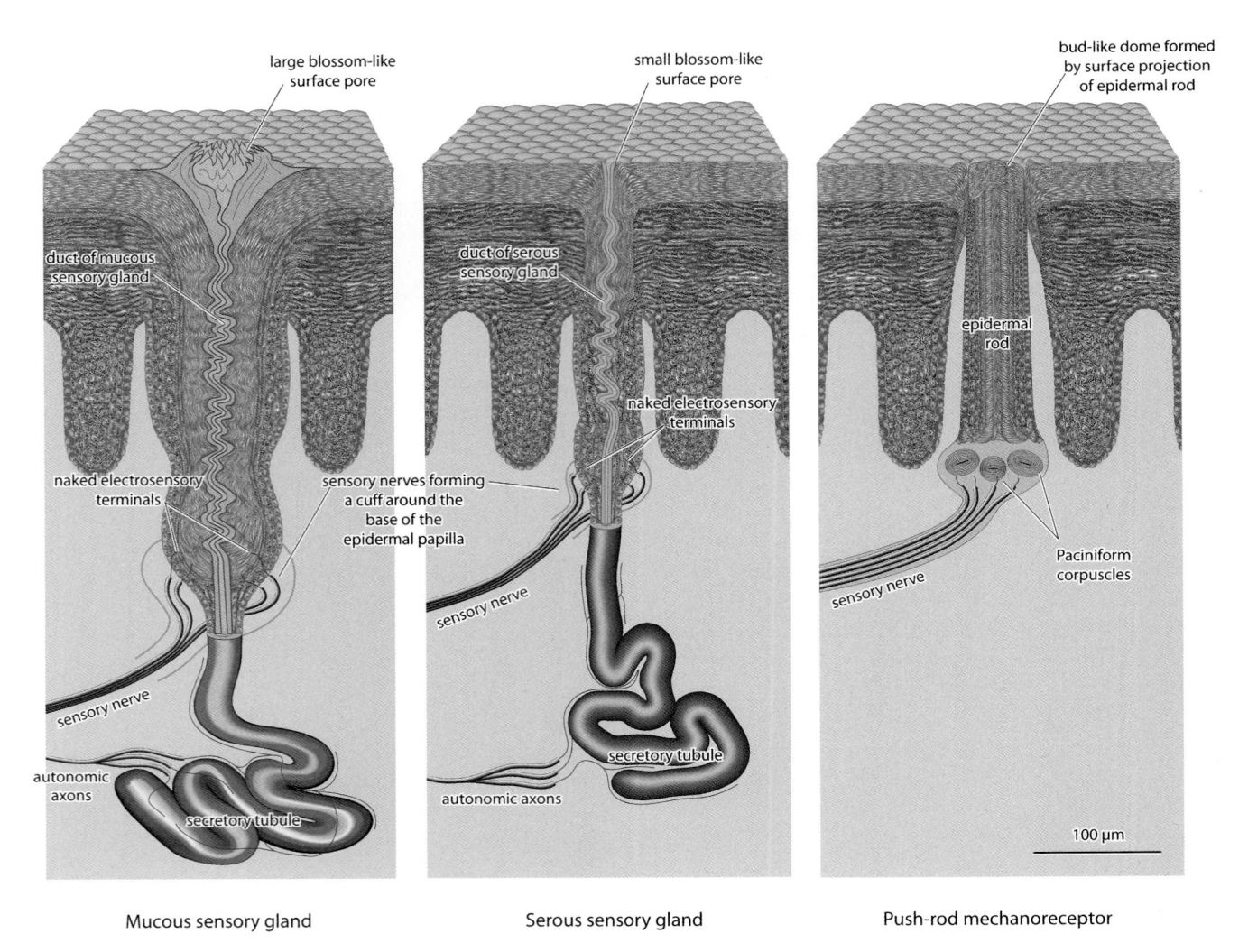

Figure 9.3: Line diagrams showing examples of sensory mucous and sensory serous glands (both putative electroreceptors) and push-rod mechanoreceptor in the bill of the platypus. The illustrations are based on figures in Andres and von Düring 1988; Proske et al. 1998; Manger and Pettigrew 1996; Proske and Gregory 2004.

layers of epidermal cells surround the peripheral vesicle chain to complete the overall shape of the rod. A group of about 12 ovoid *Merkel cells* is also incorporated into the basal part of the rod and a nerve plate process covers the dermal surface of each Merkel cell. Finally, three to six Paciniform corpuscles are located in the dermis beneath the rod. Each of these is 60 μm by 40 μm in size and consists of 13 lamellations or connective tissue layers. These corpuscles are somewhat smaller and less lamellated than true Pacinian corpuscles (which may reach 1 mm diameter and 60 lamellations in many therians).

When the push-rod complex is viewed from the bill surface, the rod looks like a bud-like dome protruding slightly from the epidermal surface and surrounded by a few concentric rings of flattened keratinocytes (Andres and von Düring 1988; Manger and Pettigrew 1996).

Physiologically, mechanoreceptors of the platypus bill include both rapidly and slowly adapting units, with a few having intermediate properties (Gregory *et al.* 1987a, 1988). Rapidly adapting receptors respond best to high frequency vibration and respond only briefly (i.e. one impulse each at on-phase and off-phase) to sustained indentation of the bill epidermis. Vibration mechanoreceptors (putative Paciniform corpuscles) respond with one impulse per indentation at vibration frequencies up to 600 Hz. Responsiveness is lost at vibration frequencies over 800 Hz.

Slowly adapting receptors have receptive fields ~2 mm in diameter and respond to indentation with a threshold of 20 μm (Gregory et al. 1988). This sort of slowly adapting receptor is known physiologically in other mammals as the cutaneous slowly adapting type I (SA I), or the Merkel receptor.

Intermediate adaptation mechanoreceptors have a relatively low mechanical threshold (25 μm indentation). Even with a maximal physical stimulus, the discharge rate is maintained for between only 1 and 2 seconds. Intermediately adapting mechanoreceptors respond poorly to vibration, giving a 1:1 response to indentation at 100 Hz and responding only intermittently to higher frequencies.

A final point to consider concerning mechanoreceptors involves the mobility of the push rod. This is a highly mobile structure compared to the corresponding structure in the echidna. It is clearly exquisitely sensitive to touch and would be at risk of damage during burrowing. Manger and colleagues (Manger et al. 1998b) have suggested that the platypus can adjust the sensitivity of the push-rod system by contracting the sphincter of actin-containing cells around the epidermal end of the rod. They suggest that when the platypus is hunting, the collar of actin-rich keratinocytes around the end of the push rod is relaxed so that the rod can move freely, but when resting in the burrow the collar of keratinocytes is contracted to prevent free movement of the rod.

Distribution of glands and push rod receptors in the platypus bill

Estimates of the numbers of the various types of receptors and glands on the upper and lower platypus bill are 30 000–40 000 mucous sensory glands, 22 500–35 000 non-sensory mucous glands, 10 000–13 500 sensory serous glands, and 46 750 push-rod mechanoreceptors (Andres and von Düring 1984, 1988; Manger and Pettigrew 1996). The mucous sensory glands are the main putative electroreceptors, while sensory serous glands have also been suggested to function as electroreceptors only at close range (Manger and Pettigrew 1996). Non-sensory mucous glands may provide for bill lubrication.

The mucous sensory glands are arranged in parasagittal stripes along the bill surfaces, on both the upper and lower surfaces of each half of the bill (Andres and von Düring 1988) in a fashion that may facilitate the rapid location of electrical signal sources when the platypus sweeps its head from side to side (Manger and Pettigrew 1996). The stripes are most prominent on the cutaneous surface of the upper bill, where the rostral three-quarters is covered by 18 to 20 stripes and they are confined to just the most rostral tip of the lingual surface of the lower bill (Manger and Pettigrew 1996). The cutaneous surface of the upper bill has the most mucous sensory glands (13 234), followed by the palatal surface of the upper bill (11 480), the cutaneous surface of the lower bill (7876) and the lingual surface of the lower bill (6164).

Serous glands are scattered throughout the whole bill surface with higher spatial densities on restricted areas such as the palatal and lip ridges, nostrils, nasopalatine duct of the vomeronasal organ, filiform border at the entrance of the cheek pouches, and the anterior segment of the tongue. This is a similar pattern to the push-rod mechanoreceptors (see below), but at a lower spatial density (Andres and von Düring 1984, 1988; Manger and Pettigrew 1996). The highest number of serous sensory glands was found on the palatal surface of the upper bill (5000) followed by the cutaneous surface of the upper bill (4358), the lingual surface of the lower bill (2118) and the cutaneous surface of the lower bill (2004). Serous sensory glands are not present on the occlusive surfaces of the bill and the depths of the grooves that drain water from the side of the mouth.

Push-rod mechanoreceptors are distributed across the four surfaces of the bill, but with most concentrations at the lateral and rostral rim (Andres and von Düring 1988; Manger and Pettigrew 1996). The cutaneous surface of the upper bill has the most push-rod mechanoreceptors (15 750), followed by the palatal surface of the upper bill (12 680), cutaneous surface of the lower bill (12 488), and lingual surface of the lower bill (5622). The mucosal surfaces where the bills came together have no push-rod mechanoreceptors, but they are found on the ridges alongside the grooves on the lingual surface of the upper bill that serve to drain water from the oral cavity (Manger and Pettigrew 1996).

Structure and function of the autonomic innervation of putative electroreceptors in the platypus

The putative electroreceptors of the platypus bill are modified eccrine glands, with pores that open onto the epidermal surface. The secretions within the duct of the gland are believed to act as a conductive pathway from the epidermal surface to the sensory nerves within the papillary portion of the gland. In electroreceptive fish the sensory glands are continuously open to the epidermal surface. Unlike fish, the semi-aquatic platypus leaves the water for a substantial part of the day (~16 hours per day, Manger *et al.* 1998b). This presumably exposes the conductive secretions of the sensory gland to the risk of drying out. The gland ducts are also vulnerable to dirt when the platypus digs its burrow or enters the muddy burrow entrance from the water (Manger *et al.* 1998b). A mechanism for controlling the exposure of the sensory gland duct contents to the desiccating effects of the air and contamination by dirt would naturally be of benefit for the platypus. Observations of the bill surface in and out of water indicate that the pores of the sensory glands are open during immersion and closed during exposure to the air, consistent with the presence of a mechanism that controls the exposure of the conductive gland duct contents to the external environment.

Opening and closing of the gland duct pores may also serve a functional role quite apart from protection of the electrosensory apparatus. Manger and colleagues have noted the importance of impedance matching between the contents of the gland duct and the varying salinity of surrounding water for optimal function of the system. They have proposed that the mechanism for closure of the duct pores may assist with active matching of the duct impedance with that of the surrounding fluid (Manger *et al.* 1998b).

Manger and colleagues went on to identify the anatomical substrates of a potential reflex mechanism for protecting the sensory gland ducts from desiccation (Manger *et al.* 1998b). They identified unmyelinated sensory axons that supply the region of the keratinocytes encircling both the sensory gland pores and mechanoreceptor rods on the bill skin. These C-type varicose sensory axons are immunoreactive for both substance P and calcitonin gene-related peptide (CGRP) and may be temperature receptors or nociceptors. Manger and colleagues propose that sensory fibres are activated by either drying or wetting of the skin (perhaps through changes in temperature). This might initiate autonomic reflexes that serve two significant effects: (1) open or close the gland duct pore; or (2) either enhance secretion while the pores are open, or reduce secretion when the pores are closed. Although it is not possible to determine whether it is immersion or drying that is the important stimulus, it is most likely that the sudden cooling of immersion is the key stimulus (Manger *et al.* 1998b).

The effector side of the reflex involves adjustment of the gland secretion rate, and opening or closure of the gland duct pore. Activity of the sensory glands is most likely controlled by a parasympathetic innervation of the glandular acini, probably involving vasoactive intestinal peptidergic (and possibly galaninergic) axons. Sympathetic axons containing catecholamines innervate the dermal blood vessels and the epidermal gland ducts in the region of the actin-rich keratinocytes at the surface. Manger and colleagues suggest that the sympathetic vasoconstrictor axons contain neuropeptide Y, whereas those supplying the gland duct and actin-rich collar around the pore do not (Manger *et al.* 1998b). They suggest that the sympathetic axons to the upper gland duct activate the actin-rich cells around the pore to close when the bill is removed from water.

The central elements to this reflex can only be deduced based on homology with therians, and need to be confirmed experimentally. It is most likely that the sensory input from immersion would be initially processed in the caudalis part of the nucleus of the trigeminal spinal tract, given that this is the part of the trigeminal sensory complex that is involved with thermoreception in placentals. Further connections probably involve the parvicellular part of the medullary reticular formation. Parasympathetic activation most likely involves preganglionic parasympathetic neurons in the pons, which send efferents from the brainstem through the facial nerve to contact postganglionic parasympathetic neurons in the pterygopalatine ganglion and on through trigeminal nerve branches to the skin surface. Sympathetic activation would most likely involve medullary reticulospinal projections from the brainstem to contact the cell bodies of preganglionic sympathetic neurons in the lateral horn of the upper thoracic spinal cord. The axons of these preganglionic sympathetic neurons project to cell bodies of postganglionic sympathetic neurons in the superior cervical ganglion, which would in turn project along branches of the external carotid artery to the bill skin.

Other authors have noted the presence of erectile tissue in the bill and beak of the monotremes (Andres *et al.* 1991; Proske and Gregory 2003), leading to the suggestion (Proske and Gregory 2003) that vascular changes in the skin may facilitate the contact between push rods and the external environment by changing the fluid content of the tissue. At present no physiological evidence for this putative role of a neurovascular apparatus has been obtained, but it may account for the rich autonomic innervation of the snout in modern monotremes.

The trigeminal nerve and ganglion of the platypus

Gross dissection of the trigeminal nerve shows that axons from the receptors of the bill form eight or nine major bundles on each side (Bohringer 1977; Manger *et al.* 1996; Manger and Pettigrew 1996). These are one or two for the ophthalmic division, four for the maxillary division and three or four for the mandibular division. The ophthalmic and maxillary divisions supply the cutaneous and palatal surfaces of the upper bill, and the maxillary division alone supplies the cutaneous and lingual surfaces of the lower bill. Cell bodies for the axons in the trigeminal nerve are located in the trigeminal ganglion, which is very large in the platypus (volume over 55 mm^3) compared to the larger-bodied short-beaked echidna (9 mm^3) (Ashwell *et al.* 2012).

The cross-sectional areas of the three divisions have been estimated as 2.45 mm^2 for the ophthalmic division, 10.38 mm^2 for the maxillary division and 4.95 mm^2 for the mandibular division (Manger and Pettigrew 1996). Early calculations of the total number of fibres in the trigeminal nerve that innervate the bill gave an estimate of 380 000 axons, ~50% of all the fibres in the trigeminal nerve (Andres and von Düring 1988), but later estimates are even higher (a total of 1 344 000 myelinated axons supplying the upper and lower bill, Manger and Pettigrew 1996). Manger and Pettigrew (1996) estimated that 640 000 axons innervate the mucous sensory glands, 175 500 axons innervate the serous sensory glands and 528 500 axons innervate the push rod mechanoreceptors.

The most parsimonious grouping of axons within the trigeminal nerve is into two size classes with mean diameters (±standard deviation) of 3.46 (±0.65) µm and 4.95 (±0.80) µm (Manger and Pettigrew 1996). The smaller size group makes up ~60% of all sensory axons, leading Manger and Pettigrew to suggest that the smaller size group might innervate electroreceptors and the larger axons might innervate mechanoreceptors, but this remains to be confirmed by correlative electrophysiological and ultrastructural analysis. Nevertheless, data on conduction velocity support this suggestion to some extent. Conduction velocity in the trigeminal nerve fibres is 21–24 m/s for mechanoreceptors, suggesting they use larger diameter or more myelinated axons, but slightly slower for electroreceptors – 19–24 m/s (Gregory *et al.* 1988).

How does the platypus use its bill in the wild?

The famous electrosensory platypus

The importance of electroreception for prey detection by the platypus has almost become an article of faith in comparative neurobiology. The notion of an electroreceptive platypus actively searching for its prey in riverbed mud has captured the imagination of the public; so much so that the animal has become an internet and children's television sensation. How does this famous status measure up scientifically? Is the bill organ capable of detecting both moving and sessile prey in the real world?

Behavioural observations

How does the platypus actually use its electrosensory capabilities in the wild? Researchers in this area have speculated that the direct current sensitivity of the platypus electrosensory apparatus may allow it to detect contact potentials between the riverbank and the water, or between the water and obstacles on the stream bottom (Gregory *et al.* 1989a). On the other hand, the optimal tuning of the electroreceptors to higher frequency signals (around 140 Hz) suggests that it is more important for the detection of free-swimming or mobile prey with rapidly changing or pulsatile electrical signals (e.g. crustaceans).

Observation of the animal in captivity emphasise the continuous sweeping of the bill from side to side (saccades) at a frequency of ~75 movements per minute. The head saccades appear to be important in allowing the platypus to locate the direction of an electrical stimulus by reorientating the head relative to the source. This reorientation is perhaps most important for presenting the more sensitive part of the bill towards the stimulus (Proske et al. 1998). When the platypus is exposed to an electrical stimulus over threshold, it also produces reflexive head movements towards the source and snaps at the electrode (Manger and Pettigrew 1995). Even when it is clear that the source is inedible, the platypus continues to show reflexive head turns towards the stimulus. Taken together, these observations suggest that there are both voluntary and involuntary (i.e. reflexive) components in the orientation of the platypus to electrical sources.

What is the sensory challenge that the wild platypus faces?

Before considering how the platypus might actually use electrosensation in the wild, it is important to be aware of the particular challenges of the natural environment. In particular, what are the electrical signals from prey that the platypus is trying to detect?

Most considerations of this question focus on crustacean prey, because it is easiest to measure electrical field strength for these animals. In the wild the platypus consumes a much wider range of prey than just crustaceans (see the cautionary note below). The electrical fields generated by crustacean prey have maximal values between 1 and 2 mV/cm (Taylor et al. 1992; Patullo and Macmillan 2004). These fields are generated when crustaceans (e.g. *Cherax destructor*) use rapid forceful contractions of their abdominal flexor and extensor muscles to swim during general locomotion and escape. Signal amplitude is correlated with crustacean size and abdominal muscle mass, so prey need to be ~6 cm in body length and have 2 g abdominal muscle mass to reach the 1 mV/cm level of signal (Patullo and Macmillan 2004). The tail flick of the crustaceans generates a fundamental frequency of 140 Hz, comparable to the optimal frequency of the platypus electroreceptive organ (Proske et al. 1998).

The electrical field strength generated by crustaceans decays with distance in a complex way that also depends on body shape (Pettigrew et al. 1998). The 1 mV/cm is maintained at about the same level for distances comparable to the length of the invertebrate (5 to 10 cm), but decays at greater distances in a complex fashion (initially by inverse square, but by inverse cube at higher distances; Pettigrew et al. 1998). To be able to hunt effectively, the platypus would need to detect the crustacean's field at ~10–20 cm. Given that the observed behavioural thresholds for the platypus are of the order of 50 µV/cm, a large shrimp would be consistently detectable at ~10 cm, but smaller prey would need to be much closer.

To be able to effectively hunt freely moving crustaceans, the platypus would need neural mechanisms for accurately assessing not only the direction of the prey, but also the range. Such a system must be able to work quickly, because free-swimming crustaceans will be actively moving or even fleeing, particularly should they become aware of a predatory platypus in their vicinity.

The threshold problem

One area of dispute in the electrosensation debate concerns the question of electrosensory thresholds (Rowe et al. 2004). Behavioural studies have reported field strength threshold values as low as 50–200 µV/cm (Scheich et al. 1986) or 20 µV/cm (Manger and Pettigrew 1996; Pettigrew 1999) for the platypus. Measured thresholds for cortical neurons are also low; for example, the field strength threshold for a response from the somatosensory cortex is of the order of 300 µV/cm (Iggo et al. 1992; Proske et al. 1992). This compares with values of 50–200 µV/cm obtained for individual cortical neurons stimulated by needle electrodes held above the surface of the bill (Manger et al. 1996). On the other hand, the thresholds of the individual trigeminal afferent fibres (in the infraorbital branch of the maxillary division of the trigeminal nerve that are believed to be responsible for electrosensation) are much higher (around 4 mV/cm; Gregory et al. 1988, 1989b). Why is there such a discrepancy (of the order of 12- to 80-fold) between behavioural and cortical thresholds on the one hand and individual fibre thresholds on the other?

The number of axons sampled in all the studies of infraorbital nerve fibres is modest (no more than a few dozen) compared to the thousands of putative electroreceptors in even one platypus bill, so perhaps the electrophysiological studies of individual infraorbital fibres have simply missed the most sensitive receptors (Proske et al. 1998). Alternatively, perhaps this discrepancy between behavioural and electrophysiological thresholds can be explained by spatial summation, i.e. the convergence of several electroreceptive trigeminal afferents onto the one central neuron (Proske et al. 1998; Pettigrew et al. 1998; Rowe et al. 2004), or by temporal summation of repeated stimulation (Fjällbrant et al. 1998).

Solving the threshold problem: spatial summation

Some spatial summation might depend on features at the level of the sensory gland itself. Proske and colleagues have noted that each sensory gland is served by an average of 16 myelinated axons. Potential gradients generated between the gland interior and the skin surface would act uniformly across the papillary epithelium so that a similar signal would be generated at each afferent ending in the gland (Proske et al. 1998). Having such a large number of parallel inputs from each gland might improve signal-to-noise ratios and allow for signal amplification.

Further spatial summation might also occur along the central trigeminal pathways, presumably in the brainstem and/or thalamus. Fjällbrant and colleagues have noted that, by signal detection theory, absolute threshold depends on the individual receptor threshold divided by the square root of the number of receptors (Green and Swets 1974; Fjällbrant et al. 1998). Given that there are 20 000 sensory mucous glands on each side of the bill and an individual receptor threshold is of the order of 2000 µV/cm, then the absolute threshold is theoretically 14.1 µV/cm for half the bill and 10 µV/cm for the whole bill (Fjällbrant et al. 1998). This summation depends on simple averaging, but more sophisticated anatomical summation may occur at the level of the somatosensory cortex (Fjällbrant et al. 1998).

A further point is that the fibre threshold in the mV/cm range was based on a one-to-one following of axon spikes to sinusoidal changes in the electrical field strength (Gregory et al. 1988, 1989a). If the central signal processor (in the brainstem, thalamus or cortex) were able to examine simultaneous input from 40 000 receptors across the bill, then a response that included only a few time-locked spikes in many thousands of axons would be detectable as a signal, yielding a potential 1000-fold improvement in threshold (Pettigrew et al. 1998).

Solving the threshold problem: temporal summation

Temporal integration of electrosensory input may also improve threshold. When a small amplitude stimulus around threshold (50 µV/cm) is presented to a platypus, often a few repetitions of the stimulus are necessary before the animal responds. The exploratory response elicited in this situation consists of voluntary exploratory movements in the direction of the source, in the form of lateral head swings and movement of the rostral half of the body towards to the source.

The lower the stimulus strength, the more repetitions of the signal are necessary to trigger the exploratory response. So 10 repetitions are needed for field strengths of 35 µV/cm and 160 repetitions are needed for field strengths of 25 µV/cm, the lowest voltage tested that gave a reliable behavioural response (Manger and Pettigrew 1995; Fjällbrant et al. 1998). These observation led Fjällbrant and colleagues to propose that the platypus can perform temporal integration or summation of several 'subthreshold' electrical stimuli to produce an active exploratory response.

The bill as a directional antenna

The most striking feature of the head saccades induced by electrical signals is that they are directional, leading towards a head and body movement in the direction of the source (Manger and Pettigrew 1995). Early studies of electrosensory threshold found considerable variation between animals (Scheich et al. 1986), but later studies showed that this variation could be explained by a pronounced directionality of the sensitivity to electrical stimulation (Manger and Pettigrew 1995). The optimal axis of electrical stimulation is ~80° from the rostral pole of the bill (i.e. to the side of the bill) and ~20° below the horizontal plane (Fjällbrant et al. 1998). The preferred axis is ~60 times more sensitive than the non-preferred axis. Most significantly for understanding the neural mechanisms of this system, the most sensitive direction is at a right angle to the parasagittal stripes of electroreceptors on the bill (Pettigrew et al. 1998).

How is directionality achieved? The platypus cannot use time-of-arrival differences in the electrical impulse (e.g. across the width of the bill), because the electrical signal propagates at light speed and differences in timing across the bill are three orders of magnitude too small to be detected by neural systems (Pettigrew *et al.* 1998).

Pettigrew and colleagues have proposed that some sort of central neural reconstruction of the isoelectric field lines over the bill achieves the directionality (Pettigrew *et al.* 1998). This is most likely achieved at the primary somatosensory cortex level. Motor mechanisms (presumably arising in the motor cortex) subsequently trigger a head saccade in a direction at right angles to the isoelectric lines, i.e. towards the source at the side of the bill. The lower threshold for field lines that run parallel to the side of the bill (and the parasagittal sensory gland stripes) can then be explained by a preferential bias to signals to the side of the bill (Pettigrew *et al.* 1998).

Directionality may also be achieved by central analysis of the decay of the electrical field strength across the bill. As noted above, the factors determining decay of the signal are complex and difficult to express in a single cubic or quadratic formula (Fjällbrant *et al.* 1998), but field strength differences are significant across the width of the bill. The physics of decay detection are complex, let alone the neurophysiology, but Fjällbrant and colleagues conclude that this form of directionality would favour the detection of weak sources at close range over strong sources at longer distances. A pattern of sensitivity that favours signals to the side of the bill also reduces the sensitivity of the system to fields induced by the platypus's own muscles (Fjällbrant *et al.* 1998; Pettigrew 1999).

The bill as an electrical and mechanical telereceptor

Early conceptions of the role of the push-rod organs have naturally focused on their role as contact tactile receptors, but more recent models of the bill emphasise the potential role of the push-rod organs as telereceptors (i.e. detectors of pressure waves in the water without any direct contact being made between prey or other solid objects and the skin). This is made possible by the exquisite sensitivity of the push-rod organs in the platypus, due to the relatively loose attachment of the rod to surrounding epidermis, as discussed above.

Electrical signals propagate at the speed of light, whereas pressure waves are substantially slower. A system for comparing input from the two stimuli could provide useful information concerning the distance to prey. The tail flick of a crustacean like *Cherax quadricarinatus* generates a mechanical disturbance in the water with a delay of ~10 ms after the electrical signal at a distance of ~15 cm, and latencies vary between 5 ms and 50 ms for distances between 5 cm and 60 cm (Pettigrew *et al.* 1998). Significantly, the mechanical disturbances of the water at these distances exceed the 20 μm displacement threshold that has been reported for the push-rod organ. Provided that the central systems of the platypus can analyse and respond to the 10 ms delay, then a system that compares these stimuli could provide an effective method for assessing range.

The potential combination of electro- and mechanosensory information has been highlighted by the presence of bimodal somatosensory cortical neurons that respond to both stimuli (Manger *et al.* 1996). Pettigrew and colleagues suggest that such a system would involve activation of cortical bimodal neurons that facilitate optimally at a variety of different time delays between electrical and mechanical signals (Fig. 9.6; Pettigrew *et al.* 1998). Selective facilitation of particular cortical neurons would provide a pathway for decision-making about range such as to direct motor systems controlling lunging head movements towards prey.

The mechano- and electrosensory systems of the platypus bill may therefore act cooperatively in judging the direction and distance of free-swimming or fleeing prey.

A cautionary note: the platypus is not such a discerning diner

Reading the discussion of sensory modelling above might leave one with the impression that the platypus has a sophisticated ability to unerringly track down moving prey and separate food from debris. The reality, however, may not be so clear-cut. Analyses of the stomach contents of platypus indicate that they are far from discerning in what they ingest. Further questions concern the magnitude of electrical signals from platypus prey and how these compare to bill electrosensitivity.

Studies that have analysed the sensitivity of the electrosensory apparatus have focused on crustaceans, but these are only a small part of the reported natural diet. Analysis of cheek pouches indicates that freshwater crustaceans make up only 12% of the winter diet (Grant and Carrick 1978). Burrell (1927) notes that natural food for the platypus must vary by season and habitat, but includes 'immature molluscs, aquatic worms, the aquatic larvae of many insects, such as dragon-flies, caddis-flies, may-flies and the like, the larvae and perfect insects of groups such as the water beetles and water-fleas – which are wholly aquatic – bottom-feeding water-bugs, and such crustacea as inhabit the bottom of streams'. Some decapod crustaceans (e.g. *Paratya australiensis*) have very high amplitude electromyograph potentials (1900 μV/cm; Taylor *et al*. 1992) and would be detectable with the platypus electrosensory apparatus, but many aquatic invertebrates do not. Earthworms, water beetles, horsefly larvae and caddis fly larvae are all unlikely to be detectable by platypuses on the basis of the muscle potentials. The field strength of electrical emissions from these animals all fall well below the observed behavioural threshold (Taylor *et al*. 1992). It may be possible that the platypus uses contact electroreception (i.e. detection of electrical signals when the prey is in contact with the bill surface) to distinguish live prey from inanimate objects, but this remains untested. Alternatively, electroreception may have evolved to serve a very specific role at critically important times of the year, when availability of electrically 'loud' prey coincides with breeding or lactation.

A further important point is that the platypus ingests its food with liberal amounts of mud, sand and even gravel, so much so as to earn the animal the epithet 'the mud-sucking platypus' in Burrell's famous doggerel. This extraneous material is not necessarily ejected from the mouth, with much of it ingested and traversing the entire gastrointestinal tract.

Two points arise from these observations. First, some of the natural food of the platypus (e.g. pelecypod molluscs, aquatic worms, some larvae) are unlikely to generate the sort of electrical signals that have been reported for crustaceans, so their detection is more likely to be based on tactile rather than electrosensation. Second, and more significantly, if the platypus has such an acute electrosensory ability, why does it ingest such large amounts of mud with its food? These observations are more consistent with the platypus adopting a strategy of swallowing mud and prey together before sifting food from substrate in the oral cavity, rather than homing in on discrete prey with a sophisticated combination of electro- and mechanoreception.

The beak receptors and trigeminal nerve of the adult echidna

Electroreception in the snout of the short-beaked echidna

The first electrophysiological study of putative electroreceptors in the short-beaked echidna was by Gregory and colleagues (Gregory *et al*. 1989b) who also conducted behavioural tests using one of their echidnas. They identified putative electroreceptors confined to the tip of the echidna beak that transmitted information centrally through axons of the infraorbital nerve (a branch of the maxillary division of the trigeminal nerve). The isolated electroreceptors were normally silent (in contrast to the tonic activity of these axons in the platypus) or had resting rates of discharge only when the skin was moistened with saline. In any event, discharge rate increased when a cathodal stimulus was applied to the site of the electroreceptor, but declined when a supramaximal stimulus was reached. Electroreceptors showed no response to anodal stimulation, but underwent a rebound response when the anodal stimulation was removed.

There is a significant difference in threshold for electrical stimuli applied in air or water, with behavioural implications for the wild animal. Electroreceptors in the snout of the echidna are poorly responsive to electrical fields in air, with thresholds in the range of 70 mV/cm to 4.0 V/cm, with an average of 1.1 V/cm (Gregory *et al*. 1989b). Threshold is always lowest when the tip of the snout has just been moistened and rises substantially as the skin dries out. Threshold is much lower when the animal either inserts its snout into water or when the snout of an anesthetised echidna is immersed in tap water. The threshold for detection of potential difference applied across water in the well of the training apparatus is between ~1.8 and 3.1 mV/cm for the behaving echidna. Thresholds for isolated electroreceptors are at ~4.5 mV/cm when the stimulus is applied underwater, with increased responsiveness up to 550 mV/cm. Threshold also varies according to the frequency of oscillating electrical fields, with the lowest threshold at about the very low frequency of 20 Hz. These observations comparing the electrical threshold in air versus water indicate that electroreception is unlikely to be of benefit to the echidna in prey detection unless the beak is inserted into sodden leaf litter, an unlikely situation in many of the environments (e.g. desert or dry sclerophyll forest) that the short-beaked echidna inhabits.

Mechanoreception in the beak of the short-beaked echidna

Electrophysiological studies of infraorbital nerve axons supplying the beak of the echidna have identified both rapidly and slowly adapting mechanoreceptors (Iggo et al. 1983, 1985; Iggo et al. 1996; Proske and Gregory 2003), with the latter being the most numerous. Rapidly adapting (i.e. vibration-sensitive) mechanoreceptors were scattered widely across the skin of the beak and responded in a sustained fashion to vibration in the 200–300 Hz range. Some were identified as Pacinian-like in their responsiveness, in that they showed sustained responsiveness over a wide range of vibration frequencies (up to 800 Hz). Others were defined as non-Pacinian-like, because they gave only intermittent responses to vibration at higher frequencies.

Slowly adapting mechanoreceptors in the beak of the echidna were classified as either SA I (Merkel type), or SA II (Ruffini type), on the basis of the regularity of their discharges in response to mechanical stimulation. SA I receptors have an indentation threshold of 4 µm and impulse discharge increases progressively up to an indentation of 40 µm. A key feature of SA I receptors is their small receptive fields (less than 100 µm diameter). SA II receptors have double the SA I receptors' responsiveness for a given skin indentation , and also have a much more regular discharge than SA I receptors, with a narrower distribution of inter-spike intervals. Both SA I and SA II receptors transmit centrally along axons with similar mean conduction velocities, but different conduction velocity ranges (ranges of 18–25 m/s for SA I, and 12–33 m/s for SA II; Iggo et al. 1985).

Iggo and colleagues (Iggo et al. 1996) were not able to definitively associate particular mechanoreceptor responses recorded electrophysiologically with the push-rod organs of the echidna snout. Nevertheless, it is most likely that SA I receptors are within the push-rod organ and that SA II receptors preferentially signal stretch of the skin (Proske and Gregory 2003).

Thermoreception in the beak of the short-beaked echidna

The first study of thermoreceptors in the beak of the echidna (Iggo et al. 1985) was undertaken before it was realised that electroreceptors were present, so some of the axons originally shown to have differing responses to changing ambient temperature are in fact electroreceptive. Iggo and colleagues identified cold receptors with small receptive fields (less than 1 mm in diameter). These were silent at 35°C but began to respond as the temperature dropped below 29°C. Warm receptors were also identified in the snout of the short-beaked echidna, again with apparently quite small receptive fields (between 1 mm and 2 mm in diameter). Warm receptors respond with a large transient increase in discharge rate within 1.5 seconds of exposure to the warm stimulus and are served by large diameter myelinated afferents. The myelination of warm receptor pathways may be beneficial, because it provides rapid warning of any elevation in ambient temperature that might cause heat stress for the echidna. The electrophysiological properties of thermoreceptors in the echidna snout generally resemble those in placentals.

Types of sensory receptors in the beak of the short-beaked echidna

The morphology and distribution of sensory receptors on the beak of the short-beaked echidna show some similarities to those in the platypus, but also some key differences (Andres et al. 1991; Manger and Hughes 1992). The most sensitive part of the snout, judged on both morphological and electrophysiological grounds, is the distal 1 cm to 2 cm.

Andres and colleagues identified several free nerve terminals and receptor complexes in the beak of the short-beaked echidna. These include two types of free nerve terminals; lamellated receptors; intraepidermal discoid, vesicle chain and cluster receptors; Merkel cell receptors; and gland duct (putative electro-) receptors (Fig. 9.4). Some of these are combined in receptor organs (e.g. the push-rod mechanoreceptor system) as described for the platypus.

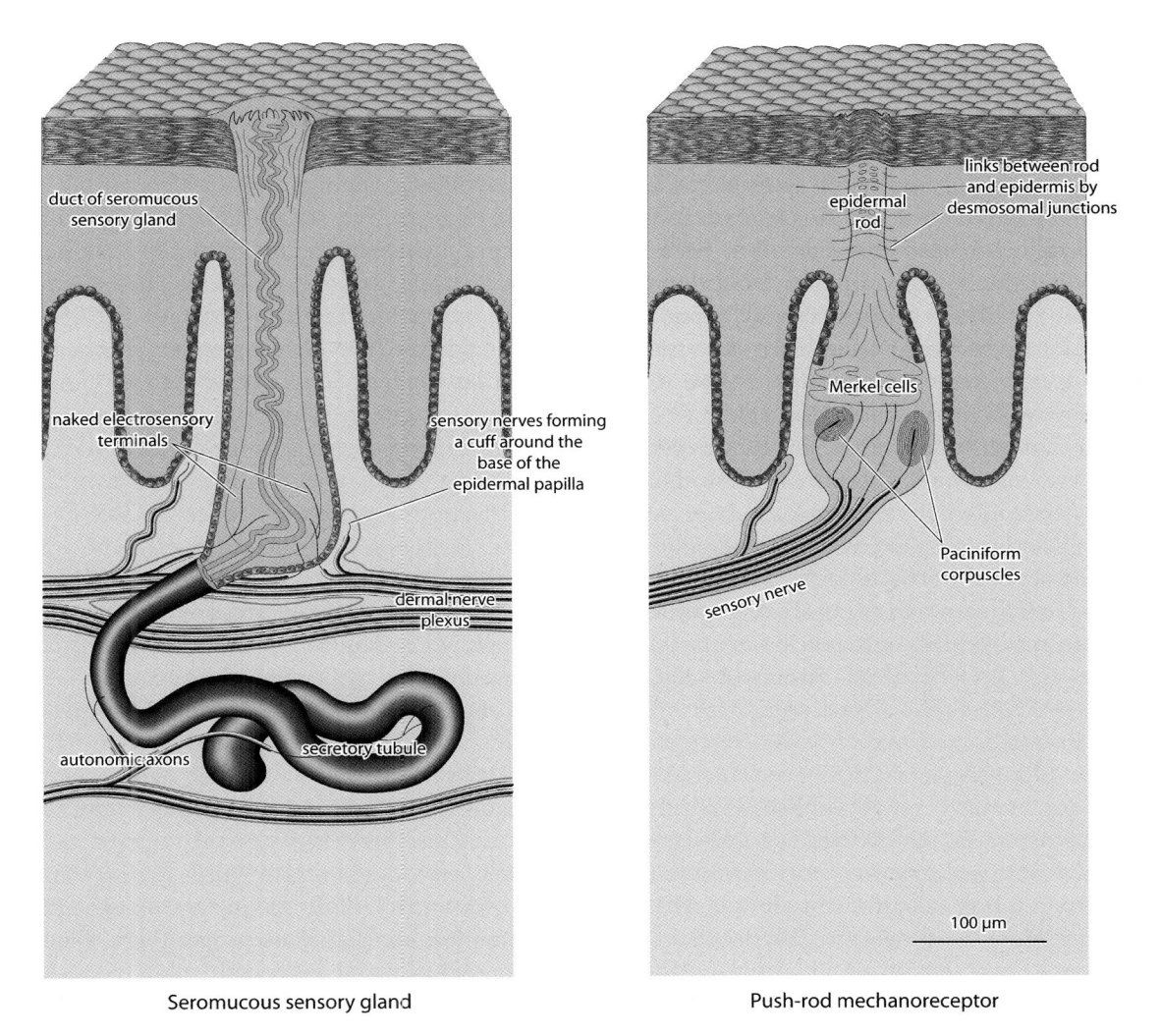

Figure 9.4: Line diagrams showing examples of sensory mucous and sensory serous glands (both putative electroreceptors) and the push-rod mechanoreceptor in the beak tip of the short-beaked echidna. The illustrations are based on figures in Gregory *et al.* 1989; Andres *et al.* 1991; Manger and Hughes 1992; Proske *et al.* 1998.

Free nerve terminals were divided into two types. One group (the first category of Andres *et al.* 1991) are supplied by thin myelinated axons with long receptor axons. This group join Remak bundles along with C fibres and terminate inconspicuously in the connective tissue compartments of the dermal reticular layer, dermal papillae and mucous membranes of the vestibulum of the mouth and nose. The second group (the second category of Andres *et al.* 1991) have a more localised distribution, are supplied by thickly myelinated axons and terminate with short receptor axons. Many are distributed to the border region between the epidermal layer and the connective tissue at the base of the dermal papillae, giving rise to the name 'borderline receptor', as was adopted by Andres and colleagues. Another type within the second category is a group of non-encapsulated Ruffini receptors, within the connective tissue of the syndesmotic joints between the bones of the snout skeleton.

Lamellate receptors include Paciniform lamellated corpuscles that are widely distributed in the skin of the snout and in the mucosa of the mouth and nasal cavity. Paciniform corpuscles are also found at the base of push-rod mechanoreceptors and in the bases of the nasal vestibular folds. Small corpuscles are 10–15 µm in diameter, while large corpuscles within the reticular layer of the lower jaw are 30–40 μm in diameter and up to 70 μm in length.

Intra-epidermal terminals include the discoid, vesicle chain and cluster receptors. Discoid receptors are found in the epidermis in the basal folds within the nasal vestibulum. Vesicle chain receptors (further divided into central and peripheral subtypes by Manger and Hughes 1992) are found exclusively within the columns of the push-rod mechanoreceptor organs (see below). Intra-epidermal cluster receptors are found within the epidermal layer of the dorsal surface of the tongue tip.

Merkel cell receptors are morphologically similar to those in other mammals. They are found at the epidermal base of the push-rod organ (see below) and in the epidermal ridges of the dorsum of the tongue.

The push-rod organs of the echidna snout have some similarities to those of the platypus, but also a key functional difference (Andres *et al.* 1991; Manger and Hughes 1992). In both species, the central element of the organ is the push rod that is composed of flattened spinous cells connected to each other by many desmosomes. The size of the push rod varies with skin region, but in short-beaked echidna it is usually ~300 μm long and 40–50 μm in diameter (Manger and Hughes 1992). A key difference from the platypus push-rod system is that the push rod of the echidna is linked to surrounding epidermal cells by desmosomal junctions, thereby limiting its mobility and sensitivity to shearing movement (Manger and Hughes 1992). An average of 10 myelinated axons arise from a nearby dermal plexus to innervate the sensory organs of the push-rod mechanoreceptor. Three types of sensory nerve endings are associated with the push-rod organ: vesicle chain (central and peripheral), Merkel cells and Paciniform corpuscles. Each push-rod is equipped with 10–24 vesicle chain receptors (depending on rod length), 18–24 Merkel receptors and several Paciniform corpuscles (Andres *et al.* 1991). There is also an important difference in the arrangement of Paciniform corpuscles at the base of the push-rod organ between platypus and short-beaked echidna. The platypus push-rod system has three Paciniform corpuscles, whereas the short-beak echidna has only two. Furthermore, the Paciniform corpuscles of the platypus are arranged in three orthogonal axes to detect movement in three spatial dimensions, whereas the Paciniform corpuscles of the echidna are randomly aligned (Manger and Hughes 1992).

Sensory and non-sensory glands have been described in the snout skin of the echidna (Andres *et al.* 1991; Manger and Hughes 1992) and these can be distinguished only by serial section histology. As in the platypus, the sensory gland apparatus is associated with an extensive dermal mucous gland ~300 μm in diameter (Manger and Hughes 1992). Axons arise from nearby dermal plexuses to innervate the modified epidermal portion of the gland by forming a sensory cuff. The sensory nerves lose their myelin sheath ~15 μm from the epidermis and, once inside the epidermis, form bulbous nerve terminals. Manger and Hughes (1992) identified several key morphological differences between echidna and platypus sensory glands. First, the epidermal pore is a quite simple structure in echidnas, whereas it is highly elaborate in the platypus. Second, myelination stops before the axon enters the epidermis in the echidna, whereas the myelination of the sensory axon continues into the epidermis in the platypus. Finally, there are only four to 10 sensory nerve terminals for each sensory mucous gland of the echidna, whereas there may be up to 30 in each platypus sensory gland.

Manger and Hughes (1992) identified a further type of sensory organ known as the innervated epidermal pit, although the detailed ultrastructure of the organ has not been described. This sensory organ has no apparent associated gland, but has a canal with a pore surrounded by keratinocytes. The organ also has an epidermal peg that is innervated by axons from a nearby dermal plexus. The epidermal pit may also serve electrosensation (Manger and Hughes 1992).

Distribution of sensory receptors in the snout of the short-beaked echidna

The push-rod and gland duct receptor systems predominate over the skin of the upper and lower lip (Andres *et al.* 1991; Manger and Hughes 1992). Mechanoreceptors are found over the entire distal 1 cm of the beak and surround the nares, and the sensory glands are located mainly on the most distal tip of the upper beak (Manger and Hughes 1992). The highest spatial density of the push-rod and gland systems is found on the underside of the upper lip and the upper surface of the lower lip, reaching spatial densities of 36–40 push-rod mechanoreceptors and seven sensory gland duct receptors per mm² (Andres *et al.* 1991). Innervated epidermal pits are concentrated on the lateral edge of the beak, immediately caudal to the opening of the nostril (Manger and Hughes 1992).

Nerve supply of the snout of the short-beaked echidna

All three divisions of the trigeminal nerve (ophthalmic, maxillary and mandibular) supply the beak of the echidna (Andres *et al.* 1991). The ophthalmic division has a large lateral ethmoidal branch that supplies both the dorsal shaft and the dorsal tip above the nares (Andres *et al.* 1991). The maxillary division has a very large infraorbital branch to the behaviourally important areas of the snout around the nares, as well as substantial buccal and palatine branches to the rim and interior of the oral cavity. Finally, the mandibular division provides an inferior alveolar branch with anterior and posterior mental branches to the lower jaw shaft and tip.

Latency measurements of electroreceptor responses in the echidna suggest much lower axonal conduction velocity in the trigeminal nerve of the echidna (18 m/s) than in the platypus (56 m/s) (Gregory *et al.* 1989a; Iggo *et al.* 1988). This may be due to the much smaller diameter of nerve fibres supplying the sensory glands in echidnas compared to the platypus.

Sensory receptors and their distribution in the snout of the long-beaked echidna

Sensory glands and push-rod mechanoreceptors have also been identified at the tip of the beak of the long-beaked echidna (*Zaglossus bruijnii*) (Manger *et al.* 1997). An early study by Kolmer (1925) identified specialised glands in the tip of the snout of *Zaglossus*. These sensory glands appear to be quite limited in their distribution, in that they are all located distal to the maxillofacial (infraorbital) foramen, i.e. within 2 cm of the tip of the beak (Manger *et al.* 1997).

The sensory gland apparatus typically includes a large dermal gland that secretes a seromucous substance and is connected to the epidermal surface by a duct. The duct is ~450 µm in length and opens at the epithelial surface through a pore 100 µm in diameter that is easily seen with a dissecting microscope. The gland is innervated at the papillary region where the duct enters an epidermal peg. Features that have been described for the papillary nerve terminals in the platypus (i.e. bulbous portions of the nerve terminals, terminal filaments and arbours of nerve terminals around the circumference of the papilla) have not been seen in *Zaglossus*. Spatial density of sensory glands is of the order of 12/mm² of skin, but might be higher at the beak tip and the total number of sensory glands has been estimated to be in the order of 3000 (i.e. more than in *Tachyglossus* but fewer than in the platypus). The question of whether sensory serous glands or non-sensory mucous glands (as described for the platypus) are present in the beak skin of the long-beaked echidna has not been resolved.

The push-rod mechanoreceptor complex has also been identified in *Zaglossus bruijnii* (Manger *et al.* 1997). In this species, the push-rod is typically 250 µm long and 45–50 µm wide, a little larger than in the short-beaked echidna but smaller than in the platypus. Like the short-beaked echidna and unlike the platypus, the push-rod of the long-beaked echidna is attached to the surrounding epidermis, implying a more restricted movement of the rod compared to the platypus. The same triad of mechanosensory structures was found associated with the push-rod complex as seen in the platypus and short-beaked echidna, i.e. the vesicle chain receptors within the rod, Merkel cells around the base and Paciniform corpuscles below the rod. The spatial density of push-rod mechanoreceptor complexes is comparable to sensory glands and is probably also confined to the skin of the distal 2 cm of the beak.

The push-rod structures are most likely mechanoreceptors based on their similarity to clearly mechanoreceptive structures in the short-beaked echidna (Gregory *et al.* 1989b). There is no direct evidence, either behavioural or electrophysiological, that the sensory glands of the beak of *Zaglossus bruijnii* are actually electrosensory. This role is presumed on the basis of purely morphological similarities between the sensory glands and the sensory mucous glands of the platypus and short-beaked echidna (Manger *et al.* 1997). Nevertheless, it is possible to make some deductions about possible uses for the sensory apparatus of the beak tip. The usual diet of the long-beaked echidna is earthworms and grubs in the leaf litter and soil of the floor of humid, montane environments of New Guinea. In such a moist environment electroreception may be of benefit in the detection of prey. In any event,

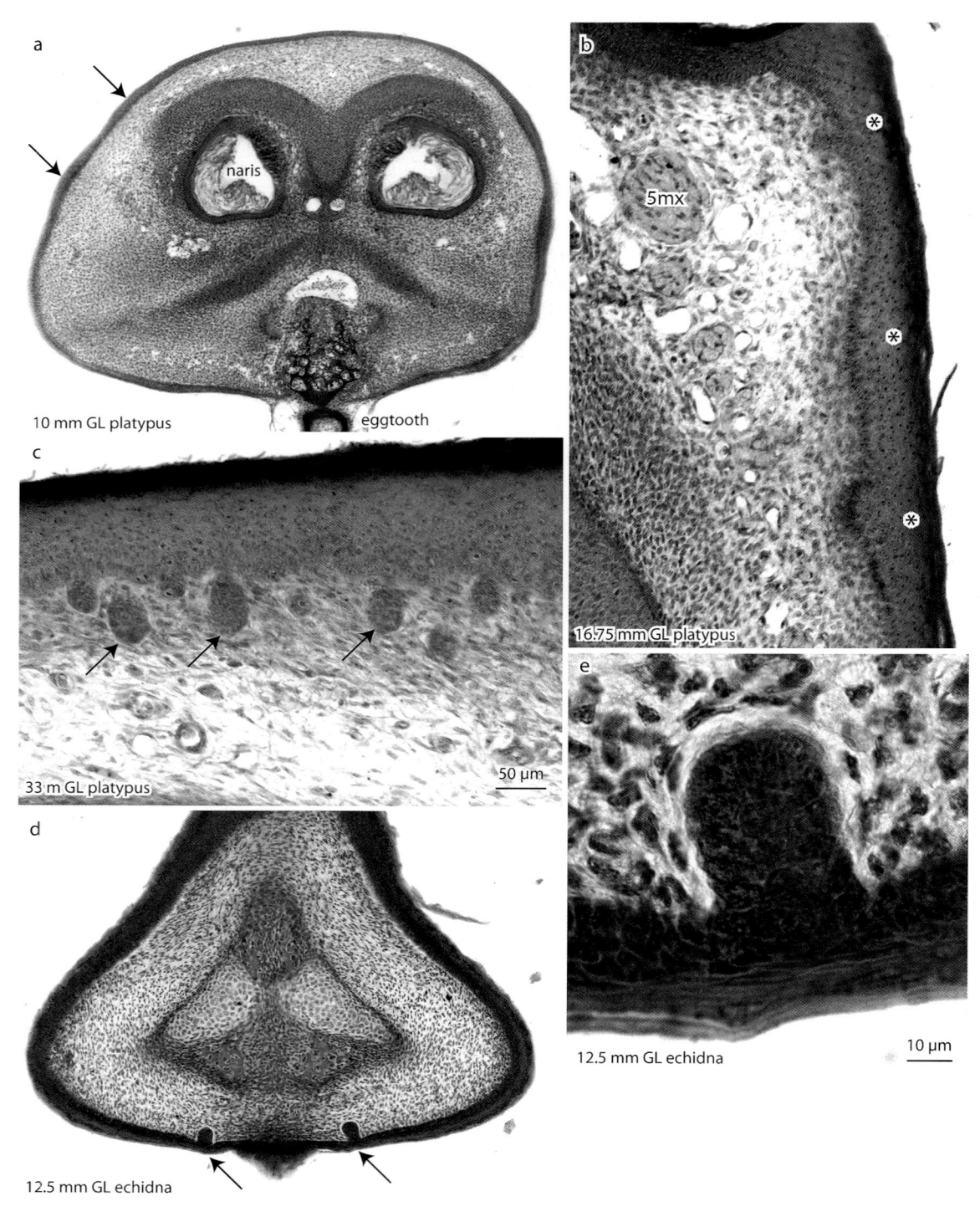

Figure 9.5: Photomicrographs showing early development of snout sensory glands of platypus and short-beaked echidna. In the platypus, the earliest precursor of sensory organs on the snout are simple epidermal ridges (arrows in a). It is not until after hatching that these begin to thicken and invade the dermis (asterisks in b) and gland primordia are not visible until the end of the first post-hatching week (arrows in c). Sensory gland development appears to occur earlier in the echidna, with epidermal specialisations visible as buds of epidermal cells (arrows in d) invading the dermis even before hatching (e – higher powered view of structures arrowed in d). 5mx – maxillary division of the trigeminal nerve.

the long beak is used as a push or thrust probe to sample the leaf litter and soil. Whether mechanoreception or electroreception is the more important sense for detecting prey remains to be determined; most likely the animal uses a combination of both.

How do the short- and long-beaked echidnas use their beaks in the wild?

The electroreceptive capabilities of the platypus have been extensively studied and modelled, but electroreception in the short-beaked echidna has received less attention and the ability is only inferred for the long-beaked echidna species on the basis of receptor morphology.

Gregory and colleagues have performed forced choice behavioural studies of short-beaked echidnas under laboratory conditions (Gregory *et al.* 1989b). Although the echidnas could detect electrical fields down to 1.8 mV/cm (similar to the threshold for individual infraorbital fibres), this was for a field applied to water in a trough, rather than the open air. This raises the question of how effective electroreception is for the wild short-beaked echidna, particularly in dry environments. Nevertheless, the close correspondence between the behavioural and individual fibre thresholds suggests that the short-beaked echidna does not have any mechanism for spatial or temporal summation. The snout tip is essentially functioning as a thrust probe sensor, with no amplification or directional capability.

So what do we know about electroreception in wild short-beaked echidnas? A study of echidnas fitted with radio-tracking devices (Augee and Gooden 1992) found that short-beaked echidnas investigated and dug up buried batteries that carried a charge significantly more often than dead batteries. The field strength generated by the buried 9V batteries used by Augee and Gooden (1992) was of the order of 15–20 mV/cm at distances of 12–14 cm. Anecdotal evidence suggests that echidnas are more active and effective in hunting for invertebrate prey after rain and in moist soil (Smith *et al.* 1989) when electroreception is likely to be more effective. Interestingly, Augee and Gooden reported in a footnote that a blind echidna from Texas, Queensland had an exceptional ability to detect charged batteries, suggesting that loss of vision had heightened its electrosensitivity.

How do field potentials in these studies compare with the observed electromyographic potentials of the invertebrates that echidnas might hunt? The limited information on this suggests that electroreception is unlikely to be of benefit in hunting at least one type of prey, i.e. *Lumbricus* spp. earthworms. The measured EMG amplitudes for earthworms are only of the order of 3 µV/cm with a frequency of only 3 Hz (Taylor *et al.* 1992), far too low to be detected by the beak of the echidna.

The electrosensory capabilities of long-beaked echidnas can only be inferred, but the moist environment occupied by *Zaglossus* species is probably more suited to the use of electroreception than that inhabited by *Tachyglossus*. Since the main dietary items for long-beaked echidnas are earthworms and grubs, and earthworms have low amplitude/low frequency EMG (Taylor *et al.* 1992), the case for active use of electroreception in prey detection by *Zaglossus* also remains disputable.

In summary, although electrosensitivity has been shown for the short-beaked echidnas, the role that it plays in actual prey detection in the wild remains unproven. Given the profound tactile sensitivity of the snout and the robust olfactory apparatus of the echidnas, it is more likely that touch and olfaction are of greater behavioural significance.

Development of bill and beak receptors and trigeminal ganglion in monotremes

The development of bill and beak receptors and the trigeminal ganglion has been analysed using the Hill and Hubrecht collection embryos and collections of post-hatching platypuses at the National Museum of Australia. Just as the adult platypus and short-beaked echidna have very distinct patterns of receptors in the bill and beak, these two species also have very distinctive developmental trajectories in both the snout receptors and the trigeminal ganglion.

Development of receptors in the platypus bill

Ashwell and colleagues (Ashwell *et al.* 2012) identified epidermal specialisations in the snout of the platypus from about the last third of incubation (Fig. 9.5a, Appendix Fig. 3). At this time, the features consist only of longitudinally running zones of epidermal thickening, which progress to ridge-like epidermal

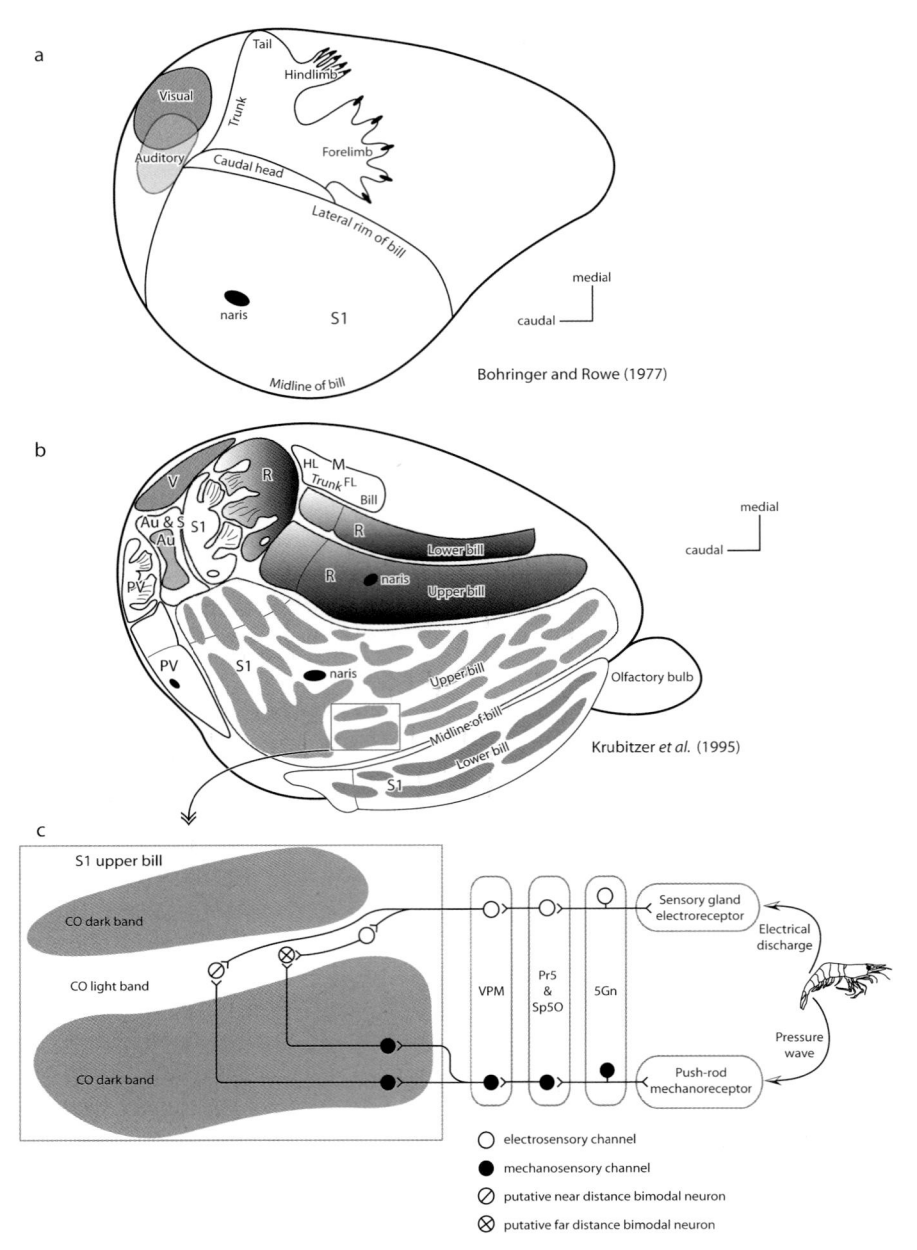

Figure 9.6: Body representations in the cerebral cortex of the platypus based on electrophysiological studies. Platypus body map on the somatosensory cortex (S1) based on the findings of (a) Bohringer and Rowe (1977) and (b) Krubitzer et al. (1995). Note the single S1 field reported by Bohringer and Rowe and the four somatosensory fields found by Krubitzer and colleagues. The dark and light zones in the bill representation of S1 are the cytochrome oxidase dark and light regions described by Krunbtizer and colleagues. The rectangular inset in (b) outlines a region of the bill representation in S1 that has been enlarged in (c). Illustration (c) is a schematic diagram to show the electrosensory and mechanosensory channels and their possible interaction in the cytochrome oxidase light regions. The hypothetical circuit diagram illustrates a possible mechanism for comparing the latency of electrical and mechanical stimuli as part of a prey ranging system (Manger et al. 1996; Pettigrew et al. 1998, Elston et al. 1999). 5Gn – trigeminal ganglion; Au – auditory cortex; Au & S – cortex responding to both auditory and somatosensory stimul; FL – forelimb; HL – hindlimb; M – manipulation somatosensory field; Pr5 – principal trigeminal nucleus; PV – parietoventral somatosensory field; R – rostral somatosensory field; S1 – primary somatosensory cortex; Sp5O – oral part of nucleus of trigeminal spinal tract; V – visual cortex; VPM – medial part of ventral posterior thalamic nucleus.

in-growths in the immediate post-hatching period (Fig. 9.5b). These in turn develop into individual finger-like epidermal invasions of the dermis at the end of the first week of post-hatching life (Fig. 9.5c). The longitudinal arrangement of these specialisations closely matches the position of the strips of sensory mucous glands of the adult bill, indicating that the striped topography of these receptors in the bill is established from the earliest time of epidermal differentiation. The specialisations are loosely clustered into longitudinal domains in the 33 mm GL platypus (around the end of the first week after hatching). One zone begins at the rostrocaudal level of the nares and runs dorsal to the nasal cavity until it reaches the large skin fold at the caudal edge of the bill. Another zone runs along the lateral edge of the upper bill, beginning from immediately rostral to the nares, but becoming discontinuous as it reaches the lateral angle of the mouth.

Quantitative estimates of sensory gland electroreceptors based on counts of pores suggest that the number of sensory glands reaches a peak at ~24 days post-hatching (Manger et al. 1998c). Manger and colleagues have reported a loss of 60% of the sensory glands between 24 days and 28 days after hatching, but independent verification by histology of the bill is necessary, given that this claim is based solely on pore counts of the bill surface. The reported drop in the sensory mucous gland population is said to coincide with the first appearance of push-rod mechanoreceptors and sensory serous glands. Innervation of the bill is undifferentiated at 28 days after hatching, but reaches the adult configuration between 3 months and 6 months post-hatching.

Development of receptors in the snout of the short-beaked echidna

The development of epidermal specialisations in the beak of the short-beaked echidna is quite different from that of the platypus (Ashwell et al. 2012, Appendix Fig. 3). Epidermal specialisations in the echidna beak appear even before hatching, as discrete epidermal pegs that penetrate to the dermis (Fig. 9.5d, e). These pegs also look more developmentally advanced than the epidermal features of the early post-hatching platypus. More significantly the epidermal specialisations of the echidna beak have a very different distribution from the platypus, in that the pegs are located only on the underside of the upper beak at this early age. During the first post-hatching week, the epidermal specialisations continue to be concentrated on the underside of the beak tip, with others appearing on the ventral surface of the lower beak. Few are seen during the first post-hatching week on the dorsolateral surface of the end of the upper beak. By the time the hatchling reaches a body length of 98 mm, the tip of the upper beak and the distal ventral surface of the lower beak are covered by putative developing electroreceptors with a few epidermal specialisations on the dorsal surface of the upper beak caudal to the nares. The presence of ducts confirms that these epidermal specialisations are eccrine members of the gland duct receptor system.

Development of the trigeminal ganglion and nerve in the platypus and echidna

The earliest outgrowth of trigeminal nerve branches into the maxillary and mandibular processes of the head is visible as soon as the ganglion is recognisable. Discrete ophthalmic, maxillary and mandibular divisions can be identified at 8.5 mm GL in the platypus and 7.3 mm GL in the short-beaked echidna (Ashwell et al. 2012, Appendix Fig. 3). Both the trigeminal ganglion and trigeminal nerve divisions are strikingly different between the platypus and echidna at the earliest age they can be recognised. For example, at around 6.0 mm GL the ganglion of the platypus has a volume of 0.0317 mm^3 compared to only 0.000 85 mm^3 for the echidna and this large difference continues throughout the entire incubation and post-hatching periods. Similarly, there is approximately a sixfold difference in the total cross-sectional area for all trigeminal divisions between the platypus and echidna at hatching, rising to more than a 15-fold difference in adulthood.

Evolutionary implications of the pattern of snout receptor development in the platypus and echidna

Ashwell and colleagues found a very different distribution of sensory gland primordia in platypus and echidna even before the time of hatching, along with species-specific growth of the trigeminal ganglion and trigeminal nerve divisions (Ashwell et al. 2012). The early ontogeny of a species-specific, spatially determined distribution of epidermal specialisations suggests that developmental processes in the two species are shaped differently from early embryonic stages to adulthood. This developmental difference produces the characteristic distribution of sensory mucous glands which is very distinct in the adults of these two species.

Ashwell and colleagues argued that the early development of a distinct pattern of peripheral trigeminal specialisations in the two monotremes is not consistent with the often-stated notion of the modern echidnas being derived from a platypus-like ancestor. Such an ancestor is supposed to have left the aquatic environment to invade a land-based insectivorous niche, thereby evolving into the modern echidna. If the common ancestor of modern monotremes were as extremely specialised as the modern platypus, then one would expect the embryos and early hatchlings of the modern echidna would show some developmental legacy of that ancestry, e.g. in the form of ganglionic hypertrophy during embryogenesis followed by pronounced developmental cell death, or share a similar pattern of distribution of electroreceptors during early development. Since the developing echidna always follows an *echidna-specific* pattern of trigeminal development, it seems unlikely that the echidna ever had a *platypus-like* ancestor.

The extreme bill and trigeminal specialisations of the platypus (i.e. the large trigeminal ganglion and trigeminal nerve divisions, and the longitudinal zonation of sensory mucous glands in the bill) are more likely to have emerged *after* the divergence of the ornithorhynchid and tachyglossid lineages. The lack of trigeminal ganglion hyperplasia or longitudinal zonation of electroreceptors at any stage during development of the echidna supports this notion. Trigeminal hyperplasia and longitudinal zonation of electroreceptors could have arisen in the ornithorhynchid lineage by the retention of trigeminal ganglion cells generated in excess during early embryonic life, perhaps occurring in concert with the progressive shift of sensory mucous glands into longitudinal arrays. This possibly requires mutations to increase the neurogenesis of profundal (trigeminal) placode and neural crest cells to boost the size of the primordial ganglion and nerve to that observed during incubation. This model of periphery-driven trigeminal evolution is consistent with the ideas of Finlay's group who claim that the periphery of the nervous system is the principal locus for producing functionally effective changes in neuroanatomy between species (Finlay *et al.* 2011).

Trigeminal sensory nuclei in monotremes

The trigeminal sensory nuclear complex of the platypus

Axons of the trigeminal nerve penetrate the pia of the brainstem and turn either rostrally to enter the principal trigeminal sensory nucleus (Pr5) or form the mesencephalic root of the trigeminal nerve (me5). Fibres which turn caudally on brainstem entry form the trigeminal spinal tract (sp5), carrying axons to the three subnuclei of the nucleus of the trigeminal spinal tract (oral part – Sp5O; interpolar part – Sp5I; and caudal part – Sp5C).

Several authors have noted unusual features of the trigeminal nuclear column in the platypus (Fig. 9.1b to f; see atlas plates Pl-Ad19 to 30) even though the same basic elements of trigeminal sensory complex are recognisable (Hines 1929; Watson *et al.* 1977). Neurons of the mesencephalic nucleus of the trigeminal nerve (Me5) are arranged into circular arcs spanning the dorsal rim of the periaqueductal grey matter (Hines 1929), with a concentration in the midline (as is seen in reptiles). Most strikingly, the hindbrain parts of the trigeminal complex are extraordinarily large in the platypus (Hines 1929; Watson *et al.* 1977; Ashwell *et al.* 2006b) and there is considerable overlap in the rostro-caudal extent of the components of the column (Watson *et al.* 1977; Ashwell *et al.* 2006b). Another key feature is the presence of distinct magno- and parvicellular components in the Pr5 cell column (Watson *et al.* 1977; Ashwell *et al.* 2006b; Fig. 9.1b; see also atlas plates Pl-Ad19 to 21).

Ashwell and colleagues (Ashwell *et al.* 2006b) did not observe any linear arrangement of parvalbumin immunoreactivity in either fibres or cell bodies in the rostral trigeminal sensory nuclei of the platypus, such as might correspond to the striped distribution of electroreceptors in the bill. They did observe a cell sparse zone separating the groups of parvalbumin immunoreactive cells in the ventromedial Pr5 into medial and lateral compartments. They suggested that these two compartments might serve the two main trigeminal divisions for which highly specialised mechano- and electroreceptor function is reported (i.e. maxillary and mandibular), but they found no chemoarchitectural evidence for any finer scale subdivision of the trigeminal nuclei.

In summary, the trigeminal sensory nuclear complex of the platypus shows several striking anatomical features that complement the pronounced trigeminal specialisations of the bill. These include a large relative size of Pr5, Sp5O and Sp5I compared to other brainstem nuclei as well as the trigeminal sensory nuclei in placentals, the high numerical density of the constituent neurons of Pr5, Sp5O and Sp5I and the intense parvalbumin immunoreactivity of the neuropil of the same nuclei (Fig. 9.1b–f). In the absence of direct unit recording from the sensory trigeminal nuclei of the playpus it is difficult to say how much these anatomical specialisations contribute to the processing of the highly discriminative mechanotactile input from the bill (Bohringer and Rowe 1977), and/or electrosensation (Scheich et al. 1986; Gregory et al. 1987b, 1988; Proske et al. 1998). The trigeminal sensory nuclear column is, however, a major site of sensory processing, based on anatomical evidence alone. The large size of the rostral trigeminal sensory nuclei stands in contrast to the unremarkable size, neuronal density and chemoarchitecture of Sp5C in the platypus (see atlas plate Pl-Ad30). Presumably the latter serves only nociceptive input from the platypus head and bill, as seen for placentals.

The sensory trigeminal nuclear complex of the short-beaked echidna

The sensory trigeminal nuclear complex of the short-beaked echidna is divisible into subnuclei along similar lines to those applied to the sensory trigeminal nuclei of therians. In fact in many respects (e.g. overall cellular and fibre architecture, parvalbumin immunoreactivity), the trigeminal nuclei of the echidna are much more like those of the laboratory rat than the platypus. An exception to this is the almost exclusively midline distribution of neurons in the mesencephalic nucleus, very like that seen in reptiles rather than therians (Abbie 1934). The mesencephalic nucleus and root of the trigeminal nerve are concerned with proprioceptive sensibility of teeth and jaw musculature (Butler and Hodos 2005). Given that the echidna is edentulous and has relatively poor jaw musculature, the large size of the mesencephalic nucleus is baffling (Hassiotis et al. 2004a).

The principal trigeminal nucleus (Pr5) lies at the level of the motor trigeminal nucleus (see atlas plate Ec-Ad23) and can be divided into dorsolateral and ventromedial components (Pr5DL, Pr5VM). The neuropil of the Pr5DL and external parts of the Pr5VM are strongly reactive for cytochrome oxidase, indicating a dense input of highly metabolically active sensory axons from the trigeminal nerve.

Starting at the caudal end of the motor trigeminal nucleus and extending to the caudal end of the facial motor nucleus (see atlas plates Ec-Ad25 and 26) are the dorsomedial spinal trigeminal nucleus (Sp5DM) and the Sp5O, which are both intensely reactive for cytochrome oxidase. As in marsupials, the fascicles of the facial nerve run either directly through Sp5DM or along its dorsal rim.

The Sp5I (see atlas plate Ec-Ad27) extends between the caudal limit of the facial nucleus and the point at which the fourth ventricle closes to the central canal. The bulk of Sp5I is strongly reactive for cytochrome oxidase, but the lateral ventral edge of the nucleus contains large numbers of poorly cytochrome oxidase reactive fascicles. It receives axonal inputs that label for the B4 isolectin derived from *Griffonia simplicifolia*, consistent with the termination of small unmyelinated axons in this region (perhaps serving dull pain and temperature from the head and neck).

The Sp5C (see atlas plate Ec-Ad30) extends from the caudal end of the fourth ventricle to at least as far as the spinomedullary junction. As in therians, it probably runs even further into the upper cervical segments, but this cannot be said with certainty without recording units from the cervical sensory column. The Sp5C receives most of the C fibre afferents which label with the isolectin B4 from *Griffonia simplicifolia*. Ashwell et al. (2006b) noted that lectin reactivity in the echidna trigeminal nuclei was distributed in discrete clusters scattered in the lateral parts of Sp5C. This pattern of labelling is quite distinct from that of rodents, where the entire crescent-shaped gelatinosus layer is densely labelled (see summary in Ashwell et al. 2006b). This discontinuous distribution of lectin labelling is similar to that for calbindin immunoreactivity in the Sp5C of the echidna and suggests that the unmyelinated C fibre afferents are irregularly distributed in the caudal parts of the Sp5 of the echidna. The functional significance of this patchy distribution of afferents remains to be determined.

In addition to the main column of the trigeminal spinal tract nuclei, the paratrigeminal nucleus (Pa5; see atlas plates Ec-Ad27 and Pl-Ad26) lies embedded within the lateral spinal trigeminal tract alongside the caudal part of Sp5I and the rostral part of Sp5C. In rodents, Pa5 has been shown to be involved in the integration of sensory information from visceral and somatosensory sources and projects to the dorsal vagal sensory complex (Menétrey and Basbaum 1987; Armstrong and Hopkins 1998). A putative peri-trigeminal nucleus (Pe5) has also been identified in the echidna alongside the ventral spinal trigeminal tract at the level of rostral Sp5C (Ashwell et al. 2006b). The presence of both these accessory trigeminal nuclei in the monotreme brainstem indicates that the entire complement of trigeminal nuclei can probably be traced to the final common ancestor for monotremes and therians in the mid-Triassic.

Reactivity for cytochrome oxidase (CO) is often used in therians to map the termination of mitochondria-rich afferents, such as would be used for discriminative touch, or perhaps electroreception. Activity of CO is strong in all the trigeminal nuclei of the echidna compared to nearby structures, even though the sensory trigeminal root and the spinal trigeminal tract are only poorly active. Significantly, reactivity for CO does not show any discrete patches like the whisker representations in the rodent trigeminal sensory complex. However, strong activity is concentrated within the cell bodies of neurons within the rostral nuclei (e.g. Sp5O). How this relates to the 'hot spot' sensitivity of the beak tip remains unknown.

Development of trigeminal sensory nuclei in monotremes

In both platypus and short-beaked echidna, large neurons of the trigeminal sensory complex begin to accumulate in the hindbrain at the beginning of the middle third of incubation (~8 mm GL, Ashwell and Hardman 2012b, Appendix Fig. 3). The spinal trigeminal tract appears alongside them at about the same time, as trigeminal afferents continue to grow into the brainstem. In both species, all the parts of the trigeminal sensory nuclear complex are very similar in appearance to the homologous nuclei of the rodent brainstem at a similar stage of hindbrain development (e.g. embryonic day 15 or 16 in the laboratory rat; Ashwell and Paxinos 2007; Ashwell and Hardman 2012b).

Striking differences between the platypus and echidna emerge during the last third of incubation and are most likely due to major differences in the expansion of the rostral rhombic lip. The rostral part of the rhombic lip (that portion which is associated with rhombomeres 2 and 3 and gives rise to small neurons of the trigeminal and cochlear nuclei) is much thicker and wider in the platypus than in the short-beaked echidna. By contrast, the (caudo)lateral rhombic lip (putatively rhombomeres 4 to 7/8 which give rise to the external granule cell layer of the cerebellum and precerebellar neurons of the medulla) is very similar in both species. There is a correspondingly larger rostral migratory stream arising from the rostral rhombic lip of the platypus, compared to the echidna. This stream carries masses of small neurons forward to the parvicellular component of the Pr5, but the caudal migratory stream is quite similar in the two species (Ashwell and Hardman 2012b; see atlas plate Pl-Ha12).

The structural differences between the trigeminal sensory nuclei of the platypus and echidna become even more pronounced during the first post-hatching week (Ashwell and Hardman 2012b). Differentiation of the large Pr5 of the platypus into distinct parts proceeds through the first post-hatching week. The lateral parvicellular region receives a dense population of cells with small darkly stained nuclei (see atlas plate Pl-Ha11) interspersed with medium-sized, palely stained nuclei. The medial magnocellular region has a more sparsely distributed population of larger, palely stained nuclei. By contrast, the trigeminal sensory nuclei of the echidna remain relatively unremarkable in size and differentiation throughout the first week of post-hatching life.

The trigeminal system probably plays a key role in the ability of the newly hatched monotreme to explore its environment. As noted in Chapter 3, juvenile monotremes are very effective at sucking, even from the time of hatching. This enhanced ability is evidenced by the milk-filled gastrointestinal tract of recently emergent monotremes (Griffiths 1968, 1978) and their capacity to repeatedly find a teat-less areola, in contrast to marsupials who can semi-permanently attach to a teat. The trigeminal nuclei of the newly hatched monotremes are in many respects similar in appearance to those of newborn marsupials, who also face significant behavioural challenges in the immediate postnatal period (e.g. climbing from the urogenital sinus to the pouch). Although behavioural studies of newly hatched monotremes are very few, Ashwell and Hardman (2012b) suggest that the functional capacities of the trigeminal sensory apparatus of the newly hatched monotremes should match those for newborn marsupials.

In summary, the key developmental mechanism underlying the greatly different sizes of the Pr5 and Sp5O in the platypus and echidna is a difference in the expansion of the rostral rhombic lip and the migration of this region's derivatives in the latter half of incubation and early post-hatching life. Ashwell and Hardman (2012b) made a prediction that the cells of the rostral migratory stream express Math-1 at some stage during their development and become glutamatergic neurons, based on observed similarities to rhombic lip derivatives in rodents (Carletti and Rossi 2008). They also note that Wnt-1 expressing cells from the rhombic lip of rodents settle in the dorsal parts of the trigeminal sensory nuclear complex (Nichols and Bruce 2006), suggesting that these are key molecular markers that should be investigated in monotreme brainstem development.

These findings also have implications for monotreme evolution. The similar development of trigeminal nuclei in the two monotremes up until the middle of incubation suggests a similar genetic control of the neurogenesis of the early generated macroneurons of the sensory nuclear complex. On the other hand, the extraordinary expansion of the rostral rhombic lip and its derivatives in the platypus during the last third of incubation suggests that the ornithorhynchid lineage has undergone major changes in the genetic control of the rhombic lip of rhombomeres 2 and 3 compared to all other mammals. Ashwell and Hardman (2012b) suggest that the expansion of the rostral rhombic lip evolved as a developmental mechanism in ornithorhynchids *after* divergence from the tachyglossid lineage.

Somatosensory receptors of the limbs and trunk

Physiology of postcranial receptors and sensory nerves

The main focus on somatosensation in the monotremes has naturally been upon the trigeminal pathways, but some data are available concerning somatosensory processing from the rest of the body. Somatosensation from glabrous (hairless) skin of the distal forelimb is particularly important in therian mammals as a means of exploring the tactile texture of the environment (Rowe *et al.* 2003). In many placentals, glabrous skin has high tactile acuity, particularly in those species that use their forelimbs to manipulate objects. Given that the echidnas use their forelimbs extensively to explore their natural environment, Rowe and co-workers were prompted to look for similarities and differences in forelimb somatosensory physiology between short-beaked echidnas and therians.

Rowe and colleagues (Mahns *et al.* 2003; Rowe *et al.* 2003, 2004) isolated single tactile sensory axons from the median and ulnar nerves in anaesthetised echidnas and used accurate, reproducible mechanical stimuli in conjunction with single unit recording to assess responses. They identified slowly adapting fibres (about half their sample of 29), which displayed responses that were maintained well into the static phase of a step indentation, as well as dynamically sensitive tactile fibres that responded only to the 'on' or 'off' component of the indentation stimulus. Table 9.2 summarises the main features of the identified somatosensory afferent types along with putative corresponding eponymous receptors. Note that the assigned receptors are speculative at this stage, but are based on presumed homology with therian receptors.

Rowe's group found wide variation in the sensitivity and responsiveness of the slowly adapting fibres, potentially reflecting differences in the types of receptors and afferent fibres. Some slowly adapting fibres (about one-eighth of all fibres studied) had very small receptive fields (less than 3 mm in diameter) and low thresholds (less than 80–120 mg weight force applied with a von Frey hair). These fibres showed a distinct onset transient response to a step indentation, followed by a sustained response to the static component of the skin indentation. The combination of the static and transient responses provided a graded sensitivity to the amplitude of indentation up to a plateau level that increased in line with vibration amplitude. Rowe's group concluded that this particular group were best adapted by their small receptive fields, high sensitivity and graded responsiveness to provide for assessment of the tactile texture of ground or other surfaces being explored by the fore-footpad.

Other slowly adapting fibres (also about one-quarter of all fibres studied) have more diffuse receptive fields and respond to much larger step displacements of skin (up to 5 mm). Rowe's group suggested that this group of fibres might serve a locomotor feedback role by signalling broad pressure on the footpad during stepping and foot placement. They also thought that this group of fibres might contribute to recognition of variations in resistive forces offered by different objects in the natural environment, such as would be helpful in controlling mechanical force when digging through soil, rock and sand.

Rowe and colleagues also mentioned a third group of slowly adapting fibres (although the number encountered was not mentioned) that are associated with the base of the powerful claws of the echidna forelimb. These show a grade response to displacement of the claw base, such as might be useful in signalling differential movement of the claw when the echidna digs through different types of substrate during its search for food or digging a burrow.

Rowe's group divided dynamically sensitive tactile fibres into two types. One type that they called rapidly adapting (about one quarter of all fibres examined) has small receptive fields (between 2 mm and 3 mm in diameter), and low thresholds to vibrating stimuli at less than 50 Hz. The other group of purely dynamically sensitive fibres have larger receptive fields (over 3 mm in diameter) and are more sensitive to a much broader range of vibration frequencies (up to 300 or 400 Hz). They thought that these fibres were functionally similar to Pacinian corpuscle receptors and their afferents in therians. The Pacinian type responded to vibrating stimuli over a very broad range of frequencies by a tight phase-locking of the impulse pattern to the vibration stimulus.

Overall, the types of slowly and rapidly adapting fibres identified by Rowe's group in the echidna forelimb match classifications in placental species (Rowe et al. 2003), leading them to conclude that somatosensory mechanisms are highly conserved across all mammals. There are two exceptions to this statement. First, Rowe and colleagues reported that slowly adapting fibres had slightly larger receptive fields and higher thresholds than comparable groups in therians. They attributed this to the thicker and more cushioned glabrous skin of the echidna and different viscoelastic properties of the forelimb skin. Second,

Table 9.2. Types of somatosensory fibres in the median and ulnar nerve of the echidna[1]

Type of fibre	Receptive field size and position	Threshold and response features	Possible behavioural role	Putative receptor[2]
Slowly adapting type 1	< 3 mm diameter on glabrous skin	Less than 80 to 120 mg weight threshold	Exploring textural detail in surfaces	? Merkel endings
Slowly adapting type 2	> 3 mm diameter on glabrous skin	Up to 5 mm indentation	Locomotor and digging feedback	? Ruffini endings
Slowly adapting type 1 (but claw associated)	Claw base	Not known	Signal claw movement during digging	? Merkel endings
Rapidly adaptive type 1	2 to 3 mm on glabrous skin	Lowest threshold (only 5 to 10 μm) below 50 Hz	Low frequency vibration assessing tactile texture	? Small lamellated corpuscles (? Meissner)
Rapidly adaptive type 2 (Pacinian)	> 3 mm on glabrous skin	Most sensitive to 50 to 400 Hz vibration	High frequency vibration assessing tactile texture	Pacinian corpuscle

1 Based on Mahns et al. (2003) and Rowe et al. (2003).
2 Based on extrapolation from therian receptors (Tracey 2004; Kaas 2012). Not yet confirmed by experiment.

Rowe's group did note that Pacinian corpuscle fibres in the echidna had a slightly narrower bandwidth of vibration sensitivity than has been reported for therians, and that peak vibration sensitivity was displaced towards slightly lower frequency values (50–200 Hz) than for placentals (200–500 Hz). They concluded that this functional difference might be due to the lower body temperature of the echidna, because the bandwidth and peak sensitivity of placental Pacinian fibres also shift down when the experimental animal's body temperature is lowered.

Spinal cord and brainstem circuitry for conducting and processing somatosensation from the body

Mammalian spinal somatosensory pathways

Sensory information from the limbs and trunk is conveyed up the spinal cord in several sensory pathways. In therians, these include the dorsal columns that serve discriminative or fine touch and vibration, the spinothalamic and spinoreticulothalamic pathways that serve simple touch, pain and temperature, and the spinocerebellar pathways that serve conscious and non-conscious proprioception (i.e. the sense of limb and joint position, and of muscle length and tension).

Somatosensory pathways in the monotremes

While something is known about the anatomy of the spinal cord and brainstem somatosensory centres in monotremes (see Chapter 5), there are very few data on the function of spinal cord pathways in the monotremes. Nevertheless, structural findings suggest that the functional capacity of these pathways is similar to that in placental mammals.

The first step in transmission of somatosensory information from the skin, joints and muscles to the brain is through the peripheral sensory nerves and dorsal roots. Myelinated axons in the first thoracic (T1) and third lumbar (L3) dorsal roots of the short-beaked echidna have mean diameters of 4.65 and 5.22 µm, respectively (Ashwell and Zhang 1997). There is a clear bimodal distribution of fibre size in the L3 dorsal root, with distinct groups at 1–5 µm and 6–12 µm, comparable to fibre calibre in placentals like the laboratory rat and domestic cat. The proportions of myelinated fibres in the dorsal roots are also similar to those reported for therians, suggesting that there are similar populations of myelinated sensory axons in the different fibre classes in this monotreme as there are in therians.

Labelling with the B4 isolectin derived from *Griffonia simplicifolia* can be used to follow the paths of smaller unmyelinated axons (i.e. C fibre nociceptors) in the dorsal horn of the spinal cord. In the short-beaked echidna, labelling of axons in the dorsal horn with this lectin is somewhat different to that observed in placentals, even though the sizes of labelled ganglion cells are similar (Ashwell and Zhang 1997). In the echidna, the isolectin-labelled axons penetrate deep into the dorsal horn, reaching Rexed's laminae R3 to R5 (a region that is usually associated with inputs from internal organs), whereas lectin-labelled C fibre afferents in placentals terminate only in the more superficial parts of the dorsal horn (Rexed's laminae R1 and R2). It may be that Ashwell and Zhang identified internal organ afferents that do not label with this lectin in placentals, or that the echidna has an unusual pattern of dorsal horn termination of C fibre afferents from the skin.

The dorsal columns are the main pathway for carrying discriminative or fine touch information up the spinal cord. The dorsal columns of the short-beaked echidna show many similarities to placentals, even those animals like the domestic cat and primates that have well-developed discriminative touch. The cross-sectional area of the dorsal column region in the echidna (2.17 mm^2) is similar to that in carnivores and primates of a similar body weight (*Felis catus* 2.46 mm^2; *Macaca fuscata* 2.57 mm^2). The mean axonal calibre of the dorsal column axons in the short-beaked echidna ranges from 2.19 µm to 3.12 µm. More than two-thirds of axons are in the small δ fibre range (2–5 µm in diameter) and 5.5% are in the larger β fibre range (6–12 µm). The slightly smaller diameter of dorsal column axons in the short-beaked echidna compared to the domestic cat suggests that conduction velocity in the echidna dorsal column pathway may be slightly slower than in the cat, even at the same body temperature. On the other hand, the number of myelinated axons in the fasciculus gracilis (the medial component of the dorsal column) of the echidna is substantial (a total of 93 000 on each side), compared to only 25 284 myelinated fibres in the domestic cat (Hwang et al. 1975). This suggests that there is a substantial flow of somatosensory information up the echidna spinal cord. Given the relatively simple uses to which the fore- and hindlimbs of the echidna are put, this is difficult to explain. What could account for such a large somatosensory pathway? Nerve branches to the panniculus carnosus are large and extensive, particularly from the brachial plexus (see Fig. 15.4), so there is probably a rich sensory input from the skin for coordination of the panniculus carnosus. The echidna is capable of some extraordinary feats with this muscular sheet and its attachments to the spines, including shimmying up drain pipes. A rich tactile and proprioceptive input from the spines and panniculus would allow fine control of the muscular sheet and provide a locomotor apparatus in addition to the limbs.

Opioids in monotremes

These peptides are used in pathways that modulate pain sensitivity, particularly in the dorsal horn of the spinal cord. Chemical circuitry of the monotreme spinal cord is largely unexplored, but some limited data are available concerning proenkephalin-derived octapeptides in the platypus (Bojnik et al. 2010). Not surprisingly, these molecules are well conserved in all terrestrial vertebrates. The platypus octapeptide is one of only five motifs seen among mammals. Curiously, the platypus motif is identical to amphibia (*Xenopus* and *Bombina*), but the mouse motif is also seen in lobe-finned fish. This probably reflects the limited range of motifs possible with only eight amino acids and gives little useful information on the evolution of opioids.

Dorsal column nuclei of monotremes

The dorsal column nuclei (gracile, cuneate and external cuneate nuclei) receive somatosensory information carried in the dorsal column pathways of the spinal cord and give rise to a pathway called the medial lemniscus, which carries this information to the somatosensory thalamus. Hines (1929) described the gracile and cuneate nuclei of the platypus, but did not note any unusual features. On the other hand, the gracile, cuneate and external cuneate nuclei are quite large in the short-beaked echidna (see atlas plates Ec-Ad29 and 30), reflecting the well-developed somatosensory pathways for the limbs (and perhaps the panniculus carnosus) of the echidna (Hassiotis et al. 2004a).

The somatosensory thalamic nuclei of the platypus and echidna

Topography of the somatosensory thalamic nucleus

The ventral posterior nucleus (sometimes called the ventrobasal complex) of the thalamus is the dedicated somatosensory relay nucleus of the thalamus. It receives the medial and trigeminal lemnisci from the dorsal column and trigeminal sensory nuclei, respectively. In therians, the ventral posterior is divided into medial and lateral parts (VPM, VPL) with VPM receiving discriminative and nociceptive tactile input from the face and head via the trigeminal nuclei and VPL receiving discriminative input from the limbs and trunk via the dorsal column nuclei (Tracey 2004).

Information is very limited concerning the structure and function of the somatosensory thalamus in the monotremes. The ventral posterior nucleus is identifiable in both platypus and short-beaked echidna on the basis of chemical and cellular architecture. In the platypus, this structure is extraordinarily large to serve the enhanced sensation from the bill, so much so that it invades the cerebral hemisphere (Mikula et al. 2008; Jones 2007).

Jones (2007) described three incompletely separated subdivisions of the monotreme ventral posterior thalamic nucleus. One subdivision was described as a region of small densely packed cells that is small at the rostral pole but expands up the lateral side and over the posterior pole of the nucleus. Jones thought that this subdivision (VMb in Fig. 9.1 g) resembled the basal ventral medial nucleus of other mammals (Jones 2007; Mikula et al. 2008; see atlas plate Pl-Ad17). In the platypus, myelinated fibre bundles traverse this subdivision, running from dorsolateral to ventromedial. The other two subdivisions are more evident in the platypus than in the echidna. These consist of a central subdivision of less densely packed neurons and a dorsal subdivision of larger, more dispersed neurons. Jones did not feel that medial and lateral subdivisions of the ventral posterior nucleus (i.e. VPM and VPL as described above) could be identified in either platypus or echidna.

Ashwell and colleagues reached slightly different conclusions in their studies of the echidna somatosensory thalamus (Ashwell and Paxinos 2005; Hassiotis et al. 2004a). They identified a VPM where neuronal spatial density is higher, as well as a more neuron sparse VPL penetrated by fibres from the internal capsule (see atlas plates Ec-Ad15 to 18). The border between these two subnuclei is, however, not clear in Nissl sections. The densest plexus of myelinated fibres was found within VPM, and the VPM stains strongly for cytochrome oxidase, in contrast to the nearby medial lemniscus that is carrying somatosensory information to the nucleus. Fibre bundles and neuronal cell bodies in the VPM and VPL are also strongly reactive for NADPH diaphorase. Consistent with the role of the ventral posterior nucleus as the major sensory relay nucleus to the isocortex, many of the VPL and VPM neurons (~17% of the total population) and lemniscal axons to the ventral posterior nucleus are strongly immunoreactive for neurofilament protein (as revealed by the SMI-32 antibody), consistent with large calibre axons rapidly carrying somatosensory information. Strongly SMI-32 immunoreactive large-bodied neurons with axons leaving the nuclear complex for the internal capsule (presumably thalamocortical axons) are also present. Immunoreactivity for calcium-binding proteins also reveals distinct classes of neurons in the ventral posterior nucleus. Of the three calcium-binding proteins studied, immunoreactivity for calbindin is found in 54% of all VPM neurons, whereas only ~7.5% of VPM neurons are immunoreactive for parvalbumin and calretinin. The functional significance of these differences is unexplored, but the findings imply different channels for sensory processing.

Connections of the monotreme somatosensory thalamus

No anatomical studies have ever been conducted of projections of the somatosensory thalamus to the isocortex in the platypus, but some data are available for the pathway in the short-beaked echidna, based on retrograde degeneration and axonal transport studies (Welker and Lende 1980; Ulinski 1984; see Chapter 6).

A factor that complicates the interpretation of these findings is that the terminology and boundaries for dorsal thalamic nuclei in the Ulinski study are rather different from those applied by Jones and colleagues (Jones 2007; Mikula et al. 2008) and Ashwell and Paxinos (2005). Ulinski places the ventral posterior nucleus much further dorsally than other authors and identifies a ventral (V) nucleus where other authors identify a ventrolateral nucleus (VL). When the tracer enzyme horseradish peroxidase was injected into an area of isocortex that covered both the rostral and caudal banks of the α sulcus (Echidna 1 of Ulinski 1984; injected in a region that corresponds to the head and forelimb somatosensory representation; Lende 1964), labelled neurons were found in the dorsal thalamus. These neurons were located in a large field that corresponds not just to the ventral posterior nucleus, but also to laterodorsal (LD), ventrolateral (VL) and posterior (Po) thalamic nuclei as delineated by Jones and colleagues (Jones 2007; Mikula et al. 2008) and Ashwell and Paxinos (2005). The extension of the injection to the rostral bank of the α sulcus may have labelled axons of passage to the motor cortex, thereby explaining the extensive spread of retrogradely labelled neurons in rostral parts of the dorsal thalamus where motor relay nuclei would be located. Injection of label into the caudal bank of the α sulcus in a region of somatosensory cortex corresponding to the tongue representation (Lende 1964) labelled neurons in the most caudal pole of the dorsal thalamus (Echidna 3R of Ulinski 1984), but not in a region congruent with the part of the dorsal thalamus that should be the ventral posterior nucleus on cyto- and chemoarchitectural grounds. These findings are difficult to interpret because of the limited number of injections (only three injections in two animals) and differences in nomenclature, highlighting the need for a more modern study that correlates thalamic nuclei with isocortical areas in the echidna.

The findings of the older study by Welker and Lende (1980) are actually much easier to interpret and more consistent with the identification of the ventral posterior nucleus in the echidna on cellular, fibre and chemical architecture grounds. The region of dorsal thalamus that contained degenerating neurons following removal of the ventrolateral somatosensory cortex by Welker and Lende (see Fig. 6.2b, f) is largely congruent with the lens-shaped structure that was identified as the ventral posterior nucleus by other authors.

Development of the monotreme somatosensory thalamus

Neurogenesis of the thalamus begins at ~8.0 mm GL in both the platypus and echidna (about a third of the way through incubation), but the distinctively large ventral posterior nucleus does not become apparent until after hatching in the platypus, when a prominent cluster of small granular cells begins to settle in the mantle region of the posterior thalamus (Ashwell 2012c; see atlas plates Pl-Ha9 and 10). Thalamocortical somatosensory fibres and the internal capsule emerge at the end of the first post-hatching week in both species and begin to expand rapidly, particularly in the platypus. The ventral posterior nucleus of the echidna shows the first signs of differentiation into a medial (VPM) and a lateral (VPL) segment at the 25 mm GL stage, but the platypus ventral posterior nucleus remains a homogeneous mass of granular cells at this time (about the end of the first post-hatching week). During the second post-hatching week, when the germinal zones of the dorsal thalamus are approaching exhaustion, the large ventral posterior nucleus of the platypus is filled with a mixture of granular neurons with dark nuclei and larger neurons with lighter nuclei. The ventral posterior nucleus of the platypus reaches a mature appearance by the fifth or sixth week after hatching.

The findings on thalamic development in the monotremes have behavioural and evolutionary implications. At hatching, the ventral posterior nucleus in both species has only a small fraction of the adult complement of neurons, does not yet receive sensory input from the brainstem or spinal cord, and has yet to extend thalamocortical axons to the isocortex. Ashwell (2012c) noted the structural similarity of the ventral posterior nucleus in the newly hatched monotreme to the limited development of dorsal thalamic nuclei in newborn marsupials. Given the limited development of the ventral posterior nucleus at hatching, the behavioural repertoire of the newly hatched monotreme is probably achieved solely with circuitry in the brainstem and spinal cord and requires no processing of information by the higher neuraxis. In other words, the large somatosensory apparatus of the monotreme thalamus and isocortex probably serves no behavioural role for the first month (at least) of post-hatching life.

The evolutionary implication is that the platypus and echidna have very different patterns of ventral posterior nucleus development and thalamic structure in the two groups diverges from the outset. At no stage of development is a large ventral posterior nucleus present in the echidna dorsal thalamus, such as might be expected if initial production of a large ventral posterior nucleus were genetically programmed in the tachyglossid lineage. The special features of the ventral posterior nucleus of the adult platypus are therefore achieved independently and with no corresponding feature in the echidna dorsal thalamus. This observation is more consistent with ornithorhynchids and tachyglossids having followed a long period of independent evolution, rather than the echidna being a recent derivative of a platypus-like ancestor.

Cortical representation and analysis of somato- and electrosensory input

Cortical somatosensory areas

There is some disagreement between early and later descriptions of the organisation of somatosensory cortex in monotremes, so the various published accounts will be examined in chronological order for each species. An important issue is that recent studies claim there are as many as four somatosensory areas and body maps within the isocortex of the platypus and echidna. This interpretation has subsequently been questioned (Rowe *et al.* 2004).

Somatosensory cortex of the platypus

Bohringer and Rowe (1977) identified only a single contralateral body representation in the isocortex of the platypus (Fig. 9.6a). They recorded cortical surface potentials following electrical or mechanical stimulation of the bill, limbs and trunk on the opposite side of the body. Responses to stimulation of the contralateral bill could be recorded over an extremely large area of the dorsolateral surface of the hemisphere. The mediolateral progression of body representation on the isocortex corresponded to the classic primary somatosensory (S1) representation of therian mammals, in that the hindlimb representation was towards the midline, and the forelimb and face were represented laterally. They found that receptive fields of cortical neurons were quite large in the tail and trunk representation (reaching an area of 15 cm²), smaller on the glabrous skin of the forelimb (0.5 µm 1.0 cm²), but less than 1 mm in diameter on the bill. Significantly, Bohringer and Rowe (1977) did not find any evidence of a second topographically organised somatosensory area (S2). They also found considerable (but not complete and congruent) overlap between motor and somatosensory cortex (see Fig. 7.3).

Krubitzer and colleagues (Krubitzer et al. 1995) confirmed the extensive area of isocortex devoted to the contralateral bill representation that Bohringer and Rowe had found, but reported the existence of as many as four topographically ordered contralateral body representations. Krubitzer and colleagues used microelectrode-based mapping techniques in conjunction with cytochrome oxidase (CO) and myelin staining. They found a large representation of cutaneous input that appears to be homologous to S1 of placentals, with the large bill representation placed laterally (Fig. 9.6b). The S1 field also included neurons responsive to low voltage electrical stimulation (i.e. electrosensory). Using cortical flatmounts, Krubitzer and colleagues identified coextensive myelin and CO-rich bands within the bill representation in the S1 field. Myelin staining identified zones of abundant myelin coated thalamocortical afferents entering the cortex, and CO reactivity demonstrated the presence of thalamocortical axons with higher than normal concentrations of mitochondria in the axon terminals (indicating higher metabolic activity). Those regions where myelin staining and CO reactivity were strongest (CO dark band) contained neurons that responded only to cutaneous stimulation, whereas regions that stained more lightly for myelin and CO (CO light band) contained bimodal neurons, i.e. they responded to both electrical and mechanical stimulation (but more vigorously to electrical stimuli) (Manger et al. 1996; Fig. 9.6b and c). The receptive fields for neurons in the myelin and CO light areas were significantly larger than those for neurons in the myelin and CO dark regions. Latencies to electrical or mechanical stimulation of the bimodal neurons were around 25 ms, but were significantly reduced when the stimuli were applied together (Manger et al. 1996).

The physiologically distinct subregions that Krubitzer and colleagues identified in platypus S1 on the basis of myelin staining and CO activity are reminiscent of the pattern found by Langner and Scheich (1986) when they used 2-deoxyglucose to map the electrosensory representation of the bill on the platypus cortex. The role of the stripes, which are also evident in parvalbumin immunoreacted sections (Fig. 9.1h), will be discussed below.

The other somatosensory regions identified in platypus cortex by Krubitzer and colleagues include the caudal or parietoventral field (PV), the rostral field (R), and the manipulation field (M). Krubitzer and colleagues reported that the PV field contained a complete body representation and was situated caudal to both the S1 and auditory cortex (Fig. 9.6b). Neurons in PV responded to cutaneous stimulation and had large receptive fields (covering as much as one-third of the body surface). As in S1, the bill had a large areal representation. The rostral somatosensory area or field contained neurons that responded most often to stimulation of deep receptors on the contralateral body (Fig. 9.6b). Receptive fields for neurons in R were also larger than for S1 and the somatotopic organisation was mirror-imaged relative to that in S1. Krubitzer and colleagues also obtained recordings from a further somatosensory field (the M field) lying rostral to R. Neurons in M respond to manipulation of joints and hard taps to body parts rather than light pressure. The internal organisation of M was less precise than that of R and several representations of a similar manipulation were seen in the one animal. Receptive fields in M were also large, but some smaller fields were found for the digits. Most neurons in the somatosensory cortex of the platypus responded transiently to the onset of repeated stimulation, but some slowly adapting neurons that responded best to sustained stimuli were also seen, particularly in the R and M fields.

Krubitzer and colleagues denoted the caudal somatosensory field as PV rather than S2, because they felt that several features of the monotreme PV were most consistent with homology to the PV region of therians rather than S2 (Krubitzer *et al.* 1995). These include: (1) the inverted representation of the body surface in the monotreme field, like the inverted representation in placental PV, but unlike the non-inverted representation seen in placental S2; and (2) the large receptive fields of neurons in the monotreme PV field, like in placental PV. However, it should be noted that in a later publication Krubitzer entertained the possible homology of the caudal monotreme field to placental S2 (Krubitzer 1998).

The role of bimodal cortical neurons in the platypus

The presence of CO dark bands serving only mechanoreception and CO light bands integrating mechano- and electroreception within S1 (Fig. 9.6c) has led to the development of physiological models to explain the role of these cortical zones and their constituent neurons. In particular, it has been proposed that the banded arrays within platypus S1 allow for the processing of temporal disparities between mechano- and electroreceptive input. Mechanical waves, in the form of displacement of water from the tail flick of prey, move through water much more slowly than the electrical field generated by the contraction of prey musculature. This disparity will increase with distance to the prey and could provide a valuable clue for the platypus as to the distance of quarry (Pettigrew *et al.* 1998; Elston *et al.* 1999). Groups of cortical neurons within the CO light bands that receive both types of input and preferentially respond to particular disparities between the detection of electrical and mechanical stimuli would be a key component in a range detection system (Fig. 9.6c). Pettigrew and colleagues (Manger *et al.* 1996; Pettigrew 1999; Pettigrew *et al.* 1998) have proposed that a system of 'near' or 'far' bimodal neurons in the CO light bands would signal the range to prey for a given point on the bill sensory organ.

Such an interaction between electrical and mechanical stimuli could be either facilitatory or inhibitory. Krubitzer and colleagues noted regions within CO light S1 where neurons responded to both cutaneous and electrical stimulation (Krubitzer *et al.* 1995), but did not provide details of the nature of this interaction. Iggo and colleagues observed inhibitory interaction at the cortical level between the two modalities (Iggo *et al.* 1992). They found that when both electrical and mechanical stimuli were applied with an inter-stimulus interval of less than 25 ms, the responses of cortical neurons generated by one stimulus were completely suppressed by the other. Although interesting, these observations are some way from demonstrating the sort of precise interaction between the two modalities that would be needed for the precise ranging of quarry.

There are also significant structural differences between pyramidal neurons in the CO dark and CO light bands of S1. Elston and colleagues showed that pyramidal neurons in the mechano- and electroresponsive CO light bands have basal dendritic fields that are wider than pyramidal neurons in purely mechanoresponsive CO dark bands (Elston *et al.* 1999). They argued that the larger basal dendritic fields might allow these neurons to integrate input from a diverse range of sources. This averaging could enhance the signal-to-noise ratio and improve spatial discrimination of stimuli on the bill, as well as analysing temporal disparity of signals in a range determination mechanism.

These models are intriguing, but there has never been any actual testing of these in either the natural or laboratory setting. At the least, a detailed study of responsiveness of neurons in the CO light bands to the temporal disparity between mechanical and electrical signals is necessary to support or refute the ranging hypothesis. Without this sort of study, the ranging model remains plausible but entirely speculative.

Somatosensory cortex of the short-beaked echidna

Early studies of somatosensory cortex in the short-beaked echidna by Lende showed that the cortex is located in the posterior part of the cortex between the α and ζ sulci (Fig. 9.7a; Lende 1964; Allison and Goff 1972; reviewed by Rowe 1990). The somatosensory cortex extends as far laterally as the rhinal fissure but the medial boundary reaches only to the midlateral surface of the cortex, where it abuts the lateral border of the visual cortex. Although shifted posteriorly, the somatosensory cortex of the echidna has a similar somatotopic representation of the contralateral head and body to that seen in placentals (Rowe 1990). The tail is represented medially with the hindlimb, trunk, forelimb and head represented progressively laterally (Lende 1964). The snout and tongue are represented most laterally, with the tongue area adjacent to the rhinal fissure, and both these body parts have an expanded area of cortex devoted to them. Lende did not find any evidence for a second somatosensory area.

A later study in the short-beaked echidna by Krubitzer and colleagues reported the existence of the same four somatosensory fields (S1, PV or C, R and M) that those authors found in the platypus (Fig. 9.7b, c; Krubitzer et al. 1995). As in the platypus, S1 of the echidna contained neurons that responded to cutaneous stimulation on the contralateral body in a map that largely corresponds with Lende's findings. Krubitzer and colleagues did not systematically investigate electrical responsiveness in S1 of the echidna, but it does not appear that the echidna has the same interdigitating subdivisions within S1 that may underlie the prey ranging system of the platypus.

Krubitzer and colleagues also identified PV, R and M somatosensory fields in the echidna cortex. The general features of these fields were broadly similar to those outlined above for the platypus. One significant difference was that they found a specialised forepaw region within the forelimb representation of PV that was not seen in the platypus. Neurons in that region responded to vibrating stimulation of the forepaw with a precise phase-locking of discharges to the sinusoidal waveform of the vibration. These neurons are probably receiving input from Pacinian or Paciniform corpuscles in the glabrous skin of the manus or forepaw. This specialised region may serve a behavioural role in the fine control of forepaw digging.

A cautionary note about reports of multiple somatic fields in monotreme cortex

It will be clear to the reader that there are significant differences in the platypus isocortical sensory maps described in the studies by Rowe and Krubitzer and their colleagues (compare Fig. 9.6a and b). Rowe and colleagues later raised a note of caution about reports of multiple body representations in monotreme cortex (Rowe et al. 2004). In Rowe's laboratory (Bohringer and Rowe 1977; Rowe 1990) only a single body representation was found on the cerebral cortex of the platypus, in contrast to the four body representations reported by Krubitzer and colleagues.

Rowe has raised several questions about the reports of multiple somatosensory fields in monotremes. First, not all sensory stimuli reported for the four somatosensory areas are convincingly tactile (i.e. from the body surface). Rowe and colleagues note that although S1 is said to respond to cutaneous sensation, area R is said to contain neurons that are most responsive to stimulation of deep receptors. Since Krubitzer and colleagues did not provide quantifiable or reproducible measures of these stimuli, Rowe felt that it is difficult to determine how Krubitzer and colleagues could clearly distinguish between deep or cutaneous neurons in the so-called different cortical representations. Therefore, the R field may actually be homologous to the deeply responsive area 2 or 3a of primate S1 (Kaas 2012). Given that it is not possible to distinguish between discrete areas on responsiveness criteria, identification of four distinct somatosensory representations therefore rests solely on the detection of four discrete and complete body maps.

The second problem raised by Rowe's group concerns a potential misinterpretation of the dermatomal distribution as translated to the cerebral cortex. The dermatomes of adjacent spinal nerves do not always serve contiguous areas of skin, particularly in the upper and lower limbs around the dorsal and ventral limb axis. Rowe has noted that the reversal of somatotopic representation that Krubitzer and colleagues have reported at the boundaries between adjacent somatosensory fields in the cortex could also be explained by a trajectory along the limb that crosses dermatomes that are not from adjacent spinal segments. A situation like this is seen in macaque S1 cortex, where representations of the postaxial and preaxial upper limb and lower limb regions are separated by representations of the more distal parts of the limbs (i.e. digits and toes, respectively) (discussed in Rowe et al. 2004).

In short, Rowe and colleagues argue that the reversals in receptive field representation that are described by Krubitzer and colleagues (1995) for their multiple somatosensory fields are also consistent with the sort of sequence of representation that would be

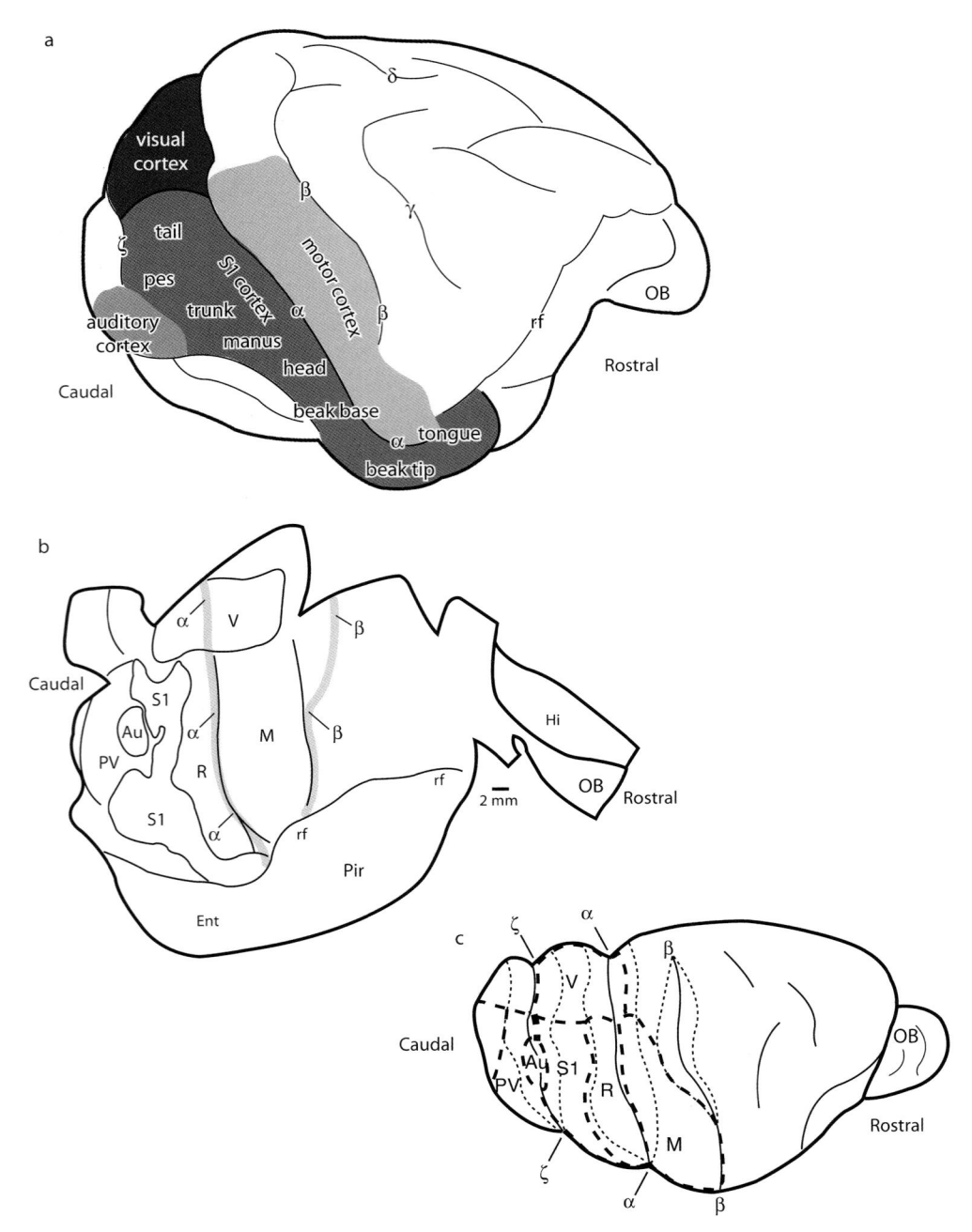

Figure 9.7: Body representation in the somatosensory cortex of the short-beaked echidna based on electrophysiological studies (a – Lende 1964; b, c – Krubitzer et al. 1995). Figure (a) shows the position of the representations of body parts on the primary somatosensory (S1) cortex as well as motor, auditory and visual cortex, as mapped by Lende, relative to major sulci indicated by Greek letters. Figure (b) shows the position of the four somatosensory fields (S1, M – manipulation cortex, R – rostral somatosensory cortex, and PV – parietoventral somatosensory cortex) as well as visual (V) and auditory areas (Au) identified by Krubitzer and colleagues on the flattened surface of the gyrencephalic echidna brain. In figure (c), the brain is intact, but the cortical sulci have been artistically opened to reveal the extent of the somatosensory areas on the cortex in the depths of the sulci. Heavy dashed lines in (c) indicate the boundaries between sensory areas; continuous thin lines indicate the surface projection of the sulci (α, β, γ, δ and ζ); and light dashed lines indicate the deepest part of each sulcus. Figures (b) and (c) have been reproduced from Hassiotis et al. (2004b) with permission from John Wiley and Sons. OB – olfactory bulb; Pir – piriform cortex; rf – rhinal fissure.

obtained if a single body map were planned according to a dermatomal trajectory. This criticism can only be satisfactorily answered by a re-examination of the issue that takes the dermatomal organisation into account in a reassessment of the monotreme cortical somatic map.

Development of somatosensory cortex

The distinctive features of monotreme isocortical development and the role of the subventricular zone are considered in the cerebral cortex chapter (Chapter 7) and will not be repeated here. A summary of the key developmental events is, however, useful in understanding the functional maturation of the somatosensory system (see Appendix Figs 1, 2 and 3). A distinct cortical plate, flanked by marginal and intermediate layers, can be seen in both species towards the end of the first post-hatching week. As noted above, the thalamocortical projection (including that to somatosensory cortex) grows towards the cortex around the end of the first post-hatching week. During the second post-hatching week, stratification begins to appear in the subventricular region and the cortical plate divides into a compact cortical zone of densely packed small cells ~30–35 μm thick, overlying a loosely packed zone ~90 μm thick. In the platypus, the subventricular zone is thickest under the trigeminal field of the primary somatosensory cortex at the lateral rim of the isocortex (Bohringer and Rowe 1977; Krubitzer et al. 1995) and is the site through which abundant axons will pass to reach the cortical plate. By 3–4.5 months post-hatching, the isocortical layering in the platypus is approaching an adult pattern of lamination. The compact cortical zone has differentiated into layers 2Cx to 4Cx and the deeper loosely packed zone is in the process of transforming into the subgranular layers (5Cx and 6Cx). The subventricular or subcortical zone persists for many months after hatching and is particularly thick beneath the somato/electrosensory cortex (Ashwell and Hardman 2012a).

The somato/electrosensory cortex of both platypus and echidna has very few postmitotic neurons at the time of birth and is much like that of marsupials in terms of its structural maturation. This immaturity indicates that whatever the behavioural abilities of newly hatched monotremes, they do not depend on cortical somatosensory function for locating the milk source or any other aspect of early post-hatching life. The somato/electrosensory cortex is unlikely to be functionally useful until many months after hatching.

Evolutionary considerations

The extraordinary specialisation of the trigeminal pathways in the platypus has attracted attention from neuroscientists for more than a century and has defined how modern science sees the group. The prevailing viewpoint has been that the remarkable trigeminal electrosensory abilities of the extant monotremes indicate that this sense was well-developed in the common ancestor of all monotremes. In this conception, the modern short-beaked echidna is seen as having lost its electrosensory capabilities as it adapted to drier terrestrial environments.

The electrosensory apparatus of the platypus is clearly a sophisticated system requiring the evolution of many anatomical and physiological specialisations at multiple levels of the trigeminal somatosensory pathway. These include: (1) a specialised parasagittal array of sensory mucous and serous glands on the bill; (2) an extraordinary enlargement of the trigeminal ganglion and nerve branches; (3) the expansion of the principal and oral spinal trigeminal nuclei; (4) the enlargement of the ventral posterior nucleus of thalamus; (5) the development of a large somatosensory cortex; and (6) the formation of bimodal/unimodal bands within S1. At present nothing is known about the genetic factors that regulate any of the developmental processes that produce these striking features. These structural changes must require the concerted action of genes that regulate (among other things) epidermal differentiation in the bill, trigeminal placode neurogenesis and migration, rhombic lip neurogenesis and migration, neurogenesis and neuronal migration from prosomere 2 (for the ventral posterior nucleus), axonal guidance of thalamocortical axons to the telencephalon, formation of cortical germinal zones (ventricular and subventricular), regulation of cortical neuron migration, and direction of competitive interactions between cortical afferents serving different modalities.

Arguably, the evolution of the platypus bill organ and its central processing systems is an evolutionary challenge comparable to the evolution of the vertebrate visual system. The emergence of genetic changes that govern just one of these specialisations while other elements in the system remain unspecialised would be of little or no benefit to the ancient platypus. The sophisticated system we see in the modern platypus probably evolved by incremental changes at all levels of the trigeminal pathway. Finlay and colleagues have argued that changes at the periphery are the major driving events in emergence of new sensory abilities (Finlay *et al.* 2011). In the case of the platypus any peripheral changes are of little benefit without profound changes in brainstem, thalamic or cortical developmental processes. Developmental studies of the trigeminal pathway (reviewed above) indicate that the unique features of the platypus trigeminal pathway have their origins in developmental events that are largely contemporaneous (i.e. all occurring around the end of incubation and the beginning of the lactational phase of development).

What do observations of trigeminal development in the platypus and echidna contribute to the argument about the evolution of electroreception? At all levels of the trigeminal pathway (bill or beak, trigeminal divisions, trigeminal ganglion, trigeminal sensory nuclei, somato/electrosensory thalamus and somato/electrosensory cortex), the platypus follows a very different developmental trajectory from the short-beaked echidna. Many of these distinctive developmental features in the two species are clearly related to the use that the adult will make of the organ, e.g. the linear 'antenna' as opposed to 'hot spot' arrangement of developing sensory glands in the snout of the platypus and echidna, respectively. The distinct anatomical features in the embryonic and early post-hatching trigeminal pathways of the platypus and echidna are more consistent with the two lineages of modern monotremes (i.e. ornithorhynchids and tachyglossids) having pursued separate paths of sensory specialisation for long periods of time, rather than the modern echidna being derived from a platypus-like ancestor that left an aquatic environment to invade a land-based insectivorous niche (Ashwell *et al.* 2012; Ashwell and Hardman 2012b).

Questions for the future

The discovery of electroreception in the trigeminal pathway of modern monotremes has provided many fascinating insights into the possible transformations the vertebrate sensory systems are capable of, but several important questions remain. These can be ordered from the periphery to the cortex:

- What is the role of sensory serous glands? Are they electrosensory and, if so, how do they differ from sensory mucous glands?

- What is the transduction mechanism that occurs at the base of the sensory mucous glands? Is it simply opening of a voltage-gated ion channel or is a more sophisticated mechanism involved?

- How is electro- and mechanosensory information processed within the trigeminal sensory nuclei?

- Is there a topographic or modality-based subdivision within the trigeminal sensory nuclei?

- How is the development of the trigeminal sensory nuclei controlled? What genetic changes have occurred in regulation of the developing rhombic lip during platypus evolution?

- What role does the enlarged ventral posterior nucleus of the platypus play in processing electrosensory information?

- What genetic changes have occurred in thalamic neurogenesis to produce the large ventral posterior nucleus of the platypus?

- How does development of cortex in the monotreme somatosensory areas differ from therians? How does the subventricular/subplate region contribute to the special features of the monotreme somatosensory cortex?

- How does the somato/electrosensory cortex of the platypus analyse decay and latency differences to determine the direction and range of prey?

- Are the multiple body representations described by Krubitzer and colleagues true whole body maps or misinterpretations of a dermatomal-based body representation?

- How is the echidna somatosensory cortex specialised for forelimb tactile sense and why does there appear to be a discrete forepaw or manus representation in this species?

10
Auditory and vestibular systems

Ken W. S. Ashwell

Summary

The middle and inner ear of the extant monotremes show anatomical features that are reminiscent of early mammals like the multituberculates, but other features of the ears of modern monotremes are clearly derived characters. Although hearing is probably not of major behavioural importance for the platypus, hearing in the short-beaked echidna may be important for detection of invertebrates like termites. In both the platypus and echidna, hearing is optimal for frequencies between 4 kHz and 5 kHz and may be superior for bone conduction through the mandible and maxilla rather than air conduction, consistent with the use of the echidna snout as a thrust probe for locating invertebrate prey. The monotremes have many of the same central components of the auditory and vestibular pathways that have been identified in therians, but some components, e.g. the superior olivary nuclear complex and the medial geniculate nucleus, are difficult to definitively identify. The auditory cortex is small in both the platypus and echidna, both in terms of absolute and relative size, but there may be two separate auditory areas, each with a tonotopic organisation. Based on the pace of structural maturation, the inner ear of the young monotreme is unlikely to be functionally capable until well into lactational phase life, although the macula of the utricle may be able to assist with orientation around the time of hatching.

Overview of the auditory and vestibular systems

The vertebrate ear is divided into three parts. The external ear consists of an external auditory canal

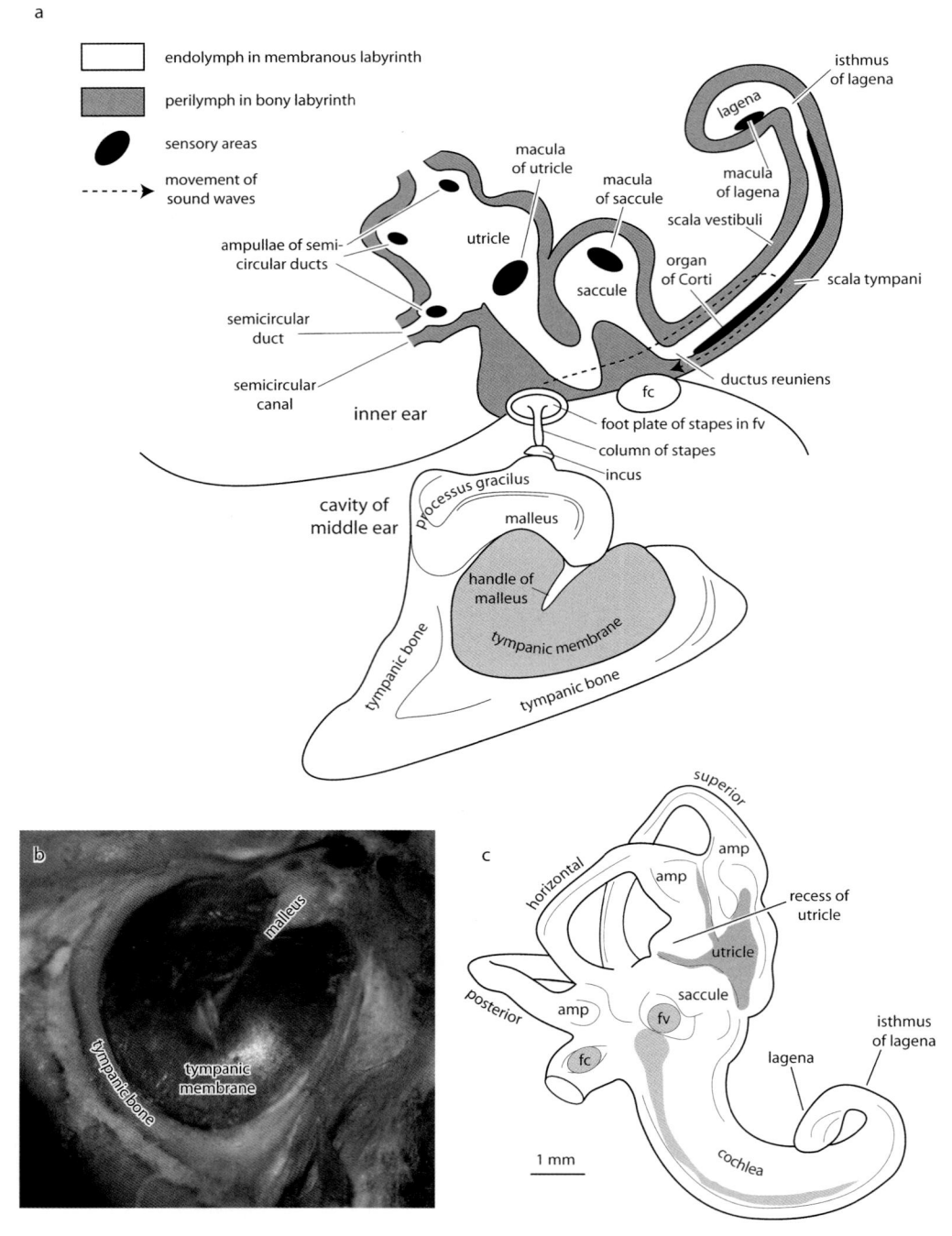

Figure 10.1: Illustrations and photographs of the monotreme ear. (a) Schematic diagram of the components of the middle and inner ear of the short-beaked echidna. Note the large malleus, diminutive incus and columellar stapes of the auditory ossicle chain. The stapes is also short relative to the diameter of its foot-plate. The cochlear duct is poorly curved and ends in a lagena, with its own sensory region, the macula of the lagena. Dashed line with arrow indicates path of pressure wave. (b) Photograph of the tympanic bone and tympanic membrane of a short-beaked echidna (AustMus specimen B1030, taken by the author). (c) Line diagram showing the major components of the osseus labyrinth of the monotreme inner ear (short-beaked echidna) adapted from a photomicrograph in Gray (1908). Note the incompletely curved cochlea; lagena at the end of the cochlea; isthmus (narrowed part) of the lagena; semicircular canals (horizontal, superior and posterior) and their respective sensory ampullae (amp); fenestra vestibuli or oval window (fv); and fenestra cochleae or round window (fc).

leading to the tympanic membrane. The wall of the external ear is often reinforced by cartilage and a tympanic bone supports and surrounds the tympanic membrane. The middle ear is an air-filled space between the tympanic membrane and the oval and round windows that, in turn, lead to the inner ear. The mammalian middle ear contains three tiny bones (auditory ossicles) in a chain (Fig. 10.1a). These three bones are the malleus (hammer), incus (anvil) and stapes (stirrup). The handle of the malleus is joined to the tympanic membrane while the foot-plate of the stapes sits over the fenestra vestibuli (oval window). The inner ear lies within the petrous bone and contains neural components concerned with auditory function (the cochlea) and vestibular (balance and acceleration) function (the vestibular apparatus including utricle, saccule and semicircular ducts). These consist anatomically of a membranous labyrinth filled with endolymph fluid (cochlear duct, utricle, saccule and semicircular ducts), suspended within the bony labyrinth filled with a fluid called perilymph.

The internal structure of the mammalian cochlea consists of an endolymph-filled cochlear duct or scala media (the auditory portion of the membranous labyrinth), flanked by the perilymph-filled scala vestibuli and scala tympani. The cochlear duct contains auditory receptors (organ of Corti) that perceive deformation of the cochlear duct as pressure waves pass from the scala vestibuli to the scala tympani (Fig. 10.1a). The organ of Corti consists of strips of hair cells and supporting cells arranged in inner and outer rows, separated by the tunnel of Corti. Stereocilia (stereovilli) of the outer hair cells are inserted into the tectorial membrane. The inner hair cells are the actual sensory cells, whereas outer hair cells function as amplifiers; sensory nerve fibres from many spiral ganglion cells are in contact with each of the inner hair cells, whereas only a few small spiral ganglion cells contact many outer hair cells.

The utricle and saccule detect linear acceleration and the gravitational field, whereas the semicircular ducts detect angular acceleration (head rotation). The three semicircular ducts are oriented at right angles to each other. In therians, the auditory and vestibular components of the vestibulocochlear nerve arise from the cochlea and vestibular apparatus, respectively.

The monotreme external and middle ear

Both groups of monotreme have the three-part structure of the typical mammalian ear as described above, but each has distinctive features. In many respects these fine scale differences reflect similarities to marsupial and/or reptilian ear structure. Figure 10.1 shows key macroscopic features of the monotreme middle and inner ear using the short-beaked echidna as an example.

The tympanic membrane of the echidna is on the ventral surface of the skull, whereas the opening of the ear faces superiorly and laterally, so the canal of the external ear is curved, with the concavity of curvature facing dorsally. The wall of the external ear has reinforcement in the form of two longitudinal strips of cartilage (Griffiths 1978). At the medial end of the external ear, the tympanic membrane is supported by a curved tympanic bone that is firmly locked to the processus gracilus of the malleus (Fig. 10.1a, b). The tympanic bone is hook-shaped in both monotremes and relatively loosely suspended from the surrounding skull. On the other hand, the malleus and incus are tightly attached to the periotic bone (Aitkin and Johnstone 1972). In fact, in both monotremes the large anterior process of the malleus is tightly bonded, even fused, to the ring of the tympanic bone (Allin 1975), so that the tympanic bone and the malleus vibrate as a single unit, a feature interpreted as retention of the primitive mammalian condition by some authors (Allin 1975). The incus is relatively small (compared to the malleus) and the stapes is a long, thin, single column of bone rather than the bicolumnar stirrup shape seen in many Eutheria. As in placentals, the stapes base is expanded into a circular plate that abuts the oval window, which is actually circular in monotremes (Gray 1908; Griffiths 1978). The auditory ossicles of the echidna are therefore so firmly joined together and to the periotic bone that they are relatively resistant to movement in response to sound, particularly from airborne vibrations (Aitkin and Johnstone 1972). Movement of the columellar stapes at 100 dB sound pressure level is very low (only 0.0014 mm/s) at 100 Hz and rises to a peak of only 0.1 mm/s at 6 kHz. This suggests that the echidna inner ear is rather poorly adapted to airborne transmission, and in fact provides the best transmission for vibrations arising through bone conduction, which would reach the middle ear along the snout and mandible.

In the platypus, the tympanic bone and tympanic membrane are also situated on the ventral surface of the skull, whereas the external orifice of the ear is situated on the dorsal surface of the head (Griffiths 1978), giving it a curved external ear much like the echidna. This means that sound waves entering the external orifice are initially directed ventrally and medially, before turning dorsally to reach the tympanic membrane, with a total external ear length of 4 cm (Grant 2007). However, there are some differences from the echidna. The cartilage of the platypus external ear is soft, in contrast to the stiff cartilage strips of the external ear of *Tachyglossus*. Gates and colleagues (Gates *et al.* 1974) also reported that the auditory ossicles of the platypus could be more easily separated from each other than is the case for the short-beaked echidna.

In summary, both monotremes have curved external auditory canals, a tympanic membrane facing ventrally and a columelliform stapes. Note that the columelliform shape of the monotreme stapes is much like that seen in most marsupials and some placentals (e.g. the pangolin). A curious feature of the monotremes is that the round window (fenestra vestibuli; for transmission of vibration from the ossicle chain to the inner ear) is oval in shape, and the oval window (fenestra cochleae; for the complementary bulging of perilymph) is round in shape (Gray 1908)!

The monotreme inner ear: cochlear part

Although it has many unusual features, the monotreme cochlea is undoubtedly mammalian in organisation (Vater *et al.* 2004), having distinct inner and outer hair cells flanking a tunnel of Corti, as well as specialised supporting cells. Ultrastructural and functional evidence also suggest the presence of a cochlear amplification mechanism.

According to Griffiths (1978), the cochlea of the short-beaked echidna is 6.0 mm long and 1.75 mm in diameter (Fig. 10.1c), but Ladhams and Pickles (1996) measured the length of the scala media at 10.6 mm and its organ of Corti at 7.6 mm. In contrast to therians, the monotreme cochlea is only poorly curved, achieving a three-quarter turn according to Griffiths (1978), although Gray (1908) and Fox and Meng (1997) reported that the extent of the curvature is larger, achieving a complete curvature of 180°. Perhaps more important than the actual angle of curvature, the direction of the cochlear turn is typical of mammals (clockwise in the right ear and counterclockwise in the left ear; Fox and Meng 1997). As observed by Luo and Ketten (1991) and Fox and Meng (1997), the membranous labyrinth in the monotremes does not coil in close correspondence with the bony labyrinth, so the basilar membrane of the scala media and bony canal do not have the same radius of curvature (Vater *et al.* 2004).

There is a lagena (dilated oval cavity) at the end of the monotreme cochlear duct (Fig. 10.1a, 10.1c) and this structure is suppled by a small nerve that joins the cochlear division of the vestibulocochlear nerve (Gray 1908; Ladhams and Pickles 1996), although its functional significance remains uncertain. Scanning electron microscopy of the basilar membrane, the width and physical properties of which determine the frequency sensitivity of the overlying hair cells in the organ of Corti, show that its width varies along the length of the cochlear duct, but not with the consistent widening with increasing distance from the vestibule that is a characteristic of the typical therian cochlea (Pickles 1992; Ladhams and Pickles 1996). The organ of Corti in the echidna is only 7.6 mm long, much shorter than any placental of a similar body weight (Vater and Kössl 2011; Fig. 10.2a): the organ of Corti of the laboratory rat, for example, is 9.5–12.0 mm long (Malmierca and Merchán 2004). The basilar membrane in the echidna is at its maximal width (374 μm) ~1.9 mm from the apex of the cochlea and tapers down to 180 μm width at 5 mm from the apex, a small variation by mammalian standards. Nevertheless, much like other mammals, the echidna organ of Corti has rows of inner and outer hair cells separated by a tunnel of Corti (Fig. 10.2b). Inner hair cells are arranged in three to five rows, but this reduces to a single irregular row at the apex and base (Chen and Anderson 1985; Ladhams and Pickles 1996). The inner hair cells have two rows of stereovilli (also known as stereocilia), or two rows with a third row of occasional stereovilli (Ladhams and Pickles 1996). Outer hair cells are mainly arranged in six rows, rising to seven or eight at the apex of the cochlea, and reducing to just two rows at the base. The regularity of their distribution decreases as one moves laterally away from the tunnel of Corti. Outer hair cells have five to seven rows of stereovilli, arranged in a straight line, a shallow 'V' or a 'W'. The tallest of the stereovilli of the outer hair cells insert into the tectorial membrane, but the stereovilli of the inner hair cells do not (Ladhams and Pickles 1996; Fig. 10.2b). Kinocilia are present on the outer hair cells of the short-beaked echidna, but not the platypus (Ladhams and Pickles 1996; but note that Chen and Anderson (1985) did not see any). Kinocilia are an early form of hair cell process that precedes the development of stereovilli and may regress in some mammals once the surrounding stereovilli bundle has matured. The functional significance of their persistence in the echidna, but not the platypus, is unknown. The pillar cells that flank the tunnel of Corti are arranged in three or four rows (Fig. 10.2b), in contrast to the two rows usually seen in placentals. This may lead to greater rigidity of the basilar membrane than in placentals. Ladhams and Pickles (1996) suggested that this might more effectively couple vertical vibrations of the fluid of the cochlear duct with vertical movement of the organ of Corti.

The organ of Corti contains active sensory cells that require a blood supply. In the short-beaked echidna, paired vessels run along the underside of the organ of Corti for about three-quarters of its length, before the lateral vessel terminates (Ladhams and Pickles 1996). In therians, only a single longitudinal vessel runs beneath the basilar membrane.

Like the echidna, the cochlea of the platypus is also poorly curved. It forms a banana-like curved tube ~6.3 mm long and 1.3 mm in diameter (Griffiths 1978), but Ladhams and Pickles (1996) measured it at 7.7 mm length. As in the short-beaked echidna, the cochlea terminates in a sensory lagena (Griffiths 1978). Fox and Meng (1997) note that the proximal cochlear duct is relatively straight with a sharp bend at the distal end, giving it something of a sickle shape. The scala media (cochlear duct) is triangular in cross-section (as in other mammals) and the organ of Corti runs along most of the length of the scala media, but at a length of only 4.4 mm it is shorter than both the echidna and all placental mammals of similar body size (Ladhams and Pickles 1996) (Fig. 10.2a). The organ of Corti sits on the basilar membrane, which, unlike in many therians (Fig. 10.2c), has only a minimal taper from the apex to the base of the cochlea (Fig. 10.2d). In fact, Ladhams and Pickles (1996) measured the width of the basilar membrane at 185 μm about halfway along the duct and a similar width (180 μm) at the apex. These measurements were made in only one specimen, but given that the width of the basilar membrane plays a major role in determining the spatial distribution of frequency sensitivity in the organ of Corti (i.e. place coding of frequencies), the observation raises questions about how the platypus analyses different pitches.

The macula lagena (the distinct sensory area at the end of the cochlear duct and separated from the organ of Corti) in both the platypus and short-beaked echidna (see Fig. 10.1a, d) consists of hair cells surmounted by a membrane with embedded otoconia crystals. Ladhams and Pickles (1996) were not able to accurately map the orientation of the stereocilia of the hair cells in the macula lagena, but they did suggest that the macula lagena can be divided into regions of opposite stereocilia polarity, much as has been seen in the maculae of the vestibular apparatus (see below). They suggested that the hair cells of the macula lagena can be divided into vestibular type I and type II hair cells, although the functions of these remain unknown.

In summary, the organ of Corti of the two monotremes shows three main differences from the organ of Corti of placentals (Ladhams and Pickles 1996). First, the cochlea (and basilar membrane) is shorter than in placentals, so much so that the basilar membrane lengths of the platypus and echidna sit far beneath the therian range for this structure at their body weight (Vater and Kössl 2011; Fig. 10.2a). Second, there are more rows of inner hair cells than in placentals (Fig. 10.2b). Finally, there are more rows of pillar cells than in placentals (Fig. 10.2b). Although the monotreme cochlea is shorter, the presence of more rows of *inner* hair cells means that the total number of inner hair cells is about the same as in placentals (Ladhams and Pickles 1996). The platypus has 1600 inner hair cells and the short-beaked echidna has 2700 compared to 1600–3000 in a range of placentals (Ladhams and Pickles 1996). By contrast, the numbers of *outer* hair cells in the two monotremes (3750 in the platypus and 5050 in the short-beaked echidna) is at the lower end of the range reported for placentals (4881–12 000; see discussion in Ladhams and Pickles 1996). Outer hair cells are part of the cochlear amplifier mechanism, so this difference has significant functional implications (see below).

a

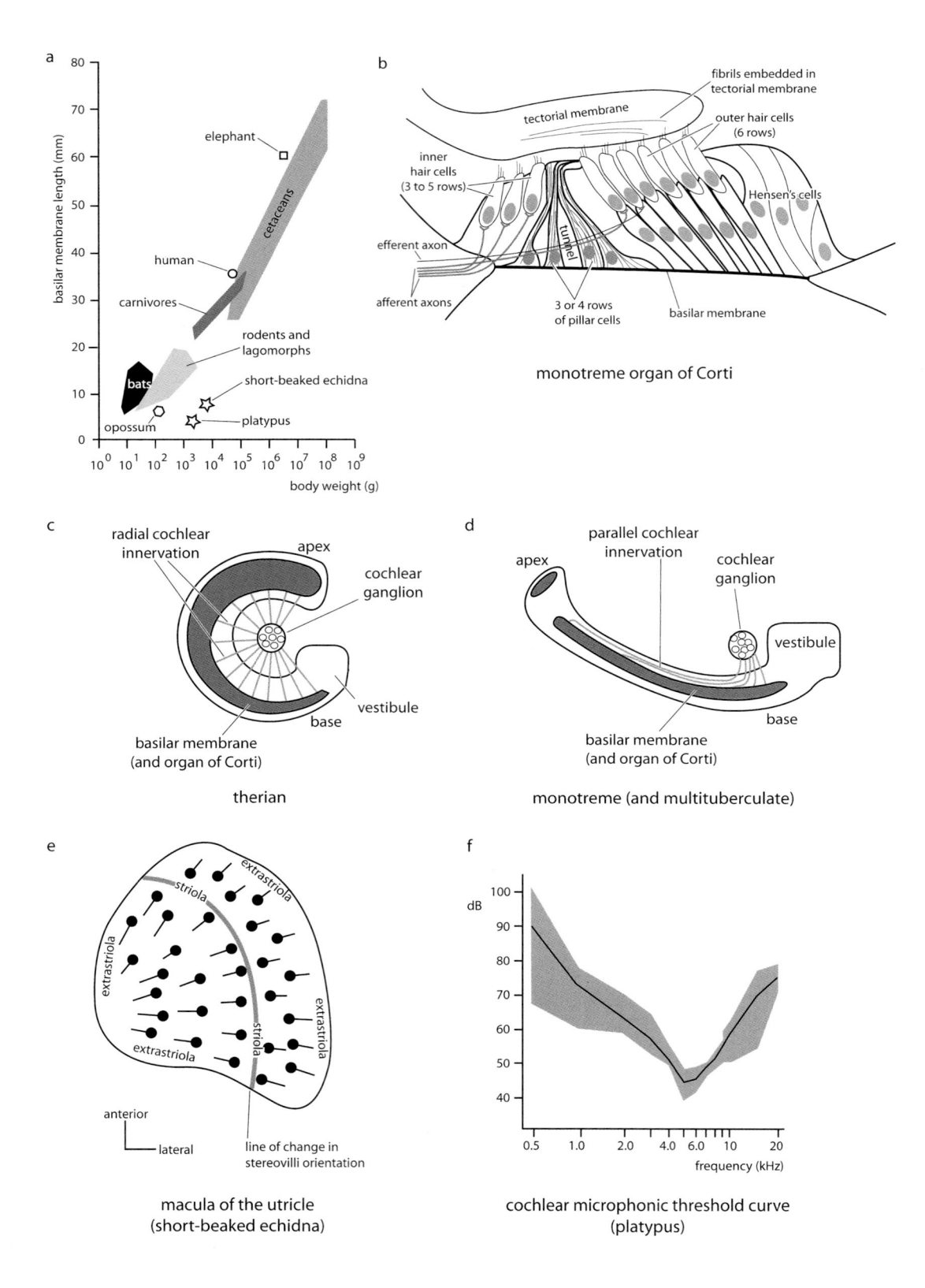

b monotreme organ of Corti

c therian

d monotreme (and multituberculate)

e macula of the utricle (short-beaked echidna)

f cochlear microphonic threshold curve (platypus)

Although the presence of multiple rows of inner hair cells is reminiscent of sauropsids, the organ of Corti in the monotremes is clearly mammalian in structure and Ladhams and Pickles (1996) reported that the ultrastructure of the inner and outer hair cells is generally similar to the matching types in the placental organ. As has already been noted, the monotreme auditory system may be more specialised for detection of ground-borne rather than airborne vibrations. In therians, the outer hair cells function as a cochlear amplifier to maximise sensitivity to high frequency sounds, so the more irregular array of outer hair cells in the monotremes may reflect a regression of the ancestral cochlea as hearing for high frequency sounds became less behaviourally important (Ladhams and Pickles 1996).

A final point to make before leaving the cochlea concerns the arrangement of cochlear nerve fibres. The cochlear nerve in adult monotremes does resemble its homologue in therians in that it passes through a cribriform plate in the floor of the internal acoustic meatus (see discussion in Fox and Meng 1997) but once within the canal there are significant differences from therians. Cochlear nerve branches radiate from the centre of the therian cochlea (Fig. 10.2c), but in monotremes these fibres enter the edge of the cochlea curvature at one point close to the base and run parallel to the turn of the cochlea to reach its distal end (Fig. 10.2d). This means that whereas cochlear fibres have a similar distance to travel to reach all points along the therian cochlear spiral, in monotremes these fibres must traverse a much greater distance to reach the apex of the cochlea compared to the base. Distance is time in nerve conduction, so the therian arrangement is clearly superior in ensuring a minimal time lapse in conveying auditory information from the cochlear apex to the brainstem. The spiral arrangement also allows more cochlear axons to be accommodated within the narrow confines of the petrosal bone (Vater et al. 2004).

The monotreme inner ear: vestibular part

In the short-beaked echidna, the size of the vestibular part of the labyrinth relative to the rest is small (3.5 mm in its longest diameter) as in other mammals, and unlike reptiles (Gray 1908). Gray also observed the presence of a recess of the utricle, a conical diverticulum arising from the utricle just where the ampulla of the horizontal canal opens into the utricle. This feature is unlike other mammals and more like that seen in reptiles and amphibians.

The posterior and horizontal semicircular canals of the echidna have smooth rounded contours, as seen in therians and quite unlike the angularity of the canals seen in the labyrinth of reptiles (Gray 1908; Griffiths 1978), but Gray (1908) reported that the superior canal of the echidna is noticeably angular

Figure 10.2: Structural and functional features of the monotreme inner ear. (a) Plot of the length of the basilar membrane (on which the organ of Corti sits) against body weight for the two monotremes and several groups of therian mammals. Animals with high frequency acuity (e.g. microchiropteran bats) have long basilar membranes for their body weight. Both the echidna and platypus have quite short basilar membranes for their body weight, consistent with poor hearing at high frequencies. The diagram is based on data in Vater and Kössl (2011). (b) Ultrastructure of the monotreme organ of Corti, as shown in a diagram of a transverse section through the organ. Note the many rows of outer hairs, the presence of fibrils embedded in the tectorial membrane, the three or four rows of pillar cells with fusion of their outwardly bending apical processes, the tunnel of Corti (tunnel) and the predominance of sensory (afferent) axons contacting inner hair cells. The diagram is based on data in Ladhams and Pickles (1996). (c) Diagrammatic representation of the radial innervation of the organ of Corti by axons from cochlear ganglion cells in therians. (d) Diagrammatic representation of the parallel innervation of the organ of Corti by axons from cochlear ganglion cells in monotremes (and probably multituberculates). Note also that the basilar membrane of the monotremes does not show the progressive increase in width along the cochlear duct, such as is characteristic of the basilar membrane in therians. The diagram has been drawn based on text and illustrations in Fox and Meng (1997). (e) Diagram of the surface of the macula of the utricle of the short-beaked echidna, showing the orientation of the hair cell processes. The dot on the end of each line indicates the position of the kinocilium relative to the stereovilli. The diagram is based on data in Jørgensen and Locket (1995). (f) Cochlear microphonic potential threshold curve for the platypus. The solid line indicates the mean intensity required to elicit a 1.0 μV cochlear microphonic potential, whereas the shaded region indicates the range. Note the peak sensitivity (i.e. lowest threshold) at 4–8 kHz. This diagram is based on data in Gates et al. (1974).

and meets with the posterior canal abruptly, a feature seen in the reptilian labyrinth. The superior canal is 4.5 mm in internal diameter and 6.0 mm in external diameter; the posterior canal is 2.5 mm in internal diameter and 4 mm externally; and the horizontal canal is 2.75 mm in internal diameter and 4 mm externally (Gray 1908). These values are broadly similar to values for rodents and carnivores of a similar body weight (Spoor and Thewissen 2008).

Two main types of cells are found in the sensory areas of the vestibular apparatus (and macula lagena). These are the sensory or hair cells, and supporting cells (Jørgensen and Locket 1995). Both cell types are visible in the echidna and overlie a basal lamina, but only the supporting cells extend from the basal lamina to the apical surface; the sensory cells have no contact with the basal lamina. There is a dense plexus of capillaries in the connective tissue underlying the basal lamina, with occasional capillary loops extending into the epithelium (Jørgensen and Locket 1995).

There are two types of vestibular hair cells in the short-beaked echidna, identified by electron microscopy mainly on the different type of innervation (Jørgensen and Locket 1995). One type is a cylindrically shaped bouton-innervated cell; the other is a bottle-shaped calyceal-innervated type. Both types of hair cells have a bundle of sensory hairs consisting of a single kinocilium and 40–60 stereovilli (stereocilia). Jørgensen and Locket (1995) thought that all of the calyceal-innervated hair cells and some of the bouton-innervated hair cells were sensory, on the basis of the presence of a moderate number of mitochondria within the axon terminals, whereas a subgroup of the boutons contains clear vesicles and may be an efferent innervation of the vestibular organ by the brainstem. The distribution of calyceal hair cells in the different sensory areas of the vestibular apparatus of the echidna is broadly similar to that in therian mammals (Jørgensen and Locket 1995).

Hair cells of the ampullae of the semicircular ducts have their hair bundles oriented in the same direction. All hair cells of the cristae of the anterior and posterior ducts have their kinocilia on the canal side, whereas the hair cells of the lateral duct crista are oriented with the kinocilia on the macular side (Jørgensen and Locket 1995). The short-beaked echidna has a distinct crista neglecta, a sensory region associated with the crista of the posterior semicircular duct, but subsumed into the posterior crista in some therians like humans.

The macula of the utricle of the short-beaked echidna has an area of 4.1–4.7 mm^3 in three echidnas (Jørgensen and Locket 1995) with ~56 350 hair cells across the entire macula. The area of the macula of the utricle is comparable in size to humans and much larger than in guinea-pigs (see Jørgensen and Locket 1995 for discussion) and the number of hair cells in the macula of the utricle is also much larger than that reported for many therians. On the other hand, the orientation of hair cell processes is broadly similar to that in therians (Fig. 10.2e). Calyceal-innervated hair cells predominate in the striola region (a specialised zone where the hair cell processes switch orientation), whereas bouton-innervated hair cells predominate outside the striola.

The vestibular branch of the vestibulocochlear nerve is formed into superior and inferior vestibular divisions. The superior division comes from the macula of the utricle and the cristae of the lateral and superior semicircular ducts, while the inferior branch comes from the macula of the saccule and the crista of the posterior semicircular duct (Jørgensen and Locket 1995). The sensory fibres from the macula lagena join the cochlear division of the vestibulocochlear nerve to enter the skull interior though the internal acoustic meatus.

Functional aspects of monotreme hearing

All species of monotremes are relatively quiet animals and it is unlikely that vocalisations play a major role in communication with conspecifics. Both Burrell and Pettigrew and colleagues noted that adult platypuses emit a tremulous snoring or growling sound, particularly when first woken from sleep (Burrell 1927; Pettigrew et al. 1998), but platypuses usually emit only faint vocalisations in the burrow. When furred platypus young are disturbed in a burrow they emit a noise like a growling puppy (Burrell 1927), but the precise role of this vocalisation is uncertain. It seems unlikely that the platypus uses audition in prey detection, but Griffiths (1978) has suggested that the optimal sensitivity of the short-beaked echidna ear to sound at 5 kHz (see below) is an adaptation to detect the sounds made by termites as they would be filtered by bone conduction through the mandible and maxilla.

As noted above, the auditory ossicle chain of the short-beaked echidna is rather stiff and firmly attached to the skull base (Aitkin and Johnstone 1972). Mills and Shepherd (2001) used distortion product otoacoustic emissions and auditory brainstem responses to assess auditory sensitivity in the short-beaked echidna. Highest sensitivity was found at 4–8 kHz, with an optimal frequency range of 1.6–13.9 kHz – a range of 3.1 octaves. Mills and Shepherd (2001) also confirmed that the short-beaked echidna does have a cochlear amplifier effect, perhaps functioning as that produced by the outer hair cells in the therian organ of Corti. They observed that the amplification effect operates to frequencies above 20 kHz, which is higher than the sensitivity of hearing function in birds or reptiles, but not as effective as in typical therians. If the echidna does posses a cochlear amplifier like therians, this would suggest that the cochlear amplifier arose before the divergence of the prototherian and therian lineages.

Nevertheless, Aitkin and Johnstone (1972) concluded that the echidna middle ear is more comparable to certain lizards than to the guinea-pig in its ability to transmit sounds at frequencies in excess of 6 kHz. It is also substantially inferior both to guinea-pigs and lizards at lower frequencies. In fact, the echidna middle ear may be better adapted for transmission of sound through the mandible than by air, a specialisation consistent with the use of the snout as a thrust probe in the search for invertebrate food.

The middle ear structures may not be as mechanically stiff in the platypus (Gates et al. 1974), but middle ear sound transmission in the platypus is nevertheless quite poor compared to many placentals and is even poorer than in the echidna. Gates and colleagues recorded cochlear microphonic potentials from the round window of the platypus in response to pure tones between 500 Hz and 20 kHz and constructed a threshold curve for the platypus (Fig. 10.2f). The optimal response is at 5 kHz with a decline in sensitivity of ~20 dB/octave for high frequency sounds and 15 dB /octave for low frequency sounds. The sensitivity at any given frequency is actually poorer than for the short-beaked echidna, perhaps because the malleus and incus are even smaller in the platypus. The question naturally arises as to how the platypus uses its (limited) auditory ability. Observations by Burrell (1927) suggested that the platypus could modify the area around its external auditory meatus to function as a directionally sensitive auricle, but this remains controversial. When the platypus submerges, it closes its external auditory meatus, so the closed air-filled tube may serve an accessory auditory function much like the swim bladder of teleost fish (Gates et al. 1974).

Evolutionary aspects of the monotreme ear

The peculiar nature of the ear of extant monotremes and its structural similarities to those of early mammals naturally raise the question of how it fits into the evolution of the mammalian ear. The tri-ossicle chain of the mammalian middle ear is derived from transformed lower jaw and jaw suspensory elements of other gnathostomes, with the malleus, incus and stapes being derived from the articular plus prearticular, quadrate and hyomandibular elements, respectively. There is considerable controversy as to exactly how many times the mammalian tri-ossicle chain evolved, with opinions ranging from once to three times (see Meng and Wyss 1995). Similarities between the middle ear structure of multituberculates and extant monotremes have led Meng and Wyss (1995) to argue for a monotreme/multituberculate pairing and a single origin of the mammalian middle ear structure. Similarities between the multituberculate and monotreme middle ear (Meng and Wyss 1995; Fox and Meng 1997; Wible and Rougier 2000; Ladevèze et al. 2010) include: (1) the dorsal position of a small incus; (2) the large ectotympanic bone; (3) the antero-medial contact between the ectotympanic and pterygoid bone; (4) horizontally positioned ectotympanic and malleus; (5) the lack of a malleus head; (6) inflation of the vestibule region of the inner ear with greater distance between the stapes foot plate and the basal turn of the cochlea; (7) the uncoiled cochlea; and (8) the presence of a lagena. On the other hand, the cribriform passage of the cochlear nerve branches in monotremes is more like the arrangement in therians than in multituberculates.

Meng and Wyss (1995) and Fox and Meng (1997) have suggested that multituberculates had a similar auditory capacity to that of modern monotremes, i.e. a relatively poor sensitivity to airborne sounds. They propose that the inflation of the vestibule with increased separation between the stapes foot-plate and the base of the cochlea better suits the

Table 10.1. Synapomorphies distinguishing therian from non-therian inner ear[1]

Synapomorphy	Functional significance
Fully coiled cochlea (more than at least 360°)	Allows the accommodation of an elongated organ of Corti and basilar membrane for a wider frequency range
Development of primary and secondary osseous spiral laminae	Supports a narrow and elongated basilar membrane for a wider frequency range
Radial cochlear nerve	Reduces impulse transmission time from apex of cochlea and allows more axons to pass through the petrosal bone
Protrusion of a narrow basilar membrane between the oval and round windows	Allows the elongation of the basilar membrane towards the base of the cochlea for perception of high frequency sounds
Formation of a perilymphatic recess that merges with the scala tympani of the cochlea	Not known
Development of a bony cochlear aqueduct and formation of a true fenestra cochleae	Not known

1 Based on Vater *et al.* (2004).

multituberculate ear to hearing by bone conduction, consistent with a fossorial lifestyle of multituberculates.

Although it is tempting to see the partially coiled cochlea of monotremes as a precursor of the highly coiled cochlea of therians, Fox and Meng (1997) have concluded that the monotreme cochlear coil is apomorphous and is an unlikely precursor of the therian coiled cochlea. There are several key derived features of the inner ear of therians (see Table 10.1; Vater *et al.* 2004) that separate them from non-therians (including modern monotremes).

Central auditory pathways

In the mammalian auditory pathway, information is passed from the spiral or cochlear ganglion to the dorsal and ventral cochlear nuclei of the brainstem and then through a complex set of pathways that incorporate relays in the superior olivary nuclear complex and the nucleus of the lateral lemniscus to reach the inferior colliculus of the midbrain. Relay neurons in the inferior colliculus project through the brachium of the inferior colliculus (inferior brachium) to the medial geniculate nucleus of the thalamus, which in turn relays auditory information to the primary auditory cortex.

The dorsal and ventral cochlear nuclei are arranged around the inferior cerebellar peduncle. In monotremes, the dorsal cochlear nucleus (DC) is unlayered, a primitive condition in which constituent neurons are scattered rather than arranged in a single layer, as in therians (Johnson *et al.* 1994). In therians, the ventral cochlear nucleus has distinct anterior and posterior components (VCA and VCP), but there is no subdivision into two parts in the platypus (Hines 1929). Central projections of the therian cochlear ganglion project into the different divisions of the cochlear nuclear complex by two principal axon collaterals. One of these is an ascending branch that terminates in the VCA; the other is a descending branch that terminates in the VCP and DC (Grothe *et al.* 2004).

Hines identified several nuclei that might be the homologue of the therian superior olivary nuclear complex (Hines 1929). This is a nucleus (or more correctly a nuclear complex) that serves diverse auditory functions including focusing sensory attention on particular frequencies and judging distance and direction of sound sources. The candidate nucleus ventrolateral to the ventral motor facial nucleus is the most likely homologue based on its apparent connections as identified in myelin stained sections. Hines was not able to identify a trapezoid nucleus (a nucleus associated with the decussating auditory axons in therians) and commented that the bulk of the axons of the trapezoid body pass dorsal and medial to the sensory trigeminal complex rather than lateral and ventral (as described for the echidna and some marsupials).

The inferior colliculus in all mammals has a central nucleus (CIC) flanked by an external cortex (ECIC) laterally and a dorsal cortex (DCIC) dorsally. These features are clearly visible in both the platypus and echidna (see for example atlas plates Pl-Ad22 and Ec-Ad22), but the functional characteristics of this structure are unknown.

Jones and colleagues have claimed that a medial geniculate nuclear complex cannot be identified in either the platypus or the short-beaked echidna (Jones 2007; Mikula et al. 2008), but Hines found the nucleus in the platypus (Hines 1929) and Ashwell and Paxinos (2005) were able to identify a poorly demarcated medial geniculate in the dorsal thalamus of the short-beaked echidna on the basis of clusters of parvalbumin immunoreactive neurons. Fibres of the brachium of the inferior colliculus can be followed entering the dorsal thalamus, but tend to disappear near the posterior pole of ventral posterior nucleus. Ashwell and Paxinos noted that the medial geniculate nucleus is not clearly defined in either Nissl stained or AChE reacted sections through the dorsal thalamus of the echidna. It has a similar cytoarchitectural appearance to the adjacent posterior thalamic nucleus and is also of a similar level of AChE reactivity. Nevertheless, they found that the medial geniculate nucleus is separated from the posterior thalamic nucleus by a band of stronger acetylcholinesterase activity. Some confirmation of the likely position of the poorly defined medial geniculate nucleus in the short-beaked echidna comes from the degeneration study of Welker and Lende (1980), which showed degeneration of neurons in the posterior lateral part of the dorsal thalamus following ablation of an area of isocortex that included the auditory cortex (see below).

The auditory cortex is located on the caudal extremity of the cerebral cortex in both the platypus and the short-beaked echidna (Lende 1964; Bohringer and Rowe 1977). In both species, the auditory region of the cerebral cortex is quite small both in absolute size (2 mm × 3 mm in the platypus, and 3 mm × 3 mm in the echidna) as well as relative to the size of the rest of the isocortex. In the platypus, a region where neurons are responsive to both auditory and somatosensory stimulation surrounds the auditory cortex (Divac 1995; Krubitzer 1998; Pettigrew et al. 1998). In the short-beaked echidna, the auditory cortex is located at the posterior pole of the cerebral cortex and extends into the depths of sulcus ζ (Lende 1964; see review in Rowe 1990; Krubitzer et al. 1995). Krubitzer and colleagues (Krubitzer et al. 1995) also found small areas of cortex between the auditory and somatosensory cortices of the echidna that respond to both auditory and tactile stimuli, suggesting a multimodal sensory role for these areas.

The cytoarchitecture of the auditory cortex in the short-beaked echidna is similar to that of the adjacent visual cortex, with small pyramidal neurons in layer 5Cx and a prominent input layer 4Cx of stellate cells. The gyrus that carries the auditory cortex has a strong band of acetylcholinesterase label in layer 4Cx as well as a rich input of myelinated axons (Krubitzer et al. 1995; Hassiotis et al. 2004b, 2005).

Krubitzer and colleagues found two tonotopic representations within the platypus auditory cortex (Krubitzer et al. 1995; Krubitzer 1998). A rostromedial field contains a tonotopic representation of frequencies from 1 to 16 kHz, with lower frequencies represented caudally and higher frequencies rostrally. A caudolateral field contains a tonotopic representation of frequencies from 3 to 13 kHz, with low frequencies rostrally and high frequencies caudally. Krubitzer proposed that the caudal field is homologous to the primary auditory cortex of therians, because of its internal organisation, strong activity for cytochrome oxidase and fibre architecture (Krubitzer 1998).

Tracing of corticocortical connections in the auditory cortex of the short-beaked echidna with the carbocyanine dye DiI has shown that the region has a columnar organisation (Dann and Buhl 1995; see Chapter 7). Anterograde and retrograde connections between cortical areas are in register, with most corticocortical connections arising from pyramidal neurons in layers 5Cx and 6Cx. Corticocortical connections are made to all six layers of the cortex.

Central vestibular pathways

The vestibular component of the vestibulocochlear nerve is distributed to the vestibular nuclei and vestibulocerebellum. Cytoarchitecturally, the vestibular nuclear complex of both the platypus and the short-beaked echidna is very similar to that in therians (Hines 1929; Ashwell et al. 2007b; Vidal and Sans 2004).

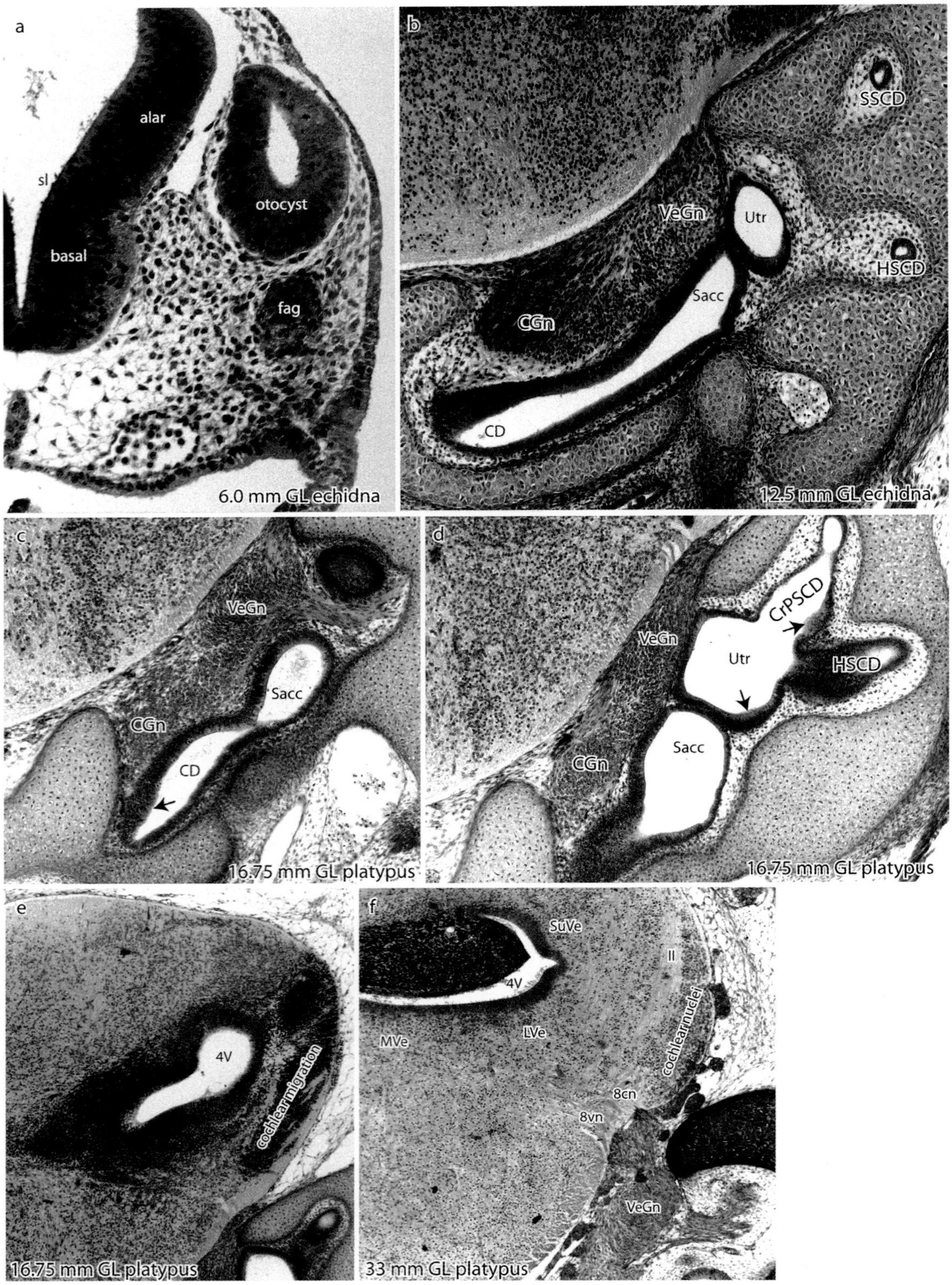
a
alar
sl
otocyst
basal
fag
6.0 mm GL echidna
b
SSCD
VeGn
Utr
HSCD
Sacc
CGn
CD
12.5 mm GL echidna
c
VeGn
Sacc
CGn
CD
16.75 mm GL platypus
d
CrPSCD
VeGn
Utr
HSCD
Sacc
CGn
16.75 mm GL platypus
e
4V
cochlear migration
16.75 mm GL platypus
f
SuVe
4V
ll
MVe
LVe
cochlear nuclei
8cn
8vn
VeGn
33 mm GL platypus

In the short-beaked echidna, the complex extends for ~6.3 mm in the rostrocaudal direction and is bounded laterally by a succession of sensory nuclei (trigeminal, dorsal cochlear and cuneate nuclei) and medially by the medial parabrachial nucleus (MPB), facial nerve fibres and nucleus of solitary tract in rostrocaudal sequence. Subnuclei of the vestibular complex include a superior vestibular nucleus (SuVe; nucleus of Bechterew) at the level of the MPB, medial and lateral vestibular nuclei (MeVe, LVe) at mid-rostrocaudal levels and a spinal vestibular nucleus (SpVe) extending from middle levels to the caudal extremity of the complex. Large neurons that would be the likely source of vestibulospinal pathways are found mainly in the LVe and SpVe (Hines 1929; Ashwell et al. 2007b).

Hines (1929) noted distribution of trigeminal nerve fibres into the superior vestibular nucleus, as well as a connection between it and the central fibre layer of the cerebellum. The LVe receives a large limb of the medial longitudinal fasciculus and this nucleus also has prominent connections with the cerebellum (Hines 1929).

Immunoreactivity for non-phosphorylated neurofilament protein (SMI-32 antibody) is concentrated in large somata in the LVe and SpVe, which are presumably cells of origin for the lateral vestibulospinal tract. Immunoreactivity to calcium-binding protein in the vestibular nuclear complex of the short-beaked echidna is similar to that in therians (e.g. rodents) with immunoreactivity for both parvalbumin and calbindin being present in vestibular afferents.

There is some limited information available concerning the projection of the vestibular nuclei to the cerebellum in the short-beaked echidna based on retrograde transport of horseradish peroxidase enzyme (Holst 1986). All four nuclei project ipsilaterally to the paravermal area of lobules 5 to 7, but the spinal and medial vestibular nuclei also project contralaterally to the same region. When the injection site in the cerebellum is slightly further rostral (i.e. the lateral parts of the rostral folium of the left paramedian lobule), a bilateral projection from the medial, spinal and superior vestibular nuclei is revealed, with most retrograde label in the ipsilateral spinal vestibular nucleus. This pattern is largely similar to the structure in the laboratory rat, where vestibulocerebellar mossy axons arise bilaterally from medial, superior and spinal vestibular nuclei in descending order of density (see review in Voogd 2004).

Development of the monotreme ear

The development of the middle and inner ear of both the platypus and echidna appears to proceed according to a similar timetable so the following description applies to both.

The tympanic bone, malleus, incus and stapes are derived from mesenchymal components of pharyngeal arches 1 and 2, while the sensorineural parts of the inner ear are derived from the otocyst and neural crest. Invagination of the otic placode ectoderm to form the otocyst occurs at ~5 mm GL, which corresponds roughly to the beginning of incubation and is contemporaneous with the closure of the anterior neuropore. The neural crest derivatives that will give rise to the facioacoustic ganglion (fag) migrate into

Figure 10.3: Development of the monotreme inner ear as shown in sections through the heads of incubation stage and hatchling specimens. The otocyst (a) forms by invagination of the otic placode epithelium around the beginning of incubation. Initially it is a teardrop shape and is accompanied by cells of the facioacoustic ganglion (fag). Elongation of the endolymphatic and cochlear ducts (CD) is followed by separation of the utricle (Utr) and saccule (Sacc) and formation of the semicircular ducts (horizontal duct – HSCD, superior duct – SSCD) so that the immediately pre-hatching embryo has all the major parts of the inner ear (b), but the sensory areas are undifferentiated at the time of birth. Separation of the ganglionic primordium into vestibular (VeGn) and cochlear (CGn) components occurs shortly before hatching (b). In the first week after hatching (c, d) the otocyst components continue to grow and the sensory regions of the cochlear duct and vestibular organs differentiate. The arrow in (c) indicates the developing organ of Corti. Arrows in (d) indicate the developing macula of the utricle (Utr) and the ampulla of the posterior semicircular duct inside the crista (CrPSCD). Neurons for the cochlear nuclei are produced by the rhombic lip and migrate into position during the first post-hatching week (e). By the second post-hatching week (f), most neurons of the vestibular and cochlear nuclei have reached their final position. 4V – fourth ventricle; 8cn – cochlear nerve axons; 8vn – vestibular nerve axons; alar – alar plate (sensory neuroepithelial zone) of the medulla; basal – basal plate (motor neuroepithelial zone) of the medulla; ll – lateral lemniscus; LVe – lateral vestibular nucleus; MVe medial vestibular nucleus; sl – sulcus limitans; SuVe – superior vestibular nucleus.

the space ventral to the otocyst at this stage (Fig. 10.3a).

The otocyst elongates and its structure becomes more elaborate during the middle third of incubation. The first two parts to develop are the endolymphatic duct that grows dorsocaudally and the cochlear duct that grows ventromedially. The utricle and saccule separate shortly afterwards and the semicircular ducts are formed by the end of the middle third of incubation. Nevertheless, the sensory surfaces of the utricle, saccule and cristae are undifferentiated and the macula lagena cannot be distinguished from the developing sensory epithelium of the cochlear duct.

During the last third of incubation (Fig. 10.3b), sensory regions begin to appear in the wall of the otocyst, including the macula lagena. The first of these to emerge and develop stereovilli is the macula of the utricle, but the maculae of the saccule, cristae ampullares and lagena do not have stereovilli until after hatching, when the sensory areas are more distinct (Fig. 10.3c, d). The ganglionic tissue separates into distinct vestibular and cochlear (spiral) ganglia during the latter part of incubation.

Development of the ossicle chain is also incomplete until well after hatching. During the incubation phase, the malleus is attached to Meckel's cartilage. It is only during the lactational phase that the malleus and incus separate from the Meckel's cartilage, allowing the dentary-squamosal suspension of the lower jaw to be established. The freeing of the malleus and incus allows them to become true ear ossicles and participate in transmission of sound across the middle ear (Griffiths 1978), but this probably does not achieve functional maturity until at least the second month after hatching.

Development of the monotreme central auditory and vestibular pathways

Generation of the neurons of the brainstem components of the auditory and vestibular pathways does not begin until the last half of incubation (around 9 mm GL in both echidna and platypus). The rhombic lip that will produce the granule cells of the cochlear nuclei expands around the end of the middle third of incubation and cochlear granule cells migrating to their final destination can be seen during the first week of post-hatching life (Fig. 10.3e). Both cochlear and vestibular nuclei become distinct at the end of the first post-hatching week (Fig. 10.3f) and ascending axons of the auditory pathway (the lateral lemniscus) appear at around this time. Vestibular neurons appear to be generated from about the middle of incubation and settle in the dorsolateral mantle layer on the floor of the fourth ventricle during the late incubation phase and first week of post-hatching life, although these conclusions are based on analysis of routine histological sections rather than DNA timed labelling experiments.

Neurogenesis of the thalamus does not begin until ~8.0 mm GL in both the platypus and echidna (Ashwell 2012c). As noted above, the medial geniculate nucleus is difficult to identify in the adult dorsal thalamus, and even more so in hatchlings, but it can be said that differentiated nuclear groups do not emerge in the dorsal thalamus until after the first week of post-hatching life. The cortical plate of the caudal cerebral cortex (where the auditory cortex is located) appears towards the end of the first post-hatching week and invasion of the plate by thalamocortical axons does not occur until the second post-hatching week. The auditory cortical plate is divided into a compact upper zone (ccz) and an underlying loosely packed zone (lpz) and cortical lamination remains undifferentiated until between 6 and 7 weeks after hatching, so the ascending auditory and vestibular projections to the cortex are unlikely to be functional until well into the second post-hatching month.

Questions for the future

Several lines of evidence, both structural and physiological, indicate that audition is a poorly developed sense in the modern monotremes. Their hearing appears to be best adapted for low frequency sounds, perhaps optimally heard through bone conduction. Hearing is unlikely to make any significant contribution to prey detection in the platypus, and is probably only of minor importance during reproduction and rearing of young. Nevertheless, questions still remain as to whether the echidna uses hearing in the detection of, or discrimination between, different types of invertebrate prey.

The middle and inner ear of the modern monotremes have been considered to be structurally and functionally similar to those of Cretaceous multituberculates. Although the modern monotreme ear bears some key features that have remained unchanged from the Cretaceous (e.g. the absence of an osseous spiral lamina in the cochlea, parallel cochlear nerve innervation, and the presence of a macula lagena), their ear should not be regarded as unmodified from the stem mammalian ear. Several lines of evidence suggest that the slight curvature of the monotreme cochlea, the enlargement of the vestibule, and the presence of a cribriform plate for the passage of vestibulocochlear nerve axons through the petrosal are all features that emerged among monotremes *after* divergence from the therian lineage (Fox and Meng 1997).

Details of the pathways and the functional anatomy of central auditory pathways in the monotremes are limited and nothing at all is known about the representation of vestibular information at the thalamic and cortical level. A key question is how auditory information is integrated with tactile or electroreceptive input to form a model of the external world. Do the monotremes have a true multimodal association cortex or is auditory sensory information poorly integrated with other modalities?

11

Chemical senses: olfactory and gustatory systems

Ken W. S. Ashwell

Summary

Olfaction (smell) and gustation (taste) are two chemical senses that allow sampling of the external environment for both food detection and selection, as well as recognition of adult conspecifics and offspring. The two modern monotreme lineages, represented by the platypus and echidnas, have strikingly different olfactory systems: small and relatively unspecialised in the poorly osmatic (some authors have said microsmatic) platypus, and complex and elaborate in the macrosmatic echidnas. Despite its reduced main olfactory system and aquatic lifestyle, the platypus has an extraordinary array of vomeronasal chemoreceptor genes, as great as any therian studied to date, but nothing is currently known about chemoreceptor genes in the echidna. Observations of the time course of development of the olfactory system in platypus and echidna indicate that olfaction is unlikely to make any significant contribution to behaviour in the first post-hatching week, because key developmental events in the central olfactory pathways do not occur until at least 6 days after hatching. The proposition that the common monotreme ancestor had a relatively unspecialised olfactory system provides the best explanation of the details of the tempo and mode of olfactory development.

Overview of mammalian olfactory systems

Olfaction is important in mammals for the location and selection of vegetable foods, the recognition of predators and tracking of prey, identifying adult members of the same species and distinguishing dependent young. Olfaction has a strong input to the amygdala for the initiation of emotional responses and to the hippocampal formation for the formation of behaviourally important memories. The olfactory system also plays a critical role in regulating

a Main Olfactory Pathway

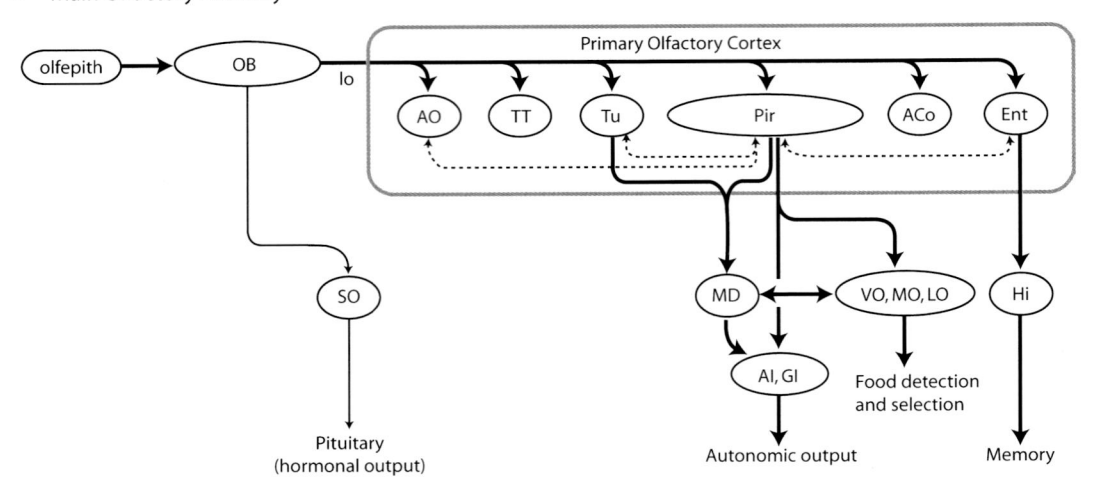

b Accessory Olfactory Pathway

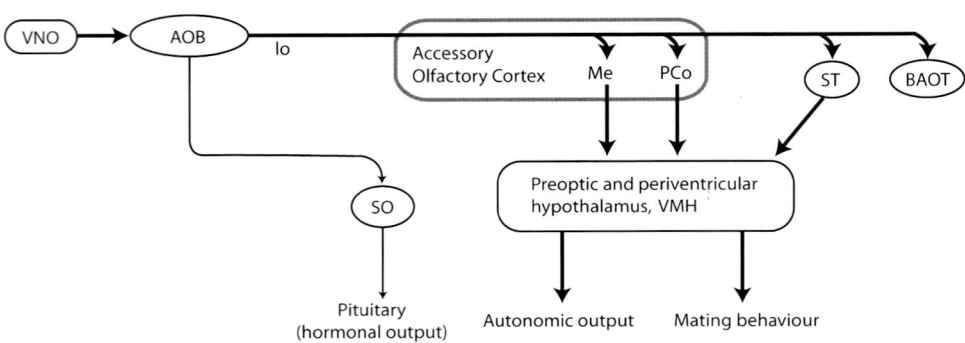

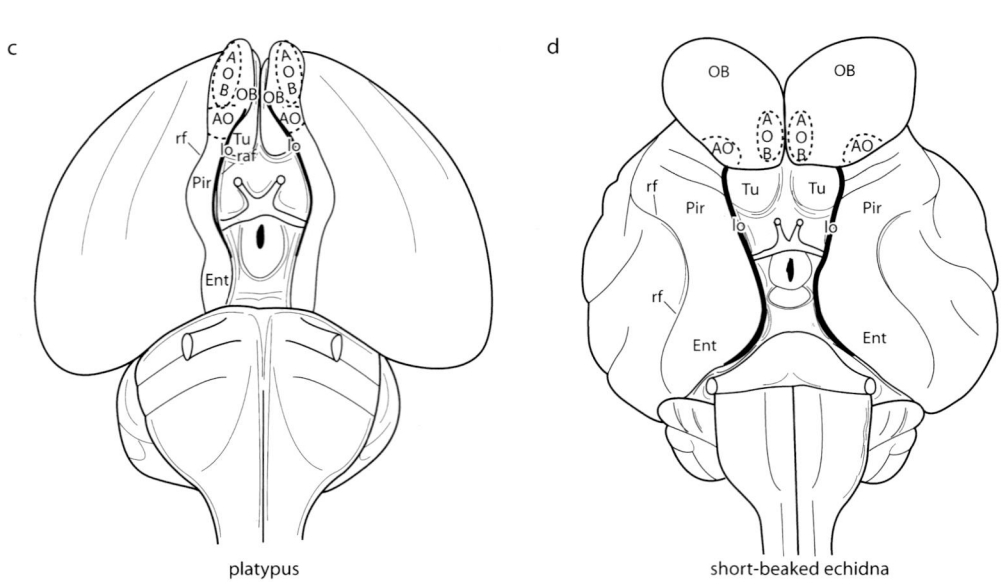

c

platypus

d

short-beaked echidna

neuroendocrine systems, particularly those concerned with the control of reproductive cycles, maternal behaviour and the feeding of young.

Two anatomical pathways serve olfaction in vertebrates and these have distinct connections for most of their central projection. The first of these is the main olfactory pathway, which is mainly concerned with the detection of a broad range of environmental odorants of relevance to food and predator detection. It begins with the main olfactory epithelium, distributed mainly across the ethmoid turbinals in the nasal cavity. The second is the accessory olfactory pathway, which is concerned with processing information from a limited range of pheromonal or intraspecific odorants of very particular behavioural importance (e.g. mate and offspring recognition). Its nasal receptor organ is the vomeronasal or Jacobson's organ. Central olfactory structures include the main and accessory olfactory bulbs, for the two channels respectively, as well as a series of allocortical and nuclear structures in the telencephalon.

The main olfactory pathway (Fig. 11.1a) proceeds by axons of mitral cells and middle or deep tufted cells of the olfactory bulb (OB). Projections pass through the lateral olfactory tract to reach the ipsilateral anterior olfactory area (AO), tenia tecta (TT), olfactory tubercle (Tu), piriform cortex (Pir), anterior amygdaloid cortex, and rostral parts of the lateral and medial entorhinal cortex (Ent) (Ashwell 2012a). Projections through the mediodorsal nucleus of the thalamus and brainstem centres, orbitofrontal cortex and hippocampus allow olfaction to influence the autonomic system, complex behaviour and memory, respectively.

The accessory olfactory pathway (Fig. 11.1b) begins at the accessory olfactory epithelium and is concerned with processing pheromone-derived information. As noted above, the accessory olfactory epithelium consists of a tubular structure in the ventral nasal septum (the vomeronasal organ of Jacobson; VNO) and projects to the accessory olfactory bulb (AOB), which is usually situated alongside the dorsocaudal edge of the main olfactory bulb. The tube of the VNO opens rostrally to the nasal or oral cavity, but the position of the opening varies substantially between mammals, depending on the behavioural role that the VNO serves. Output from the AOB follows a pathway that is segregated from that arising from the main bulb. The AOB has direct projections that reflect its critical role in regulation of emotions, mating behaviour and reproductive cycles. Its major output in well-studied therians like rodents has been found to be through the lateral olfactory tract to the medial and posterior cortical amygdala (Me and PCo, respectively), bed nuclei of the stria terminalis (ST), the bed nucleus of the accessory olfactory tract (BAOT) as well as a projection to the supraoptic nucleus (SO) for regulation of pituitary hormone production (Meisami and Bhatnagar 1998; Shipley *et al.* 2004). The AOB is often regressed in microsmatic aquatic mammals (Meisami and Bhatnagar 1998).

Figure 11.1: Key features of the main and accessory olfactory pathways in therians and mnotremes. Diagrams (a) and (b) show the main and accessory olfactory pathways, respectively, as they have been revealed in therian mammals (diagrams based on the reviews of rodent and primate olfactory pathways; Shipley *et al.* 2004; van Hartevelt and Kringelbach 2012). Since all monotremes studied to date have the main nuclear and cortical elements of these pathways (c, d), it is likely that the main features of the pathways are also present in the monotremes, but this remains to be studied by neural-tracing techniques. Both main and accessory olfactory pathways exert behavioural influences through initial connections with other forebrain structures. In the case of the main pathway, this is principally through the mediodorsal thalamic nucleus (MD), entorhinal cortex (Ent) and hippocampus (Hi). The accessory olfactory pathway exerts its main effects on behaviour through the medial and cortical parts of the amygdala, bed nuclei of the stria terminalis (ST) and bed nuclei of the accessory olfactory tract (BAOT), and thence through the hypothalamus. Note that the BAOT has not been definitively identified in the platypus. Dashed lines in (a) indicate interconnections between components of the primary olfactory cortex. Dashed lines in (c) and (d) indicate the positions of deeper structures not lying at the ventral surface. ACo – anterior cortical amygdala; AI – agranular insular cortex; AO – anterior olfactory region; AOB – accessory olfactory bulb; GI – granular insular cortex; lo – lateral olfactory tract; LO – lateral orbital cortex; Me – medial amygdala; MO – medial orbital cortex; OB – main olfactory bulb; olfepith – olfactory epithelium; PCo – posterior cortical amygdala; Pir – piriform cortex; raf – rhinal arcuate fissure; rf – rhinal fissure; SO – supraoptic nucleus of the hypothalamus; TT – tenia tecta; Tu – olfactory tubercle; VMH – ventromedial nucleus of the hypothalamus; VNO – vomeronasal organ; VO – ventral orbital cortex.

Figures 11.1c and d show the positions of the main components of the main and accessory olfactory pathways as they can be seen in a view of the ventral surface of the brains of the platypus and short-beaked echidna.

Functional importance of olfaction for monotremes

Olfaction is unlikely to make a significant contribution to prey detection in the poorly osmatic platypus because the narial valves close off the nasal cavity during diving (Griffiths 1978). Despite being only semi-aquatic, the platypus main olfactory system is reduced to a degree only exceeded among placental mammals by the fully aquatic whales (Pihlström 2008). Nevertheless, the accessory olfactory system is likely to be important in courtship, in recognition of young and perhaps for induction of lactation and milk ejection (Griffiths 1978).

No physiological or behavioural studies of olfaction have been carried out in echidnas, but several lines of evidence argue for the behavioural importance of this sense in tachyglossids. Given the poor vision of echidnas, the ease with which solitary echidnas gather to form trains during the mating season strongly suggests that olfactory signals are critically important for finding mates and assessing the reproductive status of conspecifics. Observations of the macrosmatic echidna in the wild suggest that olfaction plays an important role in detecting the ants, termites and other invertebrates that form their diet.

A recent study of the short-beaked echidna has found large sterol esters in the cloacal secretions of both sexes (Harris *et al.* 2012). These compounds comprise a sterol and a saturated or unsaturated long chain fatty acid of up to 34 carbon atoms in length. These sorts of compounds are used as scent markers by therians, so it seems likely that echidnas are able to similarly mark territory. Significantly, there are gender differences in cloacal secretions during the mating season, supporting speculation in the literature that echidnas use olfactory cues for mate attraction (Harris *et al.* 2012).

Observations by Nicol's group in Tasmania suggest that female echidnas begin to produce an attractant for males even before they emerge from hibernation (Morrow *et al.* 2009; Morrow and Nicol 2009). Rather than this being a female-specific pheromone, it is likely that the female chemical signal is made up of several compounds that encode information on sex, reproductive status and perhaps identity (see discussion in Harris *et al.* 2012). Male echidnas also produce secretions from the region of the hindlimb spur, which may serve male-specific functions (Harris *et al.* 2012).

The suggestion by Griffiths (1978) that newly hatched monotremes use olfaction to orientate to the milk source is not supported by analysis of the central olfactory connections around the time of hatching (see below).

Nasal structure in the monotremes

The nasal cavity of the platypus has a relatively simple internal structure (Fig. 11.2). There is a significant and complex maxilloturbinal in the platypus (Fig. 11.2a); but, unlike the echidnas, the nasal and ethmoidal turbinals are tiny (Fig. 11.2b) (Griffiths 1978). Consistent with the size of the central components of the accessory olfactory pathway in the platypus discussed below, the peripheral receptor organ for that pathway (the vomeronasal or Jacobson's organ) is significant in size (Fig. 11.2c) (Elliot Smith 1896a). It consists of a pair of pouches that open into the front of the oral cavity, but not the nasal cavity, an arrangement more like the reptilian vomeronasal organ than that in therians. Griffiths (1978) suggested that it may serve to smell food in the mouth, but this has never been tested. It is certainly likely to play a key role in courtship and care of the young, but behavioural observations to confirm this are lacking.

Several authors have reported the absence of a perforated ethmoid bone (lamina cribrosa) in the platypus, whether at embryonic, hatchling or adult stages (Paulli 1900; de Beer and Fell 1936). However, Zeller (1988, 1989) identified a bar of cartilage in the skulls of hatchling platypuses (at 180 and 333 mm body length) that he considered to be part of a transient lamina cribrosa. This structure never ossifies to become the cribriform plate seen in echidnas and most therians. Most olfactory nerve axons therefore pass medial and dorsal to the cartilage, rather than through a perforated structure. Zeller argued that the common ancestor of the platypus and echidnas probably possessed a lamina cribrosa, which could therefore be traced back to the ancestral mammals. He also emphasised that the lack of a lamina cribrosa in the platypus cannot be regarded as a reptilian character.

In contrast to the platypus, the olfactory apparatuses of the short- and long-beaked echidnas are strikingly large and complex. The tachyglossid nasal cavity is extremely intricate in internal structure, with seven vertical endoturbinals suspended from the lamina cribrosa of the ethmoid (cribriform plate) (Fig. 11.2d, e, f; refer also to Fig. 4.6e), many ectoturbinals, and sets of naso- and maxilloturbinals (Griffiths 1978). The resulting surface area available for olfaction must consequently be enormous, although it has never been accurately quantified. The cribriform plate of the short-beaked echidna is very large (Kuhn 1971) and has an area of around 300 mm^2 for both sides together (a range of 236–360 mm^2). In the long-beaked echidnas this structure is even larger: *Z. bartoni* (*bartoni*) – 384 mm^2; *Z. bartoni* (*diamondi*) – 440 mm^2; and *Z. bruijnii* – 326 mm^2 (unpublished observations). In all these species, the cribriform plate is pierced by many axons from olfactory receptor cells distributed across the olfactory epithelium. The report by Pihlström and colleagues (Pihlström *et al.* 2005) that the cribriform plate of *Tachyglossus aculeatus* has an area of only 98 mm^2 for both sides is a serious underestimate based on a single museum specimen. As in the platypus, the vomeronasal organ opens into the anterior part of the oral cavity (Griffiths 1978).

Olfactory receptor gene repertoire in monotremes

Both the main and accessory (vomeronasal) olfactory systems have very distinct sets of chemoreceptors and the relative importance of receptor families is directly related to habitat (Hayden *et al.* 2010). In the main olfactory system, these are odorant receptors (OR) and trace amine-associated receptors (TAAR). The OR are further subdivided into class I, which is more prevalent in aquatic vertebrates, and class II, which is dominant in terrestrial vertebrates (Freitag *et al.* 1998). In the accessory system there are type one and two vomeronasal receptors (V1R, V2R), as well as formyl peptide receptor-like proteins (FPR). The V1R family detect airborne volatiles, whereas V2R detect waterborne chemicals. Both VR families have a common receptor activation cascade involving transient receptor family potential cascade, subfamily C, member 2 (TRPC2) – a cascade that is present in the platypus (Frankenberg *et al.* 2011).

Despite the generally reduced nature of platypus olfaction, both the complexity and receptor repertoire of the accessory olfactory system in the platypus is impressive (Shi and Zhang 2007; Grus *et al.* 2007). Grus and co-workers identified an extraordinary 270 intact genes and 579 pseudogenes in the platypus V1R family; and 15 intact genes, 55 potentially intact genes, and 57 pseudogenes in the V2R (Grus *et al.* 2007). The platypus V1R family has 83 more intact genes than the largest eutherian V1R family (only 187 in the laboratory mouse; Shi *et al.* 2005; Young *et al.* 2010). Grus and co-workers identified significant chromosomal clustering of the V1R genes, indicating that the expansion of the V1R repertoire in the platypus is likely to have arisen by tandem gene duplication.

A shift of vomeronasal receptors from V2R to V1R is thought to have occurred during the emergence of vertebrates onto land, but despite the return of the platypus to an aquatic lifestyle, there has not been any retrograde shift (Grus *et al.* 2007). Instead, phylogenetic analysis indicates that the V1R repertoire has expanded strikingly in the platypus lineage since divergence from therians, whereas the V2R repertoire has undergone only a moderate expansion.

The numbers of intact OR and TAAR genes are much more modest in the platypus (261 and four, respectively) compared to 388 and six in humans, 876 and two in the domestic dog, and 1037 and 15 in the laboratory mouse. In fact as a mammal, the platypus has a relatively small repertoire of main olfactory pathway receptors (Niimura and Nei 2007), with pseudogenes making up more than 50% of OR genes. As mentioned above, OR can be divided into class I, more common in aquatic vertebrates, and class II, dominant in terrestrial vertebrates. Grus and co-workers (Grus *et al.* 2007) also observed that the aquatic platypus has a proportion of class I OR (11.5%) similar to the laboratory mouse (11%), suggesting that, despite its aquatic lifestyle, the platypus lineage has not undergone any change in the proportion of OR classes.

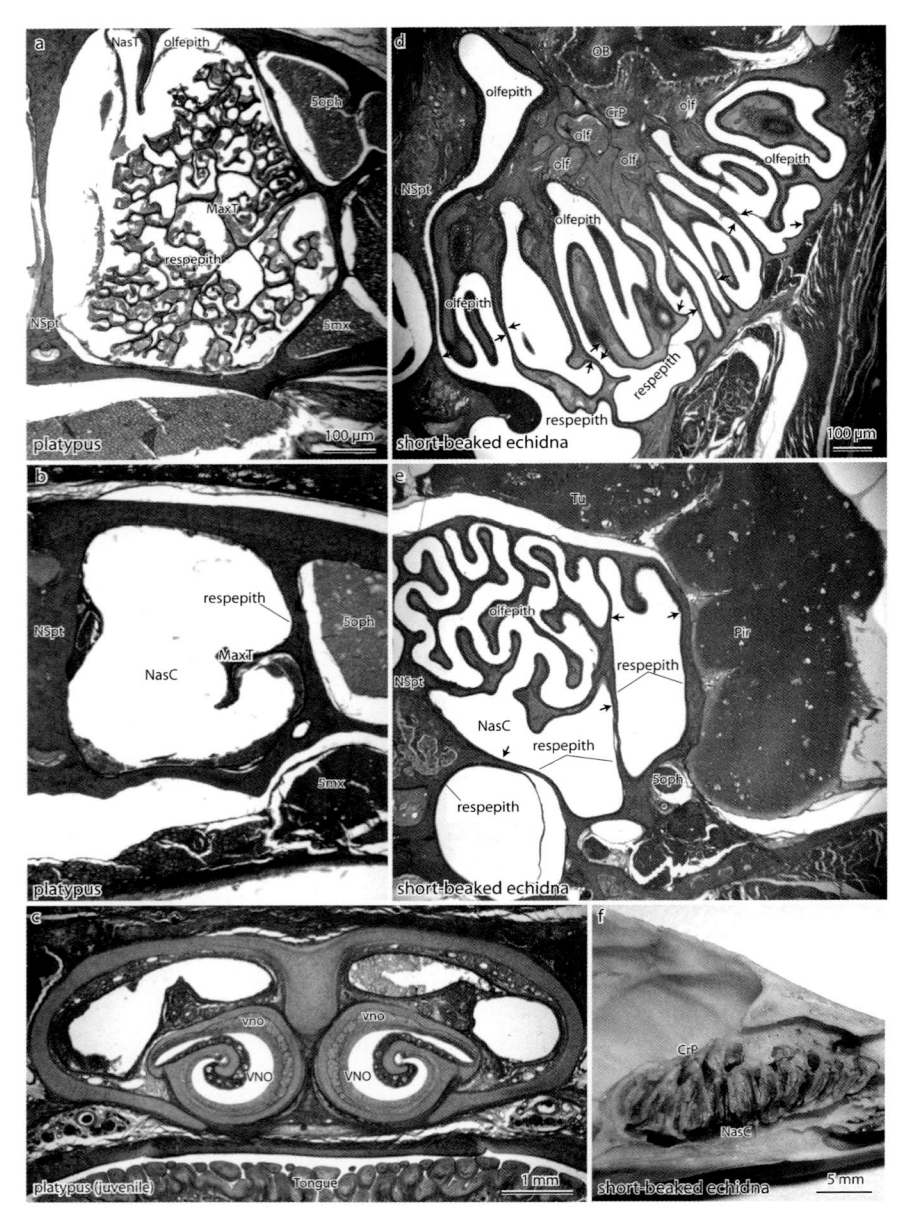

Figure 11.2: The internal structure of the nasal cavity in the platypus (a, b, c) and short-beaked echidna (d, e, f) as shown in photomicrographs. Panels (a) and (b) show frontal sections through the middle of the nasal cavity (NasC) in the two species, while panels (d) and (e) show frontal sections through the more caudal part of the nasal cavity in the two. Scale bars in (a) and (b) also apply to (c) and (d), respectively. Note the very different internal structure of the turbinates in the two species: a complex maxilloturbinate (MaxT) with no olfactory epithelium in the platypus (a) and multiple dependent plates of bone attached to the cribriform plate (CrP) of the ethmoid bone and covered with olfactory epithelium in the echidna (d, e). The caudal nasal cavity is very simple in the platypus (b), but complex in the echidna. Arrows in (d) and (e) mark the boundary between thick olfactory epithelium (olfepith) in the dorsal nasal cavity and thinner respiratory epithelium (respepith) in the ventral nasal cavity. Panel (c) shows the vomeronasal organ (VNO) and vomeronasal nerve axons (vno) in a juvenile platypus (333 mm body length), while panel (f) shows a view of the medial aspect of the left nasal cavity in a dry echidna skull. Note the presence of abundant vertical plates dependent from the cribriform plate of the ethmoid bone. 5mx – maxillary division of trigeminal nerve; 5oph – ophthalmic division of trigeminal nerve; NasC – nasal cavity; NasT – nasal turbinate; NSpt – nasal septum; OB – olfactory bulb; olf – olfactory nerve axons; Pir – piriform lobe; Tu – olfactory tubercle.

Although the echidnas are undoubtedly more olfactorily proficient than the platypus, currently nothing is known about olfactory gene repertoires in those species. Glusman and colleagues and Kishida (Glusman et al. 2000; Kishida 2008) have suggested that the common ancestral mammal had a small repertoire of OR sequences, and that expansion followed independently in all three mammalian subclasses. By extension, one would expect an even greater expansion in the tachyglossids than has been found for the platypus.

The main olfactory bulb in monotremes

The main olfactory bulb (OB) is derived from the rostral part of the pallium of the telencephalon and has a layered structure like other pallial areas. The core of the bulb contains the olfactory vesicle, a CSF fluid-filled space that is an extension of the lateral ventricle. Seven layers can be recognised in the therian OB. From internal to external these are the subependymal layer (SubE); granule cell layer (GrO); internal plexiform layer (IPl); mitral cell layer (Mi); external plexiform layer (EPl); glomerular layer (Gl); and olfactory nerve layer (ON).

The size of the olfactory bulb in the platypus is much smaller than that in the short-beaked echidna, by a factor of almost nine (see Appendix Tables 4 and 5; ventral surface area of both platypus olfactory bulbs of 25.9 ± 3.9 mm^2 compared to 215.1 ± 45.4 mm^2 for the echidna). In the platypus, the main olfactory bulb is similar in size to the accessory olfactory bulb, but in the short-beaked echidna the accessory olfactory bulb is only a few per cent of the volume of the main olfactory bulb (Ashwell 2006a). The most striking feature of the main olfactory bulb of the echidna is the extraordinary degree of folding of all layers down to the granule cells, increasing the surface area of the main bulb by 21–39%. This increases the available surface area for input from olfactory nerve axons and accommodates more olfactory glomeruli, where synaptic connections are made by olfactory nerve axons on the bulb output cells. By contrast, the main bulb of the platypus is quite smooth and slightly convex (Elliot Smith 1896a). The volume of the combined main and accessory olfactory bulbs of the short-beaked echidna reaches 630 mm^3 in some animals and is comparable to olfactorily specialised therians (Ashwell 2006a).

In most respects, the layering of the main olfactory bulb in monotremes is similar to that in therians, but Switzer and Johnson (1977) noted that the large output neurons of the bulb (mitral and tufted neurons) are dispersed through the external plexiform layer of the bulb in the monotremes, rather than being concentrated in a single layer, as in therians. Despite the diffuse distribution of large neurons through the plexiform layer, it is nevertheless possible to distinguish internal and external plexiform layers in both platypus and short-beaked echidna on the basis of differences in glial cell density and immunoreactivity for the calcium-binding protein, calretinin (Ashwell 2006a).

In many respects the chemoarchitecture of monotreme olfactory bulb neurons is similar to that in therians (e.g. for tyrosine hydroxylase and neuropeptide Y immunoreactive neurons, and NADPH diaphorase enzyme reactive neurons), but there are a few notable exceptions. First, the output neurons of the olfactory bulb (mitral and tufted neurons) are immunoreactive for non-phosphorylated neurofilament protein (SMI-32) in therians, but this is not the case in the short-beaked echidna (Ashwell 2006a). Immunoreactivity for non-phosphorylated neurofilament protein is usually a hallmark of large calibre axons running for long distances, so its absence from the robust output from the large olfactory bulb of the echidna is puzzling.

Second, there is a rather different pattern of immunoreactivity to calcium-binding proteins in the monotreme main bulb compared to therians. Immunoreactivity for parvalbumin is present in a wide range of olfactory bulb neurons in therians (e.g. parvalbumin is found in van Gehuchten cells and neurons in the external plexiform layer of marsupials – Jia and Halpern 2004; and in periglomerular, van Gehuchten, middle tufted, satellite, horizontal, pyriform and fusiform deep short-axon cells in some placentals – Kosaka et al. 1994; Kakuta et al. 1998; Briñón et al. 2001; Toida et al. 1996; Crespo et al. 2001). Although calretinin is abundantly present in periglomerular and granule cells of the monotreme olfactory bulb, only occasional neurons immunoreactive for parvalbumin or calbindin are found (Ashwell 2006a). In fact, the only parvalbumin immunoreactive neurons present in the monotreme olfactory bulb appear to be homologous to the deep multipolar short-axon cells seen in some primates (Alonso *et al.* 2001; Ashwell 2006a). Similarly, the rich array of calbindin immunoreactive neurons seen in the main olfactory bulb of therians is not present in either monotreme. The only calbindin immunroeactive neurons found in the echidna olfactory bulb appear to be homologous to the deep short-axon or deep stellate cells of the olfactory bulb of some placentals (Ashwell 2006a).

The size, folding and similar chemical features of the main bulb in the echidna strongly support the contention that the short-beaked echidna is likely to have an olfactory apparatus as effective as that in olfactorily specialised therians. There is clearly good reason to expect a major behavioural role for olfaction in this species.

The accessory olfactory bulb in monotremes

In the short-beaked echidna, the accessory olfactory bulb occupies only a small proportion of the total bulb, whereas in the platypus the accessory olfactory bulb is as large as the main bulb. In both platypus and short-beaked echidna, the cell bodies of output neurons of the accessory olfactory bulb are dispersed through the plexiform layer, as described above for the main bulb, but there is no evidence for the assertion that the monotreme accessory olfactory bulb is 'reptilian' in structure (Meisami and Bhatnagar 1998). The observations made above for the chemoarchitecture of the main bulb are also applicable to the accessory bulb.

The olfactory tubercle in monotremes

The olfactory tubercle of therians is a three-layered region on the underside of the forebrain that allows olfactory information to gain access to the ventral striatum to influence motivation and behaviour (Shipley *et al.* 2004). The tubercle has structural features that are reminiscent of both layered (i.e. pallial or cortical) and homogeneous (i.e. subpallial striatal) parts of the forebrain. Axons of mitral and tufted output cells of the main olfactory bulb terminate in the olfactory tubercle, just as much as the anterior olfactory area and primary olfactory cortex, so it is a major region by which odorants can influence behaviour. The three layers of the tubercle are (from outside to inside) a relatively cell-sparse superficial plexiform or molecular layer; an underlying cortical (or dense cell) layer; and a deep polymorphic cell layer (Shipley *et al.* 2004). Apical dendrites of the pyramidal-like neurons in the cortical layer extend into the molecular layer, much like neurons in the hippocampus or piriform cortex extend into the overlying white matter, but the dendrites of neurons of the deeper polymorphic layer have no particular orientation and look more like radially branching striatal neurons (e.g. of the caudate or putamen). These features of cellular architecture and the proximity of the tubercle to the ventral striatum have led some to argue that the olfactory tubercle is a component of the striatal complex, but its precise role remains elusive.

The olfactory tubercle also has some unusual cytoarchitectural features embedded within it. These are the granule cell clusters (sometimes called islands of Calleja). The role of these cell groups in therians remains the subject of debate, with some researchers suggesting that the clusters are part of the striatopallidal system (Ribak and Fallon 1982; Fallon 1983; Fallon *et al.* 1983) and others arguing for a neurosecretory role (Millhouse 1987). Other special cell groups in the tubercle are the cap regions that overlie folds in the cortical or dense cell layer.

The structure of the olfactory tubercle is very different between the platypus and short-beaked echidna (Ashwell 2006b). In the platypus, the olfactory tubercle is very small (less than 1.5 mm^2 in area on each side), making up only 2% of the volume of the olfactory cortex (Ashwell and Phillips 2006), but it has clearly defined superficial molecular and dense cell layers. Despite the clear layering of the superficial tubercle, the platypus has no obvious border between the deep polymorphic layer of the olfactory tubercle and the underlying ventral striatum. There are also none of the special cellular features of the tubercle: no granule cell clusters, no cap regions and no cell bridges joining the tubercle to the striatum.

The olfactory tubercle of the short-beaked echidna is somewhat larger (at least 10 mm^2 in area on each side) than in the platypus (Ashwell 2006b), but this is quite small given the huge size of the olfactory bulb and piriform lobe in this species, so the tubercle makes up only 7% of the volume of the olfactory cortex volume (Ashwell and Phillips 2006;

see also below). More significantly, the olfactory tubercle in the echidna lacks the clear layering of therians and even the platypus, such that the dense cell layer is very irregular and discontinuous (see atlas plate Ec-Ad7). Granule cell clusters are visible, but are much less striking than those in olfactory specialised placentals. The small size and poor lamination of the olfactory tubercle in the short-beaked echidna is therefore very surprising given the extensive size of the olfactory bulb and piriform cortex in echidnas.

Chemically, the olfactory tubercle in monotremes has some similarities to, but also some potentially significant differences from, therians. The presence of strong enzyme reactivity for acetylcholinesterase defines the olfactory tubercle as much in the two monotremes as in therians, reflecting its striatal affiliations and the use of acetylcholine as a neurotransmitter in the region. Likewise, immunoreactivity for calcium-binding proteins is broadly similar in the monotremes compared to therians. The major difference in tubercle chemoarchitecture between the monotremes and therians concerns activity for the enzyme involved in nitric oxide signalling, NADPH diaphorase. In therians, activity for this enzyme is intense in the terminal fields of nitric oxide containing axons within the granule cell clusters, but despite the presence of granule cell clusters in the tubercle of the echidna, there is no NADPH diaphorase activity in the echidna tubercle. This suggests that there may be significant functional differences in the use of nitric oxide by neurons projecting into the echidna and therian tubercles.

The cytology of the olfactory tubercle of the short-beaked echidna has been studied with Golgi impregnations (Ashwell 2006b). Although there are some neurons in the echidna olfactory tubercle with pyramidal morphology, the bulk of impregnated neurons resemble the medium-sized densely spined neurons of rodent olfactory tubercle. Large spine-poor cells similar to those in the laboratory rat are also present. The low spatial frequency of pyramidal neurons in the olfactory tubercle of the short-beaked echidna is entirely consistent with the region being an extension of the ventral striatum to the pial surface.

All in all, the small relative size of the tubercle in both platypus and echidna and the poor lamination in the echidna suggest that olfactory input to the ventral striatum in both these species may be less significant than in therians (Ashwell 2006b). However, it may be that the olfactory input to the striatum is indirect, i.e. from the piriform lobe rather than directly from the lateral olfactory stria, but this remains to be studied.

The anterior olfactory region of monotremes

The anterior olfactory region (also called anterior olfactory nucleus) is a layered structure around the rim of the deep white matter that connects the olfactory bulb with the rest of the forebrain. The laminated nature of the region (due to its origin from the pallium of the embryonic telencephalon) means that the older term 'nucleus' is not appropriate for this structure. In therians, the anterior olfactory region is divided into a part within the olfactory bulb (intrabulbar part), four cardinal subdivisions around the olfactory peduncle and a posterior part that merges with the septum. In macrosmatic (olfactorily proficient) therians the anterior olfactory region is much larger than in microsmatic therians.

The anterior olfactory region differs significantly between the platypus and short-beaked echidna, in ways that are puzzling given the structure of other parts of the olfactory system in these two species (Ashwell and Phillips 2006). Surprisingly, the poorly osmatic platypus has an anterior olfactory region very similar in structure to that in rodents and other macrosmatic placentals, albeit smaller in absolute size. The anterior olfactory region of the platypus has the same four cardinal segments of rodents as well as a sharply defined border with the more caudal piriform cortex.

By contrast, the anterior olfactory region of the short-beaked echidna is recognisable only as dorsal and ventral segments adjacent to the olfactory bulb (AOD and AOV; see atlas plate Ec-Ad02). Further posteriorly, the region is difficult to recognise and it is more accurate to say that the piriform cortex has a rostral extension that has incorporated the anterior olfactory region. In fact, the anterior olfactory region in the echidna is more like a transitional zone than a component of the olfactory pathway in its own right.

The chemoarchitecture of the platypus anterior olfactory region shows some similarities and differences from therians. The platypus anterior olfactory region has very little immunoreactivity for parvalbumin (Ashwell and Phillips 2006), similar to the hedgehog tenrec but very different from the laboratory rat. On the other hand, the pattern of immunoreactivity for calbindin and calretinin in the platypus is very similar to that in rodents.

The piriform cortex of monotremes

The piriform cortex is the main target for outflow from the olfactory bulb and is the largest part of the primary olfactory cortex. In the platypus it is a very modestly sized region, but it is large and complex in the short-beaked echidna, reaching a volume of 1340 mm^2 in some animals. This is comparable to olfactorily specialised placentals like lipotyphlids, rodents, lagomorphs and ungulates and about twice the volume of the region in prosimians of similar body weight (Ashwell and Phillips 2006).

The structure of the piriform cortex differs significantly between the two groups of monotremes (Ashwell and Phillips 2006). The piriform cortex of the poorly osmatic platypus has a very similar laminar structure to that in a wide variety of therians, including those with quite differentiated olfactory systems. The platypus piriform cortex has a distinct superficial plexiform layer: layer Pir1 with olfactory recipient (1a) and association (1b) layers, a middle layer of densely packed neurons (layer Pir2) and a deeper layer of more loosely packed neurons (layer Pir3); see atlas plate Pl-Ad4&5. By contrast, the short-beaked echidna has a much more complex piriform cortex, with substantial regional variation of the laminar structure and chemical architecture around the circumference of the piriform lobe.

Although the piriform cortex of the macrosmatic echidna has a superficial plexiform layer with Pir1a and 1b subzones like the platypus, the deeper layer Pir2 is much more complex and can be subdivided into a slightly less dense upper layer Pir2a and an underlying denser layer Pir2b. The next layer (Pir3) differs substantially in appearance in different regions of the piriform lobe, suggesting regional differences in neuronal size and packing that may reflect functional subdivisions. Finally the deepest layer (Pir4) also shows regional variation. It is uncertain as to whether layer Pir4 is homologous to the endopiriform cortex of therians, but Ashwell and Phillips (2006) concluded that there is also an underlying endopiriform nucleus made up of scattered neurons within the white matter core of the piriform lobe.

Immunoreactivity for calcium-binding proteins reveals both layer- and region-specific differences in the distribution of labelled neurons and axons across the piriform lobe in the echidna (Ashwell and Phillips 2006; see PirA1 to PirA4 in atlas plate Ec-Ad6). Although layer-specific patterns of calcium-binding protein immunoreactivity are seen in therians, the regional variation that allows the piriform lobe of the short-beaked echidna to be divided into four regions appears to be unique among mammals and also points to functional subdivisions within the olfactory cortex (Ashwell and Phillips 2006).

Limited data are available on neuronal shape in the piriform lobe, based on Golgi impregnations (Ashwell and Phillips 2006). Pyramidal neurons predominate in layers Pir2 and upper Pir3 of the echidna piriform lobe, but the deeper layers are dominated by non-pyramidal neurons. As in therians, the main cell type of piriform cortex is the pyramidal neuron, but pyramidal neurons in the echidna piriform lobe have shorter apical dendrites and fewer basal and apical dendrites than their counterparts in placental piriform cortex. Dendritic spines are also either absent from, or very rare on, the apical dendrites of pyramidal neurons in the echidna piriform lobe and spines are at a lower spatial density on the secondary side dendrites (Ashwell and Phillips 2006). This could have functional ramifications, suggesting that there may be a higher proportion of synaptic input onto dendritic shafts of pyramidal neurons rather than (the potentially modifiable) dendritic spines in the echidna piriform lobe.

Several types of non-pyramidal neurons in the echidna piriform lobe appear to be homologous to non-pyramidal neurons in therians. Multipolar neurons in Pir2 of the echidna piriform lobe are similar in appearance to profuse spiny neurons of Virginia opossum piriform cortex or stellate cells of deep piriform cortex in some rodents (see discussion in Ashwell and Phillips 2006). Horizontal cells can be identified in the superficial and deep layers of the echidna piriform lobe, but, unlike horizontal cells in

Pir1 of the piriform cortex of the Virginia opossum, they do not have spiny cell bodies. The vertically oriented bipolar cells seen occasionally in the piriform cortex of the echidna may be homologous to small cells with radially oriented processes seen in the Virginia opossum or fusiform cells seen in some rodents (Ashwell and Phillips 2006).

The entorhinal cortex of monotremes

Although situated far to the back of the brain, the entorhinal cortex also receives a substantial output from the main olfactory bulb. The entorhinal cortex, in turn, projects to the dentate gyrus and cornu Ammonis of the hippocampal formation, allowing olfaction to directly influence memory, and perhaps allowing the recall of olfactory memories formed by association with other events (Shipley et al. 2004). In therians, the piriform cortex also has its own projection to the entorhinal cortex.

The hallmark cytoarchitectural feature of the entorhinal cortex in therians is the presence of a relatively cell free layer (the lamina dissecans) between superficial and deep neuronal layers. This is clearly visible in the entorhinal cortex of both platypus and short-beaked echidna (see atlas plates Pl-Ad18 and Ec-Ad19; Hassiotis et al. 2004b). In contrast to many therian brains, where there is a sharp demarcation into a medial entorhinal cortex (with continuous layer 2 and poorly developed layer 3) and a lateral entorhinal cortex (with layer 2 neurons clustered into islands), the entorhinal cortex of the short-beaked echidna shows a very similar laminar pattern around its circumference (Hassiotis et al. 2004b). The functional significance of this poor cytoarchitectural differentiation remains to be explored.

Is the large frontal lobe of the echidnas a homologue of therian orbitofrontal cortex?

The large frontal cortex rostral to the identified motor and somatosensory areas of the echidnas has always been a mystery. Given the large size of the olfactory bulb and primary olfactory cortex in both short- and long-beaked echidnas, olfaction clearly plays a major role in echidna behaviour, most likely for prey detection and food selection, and/or the formation of olfactory maps of the home range.

In therians, the orbitofrontal cortex on the underside of the frontal lobe receives a rich input from the primary olfactory cortex either through a direct connection, or by intervening connections in the hypothalamus and the mediodorsal nucleus of the thalamus (see review in van Hartevelt and Kringelbach 2012). Functional magnetic resonance imaging and tract-tracing studies in rodents and primates have shown that the orbitofrontal cortex integrates olfactory information with other sensory input, in particular gustation (taste). In fact, it is the first place in the pallium where olfaction and taste converge and is probably the brain region that underlies odour discrimination and the sensation of flavour. Other sensory input to the therian orbitofrontal cortex includes somatosensory input from those parts of the primary somatosensory cortex that are concerned with the hand and mouth (such as would be activated during feeding and the 'mouth-feel' of food), as well as visual input from parietal and inferior temporal visual association cortex (such as would activate when palatable food is seen and recognised). These corticocortical connections have led to the suggestion that the orbitofrontal cortex and its connections form an *orbitofrontal network* that integrates and analyses food-related sensory information (Carmichael and Price 1996; van Hartevelt and Kringelbach 2012).

What basis is there for arguing that at least some parts of the large frontal lobe of the echidnas (e.g. Fr1, Fr2 and Fr3) are homologous to the orbitofrontal cortex of therians as a chemical sense integration centre? Topography of the isocortex is the first line of argument, in that the Fr3 and perhaps Fr1 and Fr2 regions occupy a position relative to the piriform cortex and olfactory bulb (Hassiotis et al. 2004b) similar to the orbitofrontal cortex in therians. Hodological (connectional) lines of argument are also applicable. Divac and co-workers (Divac et al. 1987a) showed that sources of afferents to the Cg and Fr cortex in the short-beaked echidna include neurons in the mediodorsal nucleus of the thalamus, piriform cortex, medial and cortical parts of the amygdala, and medial and lateral entorhinal cortex. These observations are entirely consistent with the large frontal cortex in the short-beaked echidna receiving a rich source of olfactory input, as would be expected if it were serving as a chemical sense integration centre. Some of the sources of afferents to the large frontal lobe that Divac and colleagues identified also appear to be in the position of the medial parvicellular extension of the ventral posteromedial thalamic nucleus (VPPC; see atlas plate Ec-Ad17), a region that serves the processing of gustatory information in other mammals, suggesting that gustatory information may also be reaching the frontal cortical fields of the echidna.

Nevertheless, our knowledge of the connections and functions of the frontal field in echidnas remains rudimentary. Gold standard evidence of its role in chemical sense integration would require more detailed connectional studies to confirm that olfactory and gustatory information both reach the region, as well as functional imaging studies to confirm the activation of the region during food detection and selection tasks.

Embryonic development of the main and accessory olfactory pathways

The components of the olfactory pathways develop from specialised ectoderm on the head (the olfactory placode), as well as from select parts of the forebrain. In the monotremes, the olfactory placode (the precursor of the olfactory epithelium) develops at around the end of the first third of incubation (i.e. 5.5–6 mm GL, or stage 15 of Werneburg and Sánchez-Villagra 2011) and is rather similar in appearance at this time in both the platypus and short-beaked echidna (Ashwell 2012b). Nostrils become defined by growth of the facial process around the placode by the time the embryo grows to 8.0 mm GL, and the vomeronasal organ primordium fully invaginates to form a tube at around the same time (stage 17 of Werneburg and Sánchez-Villagra 2011). Both the main olfactory epithelium and vomeronasal organ give rise to axons that reach the region of the olfactory bulb (OB) very soon after this (by 8.5 mm GL), but the olfactory bulb primordium at this stage is still very simple in structure and consists of only a 100 µm thick neuroepithelium to generate the neurons of the bulb and a very thin unlayered 80 µm thick mantle layer outside that.

The olfactory system remains undifferentiated in structure in both monotremes through the last third of incubation, although some of the distinctive anatomical features of the echidna olfactory system begin to emerge close to hatching (Fig. 11.3, Appendix Fig. 1). The first of these concerns the internal complexity of the nasal cavity. Whereas the internal surface of the nasal cavity continues to be relatively smooth in the platypus during late incubation, folding of the nasal epithelium becomes more pronounced in the echidna close to hatching. Also, the deep fissure that will separate the olfactory bulb from the rest of the forebrain in the mature echidna brain (the rhinal fissure) first appears a few days before hatching (Fig. 11.3a, b).

How much olfactory function is available to the newly hatched monotreme?

Griffiths (1978) suggested that the newly hatched echidna is able to use olfaction to navigate to behaviourally important features of its environment, e.g. the milk patch of the maternal abdomen. This line of argument was based on the presence of olfactory receptor neurons in the nasal epithelium, with axons running from the epithelium to the olfactory bulb (Griffiths et al. 1969). However, this does not take into account the relative maturity (or rather immaturity) of the central components of both the main and accessory olfactory systems. Without these being sufficiently mature and directly or indirectly connected with the brainstem, there is no pathway by which olfactory information may be used to control or modify behaviour.

Schneider (2011) argued for a role for olfaction in the early post-hatching life of the platypus, based on the presence of olfactory nerve bundles and a three-layered main olfactory bulb, but Ashwell (2012b) had a different viewpoint. Ashwell (2012b) argued that, at the time of hatching, the structure of the olfactory system in both platypus and short-beaked echidna is too structurally immature to serve any significant function. This state of immaturity at hatching is very much like that in diprotodontid marsupials at birth (Ashwell et al. 2008). Although the main and accessory pathway olfactory epithelia are in place at hatching and olfactory nerve bundles can be seen traversing the distance to the olfactory forebrain, there are no functional Bowman's glands beneath the olfactory epithelium (Fig. 11.3a). These are important for providing a fluid phase to bring odorants to the olfactory receptor neurons, but secretory acini and patent ducts do not appear until late in the first post-hatching week.

Of equal functional significance, the central components of both the main and accessory olfactory pathways have not progressed significantly beyond the neural tube stage by the time of hatching in either monotreme. The olfactory bulb of the newly hatched monotreme has a very simple structure. Although a few postmitotic cells (presumably early generated output neurons like the mitral and tufted cells) have accumulated in the mantle region of the bulb primordium by the time of hatching, the layers that underlie the functional anatomy of the bulb are yet to appear. The glomeruli where olfactory receptor axons contact projection neuron dendrites as well as the plexiform layers where interactions between bulbar interneurons occur have also not yet developed. Furthermore, intrinsic microneurons like the granule cells of the bulb have not been generated by this time.

Finally, the main output from the olfactory bulb to more caudal levels of the neuraxis (through the lateral olfactory pathway) that is essential to mediate behavioural responses to olfactory stimuli does not emerge until well after hatching. Taken together, these observations indicate that the monotreme olfactory system is far from structurally mature at birth and the hatchling is unlikely to engage in olfaction-mediated behaviour.

Post-hatching development of the monotreme olfactory system

During the first week after hatching, the extent of folding of the lateral and dorsal walls of the echidna nasal cavity is clearly more advanced than in the platypus, but the morphology of vomeronasal (Fig. 11.3c) and main olfactory epithelium (Fig. 11.3d) is similar in both species. At the end of the first post-hatching week in the platypus, lamination begins to appear in the main bulb and a distinct accessory olfactory bulb is seen for the first time (Fig. 11.3e). Surprisingly, the mitral cell layer is better defined at these early stages, particularly in the accessory olfactory bulb of the platypus, than it will be in the adult (see above). This suggests that the dispersal of output neurons observed by Switzer and Johnson (1977) in the adult bulb is most likely the result of nerve fibre development displacing neurons during post-hatching life, rather than a primary problem with migration. Primary olfactory cortex is still undifferentiated at this time (Fig. 11.3f).

Development of the olfactory system after the first week can be followed only in the platypus, due to limited specimen availability (Ashwell 2012b). During the period from 2–6 weeks after hatching, differentiation within the main and accessory olfactory bulbs proceeds rapidly, so that by the middle of the second post-hatching month all the layers of the mature main and accessory bulbs can be identified and glomeruli surround the circumference of both bulb components.

The non-olfactory maxilloturbinate of the platypus expands by a tree-like branching process during the second to sixth month of post-hatching life. This structure will warm and moisten inspired air when the young platypus leaves the burrow and does not appear to carry olfactory epithelium. Central olfactory nuclei and the piriform cortex are essentially mature in appearance by the middle of the second month of post-hatching life.

The evolution of olfaction in monotremes

The platypus and echidna clearly have very different olfactory systems. What do observations in extant monotremes tell us about the structure and function of the main and accessory olfactory systems in the ancestral monotreme, and about the evolution of mammalian olfaction in general?

Putting aside the absence of the mitral cell monolayer in monotremes, the similarity of the platypus olfactory systems (i.e. similar cellular and chemical architecture of the olfactory bulb, anterior olfactory region, tubercle, piriform and entorhinal cortex) to the olfactory systems in therians strongly suggests that the key features of the mammalian olfactory system emerged before the divergence of prototherian and therian lineages.

What do these observations of monotreme olfactory anatomy say about the time course of evolution of the two monotreme lineages? The endocranial morphology of the fossil platypus *Obdurodon dicksoni* (Macrini *et al.* 2006) indicates that the small size of the olfactory bulb has been a feature of the ornithorhynchids since the middle Miocene. In fact the size of the olfactory bulbs as a proportion of the total endocranial space in *Obdurodon* (1.90%) is comparable

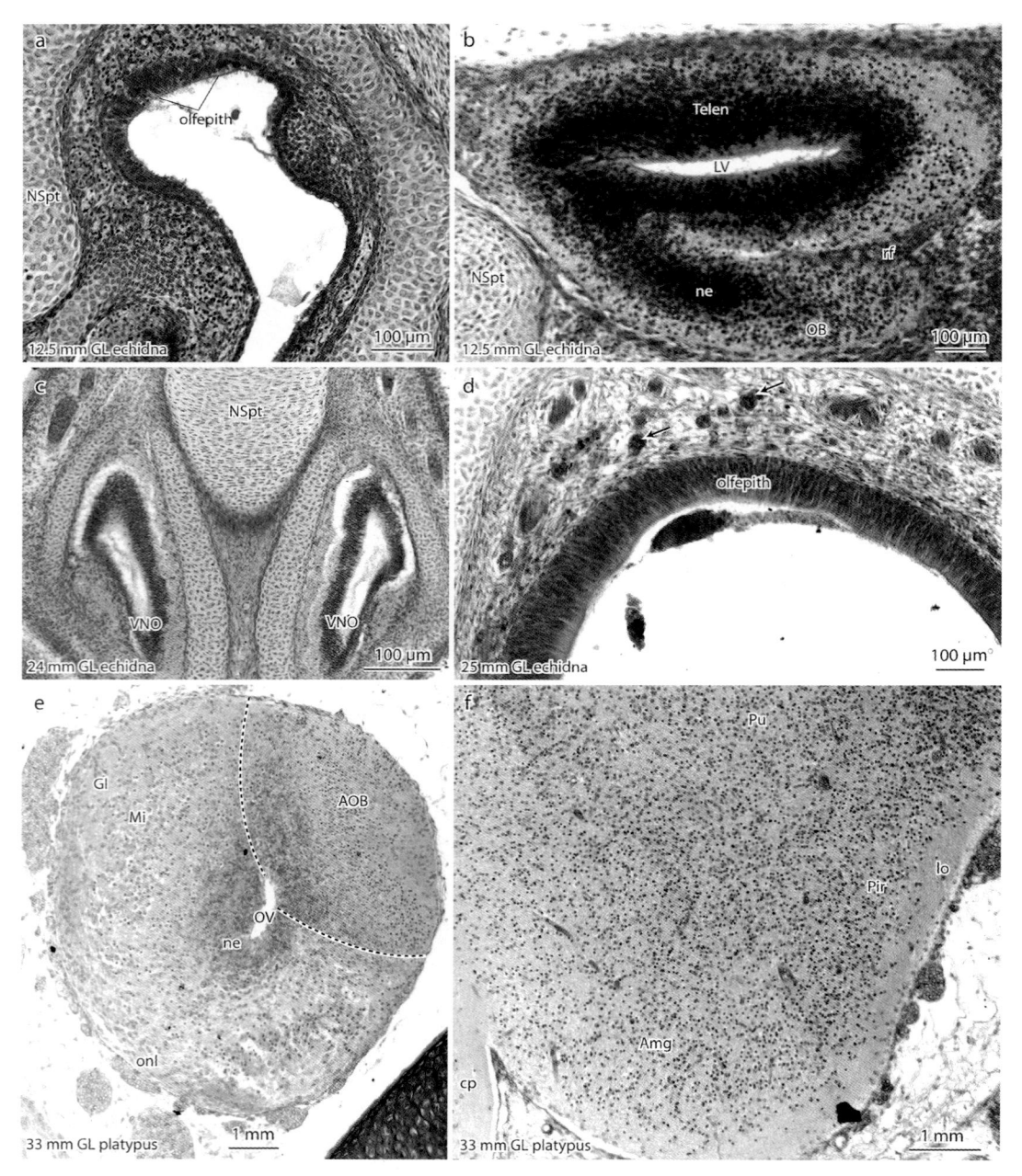

Figure 11.3: Key events in the development of the olfactory system in the monotremes. Immediately before hatching, the olfactory epithelium (olfepith) and olfactory bulb (OB) are rudimentary in the short-beaked echidna (a, b). There are no submucosal glands and the olfactory epithelium remains thin. In the brain, the bulb shows no differentiation into main or accessory bulbs and lamination is absent. There are also no output neurons (mitral and tufted cells) visible. Towards the end of the first week after hatching, the vomeronasal organ (VNO, c) and main olfactory epithelium (d) in the short-beaked echidna show increasing signs of maturation, with the emergence of Bowman's glands in the submucosa (arrows in d). Nevertheless, even at the end of the first postnatal week, the main and accessory parts of the olfactory bulb are still poorly differentiated in the platypus (e) and central olfactory structures remain undifferentiated (f). Amg – amygdala; AOB – accessory olfactory bulb; cp – cerebral peduncle; Gl – glomerular layer of main olfactory bulb; lo – lateral olfactory tract; LV – lateral ventricle; Mi – mitral cell layer of main olfactory bulb; ne – neuroepithelium; NSpt – nasal septum; onl – olfactory nerve fibre layer; OV – olfactory ventricle; Pir – piriform lobe; Pu – putamen; rf – rhinal fissure; Telen – telencephalic vesicle.

to the proportion in juvenile *Ornithorhynchus* (1.8%), but slightly larger than in the adult *Ornithorhynchus* (0.95%). It has been argued that the echidnas are relatively recent offshoots (19–48 mya) of the platypus line, and that the common ancestor for all modern monotremes was an aquatic mammal somewhat like the modern platypus (Pettigrew 1999; Phillips *et al.* 2009). If this were the case, then one would have to posit that the strikingly complex olfactory system of the echidna has evolved in that relatively short period, producing an olfactory system that rivals the most nasally proficient therians, all from the simple platypus model.

Ashwell (2012b) argued on the basis of developmental observations that the common ancestor for modern monotremes was somewhat more olfactorily proficient and somewhat less trigeminally specialised than the platypus. Two points were noted: (1) the elaboration of the dorsal and lateral walls of the nasal cavity in the two species follows quite different pathways from the time of hatching; and (2) the pronounced folding of the rostral olfactory forebrain seen in adult echidnas appears even before hatching, at a time when the platypus olfactory forebrain is quite smooth. These observations suggest that the olfactory systems in the two species follow very different developmental trajectories even from the time of incubation. Certainly this could arise from mutations among the ancestral tachyglossids that caused precocious elaboration of the nasal epithelium and central olfactory centres, but a more parsimonious explanation would be that the common ancestor of both modern monotreme lineages was relatively generalised with respect to both trigeminal electroreception and olfaction (Ashwell 2012a, b). In addition, the platypus lacks a lamina cribrosa, and yet the echidna, supposedly its relatively recent descendant, has a large and fully developed cribriform plate. This is more easily explained by the presence of the lamina cribrosa in the ancestral monotreme with the involution of this structure only in the ornithorhynchid lineage.

The sense of taste (gustation) in monotremes

Selection of foods by mammals is mainly based on olfactory input, but the assessment of the palatability of food that has been taken into the mouth depends on integrated information from both taste (or gustatory) and olfactory receptors. Very little information is currently available concerning functional aspects of taste in any of the monotremes, but something is known about the anatomy of the peripheral gustatory system.

Taste depends on taste receptors located on the dorsal surface and side of the tongue and the palate. The posterior tongue of the platypus has two small slits on its dorsal surface that lead into deep-seated circumvallate papillae with taste receptors along the walls of the slits (Griffiths 1978). Foliate papillae with taste receptors are also found on the lateral border of the tongue of the platypus (Griffiths 1978). The arrangement of taste receptors in the short-beaked echidna is broadly similar to that for the platypus. The dorsum of the tongue has two slits forming a 'V'-shaped cleft that leads to deeply placed circumvallate papillae with taste buds (see discussion in Griffiths 1978). Foliate papillae with taste buds are also located on the side of the tongue.

Some limited information is available concerning taste receptor genes in the platypus (Shi and Zhang 2006). Grus and colleagues reported that preliminary screening of the platypus genome sequence for the T1R (sweet and umami receptors) and T2R (bitter receptors) families indicates a reduced T2R repertoire in the platypus compared to other mammals (unpublished data mentioned in Grus *et al.* 2007; Shi and Zhang 2006). Whether this has any functional importance for the platypus remains to be determined. It may be that the T2R family has become reduced because the natural environment of the platypus has relatively few bitter tastants.

Taste information in therians is initially processed in the rostral part of the nucleus of the solitary tract, before being transmitted to more rostral parts of the neuraxis (parabrachial nuclei of the brainstem, parvicellular extension of the ventral posteromedial thalamic nucleus – VPPC, and eventually the dorsal insular cortex). Gustation and olfaction are brought together in the orbitofrontal cortex for an integrated perception of the palatability of food.

Information about the central connections of gustatory pathways in monotremes is at best fragmentary. There may be a VPPC homologue in the echidna and it does have connections with frontal cortical areas (see atlas plates Ec-Ad17 and 18, and below), but nothing is currently known about more caudal gustatory fields that would be in proximity to the somatosensory cortex. Certainly there are insular cortical regions in both platypus and short-beaked echidna, but the connections of these putative gustatory regions remain unexplored.

Questions for the future

There are several major deficiencies in our knowledge of monotreme olfaction. The first of these is that we simply have no good data on the olfactory acuity of either the platypus or echidnas. It is highly likely that echidnas use olfaction extensively in prey detection, mate selection and offspring recognition, but this is merely an observational conclusion rather than based on sound information about how well echidnas detect olfactory cues. Similarly, the accessory olfactory pathway is probably important for conspecific recognition in the platypus, particularly given the extensive vomeronasal receptor repertoire in that species, but this has not been conclusively demonstrated by experimental study.

The second area concerns the chemoreceptor families of the main and accessory olfactory epithelia. The platypus has an extraordinary array of genes associated with chemoreceptors in the accessory olfactory pathway, remarkable considering its aquatic lifestyle. Clearly a priority would be to establish the size of the chemoreceptor repertoire for the main and accessory olfactory pathways in the short-beaked echidna. It would also be of interest to know the relative sizes of the V1R and V2R repertoires, given the claims of relatively recent divergence between the ornithorhynchid and tachyglossid lineages (see Chapter 14 for further discussion on this point).

The third area deserving attention is the role of the structurally diverse regions of the piriform lobe in the short- and long-beaked echidnas. In therians, the piriform cortex has been divided into anterior and posterior regions on the basis of anatomical physiological and functional differences (van Hartevelt and Kringelbach 2012), but it is impossible to say at present whether the therian subdivisions have any homology with the anatomical parcellation of the monotreme piriform cortex.

The fourth area of interest is the possible role of the large *prefrontal* cortex of both short- and long-beaked echidnas as an integration centre for behaviourally important chemosensory information. Tackling this question would require much better data on connections between subcortical structures such as the gustatory VPPC nucleus of the thalamus (or its homologue) and the primary olfactory cortex, with the subregions of the frontal lobe. The ideal approach, of course, would be to undertake functional MRI imaging of the olfactory and frontal cortical areas in conscious echidnas during olfactory tasks, a prodigious challenge in a wild and uncooperative animal.

12

The hypothalamus, neuroendocrine interface and autonomic regulation

Ken W. S. Ashwell

Summary

Monotremes are characterised by their maintenance of a relatively low body temperature, a distinctive mode of reproduction (Griffiths 1999) and an unusual mammary gland structure and lactation. There are also some significant differences between the extant monotremes in hibernation, energy metabolism and thermoregulation. All of these imply some distinctive structural and functional features of the hypothalamus and its interaction with the endocrine and autonomic nervous system, and yet very few studies have focused on monotreme hypothalamic structure. Broadly speaking, the monotreme hypothalamus and pituitary are morphologically similar to those in therians and use structurally similar or identical hormones, but less is known about the details of hypothalamic function in either the platypus or the echidnas. Hypothalamic and pituitary development mainly occurs after hatching and only the lateral hypothalamus is sufficiently mature at the time of emergence from the egg to contribute to homeostatic mechanisms in the newly hatched. Data on the autonomic nervous system of monotremes are limited, but the monotreme gut has some unusual anatomical and physiological features, which appear to be reflected in enteric innervation and function.

Overview of the mammalian neuroendocrine interface

The interaction between the central nervous and endocrine systems primarily occurs at the hypothalamo-pituitary axis, but other parts of the forebrain and brainstem also play significant roles in regulating neuroendocrine activity.

The hypothalamus plays a central role in pathways concerned with autonomic, endocrine, and emotional functions (Saper 2012). The general function of the hypothalamus is to maintain the internal environment of the body (i.e. body temperature, blood gases, nutrition, blood pressure) within an optimal range, a process known as homeostasis, but its specific functions go beyond simply maintaining a constant internal state. The hypothalamus also plays a critical role in controlling reproductive cycles and mating behaviour, regulating sleep and mediating the physiological expression of emotional states (e.g. rage, fear and anxiety).

The position and connections of the hypothalamus are central to its ability to perform these functions (Saper 2012). The inputs to the hypothalamus can be divided into two groups: (1) descending control from parts of the forebrain that regulate behaviour; and (2) ascending sensory inputs that allow the hypothalamus to assess the physiological state of the internal environment and monitor changes in the external environment. The hypothalamus is influenced by input from a wide variety of sites in the brain. The main forebrain inputs to the hypothalamus are from limbic system regions like the septal nuclei and ventral striatum (nucleus accumbens), the hippocampal formation, the amygdala and selected regions of the limbic association cortex (insular and orbitofrontal cortex). Inputs from the retina are also essential for the control of diurnal and seasonal physiological rhythms.

Hypothalamic outputs are to parts of the forebrain concerned with emotions and behaviour, to the pituitary gland, and to the autonomic and respiratory centres in the brainstem. The hypothalamus has projections to the septal nuclei, hippocampus and amygdala by the same pathways that bring input from those structures to the hypothalamus. Outputs from the hypothalamus to the cerebral cortex are more widespread than the inputs, reflecting the broad effect of the hypothalamus on cerebral function in a wide range of functional areas. Particularly important pathways from the hypothalamus to the cortex are the histaminergic projections from the tuberomammillary nucleus and the orexinergic projection from the tuberal and posterior hypothalamus, because they play key roles in appetitive behaviour and sleep (Saper 2012; see discussion in Chapter 13).

The hypothalamus influences the endocrine system by its output through the anterior and posterior pituitary gland (Saper 2012). The parvicellular neurosecretory system consists of small neurons in the arcuate nucleus and nearby third ventricular wall of the hypothalamus (including some small neurons in the paraventricular nucleus). These neurons produce hypothalamic releasing and inhibitory hormones that are secreted into vessels of the hypothalamo-hypophyseal portal system to regulate the production of anterior pituitary (adenohypophyseal) hormones (e.g. ACTH – adrenocorticotrophic hormone, GH – growth hormone, TSH – thyroid stimulating hormone, PRL – prolactin, FSH – follicle stimulating hormone, and LH – luteinising hormone). The magnocellular neurosecretory system is made up of large neurons in the supraoptic and paraventricular nuclei of the hypothalamus (as well as some neurons distributed in an arc between the two) that project axons through the hypothalamo-neurohypophyseal tract to the posterior pituitary (neurohypophysis). These axon terminals release oxytocin (OX) and antidiuretic hormone (ADH or vasopressin) into a capillary plexuses within the posterior pituitary gland for distribution via the systemic circulation to the body.

There are three aspects of monotreme metabolism that have a direct bearing on the neuroendocrine axis. These are the monotreme response to stressful experiences, particularly during mating, lactation and capture; the regulation of body temperature during active times of the year and under temperature extremes; and the control of hibernation.

Energy mobilisation and stress in monotremes

Monotremes in the wild routinely deal with seasonal fluctuations in water and air temperatures, and the metabolic demands of reproduction. Seasonal fluctuation in thermoregulation, reproduction, lactation and locomotion requires the mobilisation of energy stores to meet short-term demands. Since mating occurs during the coolest months of late winter and early spring, the metabolic demands of thermoregulation, locomotion and mating usually coincide (see below). Glucocorticoids released as a result of activation of the hypothalamo-pituitary-adrenocortical (HPA) axis perform a major role in mobilising energy stores for seasonal demands and in physiologically stressful conditions. The glucocorticoids mobilise amino acids from cells, stimulate glucose production (gluconeogenesis) in the liver from the mobilised amino acids, and muster fat reserves for energy. Stores of carbohydrate as glycogen in the liver may also be mobilised in the short-term by release of adrenal medullary hormone (i.e. adrenaline). The monotremes appear to be more dependent on mobilisation of fat stores, rather than carbohydrate stores like liver glycogen, during stressful situations (McDonald et al. 1992), so the HPA axis is particularly important in these mammals.

The activity of the monotreme HPA axis is not only exquisitely sensitive to stressful events in the short term, but also fluctuates with the season (Handasyde et al. 1992, 2003). Plasma glucocorticoid concentrations in the blood of platypuses rise within 30 minutes of capture in gill-nets. This effect amounts to approximately a doubling of glucocorticoid concentration in both sexes, whether in or out of the breeding season (Handasyde et al. 2003). There is also a seasonal variation in glucocorticoid concentration in both sexes, rising from around 100–150 mM in the non-breeding season (October to June) to around 250 mM in the breeding season (July to September). Observations of seasonal variation in plasma glucocorticoid concentration correlate closely with estimates of tail fat stores in both sexes, which decline during the mating season (Handasyde et al. 2003). Feedback control of glucocorticoid secretion in the platypus is regulated by glucocorticoid inhibition of ACTH production by the anterior pituitary, as in therians (McDonald et al. 1992), but the details of regulation of the HPA axis in these mammals, particularly in the hypothalamus, are not well understood.

Thermoregulation in monotremes

Monotremes have low body temperatures (T_b) at rest (around 32°C in the platypus, 30–32°C in the short-beaked echidna and 31°C in the long-beaked echidna; Schmidt-Nielsen et al. 1966; Grigg et al. 1992; Grigg et al. 2003). Although all living monotremes have similar average body temperatures, their responses to changing ambient temperature (T_a) are very different. The platypus is effectively homeothermic over a range of temperatures, but body temperature in the tachyglossids is often labile, rising as a result of activity and declining during inactivity (Brice 2009).

Metabolic activity of the tachyglossids (0.98 W/kg$^{0.75}$ for *Tachyglossus aculeatus* and 0.86 W/kg$^{0.75}$ for *Zaglossus bruijnii*) is also much lower than in the platypus (2.21 W/kg$^{0.75}$) (Dawson et al. 1978; Grant and Dawson 1978).

Remarkably, the platypus maintains its body temperature more tightly than some semi-aquatic placentals (Brice 2009). In fact, the platypus is able to maintain its body temperature over a range of environmental temperatures (T_a) from 5 to 32°C. In cold conditions the platypus can increase metabolic heat production as effectively as many therians. It is also able to modify the thermal conductance of its skin when immersed in water, apparently through a rete mirabile counter-current heat exchange system in the pelvis and non-propulsive hindlimbs (Grant 2007). In conditions of elevated ambient temperature, sweat glands in the glabrous skin of the bill and associated with hair follicles are able to provide a sweating response when T_a rises above 30°C. Nevertheless, it is probably only rarely that the platypus is exposed to these temperatures, because burrow temperatures in most of the range of the platypus would rarely rise above 25°C. However, platypuses in tropical Queensland may be exposed to higher burrow temperatures at the height of summer.

In contrast to the platypus, body temperature in the tachyglossids is subject to diurnal variation of as much as 6°C. This daily heterothermy is particularly striking in the short-beaked echidna in cool environmental conditions (5 to 15°C; Grigg et al. 1992), but has also been observed in the much larger long-beaked echidna (Grigg et al. 2003). Minimum body temperature is found in early to midmorning, immediately before emergence from the overnight retreat. Tachyglossids are also poorer than the platypus at maintaining a stable T_b when T_a is elevated above 34°C (Brice 2009). Neither short- nor long-beaked echidnas have a panting response to heat and they do not spread saliva to facilitate evaporative cooling. Furthermore, short-beaked echidnas do not have sweat glands, although apocrine sweat glands have been reported in the long-beaked echidna. The only mechanism for heat tolerance in the short-beaked echidna appears to be a cyclical storage of heat followed by avoidance of heat sources, because there is no effective evaporative cooling mechanism in this species (see review in Brice 2009). Short-beaked echidnas appear to rely mainly on behavioural thermoregulation, i.e. by using burrows, caves, shade, or bodies of water to minimise exposure to high temperatures during the day. The short-beaked echidna is therefore diurnally active during cold weather, but nocturnal at the height of summer. By contrast, one species of long-beaked echidna, Zaglossus bartoni, when exposed to T_a above 25°C was able to increase its evaporative heat loss, but did not change its respiratory minute volume (Dawson et al. 1978) indicating an ability to use sweat, but not panting, to thermoregulate.

While the phenomenology of monotreme thermoregulation is well documented, there is a yawning gulf when it comes to understanding how T_b is controlled by the monotreme central nervous system. Baird and colleagues (Baird et al. 1974) have shown that injection of histamine and acetylcholine into the lateral ventricles of short-beaked echidnas induces similar changes in body temperatures and thermoregulatory responses to those induced in placentals, but intraventricular injection of noradrenaline does not. By contrast, whereas intraventricular injection of the prostaglandin E series induces hyperthermia in many placentals (rabbits, ungulates, primates), these compounds actually reduce heat production and decrease vasomotor tone when injected into the lateral ventricles of some but not all echidnas (Baird et al. 1974). The conclusions we can reach from these experiments are limited, but it is clear that a periventricular neural structure (presumably the hypothalamus) plays an active role in thermoregulation in the echidna, probably involving serotonin and acetylcholine as neurotransmitters in the pathways. How hypothalamic function differs between the protoendothermic echidnas and the classically thermoregulating platypus is unknown.

Hibernation in monotremes

Routine periods of both short-term (daily) and long-term torpor (hibernation) have been observed in short-beaked echidnas, but not in the platypus, and perhaps not in the long-beaked echidna (Grigg et al. 2003). Hibernation has been demonstrated in the short-beaked echidna through much of its geographical range (Green et al. 1992; Nicol and Andersen 1996, 2000; Rismiller and McKelvey 1996; Beard and Grigg 2000; Grigg and Beard 2000; Kuchel 2003), from Tasmania, through mainland alpine, semi-arid and arid environments. There is significant geographical variation in the timing of torpor; whereas echidnas from alpine conditions show a high degree of conformity in their hibernation behaviour, echidnas from more temperate conditions (e.g. Kangaroo Island in South Australia) manifest a higher degree of individualism in the hibernation behaviour (Rismiller and McKelvey 1996).

The short-beaked echidna exhibits similar patterns in T_b to those seen in therian hibernators (Brice 2009), including inexplicable periodic arousals during which the animal may change its place of hibernation. The short-beaked echidna has been said to be a spontaneous hibernator, in that it primarily enters torpor in response to a persistent circannual rhythm or photoperiod cue (Nicol and Andersen 1996), rather than in response to individual food or water shortage. Nevertheless, individual energy shortage or excess may influence the time of onset and duration of hibernation. Furthermore, warm climate echidnas do not enter hibernation at all.

Some peculiar aspects of echidna hibernation, particularly those concerning the coordination of hibernation and mating, have been observed. For example, mature individuals may emerge from hibernation during the coldest months to breed (Nicol and Andersen 1996, 2000), and females may re-enter torpor after they have mated (Morrow and Nicol 2009). This interruption of hibernation for mating presumably confers some advantages in that it allows the hatching and lactational phase development of young to be coordinated with periods of maximum resource availability. Nicol and Andersen (1996, 2000) have suggested that hibernation is used in response to a relative rather than an absolute energy shortage, explaining why some echidnas hibernate at a particular time of the year, while others do not. Alternatively, some animals that have achieved target energy stores for mating may use hibernation to avoid exposure to predators and reduce energy costs (Grigg and Beard 2000; Brice 2009).

Hibernation is an active process, characterised by specific changes in physiological measures that have in common the effect of reducing metabolic activity. Echidnas in hibernation show changes in heart rate, lung ventilation, blood gases, blood flow, blood viscosity, tissue metabolism and body temperature

(Grigg and Beard 1996; Nicol 1992; Nicol et al. 1992; Nicol and Andersen 2003). Hibernation in short-beaked echidnas is a complex physiological process that involves highly coordinated and orchestrated changes in endocrine, cardiovascular, respiratory, haematopoietic and autonomic systems. Presumably these are centrally regulated, either directly or indirectly, by the hypothalamus, but the details remain speculative. The onset of hibernation in the echidna is only loosely linked with seasonal change, so a change in photoperiod is unlikely to be the necessary and sufficient stimulus. If the role of hibernation is to conserve energy rather than avoid cold, then one would expect that some form of feedback from fat stores, perhaps by leptin, may be the important trigger to initiate hibernation, but nothing is currently known about leptin (or its equivalent) in any monotreme.

Monotreme circadian activity

In therians, the key elements of the circuit that regulates circadian rhythms are the directly photosensitive retinal ganglion cells that project through the optic nerve to the hypothalamus (retinohypothalamic tract), a pacemaker in the suprachiasmatic nucleus of the hypothalamus, a projection from the hypothalamus to autonomic centres in the brainstem and thoracic spinal cord, the sympathetic outflow from the superior cervical ganglion to the pineal gland, and the pineal gland acting as a neuroendocrine transducer through its production of melatonin.

Under laboratory conditions, platypuses show an entrainment of their activity to light/dark cycles, reflecting their predominantly nocturnal niche. Although most platypus activity is during the dark phase, there are also irregular bouts of light phase activity (Francis et al. 1999). The daily cycle of platypus behaviour shows one of the main criteria for establishing the existence of an endogenous circadian rhythm, namely the persistence of the rhythm under constant environmental conditions, i.e. continued darkness. The timing of the start and finish of activity are coupled, such that early onset of activity is paired with early offset.

The evidence for the presence of the key elements of the circadian rhythm circuitry in monotremes is limited. Although Hines did not illustrate a suprachiasmatic nucleus in the platypus hypothalamus (Hines 1929), this nucleus is clearly present in the hypothalamus of the short-beaked echidna (Fig. 12.1; Ashwell et al. 2006a). The pineal gland has also been studied in the monotremes (see below; Kenny and Scheelings 1979), but no details of the endocrine or axonal connections between these elements in monotremes are available.

Monotreme reproductive cycles

Both the platypus and echidnas are seasonal breeders (Griffiths 1978). Reproductive activity in the platypus begins in winter and finishes in late spring (Temple-Smith 1973). These behavioural changes are accompanied by seasonal changes in hormones: progesterone blood levels in female platypuses rise from August and return to baseline levels around the end of spring (Handasyde et al. 1992). In female monotremes, the elevated progesterone not only accompanies ovulation, but probably also prepares the uterine wall for the egg by inducing maturation of the uterine glands, and maintains uterine secretions during gestation (Nicol et al. 2005).

Male platypuses also undergo an annual cycle of testicular involution and recrudescence (Temple-Smith 1973; Carrick and Hughes 1978). Concentrations of testosterone in the testicular venous blood of male platypuses peak in the breeding season (reaching 400 ng/mL in July to September) when the testes are large, and decline to as little as 10 ng/mL when the testes involute (McFarlane and Carrick 1992).

Reproductive activity of the short-beaked echidna is also strongly seasonal. In this species, reproduction usually follows a variable period of inactivity. In echidnas from warmer climates (e.g. Kangaroo Island in South Australia), this period is only a few days of torpor (Rismiller and McKelvey 1996), but in alpine habitats or higher latitudes, echidnas mate after emerging from several months of hibernation (Nicol et al. 2005). Plasma progesterone and testosterone levels in the short-beaked echidna are about half those in the platypus, but undergo the same seasonal changes. Female echidnas have elevated progesterone during the June to August period (0.17 ng/mL), but the highest progesterone is found in pregnant echidnas. Male echidnas engaging in mating trains during June and August have elevated testosterone

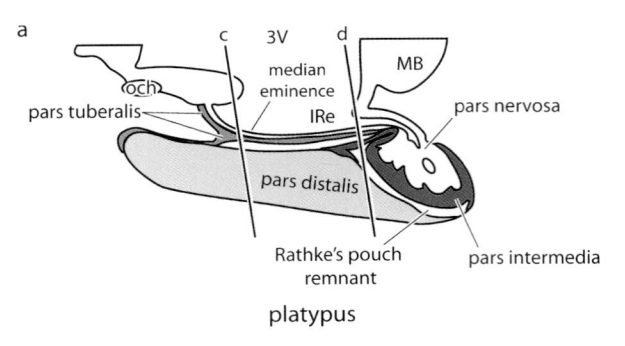

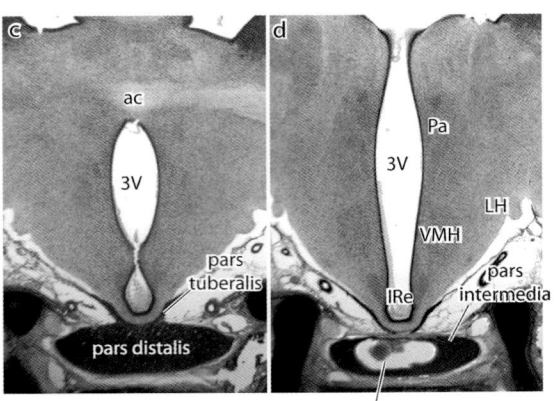

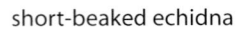

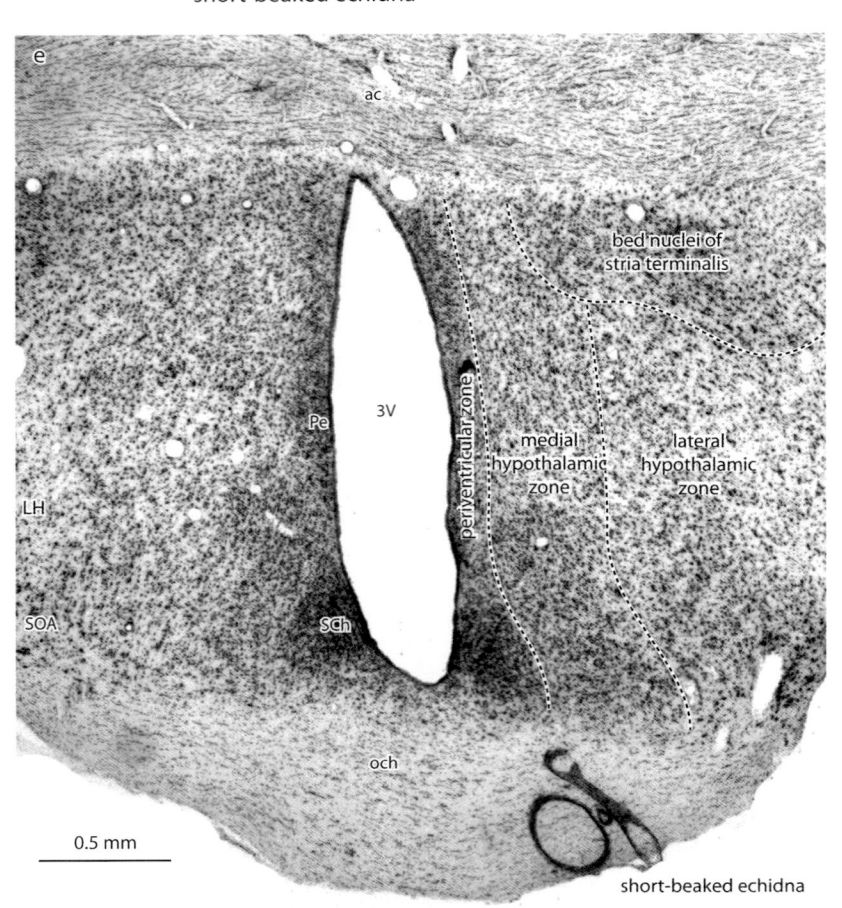

256 NEUROBIOLOGY OF MONOTREMES

levels (3.7 ng/mL), in contrast to values around 0.1 ng/mL for months other than June to August.

The seasonal control of reproduction in the monotremes suggests that changes in either temperature or photoperiod are important in controlling hypothalamic function. The environmental cue that synchronises the reproductive life history of male and female antechinus is a specific rate of change of the photoperiod (McAllan *et al.* 1991) and this could also be responsible for the observed changes in the monotremes. If this is the case, then the neural elements that would probably be involved include the suprachiasmatic nucleus, which receives the retinohypothalamic tract; the pineal gland, producing melatonin and acting as either a pacemaker or neuroendocrine transducer; secretory regions in the hypothalamus, to mediate regulatory hormone production; and a variety of forebrain nuclei (septal nuclei, amygdala, bed nuclei of the stria terminalis and limbic cortex) and hindbrain output (periaqueductal grey, nucleus retroambiguus) for behavioural effects (Ashwell 2010b).

Monotreme lactation

All the extant mammals have a complex lactational system, suggesting that the mechanism of lactation has a long evolutionary history and probably emerged in the cynodont lineage towards the end of the Triassic (~200 mya) (Lefèvre *et al.* 2010). Comparison of the casein gene cluster of the platypus with other mammalian genomes has highlighted the highly conserved nature of the casein genes in all mammals and suggests that the expression of caseins and whey proteins in the mammary gland predates the common ancestor of all living mammals. The most conserved proteins (milk fat globule) are associated with the secretory process, whereas the most divergent are associated with the nutritional and immunological components of milk (Lefèvre *et al.* 2009, 2010).

Oftedal (2002) has proposed that the mammary gland is derived from an ancestral apocrine-like gland that was associated with hair follicles. This is reflected in the structure of the monotreme mammary gland, where there is no nipple (presumably the ancestral condition) and the mammary gland ducts open directly and independently onto the areola, each in association with a hair follicle (Lefèvre *et al.* 2010). Several hypotheses have been proposed to explain the role of the protolacteal secretions of the earliest mammary gland (Lefèvre *et al.* 2010). These are: (1) thermoregulation of eggs and hatchlings through evaporative cooling; (2) preventing desiccation of leathery-shelled eggs; (3) providing pheromones for communication between mother and young; (4) supplying a protective mechanism for keeping the young close to the mother regardless of nutrition; and (5) providing antibacterial protection of the young through immunoglobulin and related secretions.

The newly hatched monotreme is highly altricial and depends completely on milk as a source of nutrition during the lactational period of development. As stated earlier, monotreme females have no teats and milk is excreted onto the areola. The areola also contains sebaceous glands, so monotreme milk may include a mixture of skin cells, immune cells, and exfoliated epithelial cells from the ducts of mammary and sebaceous glands. As in other mammals, the alveoli of the monotreme mammary gland are surrounded by a network of myoepithelial cells that contain contractile myofilaments running parallel to the long axis of the cell. These cells contract in response

Figure 12.1: Schematic representation of the morphology of the pituitary in the platypus (a) and the short-beaked echidna (b). These diagrams are based on illustrations reproduced in Griffiths (1978) for the platypus and Hanström (1954) for the short-beaked echidna. Note the pars tuberalis in contact with the median eminence of the hypothalamus and extending as far anteriorly as the optic chiasm. Note also the large infundibular recess of the third ventricle (IRe) in the echidna. The lines in (a) that have been labelled (c) and (d) show the approximate positions of the photomicrographs. Photomicrographs of the hypothalamus and pituitary are shown at the level of the anterior commissure (c) and the caudal infundibular recess (d). The photomicrograph (e) shows a transverse section though the hypothalamus of an adult short-beaked echidna at the level of the optic chiasm, illustrating the three mediolateral zones of the hypothalamus. 3V – third ventricle; ac – anterior commissure; LH – lateral hypothalamus; MB – mammillary body; och – optic chiasm; Pa – paraventricular nucleus of the hypothalamus; Pe – periventricular nucleus of the hypothalamus; SCh – suprachiasmatic nucleus; SOA – supraoptic nucleus (anterior part); VMH – ventromedial nucleus of the hypothalamus.

to circulating oxytocin (the let-down reflex) to eject milk from ducts onto the surface of the areola (Griffiths 1978).

It remains controversial as to whether monotreme milk changes in composition during the lactation period, as is seen in marsupials (asynchronous concurrent lactation) (Lefèvre et al. 2010). Nevertheless, the production of milk by monotreme mammary glands can be prodigious and the young of larger mothers can grow rapidly. Weight gain of young short-beaked echidnas varies with maternal body weight, ranging from 4 g/day for the offspring of small (1 kg body weight) females to 12 g/day for the offspring of larger (up to 5 kg body weight) females (Green et al. 1985), highlighting the importance of maternal fat stores in reproductive success.

Monotreme milk cells express all types of caseins and casein variants, many similar to those reported for other mammals. Milk proteins that monotremes share with both marsupials and placentals include: α-casein (CSN1), β-casein (CSN2), κ-casein (CSN3), β-lactoglobulin, α-lactalbumin and whey acidic protein. Milk proteins found in monotremes, but not other mammals, include casein variant CSN2b and C8orf58 (Lefèvre et al. 2010). The milks of platypuses and short-beaked echidnas have some differences at the time of hatching. The milk of the echidna has a pink colouration due to the presence of large amounts of the iron-binding protein lactotransferrin (Griffiths 1978), whereas that of the platypus is creamy white. Platypus milk also has seven unsaturated and polyunsaturated fats that are not found in *Tachyglossus* milk (Griffiths et al. 1973). The functional significance of these differences is unknown at present.

Structure of the monotreme pituitary

The pituitary gland is a pear-shaped body connected to the hypothalamus by the infundibular stalk and median eminence. As in all tetrapods, the monotreme pituitary gland is divided into an anterior part (adenohypophysis) to the front and a posterior part (neurohypophysis or pars nervosa) to the back. The adenophypophysis is further divided into three parts: a large pars distalis, a pars tuberalis over the dorsal surface of the pars distalis and extending onto the infundibular stalk and median eminence of the hypothalamus, and a pars intermedia intervening between the pars distalis and the pars nervosa (Fig. 12.1).

Early descriptions of the pituitary gland of the platypus portrayed the gland as reptilian in appearance, in that the pars distalis is elongated and lies parallel to the elevation on the underside of the hypothalamus known as the median eminence (see discussion in Griffths 1978). The pars nervosa also has an infundibular cavity, as seen in sauropsids, and there is a well-developed hypophyseal cavity (a remnant of the embryonic cavity of Rathke's pouch; see development of pituitary below) between the pars distalis and the pars intermedia (Griffiths 1978). Routine histology of the platypus pituitary has not revealed any differences from the therian pituitary gland, but no ultrastructural studies have been performed as yet. The rostral part of the pars distalis has predominantly acidophils cells (i.e. cells that stain for acidic dyes) and the caudal part has mainly basophils (cells that stain for basic dyes). In therians, acidophil cells secrete GH and PRL, whereas basophil cells produce gonadotropins, TSH and ACTH, but a modern study of the molecular biology of the monotreme anterior pituitary and its hormone secretion is clearly needed.

The arrangement of the main parts of the pituitary gland in the short-beaked echidna is similar to that of the platypus described above (Augee et al. 1971b), but with some minor differences. These are first that a downward projection from the pars tuberalis makes contact with the pars distalis in both monotremes, but this is not so extensive in the echidna (compare Fig. 12.1a and b); and second that the pars nervosa of the echidna has a more sacculate appearance because of the extension of a large infundibular recess of the third ventricle into the pars nervosa. This extension of the third ventricle is lined with ependymal cells, with the processes of the pituicytes and ependyma intermingling (Griffiths 1968). In contrast to the platypus, Fink et al. (1975) found that the rostral part of the pars distalis of the echidna has mainly basophils, whereas the caudal part of the pars distalis has mainly acidophils. Fink and colleagues (Fink et al. 1975) identified three types of secretory cell based on ultrastructure: type I with granules 100–200 nm in diameter, type II with granules 200–300 nm in diameter, and type III with granules 300–400 nm in diameter. Type II predominate in the rostral part of the gland and type III in the caudal part, but both are found throughout the pars distalis. Type I is relatively rare and cells of this type are scattered throughout the gland. The median eminence of the echidna is completely free of blood vessels, because the hypothalamo-hypophyseal portal system, which carries releasing and inhibitory factors from the hypothalamus to the anterior pituitary, runs only across the external surface of the median eminence (Smith *et al.* 1970, 1971), but the capillaries of the portal system are fenestrated for the passage of large molecules, as in therians (Smith *et al.* 1970). Axons of the hypothalamo-neurohypophyseal tract contain large secretory granules (170 to 240 nm in diameter; Smith 1971), presumably containing oxytocin and vasopressin (antidiuretic hormone) from the magnocellular neurosecretory nuclei of the hypothalamus, both bound to neurophysin carrier molecules.

No data are available concerning the structure of the anterior or posterior pituitary in *Zaglossus*.

Posterior pituitary hormones in the monotremes

Neurohypophyseal hormones have been isolated from the platypus (Chauvet *et al.* 1985; Feakes *et al.* 1950) and the platypus genome has been analysed for genes coding oxytocin and vasopressin (Wallis 2012). Chauvet and colleagues (Chauvet *et al.* 1985) identified a vasopressor hormone with an amino acid composition identical to arginine vasopressin used by many placentals (which stands as an interesting contrast with the use of phenypressin by macropod marsupials, Chauvet *et al.* 1980). An oxytocinergic hormone that resembles oxytocin was also isolated, but Chauvet and colleagues were unable to obtain a correct amino acid composition. The vasopressor peptide was present in a concentration four times higher than the oxytocinergic hormone. Chauvet and colleagues concluded that the presence of the same antidiuretic hormone in the monotremes and placentals suggests that the transformation of the reptilian vasotocin into vasopressin occurred very early in mammalian history, perhaps among therapsids. Wallis (2012) has confirmed that the platypus has genes for oxytocin and arginine vasopressin, arranged in a tail-to-tail sequence as in placentals, rather than the three to four genes for neurohypophyseal hormones seen in marsupials.

Monotreme hypothalamus

Overview of topography of the monotreme hypothalamus

The tetrapod hypothalamus is traditionally divided into a series of mediolateral zones: periventricular, medial and lateral (Saper 2012; see Fig. 12.1e). These mediolateral divisions reflect the sequence of hypothalamic neurogenesis during postembryonic life, in that the order of production of hypothalamic neurons in all mammals (including monotremes; see below) is largely from lateral to medial. Furthermore, the consistent position of major fibre bundles like the fornix and mammillothalamic tract clearly demarcate medial and lateral zones in all mammals. The subdivision of the hypothalamus in rostrocaudal and dorsoventral directions is far more contentious, because there is ongoing debate about both the neuromeric derivation of this part of the brain and the direction of developmental axes in the embryonic hypothalamus. The traditional subdivision of the mature mammalian hypothalamus into three or four antero-posterior divisions (i.e. preoptic, anterior, tuberal and mammillary regions) is serviceable for most descriptions of the adult hypothalamus, but the reader should note that this simple division is questioned by modern neuromeric models of the hypothalamus (Puelles *et al.* 2012a) that are based on gene expression during embryonic life (see discussion in Chapter 4 and monotreme hypothalamic development below).

The monotreme hypothalamus has the same broad division into both mediolateral zones and anteroposterior regions that has traditionally been used in descriptions of therian hypothalamus (Krieg 1932; Simerly 2004; Fig. 12.1e). In fact, all of the major nuclei that can be identified in therian hypothalamus can also be found in the hypothalamus of both the platypus and short-beaked echidna (Ashwell *et al.* 2006a). Furthermore, cyto- and chemoarchitectural features of the monotreme hypothalamus are also broadly similar to those in other mammals, but with some important, unique features that will be discussed below.

Serotonergic neurons of the monotreme hypothalamus

The first peculiar feature of the monotreme hypothalamus concerns the presence of serotonin-containing neurons. Serotonergic neurons have not been reported in the hypothalamus of therian mammals, but they have been identified in many other vertebrates (Smeets and Steinbusch 1988), and some hypothalamic neurons in rodents may have the ability to accumulate serotonin even if they do not habitually use serotinin as a neurotransmitter (Fuxe and Ungerstedt 1968). In contrast to placentals and marsupials, Manger and colleagues identified two groups of serotonergic neurons in the hypothalamus of both the platypus and short-beaked echidna (Manger et al. 2002c). Hypothalamic serotonergic neurons in both the monotremes are substantially smaller (cell body area of 150–220 µm^2) than most serotonergic neurons in the monotreme brainstem (cell body area of 200–400 µm^2), suggesting that the hypothalamic serotonergic neurons don't engage in long-distance connections.

The first group of serotonergic neurons is closely associated with the ependymal lining of the third ventricle and spans a rostrocaudal extent of ~1.5 mm through the periventricular zone of the hypothalamus (Manger et al. 2002c). Within this periventricular group there are two types of serotonergic neurons. The first of these lie very close to the lining of the third ventricle and have small circular cell bodies with a club-like process that extends to the ventricular wall, suggesting that they may be able to sample the cerebrospinal fluid. An axon emerges from the other pole to run laterally away from the ventricle. Neurons of the second type are arranged in an arc around the first type and have cell bodies a little further away from the ependyma. Neurons of this latter type have small oval cell bodies with two primary dendrites. The first of these dendrites runs to the neurons associated with the ependyma and the second projects away from the third ventricle. Axons arising from the cell bodies of both of these periventricular neurons bundle to form a fasciculus that runs through the zona incerta of the prethalamus and mingle with the serotonergic axons from the brainstem.

The second group of serotonergic neurons is closely associated with the infundibular recess of the third ventricle. These neurons have a similar morphology to the first type of periventricular serotonergic neuron described above, but are sufficiently separated from them as to be assigned into a separate group.

The functional significance of both these groups of serotonergic neurons remains unknown, although hypotheses linking these cells with peculiar features of the monotreme hypothalamus (e.g. thermoregulation and reproduction) are worth testing. The arrangement of both these groups is highly suggestive of some role in sampling cerebrospinal fluid and transferring information to nearby hypothalamic nuclei. Neurons with this function could play a role in regulation of circannual rhythms of hibernation and reproduction through melatonin secretion, or thermoregulation through secreted neuroactive amines.

Catecholaminergic neurons of the monotreme hypothalamus

Several clusters of catecholaminergic neurons can be identified in the monotreme preoptic area and hypothalamus with the aid of immunohistochemistry for the synthetic enzyme tyrosine hydroxylase (TH) (Manger et al. 2002b; Ashwell et al. 2006a). The positions of these groups and the morphology of the constituent neurons are broadly similar to that described for other mammals, underlining the evolutionary conservatism of the catecholaminergic system.

Hypothalamic catecholaminergic groups include five clusters of TH immunoreactive neurons. These are the A15 division in the preoptic region, further divided into dorsal and ventral components; the A14 periventricular division, which occupies most of the rostrocaudal extent of the hypothalamus and extends into the thalamus; the A13 group, which Manger and colleagues (2002b) assigned to the zona incerta, but actually begins in the lateral preoptic area level with the anterior commissure; the A12 'tuberal' group, which Manger and colleagues identified as extending from the optic chiasm (actually the anterior region of the hypothalamus) to the true tuberal region of the hypothalamus; and the A11 group, occupying the region of the ventromedial hypothalamus and extending into the lateral hypothalamus.

The TH immunoreactive neurons in the hypothalamus of the echidna tend to be ~20% larger than those in the platypus (Manger et al. 2002b), but this probably reflects nothing more than the larger size of the echidna brain and the need for a larger neuronal cell body to support the longer and larger diameter axons to reach distant sites within that larger brain.

Neurons immunoreactive for TH also contain non-catecholamine neurotransmitters. In the context of the hypothalamus, tyrosine hydroxylase immunoreactive neurons have been implicated in a wide range of behaviours, from erection to circadian rhythms, blood pressure regulation, stress and appetitive control (Brooks et al. 2011; Gogebakan et al. 2012; Yeh et al. 2011; Zhao et al. 2011). This aspect of catecholaminergic neurons has yet to be investigated in the monotremes.

Are cholinergic neurons absent from the monotreme hypothalamus?

Although cholinergic neurons (i.e. those that use acetylcholine – ACh – as their neurotransmitter) have been identified in the hypothalamus of placentals (rodents, primates and carnivores), Manger and colleagues (Manger et al. 2002a) were unable to identify cholinergic neurons in the hypothalamus of either the platypus or short-beaked echidna by using immunohistochemistry to choline acetyl transferase (ChAT), the synthetic enzyme for ACh. This is despite the presence of strongly immunoreactive axons in the fasciculus retroflexus, a fibre bundle that passes through the caudal hypothalamus.

Cholinergic neurons have been identified in the supraoptic and infundibular regions of the hypothalamus of many other tetrapod species (Ekström 1987; Marín et al. 1997; Mason et al. 1983; Medina and Reiner 1994; Medina et al. 1993; Powers and Reiner 1993; Tago et al. 1987), so, on the basis of these findings, monotremes might appear to be unique among tetrapods. Manger and colleagues assigned a functional significance to this supposed hypothalamic hiatus in the rostrocaudal sequence of cholinergic neurons, suggesting that the apparent absence of hypothalamic cholinergic neurons (as identified by their techniques) may be functionally linked with the report by the same authors that monotremes lack the two-stage pattern of sleep alternation (i.e. between slow wave and rapid eye movement sleep, but see Chapter 13 on sleep for a critical appraisal of this claim).

However, there is more to be said about the cholinergic question. Although Manger and colleagues were unable to identify hypothalamic cholinergic neurons with ChAT immunohistochemistry, enzyme histochemistry for the degradative enzyme for ACh (acetylcholinesterase – AChE) reveals AChE reactive neuronal somata in exactly the same hypothalamic regions where these neurons are found in other mammals (Mason et al. 1983), i.e. the supraoptic and tuberal regions (Ashwell et al. 2006a). Enzyme histochemistry for AChE is not as specific for cholinergic neurons as ChAT immunohistochemistry, but the AChE findings suggest that cholinergic pathways in monotremes are not as unusual as Manger et al. (2002a) have suggested.

Supraoptic and paraventricular nuclei in the monotreme hypothalamus

Several unique aspects of monotreme biology are controlled at least in part by the hypothalamo-neurohypophyseal axis. These include egg-laying, whereby the embryo is delivered from the reproductive tract in a leathery-skinned egg, from which it emerges as an immature hatchling; and the relatively primitive form of lactation used by monotremes, whereby milk is secreted onto the mother's skin at the teat-less areola. In other words, both parturition and the milk ejection reflex, which are intimately controlled by oxytocinergic neurons of the hypothalamo-neurohypophyseal pathway in therians, are remarkably different in monotremes compared to therians. These obvious behavioural and functional differences naturally raise the question of how the supraoptic and paraventricular nuclei that contain neurons that contribute axons to the hypothalamo-neurohypohyseal tract compare in monotremes and therians. Ashwell and colleagues (Ashwell et al. 2006a) have analysed the cyto- and chemoarchitecture, as well as the connections to the pituitary, of the supraoptic and paraventricular nuclei in the platypus and short-beaked echidna.

In laboratory rodents, most of the supraoptic nucleus (SO) lies adjacent to the lateral border of the optic tract, forming an anterior compartment of the supraoptic nucleus (SOA). At caudal levels of the supraoptic nucleus, a very thin layer of morphologically similar neurons lies medial to the optic tract in a group known as the retrochiasmatic continuation of SO (SOR) (Armstrong 2004; Watson and Qi 2012). The same topography of SO has also been found in other placentals like humans, in which most SO neurons are located dorsolateral to the optic tract with only a few neurons spilling medial to the tract (Saper 2012). In both the platypus and short-beaked echidna, the broad topography and chemoarchitecture of the supraoptic nucleus is similar to those in therians, but with some key differences (Ashwell *et al.* 2006a). First, supraoptic neurons appear to be more loosely clustered in the monotremes than in rodents and primates, with no clearly defined nuclear boundary to either the anterior or retrochiasmatic parts of the supraoptic nucleus. Second, in contrast to most therians that have been studied, the retrochiasmatic part of the monotreme SO (SOR; see atlas plates Ec-Ad12 and 13, Pl-Ad12) appears to be more extensive than the SOA. The SOR contains the bulk of magnocellular neurons, particularly in the platypus, where the SOA appears to be very small. Finally, the low neuronal density in the SOA and SOR of the two monotremes indicates that the absolute size of the neuronal populations in SO components is quite low relative to these nuclei in therians.

The paraventricular nucleus of the hypothalamus (Pa) lies alongside the third ventricle and plays a central role in the regulation of the neuroendocrine and autonomic systems. In particular, it is important in the control of pituitary-adrenocortical activity in response to stress, the regulation of cardiovascular activity, the maintenance of body fluid homeostasis, the milk ejection reflex (through its contribution to the hypothalamo-neurohypophyseal tract), thyroid hormone secretion, food intake, and pineal melatonin synthesis (Ciriello and Calaresu 1980; Swanson *et al.* 1986; Kiss 1988; Kiss *et al.* 1991; Dawson *et al.* 1998). The paraventricular nucleus is broadly divided into two cellular assemblages: magnocellular and parvicellular neuronal groups (Gurdjian 1927; Krieg 1932). Magnocellular neurons contribute to the pathways releasing oxytocin and antidiuretic hormone/vasopressin at the posterior pituitary. Parvicellular neurosecretory neurons of the paraventricular nucleus project axons to the external zone of the median eminence where they discharge releasing factors into the hypothalamo-hypophyseal portal system (Watson and Qi 2012).

In placental mammals, the paraventricular nucleus is subdivided on the basis of cell size and topography, peptide localisation and projection sites (see review in Watson and Qi 2012), with multiple subnuclei recognised in many experimental animals, but this subdivision is not supported for the monotreme hypothalamus, which seems to have a simpler internal structure. Ashwell and colleagues (Ashwell *et al.* 2006a) found that the Pa in the short-beaked echidna is relatively small and homogeneous. As in therians, the Pa of the echidna consists of a wedge-shaped region alongside the third ventricle, with a mixture of different-sized neurons. Occasional magnocellular neurons (cell body diameter greater than 15 µm) are also scattered in the surrounding medial hypothalamus, but these do not appear to be grouped into distinct subnuclei. From its position and cytoarchitecture, the Pa of the echidna appears to correspond most closely to the rostral magnocellular subnucleus of the Pa of the laboratory rat (Swanson and Kuypers 1980) and the magnocellular part of the paraventricular nucleus of diprotodont marsupial hypothalamus (Cheng *et al.* 2003). Nevertheless, the echidna Pa appears to be quite small, given both the brain and body size of this mammal, when compared to the paraventricular nuclei of therians of similar brain and body size. For example, Ashwell and colleagues (Ashwell *et al.* 2006a) noted that adult echidnas have body weights of 2–5 kg and brain weights of 20–23 g, but the Pa of the echidna is no larger than that of a rat with a brain and body weight as little as one-tenth that of the echidna, and much smaller than that of the diprotodont marsupial *Macropus eugenii*, which has a similar body weight and a lower brain weight (Cheng *et al.* 2003).

The Pa of the platypus also appears to be quite small, although it completely fills the region between the third ventricle and the column of the fornix. As for the echidna, Ashwell and colleagues (Ashwell *et al.* 2006a) were unable to identify any subnuclear divisions in this structure and concluded that the single wedge-shaped nucleus probably corresponds to the medial magnocellular parts of therian Pa. Surprisingly, the Pa of the platypus is comparable in size to that of the echidna, even though the body and brain weights of the platypus are substantially less.

Distribution of oxytocinergic and neurophysin immunoreactive neurons in the monotreme hypothalamus

The magnocellular neurosecretory cells of the therian Pa and SO synthesise oxytocin and vasopressin. Each of these hormones is bound to a specific type of neurophysin (NPH) for transport along unmyelinated axons of the hypothalamo-neurohypophyseal tract to the posterior pituitary, where the hormones are released into the circulatory system.

Ashwell and colleagues (Ashwell et al. 2006a) used polyclonal antisera to oxytocin and to both types of neurophysin together to map the distribution of oxytocin and NPH immunoreactive neurons in the hypothalamus of both the platypus and short-beaked echidna. They found some similarities in the distribution of these antigens in the monotreme hypothalamus to the hypothalamus of therians, but also some differences. The similarities involve the broad distribution, in that most NPH and oxytocin immunoreactive neurons in the two monotremes are located in the supraoptic and paraventricular nuclei, and in the lateral hypothalamic area, as would be found in therians.

On the other hand, NPH and oxytocin immunoreactive neurons in the monotreme hypothalamus are much more loosely distributed than similar neurons in the therian hypothalamus (Kawata and Sano 1982; Schimchowitsch et al. 1983, 1989; Iqbal and Jacobson 1995; Choy and Watkins 1979; Mai et al. 1997; Cheng et al. 2003). Consistent with the diffuse nature of SOA and SOR as described above, there is also a general lack of clustering of NPH and oxytocin immunoreactive neurons in the monotreme supraoptic nuclei. Furthermore, neither monotreme has aggregations of NPH immunoreactive or oxytocin immunoreactive neurons into a nucleus circularis, a small nucleus found in the lateral hypothalamus in some rodents (Armstrong 2004) and the domestic cat (Caverson et al. 1987). However, the nucleus circularis has not been found in the hypothalamus of a diprotodont marsupial (the tammar wallaby; Cheng et al. 2003) and no nucleus circularis has been found in the laboratory rabbit (Schimchowitsch et al. 1989) or Macaca (Caffé et al. 1989; Paxinos et al. 2000).

The monotreme hypothalamo-neurohypophyseal system may have significant differences from therians

Two lines of evidence suggest that the hypothalamo-neurohypophyseal system in monotremes may be significantly different from that in therians (Ashwell et al. 2006a), perhaps reflecting the different neuroendocrine physiology of monotremes and therians.

The first of these concerns the distribution of neurons immunoreactive for non-phosphorylated neurofilament protein (SMI-32 antibody). In the short-beaked echidna, SMI-32 immunoreactivity labels a tract arising from the lateral hypothalamus and running through the inner portion of the median eminence towards the pituitary stalk. The position of this tract within the median eminence is similar to the placental hypothalamo-neurohypophyseal tract (Lu Qui and Fox 1976; Lechan et al. 1980; Mai et al. 1997). On the other hand, the distribution of SMI-32 immunoreactive somata in the echidna hypothalamus differs from the topography of hypothalamo-neurohypophyseal tract neurons in therians. In the short-beaked echidna, SMI-32 immunoreactive neurons are mainly seen within the supraoptic nucleus and lateral hypothalamus; no SMI-32 immunoreactive neurons are found within the Pa (Ashwell et al. 2006a). This is in contrast to therian mammals, where there are large populations of SMI-32 immunoreactive neurons in the Pa (Koutcherov et al. 2000; Cheng et al. 2003).

The second line of evidence comes from direct labelling of the tract itself. In therians, the hypothalamo-neurohypophyseal tract arises from magnocellular neurons within both the supraoptic and paraventricular nuclei, as well as within accessory cell groups distributed through the lateral hypothalamic area (Sherlock et al. 1975; Lu Qui and Fox 1976; Watkins and Choy 1977; Choy and Watkins 1977; Mai et al. 1997). Ashwell and colleagues used the carbocyanine dye DiI to retrogradely trace neurons connected to the pituitary stalk by the hypothalamo-neurohypophyseal tract in the short-beaked echidna (Ashwell et al. 2006a). They were able to retrogradely fill neurons in the supraoptic nucleus and lateral hypothalamus, but surprisingly no neurons were retrogradely labelled in the Pa, even though the distance from the Pa to the point of DiI insertion was less than that between the lateral hypothalamus and the insertion site.

These two lines of evidence suggest that, in the short-beaked echidna at least, the hypothalamoneurohypophyseal tract arises exclusively from the lateral hypothalamus and SOR, rather than the Pa. The NPH and oxytocin immunoreactive neurons within the echidna Pa may be engaged in other pathways. These might be like the pathways of NPH immunoreactive neurons of the human hypothalamus to the brainstem and spinal cord (Unger and Lange 1991; Ashwell et al. 2006a).

Evolution of the milk ejection reflex and the hypothalamo-neurohypophyseal system

Since the unique neuroendocrine physiology of monotremes mainly concerns lactation, it is worth giving some consideration to how their mammary glands differ from therians, how this may reflect the evolution of mammary glands in general and whether this difference explains the unusual structural features of the hypothalamo-neurohypophyseal system noted above.

Oftedal (2002) has proposed that the mammary gland is derived from an ancestral apocrine gland that was associated with hair follicles in Triassic therapsids. This arrangement is still found in the developing monotreme mammary gland, where the primary primordia of the mammary gland show a vigorous growth of hair, even more than the surrounding skin (Griffiths 1978). In the adult monotreme, each nipple-less mammary area consists of 100–150 mammo-pilo-sebaceous units, with each unit containing a mammary lobule, a mammary hair and one or more sebaceous glands. The mammary glands of prenatal marsupials undergo a similar sequence of events during development, with the presence of true mammary hairs that are shed during epidermal eversion to form the nipple. Curiously, mammary hairs do not develop in the primary primordia of placental mammary glands (Griffiths 1978).

Oftedal suggested that the monotreme arrangement of mammo-pilo-sebaceous units opening onto a nipple-less areola was present in early Triassic therapsids and that this initial arrangement functioned (as it does in modern monotremes) to provide moisture and other constituents to permeable leathery-skinned eggs (Oftedal 2002). Ashwell and colleagues (Ashwell et al. 2006a) argued that if the initial role of the monotreme mammary apparatus were to provide moisture to a newly laid leathery egg, then one would expect that oxytocin secretion would be continuous throughout the incubation phase of monotreme development. After hatching, lactation might be more episodic in response to stimulation by the infant monotreme, but this has never been studied in detail in either monotreme under natural conditions. By contrast, therian mammary glands do not serve a primary function of maintaining a moist environment for an egg and are exclusively adapted to provide a surge of oxytocin for parturition and the lower level, but episodic, release of oxytocin during the therian milk ejection reflex.

This functional difference encapsulated in the above hypothesis may provide an explanation for the profound differences in size of the supraoptic nuclei between the monotremes and some therian mammals. Ashwell and colleagues (Ashwell et al. 2006a) argued that those nuclei contributing to the hypothalamo-neurohypophyseal tract became enlarged in therians as an adaptation for rapid delivery of bursts of oxytocin, particularly during parturition, but also during the milk ejection reflex (Tindal et al. 1963; Folley and Knaggs 1966), whereas in monotremes those nuclei have remained small because only a few neurons are necessary to meet the steady demand for oxytocin during incubation of the leathery egg. This would predict that the quantity of oxytocin released from the monotreme neurohypophysis during egg-laying is substantially lower than that released during therian parturition, so that oxytocinergic neuronal populations in the magnocellular secretory nuclei need not be as large as in therians. This hypothesis would also predict that oxytocin levels during lactation are substantially lower in monotremes than in therians, but nothing is currently known about the changes in levels of oxytocin during lactation in monotremes.

Development of the monotreme pituitary

The pituitary develops from the apposition and fusion of two developmental structures. One of these is an ectodermal pocket (Rathke's pouch) that grows up from the roof of the oral cavity. The other is an infundibular process (hence neurectodermal in nature) that grows down from the caudal midline of the prosencephalon. In the first third of incubation, the pituitary is only a rudimentary structure in the monotreme embryo. Rathke's pouch is a relatively shallow extension of the roof of the oral cavity (20–30 µm from orifice to fundus) and the infundibular recess is only a thin-walled, shallow depression.

In the middle third of incubation, Rathke's pouch and the infundibular process both elongate rapidly (Ashwell 2012e; Fig. 12.2a; Appendix Fig. 1). The posterior pituitary is still thin-walled and immature during this stage, but the pars distalis of the anterior pituitary of the echidna has begun to expand rapidly and is invaded by vasculature. The pituitary of the echidna appears to be slightly advanced relative to the platypus during this period, in that the pars distalis of the echidna has a thicker wall, suggesting that more glandular precursor cells have been generated. The platypus anterior pituitary gland remains quite simple in structure as the embryo enters the last third of incubation. Although Rathke's pouch in the platypus has elongated to a length of 400 µm, the walls of both Rathke's pouch and the infundibular recess in the platypus embryo are thin and there is no development of the pars distalis or tuberalis.

The pituitary is still very immature in both monotremes in the days leading up to hatching (Ashwell 2012e; Fig. 12.2b). The echidna pars distalis at the immediately pre-hatching age is not greatly different in appearance from that seen in the mid-incubation echidna, but the posterior pituitary has developed a lobar structure, something not seen at earlier ages. Nevertheless, there is no evidence of fibre tracts growing around the median eminence and infundibular stalk, such as would be expected if the hypothalamo-neurohypohyseal tract had developed and were preparing for secretion. This suggests that neither the adeno- nor neurohypophysis is functional at the time of hatching, but physiological confirmation of this is needed. The state of development of pituitary structure in echidna hatchlings is quite similar to that described for newborn Australian marsupials like the bandicoot and tammar wallaby (Cheng et al. 2002; Hall and Hughes 1985) and didelphids like the short-tailed opossum (Gasse and Meyer 1995). The epithelial cells of the pars anterior are similarly limited in differentiation at birth in the platypus (Fig. 12.2c): Rathke's pouch has a transverse slit, and the posterior pituitary is small and does not appear to receive a hypothalamo-neurohypophyseal tract.

Cords of epithelial cells appear in the monotreme anterior pituitary at the end of the first week after hatching (Fig. 12.2d) and the putative hypothalamo-neurophypophyseal tract is seen for the first time only during the second week of post-hatching life. The anterior pituitary has reached a relatively mature appearance by the end of the first post-hatching month, but the cavity of Rathke's pouch remains large and will form the hypophyseal cavity of the adult gland (see Fig. 12.1). Invasion of the posterior pituitary by the putative hypothalamo-neurohypophyseal tract expands the tissue of the neurohypophysis during the first month after hatching. Even after pituitary development is complete, the extension of the infundibular recess of the third ventricle into the neurohypophysis persists as an embryonic remnant into adult life.

Development of the monotreme hypothalamus

Hypothalamic development begins during incubation, but the bulk of differentiation is a post-hatching process (see Appendix Fig. 1). During the first third of incubation, the wall of the hypothalamic part of the forebrain vesicle is thin and no postmitotic cells have settled outside the neuroepithelium, but putative zones within the developing hypothalamus can be identified on the basis of their position relative to the optic stalk (Ashwell 2012e). In the middle third of incubation, postmitotic cells begin to settle in large numbers in the mantle layer outside the neuroepithelium. Neurogenesis in the monotreme hypothalamus follows the same lateral-to-medial neurogenetic gradient observed in the therian hypothalamus (Altman and Bayer 1986; Cheng et al. 2002) so the first neurons to appear are those destined for the nuclei of the lateral zone (i.e. the lateral hypothalamus, LH). At about the same time as the first neurons settle outside the neuroepithelium, a diffuse zone of fibres (probably the medial forebrain bundle) can be found running along the curved rostrocaudal axis of the hypothalamus.

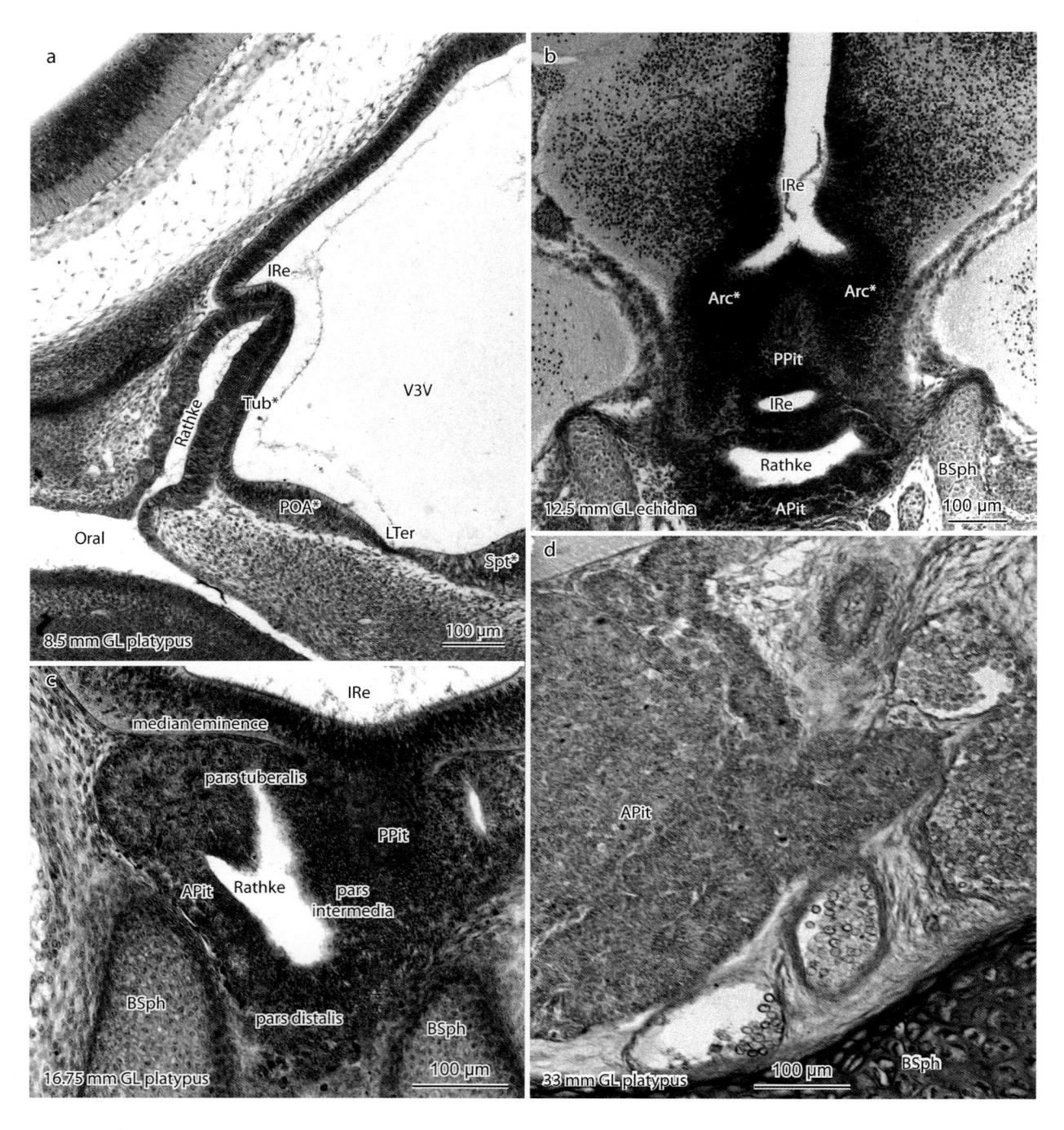

Figure 12.2: Photomicrographs illustrating the development of the pituitary and tuberal hypothalamus in the platypus and echidna. In the sagittally sectioned platypus embryo at 8.5 mm GL (a), the infundibular process has descended into contact with the wall of Rathke's pouch. Immediately before the time of hatching (b), distinct anterior and posterior pituitary components have emerged. Even in the first week after hatching (c), the anterior pituitary remains undifferentiated, although components of the anterior pituitary (pars tuberalis, pars intermedia and pars distalis) can be distinguished. Cords of cells in the anterior pituitary can be seen by the second week after hatching (d). APit – anterior pituitary; Arc* – arcuate area primordium; BSph – basisphenoid bone; IRe – infundibular recess; LTer – lamina terminalis; Oral – oral cavity; POA* – preoptic area primordium; PPit – posterior pituitary; Rathke – Rathke's pouch; Spt* – septal primordium; Tub* – tuberal area primordium; V3V – ventral third ventricle.

Leading up to the critical event of hatching, the nuclei of the lateral zone (e.g. LH; lateral preoptic area) become more defined, and components of the medial zone (e.g. anterior hypothalamic nucleus, dorsomedial hypothalamic nucleus, ventromedial hypothalamic nucleus) begin to settle in the mantle zone. Cytoarchitectural characteristics of nuclei in the lateral hypothalamic zone, e.g. the large neurons of the magnocellular nucleus of the lateral hypothalamus, also emerge in the last few days before hatching. However, the hypothalamus is still very structurally immature at hatching. Most importantly, the newly hatched monotreme does not have the full complement of hypothalamic neurons; in particular, some medial hypothalamic zone and most (if not all) periventricular zone nuclei (e.g. paraventricular and arcuate) are yet to be generated from the hypothalamic neuroepithelium. Just as significantly, several key fibre tracts (e.g. the fornix and the hypothalamo-neurohypophyseal tract) are not present in the peri-hatching period (Ashwell 2012e). This indicates that the newly hatched monotreme probably relies on the lateral hypothalamus to cope with the challenges of post-hatching life, e.g. controlling key homeostatic processes like lung ventilation, cardiovascular regulation and gastrointestinal function. In therians, the lateral hypothalamus is known to function as a feeding regulatory centre (Saper 2012; Simerly 2004) and has connections with cardiorespiratory centres in the brainstem, so the early differentiation of the lateral hypothalamic zone in newly hatched monotremes is entirely consistent with a potential role in early regulation of feeding and rudimentary cardiorespiratory control. Monotremes share the mode and tempo of hypothalamic development with marsupials, so sophisticated control of respiratory function in response to fluctuations in environmental O_2 and CO_2 is, as in marsupials, unlikely to be available until after the end of the first week (Baudinette et al. 1988).

Neurogenesis of hypothalamic neurons continues into the first few weeks after hatching. For example, the paraventricular nucleus, which plays a critical role in many magnocellular and parvicellular neurosecretory pathways, does not emerge until the first week after hatching, and periventricular hypothalamic components are not visible until week two. Differentiation of the mammillary and retromammillary regions also continues into the second week after hatching. Judging by the slow pace of post-hatching maturation of the nuclei of the medial and periventricular hypothalamus, preoptic and paraventricular nuclei, which are all involved in mediating adaptive responses to stressful environmental stimuli (Simerly 2004), stress-response systems are unlikely to be functionally active in the young monotreme until several weeks after birth. The cytoarchitecture of the monotreme hypothalamus is essentially mature by the fifth week after hatching.

Monotreme autonomic nervous system

This is a vast and unexplored area of monotreme neurobiology. Certainly, the major elements of the autonomic nervous system (e.g. sympathetic trunk, splanchnic nerves, enteric nervous system) are identifiable in both platypus and echidna, but very little is known of how the detailed structure and function of the autonomic nervous system in monotremes compares with those of therians.

One area of particular interest in the monotreme autonomic nervous system concerns the control of the sensory apparatus of the bill. There is no doubt that the autonomic control of the platypus bill is sophisticated, allowing opening of the ducts of the mucous sensory glands (putative electroreceptors) when the bill is immersed in water (see Chapter 9 for details), but nothing is known about the central control of this reflex. The response could be mediated purely at the level of the brainstem reticular formation, through a pathway that involves thermal information from the bill skin entering the caudalis part of the nucleus of the spinal trigeminal tract (Sp5C) and being relayed through parvi- and magnocellular reticular formation nuclei to autonomic centres in the medulla and thoracic spinal cord. Alternatively, the reflex could involve the hypothalamus in addition to the brainstem, as part of a more integrated response to diving.

Monotremes have distinctive cardiorespiratory responses to some of their behaviour (e.g. diving in the platypus, burrowing in both platypus and short-beaked echidna, and hibernation in the short-beaked echidna; Augee et al. 1971a; Bethge et al. 2003; Frappell 2003; Nicol and Andersen 2003). Although the phenomenology of these behaviours has been studied in some detail, nothing is known about their central regulation.

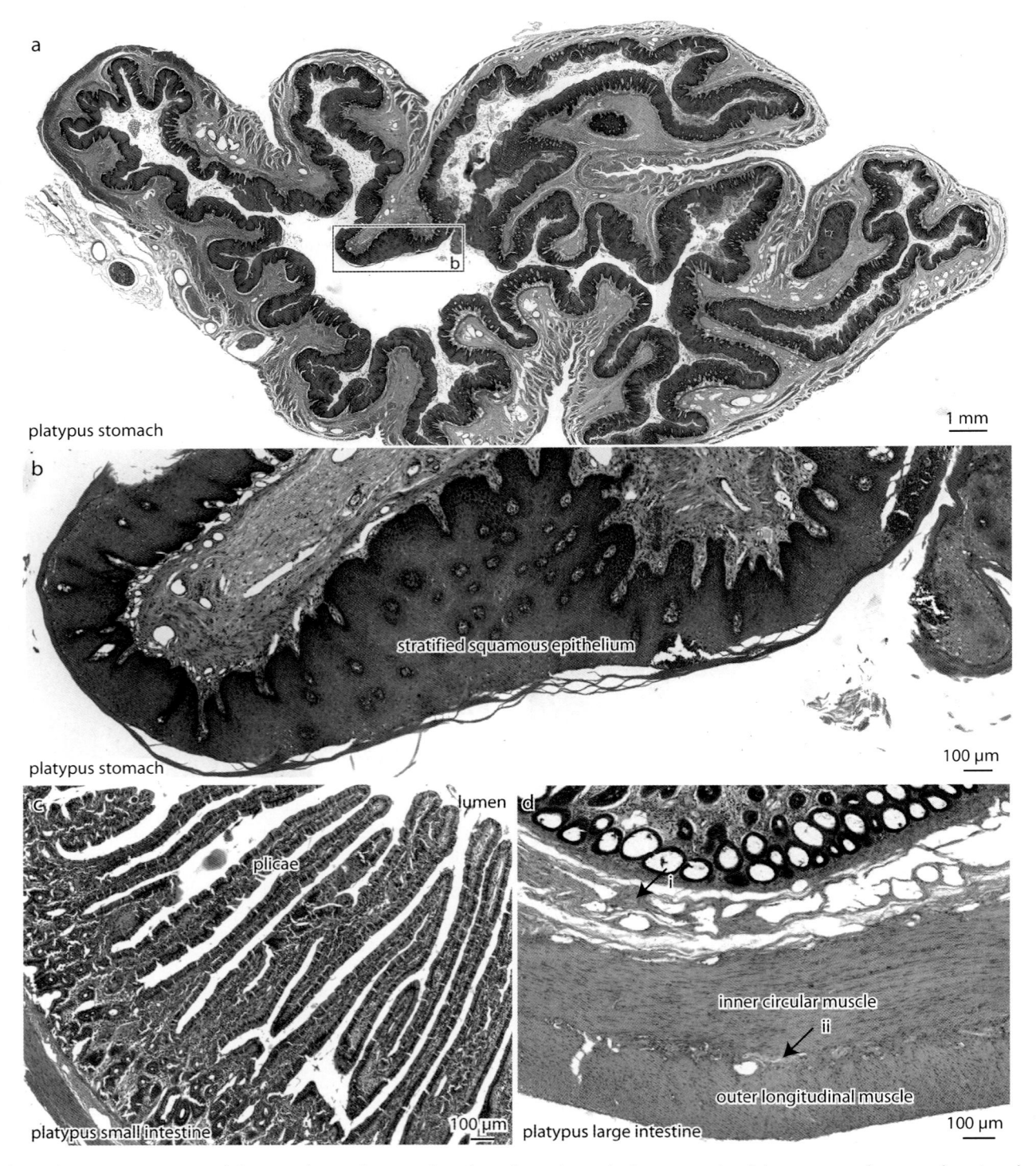

Figure 12.3: Images (a) and (b) are photomicrographs of sections through the stomach of the mature platypus showing the stratified squamous epithelium at low (a) and higher power (b). The platypus small intestine has plicae, but no villi (c). Components of the enteric nervous system are concentrated in the submucosa (submucosal or Meissner's plexus, arrow i in d) and between the circular and longitudinal muscle layers of the external muscular layer (Auerbach's plexus, arrow ii in d). Note that the external muscular layer of the platypus small intestine (lower left of c) is much thinner than the external muscular layer of the large intestine (bottom of d).

The monotreme enteric nervous system

Neurons of the enteric nervous system are embedded within the wall of the gut and comprise a significantly large population given that the gut of the echidna reaches up to 4 m in length (Griffiths 1968). These enteric neurons control smooth muscle, glands, absorptive cells, and blood vessels in the gut. The myenteric ganglion cells are primarily concerned with control of gut peristalsis through the external muscular layer of the gut wall, whereas the submucosal ganglion cells are involved in the control of the submucosal and mucosal glands, the movements of the muscularis mucosa and the regulation of water and ion transport across the gut epithelium. Study of the enteric nervous system of the monotremes has been limited, but two papers (Yamada and Krause 1983; Keast 1993) have examined the innervation of the gut and the distribution in the gut wall of neuroactive peptides and catecholamines.

Before considering the innervation of the monotreme gut, there are several features of the gut anatomy of the platypus and echidna that need to be considered. First, both species are unusual among mammals in that the stomachs of adults have no glandular cells, but are instead lined by stratified squamous epithelium (Krause and Leeson 1974; Griffiths 1978). This gives the stomach a histological appearance much like the oesophagus (Fig. 12.3a, b). Given that there are no glands, whether fundic, cardiac or pyloric, it is not surprising that the stomach contents of the short-beaked echidna are at neutral pH (pH of 7.3 between meals and never dropping below 6.2 even with ingestion of termites; Griffiths 1968). This may be a general feature of insectivorous mammals since some insectivorous placental mammals also have cornified, stratified, squamous epithelium in their stomachs (Griffiths 1968). It should be noted, however, that the stomach of the juvenile (i.e. lactational phase) echidna has a simple columnar epithelium that persists until the young reaches 850 g or 25.5 cm in body length (Griffiths 1968).

Second, the small and large intestine of the platypus have folds (plicae circulares), but no finger-like projections (true villi), even though the mucosa is quite thick and has a highly folded surface (Fig. 12.3c, d; Krause 1971). The intestinal epithelium at all levels of the bowel has a discontinuous layer of columnar cells sitting on an unusually thickened (up to 35 μm) basement membrane (Atkins and Krause 1971). The small intestine of the echidna, on the other hand, does have true villi.

Third, the transit time of the platypus gut may be much faster than for the echidna (Griffiths 1978; Keast 1993). As Keast has observed, a faster transit time along the gut tube would require a more rapid and efficient transportation of ions, water and nutrients across the mucosa, and cause greater epithelial damage due to the need to shift solid faeces rapidly. The greater damage to the epithelium might require more trophic support to maintain intestinal epithelial replenishment.

Innervation of the gut mucosa in both the platypus and echidna is mainly by peptidergic (predominantly substance P, but also neuropeptide Y, gastrin-releasing peptide, enkephalin, galanin, calcitonin gene-related peptide – CGRP, and vasoactive intestinal peptide – VIP) and tyrosine hydroxylase (TH) immunoreactive axons (Yamada and Krause 1983; Keast 1993; Fig. 12.4). In the echidna, the stomach is more densely innervated than the oesophagus; in the platypus, the oesophagus is more densely innervated than the stomach. The echidna has a dense nerve supply to the glands and villi of its intestine; but in the intestines of the platypus where there are no villi, dense bundles of axons run between the bases of the intestinal glands, divide into finer bundles and eventually single axons near the gland openings before penetrating the thick basement membrane to supply the mucosa (Keast 1993).

Peptidergic and TH immunoreactive (catecholaminergic) nerve terminals have been found in the submucous plexus of both platypus and echidna. Cell bodies containing substance P, CGRP, galanin and neuropetide Y are present in the submucous ganglia, with substance P being the main type (Keast 1993). In both species, the submucous ganglion cells are embedded within two interconnected plexuses: an outer plexus close to the inner circular layer of the muscular zone and an inner plexus closer to the muscularis mucosae. The high concentrations of substance P, motilin and serotonin in the submucosa of the platypus lend support to the argument that the smooth muscle beneath the mucosal folds produces movement of the epithelium to contact the gut contents (Yamada and Krause 1983).

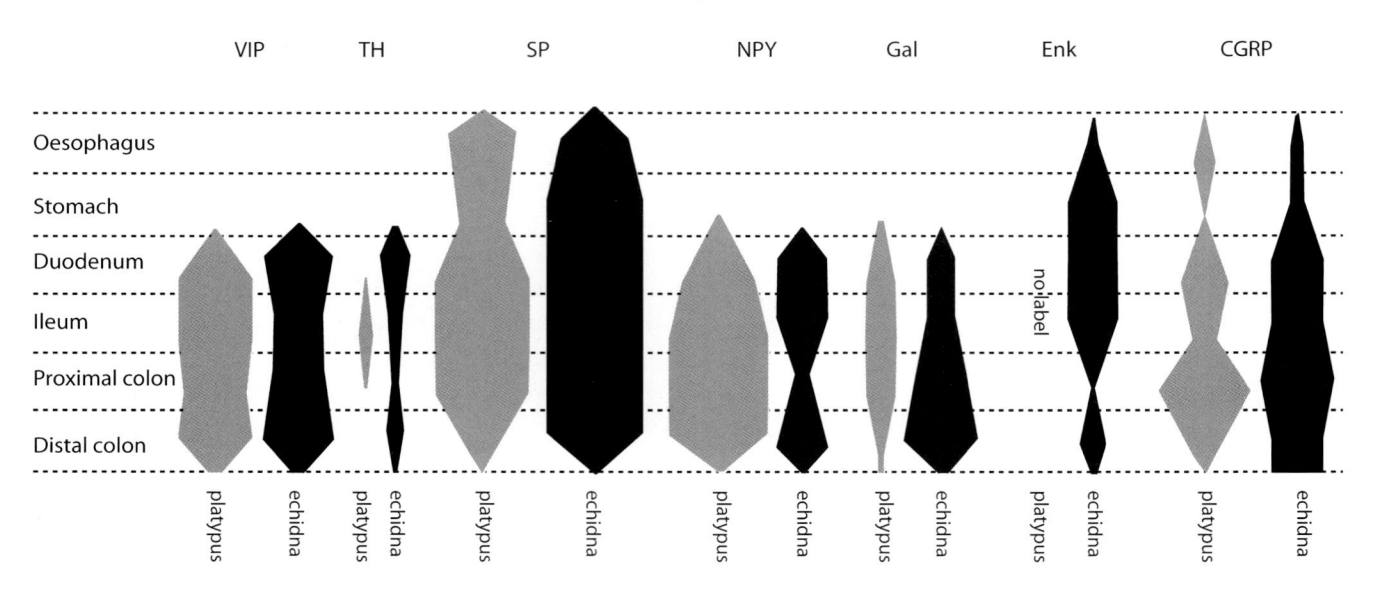

a) Neuropeptide and TH distribution in mucosa of monotreme gut

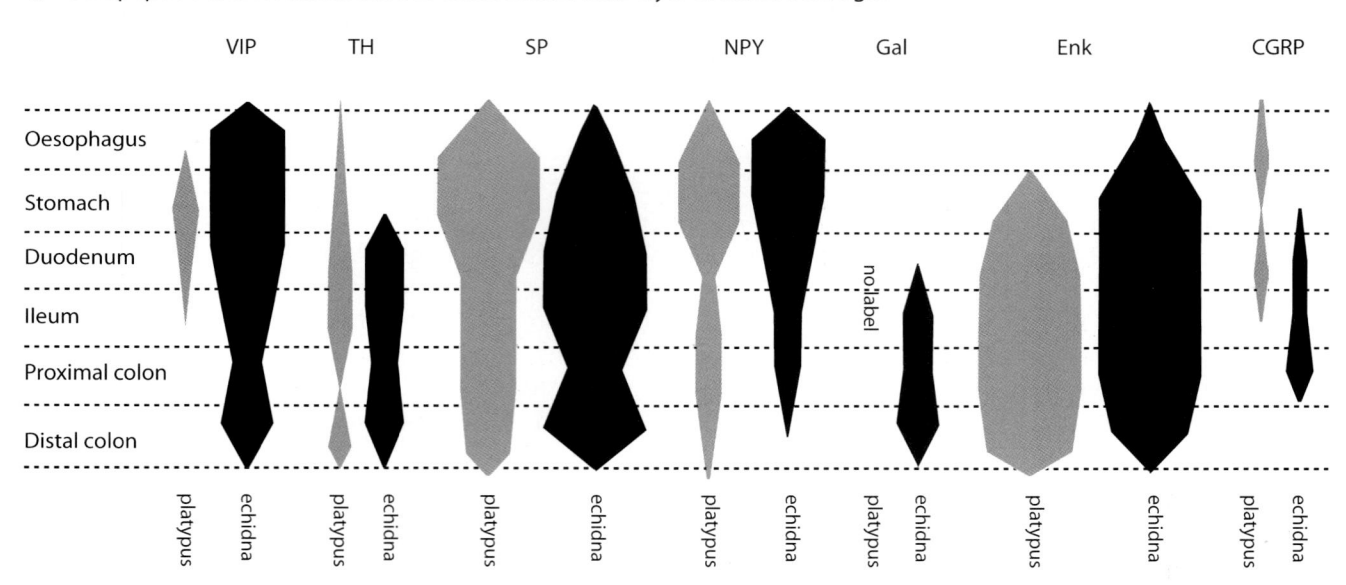

Figure 12.4: Graphs showing the distribution of neuropeptides and tyrosine hydroxylase (TH) in (a) the axons in the mucosa of the monotreme gut, and (b) the axons of the external muscular layer of the monotreme gut. This illustration is based on data published in Keast (1993). The platypus distribution pattern is shown in a series of grey polygons and that of the short-beaked echidna is shown in black polygons, with the width of each polygon at any point along the gut denoting the relative number of labelled axons. CGRP – calcitonin gene-related peptide; Enk – enkephalin; Gal – galanin; NPY – neuropeptide Y; SP – substance P; VIP – vasoactive intestinal peptide.

Enkephalin, neuropeptide Y and substance P are the main peptidergic axons supplying the external muscular layer in both species, but VIP is prevalent in the echidna. The myenteric ganglia embedded within the external muscular layer contain substance P and enkephalin and there are many axons

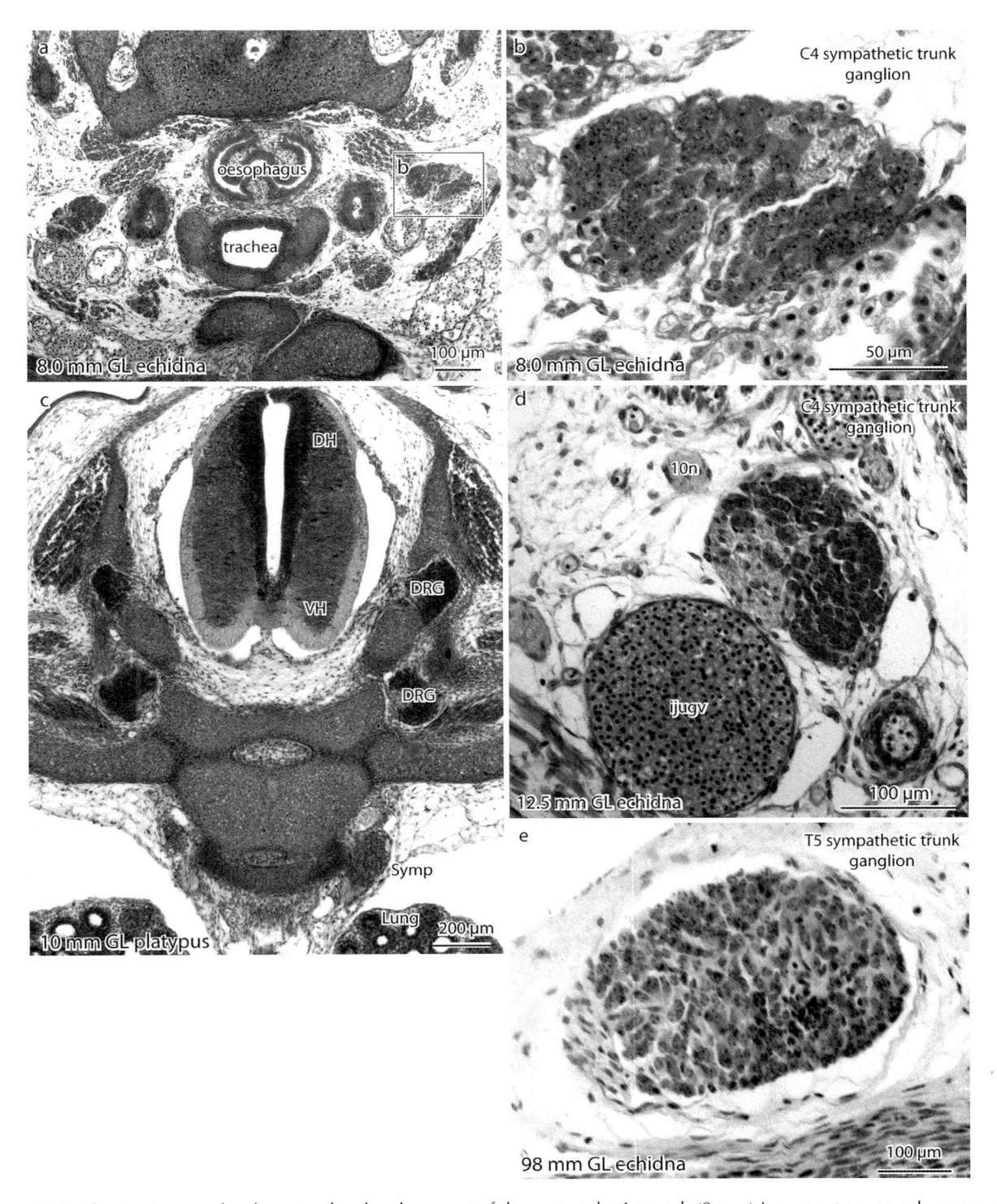

Figure 12.5: Photomicrographs showing the development of the sympathetic trunk (Symp) in monotreme embryos and hatchlings. Sympathetic trunk ganglion cells migrate into position before the middle of the incubation phase in both the short-beaked echidna (a, b) and platypus (c). Inset in (a) is shown enlarged as (b). By the last third of incubation the sympathetic trunk ganglia are connected by nerve fibres (d). By the second month after hatching (e), the sympathetic ganglia are cytologically mature. 10n – vagus nerve; DH – dorsal horn; DRG – dorsal root ganglia; ijugv – internal jugular vein; VH – ventral horn.

immunoreactive for a range of peptides and catecholamines. Keast (1993) noticed some possibly significant regional differences in peptide distribution. For example, in the echidna, enkephalinergic axons are found only in the stomach and NPY only in the oesophagus.

Overall, the range of neuroactive peptides in the monotreme gut is similar to that in placentals (Keast *et al.* 1985) and Keast observed several key similarities to therian enteric nervous system: (1) the prominence of substance P, VIP and NPY in nerve terminals of the monotreme gut; (2) the abundance of enkephalin and substance P neurons in the external muscular layer; and (3) that many vascular catecholaminergic (either dopaminergic or noradrenergic) axons also contain neuropeptide Y.

Nevertheless, Keast (1993) also noted some peculiarities of monotreme gut innervation. First, the neurotransmitters cholecystokinin and somatostatin are apparently absent from the monotreme enteric nervous system, although the antibodies used by Keast may not have recognised these if monotremes use a peculiar variant, because Yamada and Krause (1983) did find some somatostatinergic neurons. Second, the presence of enkephalinergic axons in the mucosa of the echidna is unusual for mammals. In therians, opioid peptides like enkephalin usually enhance the intestinal absorption of ions and water, but are commonly found only in the submucosa. The presence of opioids in the mucosa of the short-beaked echidna suggests that there may be direct action of opioids on the intestinal epithelium in this species. Third, there are very few CGRP, galanin, neuropeptide Y and VIP axons in the external muscular layer of the monotremes (particularly the platypus) (Keast 1993), even though they are present in the cell bodies of the myenteric ganglia. Keast suggested that the neurons containing these substances might be interneurons serving local circuits, rather than motor neurons driving the muscle itself.

The more rapid transit of gut contents through the intestines of the platypus may be facilitated by the dense nerve supply to the subepithelial region in that species. Keast noted that innervation of the intestinal mucosa in the platypus is concentrated at its superficial layer, with varicose axons running immediately beneath the thick basement membrane before penetrating through to the lumen.

Nevertheless, it is unknown whether this dense innervation is sensory, trophic or secretomotor in function (Keast 1993).

Development of the monotreme autonomic and enteric nervous system

The autonomic nervous system is derived from neural crest cells that migrate from the rim of the folding neural tube into the body cavity wall and developing gut tube. In both the platypus and echidna, this migration is a relatively early event, as demonstrated by the presence of autonomic ganglion cells in the dorsal body wall as early as the middle third of gestation (Fig. 12.5a, b). In both species, the sympathetic trunk is fully formed before hatching and axon bundles join the individual ganglia (Fig. 12.5c, d). The carotid body (a structure that detects oxygen concentration in arterial blood) develops from the second pharyngeal pouch and is identifiable at stage 40 in monotremes (an embryo ~7 mm GL). It reaches its final position well before hatching (Griffiths 1968).

Questions for the future

There is a huge disconnection between our knowledge of the behavioural phenomenology and metabolism of the monotremes and our understanding of the physiological mechanisms that underlie that behaviour and metabolism. Some of the key questions that remain include:

- How does the regulation of reproductive cycles in monotremes compare with that in therians? Although similar hormones appear to be involved, nothing is currently known about the specifics of central regulation of reproductive cycles in female monotremes or the details of the physiological interaction between hibernation and reproduction.

- How are the daily and seasonal cycles of the monotremes controlled by the hypothalamus and pineal gland? Are these species sensitive to the rate of change in photoperiod, some other seasonal clue, or endogenous signals from the body fat stores?

- How is metabolism controlled by the hypothalamus and autonomic nervous system during echidna hibernation? Does the central control of hibernation in echidnas differ from hibernating therians?

- What structural and functional differences in the hypothalamus and brainstem underlie the very different responses of the platypus and long-beaked echidna to changes in ambient environmental temperature?

- How does the innervation of the monotreme gut serve the specific needs of the two living groups?

- Why does the platypus have no true intestinal villi, but the echidna does, and how did these specific features of the gut evolve in the two living families?

- What does the enteric nervous system of monotremes tell us about the innervation of the gut in the common ancestor of therians and prototherians?

13

Monotremes and the evolution of sleep

Ken W. S. Ashwell

Summary

Sleep is an active process that requires the engagement of brainstem, hypothalamic, thalamic and cortical neural systems for the full physiological expression of the different sleep states. In mammals and birds, sleep involves cycling between slow wave sleep and paradoxical or rapid eye movement (REM) sleep states. The sleep physiology of monotremes is especially important for understanding the evolution of REM sleep among mammals in particular and amniotes in general. The presence of therian-like REM sleep in monotremes would imply that REM sleep (or at least some variant of it) was present in the common ancestor for therians and prototherians. If some variant of REM sleep is present not just in birds and mammals but in all amniotes, then the emergence of some form of REM sleep may have predated the common ancestor of all extant amniotes. Nevertheless, the nature or even existence of paradoxical sleep among the monotremes remains controversial, with some reporting that the monotremes exhibit a sleep state that has the brainstem, but not forebrain, components of paradoxical or REM sleep; whereas other reports find that echidnas have entirely typical features of therian REM sleep.

What is sleep? What is wakefulness?

Sleep is an active process and is found in a wide variety of vertebrates from fish through to amphibians, reptiles, birds and mammals, although features of sleep differ between vertebrates. Sleep differs from coma in that the sleeper can be aroused from sleep, and sleep plays an important part in normal brain function, particularly in learning and memory. Nevertheless, sleep is still a poorly understood type of brain activity.

Sleep is composed of behavioural, physiological/ metabolic and electrophysiological (i.e.

electroencephalographic – EEG, electrooculographic – EOG, and electromyographic – EMG) elements and has the additional feature of showing homeostasis (i.e. sleep rebound after deprivation). Table 13.1 lists key features of sleep in each of these categories (McNamara *et al.* 2010) along with a correlation with reported features in monotremes (see below). Sleep is characterised in most mammals by reduced motor activity or a recumbent posture, a raised threshold to sensory stimulation and the inclusion of distinctive sleep states (i.e. rapid eye movement sleep – REM sleep).

Table 13.1. Criteria for the definition of sleep contrasted with features of monotreme sleep

Criteria group	Platypus	Short-beaked echidna
Behavioural • Typical body posture • Specific sleeping site • Behavioural rituals before sleep • Physical quiescence • Elevated threshold for arousal and reactivity • Rapid state reversibility • Circadian cycle of rest/activity cycles • Hibernation/torpor	*Behavioural* • Typical body postures (Siegel *et al.* 1998) • Sleeps in burrow (Burrell 1927) • Physical quiescence • Elevated threshold for arousal and reactivity (Burrell 1927; Siegel *et al.* 1998) • Rapid state reversibility • Circadian cycle (Siegel *et al.* 1999)	*Behavioural* • Typical body postures (Siegel *et al.* 1996; Nicol *et al.* 2000) • Sleeps in burrow or in defensive posture • Physical quiescence • Elevated threshold for arousal during slow wave sleep (Allison *et al.* 1972) • Rapid state reversibility • Circadian cycle (Siegel *et al.* 1996; Nicol *et al.* 2000) • Hibernation (Grigg *et al.* 1989; Nicol and Andersen 2002)
Physiological and metabolic • REM: unstable heart rate and breathing, unstable thermoregulation • Non-REM: reduced metabolic processes, reduced body temperature	*Physiological and metabolic* • REM: reduced R wave to R wave dispersion on ECG, difficult to rouse (Siegel *et al.* 1999) • Non-REM: beat-to-beat variation in R to R period on ECG (Siegel *et al.* 1999)	*Physiological and metabolic* • REM: cardiorespiratory variability (Nicol *et al.* 2000)
Electroencephalography • REM: low voltage fast waves (paradoxical or active sleep) • Non-REM: high voltage slow waves (quiet sleep)	*Electroencephalography* • REM: Moderate or high voltage 1 to 6 Hz waves (Siegel *et al.* 1999) • Non-REM: moderate or high voltage slow waves interspersed among low voltage fast waves	*Electroencephalography* • Single sleep state with cortical slow waves simultaneous with REM like brainstem state (Siegel *et al.* 1996) • Slow wave sleep transitioning to REM sleep with moderate voltage (Nicol *et al.* 2000)
Electromyography • Progressive loss of muscle tone from wake state through non-REM to REM	*Electromyography* • Twitching of neck and jaw muscles during REM sleep (Siegel *et al.* 1999) • Quiescent during non-REM sleep	*Electromyography* • Complete atonia during sleep (Siegel *et al.* 1996) • EMG increases during transition to slow wave sleep and drops during REM sleep (Nicol *et al.* 2000)
Electrooculography • REM: rapid eye movements • Non-REM: eye movements are slow rolling, or absent	*Electrooculography* • REM: rapid eye movements during more than 56% of sleep time (Siegel *et al.* 1999) • Non-REM: eye movements small or absent	*Electrooculography* • Eye movements during REM sleep (Nicol *et al.* 2000) • No eye movements during sleep (Siegel *et al.* 1996, 1998)
Homeostatic regulation (rebound after deprivation) • Enhancement of sleep time • Intensification of sleep process	*Homeostatic regulation (rebound after deprivation)* • No data	*Homeostatic regulation (rebound after deprivation)* • No rebound after deprivation (Allison *et al.* 1972)

Just as important as what characterises sleep is the question: what is wakefulness? Wakefulness is a state of active and deliberate interaction with the environment. The maintenance of waking behaviour requires that the neural gates in the dorsal thalamus (which control the flow of information to and from the cerebral cortex) remain open. The awake or alert brain must also be tuned to behaviourally important sensory stimuli and kept activated, and the neurotransmitter input to the cerebral cortex or pallium must be compatible with the processing and recording of information.

The adaptive significance of sleep remains an elusive question in modern neuroscience. There is a remarkable similarity in the physiology of sleep states in birds and mammals. Both groups of homeothermic amniotes exhibit features of non-REM sleep and REM sleep, even though other amniotes do not show REM sleep (see below). This has led to the suggestion that REM sleep has evolved independently in birds and mammals and may be linked with other features that they share: high resting metabolic rate, relatively constant internal body temperature, high encephalisation and advanced cognitive function, and robust circuitry for transfer of information within and/or between the cerebral hemispheres. These observations suggest that sleep may either be an emergent feature of advanced nervous systems, or serve an important role in the maintenance of those complex nervous systems (Tononi and Cirelli 2006; Vyazovskiy et al. 2009). Sleep may serve to conserve energy in animals during parts of the diurnal cycle when they are at a sensory disadvantage compared to predators or competing species, or when it is simply uneconomical or physically hazardous to be active and seeking food. Alternatively, sleep may serve an active process in maintenance of the forebrain circuitry of advanced homeotherms (see below).

The physiology of sleep

Phenomenology

Sleep is divided into two types based on the pattern of electrical activity in the cerebral cortex, and associated eye movements (Hobson and Pace-Schott 2003). Non-rapid eye movement sleep (non-REM sleep) usually occurs at sleep onset and includes several stages during which the waves of brain activity on the electroencephalogram (EEG) become progressively slower. The eventual state of non-REM sleep is slow wave sleep, when the EEG is dominated by slow waves of less than 4 cycles per second (delta waves). Muscle tone and cerebral blood flow are reduced, parasympathetic activity is increased, and heart rate and breathing are slow and steady. Slow wave sleep owes its electrical features to rhythmical activity of nerve cells in the dorsal thalamus that interact with the cerebral cortex to produce the slow rhythm. This probably involves cyclical activity in reciprocal connections between the cortex and the sensory and association nuclei of the dorsal thalamus.

Non-REM sleep is followed by REM sleep, also called paradoxical sleep, dream sleep, desynchronised sleep or active sleep (Siegel et al. 1998). REM sleep is named because of the presence of rapid eye movements during this stage. In humans, REM sleep is characterised by vivid dreams; the cortical EEG is of low voltage and is similar in appearance to the waking state. The hippocampal theta rhythm is also present in REM sleep (as in the awake state). Brain metabolic activity is very high in this state, but the homeostatic regulation of ventilation and thermoregulation is less effective, with the result that the core temperature of the body slowly drifts towards the ambient temperature. Muscle tone in most skeletal muscles is abolished. Penile erection also occurs, but is not necessarily linked with dream content.

Neural activity during sleep states

Non-REM sleep is actively induced by particular groups of neurons, including preoptic area nuclei that project to the hypothalamus and brainstem. These in turn control the thalamic mechanisms that synchronise neuronal populations throughout the brain and generate the regular, slow, delta wave activity on the EEG.

REM sleep involves a very different pattern of neuronal activity from non-REM sleep (Hobson and Pace-Schott 2003). This type of sleep is characterised by rapid bursts of activity in the brainstem and cortex, with a pattern of discharge that resembles the waking state more than non-REM sleep. In fact, activity in the cortex during REM sleep may even exceed that during the waking state. Nevertheless, sensory input to the cortex is reduced during REM sleep and awareness of the external environment is lost. The vivid experiences of the dream state that accompanies REM sleep (at least in humans, but probably also in other mammals) appear to be generated entirely from sensory information stored within sensory association cortex.

REM sleep is generated by neurons in the pontine reticular formation and two groups of reticular formation neurons have been identified on the basis of their activity during sleep (Hobson and Pace-Schott 2003). One type (REM sleep-on cell) is selectively active during REM sleep, whereas the other type (REM sleep-off cell) is selectively inactive during REM sleep. The REM sleep-on cells consist of assorted cholinergic, GABAergic and glutamatergic neurons in the pons. The REM sleep-off cells include noradrenergic cells in the locus coeruleus and serotonergic cells in the raphe nuclei. Other premotor cells in the brainstem reticular formation are active during both waking and REM sleep (REM-waking active cells) and are responsible for the twitching movements of REM sleep, but overall muscle tone is reduced in REM sleep, presumably to prevent injury that might be caused by sporadic output to motor circuits and the spinal cord during REM sleep.

Regions of the human brain that are specifically increased in activity during REM sleep include the pontine tegmentum, amygdala, parahippocampal gyrus and anterior cingulate gyrus. By contrast, the dorsolateral prefrontal cortex and posterior cingulate cortex are deactivated during REM sleep (see review in Hobson and Pace-Schott 2003). These observations are consistent with the notion that the cognitive features of dreams result from simultaneous activation of the limbic brain combined with deactivation of executive or 'planning' regions of the cortex that serve working memory and logic.

What is the function of sleep?

Sleep is undoubtedly essential and complete sleep deprivation can cause death even faster than food deprivation, but the precise role of the two main phases of sleep remains obscure. Sleep is thought to serve an anabolic and actively conservative function that is in some way linked with the complex nature of the advanced brains of homeotherms.

Sleep deprivation has serious consequences for function during the waking state. In rodents, prolonged sleep deprivation (up to 6 weeks) leads to a sequence of adverse effects, beginning with inability to maintain body weight (despite an initial over-eating), followed by inability to control body temperature, and ultimately immunodeficiency leading to death from overwhelming sepsis. This would suggest that sleep is critically important for hypothalamic control of caloric balance, thermoregulation and neuro-immune function (Rechtschaffen and Bergmann 2002; Rechtschaffen et al. 2002).

A non-fatal consequence of sleep deprivation is an impaired ability to retain learned tasks. Early experimental studies in rodents suggested that REM sleep plays an important role in the retention of procedural memory (i.e. learning of sensorimotor tasks). On the other hand, slow wave sleep appears to be more important for the retention of declarative memory, such as involves recall of specific events or facts. More recent conceptions emphasise that both REM and slow wave sleep are important (perhaps in different ways) for declarative, procedural and emotional memory (Hobson and Pace-Schott 2003).

In the evolutionary context, sleep may act to conserve energy and ensure survival, particularly when the ambient temperature is low and predators are active. This is supported by the simple observation that sleep in most diurnally active mammals and birds occurs either when they are at a sensory disadvantage that would impair food gathering and render them vulnerable to predators, and/or the environmental temperature is low. Sleeping with conspecifics may also provide safety from predators and conserve energy in social animals.

Sleep in the platypus

The only sleep study ever conducted in the platypus was by Siegel and colleagues (Siegel et al. 1999). Four captive platypuses from south-east Queensland were studied under laboratory conditions in a special purpose environment. Siegel and colleagues identified five sleep-waking states on the basis of electroencephalographic (EEG), electrooculographic (EOG) and electromyographic (EMG) recordings. These include: (1) a waking state; (2) quiet sleep with moderate voltage EEG; (3) quiet sleep with high voltage EEG; (4) REM sleep with moderate voltage EEG; and (5) REM sleep with high voltage EEG. Platypuses spend 40% of their time awake, 26% in quiet sleep and 34% in REM sleep. In the laboratory setting, platypuses are awake mainly during the last 4 to 5 hours of the light period and the first 7 or 8 hours of the dark period.

Quiet sleep occupies ~44% of total sleep time. Most of the time spent in quiet sleep is in quiet sleep with moderate voltage EEG (34% of total sleep time) compared to only 10% of sleep time spent in quiet sleep with high voltage EEG. Features of EEG, EOG and EMG during quiet state sleep in the platypus all resemble non-REM sleep in therians.

The platypus spends more time in REM sleep than any other mammal (Siegel et al. 1999): the REM sleep state occupies more than 56% of the total sleep time and can be identified on the basis of eye movements shown on the EOG and accompanying head and neck movements (mastication-like twitching of the bill and side-to-side movements of the head). More than half the sleep time is spent in REM sleep with moderate voltage EEG and a little less than 6% of sleep time is spent in REM sleep with high voltage EEG. REM sleep occurs while the platypuses are immobile and in a curled or prone sleeping position. Nevertheless, adult platypuses in REM sleep show a combination of eye, head and neck twitching that suggests phasic activity in the reticular formation of the brainstem. Platypuses are difficult to rouse during REM sleep, requiring pressure stimulus to the midline of the neck with more than 20 g von Frey hairs (and often up to 50 g) to be roused. Overall, REM sleep in the platypus closely resembles REM sleep in other mammals, but with one significant difference: the REM state occurs while the EEG is moderate to high in voltage, not low in voltage as in placentals. The most common pattern of sleep progression observed in the platypus is from (1) wakefulness, to (2) quiet sleep with moderate voltage EEG, to (3) REM sleep with moderate voltage EEG, before returning to (4) wakefulness. In contrast to adult placentals and marsupials, REM sleep in the platypus can actually begin directly from the waking state.

Behavioural observations of platypuses in the wild are also consistent with this species exhibiting true REM sleep. Burrell (1927) noted that when a platypus burrow was opened during daylight hours, the inhabitant was often difficult to rouse, similar to the difficulty of rousing therians from REM sleep (Siegel et al. 1998). Behavioural observations also suggest that there are developmental differences in muscle twitching behaviour during sleep, in that observations of juvenile platypus have noted swimming movements of the forepaws during sleep, suggestive of dreaming, whereas adults do not usually show this activity (see discussion in Siegel et al. 1998).

Sleep in the short-beaked echidna

Some controversy surrounds the question of whether the short-beaked echidna shows REM sleep. An early study by Allison and colleagues (Allison et al. 1972) found that the short-beaked echidna had a state of non-REM sleep characterised by a high voltage EEG, but that this species did not have a desynchronised EEG during REM sleep. They did report some 'REM sleep-like' states, but these were not associated with either rapid eye movements or heightened arousal thresholds. These observations led them to conclude that the short-beaked echidna does not have REM sleep, but it is probably significant that their subjects had been shipped to the USA and exposed to unrecorded ambient temperatures, and so may not represent sleep in the natural environment (Nicol et al. 2000).

Siegel and colleagues (Siegel et al. 1996) did not attempt to re-examine the conclusion by Allison and colleagues that the short-beaked echidna did not show a REM sleep state based on the absence of eye movements, because they argued that eye movements are an unreliable indicator of REM sleep in mammals (like the echidna) that have limited eye mobility anyway. Instead, they addressed the question of brainstem activity during sleep states in echidnas. They used indwelling electrodes in the pontine tegmentum and midbrain reticular formation to monitor brainstem activity during waking and sleeping. They concluded that, despite the reported absence of REM sleep in echidnas, this species shows increased variability in activity of the brainstem reticular formation (a feature of REM sleep) at a time in the sleep cycle when EEG cortical activity is synchronised (a feature of slow wave sleep). This juxtaposition of physiological features led them to conclude that the echidna combines REM and non-REM features in a single sleep state. They concluded that the echidna does not show the typical features of mammalian REM sleep, i.e. low voltage EEG, increased brainstem discharge rate and variability, and synchrony of firing of brainstem reticular neurons.

By contrast to the above studies, Nicol and colleagues (Berger *et al.* 1995; Nicol *et al.* 2000) found typical mammalian cycles of slow wave sleep and REM sleep in short-beaked echidnas. The active waking state (W) was characterised by low voltage desynchronised EEG, and phasic EMG and EOG traces. The quiet state (Q) was characterised by low voltage desynchronised EEG, low non-phasic EMG and quiescent EOG. The slow wave sleep state (SWS) featured a high voltage slow wave EEG, with periodic 6–8 Hz spindles that were not seen by Allison and colleagues (Allison *et al.* 1972). Nicol and colleagues also reported a REM sleep-like state in echidnas occupying ~15.5% of total sleep time and characterised by a desynchronised EEG with low voltage waves, a reduced EMG, rapid eye movements and occasional muscle twitches. Non-REM slow wave sleep always preceded REM sleep and most of the episodes of REM sleep that they observed were terminated by waking. Significantly, deviations of body temperature away from the thermoneutral zone around 25°C depressed all nocturnal measures of REM sleep, without affecting other states. Age also reduced the percentage of time spent in REM sleep, from 6.2% in juveniles to 2.4% in adults. There were no peaks of theta wave activity in any of the observed states, a feature that Nicol and colleagues likened to avian sleep.

Is sleep in the two monotremes really different from each other and therians?

What is it about monotreme sleep that is like therian sleep and what is different? Both monotremes have brainstem activation in conjunction with a sleep state with high voltage forebrain EEG activity, but there are significant differences between the two species. The platypus spends more time in REM sleep than any other mammal (more than 8 hours per day), whereas the question of whether the short-beaked echidna has true REM sleep remains controversial. Siegel and colleagues (Siegel *et al.* 1998) noted that although the platypus exhibits muscle twitching during its REM sleep state (even more than many therians), the echidna does not. This may reflect the very different environments that each species sleeps in: twitching during sleep would make an echidna sleeping on the surface of the ground vulnerable, because twitching would be both audible and visible to predators; whereas the platypus always sleeps safely within a burrow and can twitch with impunity. Siegel and colleagues speculate that this behavioural difference between the species may have a physiological basis: brainstem reticular formation neurons in the echidna may have decreased neighbour-to-neighbour coupling when they enter the 'bursty' firing pattern of REM-like sleep, thereby reducing the vigour of each muscle twitch. This explanation ignores the points that: (1) the echidna's natural defences are equally effective whether awake or asleep; (2) slight movement of spines during sleep would be a trivial signal compared to the noise and motion generated by the echidna's natural daytime activity; and (3) echidnas also use burrows.

How do we reconcile the conflicting reports concerning sleep in the short-beaked echidna? As Nicol and colleagues have observed (Nicol *et al.* 2000), echidnas have exactly the same brainstem structures that mediate REM sleep in therians and the platypus (see below), so functional sleep differences between these species are inexplicable on anatomical grounds. Siegel and colleagues raised several questions about the methodology used by Nicol and colleagues in an abstract version of their study (Berger *et al.* 1995). Siegel *et al.* (1996) noted that Nicol and colleagues did not test arousal thresholds during the putative REM sleep state, did not use indwelling electrodes in the allocortex, and did not apply a sleep deprivation test to assess the REM sleep state. Siegel and colleagues also suggested that the EOG observation of rapid eye movements by Nicol and colleagues may be artefacts, in that the widely placed bilateral needle electrodes used by Berger *et al.* (1995) would have picked up movement artefacts in the waking state, rather than eye movements. In other words, the REM state detected by Nicol and colleagues may have been a quiescent and awake state.

On the other hand, Nicol and colleagues (Nicol *et al.* 2000) argue that the failure of Allison and colleagues to identify REM sleep is due to observations being conducted at low ambient temperatures, when REM sleep might have been depressed. They also argued that the stringent scoring criteria used by

Nicol and colleagues ensured that true paradoxical sleep could be clearly distinguished from the quiescent waking state.

The anatomy of sleep circuitry

Sleep is an active brain process, not simply the absence of waking, so activity in key neural circuits is central to generation of the different substages of sleep. Initiation and regulation of different sleep states requires the presence and activity of chemically distinct neuronal groups in the hypothalamus and brainstem (Fig. 13.1). Some of the latter are part of the generic reticular formation of the brainstem, including sensorimotor and modulatory neuronal groups. Some modulatory neurons within the reticular formation have specific chemical signatures, i.e. noradrenergic, serotonergic and cholinergic cell groups, and have highly divergent projections throughout the brain.

Hypothalamic neurons in several nuclei play a key role in the regulation of the sleep and waking (Hobson and Pace-Schott 2003). These include the ventrolateral preoptic area (VLPO), which may be the site of

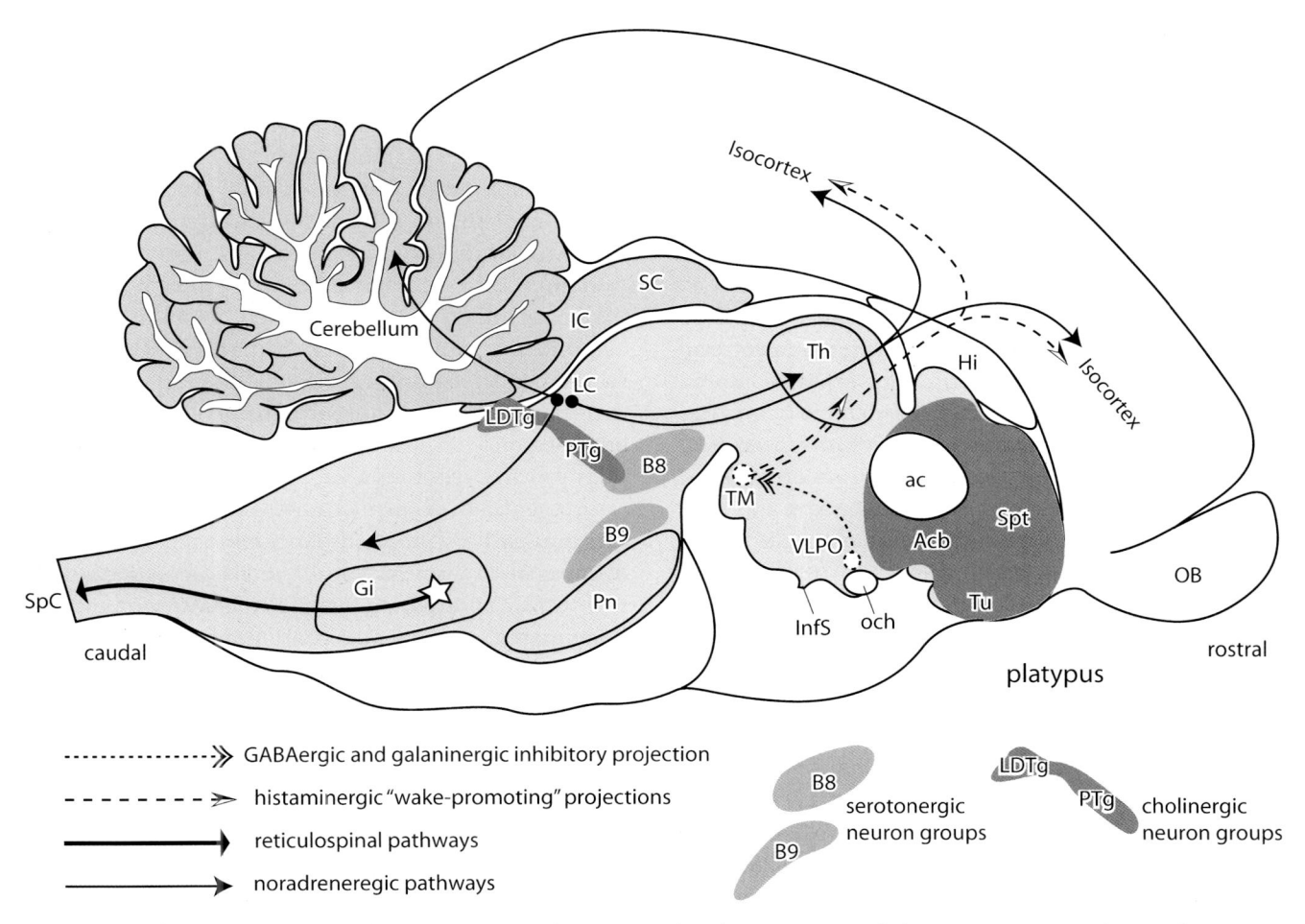

Figure 13.1: Diagram illustrating the major neuronal groups and pathways activated during non-REM and REM sleep, using a sagittal view of the platypus brain. Sleep depends on the integrated functioning of a wide variety of chemically distinct neuronal groups. The illustrated pathways represent hypotheses to be tested in future studies of monotreme neuroanatomy and sleep physiology. ac – anterior commissure; Acb – nucleus accumbens; Gi – gigantocellular reticular nucleus; Hi – hippocampus; IC – inferior colliculus; InfS – infundibular stalk; LC – locus coeruleus; LDTg – laterodorsal tegmental nucleus; OB – olfactory bulb; och – optic chiasm; Pn – pontine nuclei; PTg – peduncular tegmental nucleus; SC – superior colliculus; SpC – spinal cord; Spt – septum; Th – thalamus; TM – tuberomammillary nucleus; Tu – olfactory tubercle; VLPO – ventrolateral preoptic nucleus.

interaction of homeostatic and circadian regulatory processes of sleep, and its GABAergic and galaninergic inhibitory projection to the tuberomammillary nucleus of the caudal hypothalamus (Fig. 13.1). The tuberomammillary nucleus contains histaminergic 'wake-promoting' projections to the dorsal thalamus and cerebral cortex.

Sensorimotor neurons of the reticular formation are also critically important for the manifestation of sleep physiology. In humans they are about 50–75 μm in diameter (magnocellular and gigantocellular reticular formations) and fire continuously at rates of up to 50 Hz. They project to the spinal cord (reticulospinal pathways) with conduction velocities in excess of 100 m/s, consistent with a role in the adjustment of posture and motor control (Hobson and Pace-Schott 2003). Sensorimotor neurons are subject to exponential recruitment, particularly during directed action in the waking state and in the generation of REM sleep. Sensorimotor neurons principally use the excitatory transmitter glutamate or the inhibitory neurotransmitter GABA (Hobson and Pace-Schott 2003).

By contrast to sensorimotor neurons, modulatory neurons of the parvicellular reticular formation are relatively small (10–25 μm in diameter) with slowly conducting (1 m/s) axons. They exhibit a regular firing pattern, consistent with a pacemaker role. The feedback inhibition and pacemaker potentials of modulatory neurons are believed to allow the production of highly synchronised output during waking and sleeping. The contrasting properties of sensorimotor and modulatory neurons of the reticular formation may be the physical basis of the reciprocal interaction that causes non-REM/REM sleep cycles (Hobson and Pace-Schott 2003).

Aminergic modulatory neuron groups in the brainstem (noradrenergic neurons of the locus coeruleus and serotonergic neurons of the pontine raphe nuclei) contain pacemaker neurons that are spontaneously active during waking. The major noradrenergic cell group in the brainstem is the locus coeruleus (Fig. 13.1), lying in the rostral pontine reticular formation and central grey matter. One subset of locus coeruleus neurons projects to sensory regions of the brainstem and spinal cord, while the other set projects widely to cerebellar cortex, dorsal thalamus and cerebral cortex. This pattern of innervation is consistent with the role of the locus coeruleus in regulation of the processing of sensory information and the integration of sensory input at the cortical level (Hobson and Pace-Schott 2003).

The other group of aminergic modulatory neurons involved in waking and sleep (serotonergic neurons of the midbrain raphe nuclei, i.e. dorsal raphe and median raphe – B8 and B9 groups; Fig. 13.1; Hobson and Pace-Schott 2003) projects rostrally, innervating nearly the whole of the forebrain, also consistent with a role in the regulation of sleep and waking.

The final set of chemically distinctive modulatory neurons in the brainstem reticular formation is the cholinergic cell groups. These are located in particular non-motor nuclei, i.e. the laterodorsal tegmental and pedunculopontine nuclei (Fig. 13.1) that project to brainstem reticular formation, dorsal thalamus, hypothalamus and basal forebrain (see Hobson and Pace-Schott 2003 for discussion). In therians, the last projection in turn activates further cholinergic connections to the limbic forebrain and isocortex.

Does the anatomy of the monotreme brainstem show any peculiarities that relate to sleep function?

Aminergic and cholinergic neuronal groups have been extensively studied in the brains of both the platypus and short-beaked echidna, but these data are confined to assessment of the distribution of the cell bodies of chemically labelled neurons and the major sites of termination of labelled axons (Manger et al. 2002a, b, c). In other words, relatively little is known about the detailed projections of fine axons of aminergic and cholinergic neurons.

That said, there is nothing particularly striking about the pattern of distribution of chemically labelled neurons in the brainstems of either the platypus or short-beaked echidna (see Chapter 5 for more details) such as would explain the observed phenomenological differences in sleep. Siegel and colleagues (Siegel et al. 1998) have observed that there are two significant questions concerning the aminergic and cholinergic cell groups that remain unanswered. The first of these is whether there is a significant ascending projection from these cells, and the second is whether the ascending projection targets the same forebrain regions as in therians. To explain the observed differences in sleep physiology between monotremes and therians (as Siegel and colleagues see it) they have proposed that in monotremes 'some critical brainstem cell group, required for the generation of the REM sleep EEG pattern and normally active during REM sleep, is either inactive or does not have the projections required for inducing EEG desynchrony'. Clearly this question warrants further experimental attention.

Evolutionary implications

The question of whether echidnas exhibit true REM sleep is critically important to ideas about the phylogeny of mammalian sleep, so the uncertainty of this academic debate is particularly frustrating. If short-beaked echidnas manifest 'true' (i.e. therian-like) REM sleep, then this type of sleep may have been present in the most recent common ancestor of all extant mammals.

Conversely, if the sleep physiology of the short-beaked echidna is not functionally identical to therian sleep, then REM sleep either arose very early in the common ancestor of all modern mammals, with subsequent loss or significant modification in the short-beaked echidna, or evolved independently in therian and ornithorhynchid lineages, but never developed fully in tachyglossids.

Whatever the outcome of further investigation of the above questions, Siegel and colleagues reached several conclusions about the evolution of sleep on the basis of their observations in platypus and echidna. They suggested that their observations that brainstem activation coincides with high voltage forebrain EEG activity (i.e. the supposed combination of more than one sleep state) support the contention that REM sleep originally evolved because of its benefits to brainstem function. They argue that forebrain aspects of REM sleep (including dreaming) are a recent development, presumably confined to the therian lineage, but readers should note that this contradicts comments by the same authors that twitching of juvenile and adult platypus may indicate dreaming (Siegel et al. 1998).

Siegel and colleagues argue that the presence of a brainstem activated, but cortex synchronised, sleep state in both the platypus and echidna suggests that that particular aspect of REM sleep is very ancient and may even predate the origin of mammals (Siegel 1995; Siegel et al. 1998). In this line of argument, they propose that the low voltage EEG of REM sleep evolved some time after the brainstem aspects of REM sleep (Siegel et al. 1998). They also argue that the most parsimonious explanation of the presence of some form of REM sleep in mammals and birds is that the common amniote ancestor of both mammals and birds had REM sleep (or some precursor state), perhaps as far back as 300–350 million years ago. However, observations on sleep states in reptiles raise questions about this hypothesis. Reports of REM sleep in reptiles are contentious and most studies do not find any strong evidence of a REM sleep-like state in reptiles (Thakkar and Datta 2010). Reports of REM sleep in reptiles may actually be detecting brief arousals or a preparatory state that normally precedes arousal. Without firm evidence of true REM state sleep in other amniotes, it is most likely that the emergence of REM sleep in birds and mammals is an example of convergent evolution in their respective advanced nervous systems.

A recent study in a paleognathic bird (Lesku et al. 2011) has suggested that the REM sleep in that group is physiologically similar to REM sleep in the platypus. Lesku and colleagues found that ostriches have a heterogeneous REM sleep state with eye closure, rapid eye movements, and reduced muscle tone, all occurring together with forebrain EEG activity that flips between REM sleep-like activation and a slow wave state. Ostriches also have prolonged periods of REM sleep, more than any other bird, just as the extended REM sleep periods of the platypus occupy far more of the daily cycle than any other mammal. Lesku and colleagues suggested that this similarity reflects the basal position of the ostrich and platypus within their respective clades. They propose that the type of REM sleep seen in both platypus and ostrich represents an early stage in the evolution of REM sleep.

Questions for the future

Some of the controversy surrounding the nature and physiology of sleep states in the monotremes arises from technical differences in the assessment of physiological phenomena during the putative REM sleep state. More advanced techniques for measuring and logging eye and head movements simultaneously with EEG, EMG and temperature monitoring in naturally behaving animals may help to resolve the contentious findings reviewed above. Further studies in monotremes using sleep deprivation (which causes REM rebound in placentals) and prolactin injection (which increases sleep in placentals) would also help to clarify how similar or different the monotreme REM state is to that of placentals.

If the REM sleep state of monotremes does turn out to be quite distinctive physiologically, then the natural question is to ask how idiosyncratic physiology is produced by the apparently unremarkable internal organisation of the aminergic and cholinergic cell groups in the monotreme brainstem. As noted earlier, the apparent similarities between therians and monotremes in brainstem chemoarchitecture may belie peculiarities of ascending aminergic and cholinergic pathways. Assessment of the detailed anatomy of ascending aminergic and cholinergic projections using advanced imaging techniques like diffusion tensor imaging may help to address this lacuna in current knowledge.

14

Reflections: monotreme neurobiology in context

Ken W. S. Ashwell

Summary

The monotremes have had a long history of evolution that makes them one of the longest surviving mammalian clades. The earliest monotremes probably possessed a suite of electro- and mechanoreceptors in their snout that allowed them to correlate electrical and mechanical inputs in the exploration of their environment and the detection of prey, as well as the developmental potential to expand the pallium into the sophisticated isocortex of modern monotremes. Although echidnas are often seen as relatively recent 'platypus offshoots', a parsimonious interpretation of the key features of monotreme neurobiology and neuroembryology is more consistent with tachyglossids and ornithorhynchids having followed a prolonged period of independent evolution. Monotremes have been used in earlier neurobiological experiments as supposedly 'basal' mammals to test hypotheses on mammalian brain evolution, but this ignores the clearly advanced features of their sensory systems and isocortex.

Monotreme origins and neurobiology

The first Mesozoic monotremes are the early Cretaceous *Steropodon galmani* (Archer *et al.* 1985) from Lightning Ridge dated to ~110 mya and the slightly older (115 mya) *Teinolophos trusleri* from coastal Victoria (Rich *et al.* 2001b; Rowe *et al.* 2008). The traditional view has been that the monotremes had an evolutionary origin quite distinct from therians and this is supported by molecular studies (see discussion in Musser 2003 and Chapter 1 of this book; Kullberg *et al.* 2008).

The remains of the early (i.e. Cretaceous) monotremes are far too fragmentary to allow many deductions concerning their paleoneurology (see Chapter 1). Rowe and colleagues have maintained that the large mandibular canal of *Teinolophos trusleri* is indicative of a platypus-like lifestyle (Rowe *et al.* 2008), perhaps with the sensitivity of the (lower) bill that a large mandibular canal would imply. The presence of sensory gland receptors and push-rod mechanoreceptors in the bills and beaks of all living monotremes strongly suggests that the earliest monotremes possessed these unique sensory structures, but it is impossible to deduce the level of specialisation of central trigeminal somatosensory nuclei and pallial regions from the available fossils. However, it is likely that the earliest monotremes had sophisticated developmental mechanisms (the subcortical developmental zone discussed in Chapter 7 and below) that allowed the expansion of the isocortex to serve the processing demands of the trigeminal input, because both the platypus and short-beaked echidna have this region in their developing forebrain.

Monotreme phylogeny: perspectives from neurobiology

The two groups of living monotremes are clearly quite different, both morphologically and behaviourally, even though they share the characteristic monotreme features of reproductive anatomy and physiology, oviparity and low body temperature. Questions that often arise in the monotreme literature concern the nature of the original or ancestral monotreme and the phylogeny of the modern monotremes. Was the ancestral monotreme like either the modern platypus or the modern echidna, or was it rather different from both of them? If the ancestral platypus were 'platypus-like', are the echidnas a relatively recent offshoot of this lineage? These questions are often couched in the context of electroreception (Pettigrew 1999), since that unusual sense has been a major focus for recent neuroscientific study of the monotremes, but we should remember that electroreception is only one of a suite of sensory abilities that the monotremes use in the exploration of their environment. Naturally these questions depend on interpretation of the decidedly fragmentary fossil record of monotremes, but in recent years genomic analysis has contributed to the debate. What can a detailed consideration of the neurobiology of the living monotremes throw on these questions? In this section we will examine the arguments surrounding monotreme phylogeny with particular reference to the often-stated assertion that the modern echidnas are relatively recent descendants of a platypus-like ancestor (making Ornithorhynchidae paraphyletic), what we may call the 'platypus-first' hypothesis. In neurological terms this hypothesis posits that the ancestral monotreme was a trigeminally specialised, electroreceptive aquatic mammal.

Are echidnas recent offshoots from the main monotreme line? The case for

The argument for this proposition is often voiced in the context of electroreception and (more recently) genetic analyses. The argument is mainly based on interpretation of: (1) the available fossil evidence; (2) the suite of physical features of modern monotremes; (3) the (perceived) regression of electroreceptive ability in the echidnas; and (4) protein/genomic analysis.

The oldest known monotreme is the early Cretaceous *Teinolophos trusleri* (121 to 112.5 mya) (Musser 2003; see Chapter 1). Early analysis considered that *Teinolophos* was basal to crown (living) monotremes (Kielen-Jaworowska *et al.* 2004), but Rowe and colleagues have argued that it is aligned with the platypuses (Rowe *et al.* 2008). Phillips and colleagues argue that modern platypus and echidna are derived with respect to the *Teinolophos* and *Steropodon* morphotype, based on a more than fivefold increase in body size and a shift to relatively weak bite force. The crown position of Early Cretaceous monotremes implies extraordinarily slow molecular evolution within living monotremes (Rowe *et al.* 2008), but Phillips and colleagues have disputed the interpretation of Rowe *et al.* because (they say) Rowe's hypothesis would require rapid molecular evolution in stem monotremes and slow molecular evolution in crown monotremes (Phillips *et al.* 2009).

No definitive echidna fossils have been found in deposits older than the middle Miocene (but note the possible tachyglossid humerus reported by Pridmore *et al.* 2005), whereas several platypus-like fossils from Miocene or older sediments (e.g. *Obdurodon dicksoni* – 20–15 mya, *Monotrematum sudamericum* – 62 mya) have features similar to the modern platypus and suggest a Cretaceous origin for the Ornithorhynchidae.

There is therefore a gap of as much as 50 million years in the first appearance of the two groups (Musser 2003). If this gap is accepted at face value, it suggests that echidnas have appeared only recently, perhaps from an ornithorhynchid.

Several aspects of echidna biology have been suggested to indicate an origin of this group from an aquatic platypus-like ancestor that had aquadynamic streamlining, dorsally projecting hindlimbs acting as rudders, and locomotion based on a long-axis rotation of an hypertrophied humerus to provide an efficient swimming stroke (Phillips et al. 2009). Phillips and colleagues noted several features of the modern echidna that they felt were potential homologies with the aquatic platypus ancestor. These include the dorsoventral flattening of the body, reversed (i.e. posteriorly pointing) hindfoot posture, and 'front-wheel drive locomotion' based on the rotation of the long axis of the humerus. They argued that each of those traits would be highly anomalous if they were derived directly from a more generalised terrestrial basal mammal morphotype. They also contend that the presence in the embryonic echidna of a marginal cartilage, like that which provides the contour of the bill in the platypus, suggests that a bill (rather than a beak) was the ancestral snout structure in the earliest monotremes.

Other arguments for 'platypus-first' are based on the neurobiology of the monotremes. Pettigrew (1999) observed that although all living monotremes have some form of sensory glands in the snout or bill, there is considerable variation in the number of these putative electroreceptors in the different species. The platypus has ~40 000 mucous sensory glands and ~14 000 serous sensory glands on the bill, but the long-beaked echidna (*Zaglossus bruijnii*) is said to have only 2000 electroreceptors and the short-beaked echidna only a few hundred. Pettigrew interpreted this difference as evidence of the regression of electrosensation among the tachyglossids as their ancestors moved into progressively drier environments where electrosensory ability would be less useful in prey detection. This regression would have been driven in part by the progressive drying of the Australian mainland from the Miocene onwards (see below).

Molecular phylogeny data often cited in this debate are based on serological and amino acid sequence data (Hope et al. 1990), an early DNA hybridisation study (Westerman and Edwards 1992) and a more recent molecular clock study using seven nuclear genes and a suite of mitochondrial genes (Phillips et al. 2009). Haemoglobin sequence data indicate that the platypus and echidnas have had a long period of independent evolution (Hope et al. 1990). Similarly, Westerman and Edwards found that their DNA hybridisation data supported relatively recent divergence of the two genera of tachyglossids (mid to late Pleistocene) and a relatively ancient dichotomy between the platypus and echidna lineage. Westerman and Edwards were cautious about assigning an actual date to this divergence, noting the rarity of fossil monotremes and the lack of evidence validating the reliability of molecular clocks for monotremes. Nevertheless, they went on to assume that the molecular clock *is* regular and concluded that platypus/echidna split was at 64 mya.

In the most recent study, Phillips and colleagues derived molecular divergence dates using relaxed clocks and three alternative calibration schemes (Phillips et al. 2009). All their analyses led them to conclude that the divergence of platypus and echidna was relatively recent (mid-Tertiary; 18.5–47.8 mya by 95% highest posterior distribution, with a median estimate of 32.1 mya). By contrast, their data suggested that *Zaglossus* and *Tachyglossus* diverged at a median estimate of 5.5 mya (1.8–10.6 mya by 95% highest posterior distribution). They argued that data suggest that the base substitution rates for the monotremes are not significantly different from those for sauropsids and therian mammals. These findings should be viewed with caution, because there are few monotreme fossils to calibrate the molecular approach. As observed by Archibald (2011), molecular studies have obtained dates of splitting for mammalian clades that are clearly refuted by the fossil evidence.

In summary, the argument for the recent emergence of echidnas from a platypus-like ancestor is based on: (1) the absence of echidna-like fossils before 13 mya, whereas platypus-like fossils are evident from at least the early Tertiary; (2) the presence of mature structural features in the echidna that are most consistent with an aquatic ancestor; (3) the apparent decrease in electroreceptive ability among the tachyglossids as they began to exploit niches in the progressively drying Australian mainland; and (4) molecular clock data that suggest a recent divergence of the echidna and platypus lineages.

Are echidnas recent offshoots from the main monotreme line? The case against

The argument that echidnas are derived from the ornithorhynchid lineage is not universally accepted. The argument against the 'platypus first' hypothesis is based on paleontological, molecular, neurological and embryological evidence.

Rowe and colleagues used high resolution computed tomography of the mandible and dentition of *Teinolophos trusleri* to expand the dataset concerning derived features of this fossil (Rowe *et al.* 2008). They noted the presence of an hypertrophied mandibular canal running along the entire length of the dentary in a position lateral to the molariform tooth roots, as well as a large medial tubercle above the canal opening on the dentary. Among living mammals, only the platypus has an hypertrophied mandibular canal along the entire dentary, because this carries the abundant axons for electrosensory information from the lower bill. In the argument of Rowe and colleagues, the echidnas have a relatively narrow mandibular canal because that is the plesiomorphic condition that is found in *Morganucodon* and all therian mammals. They went on to state: 'Electroreception therefore appears to be an apomorphic characteristic of Monotremata, whereas the evolution of a specialised duckbill for high resolution aquatic electroreception is unique to the platypus clade. *Teinolophos* preserves the oldest evidence of a duckbill in its hypertrophied mandibular canal.' Since the specimen of *Steropodon* also includes the edge of a large mandibular canal, Rowe and colleagues consider *Steropodon* as a (derived) ornithorhynchid.

Rowe also argued that the low diversity of modern monotremes as compared to therians (five species as against 5362) is consistent with the slow rate of diversification that they envisage for the monotreme clade (Rowe *et al.* 2008). When they applied a relaxed molecular clock method to previously published amino acid and DNA datasets, they obtained earlier dates for the divergence of the platypus and echidna lineages than previously reported (point estimates of 88.9 mya, 79.5 mya and 63.7 mya depending on the molecular set). Taken together, these paleontological and molecular data have been used to argue for a prolonged period of separation of echidnas and platypus and the early emergence of derived morphotypes (i.e. platypus and echidna body plans) in both lineages (Rowe *et al.* 2008).

Even if one puts aside the arguments based on mandibular canal size, the paleontological evidence that supports the 'platypus first' hypothesis is logically weak because it is mainly based on an absence of evidence, i.e. the lack of tachyglossid fossils from Cretaceous or early Tertiary sediments. Camens (2010) has refuted the arguments of Phillips and colleagues, noting that the small size and fossorial habits of modern terrestrial tachyglossids make them unlikely candidates for fossilisation, thereby explaining their rarity in the fossil record. The arguments for the primacy of the platypus lineage that Phillips and colleagues have used were based on body shape, foot posture, 'front-wheel drive' and humeral rotation. Camens rejected these arguments, pointing out that flattened body shape, front-wheel drive locomotion and humeral rotation are all to be expected from the fossorial lifestyle seen in *both* the platypus and echidnas. As Camens has pointed out, the depauperate fossil record for tachyglossids is not evidence of their absence; rather, the lack of a marsupial myrmecophage from the Tertiary fossil record may indicate that that niche was occupied since the early Tertiary by an echidna-like monotreme.

The focus on electroreception as a monotreme characterisic has led to the literature ignoring several highly derived features of the echidna nervous system that are difficult to explain if a platypus-like monotreme were its ancestor. The specialised or derived features of the nervous systems of platypus and echidnas are summarised in Table 14.1. Essentially, the platypus shows striking specialisation of the trigeminal somatosensory pathways: it has extraordinary enlargement of the snout sensory glands, trigeminal ganglion, trigeminal sensory nuclei and forebrain somatosensory circuitry. Relative to the echidnas, this difference approaches several orders of magnitude in the case of the sensory glands and is around one order of magnitude for the trigeminal ganglion. By contrast, the short-beaked echidna shows striking specialisation of the olfactory pathway: it has a large and complex olfactory area,

Table 14.1. Comparison of neurological specialisations in platypus and echidna

Neural system	Platypus	Echidnas
Sensory glands on the bill	More than 40 000 glands arranged in highly ordered strips parallel to the long axis of the bill.	Only a few hundred (short-beaked) to few thousand (long-beaked) glands forming a sensitive spot on beak tip.
Trigeminal nerve divisions	Extraordinarily large cross-sectional area of all three divisions.	Cross-sectional area of all divisions is similar to that of therian trigeminal specialists (rodents and macropods).
Trigeminal ganglion	Extraordinarily large.	Similar in size and organisation to those in therian trigeminal specialists (e.g. rodents and macropods).
Trigeminal sensory nuclei in brainstem	Extraordinarily large Pr5 and Sp5O, and Pr5 shows subdivision into magno- and parvicellular components.	Similar in size and specialisation to therian trigeminal sensory nuclei.
Somatosensory thalamus	VP is greatly enlarged and protrudes into telencephalon.	VP is similar in size and shape to the VP of therian trigeminal specialists (e.g. rodents and macropods).
Somatosensory cortex	Greatly expanded S1 representation of the bill with parcellation of S1 into bimodal and unimodal bands.	Moderately sized S1 with expanded representation of the beak and forelimb, but no evidence of parcellation into bimodal and unimodal bands.
Olfactory epithelium	Rudimentary, poorly developed olfactory epithelium, simple nasal turbinate architecture. No cribriform plate.	Very high surface area with olfactory epithelium carried on a complex array of vertical bony plates. Cribriform plate present.
Olfactory bulb	Small and simple convex structure.	Large and highly gyrified olfactory bulb.
Primary and association olfactory cortex (AO, Pir, Ent)	Small and simple in structure.	Large and complex laminated structure. Evidence for regional specialisation.
Isocortex	Thick, but lissencephalic. No apparent association or prefrontal cortex.	Highly gyrified with a specialised frontal cortex, possibly homologous to therian dorsal prefrontal or orbitofrontal cortex.

large and gyrified olfactory bulb and highly derived olfactory cortical areas. Although detailed analysis of the central nervous system of the long-beaked echidna has never been undertaken, gross features of its nasal cavity and brain suggest that members of this genus share the same olfactory specialisations of the short-beaked echidna. In each of their respective sensory specialisations (trigeminal for the platypus and olfactory for the echidnas) these monotremes lie at the limits of the morphological range seen in therians. If one were to accept that the modern echidnas are derived from a trigeminally specialised platypus-like ancestor, then one would have to postulate that the ancestral echidna progressively lost much of the electrosensory ability, reversing the extraordinary anatomical specialisations of the trigeminal pathways, with a deletion or reversal of the developmental mechanisms that produce them. This would have to occur simultaneously with the redevelopment of the olfactory system from something like the rudimentary olfactory system present in the modern platypus to arrive at the highly specialised system in the modern echidnas. While this is possible, a more parsimonious interpretation of the respective specialisations would be that both the modern platypus and echidnas are derived from a common ancestor that was somewhat less specialised in either sensory system.

Differences in the neurodevelopmental processes in the platypus and short-beaked echidna also have bearing on this issue. In a series of studies of the neuroembryology of monotremes using the Hill and Hubrecht collections at Berlin, Ashwell and colleagues compared the development of the components of the somatosensory and olfactory pathways and the cerebral cortex (Ashwell 2012b; Ashwell and Hardman 2012a, b; Ashwell et al. 2012). The underlying rationale for this line of investigation was that if the modern echidna were derived from a platypus-like ancestor (with extreme trigeminal specialisation, rudimentary olfaction and a lissencephalic cortex), then the early development of the somatosensory and olfactory systems in the echidna should follow similar developmental trajectories to those systems in the platypus, perhaps with regression of those elements that are not present in the adult echidna. If, on the other hand, the tachyglossids and ornithorhynchids had pursued separate neural evolution for a large portion of the time span since their divergence from therians, then one would expect the unique features of each monotreme clade to emerge independently in each species during ontogeny. Although both the platypus and short-beaked echidna have a similar subventricular/subplate zone beneath the developing isocortex, each species has a very distinctive pattern of sensory development (Table 14.2). Characteristic features of the trigeminal system in the platypus – linear arrangement of sensory glands on the bill, large trigeminal ganglion, hypertrophied trigeminal sensory nuclei and large VP(M) nucleus of the thalamus – are never seen at any stage in the developing echidna. On the other hand, the olfactory specialisations of the short-beaked echidna (multiple vertical plates of olfactory epithelium and enlarged central olfactory regions) are never seen at any stage in the developing platypus. Finally, gyrencephaly (cortical

Table 14.2. Comparison of neuroembryology of platypus and short-beaked echidna

Feature of development	Platypus	Short-beaked echidna
Sensory gland development in snout	Highly ordered linear arrangement from time of first appearance of epidermal specialisation.	Spot arrangement from earliest time of appearance of epidermal specialisation.
Trigeminal ganglion	Very large from time of first appearance.	Modest in size throughout entire development. No evidence of major developmental regression.
Trigeminal sensory nuclei	Very large rostral rhombic lip produces many granular cells for rostral trigeminal sensory nuclei.	Rhombic lip is similar in size to therians. Trigeminal sensory nuclei are always modest in size, with no evidence of major developmental regression.
Somatosensory thalamus	Neurogenesis of large numbers of microneurons from prosomere 2 for a large VP nucleus of the thalamus.	Neurogenesis from prosomere 2 is similar to therians. VP is always modest in size and does not regress during development.
Somatosensory cortex	Large putative subventricular/subplate developmental zone for generation of microneurons for S1 and guidance of abundant thalamocortical axons.	Putative subventricular/subplate zone is present below S1 isocortex, but it is smaller than in platypus.
Olfactory epithelium	Simple folding of caudal nasal epithelium.	Complex folding of olfactory epithelium sheets during peri-hatching period.
Olfactory cortex	Always simple and unfolded during development.	Begins to fold around the time of hatching. Develops complex lamination during the lactational phase of development.
Folding of cortex	Cortex is always simple and lissencephalic throughout development.	Gyrification begins before hatching and progresses through the lactational phase of development.

folding) is seen in the developing echidna as an early event (even before hatching). Although it is possible that major changes in the timing of activation of developmental genes could produce these different patterns of neural ontogeny, the most parsimonious interpretation of the neuroembryological findings is that the platypus and short-beaked echidna had a much less specialised ancestor and have followed a prolonged period of separate evolution.

Concluding remarks on the 'platypus first' hypothesis

This issue is unlikely to be definitively settled unless or until further tachyglossid fossils from the late Cretaceous and/or early Tertiary come to light (see also Chapter 1). In the meantime, it seems unwise to dogmatically insist that the tachyglossids are derivatives of an aquatic ancestor. As Musser has observed, 'highly specialised animals seldom give rise to new taxa, particularly those that then develop novel but dissimilar specialisations of their own' (Musser 2003). The most parsimonious interpretation of the available evidence is that the ancestral monotreme had a moderate degree of trigeminal specialisation, probably with some degree of electroreception. This putative ancestor probably had the same anatomical elements of the olfactory pathway seen in all other mammals (i.e. olfactory bulb, anterior olfactory area, primary and association olfactory cortex), but with no

Table 14.3. Primitive and derived characters of the platypus nervous system

Primitive characters	Derived characters
• Freely anastomosing arteriovenous nets in neural tissue[1] • Only partial coiling of cochlear duct[2] • Presence of macula lagena in cochlear duct[2] • Unlaminated dorsal cochlear nucleus[1] • Facial nerve passes dorsal to the trigeminal sensory nuclei[1] • External cuneate nucleus not clearly separated from gracile and cuneate nuclei[1] • 'Medial' accessory inferior olivary nucleus lies in ventral position[1] • Presence of oil droplets in retinal cone photoreceptors[1] • Oil droplets in only one of a pair of cone photoreceptors[1] • Optic tract passes over the superficial grey layer of superior colliculus[1] • Retinotectal projection is almost exclusively contralateral[1] • Presence of main and accessory olfactory formations[1] • Scattered distribution of output cells (mitral, tufted) of olfactory bulb[1] • Dorsal lateral olfactory tract passes under the accessory olfactory formation[1] • Claustrum and extreme capsule not clearly demarcated[1] • Lissencephaly (flattened cortex)[1] • Absence of a corpus callosum[1] • Absence of a fasciculus aberrrans[1] • Absence of koniocortex specialisation into barrels[1] • Complete body representation on dorsal thalamic and cortical somatosensory areas[1] • Completely contralateral projection of oral and perioral input to somatosensory thalamus and cortex[1] • Anterior representation of digits in S1[1]	• Presence of sensory mucous and serous glands on bill[2] • Longitudinal striped distribution of sensory glands on bill[2] • Presence of push-rod mechanoreceptor organs on bill[2] • Bifurcated optic tract in dorsal thalamus[1] • Presence of distinct magno- and parvicellular regions in principal trigeminal nucleus[2] • Expansion of rostral rhombic lip during brainstem development[2] • Additional body representations in somatosensory cortex (PV) (controversial!)[2] • Bimodal and unimodal bands within S1 bill representation[2] • Presence of S2 in somatosensory cortex? (controversial!)[2] • Presence of subventricular zone/subplate in developing isocortex[2]

1 Reviewed by Johnson et al. (1982a); Johnson et al. (1982b); Johnson et al. (1994).
2 See relevant chapters in this book for review of literature.

particular degree of chemosensory specialisation. It is most likely that the two monotreme clades developed their respective sensory specialisations by elaboration on this basic sensory plan.

Primitive or just different?

Monotremes have traditionally been regarded as primitive or prototypical, so much so that they have often been sought out as test cases for theories of neural evolution (Lende 1964; Krubitzer *et al*. 1995). As Gould has noted, preconceptions of primitiveness masquerading as theory may hinder unbiased interpretation of evidence and render even obvious facts invisible (Gould 1992).

Of course, no living animal can be declared primitive, but most living animals (including humans!) exhibit characters that might be called primitive because they are reminiscent of the ancestral condition.

As for many mammals, the living monotremes show a mosaic of primitive and advanced characters. Non-neural plesiomorphic features of the living monotremes include egg-laying, a septomaxillary bone, the presence of an ectopterygoid in the skull, and the anatomy of the shoulder girdle (Musser 2003). There are also many neural plesiomorphic features (summarised in Tables 14.3 and 14.4). Although these are useful in phylogenetic analysis (Johnson *et al*. 1982a, b; Johnson *et al*. 1994), the presence of many of these primitive features in the monotreme nervous system should not be seen as indicative of an unsophisticated or poorly functional nervous system (Gould 1992). Some of these primitive characters do have an association with limited neural function, e.g. the limited coiling of the cochlear duct and the poor high frequency acuity of the hearing of living monotremes. On the other hand, many of these anatomical features (e.g. the lack of the mitral cell monolayer in the olfactory bulb; Switzer and Johnson 1977) do not necessarily imply less effective neural function, simply a different way to achieve the same result. Yet other characters, e.g. the presence of the macula lagena in the end of the cochlear duct, have functional implications that we do not yet understand.

Despite the many plesiomorphic features of the nervous systems of the living monotremes, it is the derived features that give a more compelling and realistic picture of the position that monotremes occupy in comparative and evolutionary neuroscience. In other words, the significance of the monotremes is defined more by their advanced sensory systems rather than their retention of primitive neural characters. Regardless of whether the platypus represents a model of the original body plan and niche of the ancestral monotreme, the extraordinary specialisation of electrosensory ability in the modern platypus is a remarkable example of sensory specialisation. The trigeminal specialisation of the platypus has required profound changes in the development of the trigeminal system at many levels: epidermal specialisation to produce the sensory gland receptors in the bill, trigeminal placode and neural crest expansion for the hypertrophied trigeminal ganglion and nerve, enhanced rhombic lip neurogenesis for the enlarged rostral trigeminal sensory nuclei, increased prosomere 2 neurogenesis for the hypertrophied ventral posterior thalamic nucleus, and a sophisticated subventricular/subplate developmental zone for the regulation of thalamocortical mapping onto the primary somatosensory cortex and the formation of the bimodal/unimodal banding.

The production of a functional trigeminal electro/mechanosensory system requires the coordinated production of neurons and circuitry at all these levels; any failure at one level would render the system ineffective. More importantly, the selection pressure to drive this sensory evolution must be free to act simultaneously at all levels of the pathway. This requires natural variation in the magnitude and duration of neurogenesis at all levels of the sensory system and a propensity for developmental plasticity that allows the matching of neural populations throughout the sensory neuron chain. Modern neuroscience is only beginning to understand the genetic regulation of the relevant developmental processes. Correlation of sequence data from the platypus genome with findings from molecular developmental studies of therians will throw light on the genetic regulation and evolutionary flexibility of these neurodevelopmental mechanisms.

The olfactory system of the monotremes has often been ignored, because of the dominant focus on electroreception in the platypus, but the echidnas clearly have highly specialised olfactory pathways with some extraordinary olfactory anatomy. The short-beaked echidna is the only mammal known that has a gyrified olfactory bulb, probably as an adaptation to expand the number of synaptic glomeruli available for analysis of the (probably large) odorant repertoire. Astonishingly, there has never been a study of the olfactory repertoire or acuity in the short-beaked echidna and nothing is known of the olfactory receptor genes. The large differentiated olfactory cortex of the short-beaked echidna also points to advanced systems for analysis of olfactory information, but the physiology of these regions remains unstudied. The potential role of the large frontal cortical fields (Fr1, Fr2 and Fr3; Hassiotis et al. 2004b) in integration of chemosensory information (putative homology with placental orbitofrontal cortex), and formation of a chemosensory model of the echidna's habitat, is also unexplored.

One derived character that has only recently come to light concerns the presence of a subcortical developmental zone beneath the developing isocortex in both the platypus and short-beaked echidna (Ashwell and Hardman 2012a; see also Chapter 7 for review). This region is present beneath the developing isocortex throughout the lactational phase of development in both species and is particularly large beneath the electroreceptive S1 cortex of the platypus. The monotreme subcortical zone has features reminiscent of the placental (or diprotodont marsupial) subventricular zone (in that it appears to be an active zone of neurogenesis for cortical microneurons) and the placental subplate (in that it is closely associated with ingrowing thalamocortical axons). The evolution of a cortical subventricular zone has been claimed to be a defining feature of advanced mammalian nervous systems, because it allows the production of large numbers of small neurons for the processing of information within cortical modules. The pallial subventricular zone has reportedly not been found in polyprotodont metatherians, so its presence within the developing cortex of monotremes, diprotodonts and placentals implies either its independent emergence as many as three times in mammalian evolution, or its presence as an unexploited potential in the developing brain of the earliest true mammal.

The possible homology of the monotreme subcortical developmental zone with the placental subplate remains unconfirmed (see Chapter 7 for discussion, and below). In placentals, the subplate serves as a support and waiting zone for developing thalamocortical axons in placentals with sophisticated cortical sensory areas (e.g. primates and carnivores). If the monotreme subcortical developmental zone does prove to be homologous to the placental subplate, then this hallmark of advanced cortical evolution would have emerged twice during mammalian evolution. It would also confirm the advanced status of the monotreme cortex.

In summary, although the plesiomorphic characters of monotremes have been emphasised in the past neuroscience literature, the modern platypus and echidnas clearly have many advanced neurological features, particularly in the respective trigeminal and olfactory sensory pathways. The group may also have independently developed sophisticated developmental mechanisms to assist connection mapping in the cerebral cortex.

The changing Australian environment and monotreme evolution

As a subclass, the monotremes have an extraordinary longevity (Musser 2003). Monotremes may have a history of as long as 200 million years (dating back to the late Triassic or early Jurassic). The fossil record indicates that Mesozoic monotremes spread from Australia through Antarctica to South America (Musser 1998; Pascual et al. 1992a, b), a range that must have incorporated a striking diversity of environments, including the polar latitudes within the Antarctic Circle. In the modern world, the platypus is confined to the river systems of the eastern seaboard of the Australian mainland plus Tasmania. On the other hand, the short-beaked echidna has a much more widespread distribution, ranging across the entire Australian mainland, and in the associated islands of Tasmania and New Guinea (a broader region also known as Greater Australia or Sahul). The long-beaked echidnas have a much more restricted distribution, being confined to the upland regions of New Guinea.

How did the changing environment of eastern Gondwana (and Greater Australia after the fragmentation) during the Mesozoic and Tertiary shape the ecology, anatomy, physiology and distribution of the modern monotremes? To begin to answer this question we need to summarise the main climatic changes of Australia during the last 200 million years (Johnson 2004, 2006).

During the Mesozoic, Australia was part of eastern Gondwana and situated closer to the pole, so that climatic conditions were cool and moist, with a pronounced seasonal variation in daylight and insolation. Marsupials probably entered Australia at the end of the Cretaceous or early in the Tertiary, so the monotremes would have faced competition from therians early in their history. In the early Tertiary, climatic conditions in Greater Australia were warmer and wetter than they are currently. Dense rainforests dominated by flowering plants and the Antarctic beech (*Nothofagus*) covered much of the Australian continent (even the now arid centre) during the Eocene (~60–37 mya). Conditions in Greater Australia began to cool and dry from ~50 mya, a trend that would eventually give rise to the aridity in the interior of modern Australia. The changed climatic conditions were largely due to the separation of the fragments of eastern Gondwana (in particular the separation of Tasmania and Antarctica) and the development of a current encircling the pole. This change in ocean currents reduced the transfer of heat between the tropics and the South Pole, so that Antarctica began to freeze and conditions in southern Australia became cooler. The *Nothofagus* forests were largely lost by the middle of the Miocene and were progressively replaced by vegetation more suited to arid conditions (open and dry forest and woodland, shrublands and grasslands). It was during this period that the dry fruited Myrtaceae (including the modern genus *Eucalyptus*) began to become more widespread. These drier conditions would have been accompanied by more frequent burning of the vegetation, and increased charcoal levels in late Tertiary deposits confirm this (Johnson 2004).

During the Pleistocene (from ~2.6 mya), global temperatures began to fluctuate as the Northern Hemisphere ice sheets advanced and retreated. For Australia, these climatic changes led to oscillations between cool and dry conditions (during glacial periods) and warm and wet conditions (during the brief interglacials). New Guinea was less susceptible to these fluctuations in rainfall and continued to be a site of mammalian biodiversity. Reductions in sea level amounting to ~130 m below the current sea level accompanied the temperature fluctuations across the Australian mainland.

It seems likely from the antiquity of clearly ornithorhynchid fossils that the platypuses found their niche as semi-aquatic insectivorous specialists at least as far back as the Cretaceous (Musser 2003) and well before they faced competition from therians. The ornithorhynchids were and are superbly adapted to their niche, with a flattened body form, wide electro- and mechanosensitive bill, and waterproof fur, but this degree of specialisation restricts their distribution to permanent river systems. This specialisation would have been ideal during the wetter and warmer periods of Australian history, but made the platypuses highly vulnerable to climatic variation as the Australian climate dried out. Permanent rivers are now largely confined to eastern Australia and the south-western corner of the continent, and the semi-aquatic lifestyle of the platypuses effectively excluded them from spreading across arid parts of the continent to new habitats in the west.

Long-beaked echidnas were widespread during the late Tertiary and are even more common as fossils from that period than their short-beaked relatives. They disappeared from Tasmania and the Australian mainland during the late Pleistocene (Murray 1978a, b; Griffiths *et al.* 1991), but survived in the isolated mountain environments of New Guinea (Baillie *et al.* 2009). The electroreceptive 'hot spot' on the end of the long beak would be an effective thrust probe tool for detecting moving prey in the leaf litter and upper soil (provided that the electromyographic potentials of prey are of sufficient strength) and/or differentiating live from dead organic material before ingestion. It is also likely that olfaction played a major role in prey detection among fossil echidnas since the nasal cavities of all echidnas (living and extinct, short- and long-beaked) are very similar in structure (see Chapters 4 and 11).

Electroreception is of little use in an arid environment. Perhaps this explains the disappearance of the long-beaked echidnas from Australia. Murray and Griffiths have suggested that long-beaked echidnas of the Australian Pleistocene had a very different diet from the short-beaked echidna (Murray 1978b; Griffiths *et al.* 1991). Griffiths and colleagues argued that the Pleistocene long-beaked echidna *Megalibgwilia* consumed large energy-rich insects such as scarab beetles and moth larvae as its major prey. Griffiths suggested that the cooling of the mainland Australian climate from 16 000 to 12 000 years ago led to a dearth of scarab beetles and a crisis for *Megalibgwilia*. The large long-beaked echidnas would have had to survive on less nutritious earthworms, ants and termites (as do modern short-beaked echidnas). This may have contributed to the extinction of large-beaked echidnas on mainland Australia, although anthropogenic overhunting is a possible contributing factor (also note the possibility of recent survival of long-beaked echidnas in the Kimberley; Helgen *et al.* 2012). Griffiths speculated that dietary changes, metabolic rate reduction and changes in skull morphology among the *Megalibgwilia* of Greater Australia led to a divergence in the group to give the modern long-beaked echidnas of New Guinea and the modern short-beaked echidnas of Australia and New Guinea. This is unlikely given that *Tachyglossus* was contemporaneous with *Megalibgwilia* during the Pleistocene in Australia.

The short-beaked echidnas are perhaps the most successful of the monotremes. Their distribution across all of Greater Australia is testimony to their ability to survive in an extraordinary range of environments, from arid zone to rainforest, and coastal plain to alpine scrubland (Augee *et al.* 2008). The progressive drying of the Australian continent, spread of grasslands and the Pleistocene extinction of herbivorous marsupial megafauna may have even benefitted the short-beaked echidnas, because their termite prey became the major 'grazers' of the Australian grasslands. The short-beaked echidnas have a suite of features that suit them well to their niche. These include powerful forelimbs with sensitive mechanosensory feedback for digging and breaking open termite mounds, a short beak permissive of low-angle tongue protrusion to catch scurrying ants and termites, a pronounced olfactory sense to detect prey and map territory, and a modest metabolic rate to conserve energy.

How did the monotremes cope with the arrival of the marsupials? The short answer is: very well. The

Table 14.4. Primitive and derived characteristics of the short-beaked echidna nervous system

Primitive characters	Derived characters
• Freely anastomosing arteriovenous nets in neural tissue[1]	• Presence of sensory mucous and serous glands on beak tip[2]
• Only partial coiling of cochlear duct[2]	• Presence of push-rod mechanoreceptor organs on beak tip[2]
• Presence of macula lagena in cochlear duct[2]	• No oil droplets in cone photoreceptors (controversial!)[2]
• Unlaminated dorsal cochlear nucleus[1]	• Bifurcated optic tract in dorsal thalamus[1]
• Facial nerve passes dorsal to the trigeminal sensory nuclei[1]	• Gyrified olfactory bulb[2]
• External cuneate nucleus not clearly separated from gracile and cuneate nuclei[1]	• Subdivisions within piriform cortex[2]
• 'Medial' accessory inferior olivary nucleus lies in ventral position[1]	• Gyrencephalic isocortex[1]
• Optic tract passes over the superficial grey layer of superior colliculus[1]	• Claustrum and extreme capsule present (controversial!)[2]
• Retinotectal projection is almost exclusively contralateral[1]	• Additional body representations in somatosensory cortex (PV) (controversial!)[2]
• Presence of main and accessory olfactory formations[1]	• Presence of S2 in somatosensory cortex? (controversial!)[2]
• Scattered distribution of output cells (mitral, tufted) of olfactory bulb[1]	• Presence of subventricular zone/subplate in developing isocortex[2]
• Dorsal lateral olfactory tract passes under the accessory olfactory formation[1]	
• Absence of a corpus callosum[1]	
• Absence of a fasciculus aberrrans[1]	
• Absence of koniocortex specialisation into barrels[1]	
• Complete body representation on dorsal thalamic and cortical somatosensory areas[1]	
• Completely contralateral projection of oral and perioral input to somatosensory thalamus and cortex[1]	
• Anterior representation of digits in S1[1]	

1 Reviewed by Johnson *et al.* (1982a); Johnson *et al.* (1982b); Johnson *et al.* (1994).
2 See relevant chapters in this book for review of literature.

marsupials certainly did not 'sweep them away' as one recent popular article on monotremes (Choi 2009) has asserted. The prolonged survival of the ornithorhynchids in an unchanged form suggests that this group so effectively filled their niche that marsupials (and later rodents) were unable to make inroads into the role of semi-aquatic insectivore. Only one Australian marsupial with a very restricted distribution (the numbat of Western Australia) has adopted an ant-eating lifestyle, suggesting that the short-beaked echidna has also effectively excluded marsupial competitors from its niche over most of Greater Australia.

How important is electroreception for the living monotremes?

Electroreception is often seen as one of the defining features of the monotremes (Pettigrew 1999; Short 2009), but how important is it for the group as a whole? Even in the platypus there are significant questions about how effective electrosensation is in the natural environment, and the importance of electrosensation for the short-beaked echidna under natural conditions is very doubtful.

The key question concerning the role of electroreception in the platypus is whether the usual prey of the platypus emits electrical fields strong enough to be detected in a real world situation. The great weakness of monotreme electroreception is that it is a passive system and requires some degree of movement by the prey before detectable fields would be generated. As has been discussed in Chapter 9, electrical field strength of some of the natural prey of the platypus (i.e. actively swimming or fleeing crustaceans) would be detectable at useful ranges in fresh water (10 to 20 cm), but electroreception cannot account for the platypus's ability to detect sedentary or slow-moving molluscs that generate only very weak electrical fields. It is likely that the platypus detects weakly electrical prey mainly by the sense of touch through the push-rod mechanoreceptor organs. Nevertheless, the electrical sense may be of benefit even in the contact situation by allowing the platypus to differentiate between live prey and detritus (Proske and Gregory 2003). Given that the platypus ingests significant quantities of mud with its prey, it may be that much of the food selection process actually takes place at the level of the palate and tongue, where sensitive mechanotransduction discriminates between soft invertebrate shells and hard debris. Some authors have speculated that the platypus can use its passive electrosensory ability to detect obstacles in its natural environment, but this has never been adequately tested with electrically neutral barriers.

The benefits of electroreception for the short-beaked echidna in the real world remain unproven. The only study of electrosensory abilities in free-ranging echidnas used very large field strengths (20 to 30 mV/cm) generated by buried 9 V batteries (Augee and Gooden 1992) and it is difficult to know how this translates to the natural environment. The threshold of 2 mV/cm found by Gregory and colleagues (Gregory *et al.* 1989b) is far higher than the field strength of many leaf-litter invertebrates and their larvae (earthworms, leeches, insects; Taylor *et al.* 1992) at even modest ranges of a few centimetres. There is a clear need for a study that investigates the ability of the short-beaked echidna to use electroreception under entirely natural conditions.

Monotremes and cortical evolution: the issue of somatosensory fields

The issue of the number of cortical somatosensory fields in the living monotremes remains contentious. This is an important question because it has a critical bearing on mammalian cortical evolution and the emergence of discrete topographically ordered somatosensory fields. If the number of somatosensory fields identified in the monotreme isocortex by Krubitzer and colleagues is confirmed and homologies between these areas and supplementary sensory cortex are established, then the antiquity of the fundamental plan of mammalian isocortex would be pushed back as far as the Early Cretaceous. Remaining questions are: (1) whether the rostral field of Krubitzer *et al.* (1995) is a truly separate topographically structured somatosensory field, or is simply the deep sensation component of the primary somatosensory cortex (S1); (2) whether the PV (or C) field identified by Krubitzer and colleagues is homologous to the PV of therians, or perhaps the secondary somatosensory field (S2); and (3) whether a mapping of the monotreme somatosensory cortex that takes dermatomal trajectories into account would lead to a revision of the reported number of topographically organised somatosensory areas. At stake is the question of how ancient the S2 field is. Krubitzer and colleagues hypothesise that the basic mammalian plan of cortical organisation included S1, PV, R as purely somatosensory areas, as well as visual, auditory and motor (or manipulation somatosensory) cortex and a limited, but variable, amount of association cortex. In neither Krubitzer's nor Bohringer and Rowe's (Bohringer and Rowe 1977; Rowe 1990) conceptions is the S2 field present in the stem mammal, so it is generally believed to have emerged only in the therian lineage.

Monotremes and cortical evolution: ontogeny of cortical circuitry components

Decades of study of cortical development in eutherians have revealed some sophisticated developmental mechanisms that produce the many morphologically and functionally diverse microneurons essential for cortical processing, and contribute to the formation of patterned connections. Key elements of isocortical development in so-called 'advanced' placentals include the two-zone or -stage pattern of neurogenesis (the ventricular germinal zone and subventricular zone) and the evolution of a transient cortical subplate to support incoming thalamocortical afferents during cortical ontogeny (see Chapter 7 for details). The substantial size of the isocortex in the living monotremes (thick, but lissencephalic in the platypus and with complex regional subdivision; and gyrencephalic and regionally complex in the echidnas) is highly suggestive of sophisticated developmental mechanisms for building the isocortex.

The evolution of a developmental subventricular zone for the cortex has been seen as an important event in mammalian brain evolution, because it allowed the prolonged and amplified neurogenesis of microneurons for a thicker and functionally more effective isocortex (Charvet and Striedter 2011; Charvet *et al.* 2009; Cheung *et al.* 2007; Kriegstein *et al.* 2006). The emergence of the telencephalic subventricular zone is seen as a critical development that allowed the expansion of the cerebral cortex from the three layers of therapsids and sauropsids to the six layers characteristic of all modern mammals (Cheung *et al.* 2007;

Cheung *et al.* 2010). It may have arisen by a delay of neurogenesis in the ventricular zone (VZ) so that the size of the proliferative cell population exceeded the growth of the ventricular surface, forcing mitotic cells to leave the VZ and form a subventricular proliferative zone outside the VZ (Charvet and Striedter 2011). Some authors have claimed that didelphid marsupials do not have an organised subventricular zone beneath the isocortex, but instead rely on the migration of microneurons from the subpallial (i.e. ganglionic eminence) subventricular zone (Abdel-Mannan *et al.* 2008; Charvet *et al.* 2009; Puzzolo and Mallamaci 2010), but this remains controversial (Cheung *et al.* 2010). It is certainly true that those therians that have very large and gyrencephalic isocortices as adults have a thick and very mitotically active subventricular zone beneath both the developing cortex and the ganglionic eminences. The identification in platypus and echidna hatchlings of a subcortical region with proliferative properties similar to the subventricular zone of advanced therians naturally raises the question of whether such a developmental mechanism evolved as an unexplored potential before the divergence of therians and prototherians, or evolved independently in the two lineages.

The other important developmental device that characterises the large and complex isocortex of so-called 'advanced' therians is the subplate. This region contains some of the earliest neurons in the cortex and provides transient targets for the developing thalamocortical projections in all sensory modalities, as well as association connections. Ingrowing thalamocortical axons in some therians spend a 'waiting period' during which they form synapses on the subplate neurons, before continuing into the cerebral cortex proper (see review in Montiel *et al.* 2011). Subplate neurons in therians have also been shown to pioneer the corticofugal pathway and are believed to play a key role in the guidance of thalamocortical afferents. The subplate neurons are thought to be important in the establishment of complex cortical functional domains such as the ocular dominance columns in carnivores and primates. Not all therians have a similar degree of subplate development: the subplate is reported to be almost undetectable in marsupials (Montiel *et al.* 2011) and to be correspondingly large in placentals with big brains (i.e. carnivores, large ungulates and primates).

Some authors have suggested that the pre- or subplate, generated in the postembryonic mammalian brain, is really the ancestral cortex of the first amniotes (Marin-Padilla 1978) and that mammals expanded the duration of cortical neurogenesis to build a six-layered isocortex proper external to it. Others argue that the subplate is a phylogenetically recent developmental device exclusive to mammals and has been elaborated hand-in-hand with the enlargement of the isocortex in therians (Molnár *et al.* 2006). Montiel and colleagues have suggested that the subplate originally derived from a very old structure in the dorsal pallium of stem amniotes, subsequently expanding in particular branches of the synapsid lineage (Montiel *et al.* 2011; Wang *et al.* 2011). The presence of a long-lasting cortical subzone that is penetrated by thalamocortical axons in the developing telencephalon of the platypus and echidna raises the possibility that the monotremes may have a subplate of their own. Nevertheless, more detailed study including analysis for molecular markers of the subplate (*Cplx*, *Ctgf*, *Moxd1*, *Nurr1* and *Tmem163*; see Wang *et al.* 2011 for review) is necessary to confirm this.

The subventricular zone and subplate are developmental hallmarks of complex mammalian nervous systems. If future studies in this important area do demonstrate that the monotremes have similar cortical developmental mechanisms to those in the so-called 'advanced' therians, then the descriptor 'primitive' sometimes applied to monotremes must be discarded.

Questions for the future

All the living monotremes clearly have highly advanced nervous systems, but our current understanding of both the phylogeny and ontogeny of their nervous system is poor. There are many questions in phylogeny, physiology and development that remain unanswered. The most important of these concern the molecular controls of the development of the distinctive anatomy and physiology of the platypus and echidnas.

- How is the distinctive pattern of sensory glands on the bill and beaks of monotremes controlled at the molecular level? What is the relationship between sensory glands of monotremes and non-sensory glands in the snout skin of other mammals?

- How is the expansion of the trigeminal placode and neural crest regulated in the platypus? Why is an expanded trigeminal ganglion seen only in the embryonic platypus, and not the embryonic echidna?

- How is the expansion of the rostral rhombic lip controlled in the platypus? How does the molecular regulation of the development of this region and its derivatives compare with placentals?

- How does the proliferative population of the putative monotreme subventricular zone compare functionally with the placental subventricular zone? What is its contribution to the neuronal populations of the monotreme isocortex? Is this zone subject to the same molecular controls as in placentals?

- Is there a true subplate in the developing monotreme cortex? If so, how does it assist the formation of thalamocortical connection patterning and does it share the molecular markers identified in the placental subplate?

- How is the bimodal/unimodal banding of the platypus S1 cortex produced during development? Is its development under purely molecular control or does it, like placental ocular dominance columns, arise by competitive interactions between axons during the lactational phase of development?

15

Atlas and tables of peripheral nervous system anatomy

Ken W. S. Ashwell and Anne M. Musser

Summary

This chapter provides an illustrative and tabular summary of the innervation of the skeletal musculature of the head, neck, limbs and trunk in the monotremes. The original data that have been brought together for this chapter are scattered throughout the literature, much of it in very old publications that are difficult to access. It is hoped that the compilation of this information in a single chapter will facilitate further studies in this area.

Innervation of the musculature of the head and neck

Cranial nerves 3, 4, 5, 6, 7, 9, 10, 11 and 12 (Fig. 15.1) all supply skeletal muscle in the head and neck. These muscle groups (see Table 15.1) include extraocular muscles for control of eye movement (oculomotor nerve – 3n), masticatory (chewing) muscles (from the first pharyngeal arch and supplied by the trigeminal nerve – 5n), facial muscles (from the second pharyngeal arch and supplied by the facial nerve – 7n), palatine and laryngeal muscles (from arches 3 to 6 and supplied by the glossopharyngeal and vagal nerves – 9n and 10n), somatic neck muscles (from paraxial mesoderm and supplied by the spinal accessory – 11n, and cervical plexus) and tongue muscles (from occipital somites and supplied by the hypoglossal nerve – 12n).

The main groups of muscles of the head and neck of the monotremes (Figs. 15.2, 15.3) are broadly similar to those in therians, but with some notable exceptions for specific muscles (Diogo et al. 2008). These include the apparent absence of one deep facial muscle (digastricus posterior) and some of the smaller superficial facial muscles of therians (occipitalis, auricularis posterior, zygomaticus major and minor, levator anguli oris facialis); as well as the absence of some

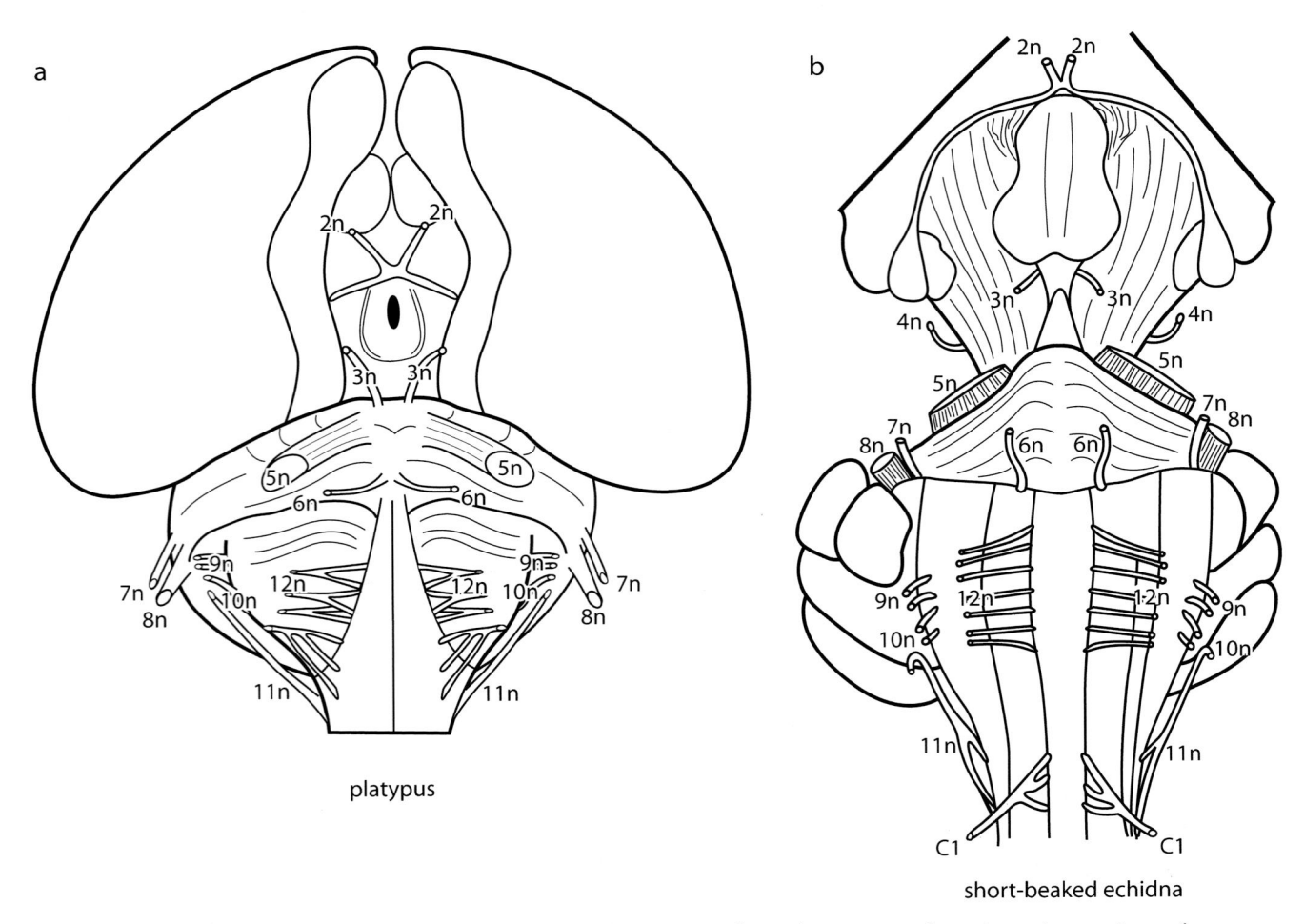

Figure 15.1: Cranial nerves arising from the monotreme brainstem. These diagrams are based on observations of preserved monotreme brains in the National Museum of Australia collection as well as illustrations in Ziehen (1897) and Abbie (1934). 2n – optic nerve; 3n – oculomotor nerve; 4n – trochlear nerve; 5n – trigeminal nerve; 6n – abducens (or abducent) nerve; 7n – facial nerve; 8n – vestibulocochlear nerve; 9n – glossopharyngeal nerve; 10n – vagus nerve; 11n – accessory nerve; 12n – hypoglossal nerve; C1 – first cervical nerve root.

palatine, pharyngeal and laryngeal muscles (levator veli palatini, discrete pharyngeal constrictors, crico-arytenoideus lateralis), and a tongue muscle (palatoglossus). Griffiths (1968, 1978) also included several extrinsic tongue muscles in the short-beaked echidna (sternoglossus superior and inferior, laryngoglossus, styloglossus, myloglossus) that were not incorporated by Diogo and colleagues in their list of lingual muscles in the platypus (Diogo *et al*. 2008). This might reflect a true difference in function (see below). Overall, the monotremes have fewer facial muscles (only 10) than many therians (rats – 20; colugos – 19; *Tupaia glis* – 21; humans – 24) and fewer laryngeal muscles (three in the platypus compared to six in modern humans; four in *Tupaia glis*, the domestic rat and the colugo) (Diogo *et al*. 2008).

The special features of the tongue and masticatory musculature of the echidnas are key to the unique feeding mechanism of this group. Protrusion of the tongue is brought about by contraction of loosely arranged circular muscles that surround a core of longitudinal muscle (Doran 1973; Griffiths 1978). Since the base of the tongue is fixed, contraction of the circular muscle rapidly ejects the tip of the tongue from the mouth. Radially arranged muscles in the distal tongue allow the tongue tip to be directed up, down and to each side while it is protruded. A vascular stiffening mechanism also contributes to tongue

Table 15.1. Cranial musculature and innervation[1]

Muscle group	Muscle name	Innervation
Extrinsic eye muscles	• Medial rectus, inferior rectus, superior rectus and inferior oblique	Oculomotor (3n)
	• Superior oblique	Trochlear nerve (4n)
	• Lateral (external) rectus	Abducens nerve (6n)
Ventral mandibular (arch 1 derivatives)	• Mylohyoideus (superficialis and profundus) • Digastricus anterior	Trigeminal: mandibular division (5man)
Adductor mandibulae (arch 1 derivatives)	• Masseter • Detrahens mandibulae • Temporalis • Pterygoideus lateralis • Pterygoideus medialis • Tensor tympani • Tensor veli palatini	Trigeminal nerve: mandibular division (5man)
Dorsomedial hyoid muscles (arch 2 derivatives)	• Styloideus • Stapedius • Platysma cervicale • Platysma myoides • External ear musculature	Facial nerve (7n)
Ventral hyoid muscles (arch 2 derivatives)	• Interhyoideus profundus • Sphincter colli superficialis • Sphincter colli profundus (in echidna only) • Cervicalis transversus • Orbicularis oculi • Nasolabialis • Buccinatorius • Orbicularis oris • Mentalis	Facial nerve (7n)
True branchial muscles (*sensu stricto*)(arch 3 derivatives)	• Stylopharyngeus • Ceratohyoideus • Subarcualis rectus III	Glossopharyngeal (9n) Glossopharyngeal (9n)? ?
True branchial muscles (other), derived from paraxial mesoderm, but perhaps some somite contribution	• Acromiotrapezius • Spinotrapezius • Dorsocutaneous • Cleidomastoideus • Sternomastoideus	Spinal accessory (11n)
Pharyngeal muscles (arches 4 to 6 derivatives)	• Constrictor pharyngis • Cricothyroideus • Palatopharyngeus	Vagus (10n)
Laryngeal muscles (arches 4 to 6 derivatives)	• Thyrocricoarytenoideus • Arytenoideus • Cricoarytenoideus posterior	Vagus (10n)
Geniohyoideus (lingual musculature)	• Geniohyoideus • Genioglossus • Intrinsic muscles of tongue • Hyoglossus	Hypoglossal (12n)
Rectus cervicis (infrahyoid musculature)	• Sternohyoideus • Omohyoideus (superficialis and profundus) • Sternothyroideus	C1, C2

1 Muscles based on review by Diogo *et al.* (2008), with additional points from Griffiths (1968, 1978).

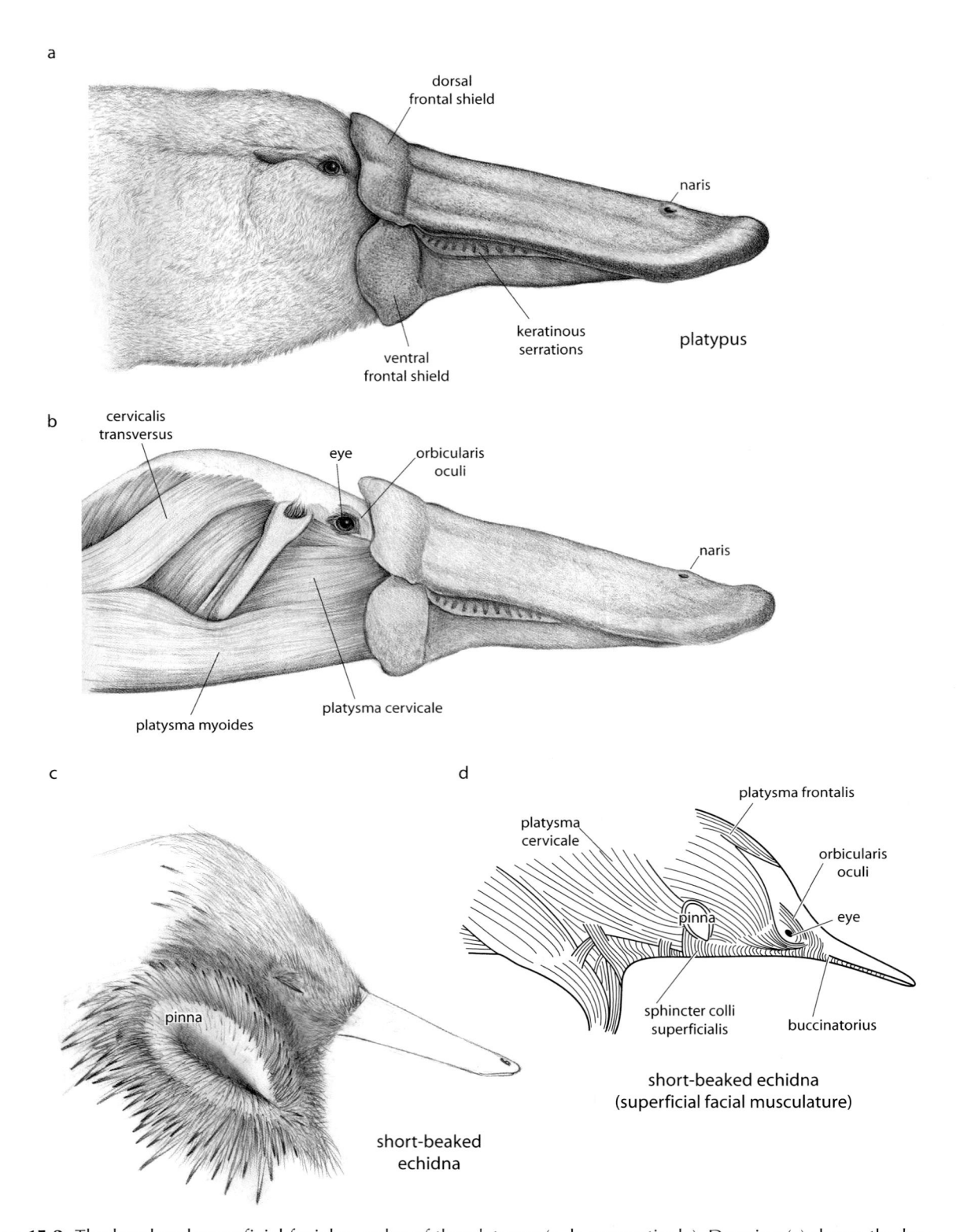

Figure 15.2: The head and superficial facial muscles of the platypus (a, b, respectively). Drawing (a) shows the key features of the platypus head and bill. Drawing (b) is a superficial dissection showing the extensive platysma. Drawing (c) shows the external features of the head of the short-beaked echidna and (d) illustrates the superficial facial muscles of the same species. Illustrations (a) to (c) are by Anne Musser. Illustration d) is based on Griffiths (1978) with modification.

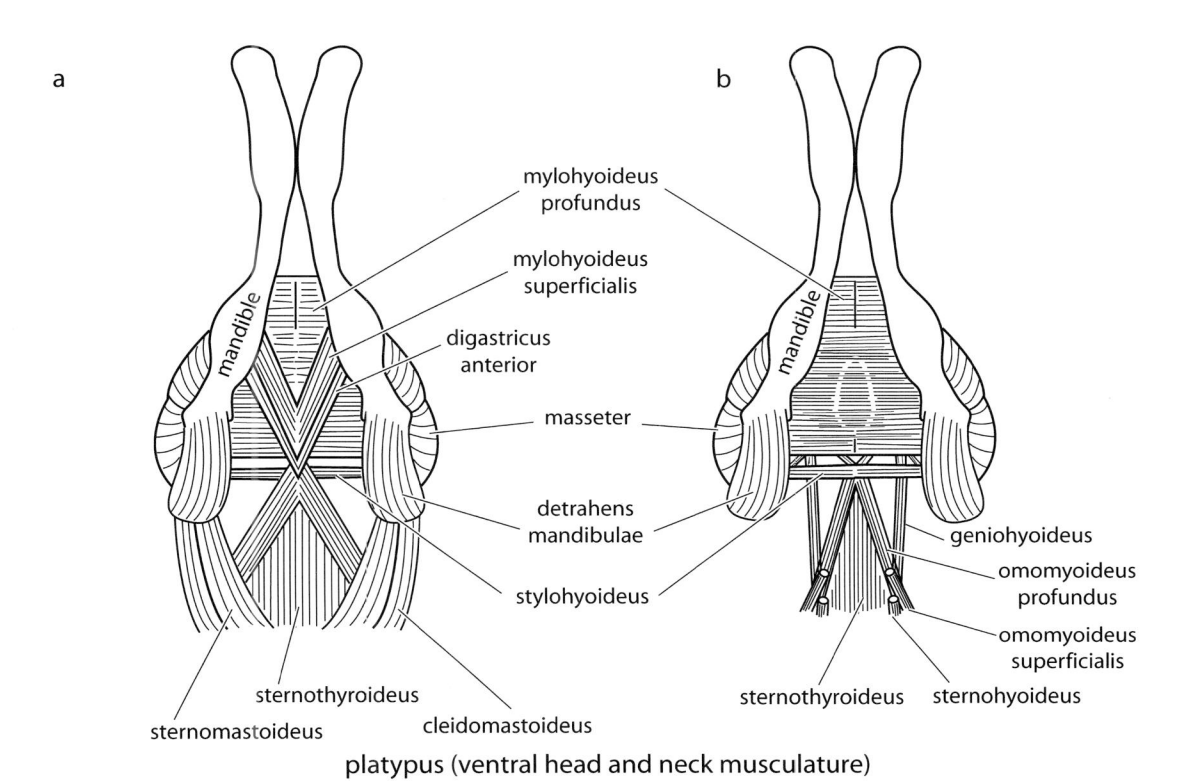

Figure 15.3: The superficial and deep muscles of the floor of the oral cavity and ventral neck of the platypus are shown in (a) and (b), respectively. The laryngeal muscles of the platypus are illustrated in (c). Finally, the superficial and deep masticatory muscles of the short-beaked echidna are illustrated in (d) and (e), respectively, with a section of the zygoma removed in (e). Illustrations (a), (b) and (c) are based on Diogo et al. (2008). The illustrations of the masticatory musculature of the echidna (d, e) are based on Murray (1981).

rigidity during these manoeuvres (Doran and Baggett 1970). Retraction of the tongue is achieved by contraction of two internal longitudinal muscles that run the length of the tongue (Griffiths 1978).

Both long- and short-beaked echidnas have keratinous spines on their tongue and palate to assist with the comminution of invertebrate prey. The tongue spines lie in opposition to similar spines on the underside of the palate, so forward and backward movement of the tongue can effectively break up chitinous body parts (Griffiths 1978). In long-beaked echidnas, the end of the tongue also carries several rows of regularly arranged horny teeth that are directed backwards and presumably assist in returning prey to the mouth (Griffiths 1968). In the platypus, keratinous pads on the oral surfaces of the upper and lower jaws, as well as backwardly directed cuticular papillae and two large forwardly directed keratinous spines on the tongue, all work together to perform a similar function (Griffiths 1978). Keratinous serrations are often visible externally (Fig. 15.2a).

Both platypus and echidnas have only a limited need to masticate, but they have retained masticatory musculature (Fig. 15.3d, e) and its innervation by the motor root of the trigeminal nerve. Murray (1981) has noted some unique features of the jaw mechanism in the echidna. The jaw of the echidna does not open and shut by hinging at the temporomandibular joints, as do other mammals. Opening of the mouth in the echidna is limited to only ~5 mm and is achieved by rotation of the spindle-like dentaries around their long axes. Medial rotation of the dentary shaft to open the mouth is achieved by the contraction of the morphological equivalent of the therian pterygoideus medialis. Conversely, contraction of the temporalis laterally rotates the dentary shaft laterally around its long axis and closes the mouth. This peculiar arrangement ensures that the grinding surfaces of the tongue dorsum and the palate can continue in contact even as the tongue is flicked in and out.

The monotremes also have some muscles that are not present in therians. In particular these include: (1) the detrahens mandibulae (supplied by 5n), which is also not present in non-mammalian tetrapods (Diogo *et al.* 2008); (2) the dorsocutaneous component of the trapezius group (supplied by 11n); and (3) the interhyoideus profundus (presumably supplied by the upper cervical plexus). The presence of the detrahens in monotremes has been used as an argument for an ancient origin: the detrahens may be a holdover from the time, far back in mammalian history, when the jaw joint was evolving from the articular-quadrate joint to the dentary-squamosal joint (Musser 2006b).

The panniculus carnosus, a sheet of muscle that covers the trunk, also has an extension into the neck region (the hypodermal slip) that runs over the deeper neck muscles (McKay 1894). It is innervated by branches of the cervical plexus (see below).

Innervation of the musculature of the pectoral girdle and forelimb

Muscles of the pectoral girdle and brachium

The pectoral girdle of the monotremes has long been recognised as incorporating many plesiomorphic features (Musser 2003), so one would expect significant differences in pectoral musculature compared to therians. In *Tachyglossus*, each pectoral girdle includes a scapula, coracoid, clavicle, pro- (or epi-) coracoid and a midline interclavicle (Griffiths 1968, 1978). A similar arrangement is present in *Ornithorhynchus*, but *Zaglossus* has an unusual arrangement of epicoracoids overlapping in the midline (right over left), with the interclavicle displaced to the right of the midline (see discussion in Griffiths 1968).

Despite these unusual skeletal features, the pectoral girdle muscles of the monotremes (Table 15.2) are surprisingly similar to those of therians, but with some significant differences. For example, McKay did not consider that either monotreme had a true pectoralis minor (McKay 1894), but both species have a pectoralis quartus that arises from the linea alba at the midline of the abdomen and inserts into the superficial part of the tendon of the pectoralis major. The deltoid has ventral acromioclavicular and dorsal scapular components in both species. Other muscles of the shoulder region in the monotremes include the rhomboideus, levator scapulae, serratus magnus, acromiotrachelien, epicoracohumeralis, subclavius (with costocoracoideus and sterno-epicoracoideus components), subscapularis, supraspinatus, infraspinatus, teres minor, teres major, trapezius, latissimus dorsi and dorso-epitrochlearis.

Table 15.2. Muscles of the pectoral girdle and brachium and their innervation[1]

Muscle group	Muscle		Innervation	Myotomal supply
Muscles inserting onto the scapula and clavicle	Trapezius	Rostral part	Accessory nerve (11n)	11n from brainstem
		Caudal part	Thoracic spinal nerves	T5 to T13 dorsal rami
	Rhomboideus		Cervical spinal nerves	Platypus: C2; Echidna: C2 and C3
	Levator scapulae		Cervical spinal nerves	C3
	Serratus magnus		Cervical spinal nerves	C5 and C6
	Subclavius	Costocoracoideus	Platypus: nerve to subclavius Echidna: anterior thoracic nerve	Platypus: C4 to C6; Echidna: C4 and C5
		Sterno-epicoracoideus		
Muscles inserting onto the humerus	Pectoralis major		Nerve to the pectoralis major	Platypus: C4 to C7; Echidna: C4 to C7
	Pectoralis quartus		External anterior thoracic nerve	C4 to C6
	Latissimus dorsi Dorso-epitrochlearis		Cervical spinal nerves	C6 to C7. Also T1 and T2 for dorso-epitrochlearis in the platypus
	Deltoideus		Axillary nerve	C4 to C6
	Epicoracohumeralis		Supracoracoid nerve	C4, C5
	Teres major		From the cord of the brachial plexus formed from C4 to C6	C4 to C6
	Teres minor		From the cord formed from C4 to C6. In the echidna there is also a supply from the axillary nerve	C4 to C6
	Supraspinatus		Supracoracoid nerve	C4, C5
	Infraspinatus		Supracoracoid nerve with some twigs from the axillary nerve	C4, C5, C6
	Subscapularis		Nerve to the subscapularis	C4 to C6
Brachial muscles	Biceps brachii		Musculocutaneous nerve	C4 to C7
	Coracobrachialis		Musculocutaneous nerve	C4 to C7
	Epicoracobrachialis		From the musculocutaneous (platypus) or from the brachial plexus cord formed by C4 to C7 (echidna)	C4 to C7
	Triceps brachii		Radial nerve (deep branch). Also called musculospiral nerve	C7

1 Based on McKay (1894). Note that there is some disagreement between authors concerning segmental innervation of some nerves (reviewed in McKay 1894). The segments provided here are a consensus and may differ between individual animals.

Brachial musculature is divided into flexor and extensor compartments, separated by neurovascular structures (brachial vessels and nerves). In the flexor compartment is the biceps brachii (with epicoracoid and coracoid heads), the coracobrachialis and the epicoracobrachialis. The extensor compartment is occupied by the triceps brachii muscle. Curiously, despite its name, the triceps brachii of the extensor compartment actually has four heads in the monotremes (scapular, external humeral, proximal

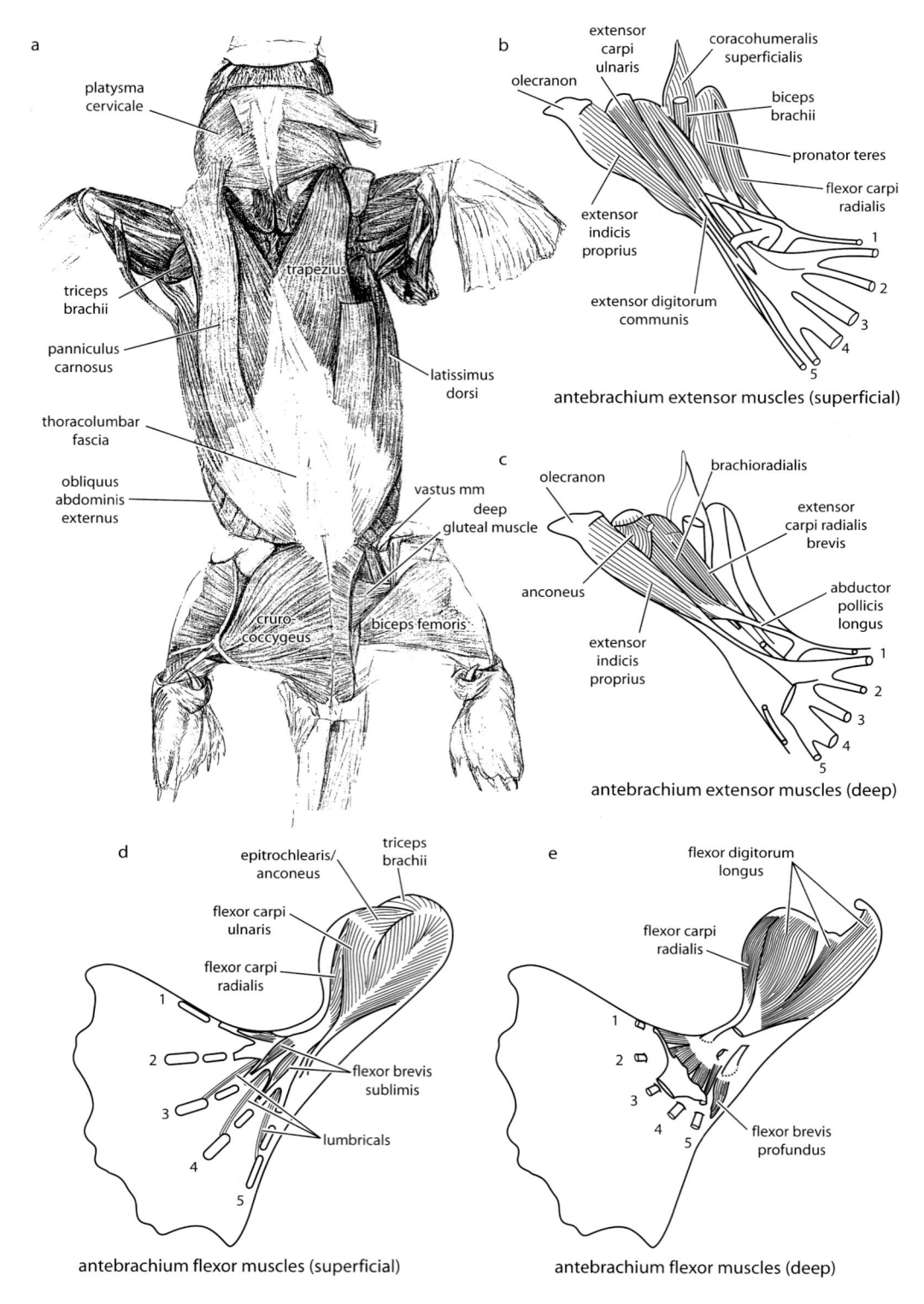

Figure 15.4: Musculature of the trunk and limbs in the platypus. Illustration (a) is by Anne Musser and shows the major superficial muscles of the limbs and trunk. Illustrations (b) and (c) show, respectively, superficial and deep extensor compartment muscles of the antebrachium (forearm). Illustrations (d) and (e) are of respective superficial and deep flexor compartment muscles. Illustrations (b) to (e) are based on illustrations in Howell (1936).

Table 15.3. Muscles of the antebrachium (forearm) and manus and their innervation[1]

Compartment group	Muscle group	Divisions	Innervation
Extensor (dorsal) compartment of forearm	Extensor humeroradialis	• Extensor humeroradialis (carpi) longus • Extensor humeroradialis (carpi) brevis • Brachioradialis	Radial nerve
	Extensor humerodorsalis	• Extensor digitorum communis in two divisions	Radial nerve
	Extensor humeroulnaris	• Extensor carpi ulnaris • (Extensor digiti quinti?) • Anconeus	Radial nerve
Flexor (ventral) compartment of forearm	Flexor humeroradialis	• Flexor carpi radialis • Pronator teres	Median nerve
	Flexor humeropalmaris	• Humeropalmar part of flexor digitorum longus	Ulnar and median nerve
	Flexor humeroulnaris	• Flexor carpi ulnaris (arising by two heads, one from the humerus, one from the ulna) • Epitrochleoanconeus	Ulnar nerve
Extensor (dorsal) compartment of manus	Ulnocarpal series	• Extensor indicis proprius • Abductor pollicis longus	Radial nerve
Flexor (ventral) compartment of manus	Ulnopalmar flexors	• Ulnopalmar part of flexor digitorum longus	Ulnar and median nerve
	Superficial podial flexors	• Flexor digitorum brevis sublimis • Lumbricals	Ulnar nerve
	Deep podial flexors	• Flexor digitorum brevis profundus (interossei)	Ulnar nerve

1 Based on Howell (1936).

internal humeral and distal internal humeral) in both monotremes (McKay 1894).

Muscles of the antebrachium and manus

The musculature of the platypus forelimb is highly specialised for the characteristic alternate rowing movements of this mammal and cannot be used as a model for deducing the primitive mammalian arrangement (Howell 1936). Howell divided musculature of the forearm and hand of the platypus into dorsal and ventral groups attaching to the elbow, and dorsal and ventral podial groups (see Table 15.3, Fig. 15.4b–d).

Howell has stressed that the brachioradialis of the platypus may not be homologous to the placental brachioradialis. There also does not appear to be a true supinator muscle, pronator quadratus, or short extensors to digits 3 to 5 in the platypus. Division of the deep podial flexors into opponens pollicis or flexor pollicis brevis is not justified.

The innervation of the muscular compartments of the forearm and manus is broadly similar to that in therians, but Howell noted that *both* the median and ulnar nerves innervate *both* the ulnopalmar and humeropalmar heads of the flexor digitorum longus (see Table 15.3). Myotomal supply for the muscles of the antebrachium and manus is unknown.

Cervical plexus

The main branches of the cervical plexus (C1 to C4) are summarised in Table 15.4. As in other mammals, C1 and C2 contribute branches to the descending part of the hypoglossal nerve to supply the infrahyoid muscles. The cervical plexus supplies the levator scapulae and rhomboideus muscles that draw the scapula rostrally and medially. The phrenic nerve (for innervation of the diaphragm) is also derived from the caudal divisions of C3 to C5, as in other mammals.

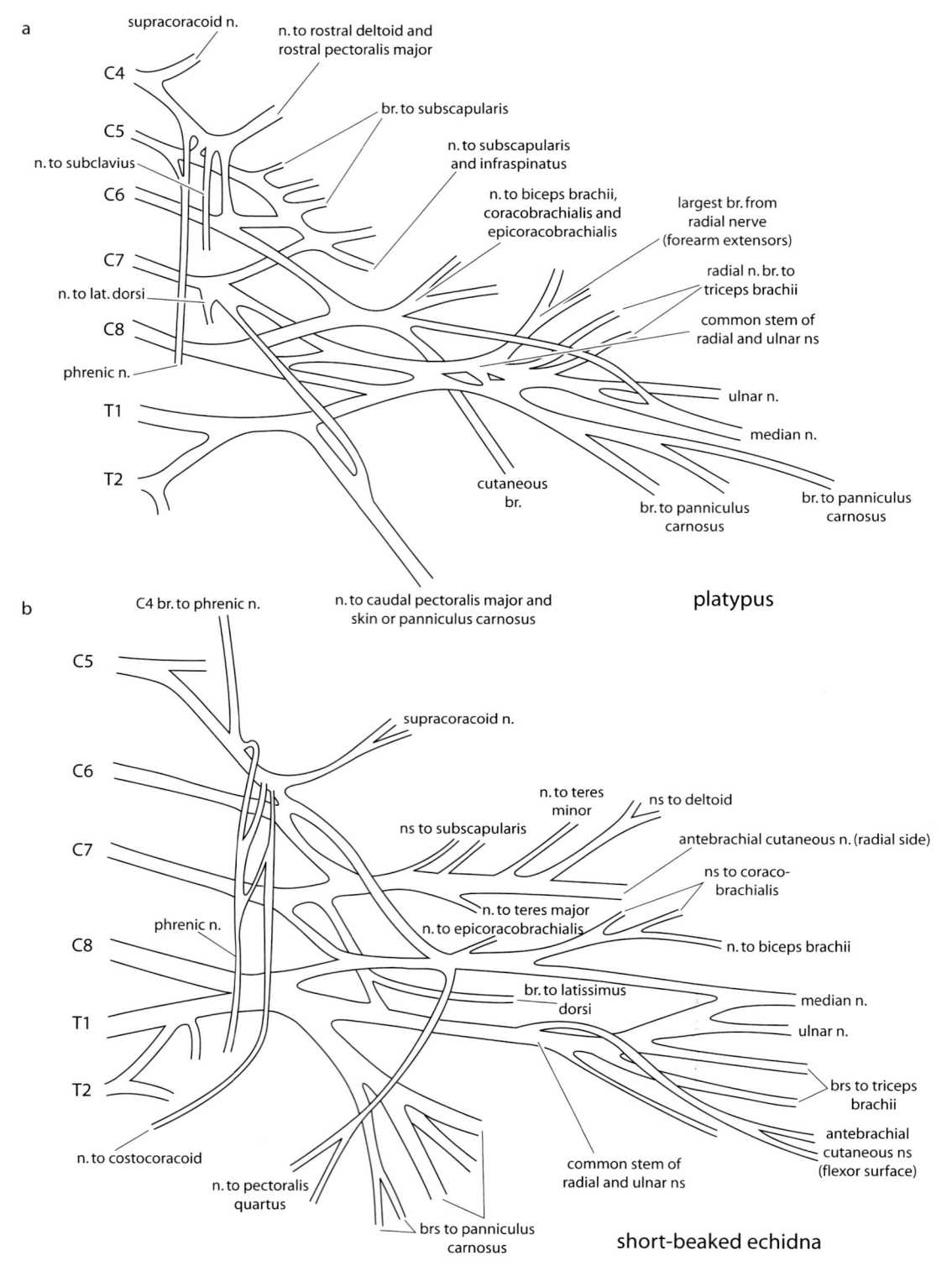

Figure 15.5: The brachial plexus of the platypus and short-beaked echidna. These line diagrams are based on illustrations and descriptions in McKay (1894) and Koizumi and Sakai (1997). Note the poor dorsoventral separation of nerves in the caudal part of the brachial plexus, the long common bundle for the radial and ulnar nerves, and the extensive nerve supply to the panniculus carnosus in both species.

Table 15.4. Branches of the cervical plexus (C1 to C4)[1]

Nerve	Cervical segments	Course and connections
Cervical 1	C1	Provides a branch to the descending part of the hypoglossal nerve.
Cervical 2	C2	Gives a branch to the descending part of the hypoglossal nerve and a branch to the spinal accessory nerve. It also has branches to the auricular region and joins with C3 to form the nerve to the rhomboideus muscle.
Cervical 3	C3	Divides into two branches. The rostral branch joins with C2 to form the nerve to the rhomboideus. The caudal branch contributes to the phrenic nerve, and supplies levator scapulae and serratus magnus.
Cervical 4	C4	Divides into two divisions. The rostral division joins with C3, while the caudal division contributes to the phrenic nerve and joins with C5.
Phrenic	C3 to C5	Takes nerve branches from C3, C4 and C5 spinal nerves as well as the nerve to subclavius.

1 Based on McKay (1894).

Brachial plexus

According to McKay (1894) the brachial plexus in the monotremes is formed from C5 to C8 and T1 spinal nerves, but Koizumi and Sakai (1997) identified contributions from C4 and T2 to the plexus. In general, brachial plexus branches (summarised in Table 15.5) by the schema of Koizumi and Sakai (1997) are one segment rostral to those of McKay (1894). Several authors have noted two distinctive features of the brachial plexus in the platypus and echidna. These are: (1) the lack of dorsal and ventral bundles in the caudal part of the plexus (C7 to T1), even though these are present in the rostral plexus where they give rise to the axillary nerve and lateral cord, respectively (Koizumi and Sakai 1997); (2) the extensive contribution of the brachial plexus to the panniculus carnosus (see below); and (3) the common origin of the radial and ulnar nerves, such that these do not separate until the distal antebrachium in some animals (McKay 1894, Koizumi and Sakai 1997, Fig. 15.5). The lack of dorsal and ventral bundles is a feature of the brachial plexus in several non-mammals, prompting Koizumi and Sakai to suggest that the lack of a dorsoventral division in the caudal monotreme brachial plexus represents an intermediate condition between therians and some non-mammals (i.e. reptiles and urodeles).

Figure 15.5 shows the branches of the brachial plexus in the platypus and echidna based on McKay 1894. After the C4 and C5 nerves have joined, they give off: (1) a branch to the phrenic nerve; (2) a supply to the costocoracoid and sterno-epicoracoid muscles; (3) the supracoracoid nerve; and (4) a dorsal cord that joins the cords from C6 and C7.

The C6 spinal nerve ventral ramus divides into two divisions, the dorsal of which joins with the dorsal cord of C4 and C5 to form a stout trunk that supplies the deltoid, the teres major, and some of the rotator cuff muscles (subscapularis, teres minor, infraspinatus), as well as providing the axillary and radial nerves.

The C7 spinal nerve ventral ramus also divides into two cords, the rostral of which joins with the dorsal branch of C6 before dividing into two parts. One of these parts joins with the ventral division of C6 and the ventral division formed by the junction of C4 and C5 to form another stout cord that supplies the epicoracobrachialis, pectoralis major, pectoralis quartus, and coracobrachialis, as well as giving off the musculocutaneous nerve. The other part of the rostral division of C7 gives off the radial (musculospiral) and ulnar nerves, as well as a contribution to the median nerve. The more caudal of the C7 divisions forms a cord with C8, T1 and T2 to form a complicated network of nerve branches over the lateral and ventral aspects of the thorax and abdomen, many of which provide extensive supply to the panniculus carnosus.

Superficial musculature of the trunk

Panniculus carnosus

The panniculus carnosus is a sheet of muscle that covers much of the trunk in both the platypus and echidnas. It develops during a period from immediately before to ~10 weeks after hatching (Fig. 15.6). In

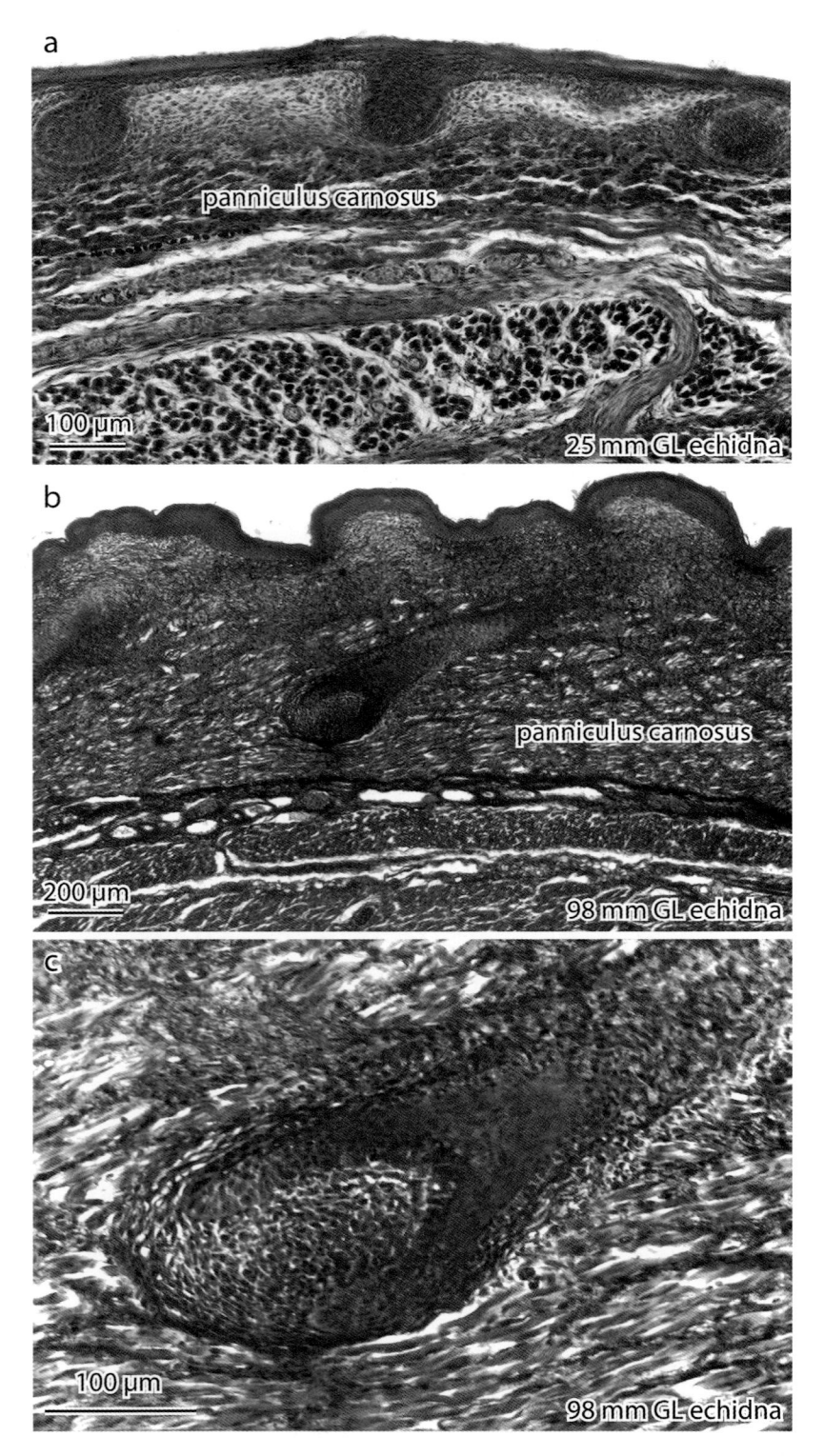

Figure 15.6: The histology and distribution of the developing panniculus carnosus. The photomicrographs show the histology of the panniculus carnosus in a 25 mm GL (a) and a 98 mm (b, c) short-beaked echidna. Photomicrograph (c) is a higher power view of the spine follicle in (b). Note that spine bases are above the panniculus at 25 mm GL, but fully embedded by 98 mm GL.

Table 15.5. Brachial plexus branches in monotremes

Nerve	Origin	Notes on course and muscles supplied
Supracoracoid nerve	C4[1] or C5[2]	Supplies epicoracohumeralis muscle (ventral branch) and supraspinatus and infraspinatus (dorsal branch)
Nerve to subscapularis	C4/5/6[1] C5/6[2]	Arises from the dorsal division of the upper brachial plexus trunk
Nerve to teres major	C5/6[1] C4/5/6[2]	Arises from the dorsal division of the upper brachial plexus trunk
Nerve to teres minor	C5/6[1] C4/5/6[2]	Arises either as a discrete nerve[2] or a branch of the axillary nerve[1]
Axillary nerve	C4/5/6[1] C6/7[2]	Branches from the dorsal division of the upper trunk[1] to supply the deltoid or teres minor.
Nerve to deltoid	C4/5/6[1] C4/5/6[2]	May be a branch of the axillary nerve[1] or a discrete branch.[2]
Nerve to infraspinatus	C4/5/6[2]	Not shown as a discrete branch by Koizumi and Sakai (1997). Arises in common with nerve to deltoid according to McKay (1894).
Superficial radial nerve	C5/C6[1,2]	Extension of the axillary nerve[1]. Continuing fibres of nerves to shoulder girdle muscles[2].
Nerve to epicoracobrachialis	C4/5/6/7[1,2]	May arise from the ventral aspect of the rostral trunk of the brachial plexus.[1,2]
Nerve to pectoralis major	C4/5/6/7[1,2]	May arise from the ventral aspect of the rostral trunk of the brachial plexus.[1,2]
Musculocutaneous nerve	C5/6[1] C4/5/6/7[2]	Branches off the ventral bundle formed by the union of the ventral divisions of the upper and middle trunks.[1]
Median nerve	C5/6[1] C6/7[2]	Branches off the ventral bundle formed by the union of the ventral divisions of the upper and middle trunks.[1] It supplies some flexor muscles of antebrachium and manus.
Nerve to dorsoepitrochlear	C7[2]	Branches off the ventral division of C7.
Radial (musculospiral) nerve	C6/7/8[1] C8[2]	Arises from the caudal half of the plexus. Usually arises in common with the ulnar nerve.[1,2] It supplies the triceps brachii and the extensor compartment muscles of antebrachium.
Ulnar nerve	C7/8[1,2]	Usually arises in common with the radial nerve.[1,2] It supplies most flexors of antebrachium and manus.

1 Based on Koizumi and Sakai (1997).
2 Based on McKay (1894).

the short-beaked echidna, the spine bases both penetrate the panniculus carnosus and receive individual muscle slips from the panniculus (McKay 1894), allowing contraction of the panniculus to move the spines and assist locomotion in confined spaces. Shortly after hatching, the bases of the developing spines are separate from the panniculus, but the spines become embedded during the second month of post-hatching life (Fig. 15.6).

In the adult short-beaked echidna, the panniculus starts from the posterovertebral region and runs ventrally, winding around the lateral aspects of the body. Further rostrally, the panniculus runs over the scapula and cervical regions and even extends onto the forearm. The panniculus also has special slips that run into both the neck (hypodermal slip) and head (dermodorsi cervicales). If one follows the panniculus of the echidna around the lateral curvature of the body, it forms a continuous sheet overlying the abdominal muscles. When the muscle fibres approach the metasternum they split into superficial and deep layers. The panniculus also projects into the forelimbs: some fibres of the superficial panniculus connect to the distal ulna close to its articulation with the carpal bones to form a dermoflexor antebrachii muscle; and the deep layer of the panniculus attaches

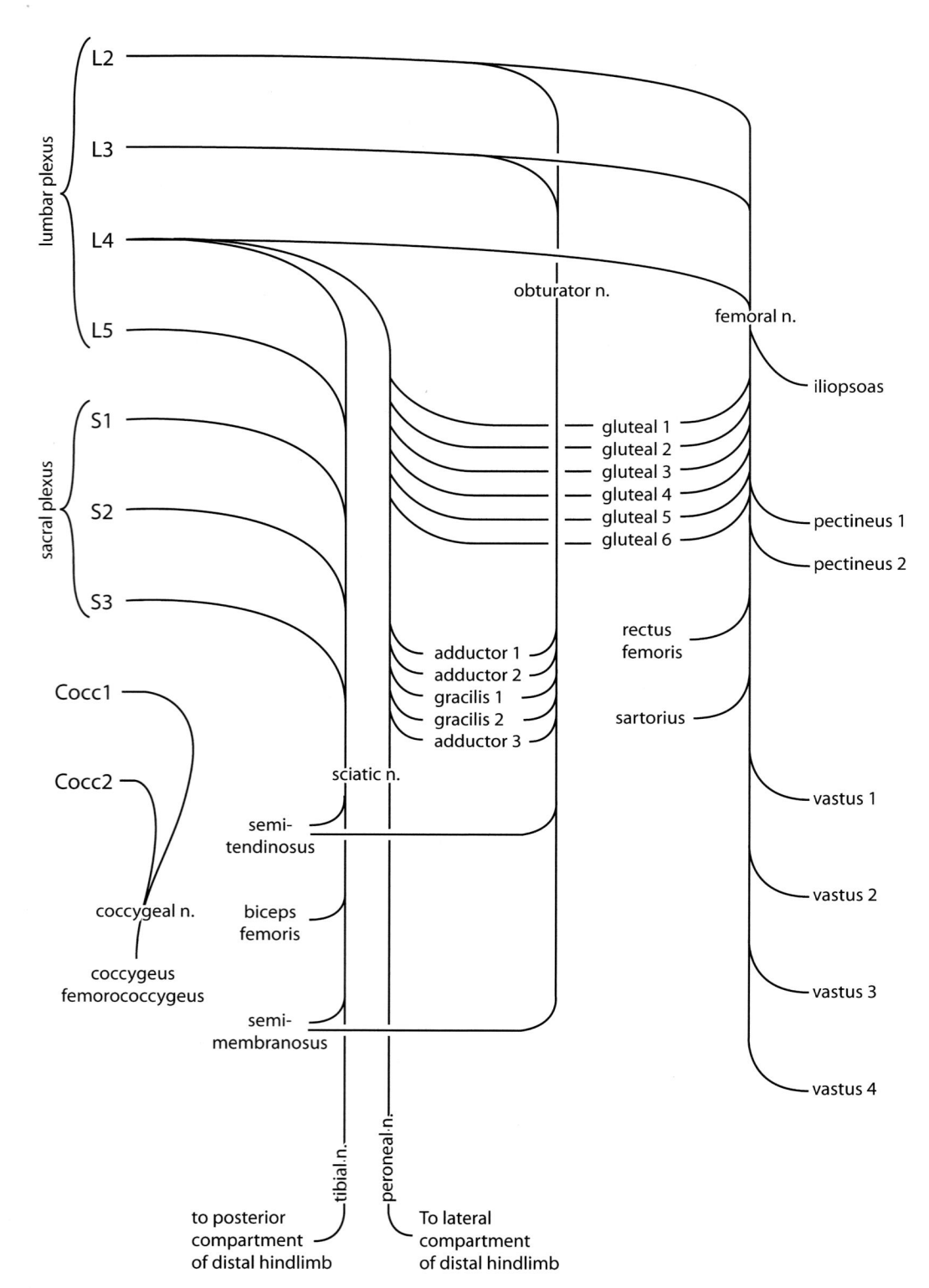

Figure 15.7: Schematic representation of hindlimb nerves in the short-beaked echidna. This diagram is based on illustrations in Ellsworth (1974). Note the contribution of both the femoral nerve and sciatic L4 branch to the supply of the gluteal musculature by multiple nerve branches. The sciatic and obturator nerve also share some nerve supply with the adductor group of muscles (including the gracilis) and the semitendinosus and semimembranosus.

to the deltopectoral ridge of the humerus as the dermoflexor brachii muscle (McKay 1894). These connections to the forelimb are most likely used to shift the panniculus rostrally, but may also assist in forelimb movements.

The panniculus carnosus of the platypus is broadly similar to that in the short-beaked echidna, except that the ventral sheet of the panniculus that covers the abdomen is not so readily separated into divisions.

Innervation of the panniculus carnosus is (presumably) segmentally based and derived from dorsal and ventral rami of the spinal nerves (including cervical spinal nerves C1 to C4, the brachial plexus – see above, and the lateral cutaneous nerve of the thorax (McKay 1894, Fig. 15.5), but this has not been tested by electromyography.

Innervation of the musculature of the pelvis and hindlimb

Muscles of the monotreme pelvis and thigh

Musculature of the monotreme hindlimb is broadly similar to that in therians. The following description of the musculature of the pelvis and gluteal region is based mainly on the findings of Ellsworth (1974) and Walter (1988) in the short-beaked echidna. Superficial muscles of the gluteal and thigh region include the gluteus maximus caudally, the gluteus medius and gluteus minimus further rostrally, as well as other much smaller unspecified gluteal muscles between the gluteus minimus and medius (Walter 1988), although Ellsworth simply delineated these as gluteus 1 to 4 (Fig. 15.4a). The biceps femoris muscle also lies at this level, emerging from under the cover of the gluteus maximus to insert into a band of crural fascia that runs from the knee to the ankle (Fig. 15.4a). In the platypus, Pearson (1926) identified a large superficial muscle extending from the sacrum to the hindlimb (crurococcygeus), which appears to correspond to gluteus maximus, and with the biceps femoris visible beneath its lateral border (Fig. 15.4c). Additional gluteal muscles are visible when the crurococcygeus is removed (Pearson 1926).

As one approaches from the lateral side of the animal, two further layers of muscle can be dissected. Deep to the gluteal and biceps femoris muscles is an intermediate layer that includes the iliacus, rectus femoris, pectineus and piriformis muscles (Ellsworth 1974; Pearson 1926). This in turn overlies a deeper layer of thigh and gluteal muscles, including psoas magnus, obturator externus, obturator intermedius, quadratus femoris, and vastus externus and internus.

In dissections from the ventral or medial aspect of the hindlimb, the most superficial muscles are the sartorius and gracilis. These overlie the medial hamstrings (semimembranosus and semitendinosus muscles) and a group of femoral adductors (brevis and longus). A pectineus muscle lies deep to the sartorius.

Muscles of the monotreme crus and pes

Extensor compartment muscles of the crus of the echidna (Walter 1988) include: (1) a superficial group made up of tibialis anticus (anterior), extensor digitorum for digit 2, extensor digitorum communis, extensor digitorum for digits 4 and 5, and peroneus longus; and (2) a deep group made up of the extensor hallucis longus. The flexor compartment of the crus and pes in the echidna includes: (1) a superficial group comprising the gastrocnemius, soleus and plantaris; and (2) a deep group made up of tibialis posticus (posterior), flexor digitorum longus, popliteus, a single lumbrical and the abductor hallucis.

Innervation of the hindlimb

The nerve supply of the monotreme hindlimb is organised much like than in therians, with distinct femoral, obturator, and sciatic nerves, and with the sciatic nerve exhibiting tibial and peroneal components (Fig. 15.7).

16

Atlas of the adult and developing brain and spinal cord of the short-beaked echidna (*Tachyglossus aculeatus aculeatus*)

Ken W. S. Ashwell and Craig D. Hardman

Introduction

The short-beaked echidna (*Tachyglossus aculeatus*) is widespread throughout continental Australia as well as on associated islands, including Tasmania and New Guinea. Several subspecies are recognised in different parts of Australia (see Chapter 2): *Tachyglossus aculeatus acanthion* (Northern Territory, northern Queensland, inland Australia and Western Australia); *Tachyglossus aculeatus aculeatus* (eastern New South Wales and Victoria, southern Queensland); *Tachyglossus aculeatus lawesii* (New Guinea lowlands); *Tachyglossus aculeatus multiaculeatus* (South Australia and Kangaroo Island); and *Tachyglossus aculeatus setosus* (Tasmania).

Adult body weight for short-beaked echidnas may reach 7 kg, but is usually between 3 kg and 5 kg. Males are ~25% larger than females and have a crural spur on the hindlimb, although this may sometimes also be present on females. The body is covered on the back and sides with spines interspersed within fur that is usually brown or black, but occasionally lighter in poorly pigmented individuals. The eyes are small and black and the external ear opening is difficult to see. The head tapers to a straight, cylindrical snout or beak that may be used as a probe or lever in the search for invertebrate food. There is a pronounced flexure of the vertebral column at the junction of the short neck with the trunk. The limbs are held horizontally away from the body and the hindlimb is rotated so as to point to the rear. Forelimbs have spatulate claws for digging, whereas the hindlimb claws are thinner. Body temperature rarely rises above 33°C and basal metabolic rate is approximately one-third that of a placental carnivore of the same body weight. Life span may be up to 45 years in the wild (Augee *et al*. 2006).

Both the cerebral cortex and the olfactory bulb are highly folded (gyrified), but the cerebral cortex is thinner than in the platypus.

Materials and methods

Animals, ethics and anaesthesia for the adult brain atlas

The four short-beaked echidnas (*Tachyglossus aculeatus aculeatus*) used in this study were obtained from Mudgee and Tumut in eastern and southern rural New South Wales, respectively. The animals were obtained from the wild under licence from the New South Wales National Parks and Wildlife Service. All procedures were performed under guidelines established by the National Health and Medical Research Council of Australia, and the Animal Care and Ethics Committee of the University Of New South Wales approved all methods for animal capture and handling. All work was supported by a grant from the Australian Research Council.

The animal depicted in the brain plates (Ec-Ad01 to 30) and mapped in the line diagrams (Echidna 96/2) was used for neurophysiological experiments on somatosensory pathways before being overdosed by intramuscular injection of pentobarbitone sodium (60 mg/kg) and transcardially perfused with normal saline at 4°C for 10 minutes, 4% paraformaldehyde in 0.1 M phosphate buffer (4°C, pH 7.4) for 20 minutes, and cold (4°C) 0.1 M phosphate buffer for 10 minutes. The animal used for the depiction of the adult spinal cord (Ec-Ad31 to 33) is different from that used for the brain, but it had been obtained and anaesthetised in a similar fashion (see details in Ashwell and Zhang 1997).

Tissue processing for the adult brains and spinal cord

The brain tissue was immersed in 30% sucrose for 24 hours at 4°C and sectioned in the frontal plane with the aid of a cryostat at a thickness of 40 mm. Ten series through the brain were collected and stained for a variety of markers (Nissl substance; enzyme histochemistry for cytochrome oxidase; NADPH-diaphorase; acetylcholinesterase; and immunohistochemistry to parvalbumin, calbindin, calretinin, tyrosine hydroxylase, neurofilament protein – SMI-32 antibody – and neuropeptide Y). Only Nissl and parvalbumin-immunoreacted sections have been depicted here. The details of stains and antibodies used are provided in Hassiotis *et al.* (2004b).

The spinal cord and dural sac of the animal used for the spinal cord atlas were removed by laminectomy and the spinal cord cut into blocks 6–10 mm in length. Four series of sections were cut in a transverse plane at a thickness of 30 μm with the aid of a cryostat and one series was stained with cresyl violet.

Imaging and delineation

Sections were digitally scanned with the aid of an Aperio ScanScope XT slide scanner in either the Histology and Microscopy Unit of the School of Medical Sciences at the University of New South Wales or the Paxinos laboratory at Neuroscience Research Australia (NeuRa). Images were cleaned and digitally enhanced (confined to adjustment of contrast and brightness for the whole image) in Adobe Photoshop CS2 and placed in Adobe Illustrator CS2. Outlines of structures (brain nuclei, fibre bundles, cortical functional areas) were drawn and labelled using a system of abbreviations derived from that used by the Paxinos group. Each plate includes a low power line diagram of a section from the right half of the brain, accompanied by a caption and one or more higher-powered images of specific regions of interest. The numbers beneath the finder diagram in the upper right-hand corner of the Ec-Ad illustrations denote the distance in millimetres from the front of the brain.

Embryos

The two embryonic specimens illustrated (Ec-InA – M154, Ec-InB – M158) are from the Hill collection in the Museum für Naturkunde, Berlin. No information is available concerning the precise site of origin of the specimens, but they probably came from eastern New South Wales or southern Queensland, since that is the likely region from which Caldwell and Hill would have collected in the late nineteenth and early twentieth centuries.

The early to middle incubation phase embryo (Ec-InA – M154, also called B119) has a greatest length of 7.5 mm, but that length was estimated from the known section thickness and the number of sections available. The embryo corresponds to stage 15 as depicted in Werneburg and Sánchez-Villagra (2011) with 26 or 27 somite pairs, four pharyngeal arches and a gut that is wide open to the yolk sac. The specimen had been embedded in paraffin and sectioned at a thickness of 10 µm. Sections were mounted on 16 slides and appear to have been stained for hematoxylin and eosin.

The late incubation specimen (Ec-InB – M158) had a greatest length of 12.5 mm (probably a few days before hatching; corresponding to stages 22 and 23 of Werneburg and Sánchez-Villagra 2011) and was apparently embedded in paraffin. The head and upper trunk was sectioned in the frontal plane at a thickness of 10 µm and the sections mounted on 27 glass slides and stained with hematoxylin and eosin. The embryo appears to have its gut still inside a physiological herniation.

Photography and delineation of embryos

The sections were photographed with the aid of a Zeiss Axioplan2 microscope fitted with an AxioCam MRc5 camera. All images were calibrated by photographing a scale bar at the same magnification. The nervous systems of both embryonic specimens were mapped using modern terminology for neuromeric subdivisions and labelled using a system of abbreviations derived from that used by the Paxinos group (Ashwell and Paxinos 2008). Features of the skeleton were based on Kuhn (1971). Primordia of brain regions were denoted by an asterisk, e.g. Cb* denotes the primordium of the cerebellum. The numbers beneath the finder diagram in the upper right-hand corner of the Ec-InB illustrations denote the distance in millimetres from the front of the brain.

Ec-Ad01

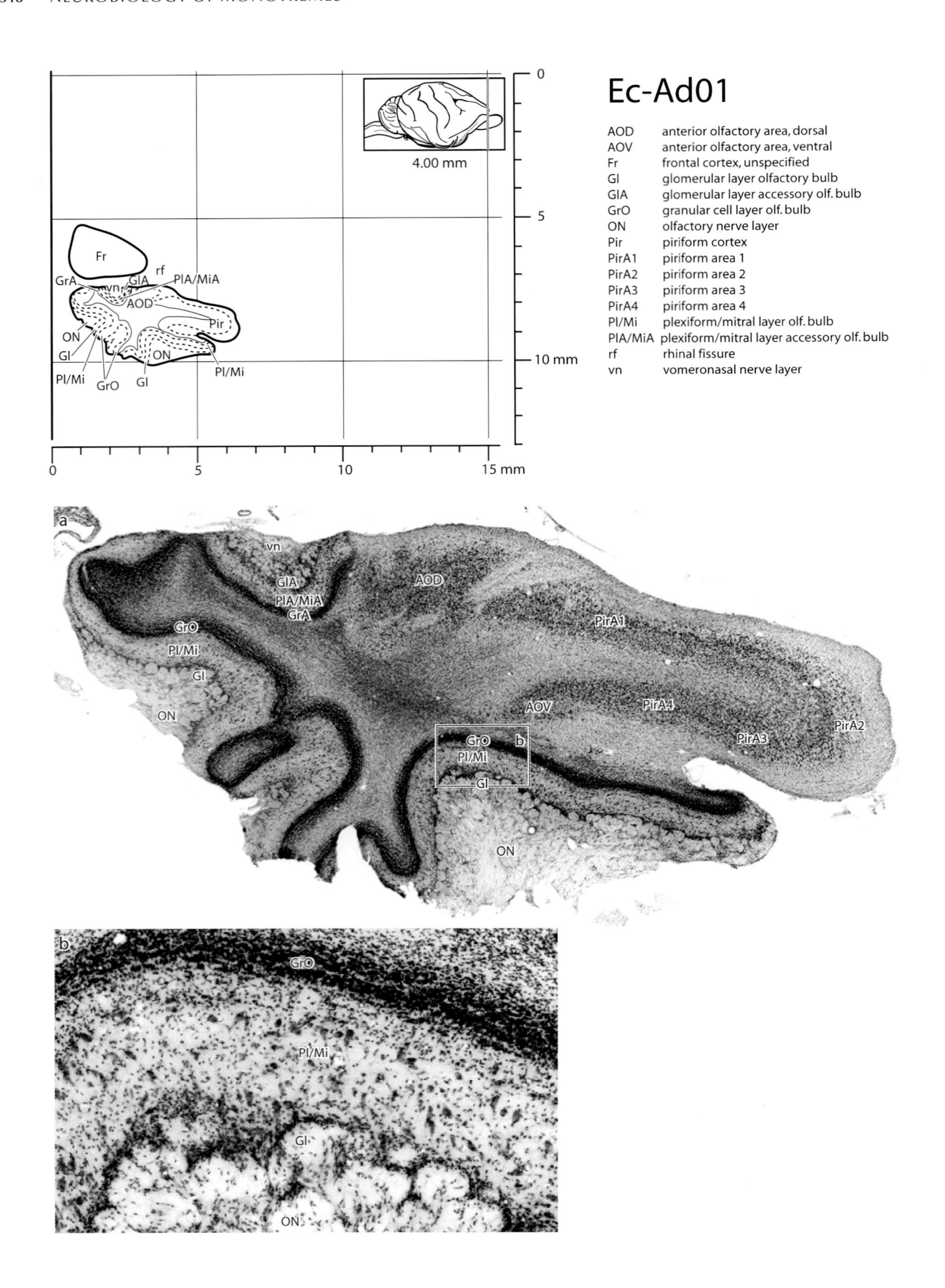

AOD	anterior olfactory area, dorsal
AOV	anterior olfactory area, ventral
Fr	frontal cortex, unspecified
Gl	glomerular layer olfactory bulb
GlA	glomerular layer accessory olf. bulb
GrO	granular cell layer olf. bulb
ON	olfactory nerve layer
Pir	piriform cortex
PirA1	piriform area 1
PirA2	piriform area 2
PirA3	piriform area 3
PirA4	piriform area 4
Pl/Mi	plexiform/mitral layer olf. bulb
PlA/MiA	plexiform/mitral layer accessory olf. bulb
rf	rhinal fissure
vn	vomeronasal nerve layer

Ec-Ad02

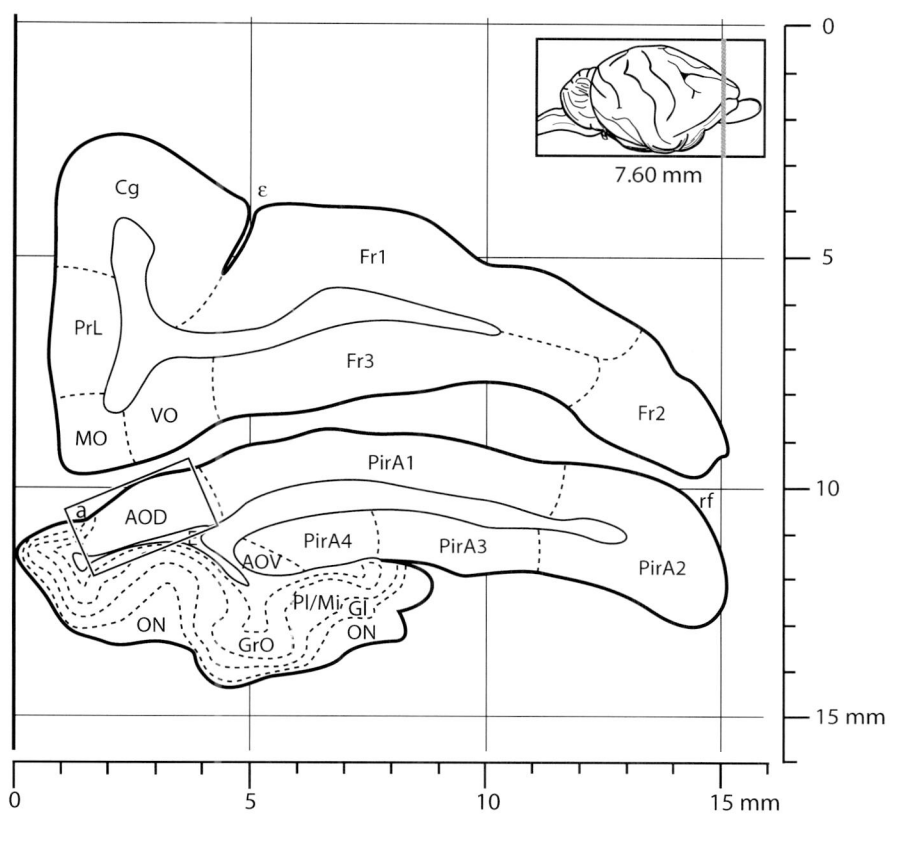

ε	sulcus ε
AOD	anterior olfactory area, dorsal
AOV	anterior olfactory area, ventral
Cg	cingulate cortex
E	ependyma and subependyma
Fr1	frontal cortex area 1
Fr2	frontal cortex area 2
Fr3	frontal cortex area 3
Gl	glomerular layer olfactory bulb
GrO	granular cell layer olfactory bulb
MO	medial orbital cortex
ON	olfactory nerve layer
PirA1	piriform area 1
PirA2	piriform area 2
PirA3	piriform area 3
PirA4	piriform area 4
Pl/Mi	plexiform/mitral layer olfactory bulb
PrL	prelimbic cortex
rf	rhinal fissure
VO	ventral orbital cortex

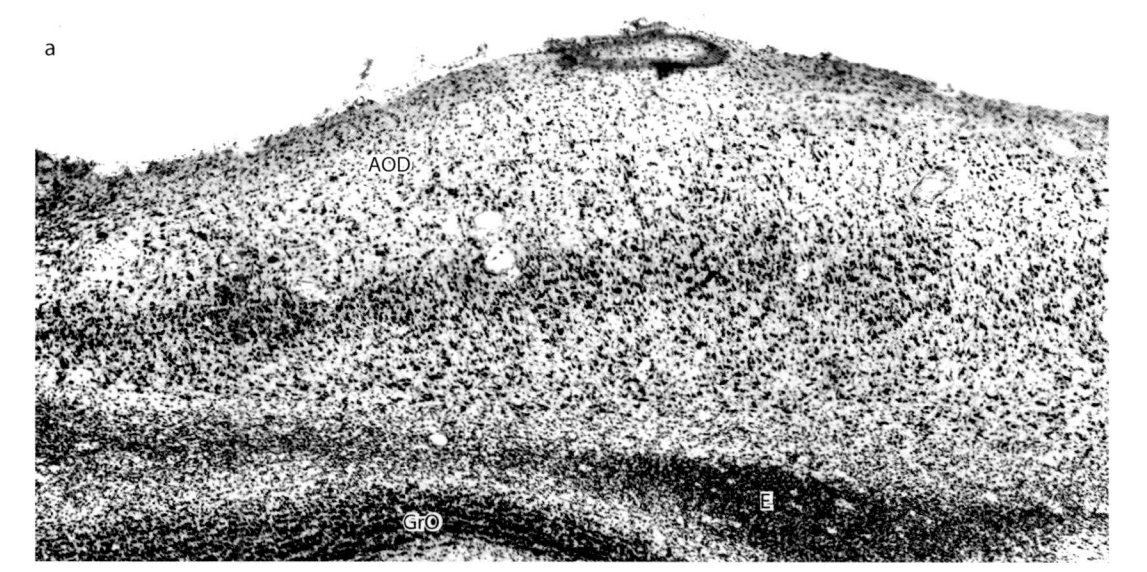

7.60 mm

Ec-Ad03

ε	sulcus ε
Cg	cingulate cortex
DTr	dorsal transition zone
E	ependyma and subependyma
Fr1	frontal cortex area 1
Fr2	frontal cortex area 2
Fr3	frontal cortex area 3
Gl	glomerular layer olfactory bulb
GrO	granular cell layer olfactory bulb
Ins	insular cortex
MO	medial orbital cortex
ON	olfactory nerve layer
PirA1	piriform area 1
PirA2	piriform area 2
PirA3	piriform area 3
PirA4	piriform area 4
Pl/Mi	plexiform/mitral layer olfactory bulb
PrL	prelimbic cortex
rf	rhinal fissure
VO	ventral orbital cortex

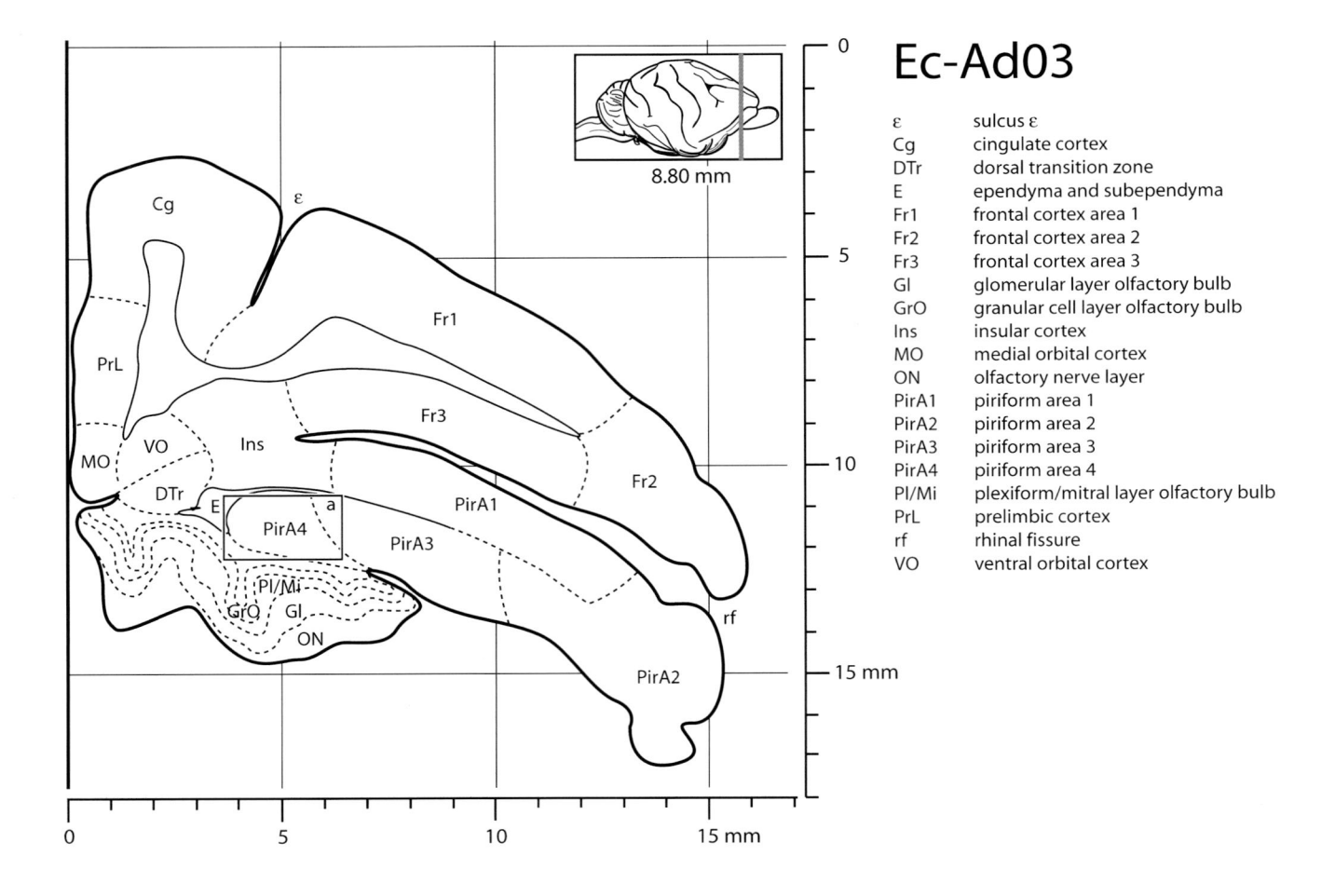

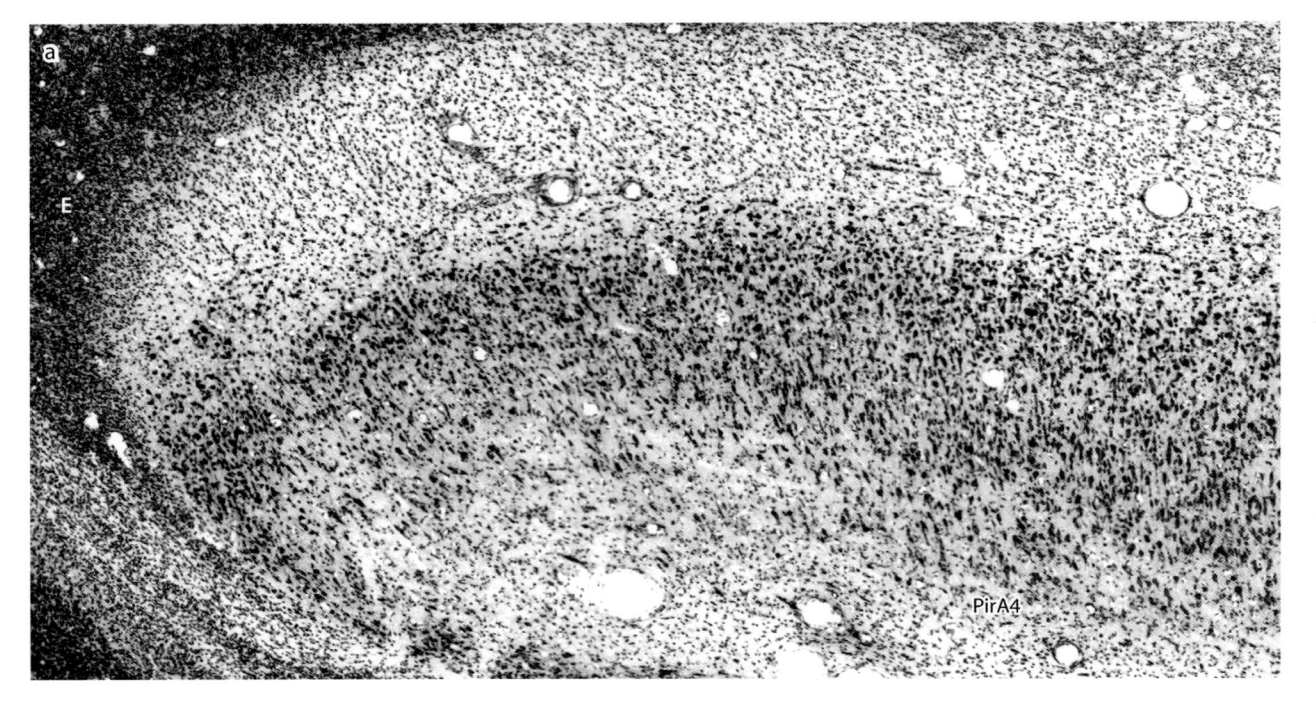

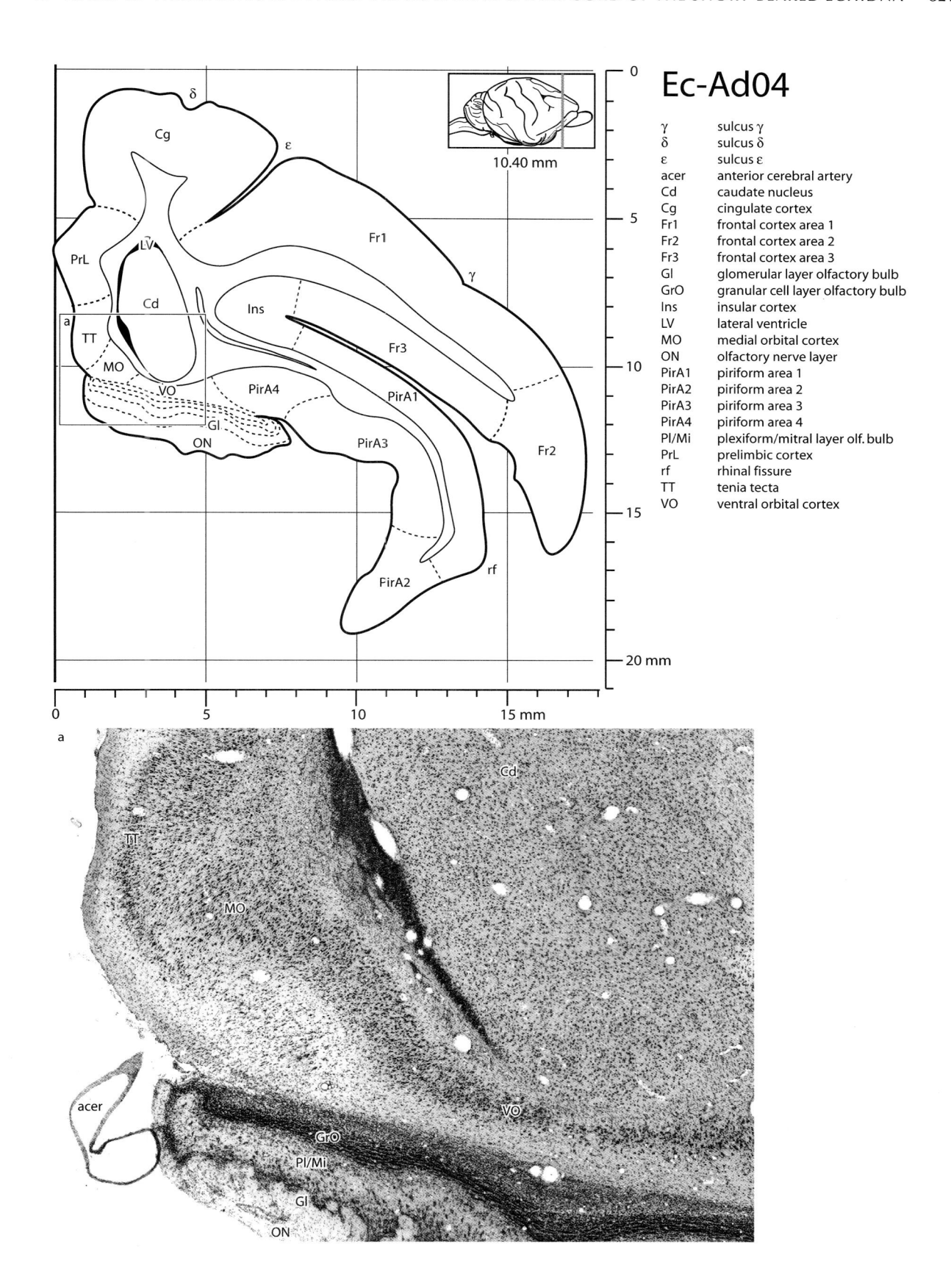

Ec-Ad04

10.40 mm

γ	sulcus γ
δ	sulcus δ
ε	sulcus ε
acer	anterior cerebral artery
Cd	caudate nucleus
Cg	cingulate cortex
Fr1	frontal cortex area 1
Fr2	frontal cortex area 2
Fr3	frontal cortex area 3
Gl	glomerular layer olfactory bulb
GrO	granular cell layer olfactory bulb
Ins	insular cortex
LV	lateral ventricle
MO	medial orbital cortex
ON	olfactory nerve layer
PirA1	piriform area 1
PirA2	piriform area 2
PirA3	piriform area 3
PirA4	piriform area 4
Pl/Mi	plexiform/mitral layer olf. bulb
PrL	prelimbic cortex
rf	rhinal fissure
TT	tenia tecta
VO	ventral orbital cortex

Ec-Ad05

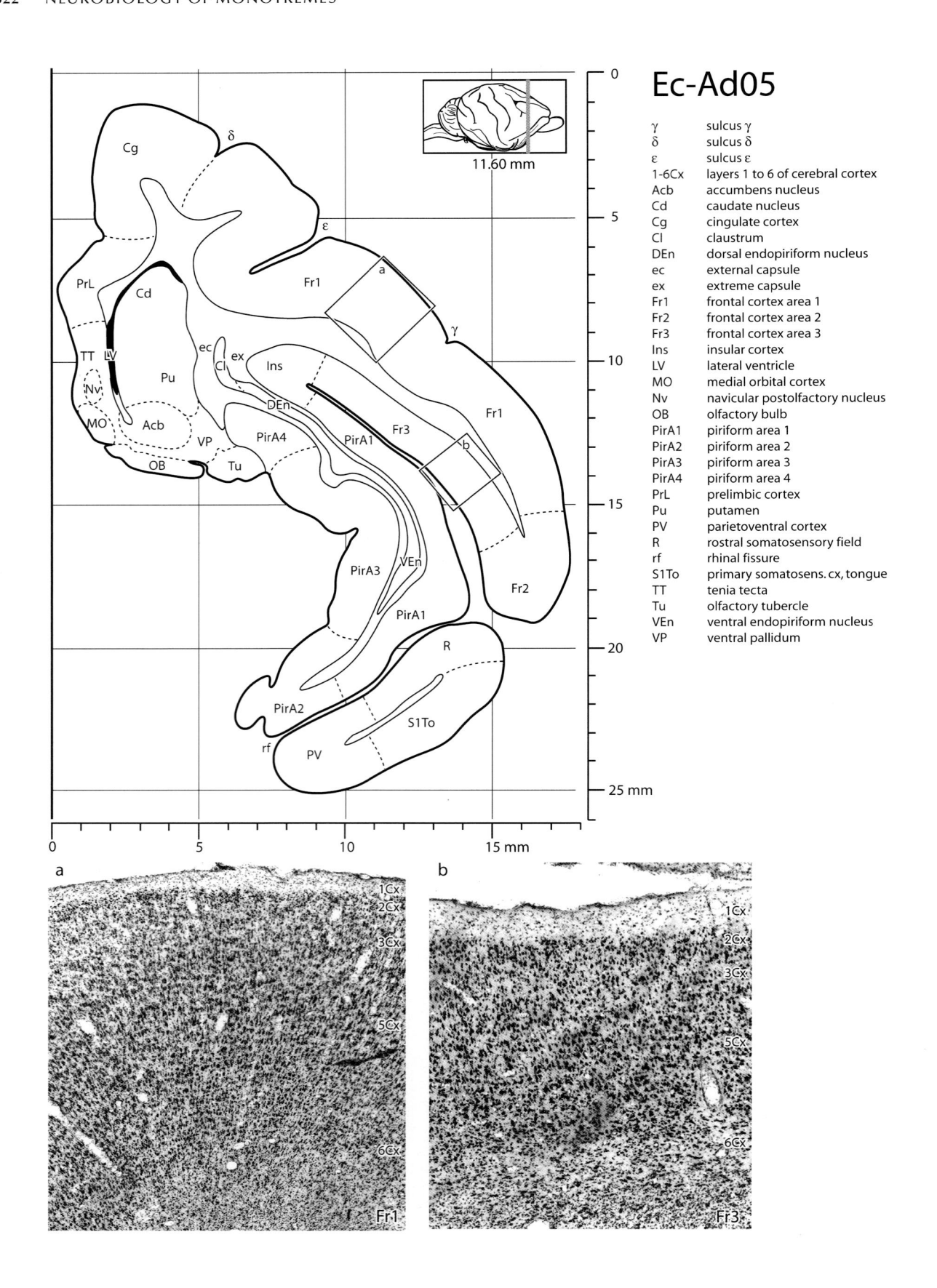

11.60 mm

γ	sulcus γ
δ	sulcus δ
ε	sulcus ε
1-6Cx	layers 1 to 6 of cerebral cortex
Acb	accumbens nucleus
Cd	caudate nucleus
Cg	cingulate cortex
Cl	claustrum
DEn	dorsal endopiriform nucleus
ec	external capsule
ex	extreme capsule
Fr1	frontal cortex area 1
Fr2	frontal cortex area 2
Fr3	frontal cortex area 3
Ins	insular cortex
LV	lateral ventricle
MO	medial orbital cortex
Nv	navicular postolfactory nucleus
OB	olfactory bulb
PirA1	piriform area 1
PirA2	piriform area 2
PirA3	piriform area 3
PirA4	piriform area 4
PrL	prelimbic cortex
Pu	putamen
PV	parietoventral cortex
R	rostral somatosensory field
rf	rhinal fissure
S1To	primary somatosens. cx, tongue
TT	tenia tecta
Tu	olfactory tubercle
VEn	ventral endopiriform nucleus
VP	ventral pallidum

Ec-Ad06

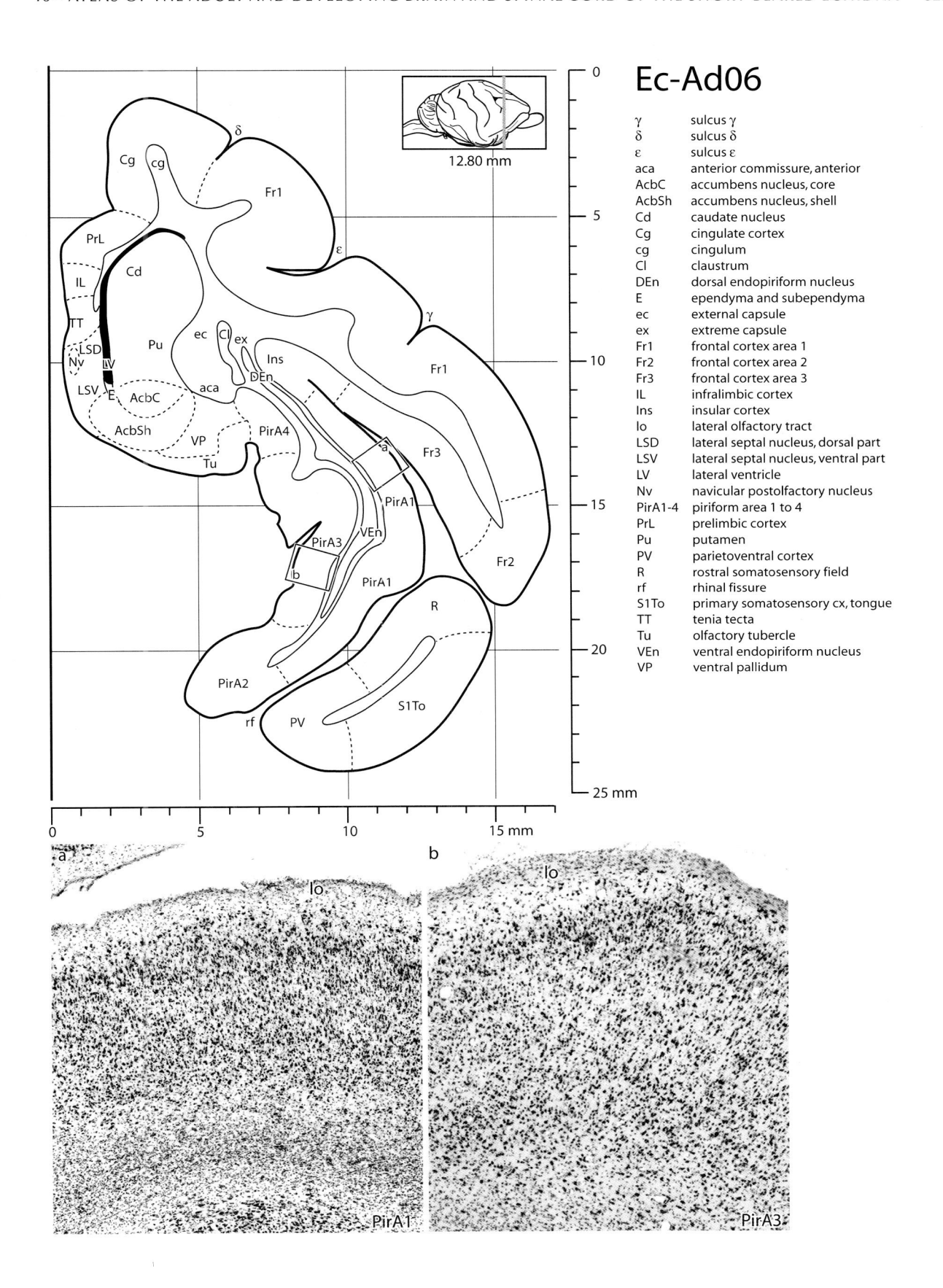

12.80 mm

γ	sulcus γ
δ	sulcus δ
ε	sulcus ε
aca	anterior commissure, anterior
AcbC	accumbens nucleus, core
AcbSh	accumbens nucleus, shell
Cd	caudate nucleus
Cg	cingulate cortex
cg	cingulum
Cl	claustrum
DEn	dorsal endopiriform nucleus
E	ependyma and subependyma
ec	external capsule
ex	extreme capsule
Fr1	frontal cortex area 1
Fr2	frontal cortex area 2
Fr3	frontal cortex area 3
IL	infralimbic cortex
Ins	insular cortex
lo	lateral olfactory tract
LSD	lateral septal nucleus, dorsal part
LSV	lateral septal nucleus, ventral part
LV	lateral ventricle
Nv	navicular postolfactory nucleus
PirA1-4	piriform area 1 to 4
PrL	prelimbic cortex
Pu	putamen
PV	parietoventral cortex
R	rostral somatosensory field
rf	rhinal fissure
S1To	primary somatosensory cx, tongue
TT	tenia tecta
Tu	olfactory tubercle
VEn	ventral endopiriform nucleus
VP	ventral pallidum

Ec-Ad07

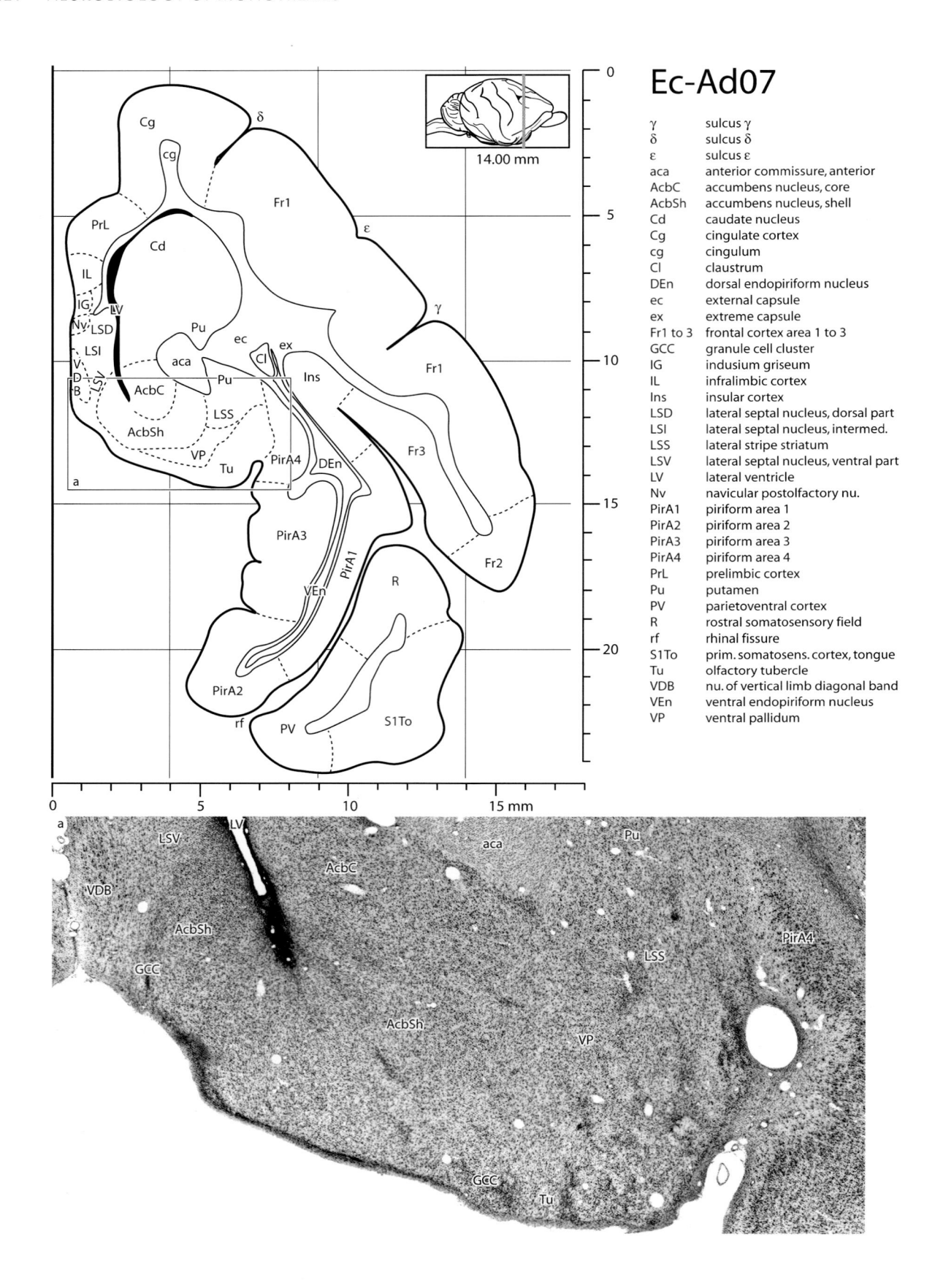

γ	sulcus γ
δ	sulcus δ
ε	sulcus ε
aca	anterior commissure, anterior
AcbC	accumbens nucleus, core
AcbSh	accumbens nucleus, shell
Cd	caudate nucleus
Cg	cingulate cortex
cg	cingulum
Cl	claustrum
DEn	dorsal endopiriform nucleus
ec	external capsule
ex	extreme capsule
Fr1 to 3	frontal cortex area 1 to 3
GCC	granule cell cluster
IG	indusium griseum
IL	infralimbic cortex
Ins	insular cortex
LSD	lateral septal nucleus, dorsal part
LSI	lateral septal nucleus, intermed.
LSS	lateral stripe striatum
LSV	lateral septal nucleus, ventral part
LV	lateral ventricle
Nv	navicular postolfactory nu.
PirA1	piriform area 1
PirA2	piriform area 2
PirA3	piriform area 3
PirA4	piriform area 4
PrL	prelimbic cortex
Pu	putamen
PV	parietoventral cortex
R	rostral somatosensory field
rf	rhinal fissure
S1To	prim. somatosens. cortex, tongue
Tu	olfactory tubercle
VDB	nu. of vertical limb diagonal band
VEn	ventral endopiriform nucleus
VP	ventral pallidum

Ec-Ad08

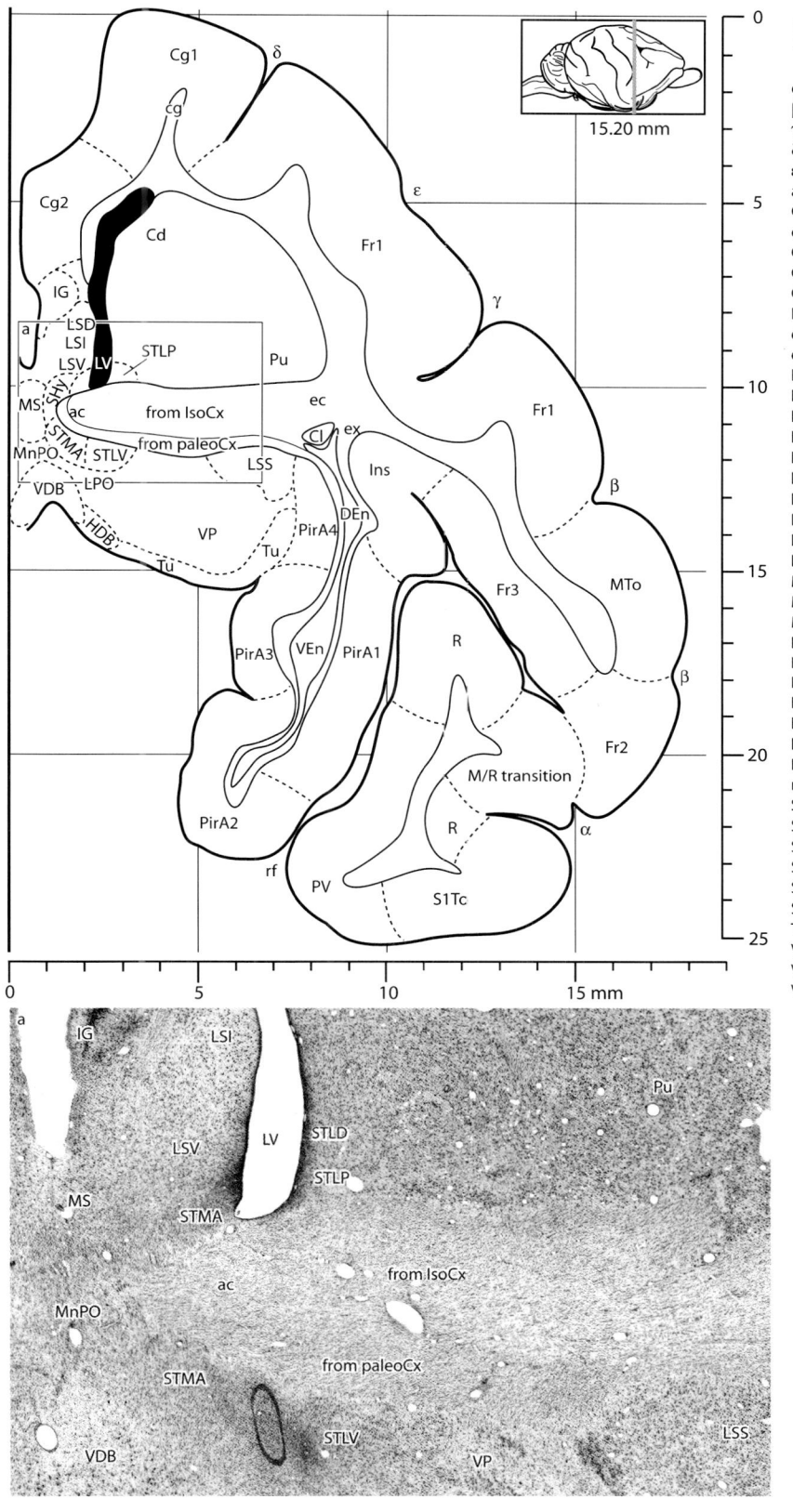

α	sulcus α
β	sulcus β
γ	sulcus γ
δ	sulcus δ
ε	sulcus ε
ac	anterior commissure
Cd	caudate nucleus
cg	cingulum
Cg1	cingulate cortex, area 1
Cg2	cingulate cortex, area 2
Cl	claustrum
DEn	dorsal endopiriform nucleus
ec	external capsule
ex	extreme capsule
Fr1 to 3	frontal cortex areas 1 to 3
HDB	nu. of horiz. limb diagonal band
IG	indusium griseum
Ins	insular cortex
LPO	lateral preoptic area
LSD	lateral septal nu., dorsal part
LSI	lateral septal nu., intermed.
LSS	lateral stripe striatum
LSV	lateral septal nu., ventral part
LV	lateral ventricle
MnPO	median preoptic nu.
MS	medial septal nu.
MTo	motor cortex tongue
PirA1	piriform area 1
PirA2	piriform area 2
PirA3	piriform area 3
PirA4	piriform area 4
Pu	putamen
PV	parietoventral cortex
R	rostral somatosensory field
rf	rhinal fissure
S1To	prim. somatosens. cx, tongue
SHy	septohypothalamic nu.
STLD	bed nu. st, lat. divn, dorsal
STLP	bed nu. st, lat. divn, posterior
STLV	bed nu. st, lat. divn, ventral
STMA	bed nu. st, med. divn, ant.
Tu	olfactory tubercle
VDB	nu. of vertical limb diag. band
VEn	ventral endopiriform nucleus
VP	ventral pallidum

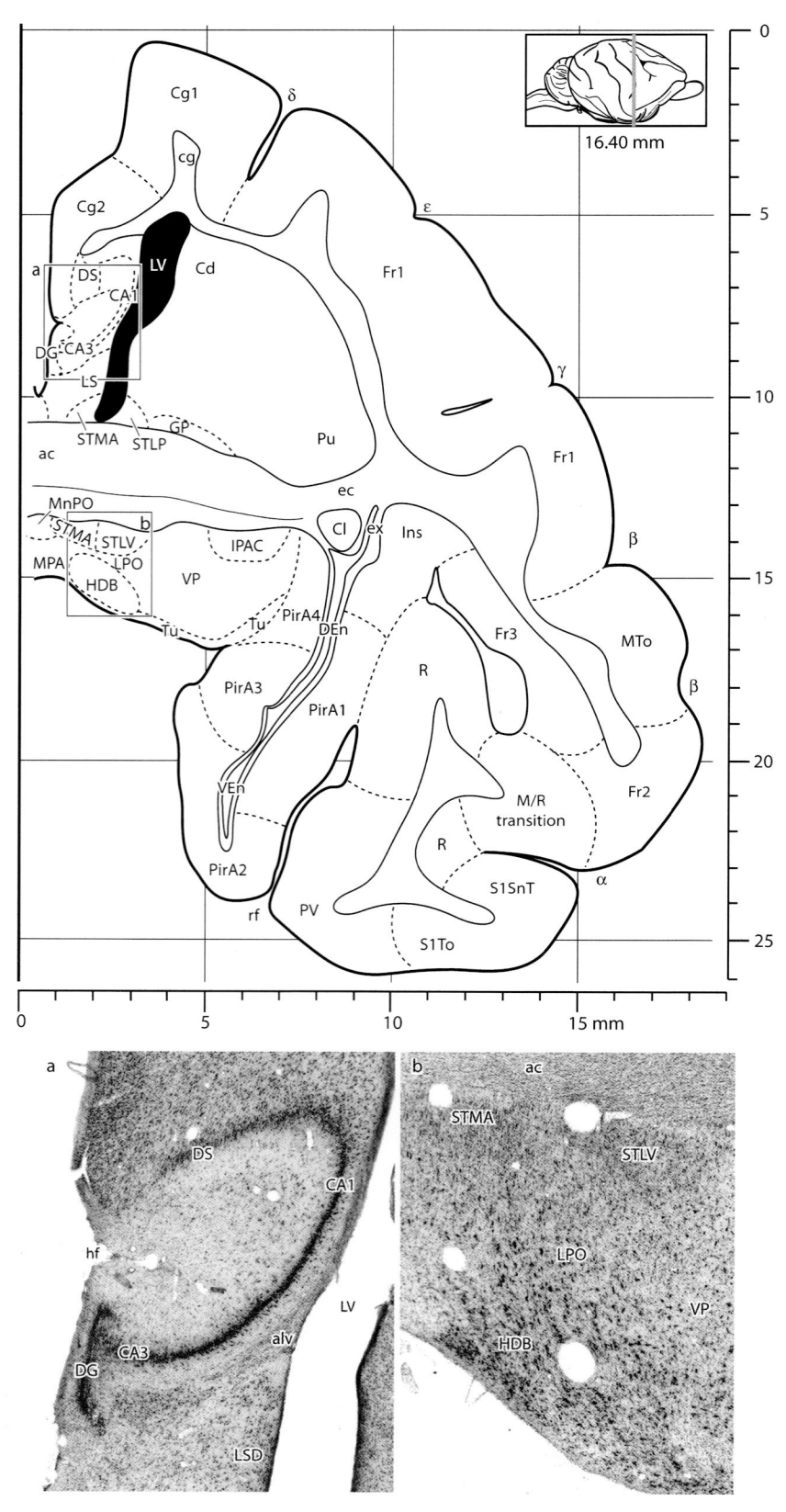

Ec-Ad09

α	sulcus α
β	sulcus β
γ	sulcus γ
δ	sulcus δ
ε	sulcus ε
ac	anterior commissure
alv	alveus of the hippocampus
CA1	field CA1 hippocampus
CA3	field CA3 hippocampus
Cd	caudate nucleus
cg	cingulum
Cg1	cingulate cortex, area 1
Cg2	cingulate cortex, area 2
Cl	claustrum
DEn	dorsal endopiriform nucleus
DG	dentate gyrus
DS	dorsal subiculum
ec	external capsule
ex	extreme capsule
Fr1 to 3	frontal cortex area 1 to 3
GP	globus pallidus
HDB	nu. of horiz. limb diag. band
hif	hippocampal fissure
Ins	insular cortex
IPAC	interstitial nu, posterior limb ac
LPO	lateral preoptic area
LS	lateral septal nu.
LSD	lateral septal nu., dorsal part
LV	lateral ventricle
M/R	M/R transition somato. cortex
MnPO	median preoptic nu.
MPA	medial preoptic area
MTo	motor cortex tongue
PirA1	piriform area 1
PirA2	piriform area 2
PirA3	piriform area 3
PirA4	piriform area 4
Pu	putamen
PV	parietoventral cortex
R	rostral somatosensory field
rf	rhinal fissure
S1SnT	prim. somatosens. cx, snout tip
S1To	prim. somatosens. cx, tongue
STLP	bed nu. st, lat. divn, posterior
STLV	bed nu. st, lat. divn, ventral
STMA	bed nu. st, med. divn, anterior
Tu	olfactory tubercle
VEn	ventral endopiriform nucleus
VP	ventral pallidum

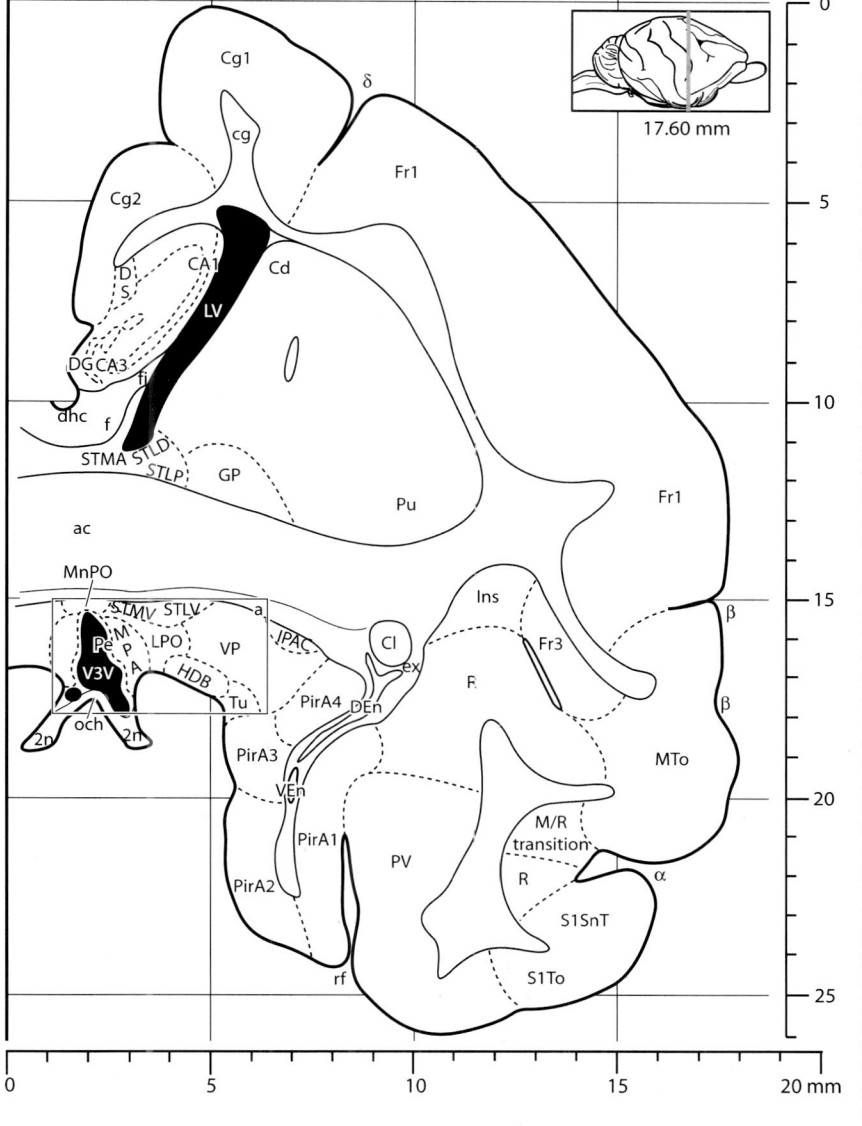

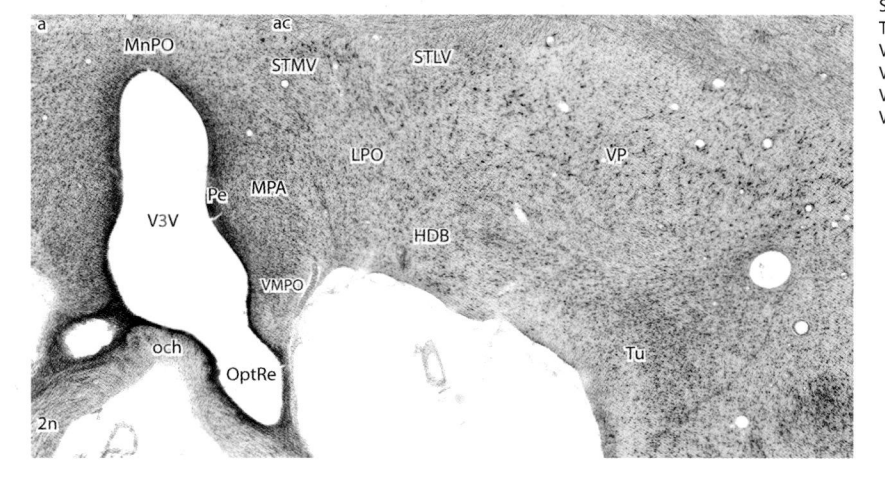

16 – ATLAS OF THE ADULT AND DEVELOPING BRAIN AND SPINAL CORD OF THE SHORT-BEAKED ECHIDNA

Ec-Ad10

α	sulcus α
β	sulcus β
δ	sulcus δ
2n	optic nerve
ac	anterior commissure
CA1	field CA1 hippocampus
CA3	field CA3 hippocampus
Cd	caudate nucleus
cg	cingulum
Cg1	cingulate cortex, area 1
Cg2	cingulate cortex, area 2
Cl	claustrum
DEn	dorsal endopiriform nucleus
DG	dentate gyrus
dhc	dorsal hippocampal commiss.
DS	dorsal subiculum
ex	extreme capsule
f	fornix
fi	fimbria of the hippocampus
Fr1	frontal cortex area 1
Fr3	frontal cortex area 3
GP	globus pallidus
HDB	nu. of horiz limb diagonal band
Ins	insular cortex
IPAC	interstitial nu. of post. limb ac
LPO	lateral preoptic area
LV	lateral ventricle
M/R	M/R transition somato. cx
MnPO	median preoptic nucleus
MPA	medial preoptic area
MTo	motor cortex tongue
och	optic chiasm
OptRe	optic recess of third ventricle
Pe	periventricular hypothal. nu.
PirA1	piriform area 1
PirA2	piriform area 2
PirA3	piriform area 3
PirA4	piriform area 4
Pu	putamen
PV	parietoventral cortex
R	rostral somatosensory field
rf	rhinal fissure
S1SnT	prim. somato. cx, snout tip
S1To	prim. somatosens. cx, tongue
STLD	bed nu. st, lat. divn, dorsal
STLP	bed nu. st, lat. divn, posterior
STLV	bed nu. st, lat. divn, ventral
STMA	bed nu. st, med. divn, anterior
STMV	bed nu. st, med. divn, ventral
Tu	olfactory tubercle
V3V	ventral third ventricle
VEn	ventral endopiriform nucleus
VMPO	ventromedial preoptic nucleus
VP	ventral pallidum

328 NEUROBIOLOGY OF MONOTREMES

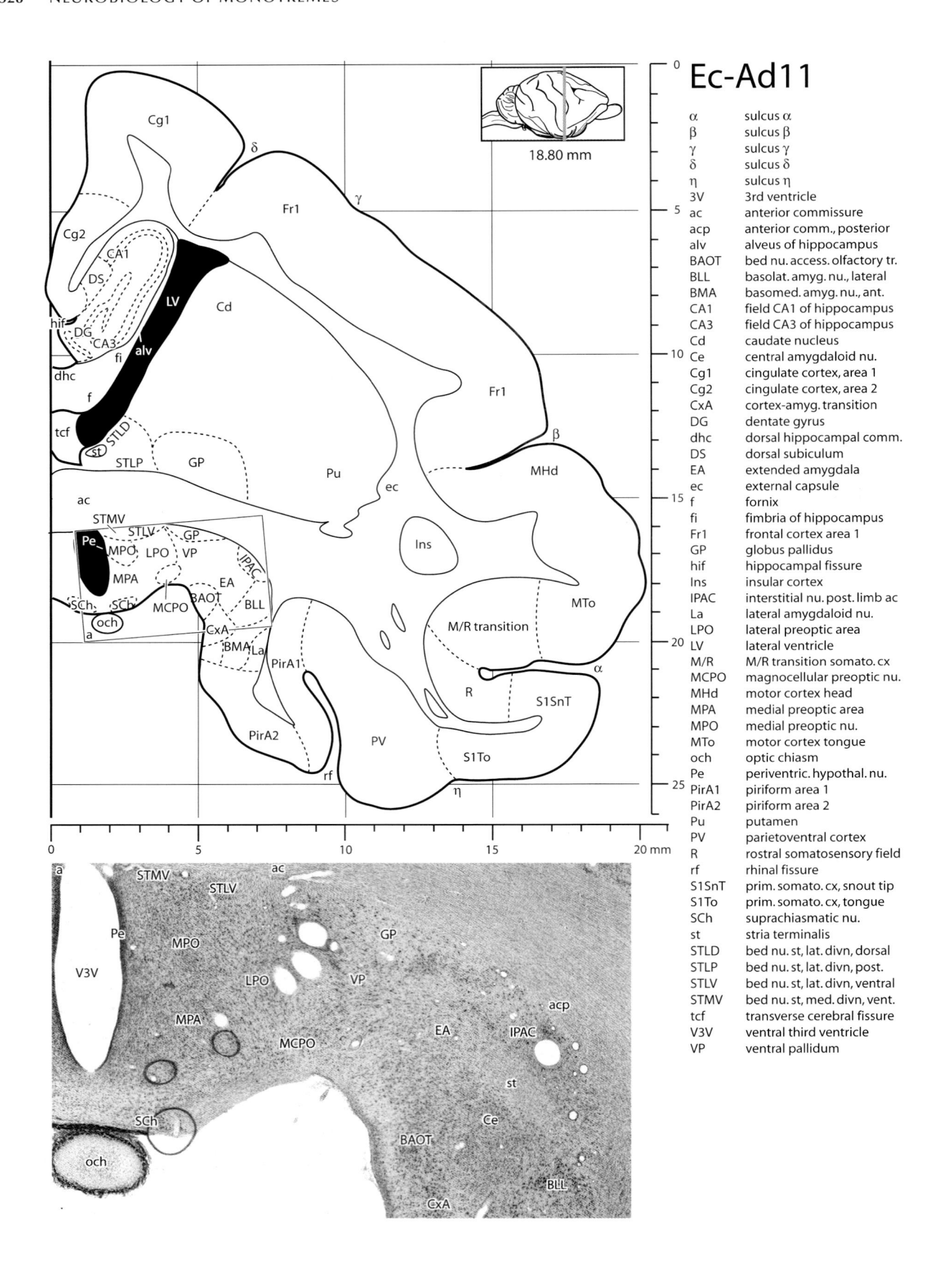

Ec-Ad11

18.80 mm

α	sulcus α
β	sulcus β
γ	sulcus γ
δ	sulcus δ
η	sulcus η
3V	3rd ventricle
ac	anterior commissure
acp	anterior comm., posterior
alv	alveus of hippocampus
BAOT	bed nu. access. olfactory tr.
BLL	basolat. amyg. nu., lateral
BMA	basomed. amyg. nu., ant.
CA1	field CA1 of hippocampus
CA3	field CA3 of hippocampus
Cd	caudate nucleus
Ce	central amygdaloid nu.
Cg1	cingulate cortex, area 1
Cg2	cingulate cortex, area 2
CxA	cortex-amyg. transition
DG	dentate gyrus
dhc	dorsal hippocampal comm.
DS	dorsal subiculum
EA	extended amygdala
ec	external capsule
f	fornix
fi	fimbria of hippocampus
Fr1	frontal cortex area 1
GP	globus pallidus
hif	hippocampal fissure
Ins	insular cortex
IPAC	interstitial nu. post. limb ac
La	lateral amygdaloid nu.
LPO	lateral preoptic area
LV	lateral ventricle
M/R	M/R transition somato. cx
MCPO	magnocellular preoptic nu.
MHd	motor cortex head
MPA	medial preoptic area
MPO	medial preoptic nu.
MTo	motor cortex tongue
och	optic chiasm
Pe	periventric. hypothal. nu.
PirA1	piriform area 1
PirA2	piriform area 2
Pu	putamen
PV	parietoventral cortex
R	rostral somatosensory field
rf	rhinal fissure
S1SnT	prim. somato. cx, snout tip
S1To	prim. somato. cx, tongue
SCh	suprachiasmatic nu.
st	stria terminalis
STLD	bed nu. st, lat. divn, dorsal
STLP	bed nu. st, lat. divn, post.
STLV	bed nu. st, lat. divn, ventral
STMV	bed nu. st, med. divn, vent.
tcf	transverse cerebral fissure
V3V	ventral third ventricle
VP	ventral pallidum

Ec-Ad12

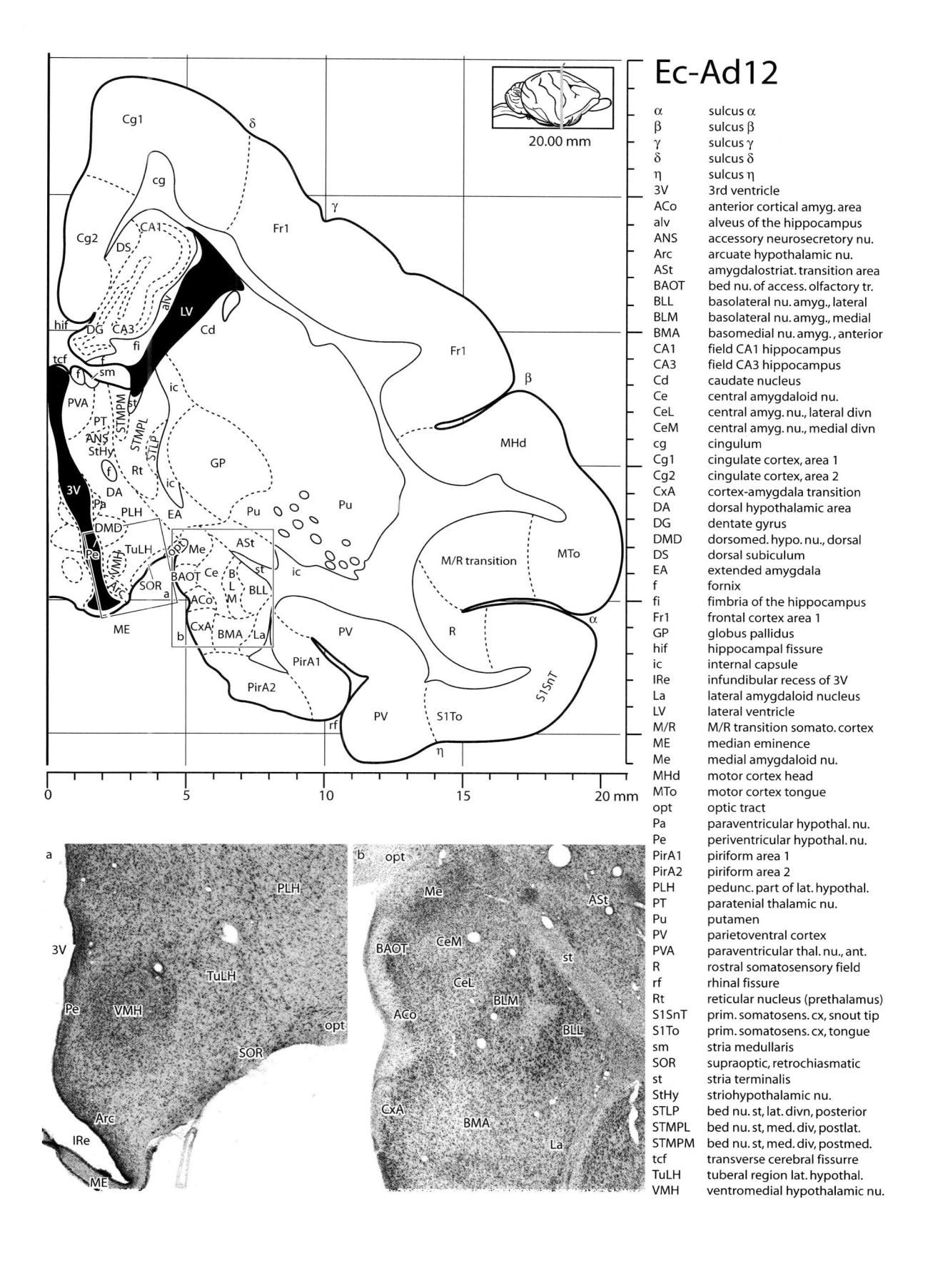

α	sulcus α
β	sulcus β
γ	sulcus γ
δ	sulcus δ
η	sulcus η
3V	3rd ventricle
ACo	anterior cortical amyg. area
alv	alveus of the hippocampus
ANS	accessory neurosecretory nu.
Arc	arcuate hypothalamic nu.
ASt	amygdalostriat. transition area
BAOT	bed nu. of access. olfactory tr.
BLL	basolateral nu. amyg., lateral
BLM	basolateral nu. amyg., medial
BMA	basomedial nu. amyg., anterior
CA1	field CA1 hippocampus
CA3	field CA3 hippocampus
Cd	caudate nucleus
Ce	central amygdaloid nu.
CeL	central amyg. nu., lateral divn
CeM	central amyg. nu., medial divn
cg	cingulum
Cg1	cingulate cortex, area 1
Cg2	cingulate cortex, area 2
CxA	cortex-amygdala transition
DA	dorsal hypothalamic area
DG	dentate gyrus
DMD	dorsomed. hypo. nu., dorsal
DS	dorsal subiculum
EA	extended amygdala
f	fornix
fi	fimbria of the hippocampus
Fr1	frontal cortex area 1
GP	globus pallidus
hif	hippocampal fissure
ic	internal capsule
IRe	infundibular recess of 3V
La	lateral amygdaloid nucleus
LV	lateral ventricle
M/R	M/R transition somato. cortex
ME	median eminence
Me	medial amygdaloid nu.
MHd	motor cortex head
MTo	motor cortex tongue
opt	optic tract
Pa	paraventricular hypothal. nu.
Pe	periventricular hypothal. nu.
PirA1	piriform area 1
PirA2	piriform area 2
PLH	pedunc. part of lat. hypothal.
PT	paratenial thalamic nu.
Pu	putamen
PV	parietoventral cortex
PVA	paraventricular thal. nu., ant.
R	rostral somatosensory field
rf	rhinal fissure
Rt	reticular nucleus (prethalamus)
S1SnT	prim. somatosens. cx, snout tip
S1To	prim. somatosens. cx, tongue
sm	stria medullaris
SOR	supraoptic, retrochiasmatic
st	stria terminalis
StHy	striohypothalamic nu.
STLP	bed nu. st, lat. divn, posterior
STMPL	bed nu. st, med. div, postlat.
STMPM	bed nu. st, med. div, postmed.
tcf	transverse cerebral fissurre
TuLH	tuberal region lat. hypothal.
VMH	ventromedial hypothalamic nu.

Ec-Ad13

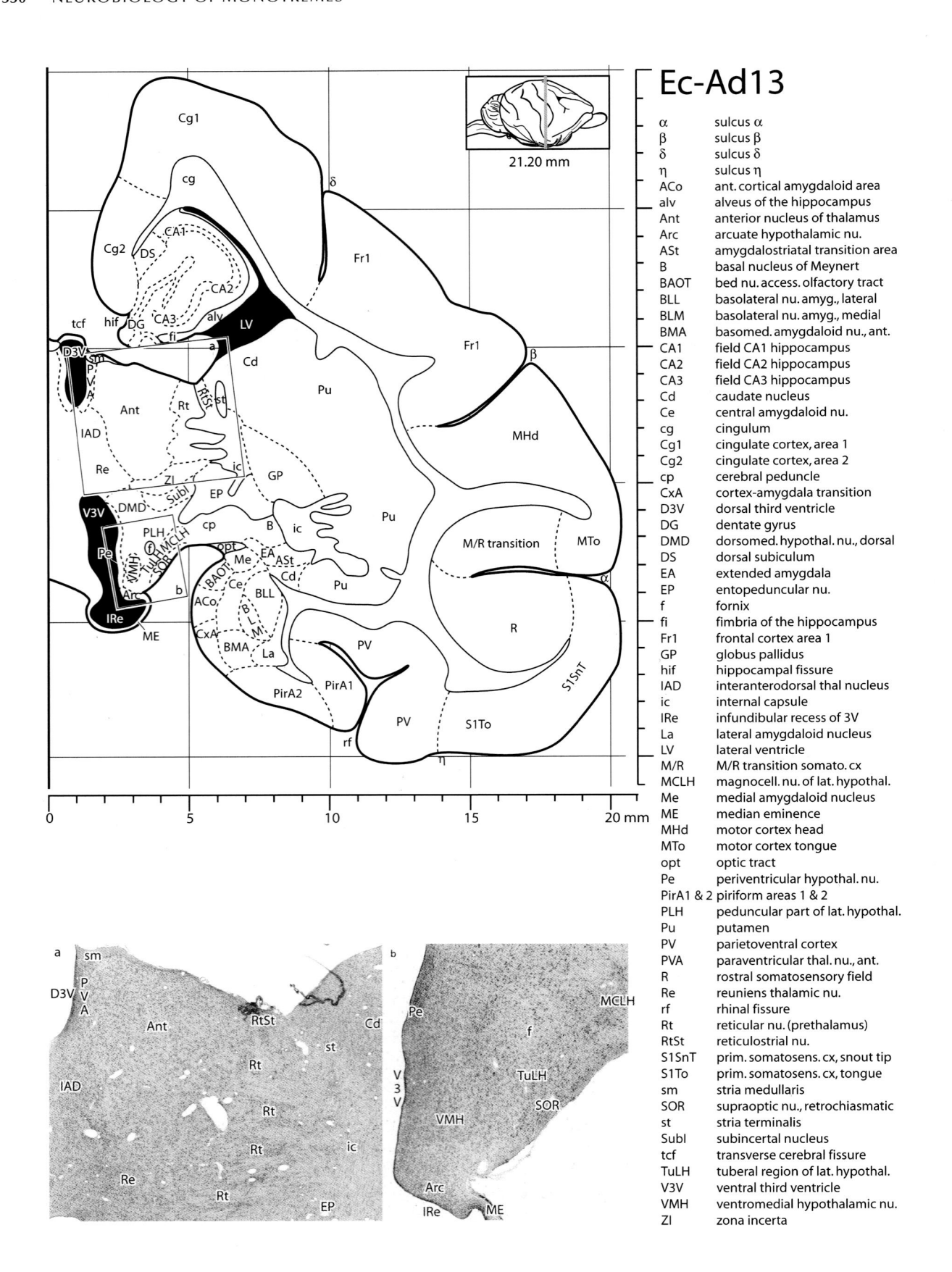

α	sulcus α
β	sulcus β
δ	sulcus δ
η	sulcus η
ACo	ant. cortical amygdaloid area
alv	alveus of the hippocampus
Ant	anterior nucleus of thalamus
Arc	arcuate hypothalamic nu.
ASt	amygdalostriatal transition area
B	basal nucleus of Meynert
BAOT	bed nu. access. olfactory tract
BLL	basolateral nu. amyg., lateral
BLM	basolateral nu. amyg., medial
BMA	basomed. amygdaloid nu., ant.
CA1	field CA1 hippocampus
CA2	field CA2 hippocampus
CA3	field CA3 hippocampus
Cd	caudate nucleus
Ce	central amygdaloid nu.
cg	cingulum
Cg1	cingulate cortex, area 1
Cg2	cingulate cortex, area 2
cp	cerebral peduncle
CxA	cortex-amygdala transition
D3V	dorsal third ventricle
DG	dentate gyrus
DMD	dorsomed. hypothal. nu., dorsal
DS	dorsal subiculum
EA	extended amygdala
EP	entopeduncular nu.
f	fornix
fi	fimbria of the hippocampus
Fr1	frontal cortex area 1
GP	globus pallidus
hif	hippocampal fissure
IAD	interanterodorsal thal nucleus
ic	internal capsule
IRe	infundibular recess of 3V
La	lateral amygdaloid nucleus
LV	lateral ventricle
M/R	M/R transition somato. cx
MCLH	magnocell. nu. of lat. hypothal.
Me	medial amygdaloid nucleus
ME	median eminence
MHd	motor cortex head
MTo	motor cortex tongue
opt	optic tract
Pe	periventricular hypothal. nu.
PirA1 & 2	piriform areas 1 & 2
PLH	peduncular part of lat. hypothal.
Pu	putamen
PV	parietoventral cortex
PVA	paraventricular thal. nu., ant.
R	rostral somatosensory field
Re	reuniens thalamic nu.
rf	rhinal fissure
Rt	reticular nu. (prethalamus)
RtSt	reticulostrial nu.
S1SnT	prim. somatosens. cx, snout tip
S1To	prim. somatosens. cx, tongue
sm	stria medullaris
SOR	supraoptic nu., retrochiasmatic
st	stria terminalis
SubI	subincertal nucleus
tcf	transverse cerebral fissure
TuLH	tuberal region of lat. hypothal.
V3V	ventral third ventricle
VMH	ventromedial hypothalamic nu.
ZI	zona incerta

21.20 mm

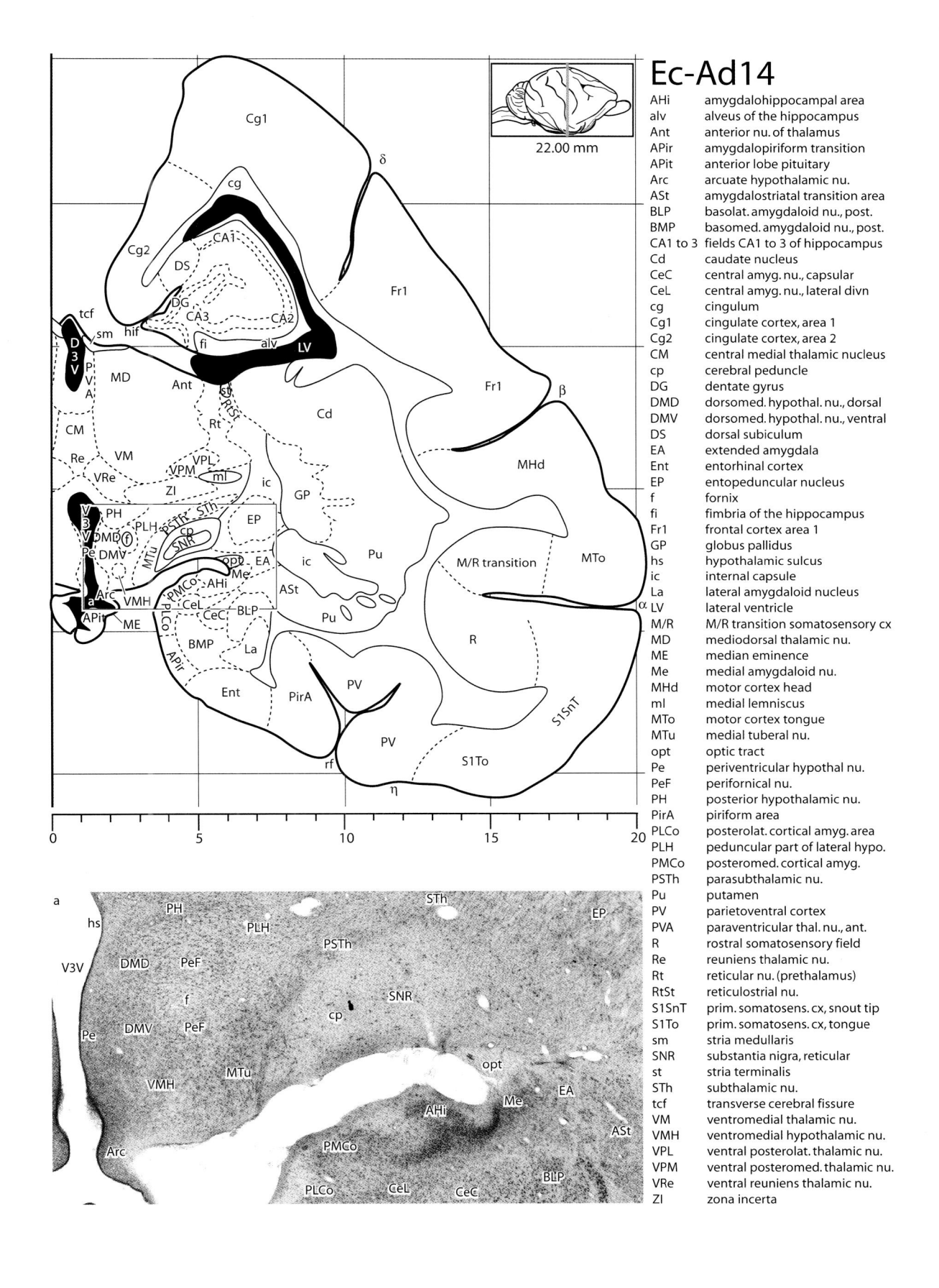

16 – ATLAS OF THE ADULT AND DEVELOPING BRAIN AND SPINAL CORD OF THE SHORT-BEAKED ECHIDNA

Ec-Ad14

AHi	amygdalohippocampal area
alv	alveus of the hippocampus
Ant	anterior nu. of thalamus
APir	amygdalopiriform transition
APit	anterior lobe pituitary
Arc	arcuate hypothalamic nu.
ASt	amygdalostriatal transition area
BLP	basolat. amygdaloid nu., post.
BMP	basomed. amygdaloid nu., post.
CA1 to 3	fields CA1 to 3 of hippocampus
Cd	caudate nucleus
CeC	central amyg. nu., capsular
CeL	central amyg. nu., lateral divn
cg	cingulum
Cg1	cingulate cortex, area 1
Cg2	cingulate cortex, area 2
CM	central medial thalamic nucleus
cp	cerebral peduncle
DG	dentate gyrus
DMD	dorsomed. hypothal. nu., dorsal
DMV	dorsomed. hypothal. nu., ventral
DS	dorsal subiculum
EA	extended amygdala
Ent	entorhinal cortex
EP	entopeduncular nucleus
f	fornix
fi	fimbria of the hippocampus
Fr1	frontal cortex area 1
GP	globus pallidus
hs	hypothalamic sulcus
ic	internal capsule
La	lateral amygdaloid nucleus
LV	lateral ventricle
M/R	M/R transition somatosensory cx
MD	mediodorsal thalamic nu.
ME	median eminence
Me	medial amygdaloid nu.
MHd	motor cortex head
ml	medial lemniscus
MTo	motor cortex tongue
MTu	medial tuberal nu.
opt	optic tract
Pe	periventricular hypothal nu.
PeF	perifornical nu.
PH	posterior hypothalamic nu.
PirA	piriform area
PLCo	posterolat. cortical amyg. area
PLH	peduncular part of lateral hypo.
PMCo	posteromed. cortical amyg.
PSTh	parasubthalamic nu.
Pu	putamen
PV	parietoventral cortex
PVA	paraventricular thal. nu., ant.
R	rostral somatosensory field
Re	reuniens thalamic nu.
Rt	reticular nu. (prethalamus)
RtSt	reticulostrial nu.
S1SnT	prim. somatosens. cx, snout tip
S1To	prim. somatosens. cx, tongue
sm	stria medullaris
SNR	substantia nigra, reticular
st	stria terminalis
STh	subthalamic nu.
tcf	transverse cerebral fissure
VM	ventromedial thalamic nu.
VMH	ventromedial hypothalamic nu.
VPL	ventral posterolat. thalamic nu.
VPM	ventral posteromed. thalamic nu.
VRe	ventral reuniens thalamic nu.
ZI	zona incerta

Ec-Ad15

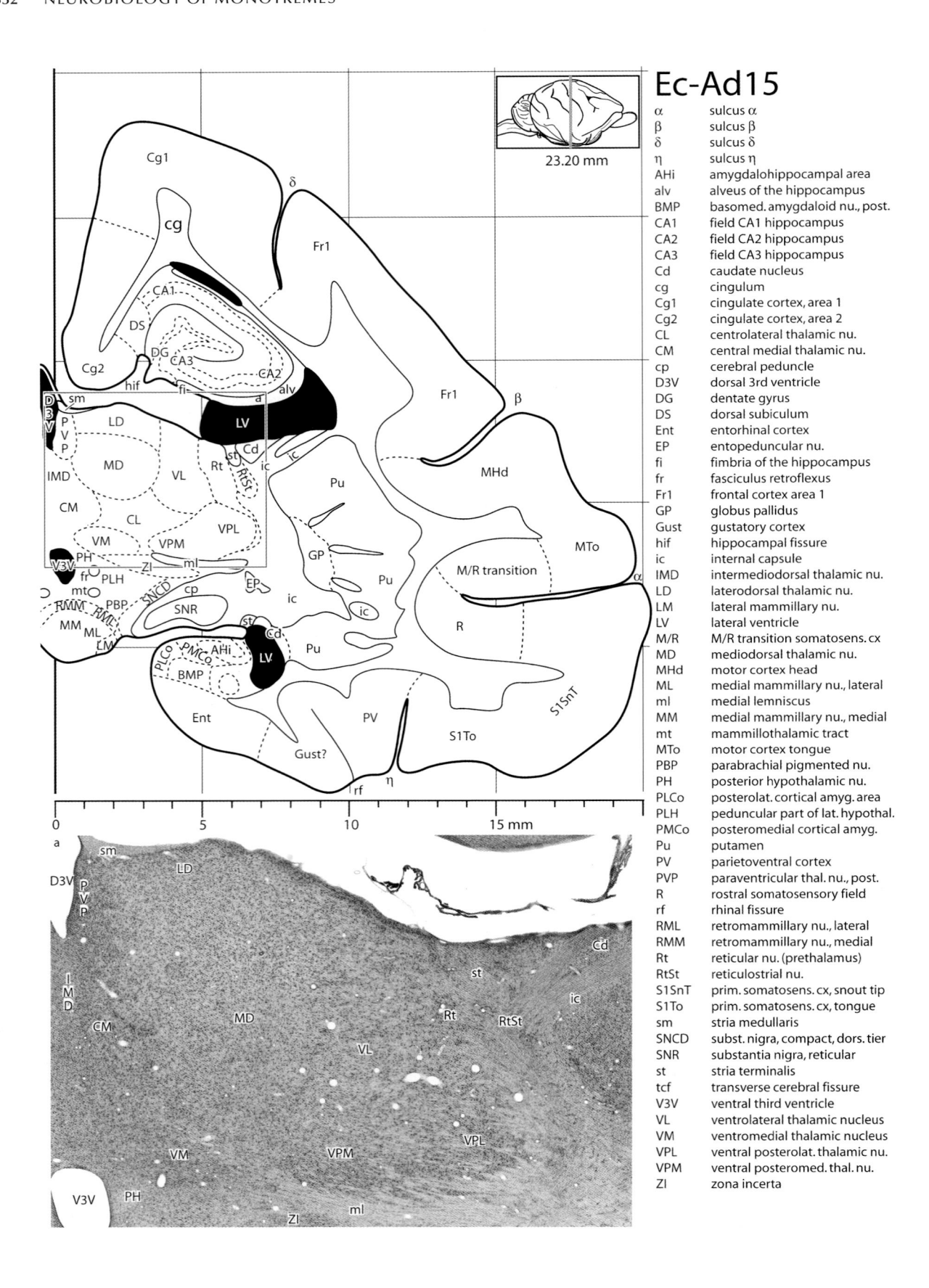

α	sulcus α
β	sulcus β
δ	sulcus δ
η	sulcus η
AHi	amygdalohippocampal area
alv	alveus of the hippocampus
BMP	basomed. amygdaloid nu., post.
CA1	field CA1 hippocampus
CA2	field CA2 hippocampus
CA3	field CA3 hippocampus
Cd	caudate nucleus
cg	cingulum
Cg1	cingulate cortex, area 1
Cg2	cingulate cortex, area 2
CL	centrolateral thalamic nu.
CM	central medial thalamic nu.
cp	cerebral peduncle
D3V	dorsal 3rd ventricle
DG	dentate gyrus
DS	dorsal subiculum
Ent	entorhinal cortex
EP	entopeduncular nu.
fi	fimbria of the hippocampus
fr	fasciculus retroflexus
Fr1	frontal cortex area 1
GP	globus pallidus
Gust	gustatory cortex
hif	hippocampal fissure
ic	internal capsule
IMD	intermediodorsal thalamic nu.
LD	laterodorsal thalamic nu.
LM	lateral mammillary nu.
LV	lateral ventricle
M/R	M/R transition somatosens. cx
MD	mediodorsal thalamic nu.
MHd	motor cortex head
ML	medial mammillary nu., lateral
ml	medial lemniscus
MM	medial mammillary nu., medial
mt	mammillothalamic tract
MTo	motor cortex tongue
PBP	parabrachial pigmented nu.
PH	posterior hypothalamic nu.
PLCo	posterolat. cortical amyg. area
PLH	peduncular part of lat. hypothal.
PMCo	posteromedial cortical amyg.
Pu	putamen
PV	parietoventral cortex
PVP	paraventricular thal. nu., post.
R	rostral somatosensory field
rf	rhinal fissure
RML	retromammillary nu., lateral
RMM	retromammillary nu., medial
Rt	reticular nu. (prethalamus)
RtSt	reticulostrial nu.
S1SnT	prim. somatosens. cx, snout tip
S1To	prim. somatosens. cx, tongue
sm	stria medullaris
SNCD	subst. nigra, compact, dors. tier
SNR	substantia nigra, reticular
st	stria terminalis
tcf	transverse cerebral fissure
V3V	ventral third ventricle
VL	ventrolateral thalamic nucleus
VM	ventromedial thalamic nucleus
VPL	ventral posterolat. thalamic nu.
VPM	ventral posteromed. thal. nu.
ZI	zona incerta

23.20 mm

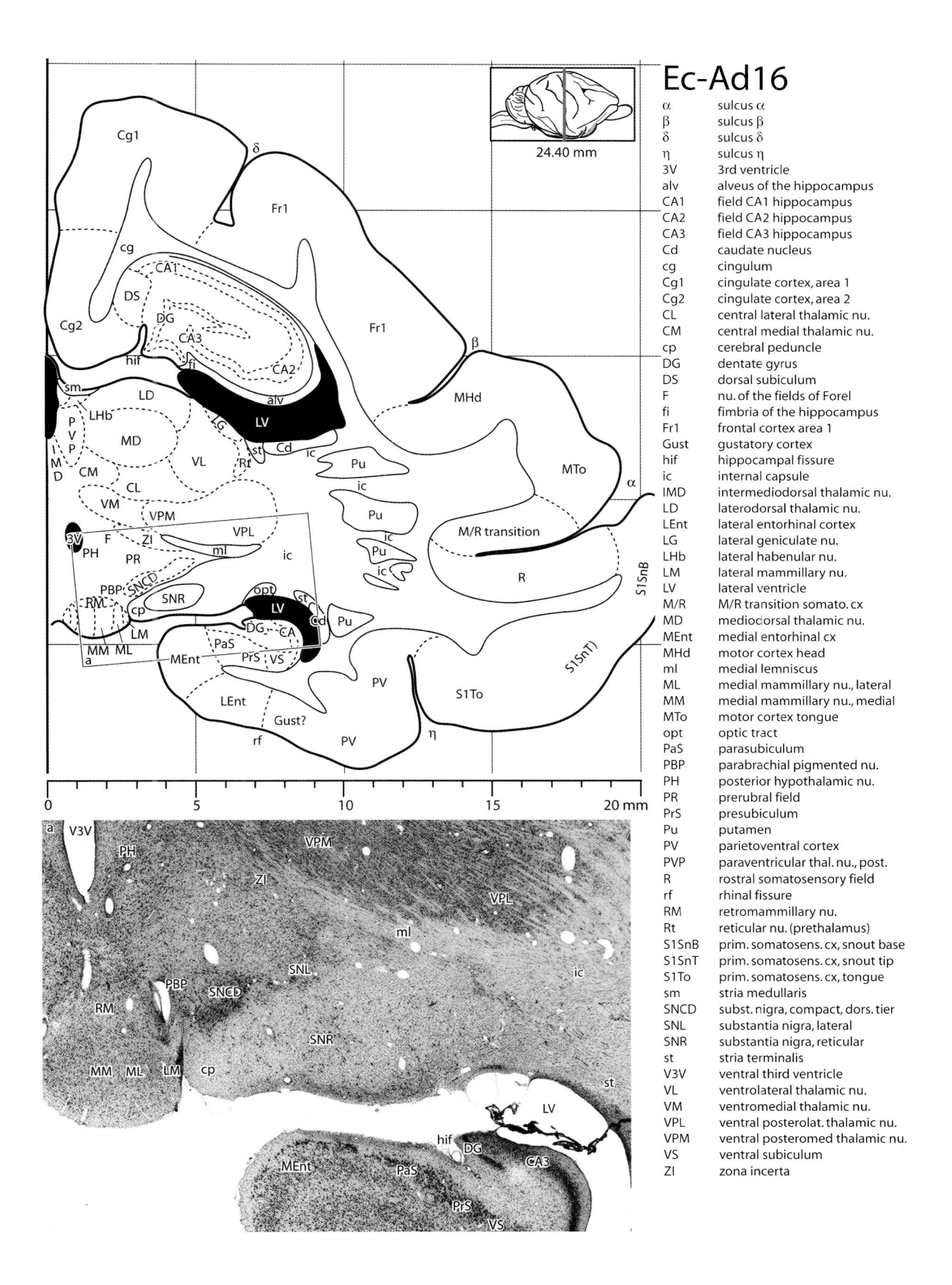

Ec-Ad16

α	sulcus α
β	sulcus β
δ	sulcus δ
η	sulcus η
3V	3rd ventricle
alv	alveus of the hippocampus
CA1	field CA1 hippocampus
CA2	field CA2 hippocampus
CA3	field CA3 hippocampus
Cd	caudate nucleus
cg	cingulum
Cg1	cingulate cortex, area 1
Cg2	cingulate cortex, area 2
CL	central lateral thalamic nu.
CM	central medial thalamic nu.
cp	cerebral peduncle
DG	dentate gyrus
DS	dorsal subiculum
F	nu. of the fields of Forel
fi	fimbria of the hippocampus
Fr1	frontal cortex area 1
Gust	gustatory cortex
hif	hippocampal fissure
ic	internal capsule
IMD	intermediodorsal thalamic nu.
LD	laterodorsal thalamic nu.
LEnt	lateral entorhinal cortex
LG	lateral geniculate nu.
LHb	lateral habenular nu.
LM	lateral mammillary nu.
LV	lateral ventricle
M/R	M/R transition somato. cx
MD	mediodorsal thalamic nu.
MEnt	medial entorhinal cx
MHd	motor cortex head
ml	medial lemniscus
ML	medial mammillary nu., lateral
MM	medial mammillary nu., medial
MTo	motor cortex tongue
opt	optic tract
PaS	parasubiculum
PBP	parabrachial pigmented nu.
PH	posterior hypothalamic nu.
PR	prerubral field
PrS	presubiculum
Pu	putamen
PV	parietoventral cortex
PVP	paraventricular thal. nu., post.
R	rostral somatosensory field
rf	rhinal fissure
RM	retromammillary nu.
Rt	reticular nu. (prethalamus)
S1SnB	prim. somatosens. cx, snout base
S1SnT	prim. somatosens. cx, snout tip
S1To	prim. somatosens. cx, tongue
sm	stria medullaris
SNCD	subst. nigra, compact, dors. tier
SNL	substantia nigra, lateral
SNR	substantia nigra, reticular
st	stria terminalis
V3V	ventral third ventricle
VL	ventrolateral thalamic nu.
VM	ventromedial thalamic nu.
VPL	ventral posterolat. thalamic nu.
VPM	ventral posteromed thalamic nu.
VS	ventral subiculum
ZI	zona incerta

Ec-Ad17

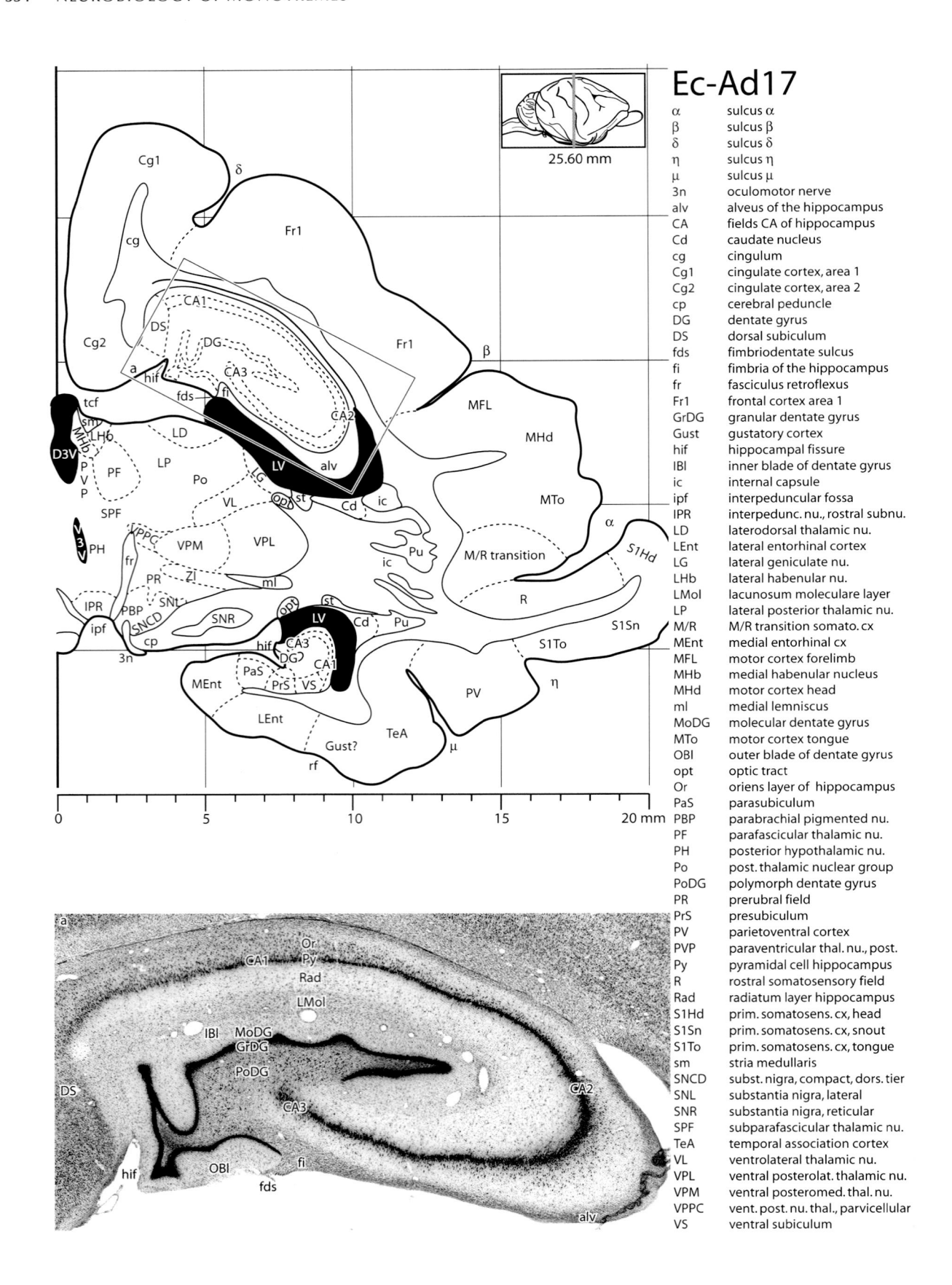

α	sulcus α
β	sulcus β
δ	sulcus δ
η	sulcus η
μ	sulcus μ
3n	oculomotor nerve
alv	alveus of the hippocampus
CA	fields CA of hippocampus
Cd	caudate nucleus
cg	cingulum
Cg1	cingulate cortex, area 1
Cg2	cingulate cortex, area 2
cp	cerebral peduncle
DG	dentate gyrus
DS	dorsal subiculum
fds	fimbriodentate sulcus
fi	fimbria of the hippocampus
fr	fasciculus retroflexus
Fr1	frontal cortex area 1
GrDG	granular dentate gyrus
Gust	gustatory cortex
hif	hippocampal fissure
IBl	inner blade of dentate gyrus
ic	internal capsule
ipf	interpeduncular fossa
IPR	interpedunc. nu., rostral subnu.
LD	laterodorsal thalamic nu.
LEnt	lateral entorhinal cortex
LG	lateral geniculate nu.
LHb	lateral habenular nu.
LMol	lacunosum moleculare layer
LP	lateral posterior thalamic nu.
M/R	M/R transition somato. cx
MEnt	medial entorhinal cx
MFL	motor cortex forelimb
MHb	medial habenular nucleus
MHd	motor cortex head
ml	medial lemniscus
MoDG	molecular dentate gyrus
MTo	motor cortex tongue
OBl	outer blade of dentate gyrus
opt	optic tract
Or	oriens layer of hippocampus
PaS	parasubiculum
PBP	parabrachial pigmented nu.
PF	parafascicular thalamic nu.
PH	posterior hypothalamic nu.
Po	post. thalamic nuclear group
PoDG	polymorph dentate gyrus
PR	prerubral field
PrS	presubiculum
PV	parietoventral cortex
PVP	paraventricular thal. nu., post.
Py	pyramidal cell hippocampus
R	rostral somatosensory field
Rad	radiatum layer hippocampus
S1Hd	prim. somatosens. cx, head
S1Sn	prim. somatosens. cx, snout
S1To	prim. somatosens. cx, tongue
sm	stria medullaris
SNCD	subst. nigra, compact, dors. tier
SNL	substantia nigra, lateral
SNR	substantia nigra, reticular
SPF	subparafascicular thalamic nu.
TeA	temporal association cortex
VL	ventrolateral thalamic nu.
VPL	ventral posterolat. thalamic nu.
VPM	ventral posteromed. thal. nu.
VPPC	vent. post. nu. thal., parvicellular
VS	ventral subiculum

Ec-Ad18

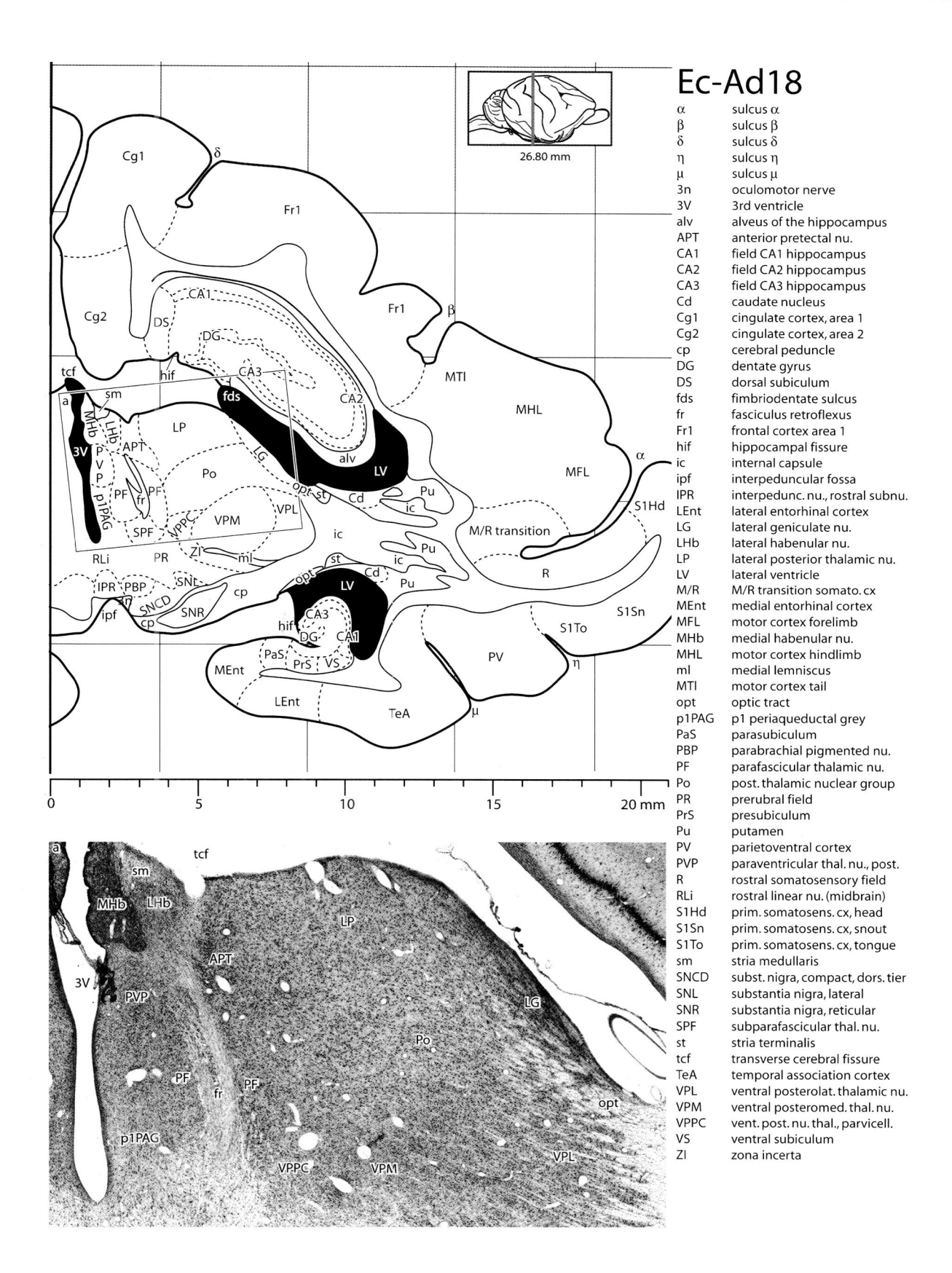

26.80 mm

α	sulcus α
β	sulcus β
δ	sulcus δ
η	sulcus η
μ	sulcus μ
3n	oculomotor nerve
3V	3rd ventricle
alv	alveus of the hippocampus
APT	anterior pretectal nu.
CA1	field CA1 hippocampus
CA2	field CA2 hippocampus
CA3	field CA3 hippocampus
Cd	caudate nucleus
Cg1	cingulate cortex, area 1
Cg2	cingulate cortex, area 2
cp	cerebral peduncle
DG	dentate gyrus
DS	dorsal subiculum
fds	fimbriodentate sulcus
fr	fasciculus retroflexus
Fr1	frontal cortex area 1
hif	hippocampal fissure
ic	internal capsule
ipf	interpeduncular fossa
IPR	interpedunc. nu., rostral subnu.
LEnt	lateral entorhinal cortex
LG	lateral geniculate nu.
LHb	lateral habenular nu.
LP	lateral posterior thalamic nu.
LV	lateral ventricle
M/R	M/R transition somato. cx
MEnt	medial entorhinal cortex
MFL	motor cortex forelimb
MHb	medial habenular nu.
MHL	motor cortex hindlimb
ml	medial lemniscus
MTl	motor cortex tail
opt	optic tract
p1PAG	p1 periaqueductal grey
PaS	parasubiculum
PBP	parabrachial pigmented nu.
PF	parafascicular thalamic nu.
Po	post. thalamic nuclear group
PR	prerubral field
PrS	presubiculum
Pu	putamen
PV	parietoventral cortex
PVP	paraventricular thal. nu., post.
R	rostral somatosensory field
RLi	rostral linear nu. (midbrain)
S1Hd	prim. somatosens. cx, head
S1Sn	prim. somatosens. cx, snout
S1To	prim. somatosens. cx, tongue
sm	stria medullaris
SNCD	subst. nigra, compact, dors. tier
SNL	substantia nigra, lateral
SNR	substantia nigra, reticular
SPF	subparafascicular thal. nu.
st	stria terminalis
tcf	transverse cerebral fissure
TeA	temporal association cortex
VPL	ventral posterolat. thalamic nu.
VPM	ventral posteromed. thal. nu.
VPPC	vent. post. nu. thal., parvicell.
VS	ventral subiculum
ZI	zona incerta

Ec-Ad19

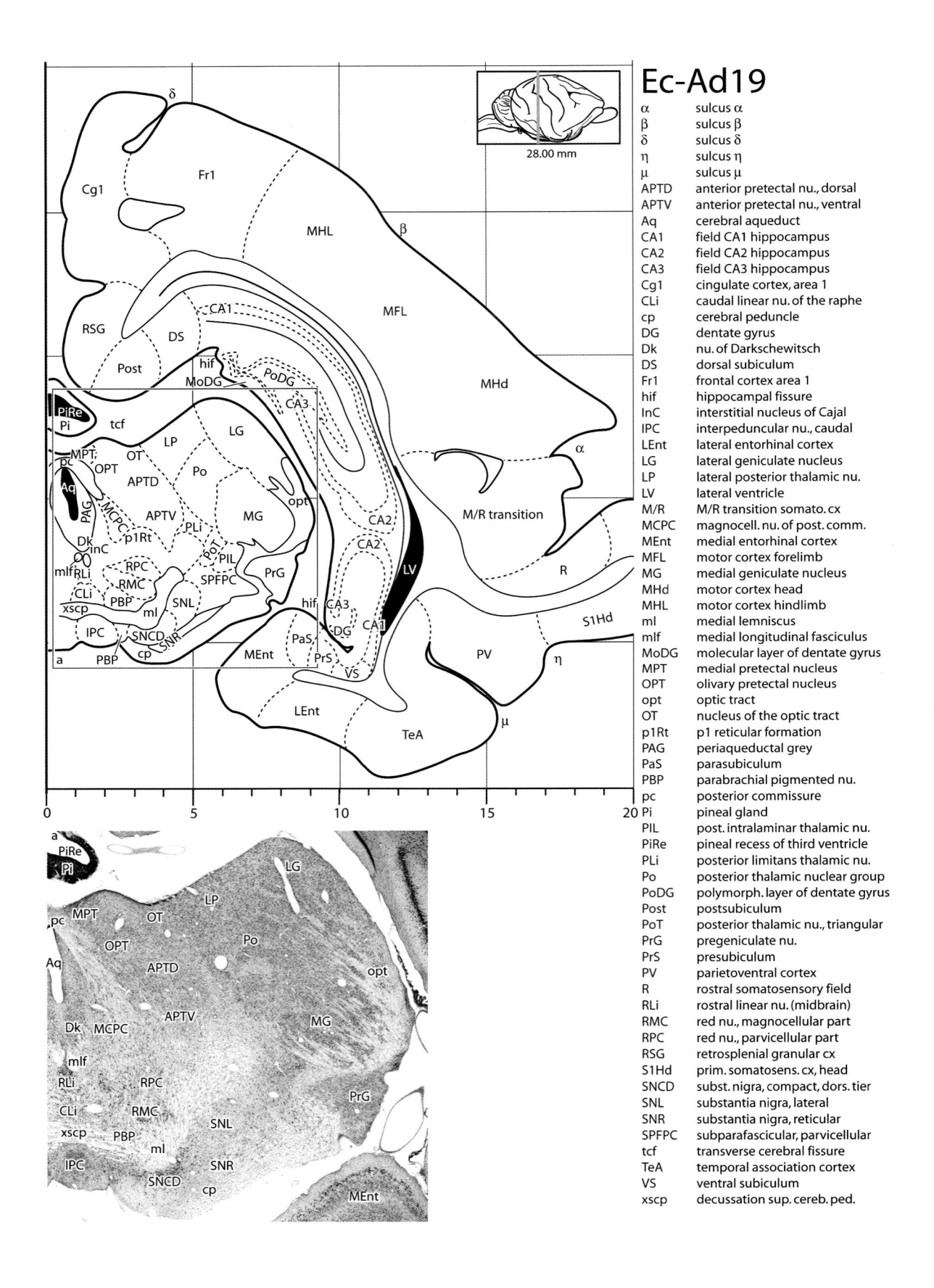

α	sulcus α
β	sulcus β
δ	sulcus δ
η	sulcus η
μ	sulcus μ
APTD	anterior pretectal nu., dorsal
APTV	anterior pretectal nu., ventral
Aq	cerebral aqueduct
CA1	field CA1 hippocampus
CA2	field CA2 hippocampus
CA3	field CA3 hippocampus
Cg1	cingulate cortex, area 1
CLi	caudal linear nu. of the raphe
cp	cerebral peduncle
DG	dentate gyrus
Dk	nu. of Darkschewitsch
DS	dorsal subiculum
Fr1	frontal cortex area 1
hif	hippocampal fissure
InC	interstitial nucleus of Cajal
IPC	interpeduncular nu., caudal
LEnt	lateral entorhinal cortex
LG	lateral geniculate nucleus
LP	lateral posterior thalamic nu.
LV	lateral ventricle
M/R	M/R transition somato. cx
MCPC	magnocell. nu. of post. comm.
MEnt	medial entorhinal cortex
MFL	motor cortex forelimb
MG	medial geniculate nucleus
MHd	motor cortex head
MHL	motor cortex hindlimb
ml	medial lemniscus
mlf	medial longitudinal fasciculus
MoDG	molecular layer of dentate gyrus
MPT	medial pretectal nucleus
OPT	olivary pretectal nucleus
opt	optic tract
OT	nucleus of the optic tract
p1Rt	p1 reticular formation
PAG	periaqueductal grey
PaS	parasubiculum
PBP	parabrachial pigmented nu.
pc	posterior commissure
Pi	pineal gland
PIL	post. intralaminar thalamic nu.
PiRe	pineal recess of third ventricle
PLi	posterior limitans thalamic nu.
Po	posterior thalamic nuclear group
PoDG	polymorph. layer of dentate gyrus
Post	postsubiculum
PoT	posterior thalamic nu., triangular
PrG	pregeniculate nu.
PrS	presubiculum
PV	parietoventral cortex
R	rostral somatosensory field
RLi	rostral linear nu. (midbrain)
RMC	red nu., magnocellular part
RPC	red nu., parvicellular part
RSG	retrosplenial granular cx
S1Hd	prim. somatosens. cx, head
SNCD	subst. nigra, compact, dors. tier
SNL	substantia nigra, lateral
SNR	substantia nigra, reticular
SPFPC	subparafascicular, parvicellular
tcf	transverse cerebral fissure
TeA	temporal association cortex
VS	ventral subiculum
xscp	decussation sup. cereb. ped.

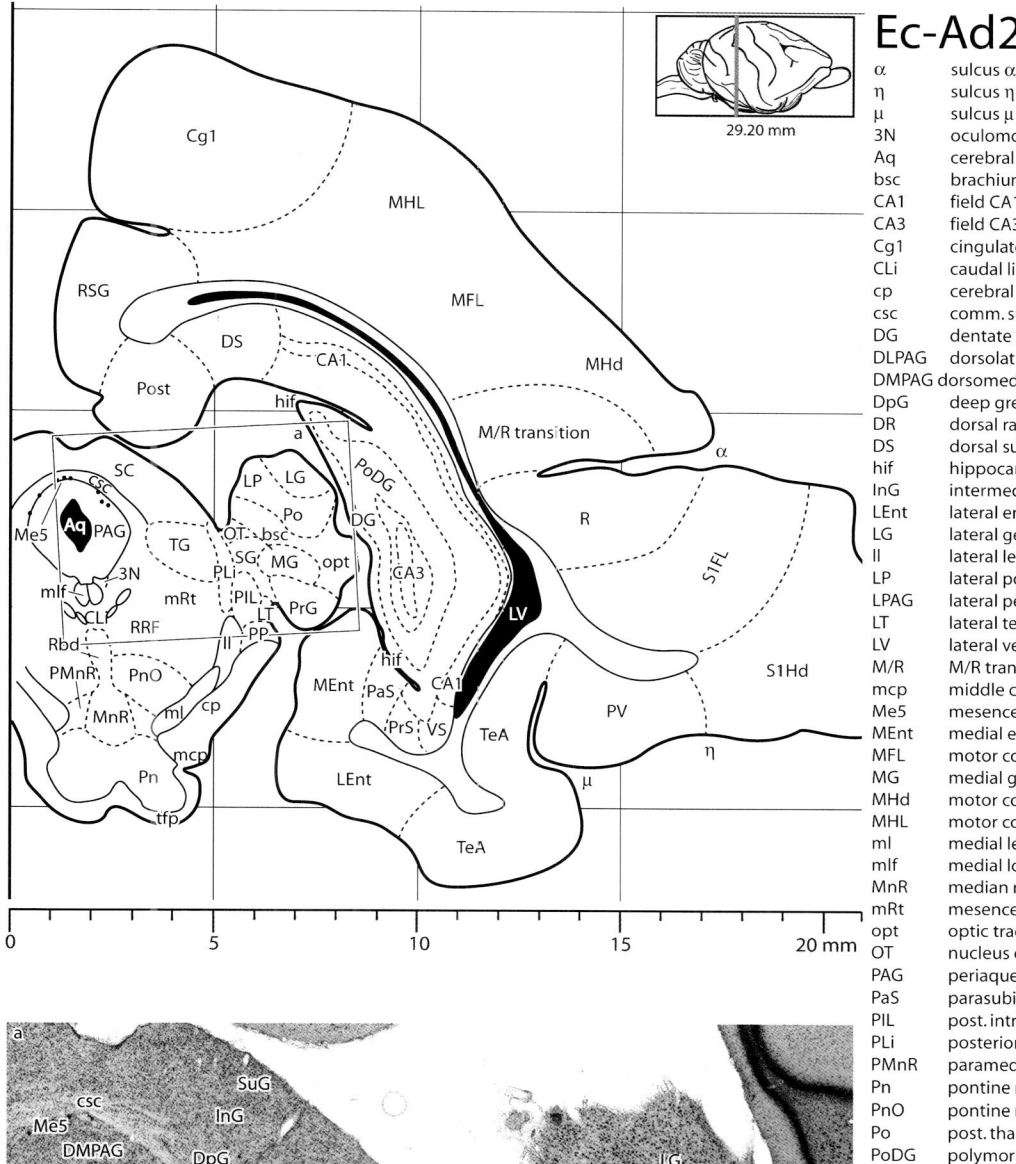

Ec-Ad20

α	sulcus α
η	sulcus η
μ	sulcus μ
3N	oculomotor nu.
Aq	cerebral aqueduct
bsc	brachium superior colliculus
CA1	field CA1 hippocampus
CA3	field CA3 hippocampus
Cg1	cingulate cortex, area 1
CLi	caudal linear nu. of the raphe
cp	cerebral peduncle
csc	comm. superior colliculus
DG	dentate gyrus
DLPAG	dorsolat. periaqueductal grey
DMPAG	dorsomed. periaqueductal grey
DpG	deep grey superior colliculus
DR	dorsal raphe nu.
DS	dorsal subiculum
hif	hippocampal fissure
InG	intermediate grey layer SC
LEnt	lateral entorhinal cortex
LG	lateral geniculate nu.
ll	lateral lemniscus
LP	lateral posterior thalamic nu.
LPAG	lateral periaqueductal grey
LT	lateral terminal nu.
LV	lateral ventricle
M/R	M/R transition somato. cx
mcp	middle cerebellar peduncle
Me5	mesencephalic trigeminal nu.
MEnt	medial entorhinal cortex
MFL	motor cortex forelimb
MG	medial geniculate nucleus
MHd	motor cortex head
MHL	motor cortex hindlimb
ml	medial lemniscus
mlf	medial longitudinal fasciculus
MnR	median raphe nucleus
mRt	mesencephalic reticular formn
opt	optic tract
OT	nucleus of the optic tract
PAG	periaqueductal grey
PaS	parasubiculum
PIL	post. intralaminar thalamic nu.
PLi	posterior limitans thalamic nu.
PMnR	paramedian raphe nu.
Pn	pontine nuclei
PnO	pontine reticular nu., oral
Po	post. thalamic nuclear group
PoDG	polymorph. dentate gyrus
Post	postsubiculum
PP	peripeduncular nu.
PrG	pregeniculate nu.
PrS	presubiculum
PV	parietoventral cortex
R	rostral somatosensory field
Rbd	rhabdoid nucleus
RRF	retrorubral field
RSG	retrosplenial granular cortex
S1FL	prim. somatosens. cx, forelimb
S1Hd	prim. somatosens. cx, head
SC	superior colliculus
SG	suprageniculate thalamic nu.
Su3	supraoculomotor PAG
SuG	superficial grey sup. colliculus
TeA	temporal association cortex
tfp	transverse fibres of pons
TG	tectal grey
VLPAG	ventrolateral periaq. grey
VS	ventral subiculum

338 NEUROBIOLOGY OF MONOTREMES

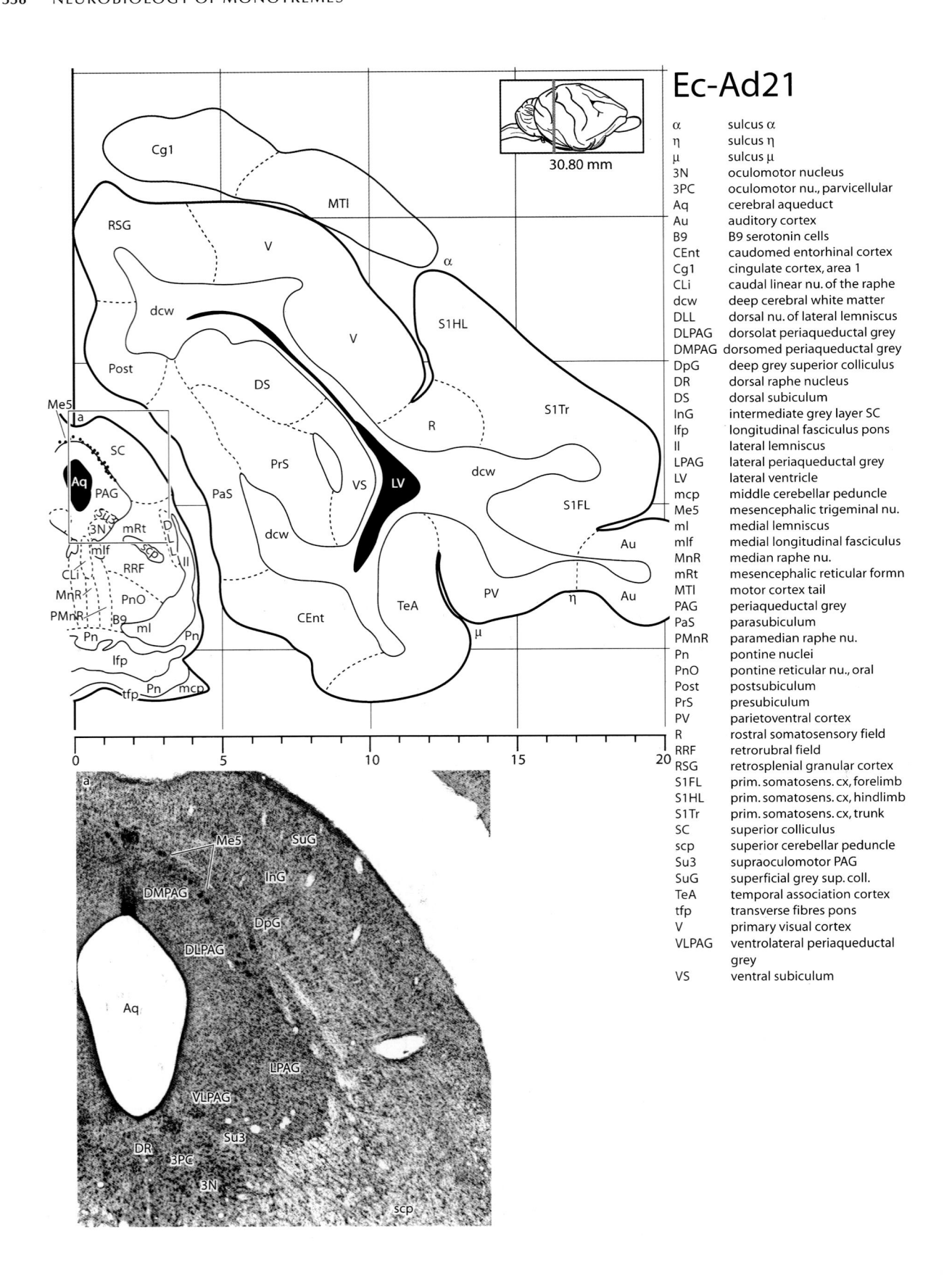

Ec-Ad21

α	sulcus α
η	sulcus η
μ	sulcus μ
3N	oculomotor nucleus
3PC	oculomotor nu., parvicellular
Aq	cerebral aqueduct
Au	auditory cortex
B9	B9 serotonin cells
CEnt	caudomed entorhinal cortex
Cg1	cingulate cortex, area 1
CLi	caudal linear nu. of the raphe
dcw	deep cerebral white matter
DLL	dorsal nu. of lateral lemniscus
DLPAG	dorsolat periaqueductal grey
DMPAG	dorsomed periaqueductal grey
DpG	deep grey superior colliculus
DR	dorsal raphe nucleus
DS	dorsal subiculum
InG	intermediate grey layer SC
lfp	longitudinal fasciculus pons
ll	lateral lemniscus
LPAG	lateral periaqueductal grey
LV	lateral ventricle
mcp	middle cerebellar peduncle
Me5	mesencephalic trigeminal nu.
ml	medial lemniscus
mlf	medial longitudinal fasciculus
MnR	median raphe nu.
mRt	mesencephalic reticular formn
MTl	motor cortex tail
PAG	periaqueductal grey
PaS	parasubiculum
PMnR	paramedian raphe nu.
Pn	pontine nuclei
PnO	pontine reticular nu., oral
Post	postsubiculum
PrS	presubiculum
PV	parietoventral cortex
R	rostral somatosensory field
RRF	retrorubral field
RSG	retrosplenial granular cortex
S1FL	prim. somatosens. cx, forelimb
S1HL	prim. somatosens. cx, hindlimb
S1Tr	prim. somatosens. cx, trunk
SC	superior colliculus
scp	superior cerebellar peduncle
Su3	supraoculomotor PAG
SuG	superficial grey sup. coll.
TeA	temporal association cortex
tfp	transverse fibres pons
V	primary visual cortex
VLPAG	ventrolateral periaqueductal grey
VS	ventral subiculum

30.80 mm

Ec-Ad22

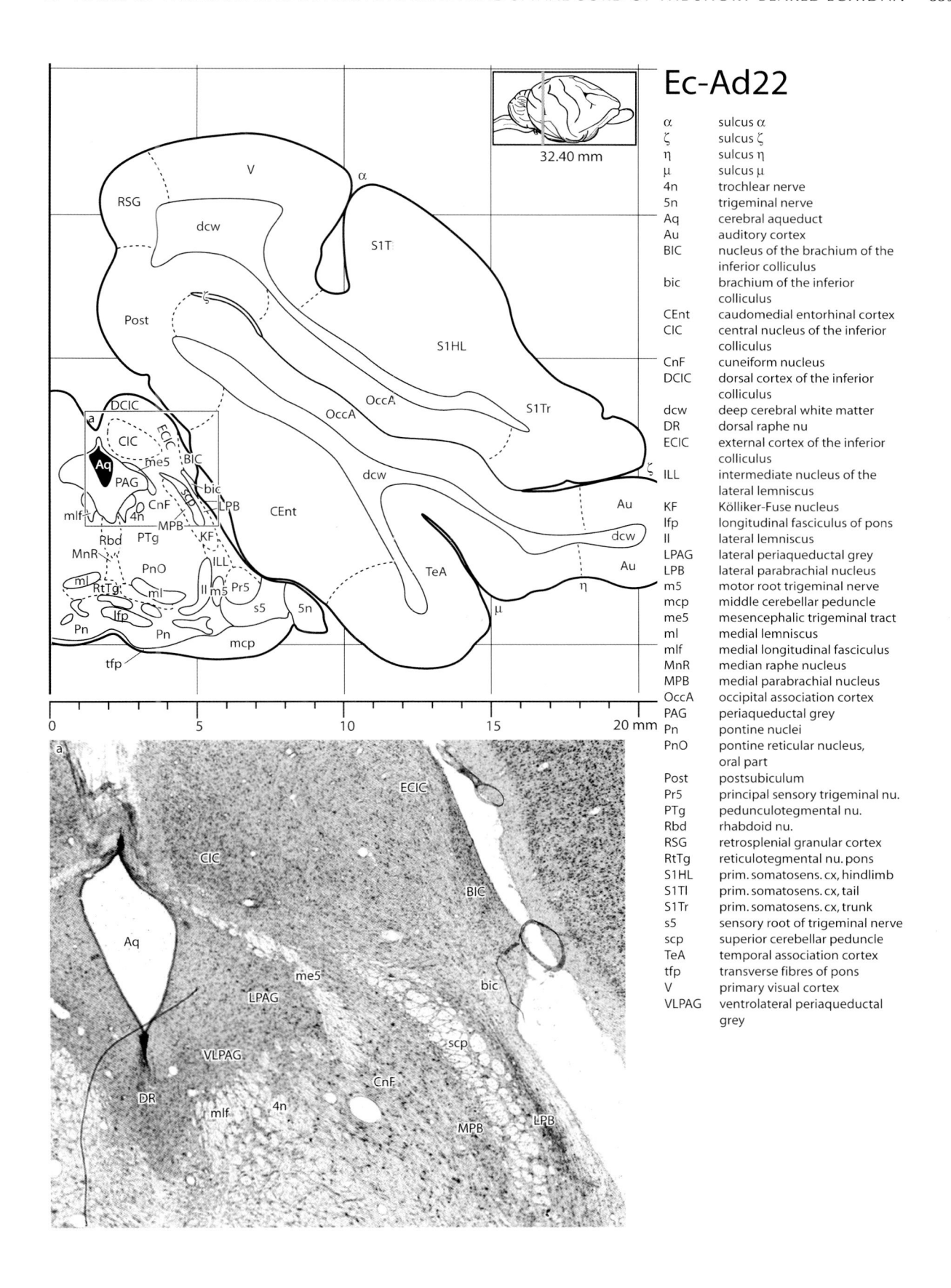

32.40 mm

α	sulcus α
ζ	sulcus ζ
η	sulcus η
μ	sulcus μ
4n	trochlear nerve
5n	trigeminal nerve
Aq	cerebral aqueduct
Au	auditory cortex
BIC	nucleus of the brachium of the inferior colliculus
bic	brachium of the inferior colliculus
CEnt	caudomedial entorhinal cortex
CIC	central nucleus of the inferior colliculus
CnF	cuneiform nucleus
DCIC	dorsal cortex of the inferior colliculus
dcw	deep cerebral white matter
DR	dorsal raphe nu
ECIC	external cortex of the inferior colliculus
ILL	intermediate nucleus of the lateral lemniscus
KF	Kölliker-Fuse nucleus
lfp	longitudinal fasciculus of pons
ll	lateral lemniscus
LPAG	lateral periaqueductal grey
LPB	lateral parabrachial nucleus
m5	motor root trigeminal nerve
mcp	middle cerebellar peduncle
me5	mesencephalic trigeminal tract
ml	medial lemniscus
mlf	medial longitudinal fasciculus
MnR	median raphe nucleus
MPB	medial parabrachial nucleus
OccA	occipital association cortex
PAG	periaqueductal grey
Pn	pontine nuclei
PnO	pontine reticular nucleus, oral part
Post	postsubiculum
Pr5	principal sensory trigeminal nu.
PTg	pedunculotegmental nu.
Rbd	rhabdoid nu.
RSG	retrosplenial granular cortex
RtTg	reticulotegmental nu. pons
S1HL	prim. somatosens. cx, hindlimb
S1Tl	prim. somatosens. cx, tail
S1Tr	prim. somatosens. cx, trunk
s5	sensory root of trigeminal nerve
scp	superior cerebellar peduncle
TeA	temporal association cortex
tfp	transverse fibres of pons
V	primary visual cortex
VLPAG	ventrolateral periaqueductal grey

Ec-Ad23

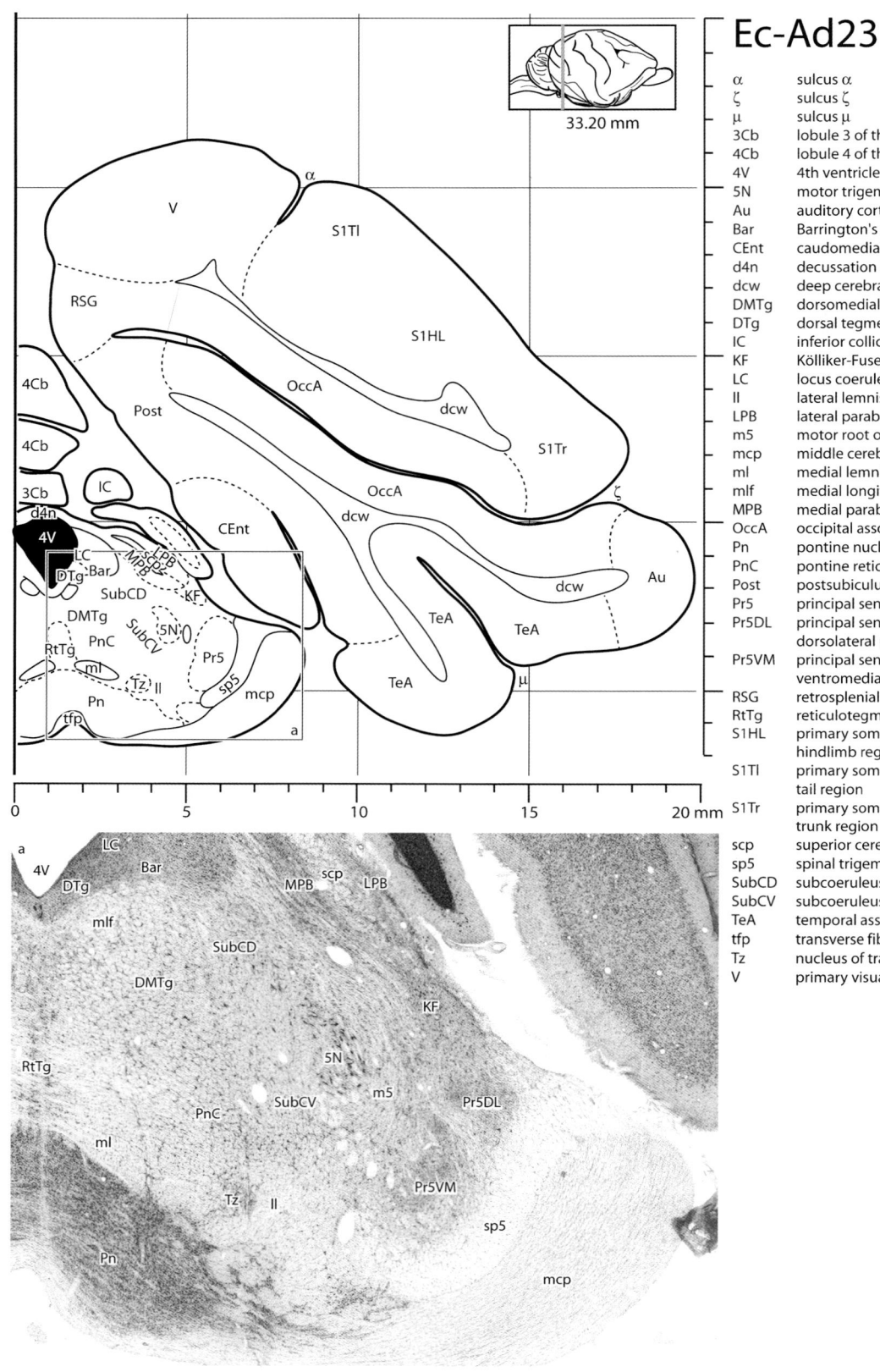

33.20 mm

α	sulcus α
ζ	sulcus ζ
μ	sulcus μ
3Cb	lobule 3 of the Cb vermis
4Cb	lobule 4 of the Cb vermis
4V	4th ventricle
5N	motor trigeminal nucleus
Au	auditory cortex
Bar	Barrington's nucleus
CEnt	caudomedial entorhinal cortex
d4n	decussation of trochlear nerve
dcw	deep cerebral white matter
DMTg	dorsomedial tegmental area
DTg	dorsal tegmental nu.
IC	inferior colliculus
KF	Kölliker-Fuse nu.
LC	locus coeruleus
ll	lateral lemniscus
LPB	lateral parabrachial nu.
m5	motor root of trigeminal nerve
mcp	middle cerebellar peduncle
ml	medial lemniscus
mlf	medial longitudinal fasciculus
MPB	medial parabrachial nu.
OccA	occipital association cortex
Pn	pontine nuclei
PnC	pontine reticular nu., caudal
Post	postsubiculum
Pr5	principal sensory trigem. nu.
Pr5DL	principal sensory trigeminal nu. dorsolateral part
Pr5VM	principal sensory trigeminal nu. ventromedial part
RSG	retrosplenial granular cortex
RtTg	reticulotegmental nu. pons
S1HL	primary somatosensory cortex hindlimb region
S1Tl	primary somatosensory cortex tail region
S1Tr	primary somatosensory cortex trunk region
scp	superior cerebellar peduncle
sp5	spinal trigeminal tract
SubCD	subcoeruleus nu., dorsal
SubCV	subcoeruleus nu., ventral
TeA	temporal association cortex
tfp	transverse fibres of pons
Tz	nucleus of trapezoid body
V	primary visual cortex

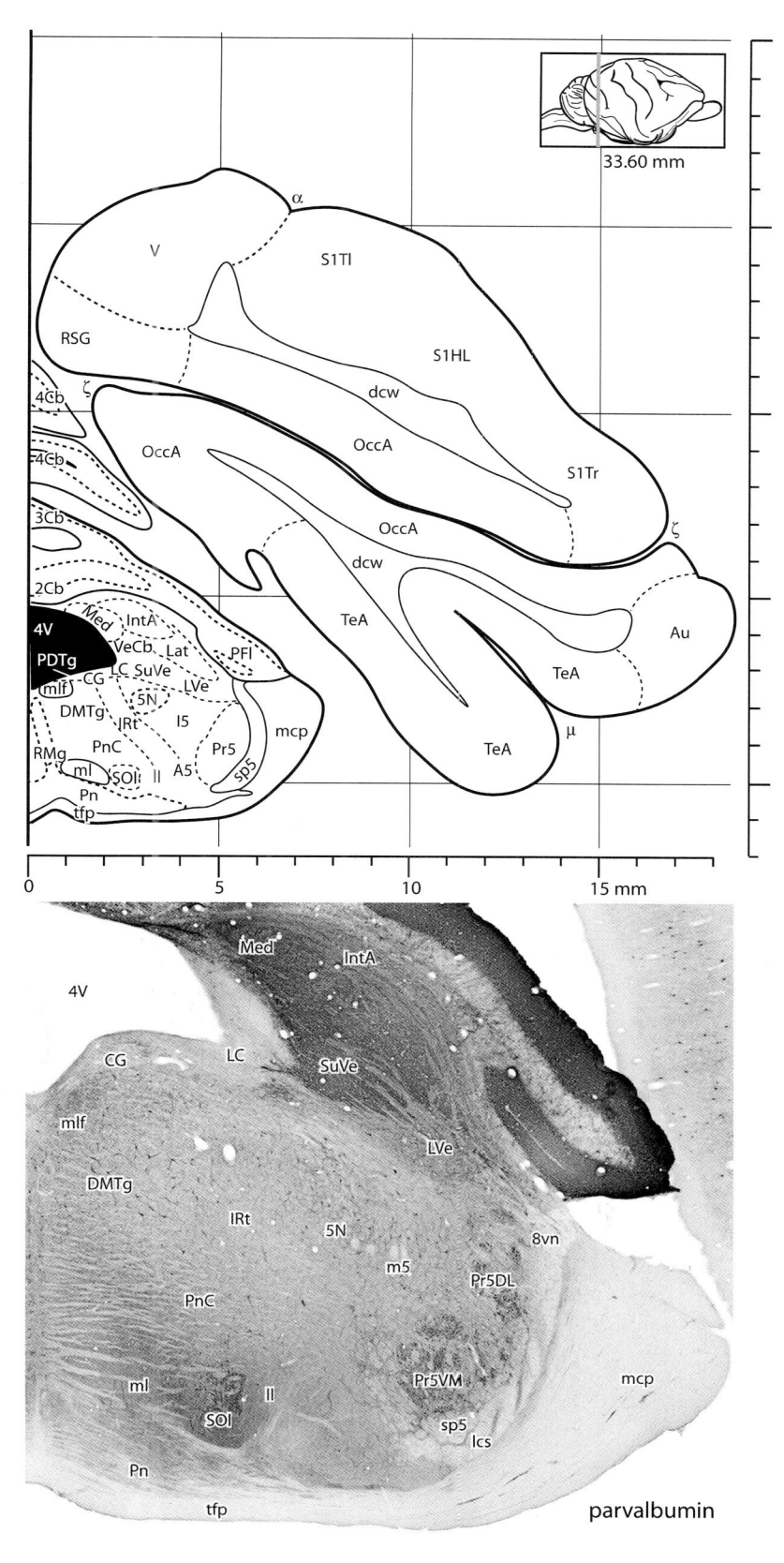

Ec-Ad24

α	sulcus α
ζ	sulcus ζ
μ	sulcus μ
2Cb	lobule 2 of the Cb vermis
3Cb	lobule 3 of the Cb vermis
4Cb	lobule 4 of the Cb vermis
4V	4th ventricle
5N	motor trigeminal nucleus
8vn	vestibular root of 8th nerve
A5	A5 noradrenaline cells
Au	auditory cortex
CG	central grey
dcw	deep cerebral white matter
DMTg	dorsomedial tegmental area
I5	intertrigeminal nu.
IntA	interposed cerebellar nu., anterior
IRt	intermediate reticular nucleus
Lat	lateral cerebellar nucleus
LC	locus coeruleus
lcs	lateral corticospinal tract
ll	lateral lemniscus
LVe	lateral vestibular nucleus
m5	motor root of trigeminal nerve
mcp	middle cerebellar peduncle
Med	medial cerebellar nucleus
ml	medial lemniscus
mlf	medial longitudinal fasciculus
OccA	occipital association cortex
PDTg	posterodorsal tegmental nucleus
PFl	paraflocculus of cerebellum
Pn	pontine nuclei
PnC	pontine reticular nucleus, caudal
Pr5	principal sensory trigeminal nu.
Pr5DL	principal sensory trigeminal nu. dorsolateral part
Pr5VM	principal sensory trigeminal nu. ventromedial part
RMg	raphe magnus nucleus
RSG	retrosplenial granular cortex
S1HL	primary somatosensory cortex hindlimb region
S1Tl	primary somatosensory cortex tail region
S1Tr	primary somatosensory cortex trunk region
SOl	superior olive
sp5	spinal trigeminal tract
SuVe	superior vestibular nucleus
TeA	temporal association cortex
tfp	transverse fibres of pons
V	primary visual cortex
VeCb	vestibulocerebellar nucleus

33.60 mm

parvalbumin

Ec-Ad25

μ	sulcus μ
ζ	sulcus ζ
1Cb	lobule 1 of the Cb vermis
2Cb	lobule 2 of the Cb vermis
3Cb	lobule 3 of the Cb vermis
4Cb	lobule 4 of the Cb vermis
10Cb	lobule 10 of the Cb vermis
4V	4th ventricle
6N	abducens nucleus
7n	facial nerve
7Nd	facial nucleus, dorsal part
7Nv	facial nucleus, ventral part
8n	vestibulocochlear nerve
8vn	vestibular root of 8th nerve
Au	auditory cortex
dcw	deep cerebral white matter
DMSp5	dorsomedial spinal trigeminal nucleus
DPGi	dorsal paragigantocell. nu.
Fl	flocculus
Gi	gigantocellular reticular nu.
I8	interstitial nucleus of 8th nerve
IntA	interposed cerebellar nu., ant.
IRt	intermediate reticular nu.
Lat	lateral cerebellar nu.
LVe	lateral vestibular nu.
Med	medial cerebellar nu.
ml	medial lemniscus
mlf	medial longitudinal fasciculus
OccA	occipital association cortex
PCRtA	parvicellular reticular nucleus alpha region
PFl	paraflocculus of cerebellum
RMg	raphe magnus nu.
RPa	raphe pallidus nu.
RSG	retrosplenial granular cortex
S1HL	primary somatosensory cortex hindlimb region
S1Tl	primary somatosensory cortex tail region
S1Tr	primary somatosensory cortex trunk region
scp	superior cerebellar peduncle
SOl	superior olive
sp5	spinal trigeminal tract
Sp5O	spinal trigeminal nucleus, oral part
SuVe	superior vestibular nucleus
TeA	temporal association cortex
V	primary visual cortex
VeCb	vestibulocerebellar nucleus

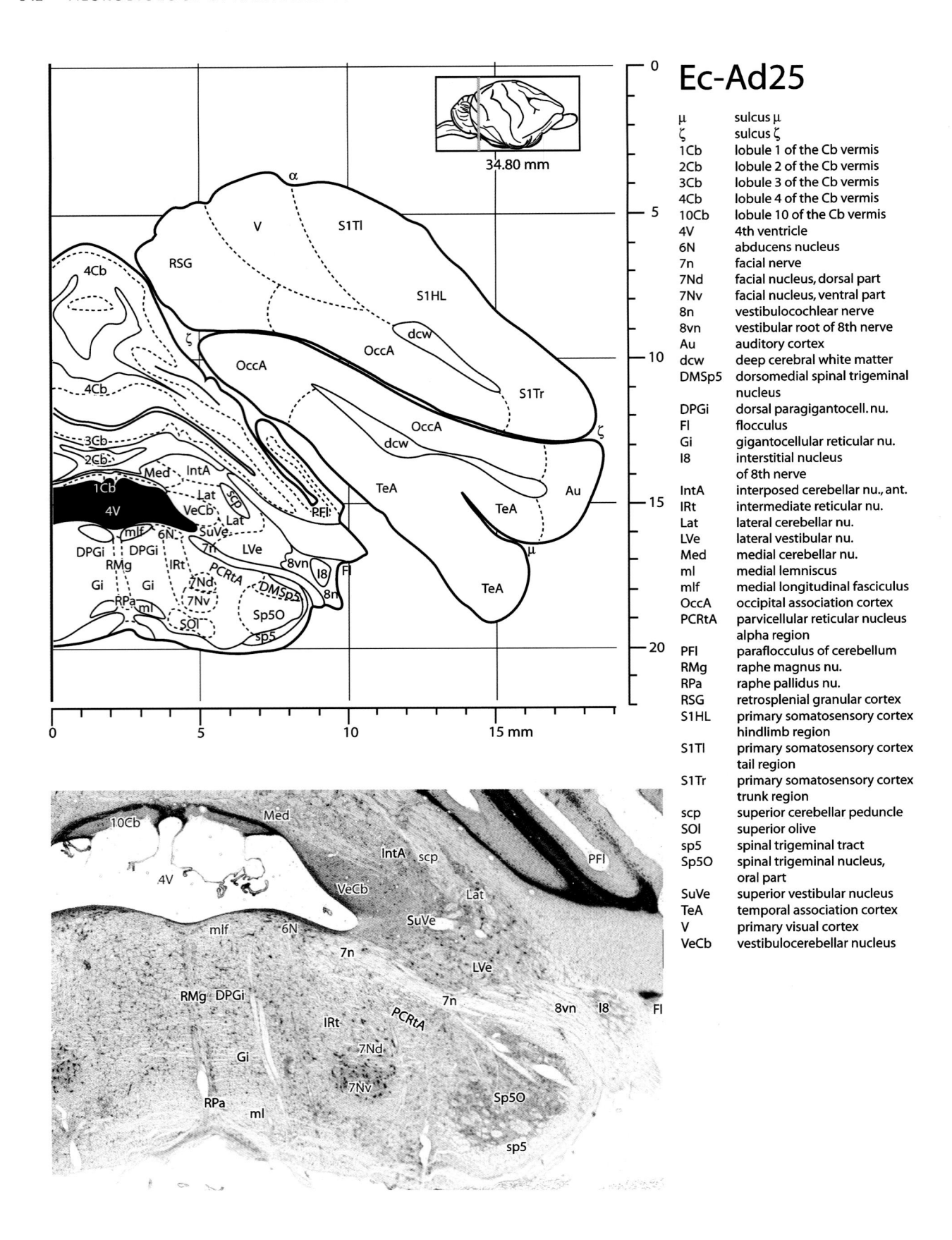

34.80 mm

Ec-Ad26

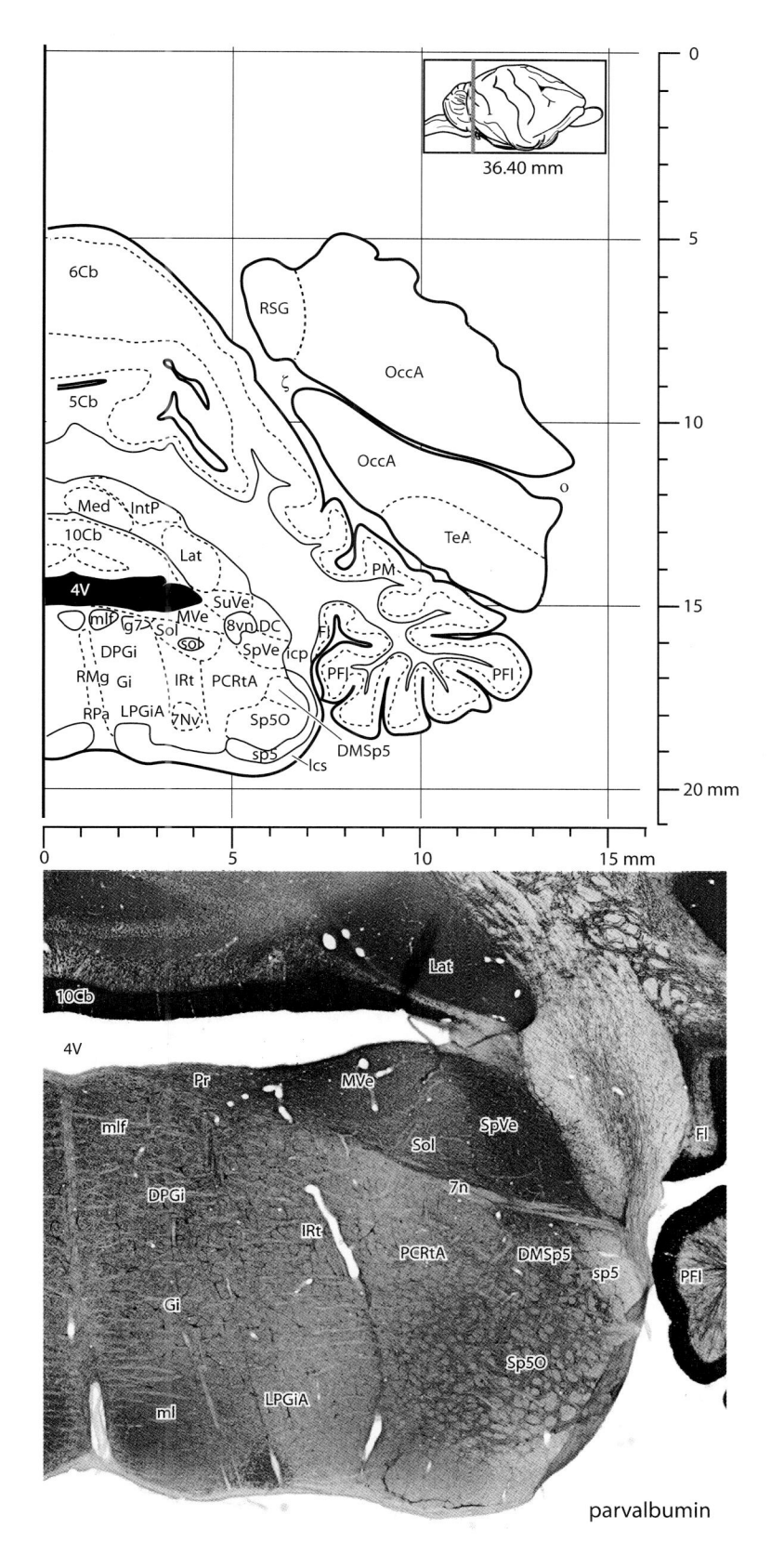

ζ	sulcus ζ
o	sulcus o
4V	4th ventricle
5Cb	lobule 5 of the Cb vermis
6Cb	lobule 6 of the Cb vermis
10Cb	lobule 10 of the Cb vermis
7n	facial nerve
7Nv	facial nucleus, ventral part
8vn	vestibular root of 8th nerve
DC	dorsal cochlear nu.
DMSp5	dorsomedial spinal trigeminal nu.
DPGi	dorsal paragigantocellular nu.
Fl	flocculus of Cb
g7	genu of the facial nerve
Gi	gigantocellular reticular nu.
icp	inferior cerebellar peduncle
IntP	interposed Cb nu., posterior
IRt	intermediate reticular nu.
Lat	lateral cerebellar nu.
lcs	lateral corticospinal tract
LPGiA	lateral paragigantocellular nu., alpha
Med	medial cerebellar nu.
ml	medial lemniscus
mlf	medial longitudinal fasciculus
MVe	medial vestibular nu.
OccA	occipital association cortex
PCRtA	parvicellular reticular nu., alpha
PFl	paraflocculus
PM	paramedian lobule of cerebellum
Pr	prepositus nu.
RMg	raphe magnus nu.
RPa	raphe pallidus nu.
RSG	retrosplenial granular cortex
Sol	solitary nucleus
sol	solitary tract
sp5	spinal trigeminal tract
Sp5O	spinal trigeminal nu., oral part
SpVe	spinal vestibular nu.
SuVe	superior vestibular nu.
TeA	temporal association cortex

parvalbumin

Ec-Ad27

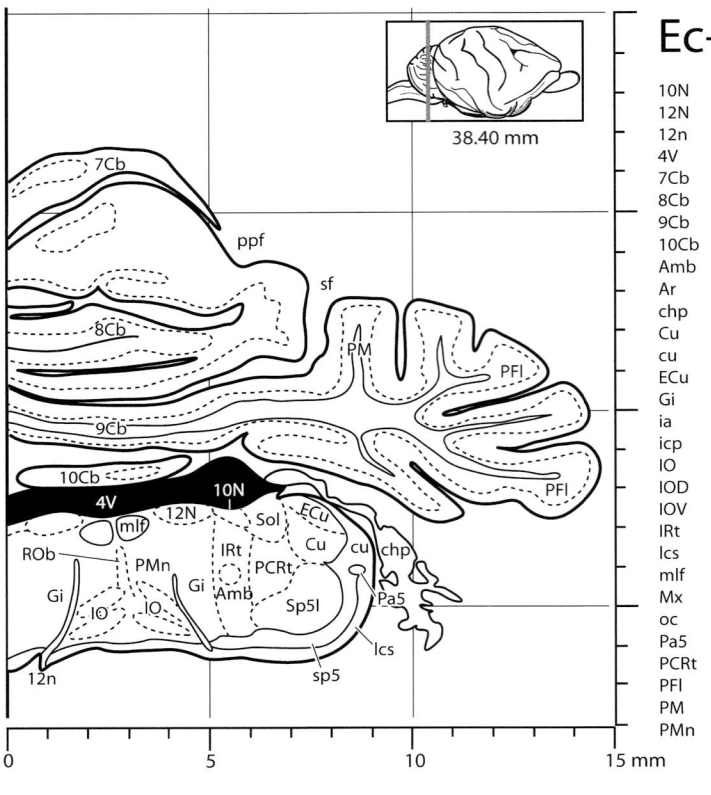

38.40 mm

10N	vagus nerve nu.	ppf	prepyramidal fissure
12N	hypoglossal nu.	ROb	raphe obscurus nu.
12n	hypoglossal nerve	sf	secondary fissure
4V	4th ventricle	Sol	solitary nucleus
7Cb	lobule 7 of Cb vermis	sol	solitary tract
8Cb	lobule 8 of Cb vermis	sp5	spinal trigeminal tract
9Cb	lobule 9 of Cb vermis	Sp5I	spinal trigeminal nu., interpolar part
10Cb	lobule 10 of Cb vermis		
Amb	ambiguus nu.		
Ar	arcuate medullary nu.		
chp	choroid plexus		
Cu	cuneate nu.		
cu	cuneate fasciculus		
ECu	external cuneate nu.		
Gi	gigantocellular reticular nu.		
ia	internal arcuate fibres		
icp	inferior cerebellar peduncle		
IO	inferior olivary nu.		
IOD	inferior olive, dorsal nu.		
IOV	inferior olive, ventral nu.		
IRt	intermediate reticular nu.		
lcs	lateral corticospinal tract		
mlf	medial longitudinal fasciculus		
Mx	matrix region of the medulla		
oc	olivocerebellar tract		
Pa5	paratrigeminal nu.		
PCRt	parvicellular reticular nu.		
PFl	paraflocculus of cerebellum		
PM	paramedian lobule of Cb		
PMn	paramedian reticular nu.		

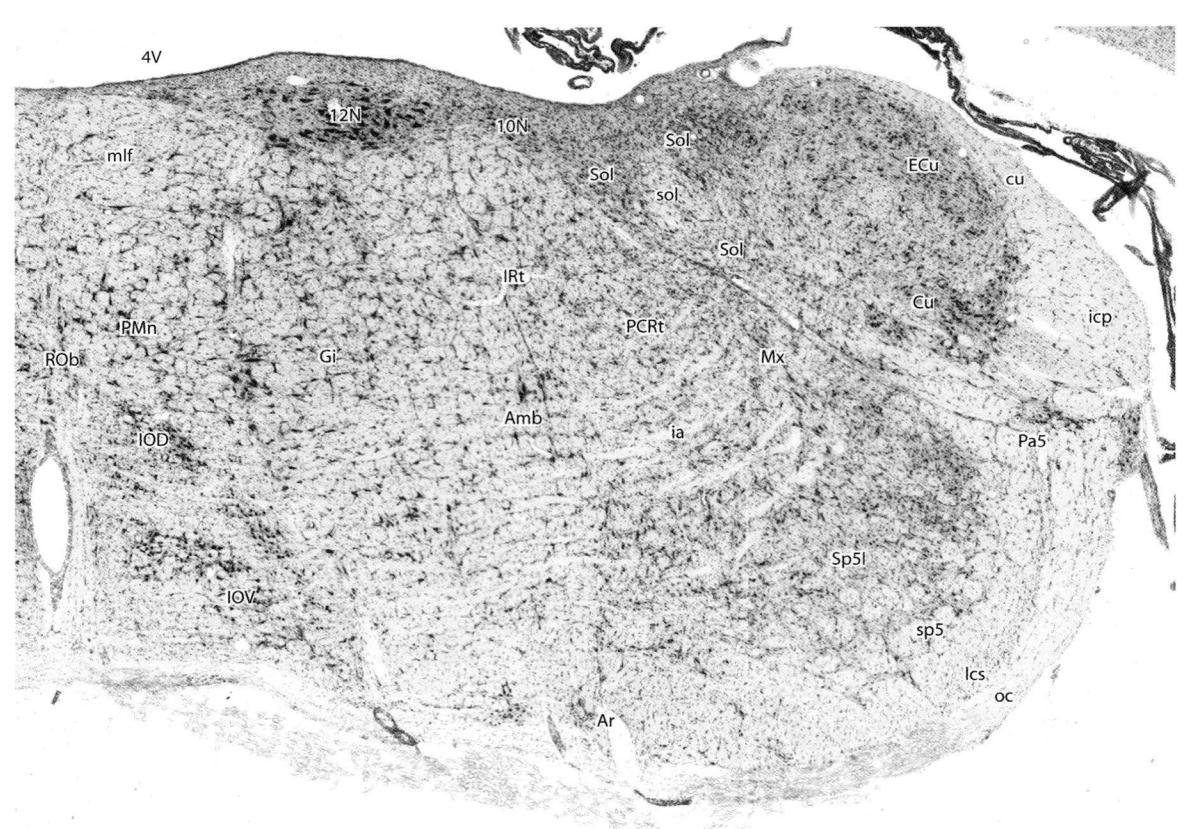

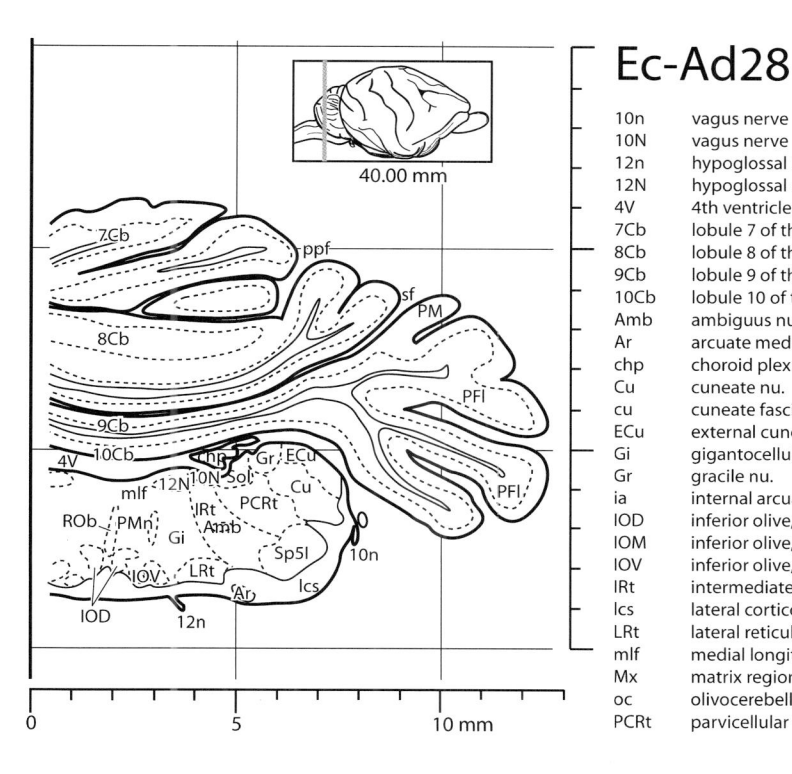

Ec-Ad28

10n	vagus nerve	PFl	paraflocculus
10N	vagus nerve nu.	PM	paramedian lobule
12n	hypoglossal nerve	PMn	paramedian reticular nu.
12N	hypoglossal nu.	ppf	prepyramidal fissure
4V	4th ventricle	ROb	raphe obscurus nu.
7Cb	lobule 7 of the Cb vermis	RPa	raphe pallidus nu.
8Cb	lobule 8 of the Cb vermis	sf	secondary fissure
9Cb	lobule 9 of the Cb vermis	Sol	solitary nucleus
10Cb	lobule 10 of the Cb vermis	sol	solitary tract
Amb	ambiguus nu.	sp5	spinal trigeminal tract
Ar	arcuate medullary nu.	Sp5I	spinal trig. nu., interpolar
chp	choroid plexus	ts	tectospinal tract
Cu	cuneate nu.		
cu	cuneate fasciculus		
ECu	external cuneate nu.		
Gi	gigantocellular reticular nu.		
Gr	gracile nu.		
ia	internal arcuate fibres		
IOD	inferior olive, dorsal nu.		
IOM	inferior olive, medial nu.		
IOV	inferior olive, ventral nu.		
IRt	intermediate reticular nu.		
lcs	lateral corticospinal tract		
LRt	lateral reticular nu.		
mlf	medial longitudinal fasciculus		
Mx	matrix region of the medulla		
oc	olivocerebellar tract		
PCRt	parvicellular reticular nu.		

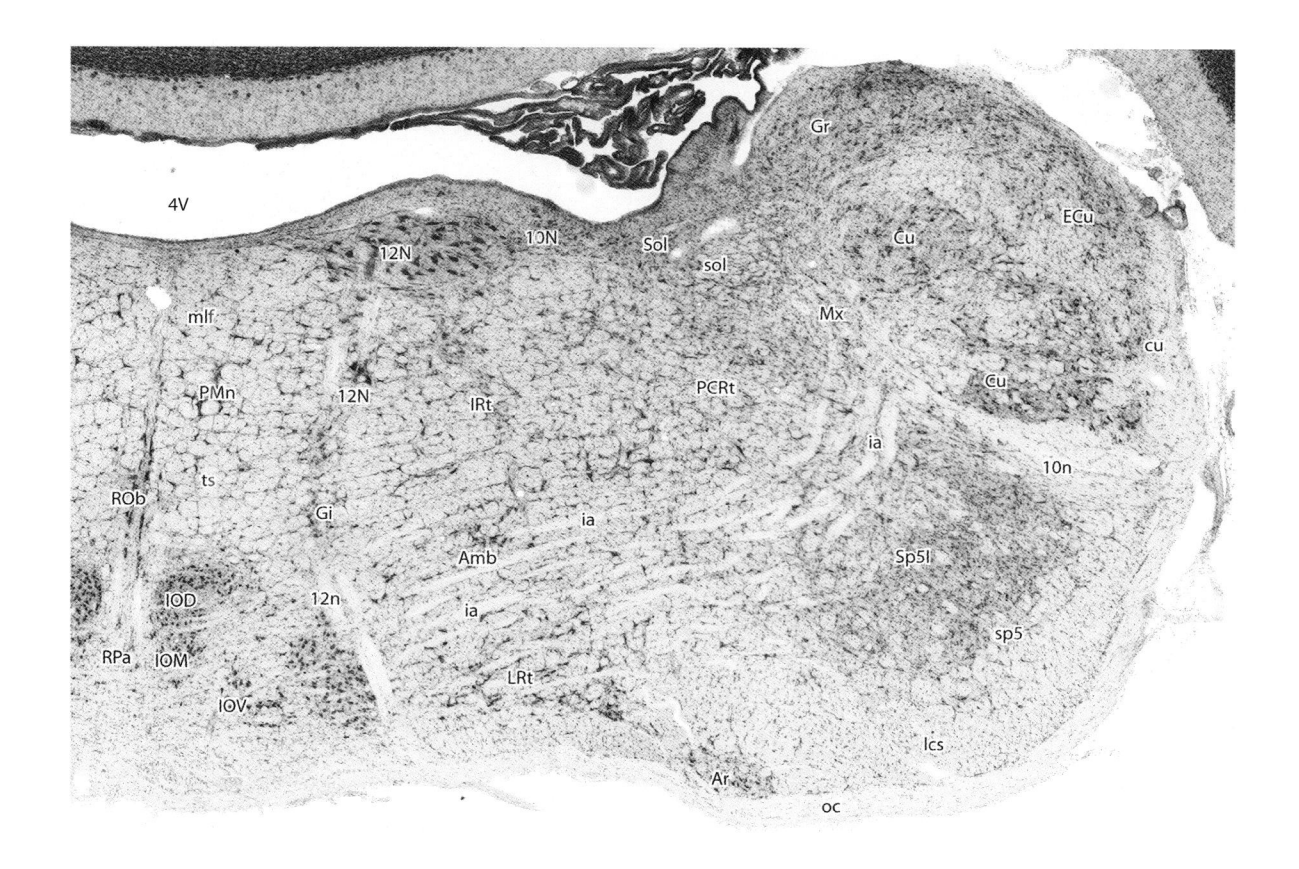

Ec-Ad29

10N	vagus nerve nucleus
12n	hypoglossal nerve
12N	hypoglossal nucleus
4V	4th ventricle
9Cb	lobule 9 of the Cb vermis
10Cb	lobule 10 of the Cb vermis
Ar	arcuate medullary nucleus
Cu	cuneate nucleus
cu	cuneate fasciculus
ECu	external cuneate nucleus
Fl	flocculus of cerebellum
Gi	gigantocellular reticular nucleus
Gr	gracile nucleus
gr	gracile fasciculus
IO	inferior olivary nucleus
IOA	inferior olive, subnu. A medial nu.
IOBe	inferior olive, beta subnucleus
IOD	inferior olive, dorsal nucleus
IRt	intermediate reticular nucleus
lcs	lateral corticospinal tract
LRt	lateral reticular nucleus
mlf	medial longitudinal fasciculus
Mx	matrix region of the medulla
PCRt	parvicellular reticular nucleus
PMn	paramedian reticular nucleus
ROb	raphe obscurus nucleus
RPa	raphe pallidus nucleus
Sol	solitary nucleus
sol	solitary tract
sp5	spinal trigeminal tract
Sp5I	spinal trigeminal nu., interpolar
ts	tectospinal tract

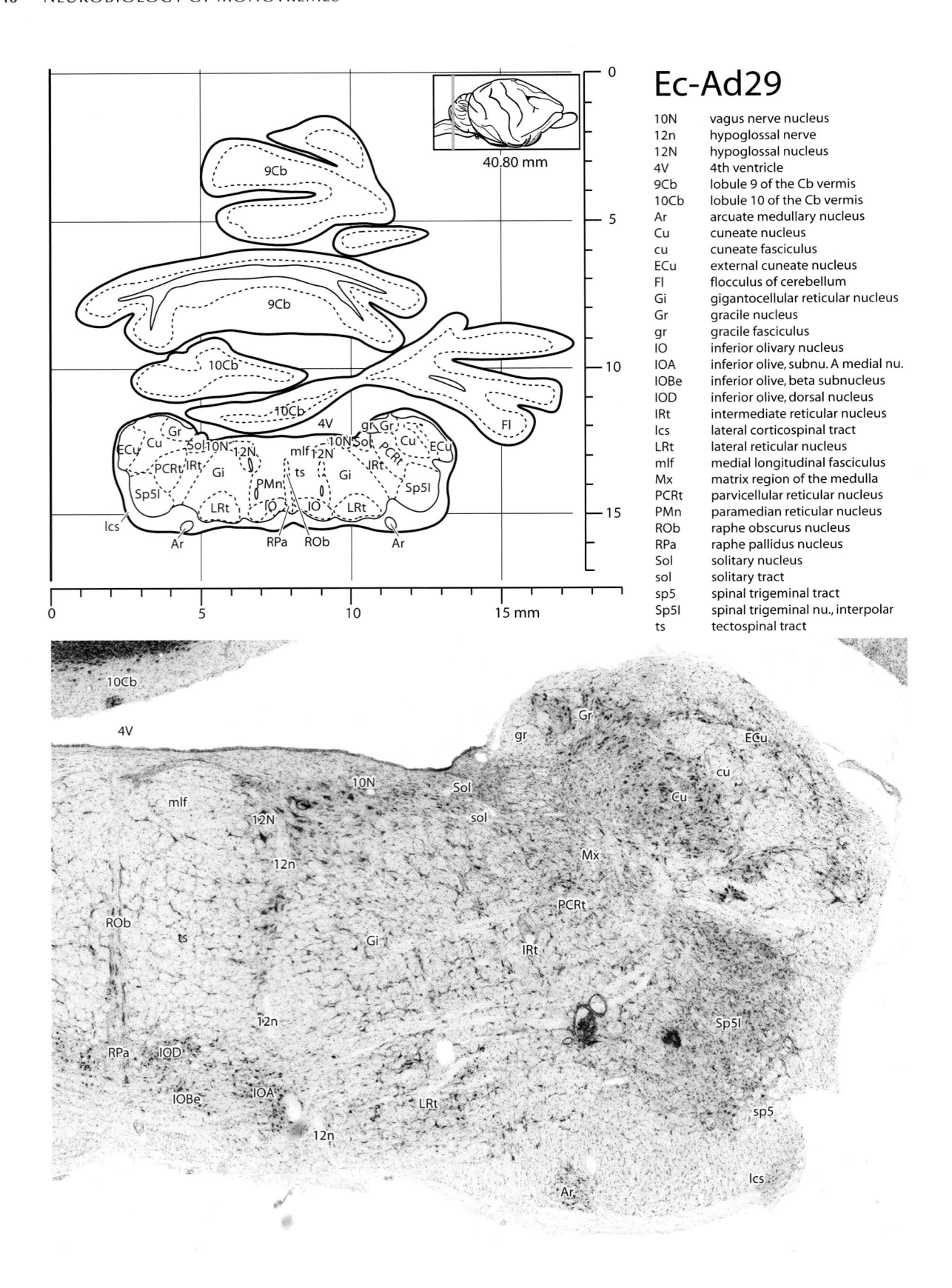

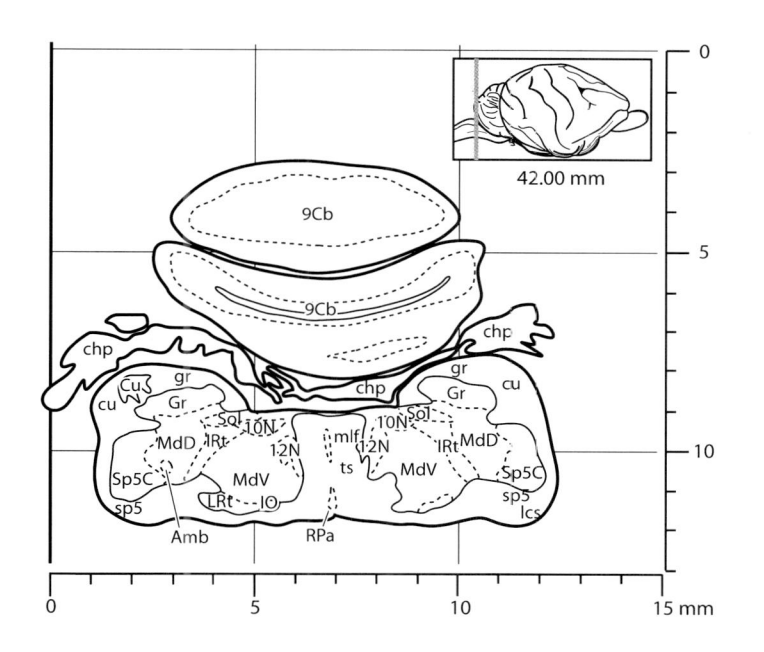

Ec-Ad30

10N	vagus nerve nucleus
12N	hypoglossal nucleus
4V	4th ventricle
9Cb	lobule 9 of the Cb vermis
Amb	ambiguus nucleus
chp	choroid plexus
cu	cuneate fasciculus
Cu	cuneate nucleus
Ge5	gelatinous layer of caudal Sp5
gr	gracile fasciculus
Gr	gracile nucleus
IO	inferior olivary nucleus
IRt	intermediate reticular nucleus
lcs	lateral corticospinal tract
LRt	lateral reticular nucleus
MdD	medullary reticular nucleus, dorsal
MdV	medullary reticular nucleus, ventral
mlf	medial longitudinal fasciculus
RPa	raphe pallidus nucleus
Sol	solitary nucleus
sol	solitary tract
sp5	spinal trigeminal tract
Sp5C	spinal trigeminal nucleus, caudal
ts	tectospinal tract

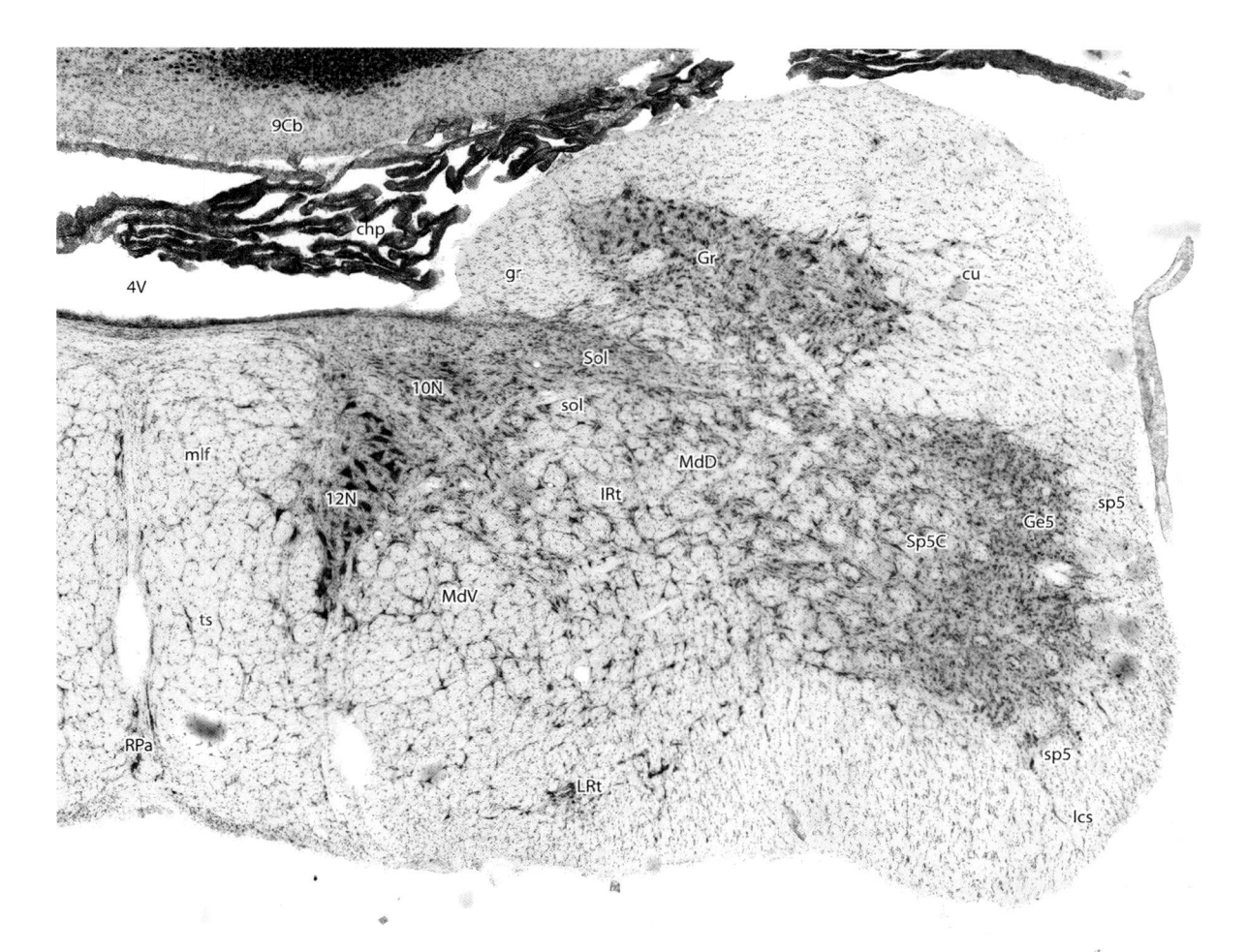

Ec-Ad31 SpC Cervical

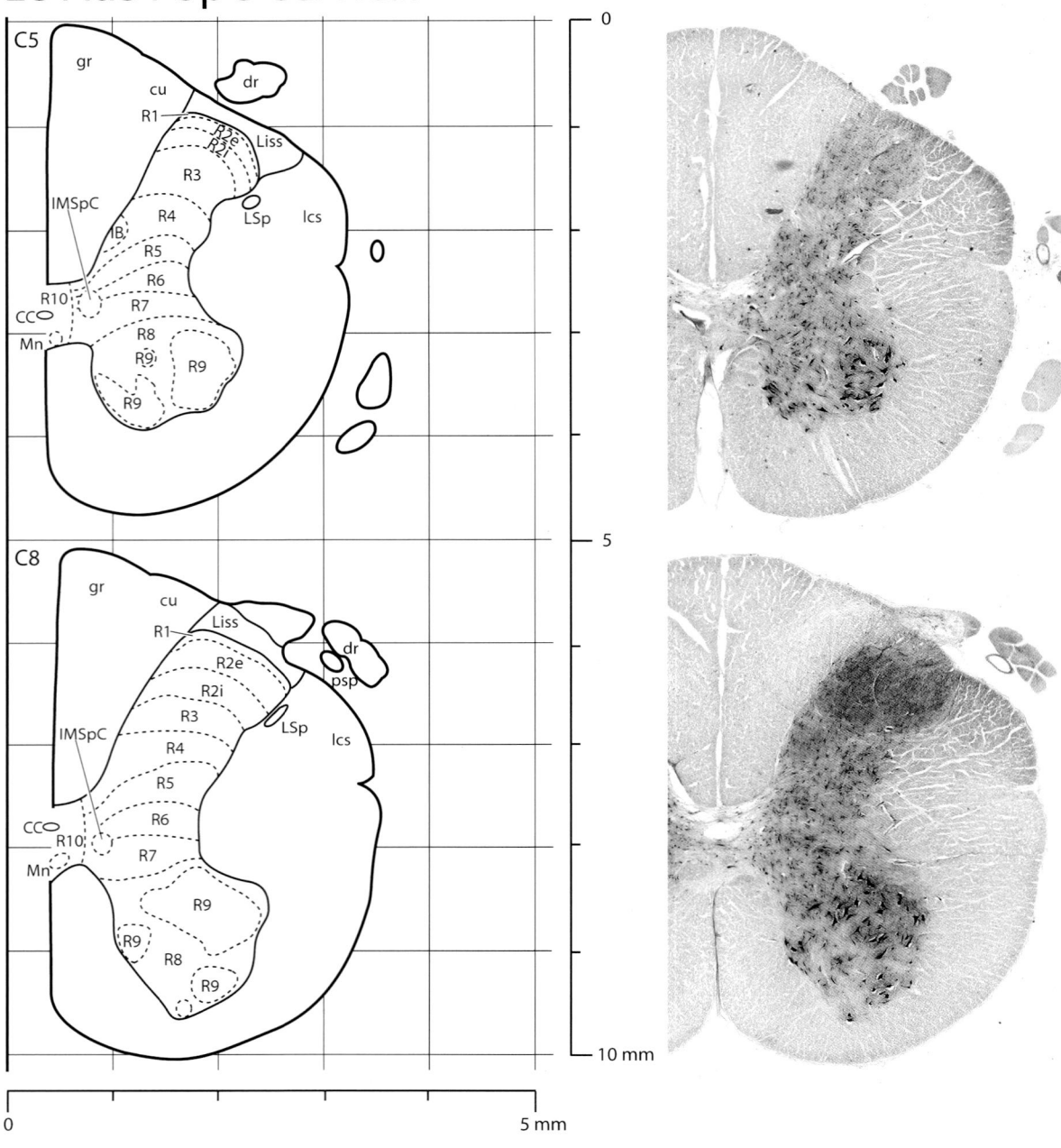

CC	central canal
cu	cuneate fasciculus
dr	dorsal root
gr	gracile fasciculus
IB	internal basilar nucleus of SpC
IMSpC	intermediomedial nucleus of SpC
lcs	lateral corticospinal tract
Liss	Lissauer's zone
LSp	lateral spinal nucleus
Mn	median nucleus of SpC
psp	posterior spinal artery
R1	Rexed's lamina 1
R2e	Rexed's lamina 2, external
R2i	Rexed's lamina 2, internal
R3	Rexed's lamina 3
R4	Rexed's lamina 4
R5	Rexed's lamina 5
R6	Rexed's lamina 6
R7	Rexed's lamina 7
R8	Rexed's lamina 8
R9	Rexed's lamina 9
R10	Rexed's lamina 10

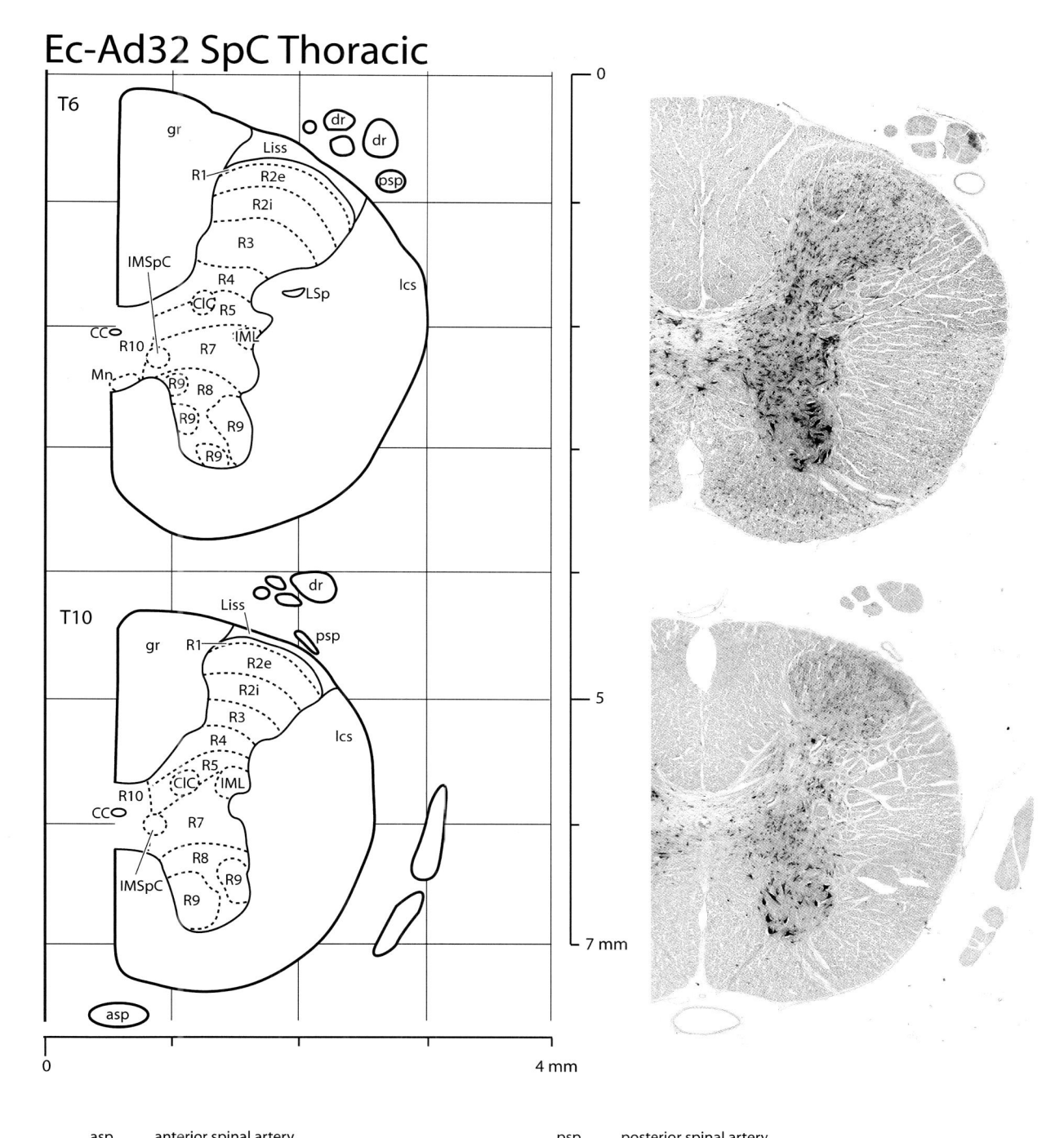

Ec-Ad32 SpC Thoracic

asp	anterior spinal artery
CC	central canal
ClC	Clarke's column
dr	dorsal root
gr	gracile fasciculus
IML	intermediolateral cell column
IMSpC	intermediomedial nu. SpC
lcs	lateral corticospinal tract
Liss	Lissauer's zone
LSp	lateral spinal nu.
Mn	median nucleus of SpC
psp	posterior spinal artery
R1	Rexed's lamina 1
R2e	Rexed's lamina 2, external
R2i	Rexed's lamina 2, internal
R3	Rexed's lamina 3
R4	Rexed's lamina 4
R5	Rexed's lamina 5
R7	Rexed's lamina 7
R8	Rexed's lamina 8
R9	Rexed's lamina 9
R10	Rexed's lamina 10

Ec-Ad33 SpC Lumbosacral

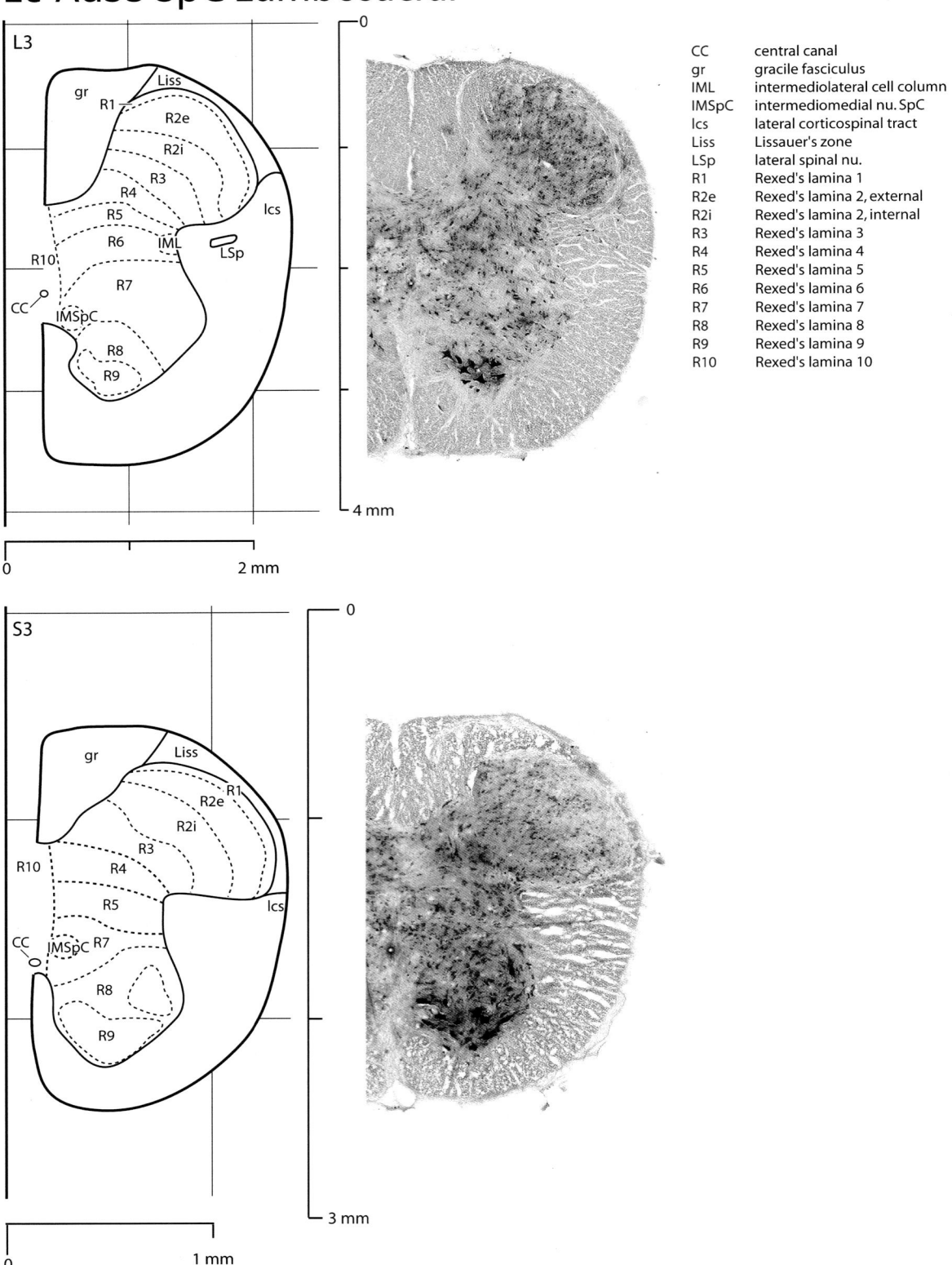

CC	central canal
gr	gracile fasciculus
IML	intermediolateral cell column
IMSpC	intermediomedial nu. SpC
lcs	lateral corticospinal tract
Liss	Lissauer's zone
LSp	lateral spinal nu.
R1	Rexed's lamina 1
R2e	Rexed's lamina 2, external
R2i	Rexed's lamina 2, internal
R3	Rexed's lamina 3
R4	Rexed's lamina 4
R5	Rexed's lamina 5
R6	Rexed's lamina 6
R7	Rexed's lamina 7
R8	Rexed's lamina 8
R9	Rexed's lamina 9
R10	Rexed's lamina 10

Ec-InA01

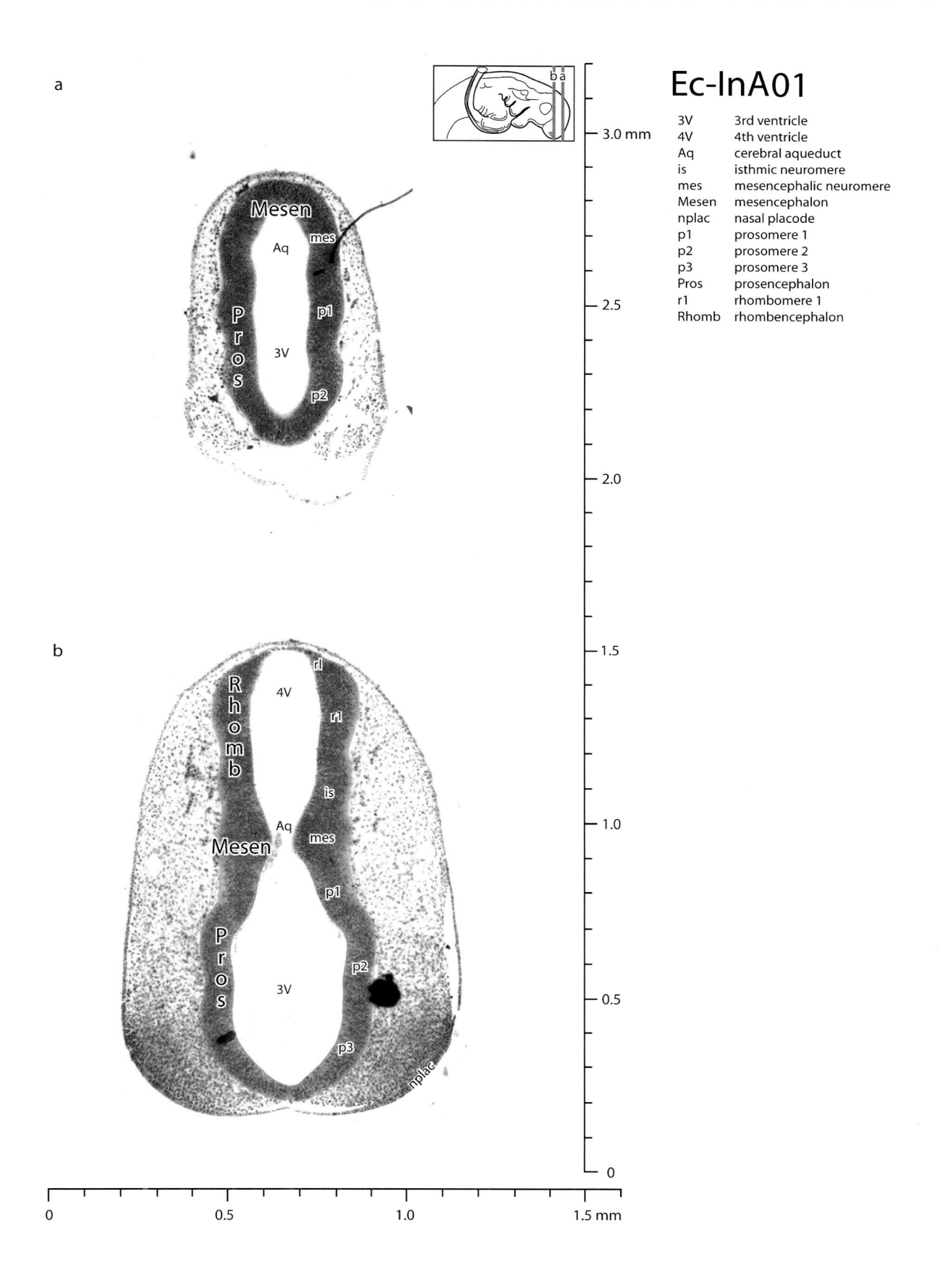

3V	3rd ventricle
4V	4th ventricle
Aq	cerebral aqueduct
is	isthmic neuromere
mes	mesencephalic neuromere
Mesen	mesencephalon
nplac	nasal placode
p1	prosomere 1
p2	prosomere 2
p3	prosomere 3
Pros	prosencephalon
r1	rhombomere 1
Rhomb	rhombencephalon

Ec-InA02

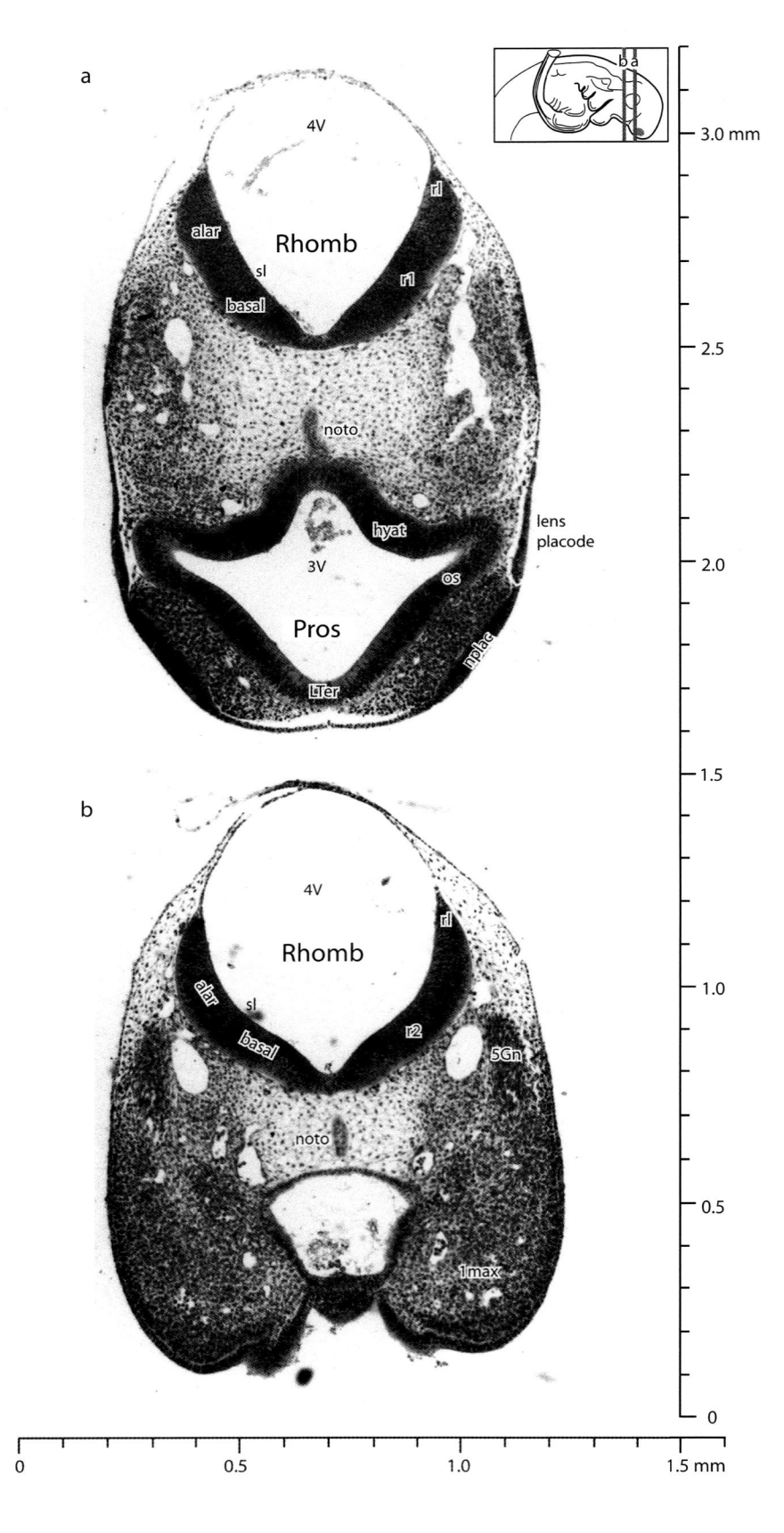

1max	maxillary process of arch 1
3V	3rd ventricle
4V	4th ventricle
5Gn	trigeminal ganglion
alar	alar plate
basal	basal plate
hyat	acroterminal hypothalamus
LTer	lamina terminalis
noto	notochord
nplac	nasal placode
os	optic stalk
Pros	prosencephalon
r1	rhombomere 1
r2	rhombomere 2
Rhomb	rhombencephalon
rl	rhombic lip
sl	sulcus limitans

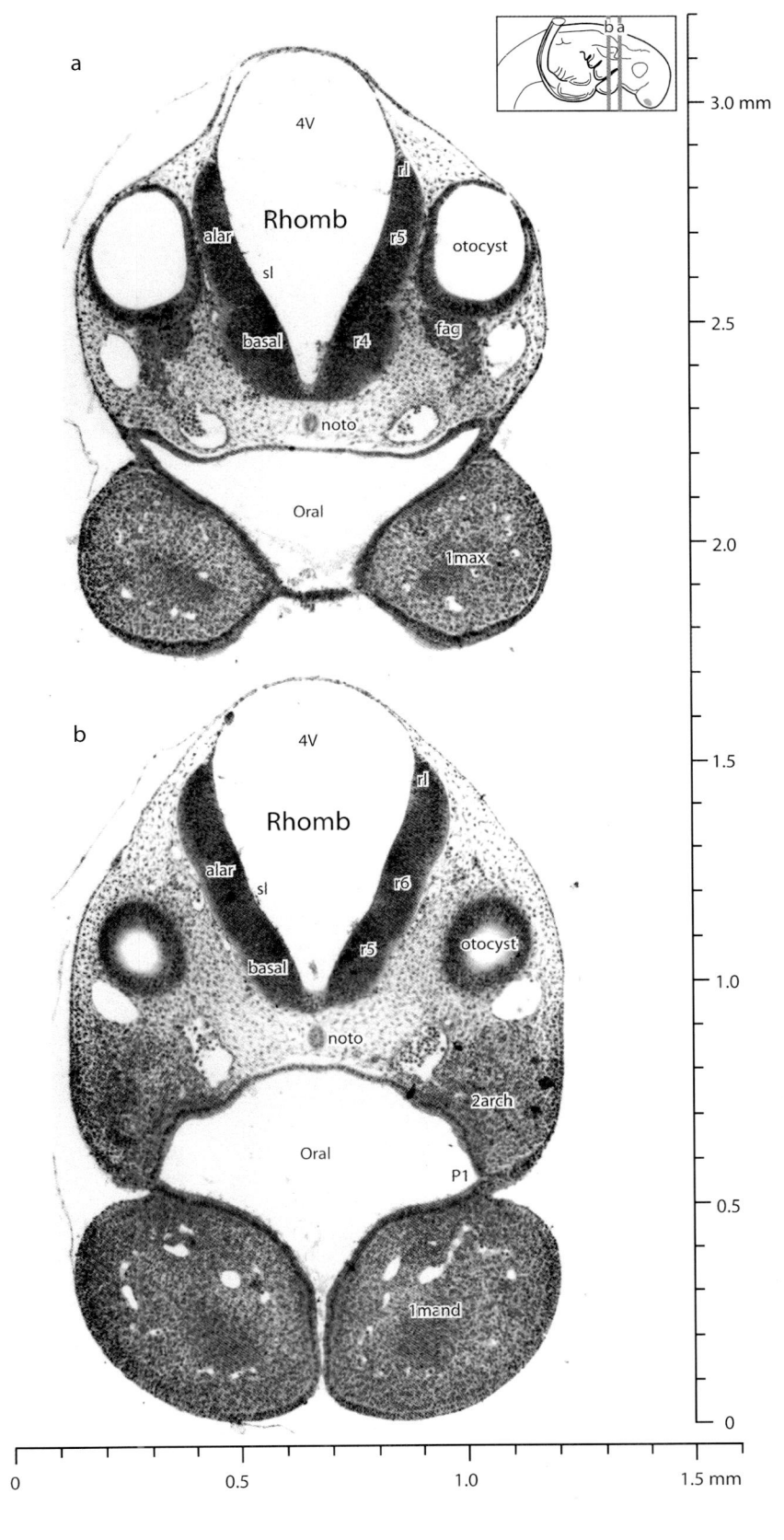

Ec-InA03

1mand	mandibular process of arch 1
1max	maxillary process of arch 1
2arch	second pharyngeal arch
4V	4th ventricle
alar	alar plate
basal	basal plate
fag	facioacoustic ganglion
noto	notochord
Oral	oral cavity
P1	pharyngeal pouch 1
r4	rhombomere 4
r5	rhombomere 5
r6	rhombomere 6
Rhomb	rhombencephalon
rl	rhombic lip
sl	sulcus limitans

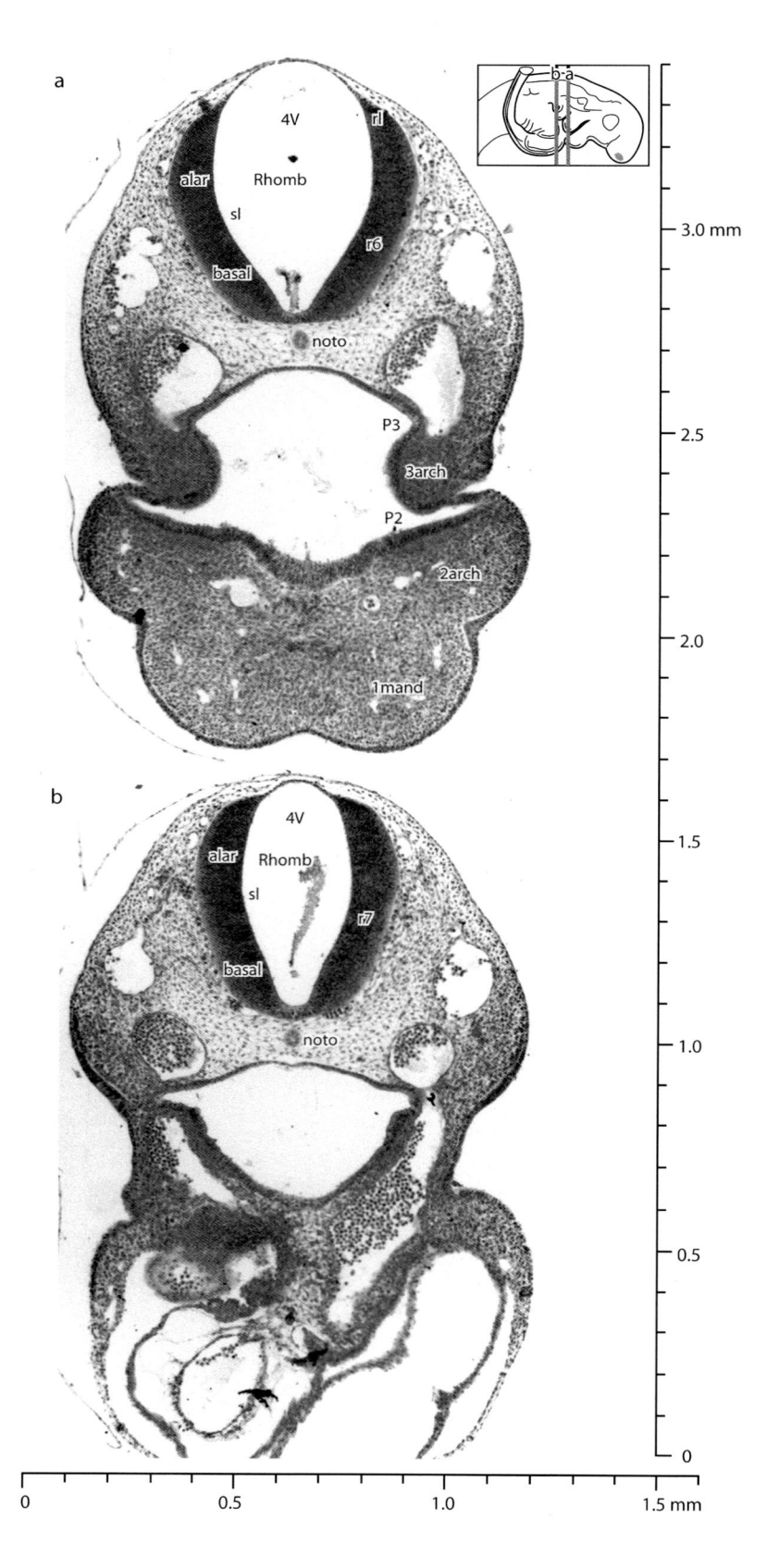

Ec-InA04

1mand	mandibular process of arch 1
2arch	second pharyngeal arch
3arch	third pharyngeal arch
4V	4th ventricle
alar	alar plate
basal	basal plate
noto	notochord
P2	pharyngeal pouch 2
P3	pharyngeal pouch 3
r6	rhombomere 6
r7	rhombomere 7
Rhomb	rhombencephalon
rl	rhombic lip
sl	sulcus limitans

Ec-InA05

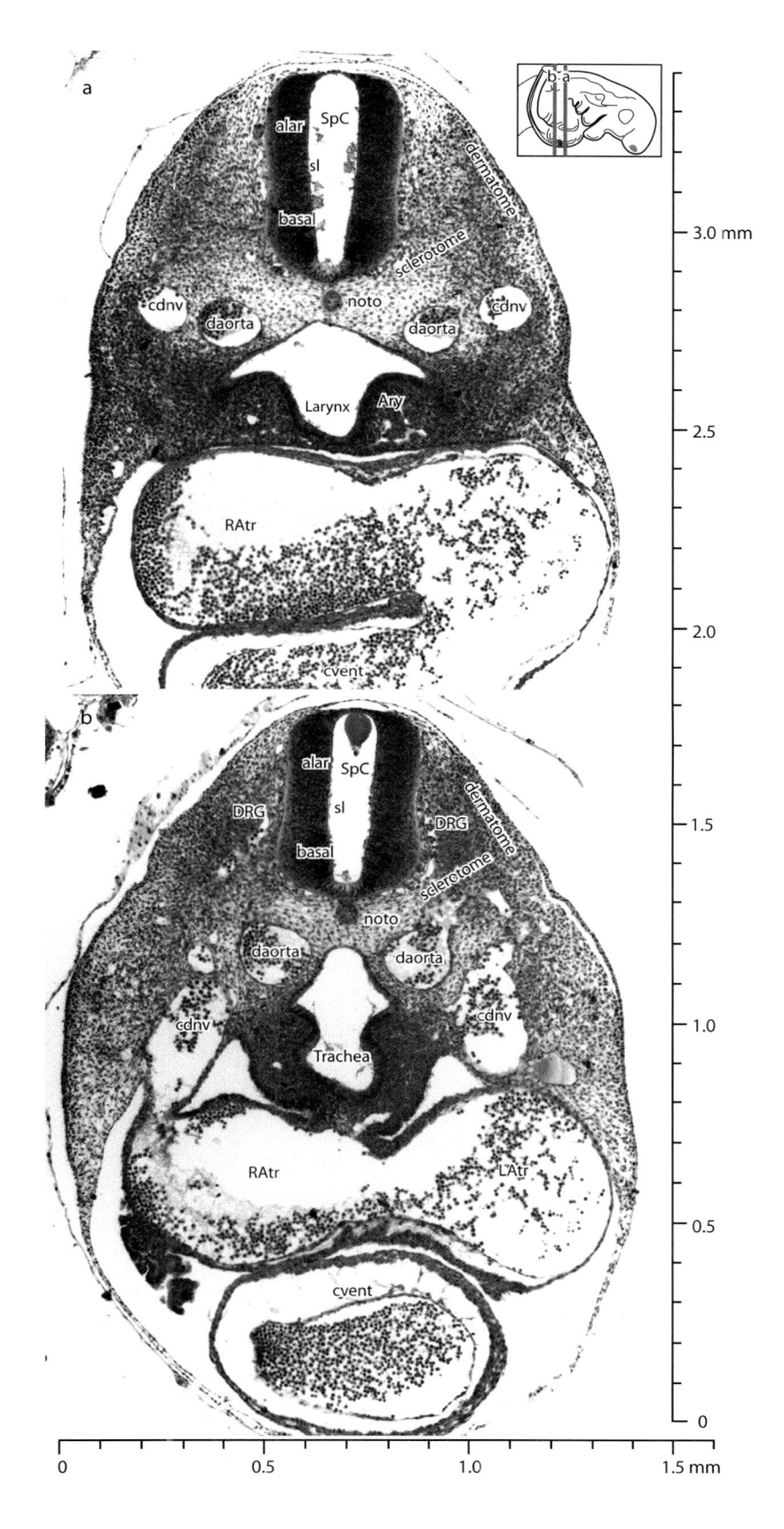

alar	alar plate
Ary	arytenoid cartilage
basal	basal plate
cdnv	cardinal vein
cvent	common ventricular chamber
daorta	dorsal aorta
DRG	dorsal root ganglion
LAtr	left atrium
noto	notochord
RAtr	right atrium
sl	sulcus limitans

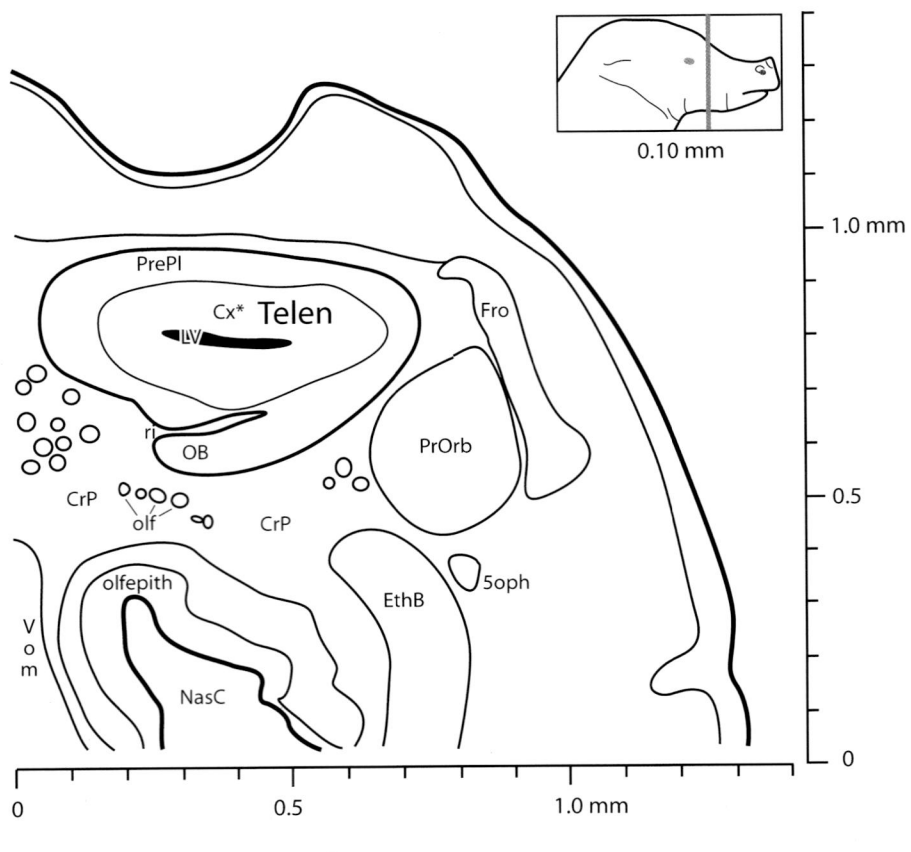

Ec-InB01

*	denotes precursor
5oph	ophthalmic division of trigeminal nerve
CrP	cribriform plate of ethmoid bone
Cx	cerebral cortex
EthB	ethmoid bone
Fro	frontal bone
LV	lateral ventricle
NasC	nasal cavity
OB	olfactory bulb
olf	olfactory nerve
olfepith	olfactory epithelium
PrePl	preplate of cortex
PrOrb	preoptic root of orbital bone
ri	rhinal incisure
Telen	telencephalon
Vom	vomer

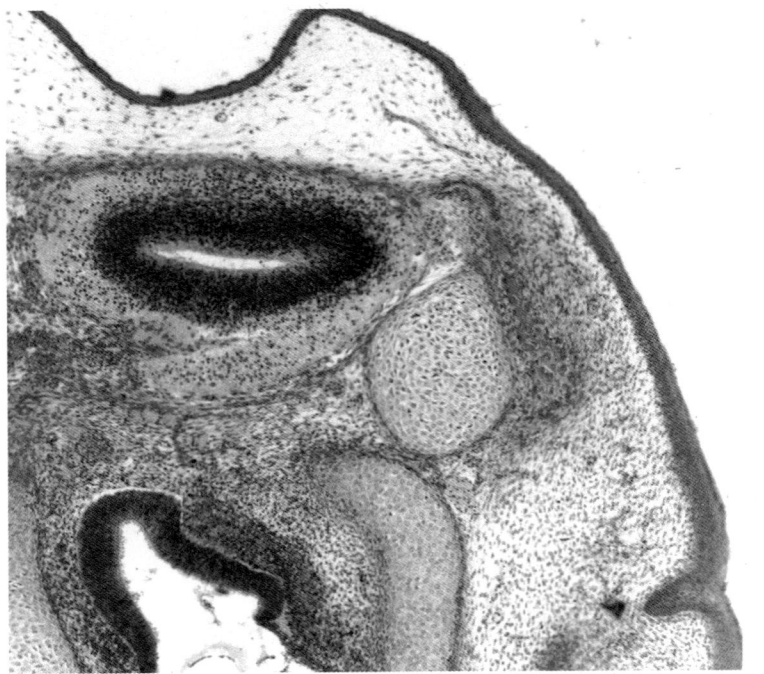

Ec-InB02

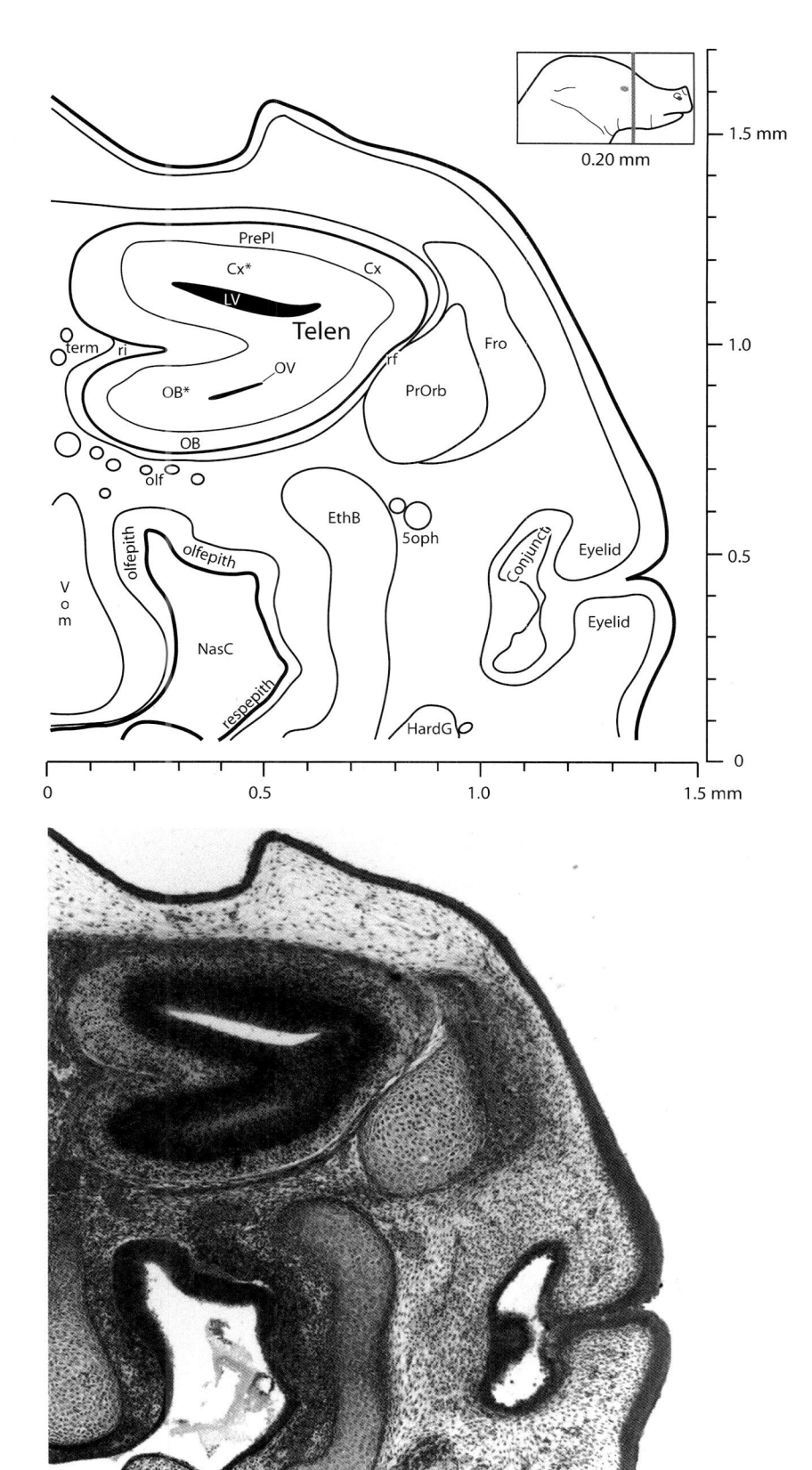

*	denotes precursor
5oph	ophthalmic division of trigeminal nerve
Conjunct	conjunctival sac
Cx	cerebral cortex
EthB	ethmoid bone
Fro	frontal bone
HardG	Harderian gland
LV	lateral ventricle
NasC	nasal cavity
OB	olfactory bulb
olf	olfactory nerve
olfepith	olfactory epithelium
OV	olfactory ventricle
PrePl	preplate of cortex
PrOrb	preoptic root orbital bone
respepith	respiratory epithelium
rf	rhinal fissure
ri	rhinal incisure
Telen	telencephalon
term	terminal nerve
Vom	vomer

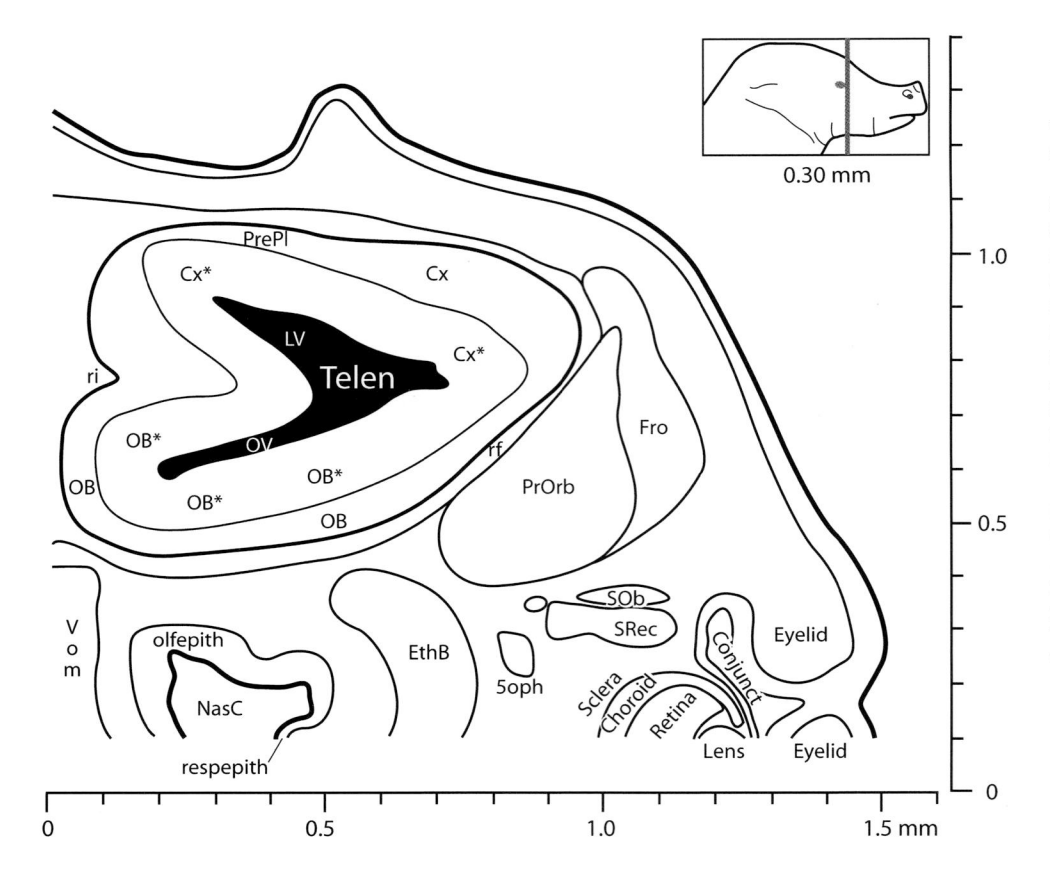

Ec-InB03

*	denotes precursor
5oph	ophthalmic division of trigeminal nerve
Conjunct	conjunctival sac
Cx	cerebral cortex
EthB	ethmoid bone
Fro	frontal bone
LV	lateral ventricle
NasC	nasal cavity
OB	olfactory bulb
olfepith	olfactory epithelium
OV	olfactory ventricle
PrePl	preplate of cortex
PrOrb	preoptic root of orbital bone
respepith	respiratory epithelium
rf	rhinal fissure
ri	rhinal incisure
SOb	superior oblique muscle
SRec	superior rectus muscle
Telen	telencephalon
Vom	vomer

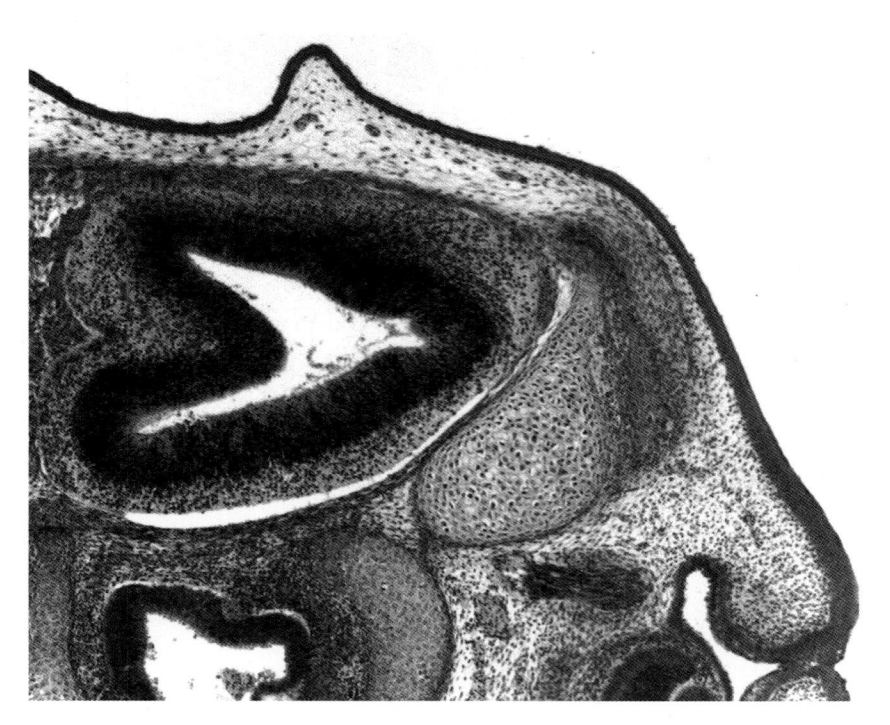

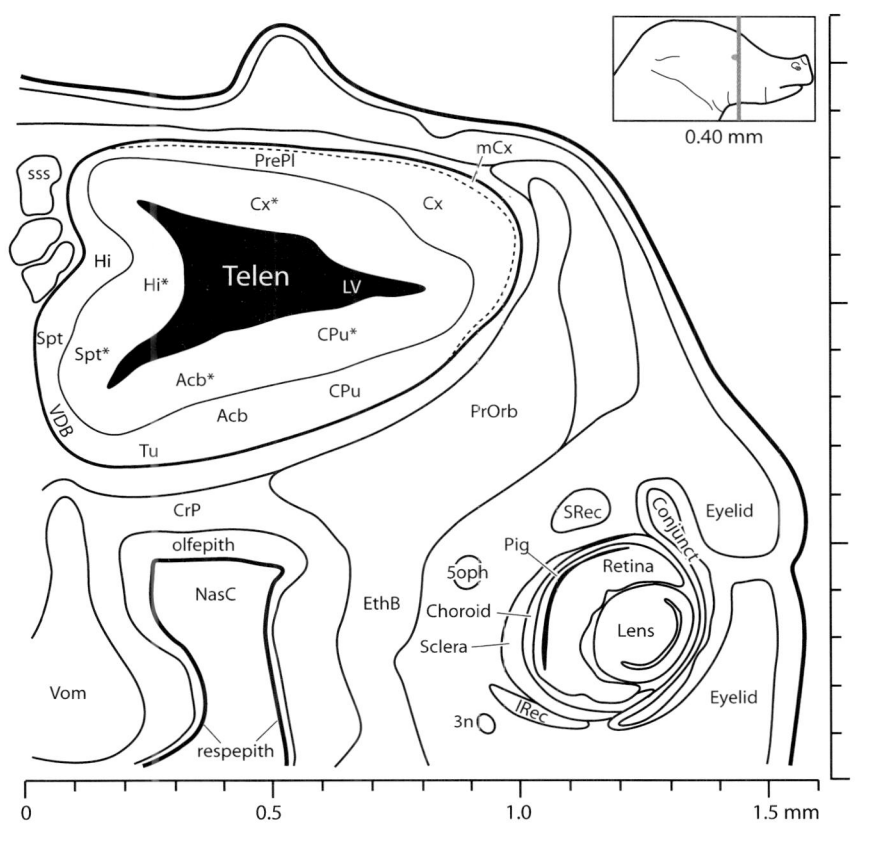

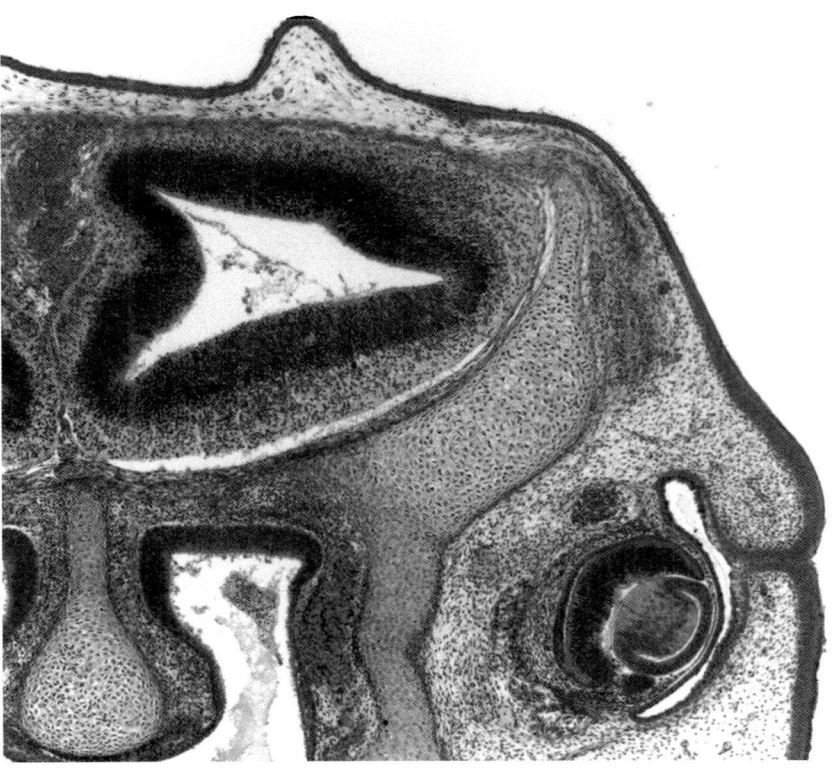

16 – ATLAS OF THE ADULT AND DEVELOPING BRAIN AND SPINAL CORD OF THE SHORT-BEAKED ECHIDNA   359

Ec-InB04

*	denotes precursor
3n	oculomotor nerve
5oph	ophthalmic division of trigeminal nerve
Acb	accumbens nu.
Conjunct	conjunctival sac
CPu	caudate putamen (striatum)
CrP	cribriform plate of ethmoid bone
Cx	cerebral cortex
EthB	ethmoid bone
Hi	hippocampus
IRec	inferior rectus muscle
LV	lateral ventricle
mCx	marginal zone of developing cortex
NasC	nasal cavity
olfepith	olfactory epithelium
Pig	pigment layer of the eye
PrePl	preplate of developing cortex
PrOrb	preoptic root of orbital bone
respepith	respiratory epithelium
Spt	septal region of brain
SRec	superior rectus muscle
sss	superior sagittal sinus
Telen	telencephalon
Tu	olfactory tubercle
VDB	nucleus of the vertical limb of the diagonal band
Vom	vomer

360 NEUROBIOLOGY OF MONOTREMES

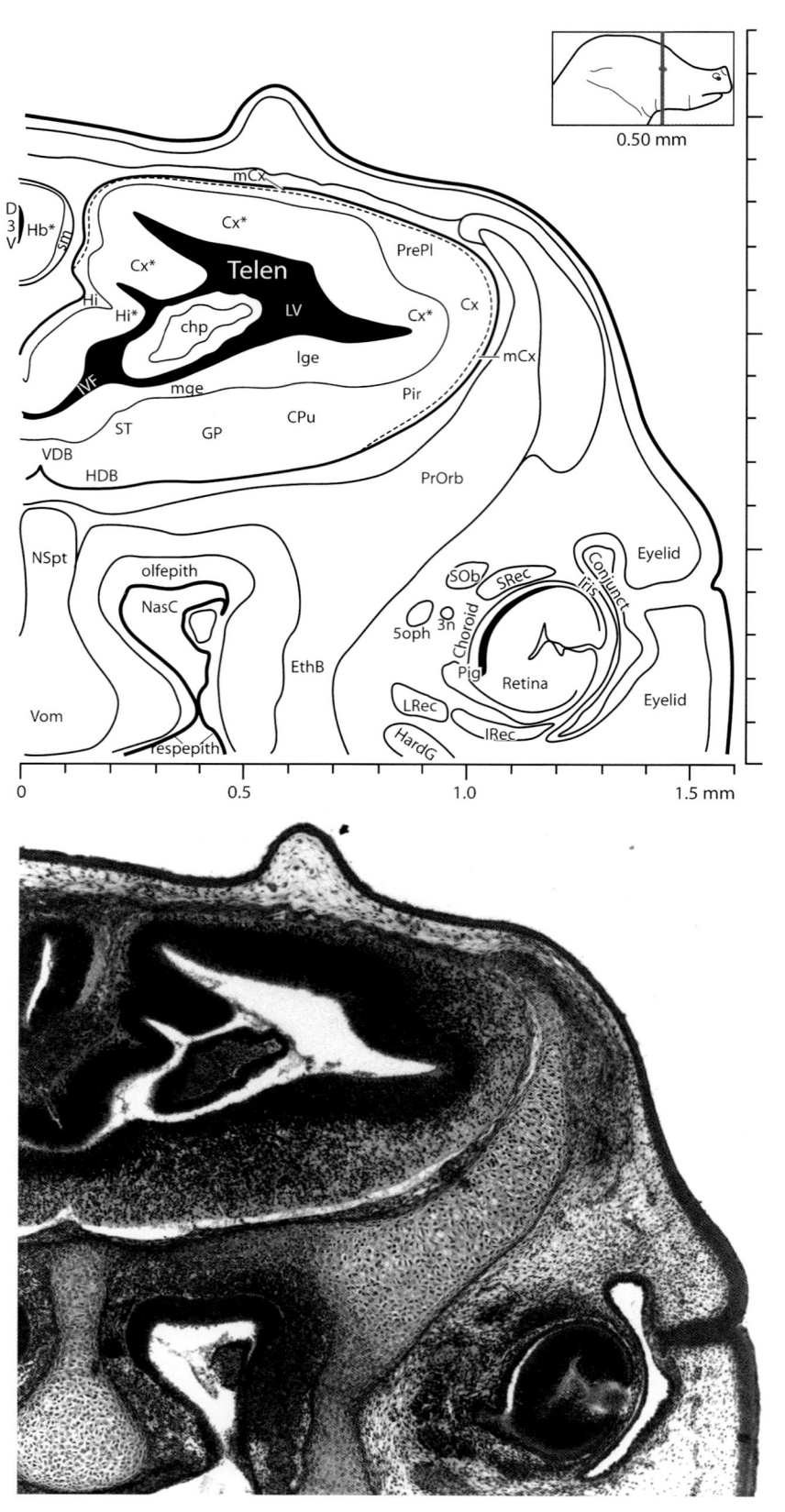

Ec-InB05

*	denotes precursor
3n	oculomotor nerve
5oph	ophthalmic division of trigeminal nerve
chp	choroid plexus
Conjunct	conjunctival sac
CPu	caudate putamen (striatum)
Cx	cerebral cortex
D3V	dorsal third ventricle
EthB	ethmoid bone
GP	globus pallidus
HardG	Harderian gland
Hb	habenular nuclei
HDB	nucleus of the horizontal limb of the diagonal band
Hi	hippocampus
IRec	inferior rectus muscle
Iris	iris
IVF	interventricular foramen
lge	lateral ganglionic eminence of subpallium
LRec	lateral rectus muscle
LV	lateral ventricle
mCx	marginal zone of developing cortex
mge	medial ganglionic eminence of subpallium
NasC	nasal cavity
NSpt	nasal septum
olfepith	olfactory epithelium
Pig	pigment layer of the eye
Pir	piriform cortex
PrePl	preplate of cortex
PrOrb	preoptic root of orbital bone
respepith	respiratory epithelium
sm	stria medullaris
SOb	superior oblique muscle
SRec	superior rectus muscle
ST	bed nucleus of stria terminalis
Telen	telencephalon
VDB	nucleus of the vertical limb of the diagonal band
Vom	vomer

16 – ATLAS OF THE ADULT AND DEVELOPING BRAIN AND SPINAL CORD OF THE SHORT-BEAKED ECHIDNA

Ec-InB06

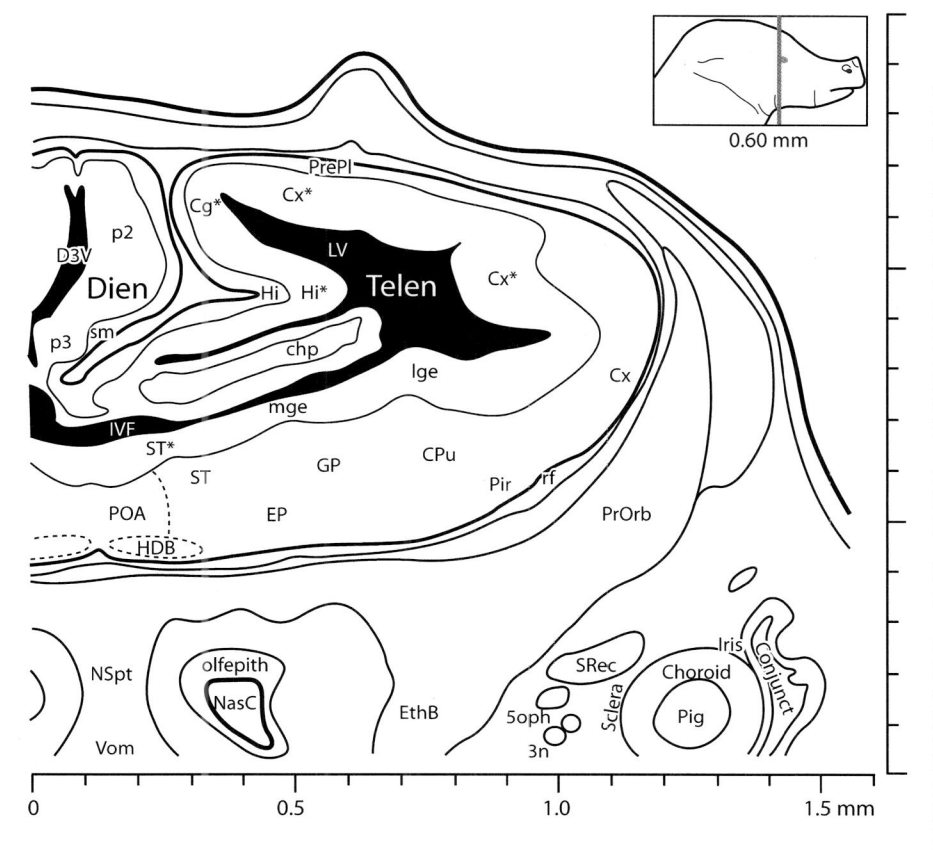

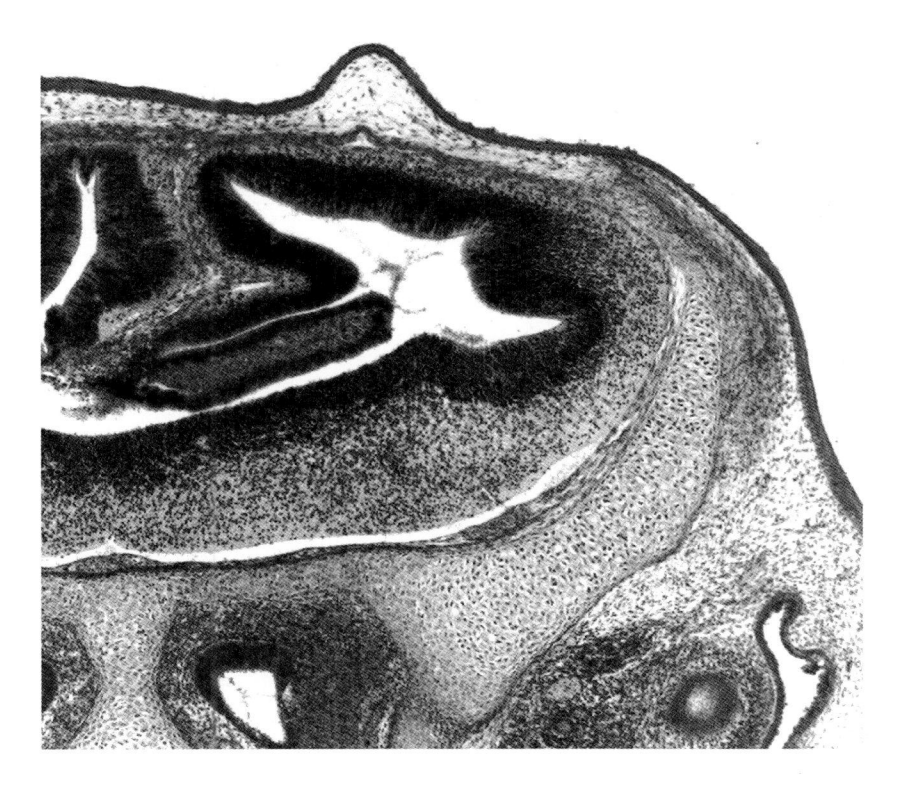

*	denotes precursor
3n	oculomotor nerve
5oph	ophthalmic division of trigeminal nerve
Cg	cingulate cortex
chp	choroid plexus
Conjunct	conjunctival sac
CPu	caudate putamen (striatum)
Cx	cerebral cortex
D3V	dorsal third ventricle
Dien	diencephalon
EP	entopeduncular nucleus
EthB	ethmoid bone
GP	globus pallidus
HDB	nucleus of the horiz.ontal limb of the diagonal band
Hi	hippocampus
IVF	interventricular foramen
lge	lateral ganglionic eminence of the subpallium
LV	lateral ventricle
mge	medial ganglionic eminence of the subpallium
NasC	nasal cavity
NSpt	nasal septum
olfepith	olfactory epithelium
p2	prosomere 2
p3	prosomere 3
Pig	pigment layer of the eye
Pir	piriform cortex
POA	preoptic area
PrePl	preplate of cortex
PrOrb	preoptic root of orbital bone
rf	rhinal fissure
sm	stria medullaris
SRec	superior rectus muscle
ST	bed nucleus of stria terminalis
Telen	telencephalon
Vom	vomer

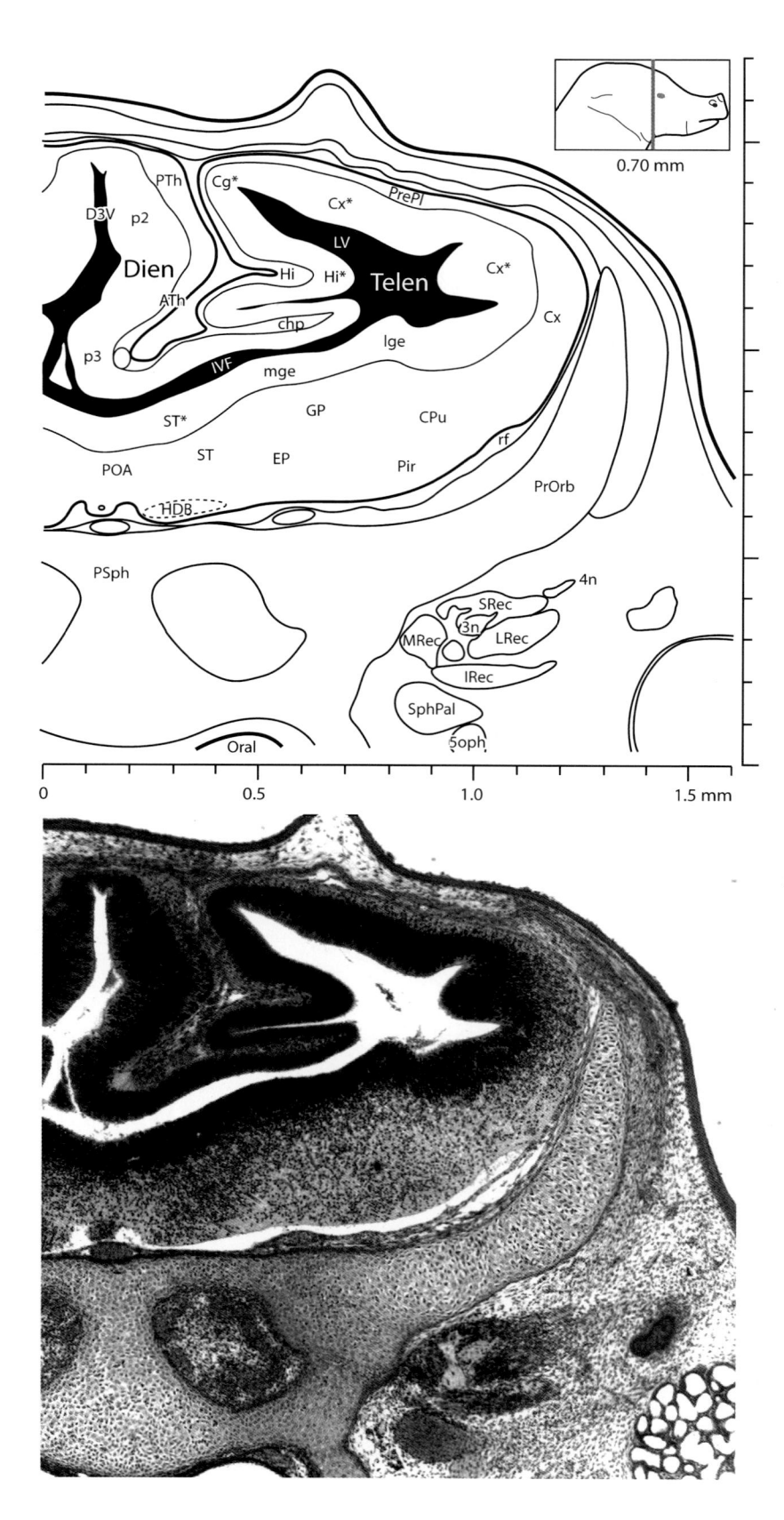

Ec-InB07

*	denotes precursor
3n	oculomotor nerve
4n	trochlear nerve
5oph	ophthalmic division of trigeminal nerve
ATh	anterior thalamic region
Cg	cingulate cortex
chp	choroid plexus
CPu	caudate putamen (striatum)
Cx	cerebral cortex
D3V	dorsal third ventricle
Dien	diencephalon
EP	entopeduncular nucleus
GP	globus pallidus
HDB	nucleus of the horizontal limb of the diagonal band
Hi	hippocampus
IRec	inferior rectus muscle
IVF	interventricular foramen
lge	lateral ganglionic eminence of the subpallium
LRec	lateral rectus muscle
LV	lateral ventricle
mge	medial ganglionic eminence of the subpallium
MRec	medial rectus muscle
Oral	oral cavity
p2	prosomere 2
p3	prosomere 3
Pir	piriform cortex
POA	preoptic area
PrePl	preplate of cortex
PrOrb	preoptic root of orbital bone
PSph	presphenoid bone
PTh	posterior thalamus
rf	rhinal fissure
SphPal	sphenopalatine ganglion
SRec	superior rectus muscle
ST	bed nucleus of stria terminalis
Telen	telencephalon

0.70 mm

0.80 mm

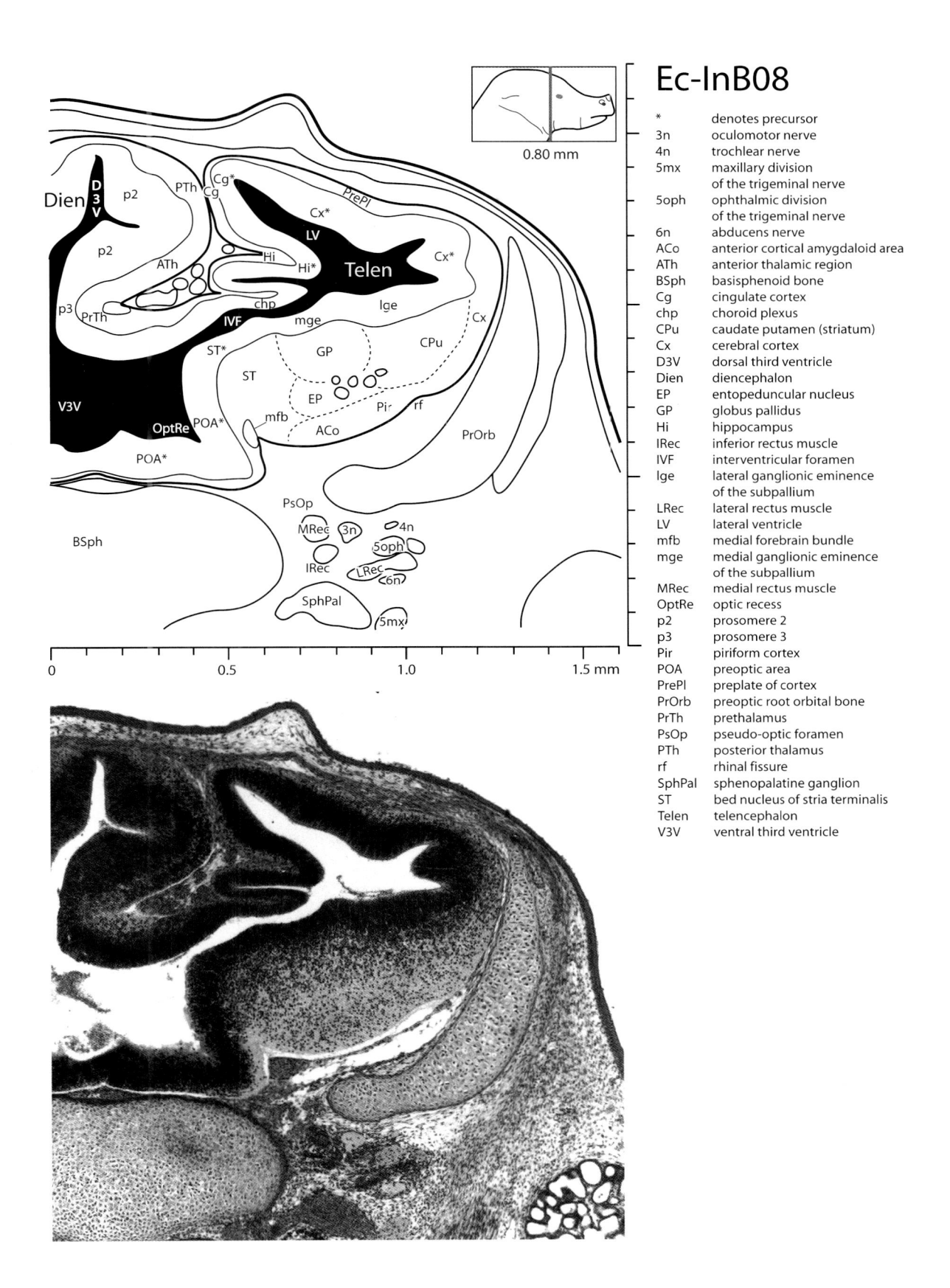

Ec-InB08

*	denotes precursor
3n	oculomotor nerve
4n	trochlear nerve
5mx	maxillary division of the trigeminal nerve
5oph	ophthalmic division of the trigeminal nerve
6n	abducens nerve
ACo	anterior cortical amygdaloid area
ATh	anterior thalamic region
BSph	basisphenoid bone
Cg	cingulate cortex
chp	choroid plexus
CPu	caudate putamen (striatum)
Cx	cerebral cortex
D3V	dorsal third ventricle
Dien	diencephalon
EP	entopeduncular nucleus
GP	globus pallidus
Hi	hippocampus
IRec	inferior rectus muscle
IVF	interventricular foramen
lge	lateral ganglionic eminence of the subpallium
LRec	lateral rectus muscle
LV	lateral ventricle
mfb	medial forebrain bundle
mge	medial ganglionic eminence of the subpallium
MRec	medial rectus muscle
OptRe	optic recess
p2	prosomere 2
p3	prosomere 3
Pir	piriform cortex
POA	preoptic area
PrePl	preplate of cortex
PrOrb	preoptic root orbital bone
PrTh	prethalamus
PsOp	pseudo-optic foramen
PTh	posterior thalamus
rf	rhinal fissure
SphPal	sphenopalatine ganglion
ST	bed nucleus of stria terminalis
Telen	telencephalon
V3V	ventral third ventricle

Ec-InB09

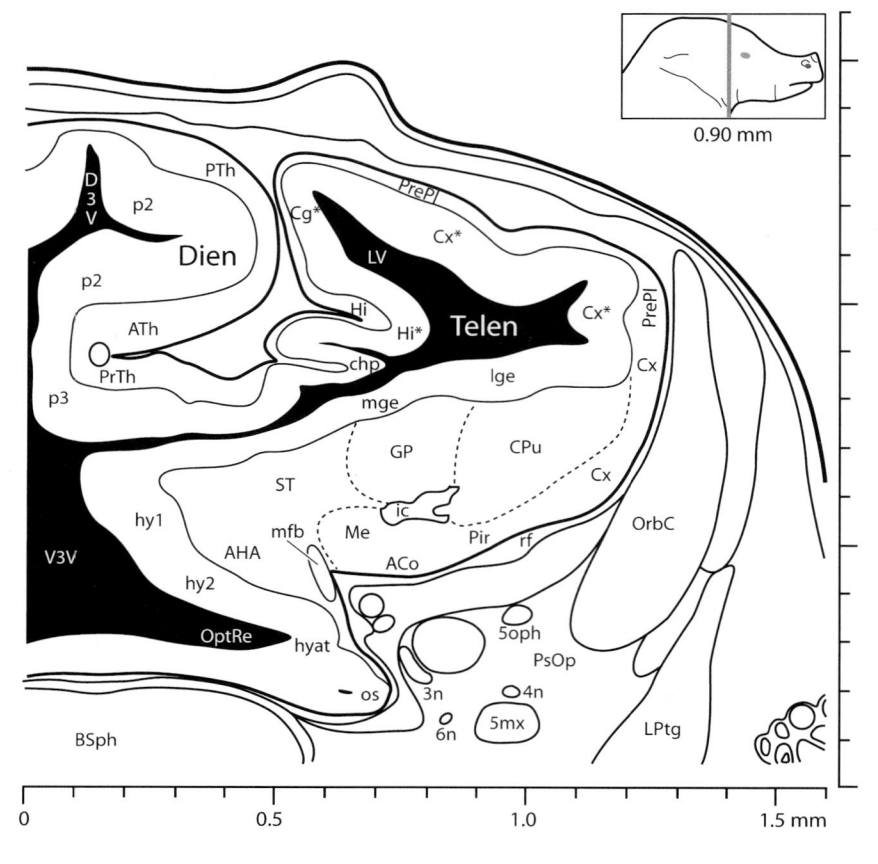

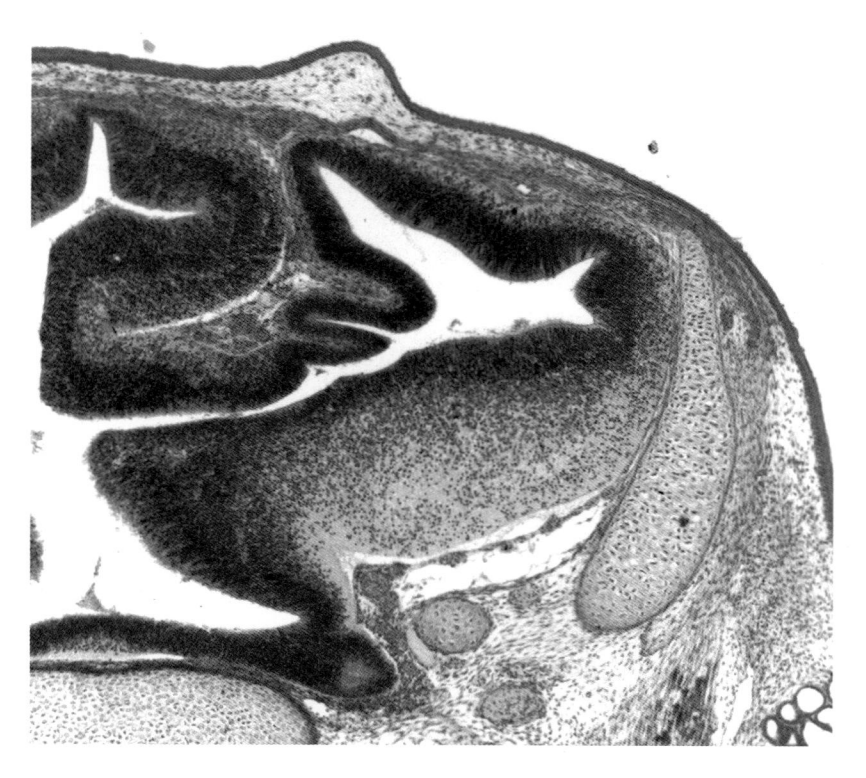

0.90 mm

*	denotes precursor
3n	oculomotor nerve
4n	trochlear nerve
5mx	maxillary division of the trigeminal nerve
5oph	ophthalmic division of the trigeminal nerve
6n	abducens nerve
ACo	anterior cortical amygdaloid area
AHA	anterior hypothalamic area, ant.
ATh	anterior thalamic region
BSph	basisphenoid bone
Cg	cingulate cortex
chp	choroid plexus
CPu	caudate putamen (striatum)
Cx	cerebral cortex
D3V	dorsal third ventricle
Dien	diencephalon
GP	globus pallidus
Hi	hippocampus
hy1	peduncular hypothalamus
hy2	tuberal hypothalamus
hyat	acroterminal hypothalamus
ic	internal capsule
lge	lateral ganglionic eminence of the subpallium
LPtg	lateral pterygoid muscle
LV	lateral ventricle
Me	medial amygdaloid nu.
mfb	medial forebrain bundle
mge	medial ganglionic eminence of the subpallium
OptRe	optic recess of the third ventricle
OrbC	orbital cartilage
os	optic stalk
p2	prosomere 2
p3	prosomere 3
Pir	piriform cortex
PrePl	preplate of cortex
PrTh	prethalamus
PsOp	pseudo-optic foramen
PTh	posterior thalamus
rf	rhinal fissure
ST	bed nu. of stria terminalis
Telen	telencephalon
V3V	ventral third ventricle

Ec-InB10

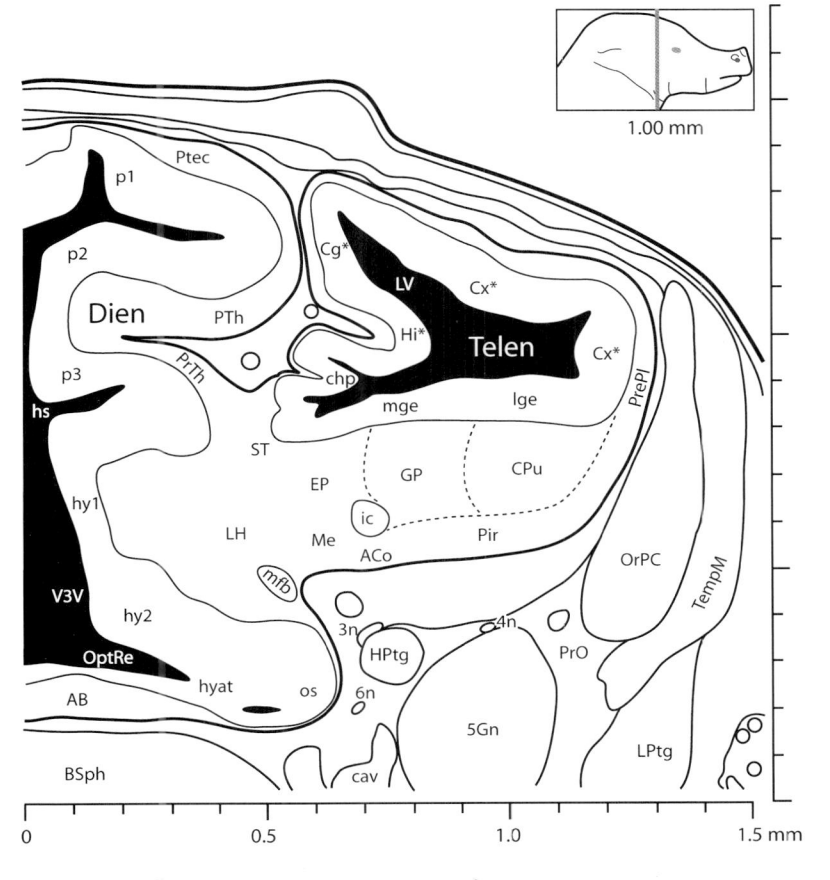

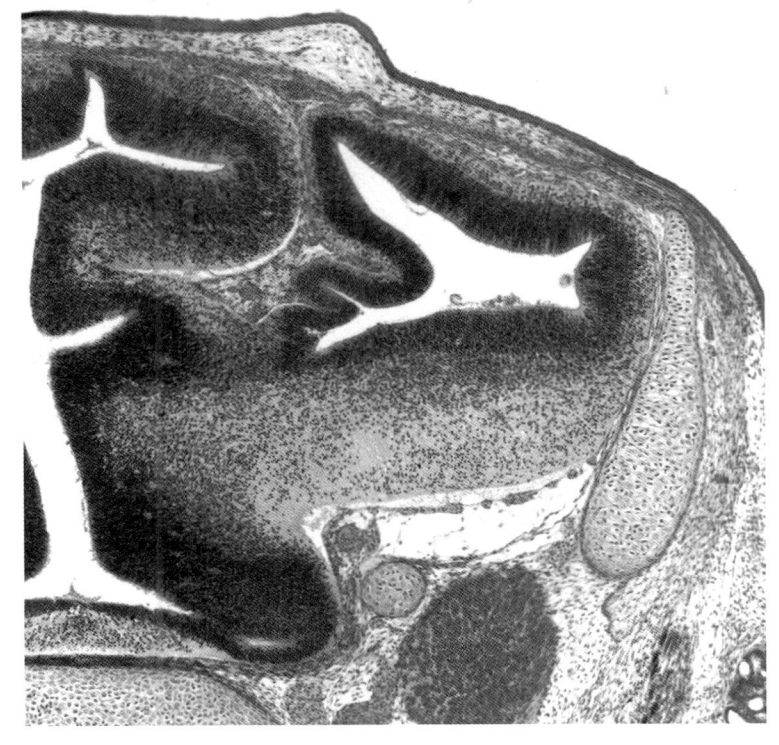

*	denotes precursor
3n	oculomotor nerve
4n	trochlear nerve
5Gn	trigeminal ganglion
6n	abducens nerve
AB	anterobasal nucleus
ACo	anterior cortical amygdaloid area
BSph	basisphenoid bone
cav	cavernous sinus
Cg	cingulate cortex
chp	choroid plexus
CPu	caudate putamen (striatum)
Cx	cerebral cortex
Dien	diencephalon
EP	entopeduncular nucleus
GP	globus pallidus
Hi	hippocampus
HPtg	hamulus of pterygoid bone
hs	hypothalamic sulcus
hy1	peduncular hypothalamus
hy2	tuberal hypothalamus
hyat	acroterminal hypothalamus
ic	internal capsule
lge	lateral ganglionic eminence of the subpallium
LH	lateral hypothalamic area
LPtg	lateral pterygoid muscle
LV	lateral ventricle
Me	medial amygdaloid nucleus
mfb	medial forebrain bundle
mge	medial ganglionic eminence of the subpallium
OptRe	optic recess of third ventricle
OrPC	orbitoparietal commissure
os	optic stalk
p1	prosomere 1
p2	prosomere 2
p3	prosomere 3
Pir	piriform cortex
PrePl	preplate of cortex
PrO	pro-otic foramen
PrTh	prethalamus
Ptec	pretectum
PTh	posterior thalamus
ST	bed nucleus of stria terminalis
Telen	telencephalon
TempM	temporalis muscle
V3V	ventral third ventricle

366 NEUROBIOLOGY OF MONOTREMES

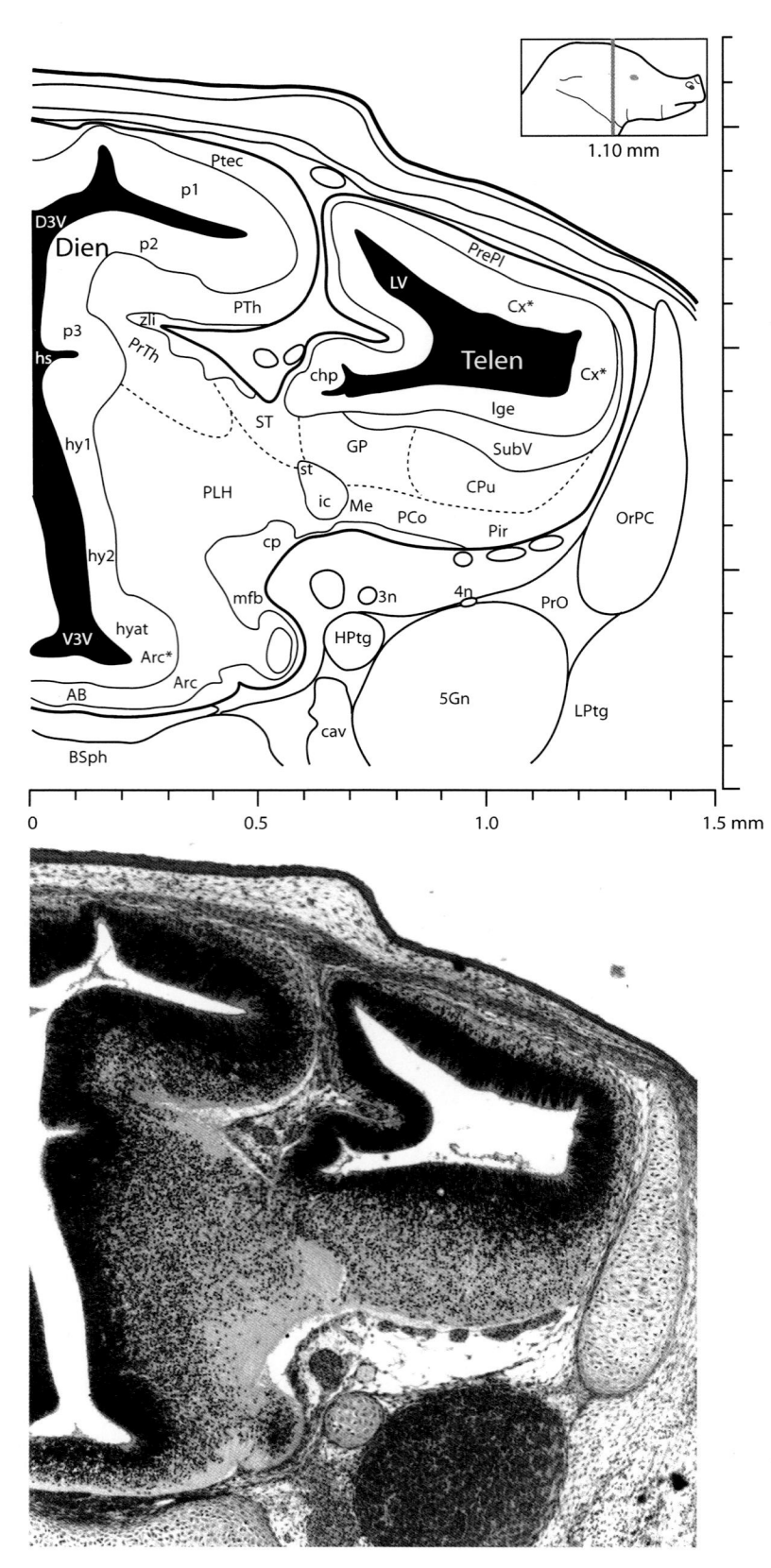

Ec-InB11

*	denotes precursor
3n	oculomotor nerve
4n	trochlear nerve
5Gn	trigeminal ganglion
AB	anterobasal nucleus
Arc	arcuate hypothalamic nucleus
BSph	basisphenoid bone
cav	cavernous sinus
chp	choroid plexus
cp	cerebral peduncle
CPu	caudate putamen (striatum)
Cx	cerebral cortex
D3V	dorsal third ventricle
Dien	diencephalon
GP	globus pallidus
HPtg	hamulus of pterygoid bone
hs	hypothalamic sulcus
hy1	peduncular hypothalamus
hy2	tuberal hypothalamus
hyat	acroterminal hypothalamus
ic	internal capsule
lge	lateral ganglionic eminence of the subpallium
LPtg	lateral pterygoid muscle
LV	lateral ventricle
Me	medial amygdaloid nucleus
mfb	medial forebrain bundle
OrPC	orbitoparietal commissure
p1	prosomere 1
p2	prosomere 2
p3	prosomere 3
PCo	posterior cortical amygdaloid area
Pir	piriform cortex
PLH	peduncular part of lateral hypothalamus
PrePl	preplate of cortex
PrO	pro-otic foramen
PrTh	prethalamus
Ptec	pretectum
PTh	posterior thalamus
st	stria terminalis
ST	bed nucleus of stria terminalis
SubV	subventricular layer of developing cortex
Telen	telencephalon
V3V	ventral third ventricle
zli	zona limitans interthalamica

Ec-InB12

*	denotes precursor
3n	oculomotor nerve
4n	trochlear nerve
5Gn	trigeminal ganglion
6n	abducens nerve
Amg	amygdala
APit	anterior lobe of the pituitary gland
Arc	arcuate hypothalamic nucleus
BSph	basisphenoid bone
cav	cavernous sinus
Ce	central amygdaloid nucleus
cp	cerebral peduncle
Cx	cerebral cortex
D3V	dorsal third ventricle
Dien	diencephalon
Hi	hippocampus
hs	hypothalamic sulcus
hy1	peduncular hypothalamus
hy2	tuberal hypothalamus
hyat	acroterminal hypothalamus
ictd	internal carotid artery
IRe	infundibular recess of the third ventricle
La	lateral amygdaloid nucleus
lge	lateral ganglionic eminence of the subpallium
LH	lateral hypothalamic area
LV	lateral ventricle
Me	medial amygdaloid nucleus
mfb	medial forebrain bundle
OrPC	orbitoparietal commissure
p1	prosomere 1
p2	prosomere 2
p3	prosomere 3
ParB	parietal bone
PLCo	posterolateral cortical amygdaloid area
PLH	peduncular part of lateral hypothalamus
PrePl	preplate of cortex
PrO	pro-otic foramen
PrTh	prethalamus
Ptec	pretectum
PTh	posterior thalamus
st	stria terminalis
SubV	subventricular layer of developing cortex
Telen	telencephalon
V3V	ventral third ventricle
zli	zona limitans interthalamica

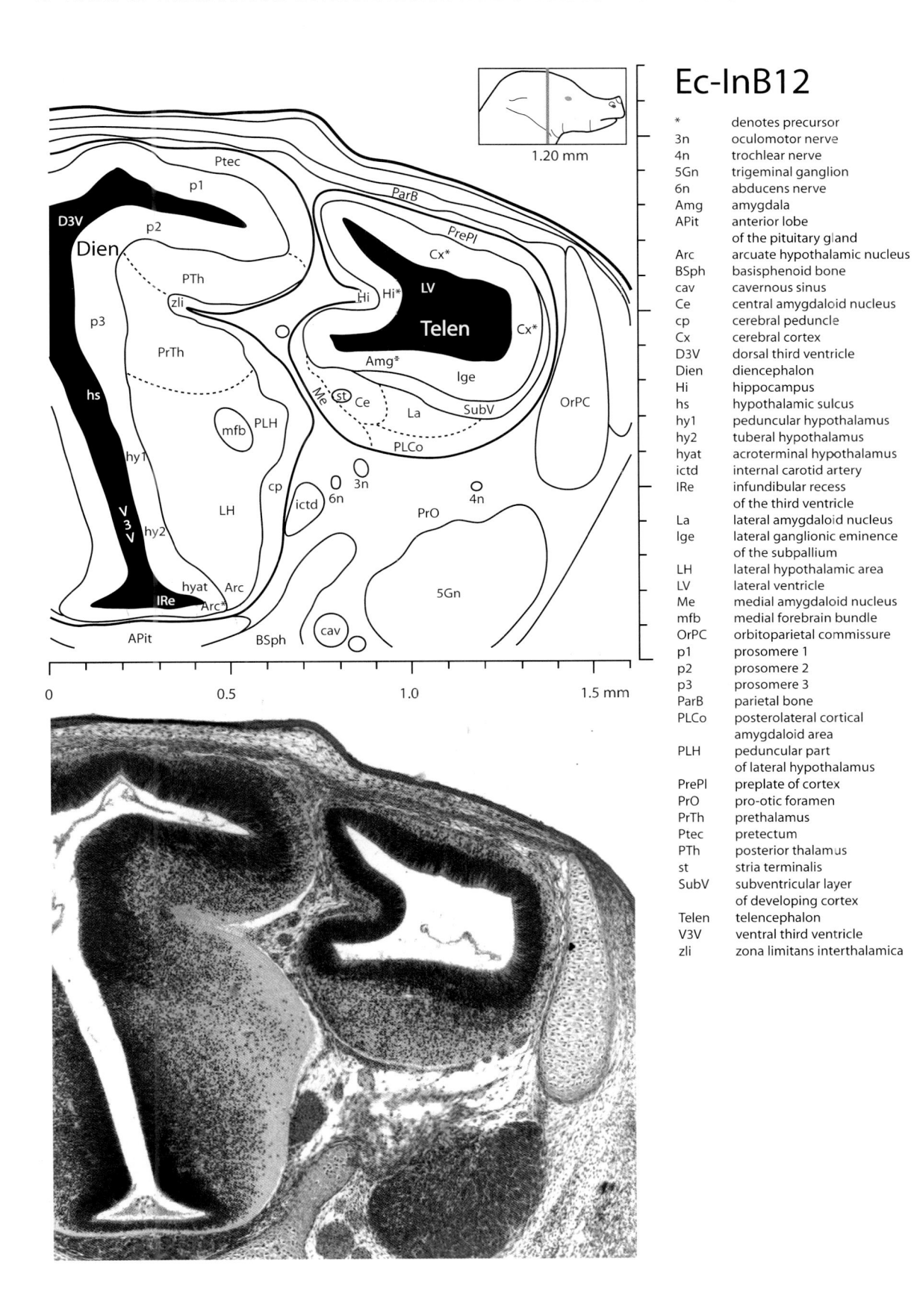

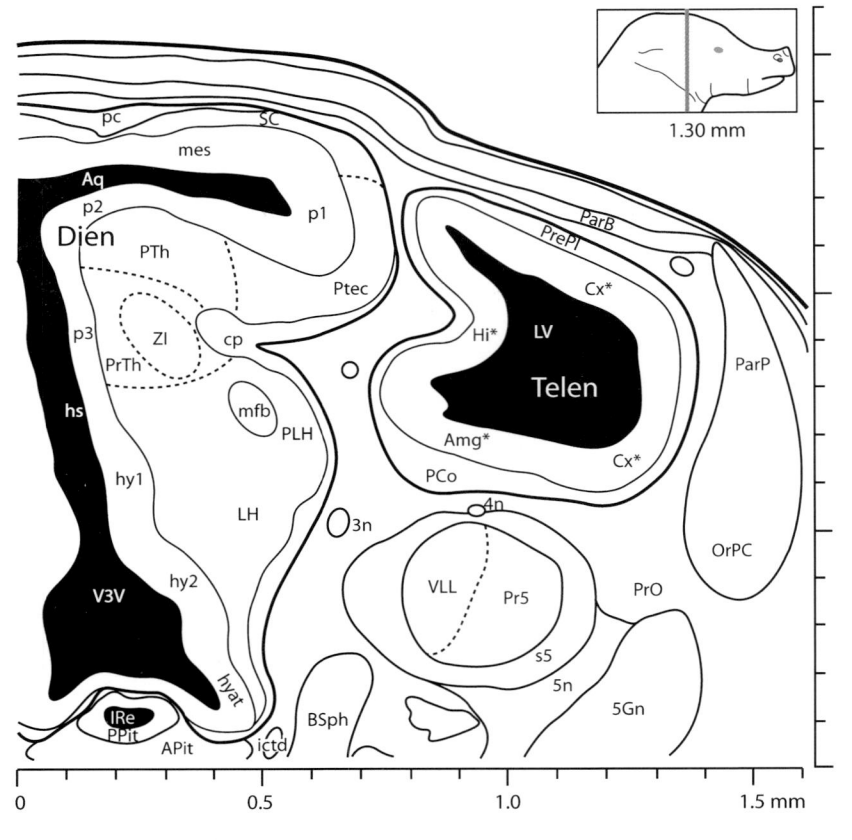

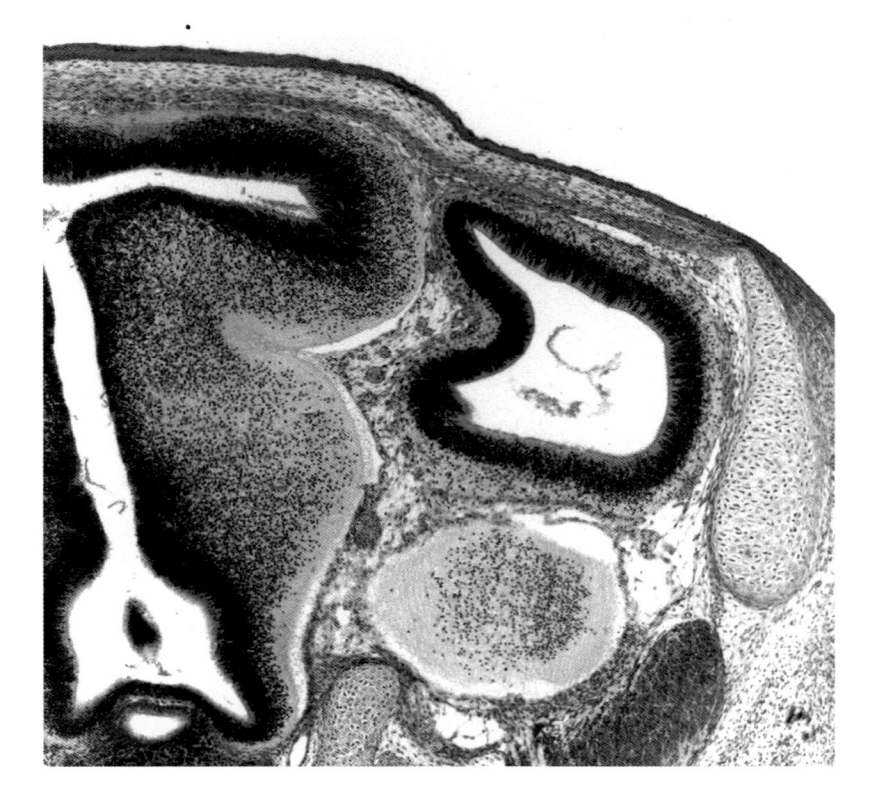

Ec-InB13

*	denotes precursor
3n	oculomotor nerve
4n	trochlear nerve
5Gn	trigeminal ganglion
5n	trigeminal nerve
Amg	amygdala
APit	anterior lobe of the pituitary gland
Aq	cerebral aqueduct
BSph	basisphenoid bone
cp	cerebral peduncle
Cx	cerebral cortex
Dien	diencephalon
Hi	hippocampus
hs	hypothalamic sulcus
hy1	peduncular hypothalamus
hy2	tuberal hypothalamus
hyat	acroterminal hypothalamus
ictd	internal carotid artery
IRe	infundibular recess of third ventricle
LH	lateral hypothalamic area
LV	lateral ventricle
mes	mesencephalic neuromere
mfb	medial forebrain bundle
OrPC	orbitoparietal commissure
p1	prosomere 1
p2	prosomere 2
p3	prosomere 3
ParB	parietal bone
ParP	parietal plate
pc	posterior commissure
PCo	posterior cortical amygdaloid area
PLH	peduncular part of lateral hypothalamus
PPit	posterior lobe of the pituitary gland
Pr5	principal sensory trigeminal nucleus
PrePl	preplate of developing cortex
PrO	pro-otic foramen
PrTh	prethalamus
Ptec	pretectum
PTh	posterior thalamus
s5	sensory root of trigeminal nerve
SC	superior colliculus
Telen	telencephalon
V3V	ventral third ventricle
VLL	ventral nucleus of lateral lemniscus
ZI	zona incerta

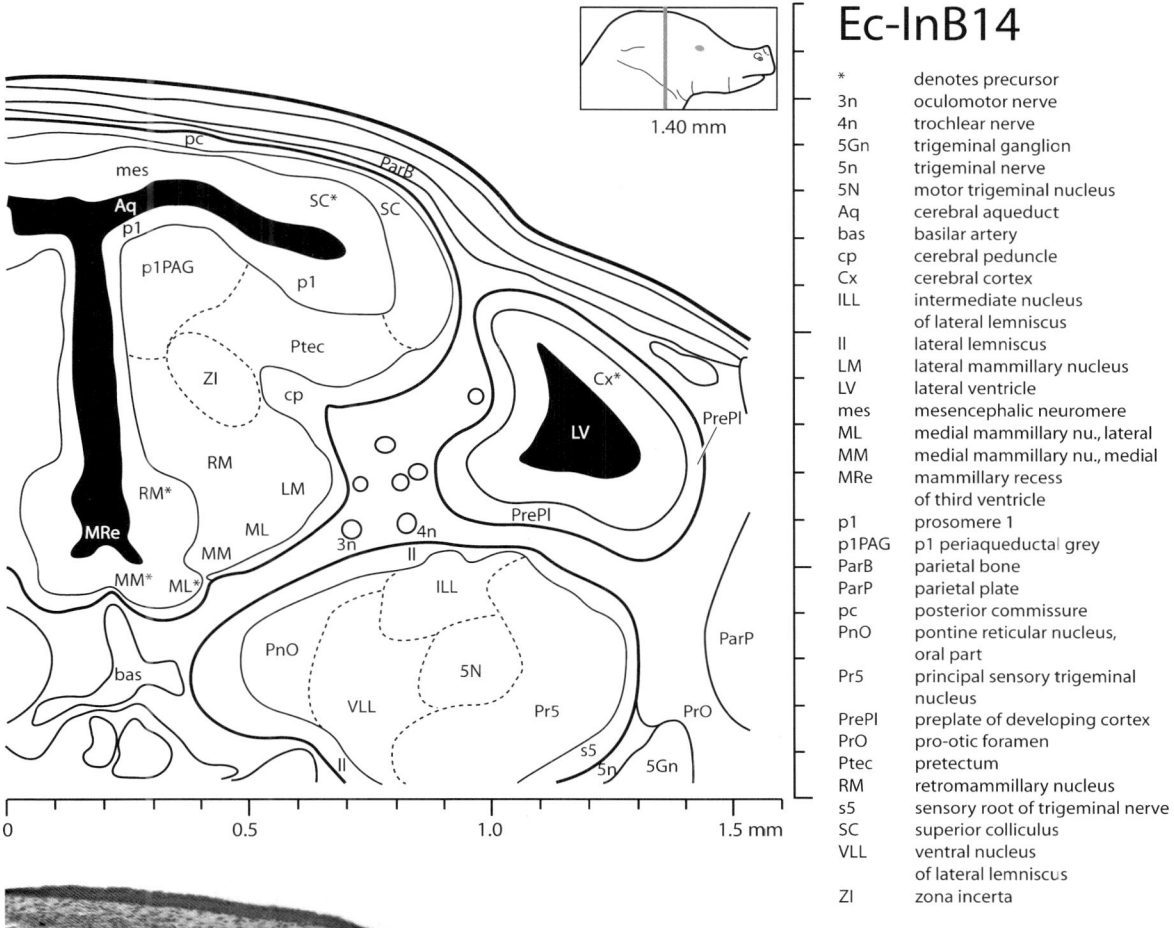

Ec-InB14

1.40 mm

*	denotes precursor
3n	oculomotor nerve
4n	trochlear nerve
5Gn	trigeminal ganglion
5n	trigeminal nerve
5N	motor trigeminal nucleus
Aq	cerebral aqueduct
bas	basilar artery
cp	cerebral peduncle
Cx	cerebral cortex
ILL	intermediate nucleus of lateral lemniscus
ll	lateral lemniscus
LM	lateral mammillary nucleus
LV	lateral ventricle
mes	mesencephalic neuromere
ML	medial mammillary nu., lateral
MM	medial mammillary nu., medial
MRe	mammillary recess of third ventricle
p1	prosomere 1
p1PAG	p1 periaqueductal grey
ParB	parietal bone
ParP	parietal plate
pc	posterior commissure
PnO	pontine reticular nucleus, oral part
Pr5	principal sensory trigeminal nucleus
PrePl	preplate of developing cortex
PrO	pro-otic foramen
Ptec	pretectum
RM	retromammillary nucleus
s5	sensory root of trigeminal nerve
SC	superior colliculus
VLL	ventral nucleus of lateral lemniscus
ZI	zona incerta

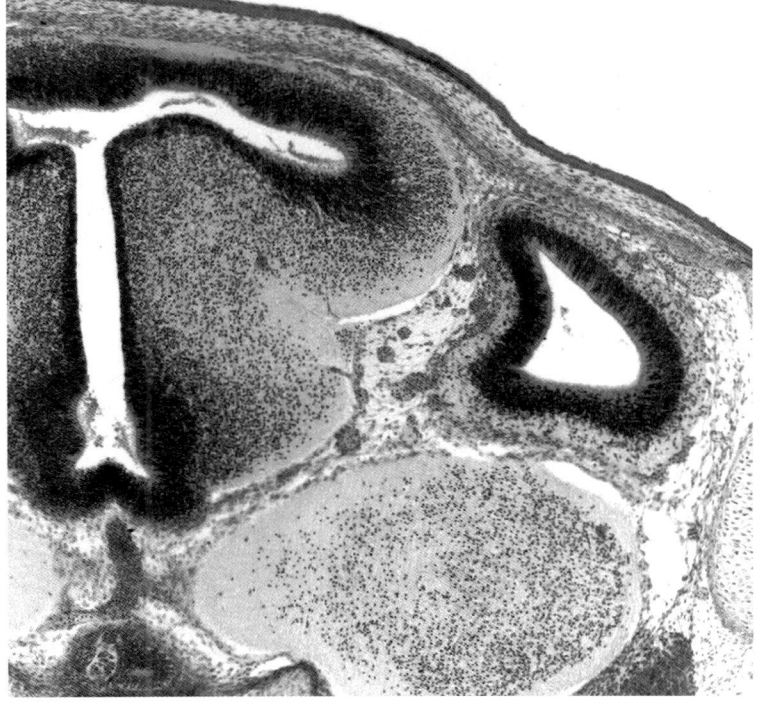

370 NEUROBIOLOGY OF MONOTREMES

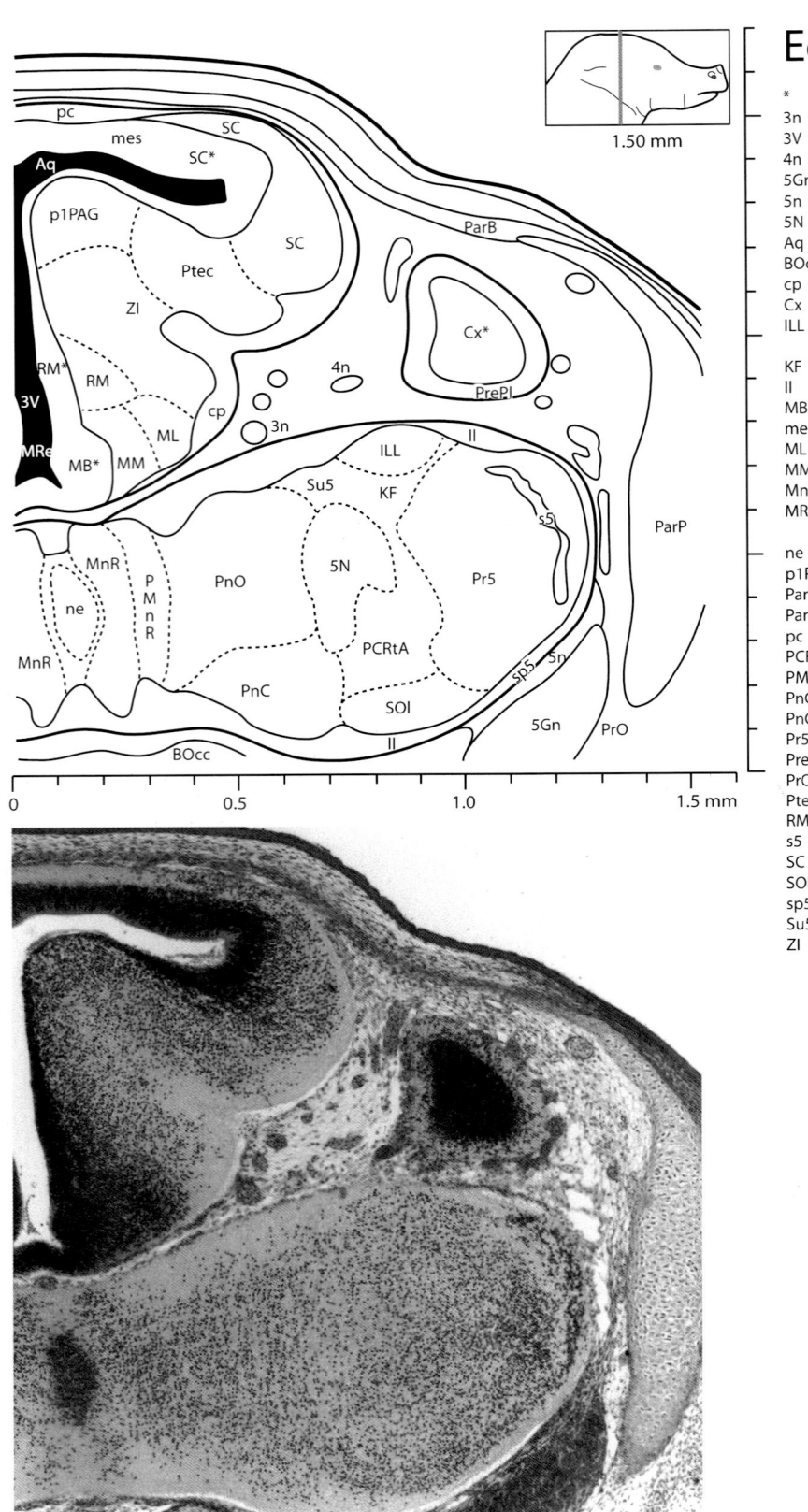

Ec-InB15

1.50 mm

*	denotes precursor
3n	oculomotor nerve
3V	3rd ventricle
4n	trochlear nerve
5Gn	trigeminal ganglion
5n	trigeminal nerve
5N	motor trigeminal nucleus
Aq	cerebral aqueduct
BOcc	basioccipital bone
cp	cerebral peduncle
Cx	cerebral cortex
ILL	intermediate nucleus of lateral lemniscus
KF	Kölliker-Fuse nu.
ll	lateral lemniscus
MB	mammillary body
mes	mesencephalic neuromere
ML	medial mammillary nu., lateral
MM	medial mammillary nu., medial
MnR	median raphe nu.
MRe	mammillary recess of third ventricle
ne	neuroepithelium
p1PAG	p1 periaqueductal grey
ParB	parietal bone
ParP	parietal plate
pc	posterior commissure
PCRtA	parvicellular reticular nu., alpha
PMnR	paramedian raphe nu.
PnC	pontine reticular nu., caudal
PnO	pontine reticular nu., oral
Pr5	principal sensory trigeminal nu.
PrePl	preplate of cortex
PrO	pro-otic foramen
Ptec	pretectum
RM	retromammillary nu.
s5	sensory root of trigeminal nerve
SC	superior colliculus
SOl	superior olive
sp5	spinal trigeminal tract
Su5	supratrigeminal nu.
ZI	zona incerta

Ec-InB16

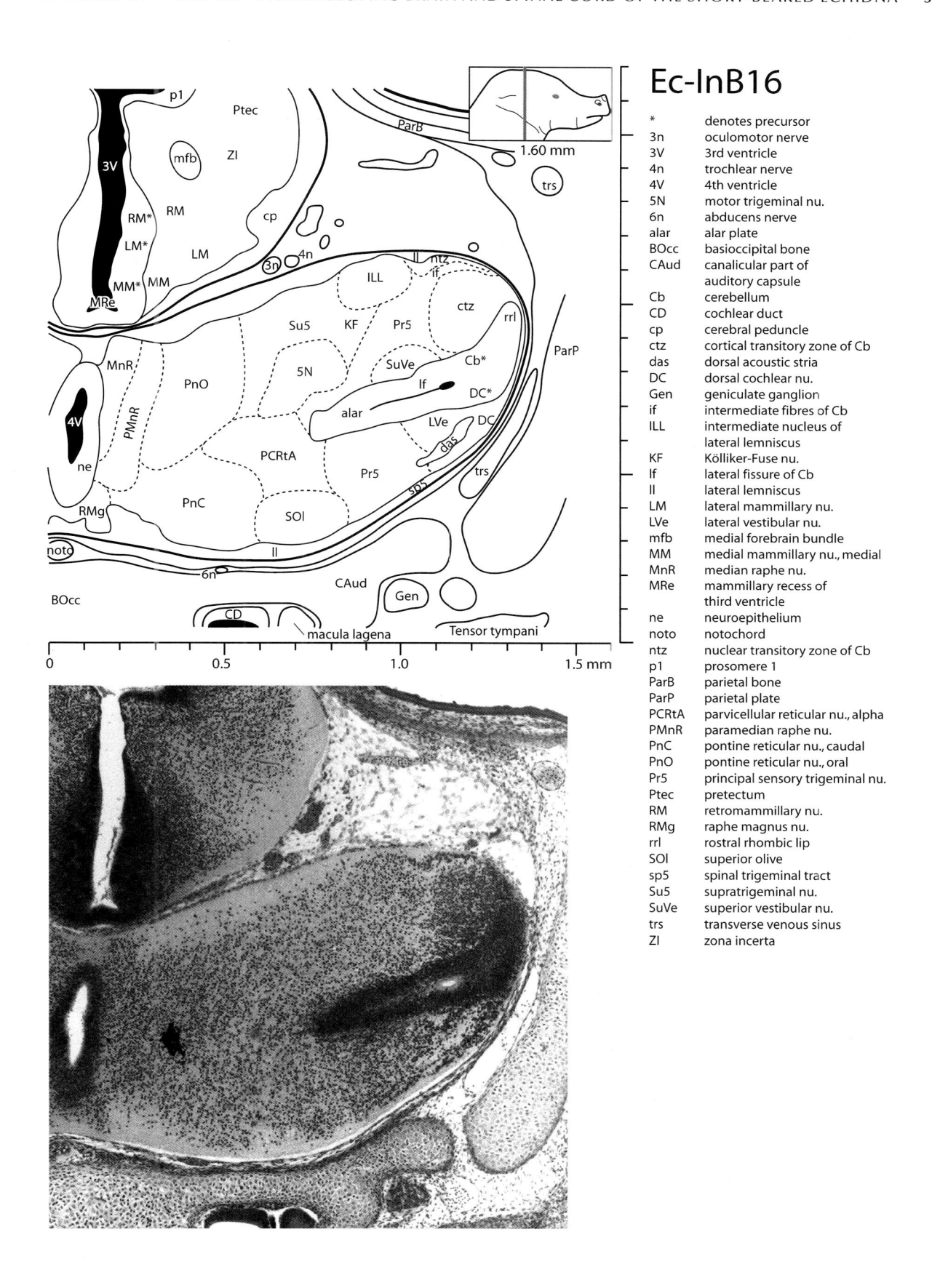

*	denotes precursor
3n	oculomotor nerve
3V	3rd ventricle
4n	trochlear nerve
4V	4th ventricle
5N	motor trigeminal nu.
6n	abducens nerve
alar	alar plate
BOcc	basioccipital bone
CAud	canalicular part of auditory capsule
Cb	cerebellum
CD	cochlear duct
cp	cerebral peduncle
ctz	cortical transitory zone of Cb
das	dorsal acoustic stria
DC	dorsal cochlear nu.
Gen	geniculate ganglion
if	intermediate fibres of Cb
ILL	intermediate nucleus of lateral lemniscus
KF	Kölliker-Fuse nu.
lf	lateral fissure of Cb
ll	lateral lemniscus
LM	lateral mammillary nu.
LVe	lateral vestibular nu.
mfb	medial forebrain bundle
MM	medial mammillary nu., medial
MnR	median raphe nu.
MRe	mammillary recess of third ventricle
ne	neuroepithelium
noto	notochord
ntz	nuclear transitory zone of Cb
p1	prosomere 1
ParB	parietal bone
ParP	parietal plate
PCRtA	parvicellular reticular nu., alpha
PMnR	paramedian raphe nu.
PnC	pontine reticular nu., caudal
PnO	pontine reticular nu., oral
Pr5	principal sensory trigeminal nu.
Ptec	pretectum
RM	retromammillary nu.
RMg	raphe magnus nu.
rrl	rostral rhombic lip
SOl	superior olive
sp5	spinal trigeminal tract
Su5	supratrigeminal nu.
SuVe	superior vestibular nu.
trs	transverse venous sinus
ZI	zona incerta

Ec-InB17

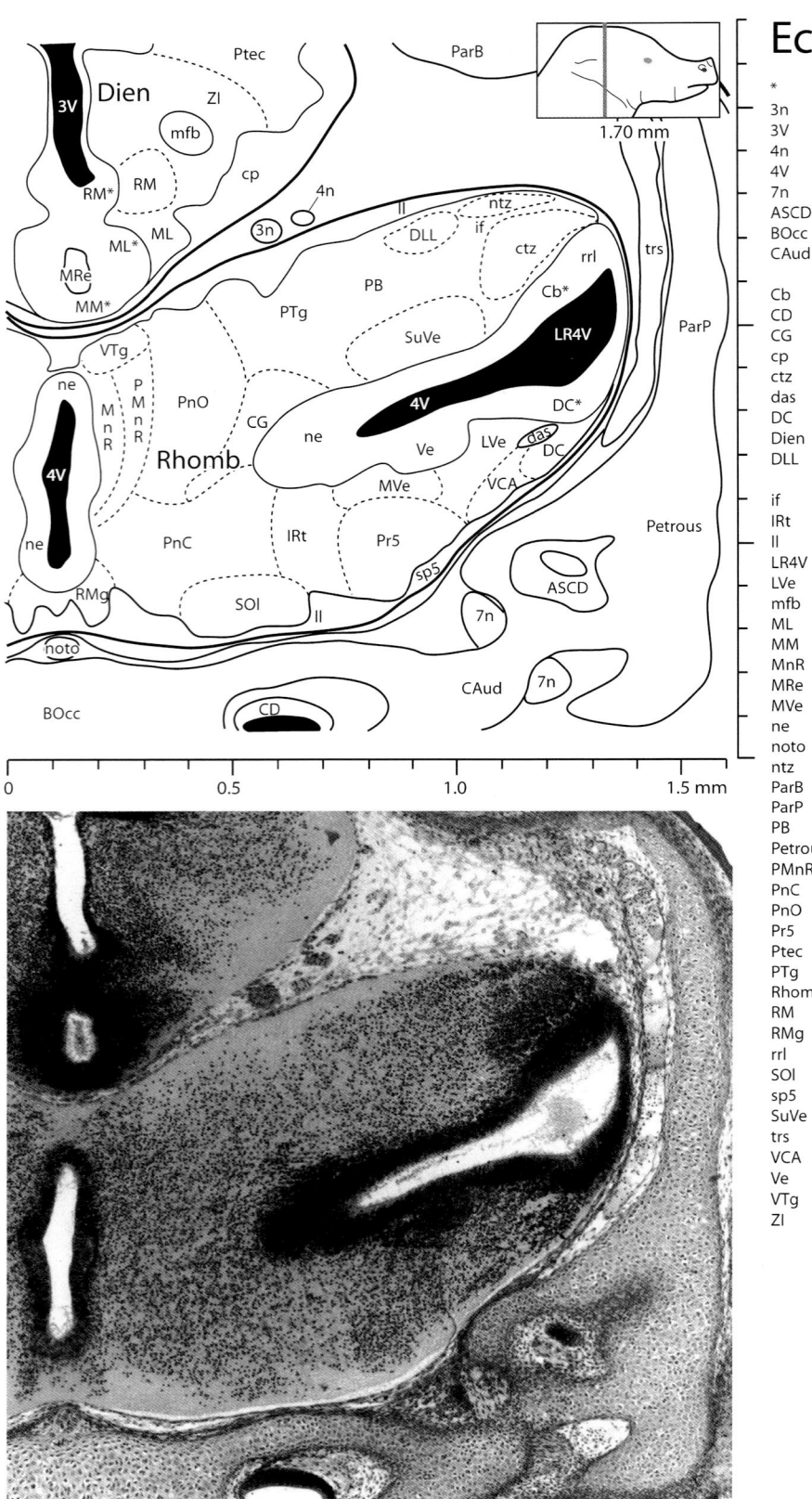

*	denotes precursor
3n	oculomotor nerve
3V	3rd ventricle
4n	trochlear nerve
4V	4th ventricle
7n	facial nerve
ASCD	anterior semicircular duct
BOcc	basioccipital bone
CAud	canalicular part of auditory capsule
Cb	cerebellum
CD	cochlear duct
CG	central grey
cp	cerebral peduncle
ctz	cortical transitory zone of Cb
das	dorsal acoustic stria
DC	dorsal cochlear nu.
Dien	diencephalon
DLL	dorsal nucleus of lateral lemniscus
if	intermediate fibres of Cb
IRt	intermediate reticular nu.
ll	lateral lemniscus
LR4V	lateral recess of 4th ventricle
LVe	lateral vestibular nu.
mfb	medial forebrain bundle
ML	medial mammillary nu., lateral
MM	medial mammillary nu., medial
MnR	median raphe nu.
MRe	mammillary recess of 3V
MVe	medial vestibular nu.
ne	neuroepithelium
noto	notochord
ntz	nuclear transitory zone of Cb
ParB	parietal bone
ParP	parietal plate
PB	parabrachial complex
Petrous	petrous temporal bone
PMnR	paramedian raphe nu.
PnC	pontine reticular nu., caudal
PnO	pontine reticular nu., oral
Pr5	principal sensory trigeminal nu.
Ptec	pretectum
PTg	pedunculotegmental nu.
Rhomb	rhombencephalon
RM	retromammillary nu.
RMg	raphe magnus nu.
rrl	rostral rhombic lip
SOl	superior olive
sp5	spinal trigeminal tract
SuVe	superior vestibular nu.
trs	transverse venous sinus
VCA	ventral cochlear nu., anterior
Ve	vestibular neuroepithelium
VTg	ventral tegmental nu.
ZI	zona incerta

Ec-InB18

*	denotes precursor
3n	oculomotor nerve
3N	oculomotor nu.
4n	trochlear nerve
5Tr	trigeminal transition zone
6N	abducens nu.
8n	vestibulocochlear nerve
alar	alar plate
ASCD	anterior semicircular duct
BOcc	basioccipital bone
CAud	canalicular part of auditory capsule
Cb	cerebellum
CG	central grey
CGn	cochlear (spiral) ganglion
cp	cerebral peduncle
crl	caudal rhombic lip
ctz	cortical transitory zone of Cb
DC	dorsal cochlear nu.
DLL	dorsal nu. lateral lemniscus
EW	Edinger-Westphal nu.
GiA	gigantocellular reticular nu., alpha
if	intermediate fibres of Cb
IRt	intermediate reticular nu.
lf	lateral fissure of Cb
ll	lateral lemniscus
LR4V	lateral recess of 4th ventricle
LVe	lateral vestibular nu.
MB	mammillary body
me5	mesencephalic trigeminal tract
Me5	mesencephalic trigeminal nu.
Mesen	mesencephalon
mlf	medial longitudinal fasciculus
MnR	median raphe nu.
mRt	mesencephalic reticular formation
MVe	medial vestibular nu.
ne	neuroepithelium
noto	notochord
ntz	nuclear transitory zone of Cb
ParB	parietal bone
PB	parabrachial complex
PCRtA	parvicellular reticular nu., alpha
PMnR	paramedian raphe nu.
PN	paranigral nu. of the VTA
PnC	pontine reticular nu., caudal
PnO	pontine reticular nu., oral
Pr5	principal sensory trigeminal nu.
PTg	pedunculotegmental nu.
Red	red nucleus
Rhomb	rhombencephalon
RMg	raphe magnus nu.
rrl	rostral rhombic lip
Sacc	saccule
SNC	substantia nigra, compact
SNR	substantia nigra, reticular
Sol	solitary nucleus
sp5	spinal trigeminal tract
SuVe	superior vestibular nu.
trs	transverse venous sinus
UMac	macula of the utricle
Utr	utricle
VCP	ventral cochlear nu., posterior
Ve	vestibular neuroepithelium
VeGn	vestibular nerve ganglion
VTA	ventral tegmental area
VTg	ventral tegmental nu.

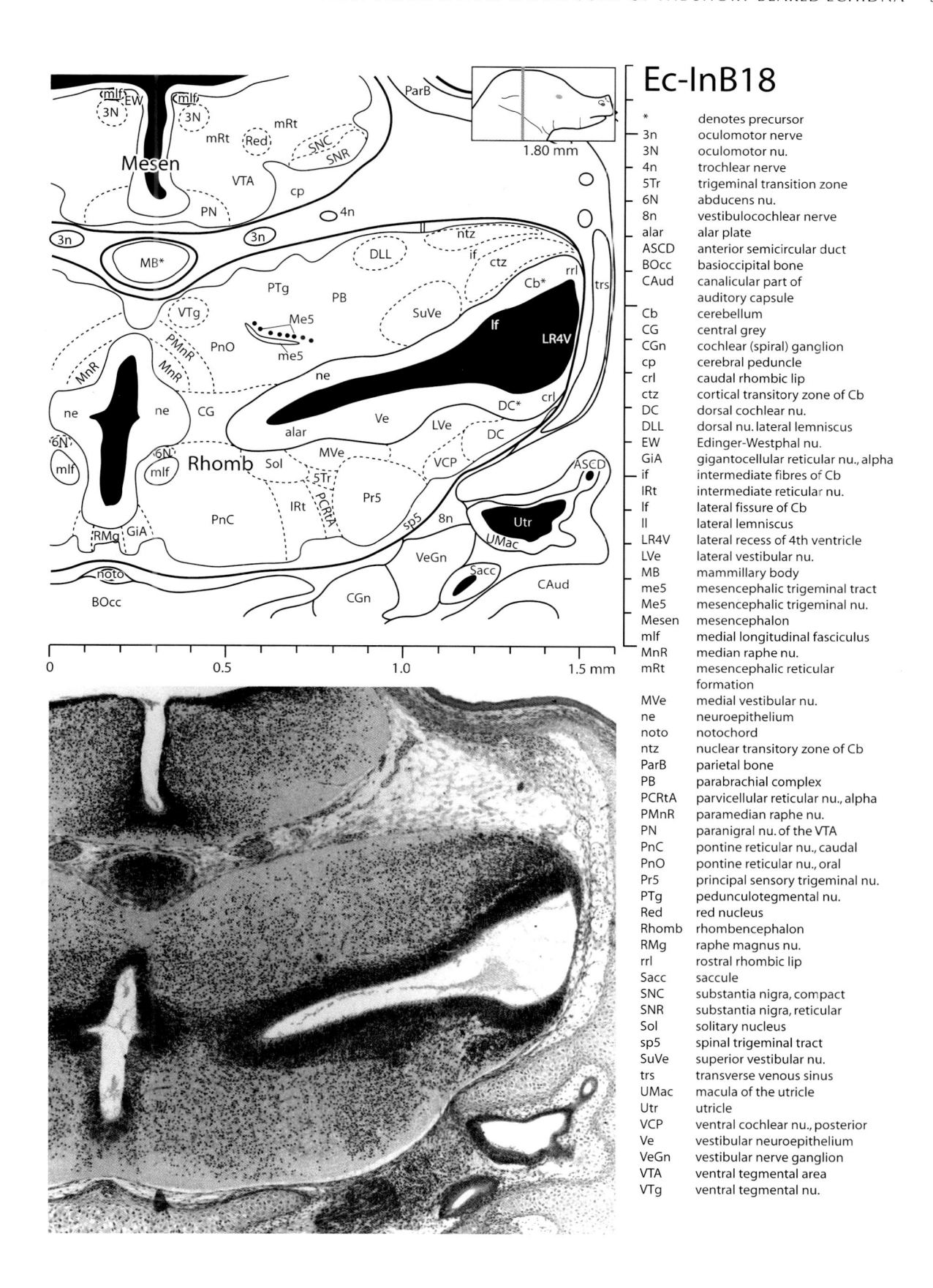

Ec-InB19

*	denotes precursor
3n	oculomotor nerve
3N	oculomotor nu.
4n	trochlear nerve
4V	4th ventricle
5Sol	trigeminal-solitary transition zone
7N	facial nu.
8n	vestibulocochlear nerve
alar	alar plate
ASCD	anterior semicircular duct
basal	basal plate
BOcc	basioccipital bone
CAud	canalicular part auditory capsule
Cb	cerebellum
CG	central grey
CGn	cochlear (spiral) ganglion
cp	cerebral peduncle
crl	caudal rhombic lip
ctz	cortical transitory zone of Cb
DC	dorsal cochlear nu.
DLL	dorsal nu. of lateral lemniscus
DMSp5	dorsomedial spinal trigem nu.
DMTg	dorsomedial tegmental area
DR	dorsal raphe nu.
EW	Edinger-Westphal nu.
Gi	gigantocellular reticular nu.
GiA	gigantocellular reticular nu., alpha
HSCD	horizontal semicircular duct
IAud	internal auditory meatus
if	intermediate fibres of Cb
IP	interpeduncular nu.
IRt	intermediate reticular nu.
lf	lateral fissure
LPGi	lateral paragigantocellular nu.
LR4V	lateral recess of 4th ventricle
LVe	lateral vestibular nu.
Me5	mesencephalic trigem nu.
me5	mesencephalic trigem tract
Mesen	mesencephalon
mlf	medial longitudinal fasciculus
mRt	mesencephalic reticular formn
MVe	medial vestibular nu.
noto	notochord
ntz	nuclear transitory zone of Cb
ParB	parietal bone
ParP	parietal plate
PB	parabrachial complex
PCRtA	parvicellular reticular nu., alpha
PN	paranigral nu. of the VTA
PPy	parapyramidal nu.
Pr	prepositus nu.
PTg	pedunculotegmental nu.
Red	red nucleus
Rhomb	rhombencephalon
RMg	raphe magnus nu.
rrl	rostral rhombic lip
sl	sulcus limitans
SNC	substantia nigra, compact
SNR	substantia nigra, reticular
Sol	solitary nucleus
Sp5O	spinal trigeminal nu., oral
SuVe	superior vestibular nu.
trs	transverse venous sinus
Utr	utricle
VC	ventral cochlear nu.
VCP	ventral cochlear nu., posterior
Ve	vestibular neuroepithelium
VeGn	vestibular nerve ganglion
VTA	ventral tegmental area
VTg	ventral tegmental nu.

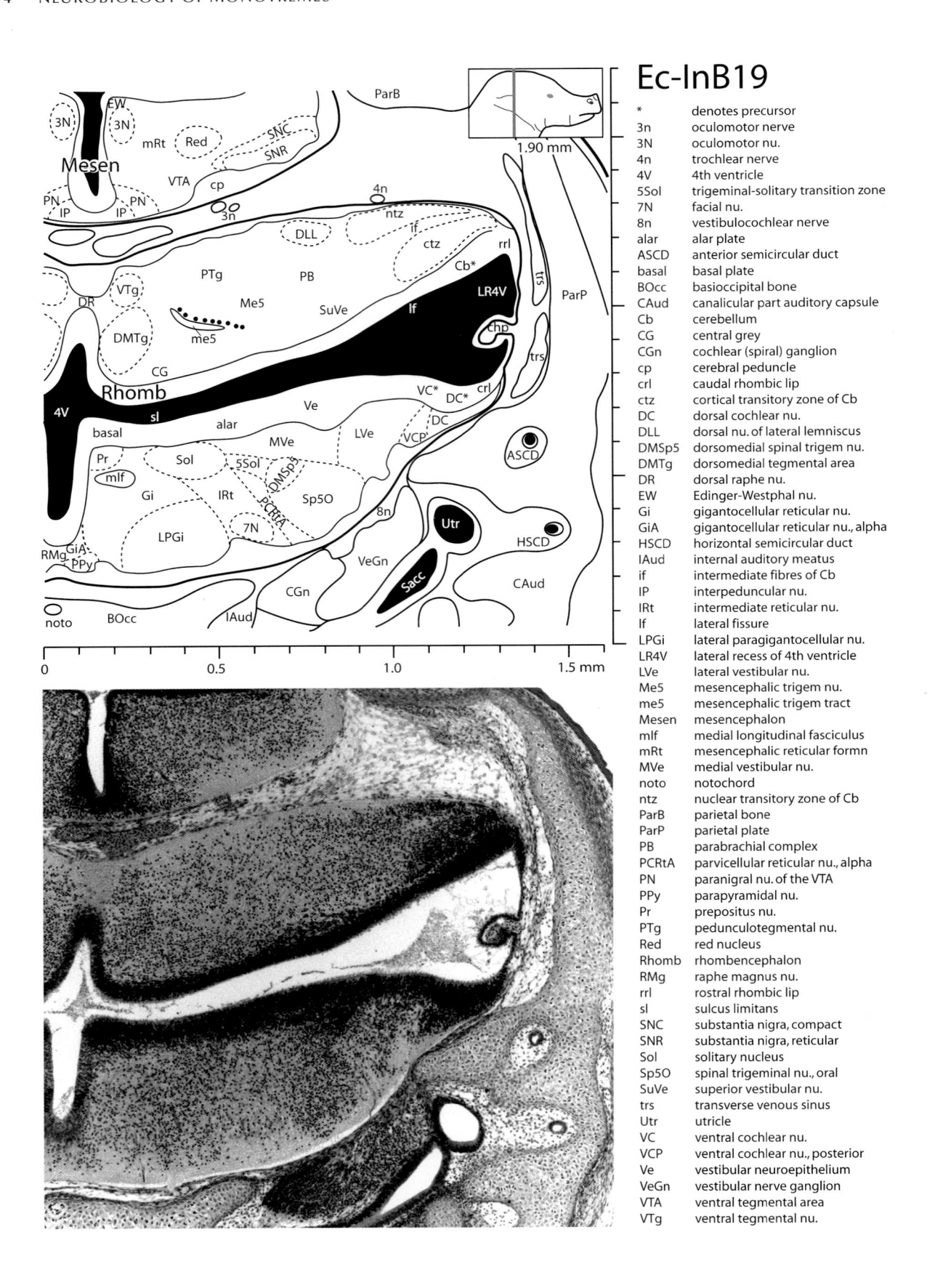

Ec-InB20

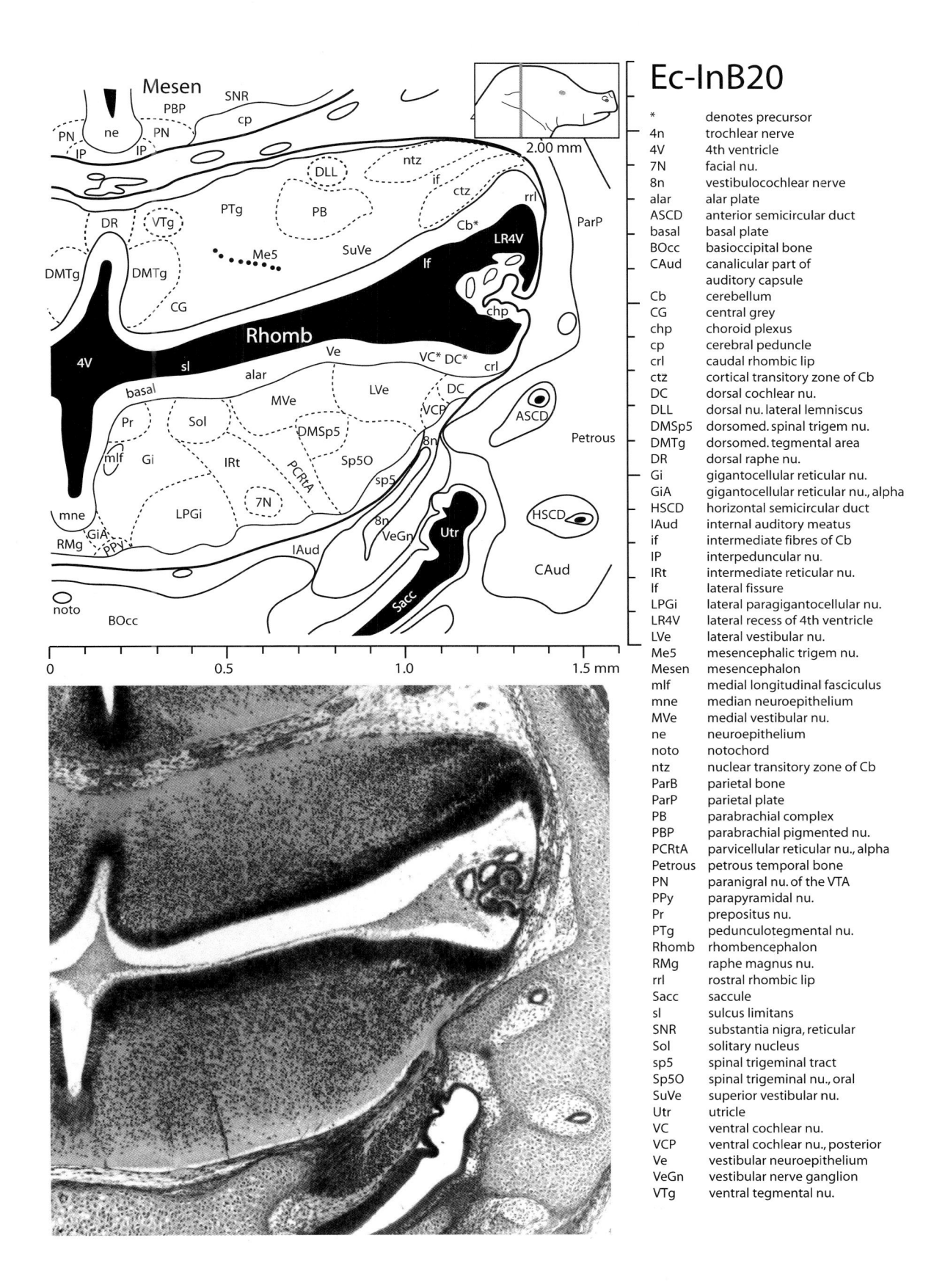

*	denotes precursor
4n	trochlear nerve
4V	4th ventricle
7N	facial nu.
8n	vestibulocochlear nerve
alar	alar plate
ASCD	anterior semicircular duct
basal	basal plate
BOcc	basioccipital bone
CAud	canalicular part of auditory capsule
Cb	cerebellum
CG	central grey
chp	choroid plexus
cp	cerebral peduncle
crl	caudal rhombic lip
ctz	cortical transitory zone of Cb
DC	dorsal cochlear nu.
DLL	dorsal nu. lateral lemniscus
DMSp5	dorsomed. spinal trigem nu.
DMTg	dorsomed. tegmental area
DR	dorsal raphe nu.
Gi	gigantocellular reticular nu.
GiA	gigantocellular reticular nu., alpha
HSCD	horizontal semicircular duct
IAud	internal auditory meatus
if	intermediate fibres of Cb
IP	interpeduncular nu.
IRt	intermediate reticular nu.
lf	lateral fissure
LPGi	lateral paragigantocellular nu.
LR4V	lateral recess of 4th ventricle
LVe	lateral vestibular nu.
Me5	mesencephalic trigem nu.
Mesen	mesencephalon
mlf	medial longitudinal fasciculus
mne	median neuroepithelium
MVe	medial vestibular nu.
ne	neuroepithelium
noto	notochord
ntz	nuclear transitory zone of Cb
ParB	parietal bone
ParP	parietal plate
PB	parabrachial complex
PBP	parabrachial pigmented nu.
PCRtA	parvicellular reticular nu., alpha
Petrous	petrous temporal bone
PN	paranigral nu. of the VTA
PPy	parapyramidal nu.
Pr	prepositus nu.
PTg	pedunculotegmental nu.
Rhomb	rhombencephalon
RMg	raphe magnus nu.
rrl	rostral rhombic lip
Sacc	saccule
sl	sulcus limitans
SNR	substantia nigra, reticular
Sol	solitary nucleus
sp5	spinal trigeminal tract
Sp5O	spinal trigeminal nu., oral
SuVe	superior vestibular nu.
Utr	utricle
VC	ventral cochlear nu.
VCP	ventral cochlear nu., posterior
Ve	vestibular neuroepithelium
VeGn	vestibular nerve ganglion
VTg	ventral tegmental nu.

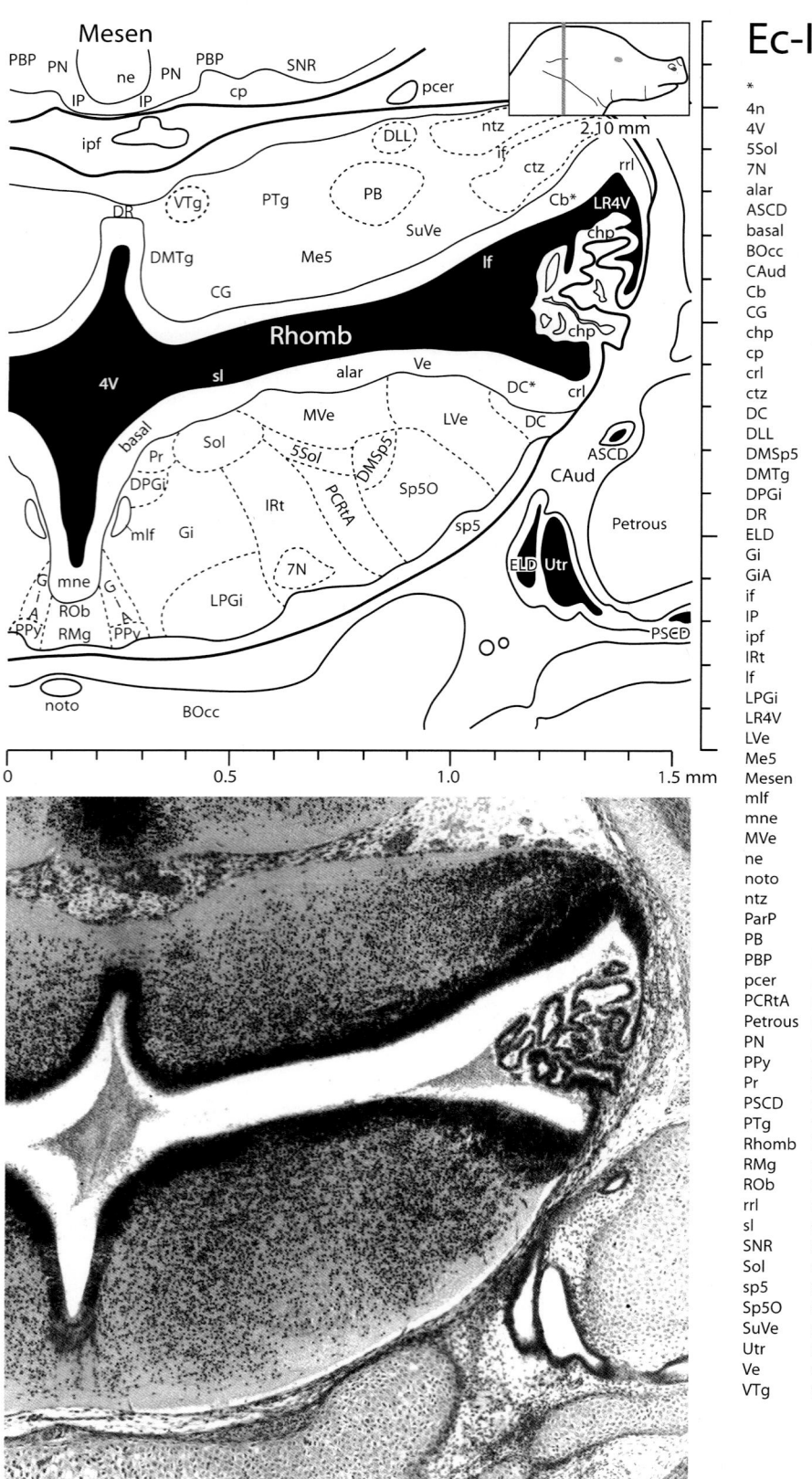

Ec-InB21

*	denotes precursor
4n	trochlear nerve
4V	4th ventricle
5Sol	trigeminal-solitary transition zone
7N	facial nu.
alar	alar plate
ASCD	anterior semicircular duct
basal	basal plate
BOcc	basioccipital bone
CAud	canalicular part of auditory capsule
Cb	cerebellum
CG	central grey
chp	choroid plexus
cp	cerebral peduncle
crl	caudal rhombic lip
ctz	cortical transitory zone of Cb
DC	dorsal cochlear nu.
DLL	dorsal nu. of lateral lemniscus
DMSp5	dorsomedial spinal trigeminal nu.
DMTg	dorsomedial tegmental area
DPGi	dorsal paragigantocellular nu.
DR	dorsal raphe nu.
ELD	endolymphatic duct
Gi	gigantocellular reticular nu.
GiA	gigantocellular reticular nu., alpha
if	intermediate fibres of cerebellum
IP	interpeduncular nu.
ipf	interpeduncular fossa
IRt	intermediate reticular nu.
lf	lateral fissure of cerebellum
LPGi	lateral paragigantocellular nu.
LR4V	lateral recess of 4th ventricle
LVe	lateral vestibular nu.
Me5	mesencephalic trigeminal nu.
Mesen	mesencephalon
mlf	medial longitudinal fasciculus
mne	median neuroepithelium
MVe	medial vestibular nu.
ne	neuroepithelium
noto	notochord
ntz	nuclear transitory zone of Cb
ParP	parietal plate
PB	parabrachial complex
PBP	parabrachial pigmented nu.
pcer	posterior cerebral artery
PCRtA	parvicellular reticular nu., alpha
Petrous	petrous temporal bone
PN	paranigral nu. of the VTA
PPy	parapyramidal nu.
Pr	prepositus nu.
PSCD	posterior semicircular duct
PTg	pedunculotegmental nu.
Rhomb	rhombencephalon
RMg	raphe magnus nu.
ROb	raphe obscurus nu.
rrl	rostral rhombic lip
sl	sulcus limitans
SNR	substantia nigra, reticular
Sol	solitary nucleus
sp5	spinal trigeminal tract
Sp5O	spinal trigeminal nu., oral
SuVe	superior vestibular nu.
Utr	utricle
Ve	vestibular neuroepithelium
VTg	ventral tegmental nu.

Ec-InB22

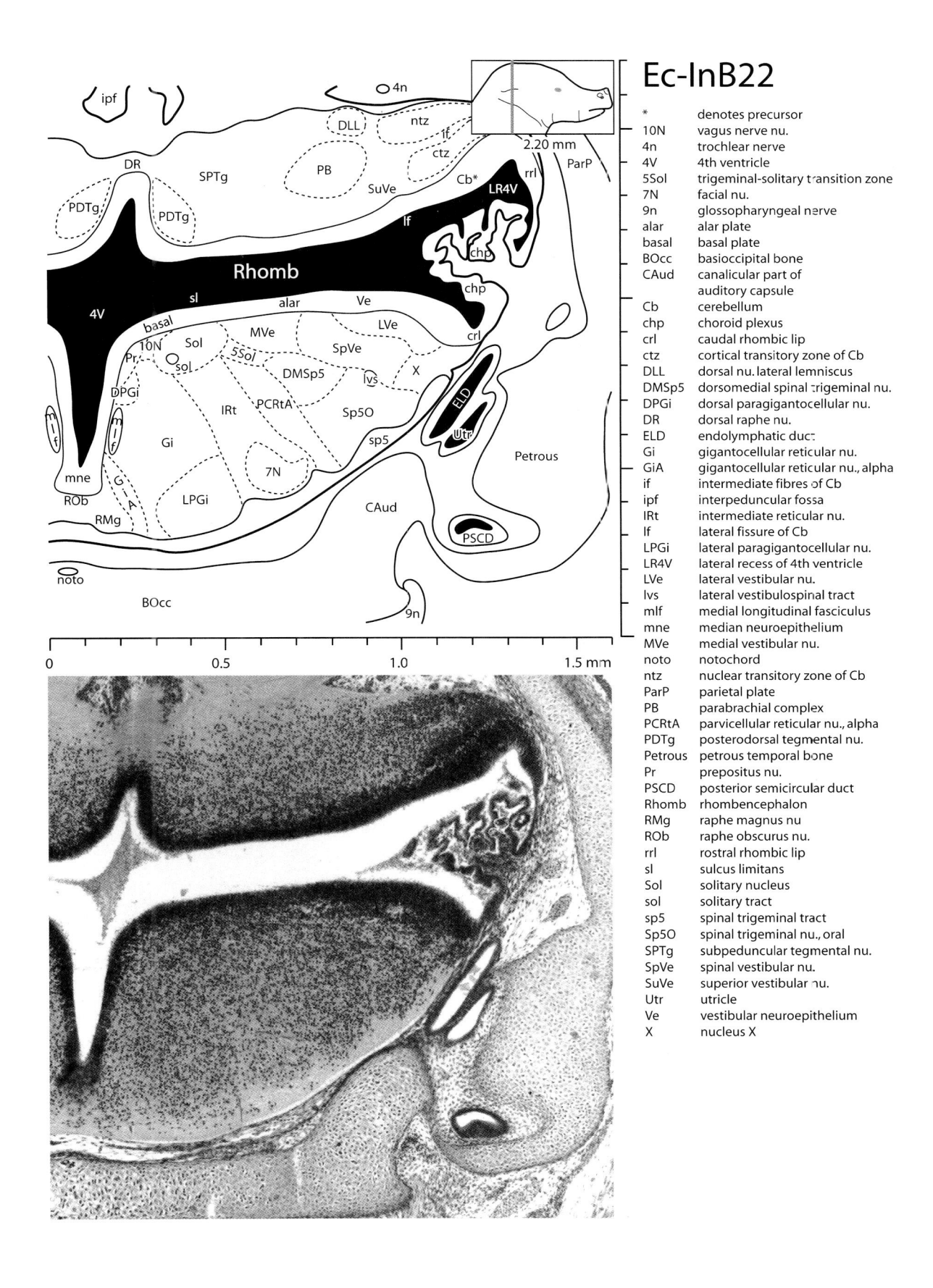

*	denotes precursor
10N	vagus nerve nu.
4n	trochlear nerve
4V	4th ventricle
5Sol	trigeminal-solitary transition zone
7N	facial nu.
9n	glossopharyngeal nerve
alar	alar plate
basal	basal plate
BOcc	basioccipital bone
CAud	canalicular part of auditory capsule
Cb	cerebellum
chp	choroid plexus
crl	caudal rhombic lip
ctz	cortical transitory zone of Cb
DLL	dorsal nu. lateral lemniscus
DMSp5	dorsomedial spinal trigeminal nu.
DPGi	dorsal paragigantocellular nu.
DR	dorsal raphe nu.
ELD	endolymphatic duct
Gi	gigantocellular reticular nu.
GiA	gigantocellular reticular nu., alpha
if	intermediate fibres of Cb
ipf	interpeduncular fossa
IRt	intermediate reticular nu.
lf	lateral fissure of Cb
LPGi	lateral paragigantocellular nu.
LR4V	lateral recess of 4th ventricle
LVe	lateral vestibular nu.
lvs	lateral vestibulospinal tract
mlf	medial longitudinal fasciculus
mne	median neuroepithelium
MVe	medial vestibular nu.
noto	notochord
ntz	nuclear transitory zone of Cb
ParP	parietal plate
PB	parabrachial complex
PCRtA	parvicellular reticular nu., alpha
PDTg	posterodorsal tegmental nu.
Petrous	petrous temporal bone
Pr	prepositus nu.
PSCD	posterior semicircular duct
Rhomb	rhombencephalon
RMg	raphe magnus nu
ROb	raphe obscurus nu.
rrl	rostral rhombic lip
sl	sulcus limitans
Sol	solitary nucleus
sol	solitary tract
sp5	spinal trigeminal tract
Sp5O	spinal trigeminal nu., oral
SPTg	subpeduncular tegmental nu.
SpVe	spinal vestibular nu.
SuVe	superior vestibular nu.
Utr	utricle
Ve	vestibular neuroepithelium
X	nucleus X

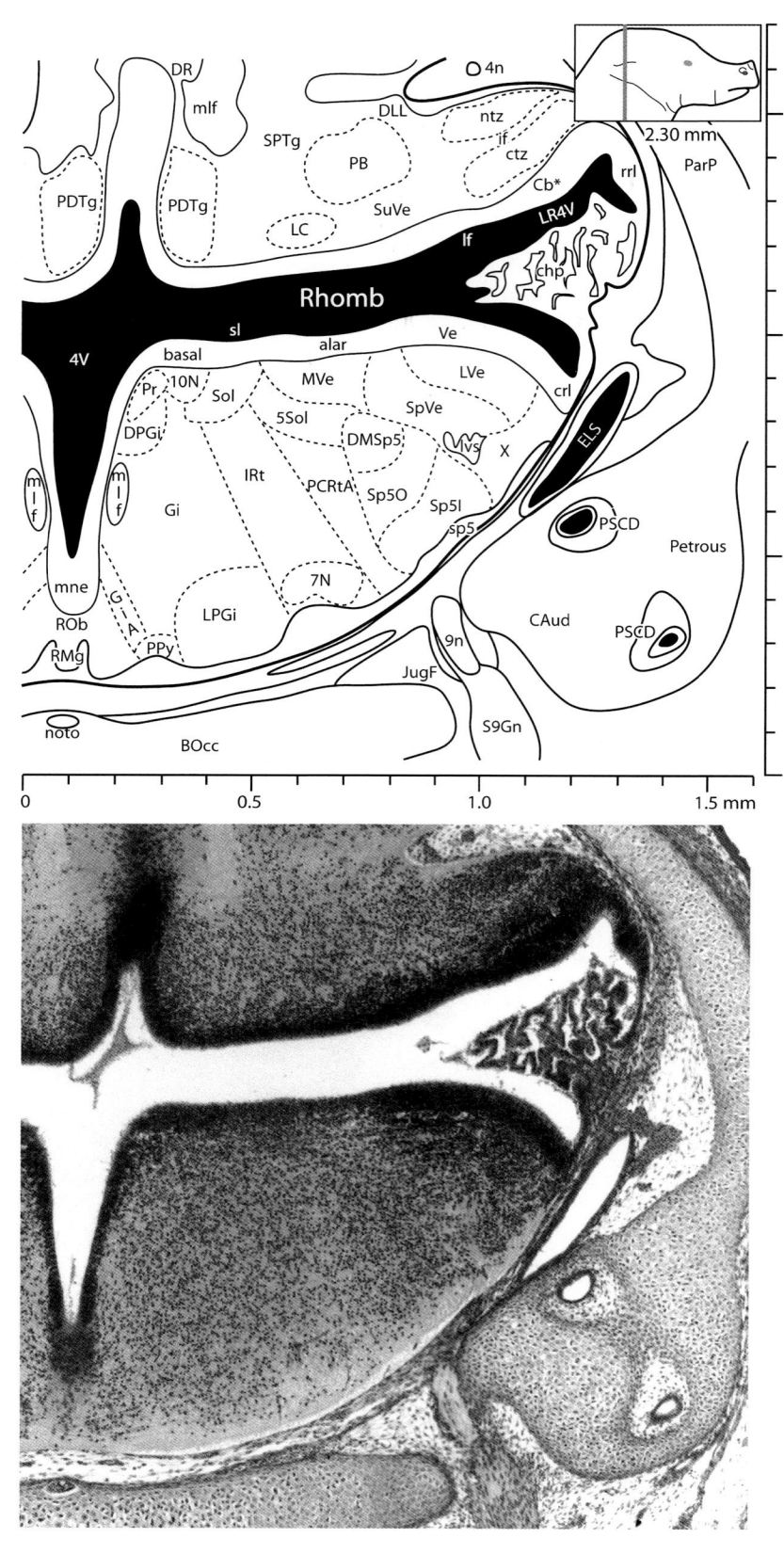

Ec-InB23

*	denotes precursor
10N	vagus nerve nu.
4n	trochlear nerve
4V	4th ventricle
5Sol	trigeminal-solitary transition zone
7N	facial nu.
9n	glossopharyngeal nerve
alar	alar plate
basal	basal plate
BOcc	basioccipital bone
CAud	canalicular part of auditory capsule
Cb	cerebellum
chp	choroid plexus
crl	caudal rhombic lip
ctz	cortical transitory zone of Cb
DLL	dorsal nu. of lateral lemniscus
DMSp5	dorsomedial spinal trigeminal nu.
DPGi	dorsal paragigantocellular nu.
DR	dorsal raphe nucleus
ELS	endolymphatic sac
Gi	gigantocellular reticular nu.
GiA	gigantocellular reticular nu., alpha
if	intermediate fibres of Cb
IRt	intermediate reticular nucleus
JugF	jugular foramen
LC	locus coeruleus
lf	lateral fissure of Cb
LPGi	lateral paragigantocellular nu.
LR4V	lateral recess of 4th ventricle
LVe	lateral vestibular nucleus
lvs	lateral vestibulospinal tract
mlf	medial longitudinal fasciculus
mne	median neuroepithelium
MVe	medial vestibular nu.
noto	notochord
ntz	nuclear transitory zone of Cb
ParP	parietal plate
PB	parabrachial complex
PCRtA	parvicellular reticular nu., alpha
PDTg	posterodorsal tegmental nu.
Petrous	petrous temporal bone
PPy	parapyramidal nu.
Pr	prepositus nu.
PSCD	posterior semicircular duct
Rhomb	rhombencephalon
RMg	raphe magnus nu.
ROb	raphe obscurus nu.
rrl	rostral rhombic lip
S9Gn	superior ganglion of the glossopharyngeal nerve
sl	sulcus limitans
Sol	solitary nucleus
sp5	spinal trigeminal tract
Sp5I	spinal trigeminal nu., interpolar
Sp5O	spinal trigeminal nu., oral
SPTg	subpeduncular tegmental nu.
SpVe	spinal vestibular nu.
SuVe	superior vestibular nu.
Ve	vestibular neuroepithelium
X	nucleus X

Ec-InB24

*	denotes precursor
10N	vagus nerve nu.
10n	vagus nerve
4n	trochlear nerve
4V	4th ventricle
5Sol	trigeminal-solitary transition zone
alar	alar plate
AmbSC	ambiguus nu., subcompact part
basal	basal plate
BOcc	basioccipital bone
CAud	canalicular part of auditory capsule
Cb	cerebellum
chp	choroid plexus
crl	caudal rhombic lip
ctz	cortical transitory zone of Cb
DMSp5	dorsomedial spinal trigeminal nu.
DPGi	dorsal paragigantocellular nu.
DR	dorsal raphe nu.
DTg	dorsal tegmental nu.
ELS	endolymphatic sac
Gi	gigantocellular reticular nu.
GiA	gigantocellular reticular nu., alpha
IC	inferior colliculus
if	intermediate fibres of Cb
IRt	intermediate reticular nu.
JugF	jugular foramen
LC	locus coeruleus
lf	lateral fissure of Cb
LPGi	lateral paragigantocellular nu.
LR4V	lateral recess of 4th ventricle
lvs	lateral vestibulospinal tract
Mesen	mesencephalon
mlf	medial longitudinal fasciculus
mne	median neuroepithelium
MVe	medial vestibular nu.
noto	notochord
ntz	nuclear transitory zone of Cb
PB	parabrachial complex
PCRt	parvicellular reticular nu.
Pr	prepositus nu.
PSCD	posterior semicircular duct
Rhomb	rhombencephalon
RMg	raphe magnus nu.
ROb	raphe obscurus nu.
rrl	rostral rhombic lip
S10Gn	superior ganglion 10n
sl	sulcus limitans
Sol	solitary nucleus
sol	solitary tract
sp5	spinal trigeminal tract
Sp5I	spinal trigeminal nu., interpolar
SpVe	spinal vestibular nu.
SuVe	superior vestibular nu.
vert	vertebral artery
X	nucleus X

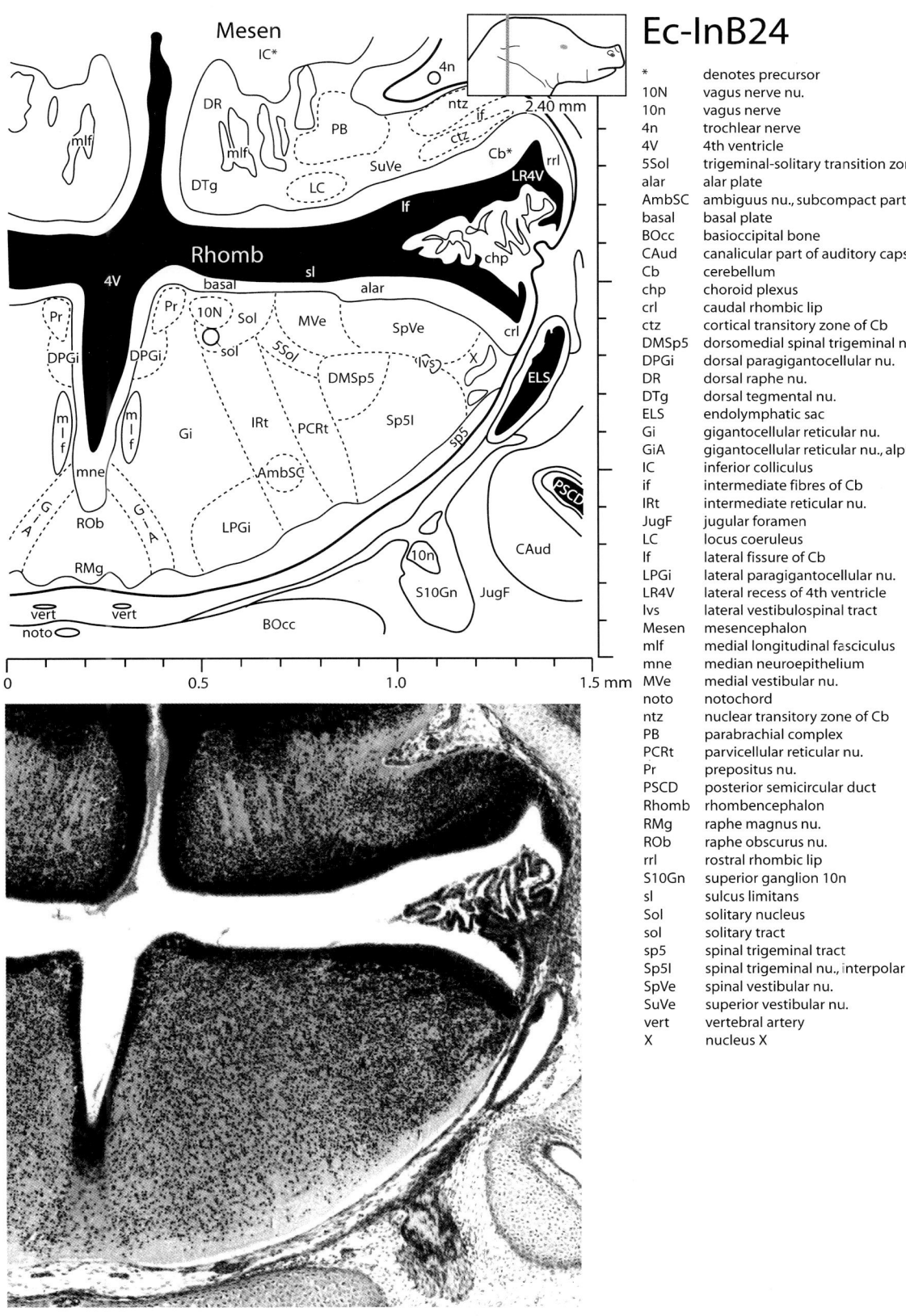

Ec-InB25

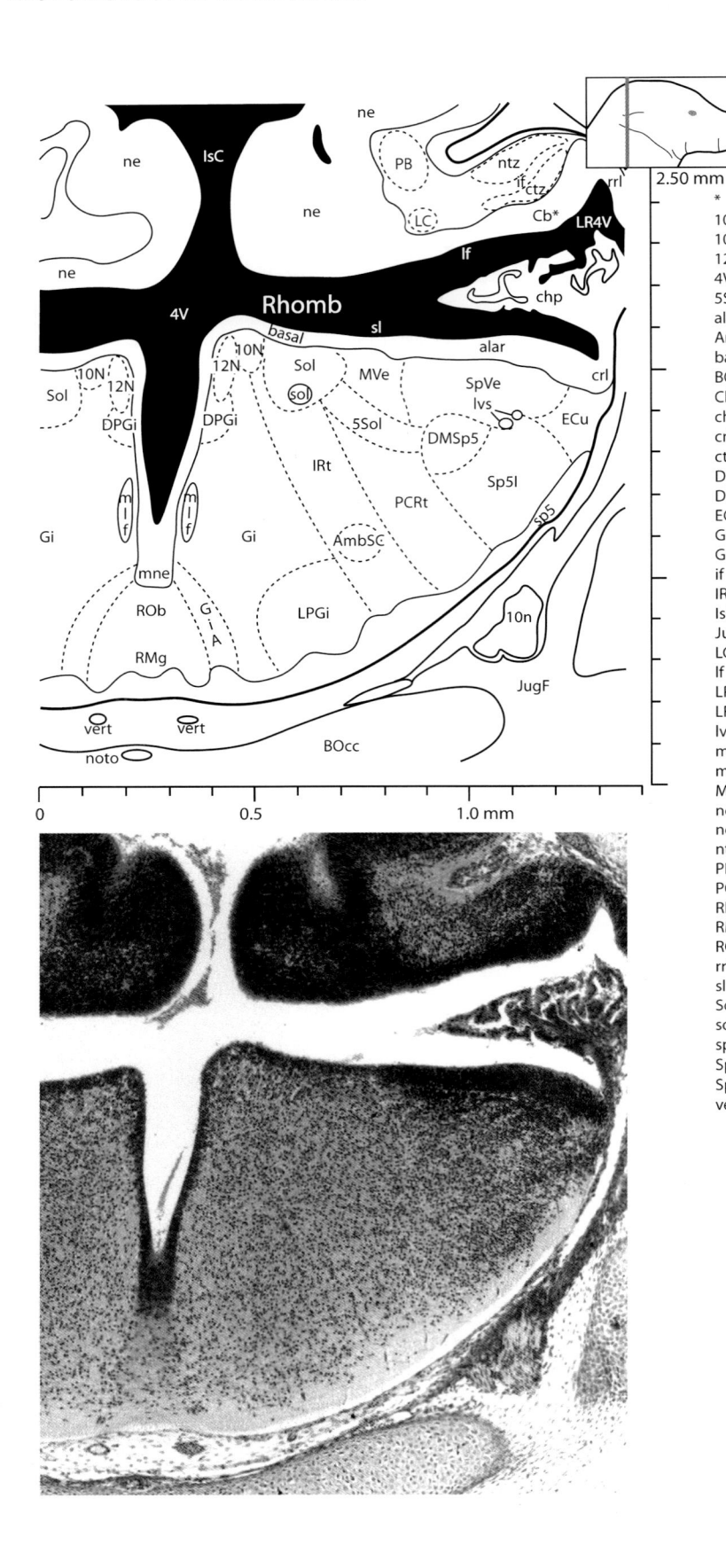

2.50 mm

*	denotes precursor
10N	vagus nerve nu.
10n	vagus nerve
12N	hypoglossal nu.
4V	4th ventricle
5Sol	trigeminal-solitary transition zone
alar	alar plate
AmbSC	ambiguus nu., subcompact
basal	basal plate
BOcc	basioccipital bone
Cb	cerebellum
chp	choroid plexus
crl	caudal rhombic lip
ctz	cortical transitory zone of Cb
DMSp5	dorsomedial spinal trigeminal nu.
DPGi	dorsal paragigantocellular nu.
ECu	external cuneate nu.
Gi	gigantocellular reticular nu.
GiA	gigantocellular reticular nu., alpha
if	intermediate fibres of Cb
IRt	intermediate reticular nu.
IsC	isthmic canal
JugF	jugular foramen
LC	locus coeruleus
lf	lateral fissure of Cb
LPGi	lateral paragigantocellular nu.
LR4V	lateral recess of 4th ventricle
lvs	lateral vestibulospinal tract
mlf	medial longitudinal fasciculus
mne	median neuroepithelium
MVe	medial vestibular nu.
ne	neuroepithelium
noto	notochord
ntz	nuclear transitory zone of Cb
PB	parabrachial complex
PCRt	parvicellular reticular nu.
Rhomb	rhombencephalon
RMg	raphe magnus nu.
ROb	raphe obscurus nu.
rrl	rostral rhombic lip
sl	sulcus limitans
Sol	solitary nucleus
sol	solitary tract
sp5	spinal trigeminal tract
Sp5I	spinal trigeminal nu., interpolar
SpVe	spinal vestibular nu.
vert	vertebral artery

Ec-InB26

*	denotes precursor
10n	vagus nerve
10N	vagus nerve nucleus
12N	hypoglossal nucleus
4V	4th ventricle
5Sol	trigeminal-solitary transition zone
alar	alar plate
AmbC	ambiguus nu., compact part
basal	basal plate
BOcc	basioccipital bone
Cb	cerebellum
chp	choroid plexus
crl	caudal rhombic lip
DMSp5	dorsomedial spinal trigeminal nu.
DPGi	dorsal paragigantocellular nu.
ECu	external cuneate nu.
ELD	endolymphatic duct
ELS	endolymphatic sac
Gi	gigantocellular reticular nu.
GiV	gigantocellular reticular nu., ventral
IRt	intermediate reticular nu.
LPGi	lateral paragigantocellular nu.
LR4V	lateral recess of 4th ventricle
lvs	lateral vestibulospinal tract
mlf	medial longitudinal fasciculus
mne	median neuroepithelium
MVe	medial vestibular nu.
noto	notochord
Occ	occipital bone
PCRt	parvicellular reticular nu.
Rhomb	rhombencephalon
RMg	raphe magnus nu.
ROb	raphe obscurus nu.
RPa	raphe pallidus nu.
S10Gn	superior ganglion of the vagus nerve
sl	sulcus limitans
Sol	solitary nucleus
sol	solitary tract
sp5	spinal trigeminal tract
Sp5I	spinal trigeminal nu., interpolar
SpVe	spinal vestibular nu.
vert	vertebral artery

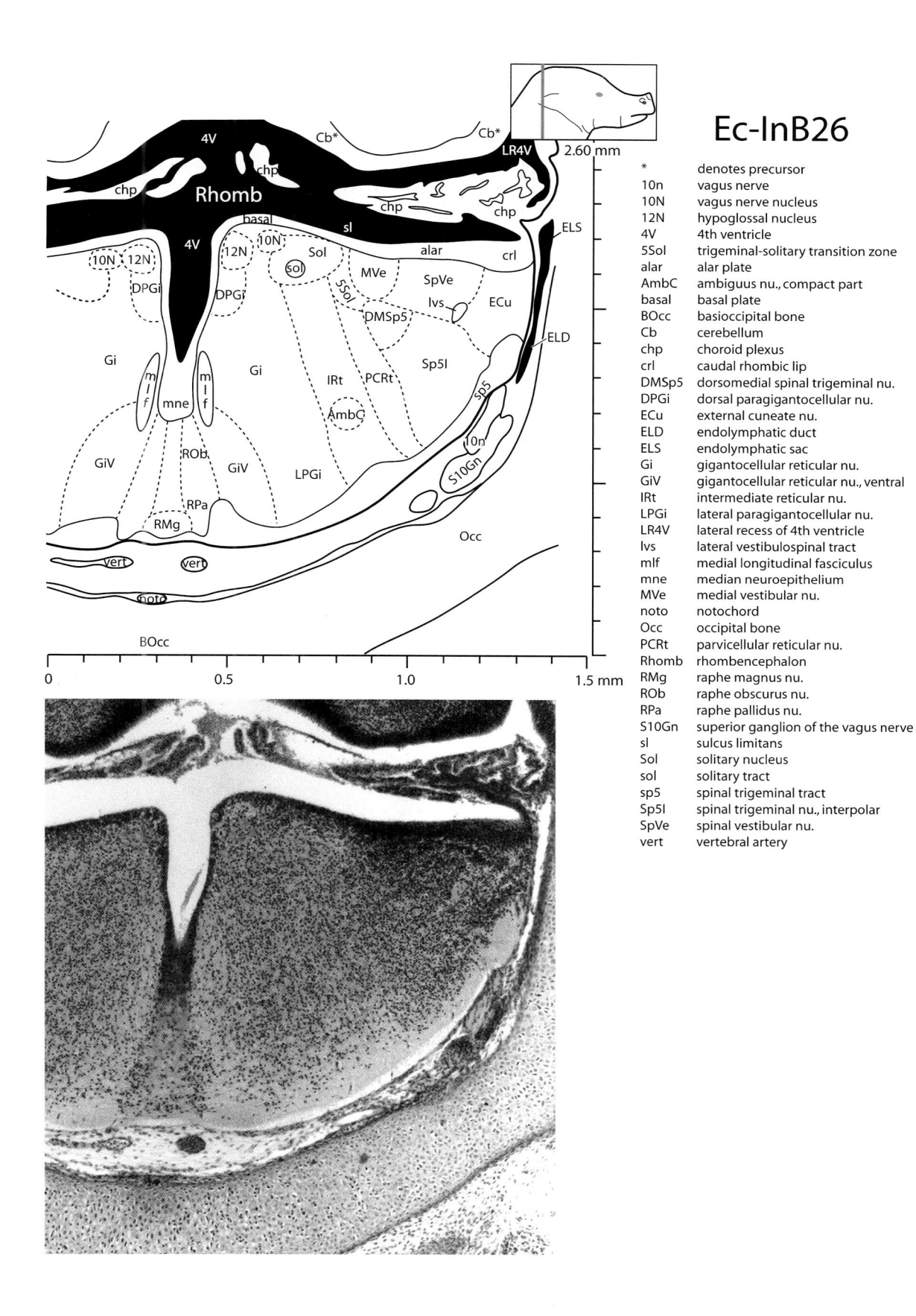

Ec-InB27

*	denotes precursor
10N	vagus nerve nu.
10n	vagus nerve
12N	hypoglossal nu.
4V	4th ventricle
5Sol	trigeminal-solitary transition zone
alar	alar plate
AmbC	ambiguus nu., compact part
basal	basal plate
Cb	cerebellum
chp	choroid plexus
crl	caudal rhombic lip
DMSp5	dorsomedial spinal trigeminal nu.
DPGi	dorsal paragigantocellular nu.
ECu	external cuneate nu.
ELS	endolymphatic sac
Gi	gigantocellular reticular nu.
GiV	gigantocellular reticular nu., ventral
IRt	intermediate reticular nu.
LPGi	lateral paragigantocellular nu.
lvs	lateral vestibulospinal tract
mlf	medial longitudinal fasciculus
mne	median neuroepithelium
MVe	medial vestibular nu.
Occ	occipital bone
PCRt	parvicellular reticular nu.
Rhomb	rhombencephalon
ROb	raphe obscurus nu.
RPa	raphe pallidus nu.
sl	sulcus limitans
sol	solitary tract
Sol	solitary nucleus
sp5	spinal trigeminal tract
Sp5I	spinal trigeminal nu., interpolar
SpVe	spinal vestibular nu.
vert	vertebral artery

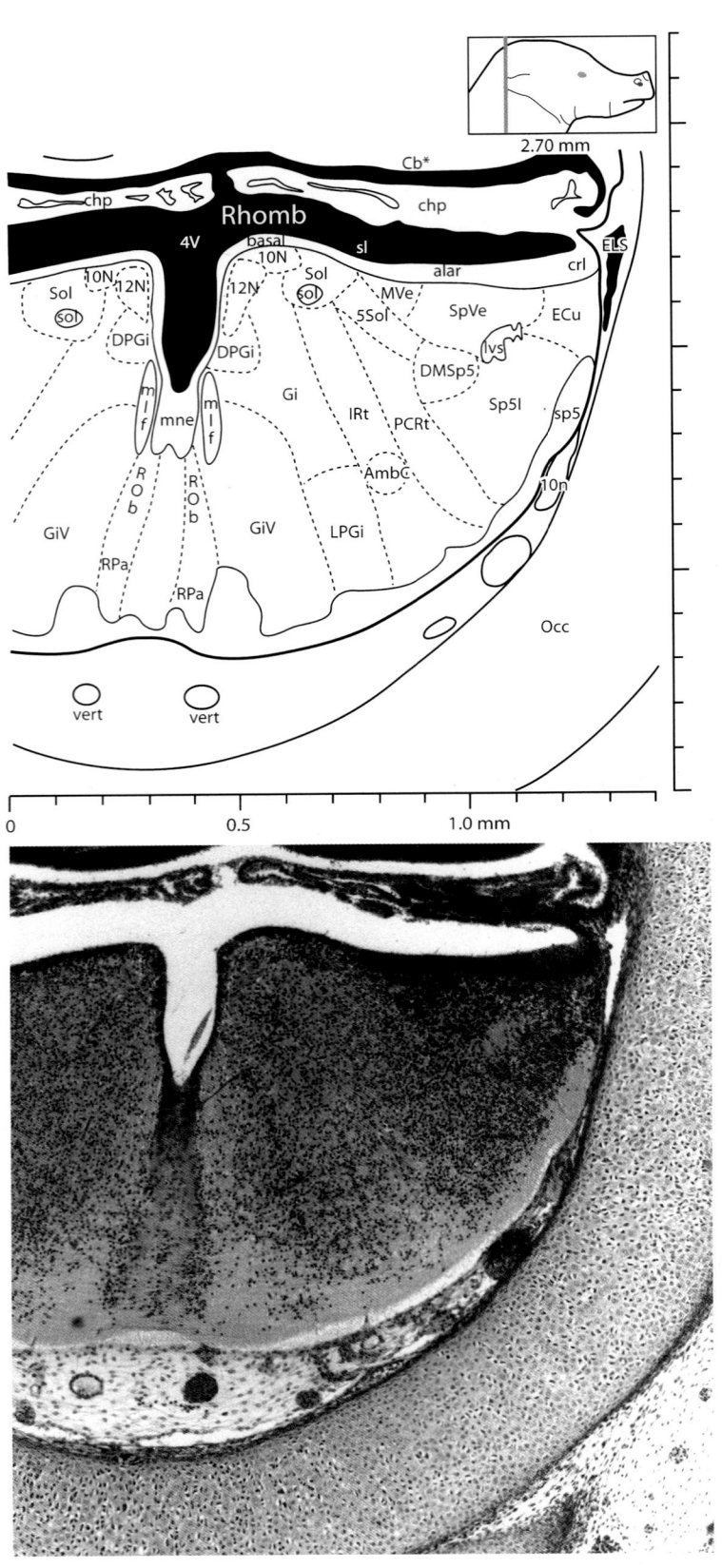

Ec-InB28

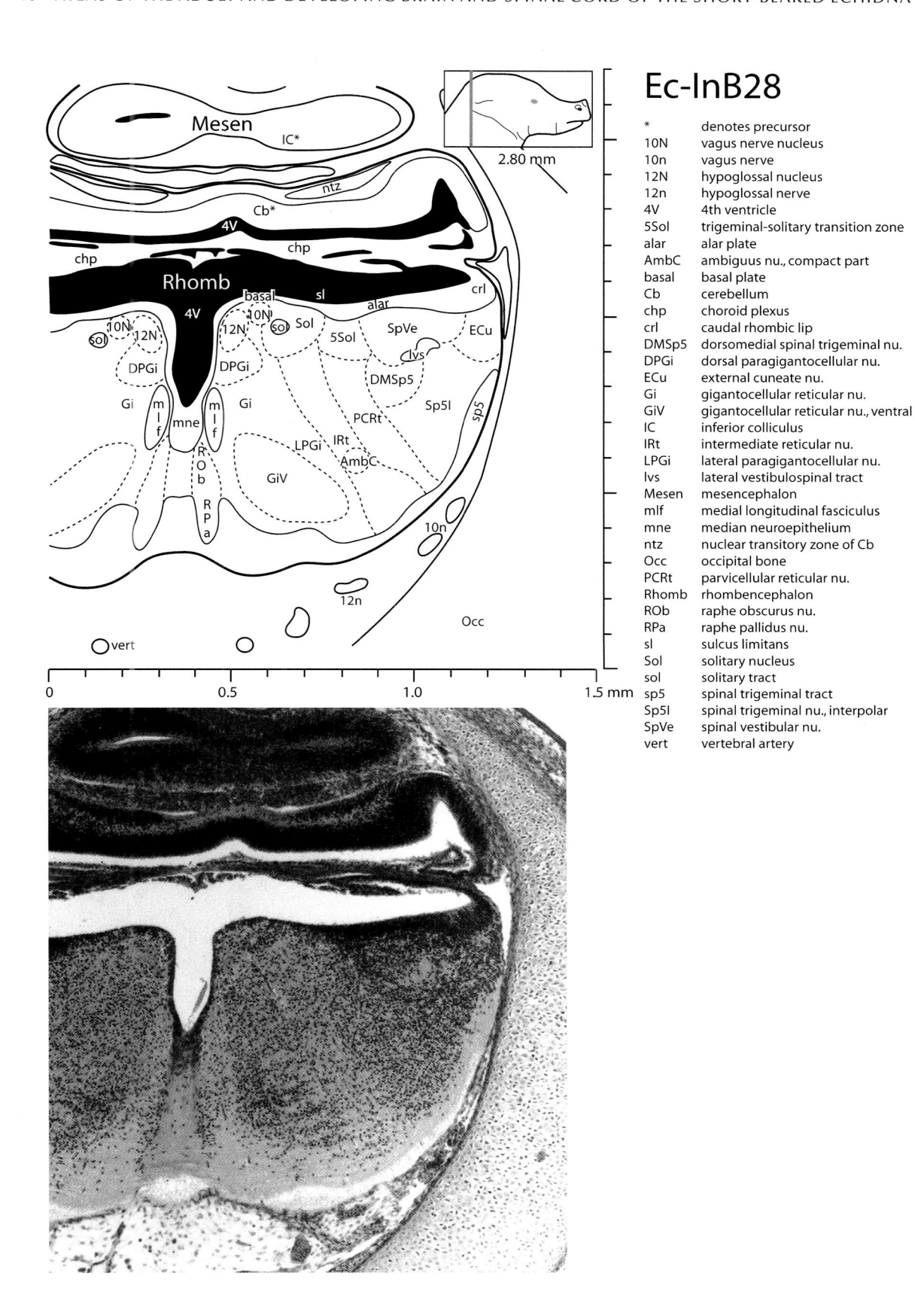

2.80 mm

*	denotes precursor
10N	vagus nerve nucleus
10n	vagus nerve
12N	hypoglossal nucleus
12n	hypoglossal nerve
4V	4th ventricle
5Sol	trigeminal-solitary transition zone
alar	alar plate
AmbC	ambiguus nu., compact part
basal	basal plate
Cb	cerebellum
chp	choroid plexus
crl	caudal rhombic lip
DMSp5	dorsomedial spinal trigeminal nu.
DPGi	dorsal paragigantocellular nu.
ECu	external cuneate nu.
Gi	gigantocellular reticular nu.
GiV	gigantocellular reticular nu., ventral
IC	inferior colliculus
IRt	intermediate reticular nu.
LPGi	lateral paragigantocellular nu.
lvs	lateral vestibulospinal tract
Mesen	mesencephalon
mlf	medial longitudinal fasciculus
mne	median neuroepithelium
ntz	nuclear transitory zone of Cb
Occ	occipital bone
PCRt	parvicellular reticular nu.
Rhomb	rhombencephalon
ROb	raphe obscurus nu.
RPa	raphe pallidus nu.
sl	sulcus limitans
Sol	solitary nucleus
sol	solitary tract
sp5	spinal trigeminal tract
Sp5I	spinal trigeminal nu., interpolar
SpVe	spinal vestibular nu.
vert	vertebral artery

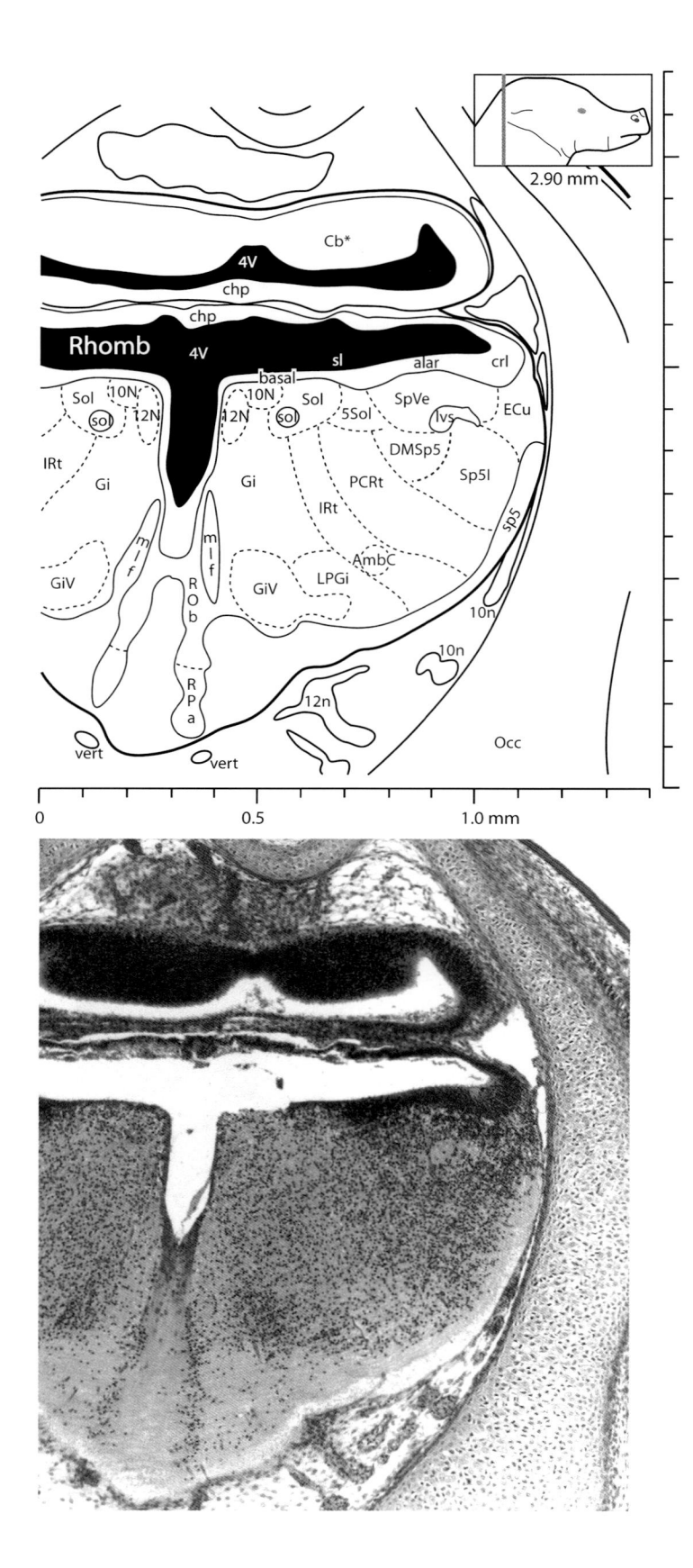

Ec-InB29

*	denotes precursor
10N	vagus nerve nu.
10n	vagus nerve
12N	hypoglossal nu.
12n	hypoglossal nerve
4V	4th ventricle
5Sol	trigeminal-solitary transition zone
alar	alar plate
AmbC	ambiguus nu., compact part
basal	basal plate
Cb	cerebellum
chp	choroid plexus
crl	caudal rhombic lip
DMSp5	dorsomedial spinal trigeminal nu.
ECu	external cuneate nu.
Gi	gigantocellular reticular nu.
GiV	gigantocellular reticular nu., ventral
IRt	intermediate reticular nu.
LPGi	lateral paragigantocellular nu.
lvs	lateral vestibulospinal tract
mlf	medial longitudinal fasciculus
Occ	occipital bone
PCRt	parvicellular reticular nu.
Rhomb	rhombencephalon
ROb	raphe obscurus nu.
RPa	raphe pallidus nu.
sl	sulcus limitans
Sol	solitary nucleus
sol	solitary tract
sp5	spinal trigeminal tract
Sp5I	spinal trigeminal nu., interpolar
SpVe	spinal vestibular nu.
vert	vertebral artery

Ec-InB30

*	denotes precursor
10n	vagus nerve
10N	vagus nerve nu.
12N	hypoglossal nu.
12n	hypoglossal nerve
4V	4th ventricle
5Sol	trigeminal-solitary transition zone
alar	alar plate
AmbC	ambiguus nu., compact part
basal	basal plate
chp	choroid plexus
crl	caudal rhombic lip
DMSp5	dorsomedial spinal trigeminal nu.
ECu	external cuneate nu.
Gi	gigantocellular reticular nu.
GiV	gigantocellular reticular nu., ventral
IRt	intermediate reticular nu.
LPGi	lateral paragigantocellular nu.
lvs	lateral vestibulospinal tract
mlf	medial longitudinal fasciculus
Occ	occipital bone
PCRt	parvicellular reticular nu.
Rhomb	rhombencephalon
sl	sulcus limitans
Sol	solitary nucleus
sol	solitary tract
sp5	spinal trigeminal tract
Sp5I	spinal trigeminal nu., interpolar
SpVe	spinal vestibular nu.
Ve	vestibular neuroepithelium
vert	vertebral artery

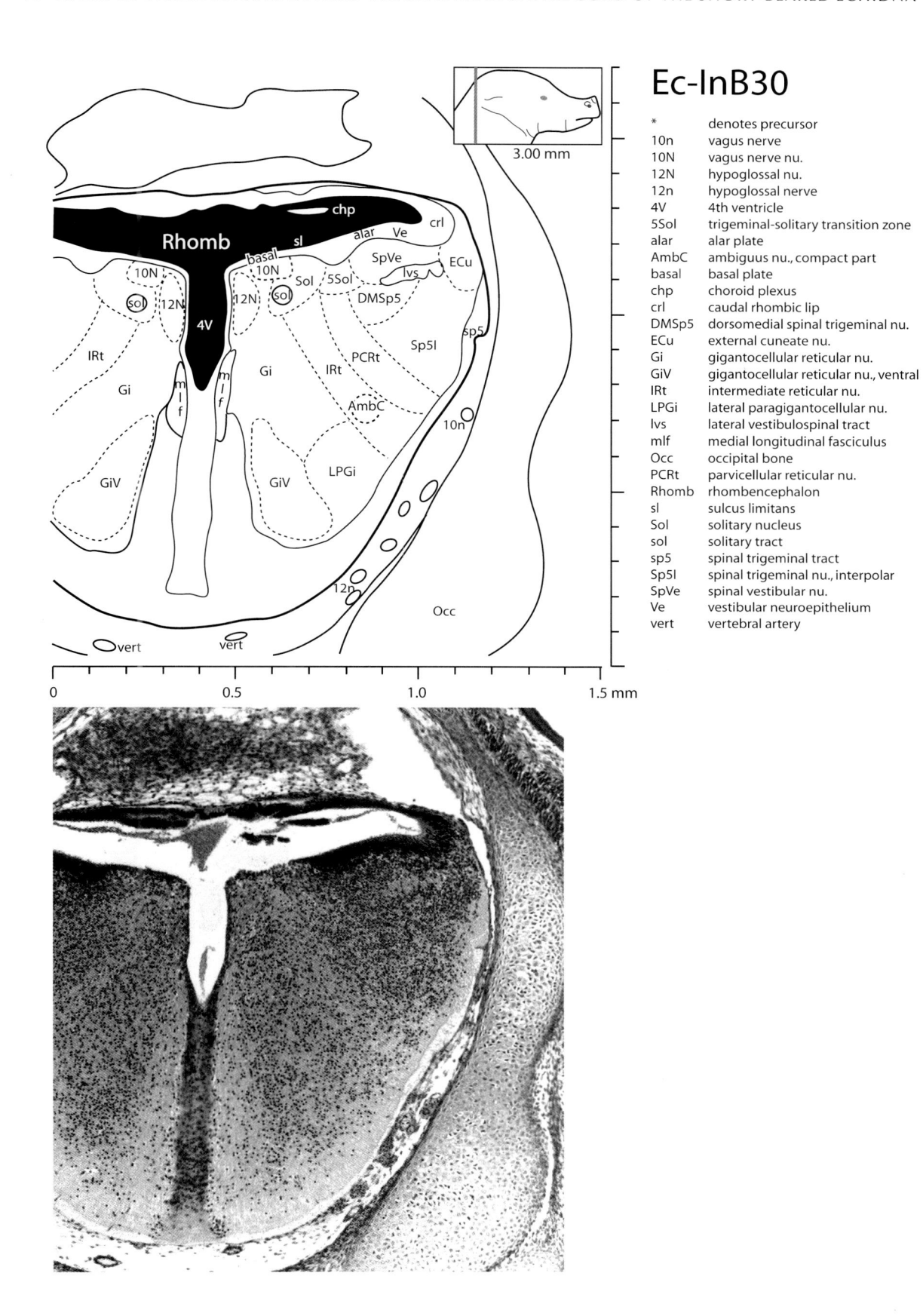

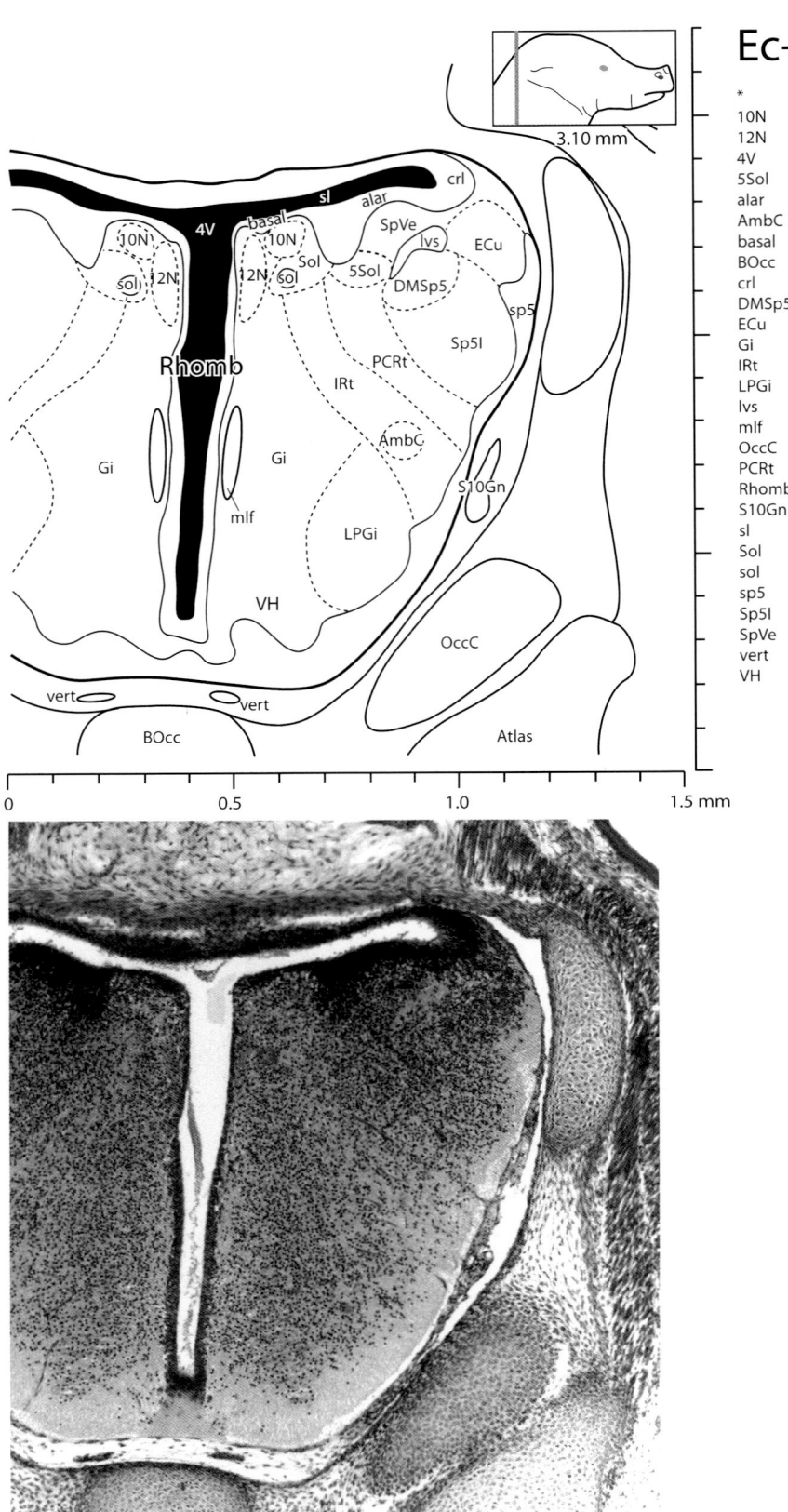

Ec-InB31

3.10 mm

*	denotes precursor
10N	vagus nerve nu.
12N	hypoglossal nu.
4V	4th ventricle
5Sol	trigeminal-solitary transition zone
alar	alar plate
AmbC	ambiguus nu., compact part
basal	basal plate
BOcc	basioccipital bone
crl	caudal rhombic lip
DMSp5	dorsomedial spinal trigeminal nu.
ECu	external cuneate nu.
Gi	gigantocellular reticular nu.
IRt	intermediate reticular nu.
LPGi	lateral paragigantocellular nu.
lvs	lateral vestibulospinal tract
mlf	medial longitudinal fasciculus
OccC	occipital condyle
PCRt	parvicellular reticular nu.
Rhomb	rhombencephalon
S10Gn	superior ganglion 10n
sl	sulcus limitans
Sol	solitary nucleus
sol	solitary tract
sp5	spinal trigeminal tract
Sp5I	spinal trigeminal nu., interpolar
SpVe	spinal vestibular nu.
vert	vertebral artery
VH	ventral horn

17

Atlas of the adult and developing brain of the platypus (*Ornithorhynchus anatinus*)

Ken W. S. Ashwell and Craig D. Hardman

Introduction

The platypus (*Ornithorhynchus anatinus*) has a natural distribution in the freshwater rivers and lakes down the eastern coast of Australia, in Tasmania and King Island in Bass Strait (see Chapter 2). Individuals have occasionally been recorded in the lower Murray River from western New South Wales into South Australia. Some platypuses have also been introduced to the western end of Kangaroo Island in South Australia and even Western Australia, although it is unknown whether these latter have survived to the present day (Grant 2007).

Adult platypuses are smaller than most people expect, weighing less than the domestic cat (Grant 2007). Large males can reach 60 cm dorsal contour length (bill tip to tail tip), but most are 40–50 cm in greatest length. Males are distinctly larger than females by a factor of 1.4–1.9 and there is, in general, a latitudinal variation in body weight, such that males from Tasmania may reach 3 kg, whereas average male weight from north Queensland is around 1 kg.

The cerebral cortex of the adult platypus is quite thick but smooth (lissencephalic), in contrast to the gyrified cerebral cortex of the echidna.

Materials and methods

Adult animal acquisition and ethical issues

The procedures outlined below conform to the guidelines established by the National Health and Medical Research Council of Australia. The platypus depicted in the following atlas was a juvenile female (291 g), which had been brought injured to the native animal veterinary clinic at Taronga Zoological Park, Sydney, and died during the course of treatment (Ashwell *et al.* 2006b).

Tissue processing

The right half of the brain was immersion fixed in 4% paraformaldehyde in 0.1 M phosphate buffer (4°C, pH 7.4) and subsequently immersed in 30% sucrose in 0.1 M phosphate buffer for 24 hours at 4°C. After this time the brain was sectioned coronally with the aid of a cryostat at a thickness of 40 μm and sections stained for Nissl substance (cresyl violet), enzyme histochemistry (acetylcholinesterase and NADPH diaphorase), and immunohistochemistry (parvalbumin, calbindin, calretinin). Only Nissl-stained sections and sections immunoreacted for parvalbumin and calbindin are illustrated in the plates. Details of processing can be found in Ashwell *et al.* (2006b).

Imaging and delineation

As for the adult echidna brain, sections were digitally scanned with the aid of Aperio ScanScope XT slide scanners in either the Paxinos laboratory of Neuroscience Research Australia or the Histology and Microscopy Unit of the School of Medical Sciences at the University of New South Wales. Images were processed using Adobe Photoshop CS2 and placed in Adobe Illustrator CS2. Structures of interest were drawn and labelled using the same system of abbreviations applied to the echidna brain. Each plate includes a low power line diagram of a section from the right half of the brain, accompanied by a caption and one or more higher-powered images of specific regions of interest.

Note that the delineation of somatosensory areas (R, S1, PV) is based on Krubitzer *et al.* (1995), whereas the motor areas are based on Bohringer and Rowe (1977). This leads to occasional disparities between the motor and somatosensory representation of some body parts, which need further studies to eliminate.

Platypus hatchling

This newly hatched specimen (Pl-Ha – M44) is from the Hill collection held in the Museum für Naturkunde in Berlin. No details are available for the site of origin of the specimen, but it was probably collected in eastern New South Wales, because that is the region from which Hill collected. Records state that the specimen was newly hatched. Its greatest length is given as 16.45 mm on the cards with the slides, but the slides themselves say the greatest length is 16.75 mm. Its dorsal contour length (i.e. the curved length around the dorsum from snout to tail tip) is listed as 28.0 mm and its head length is given as 6.0 mm. A hatchling of this size should be only a day or two old, but the precise age is not known.

The head of the specimen had been embedded in paraffin and sectioned in the frontal plane at a thickness of 10 μm. Sections had been mounted on 23 slides. The sections appear to have been subsequently stained with hematoxylin and eosin.

Photography and delineation of the nervous system of the platypus hatchling

Sections through this specimen were photographed with the aid of a Zeiss Axioplan2 microscope fitted with an AxioCam MRc5 camera. The images were cleaned and adjusted for contrast and brightness using Adobe Photoshop CS2. All images were calibrated by photographing a scale bar at the same magnification. The brain and associated nerves were mapped in Adobe Illustrator CS2. Nomenclature and abbreviations were derived from the system used by the Paxinos group (Ashwell and Paxinos 2008). Features of the skeleton were based on de Beer and Fell (1936) and Zeller (1989). Primordia of brain regions were denoted by an asterisk, e.g. Cx* denotes the primordium of the cerebral cortex.

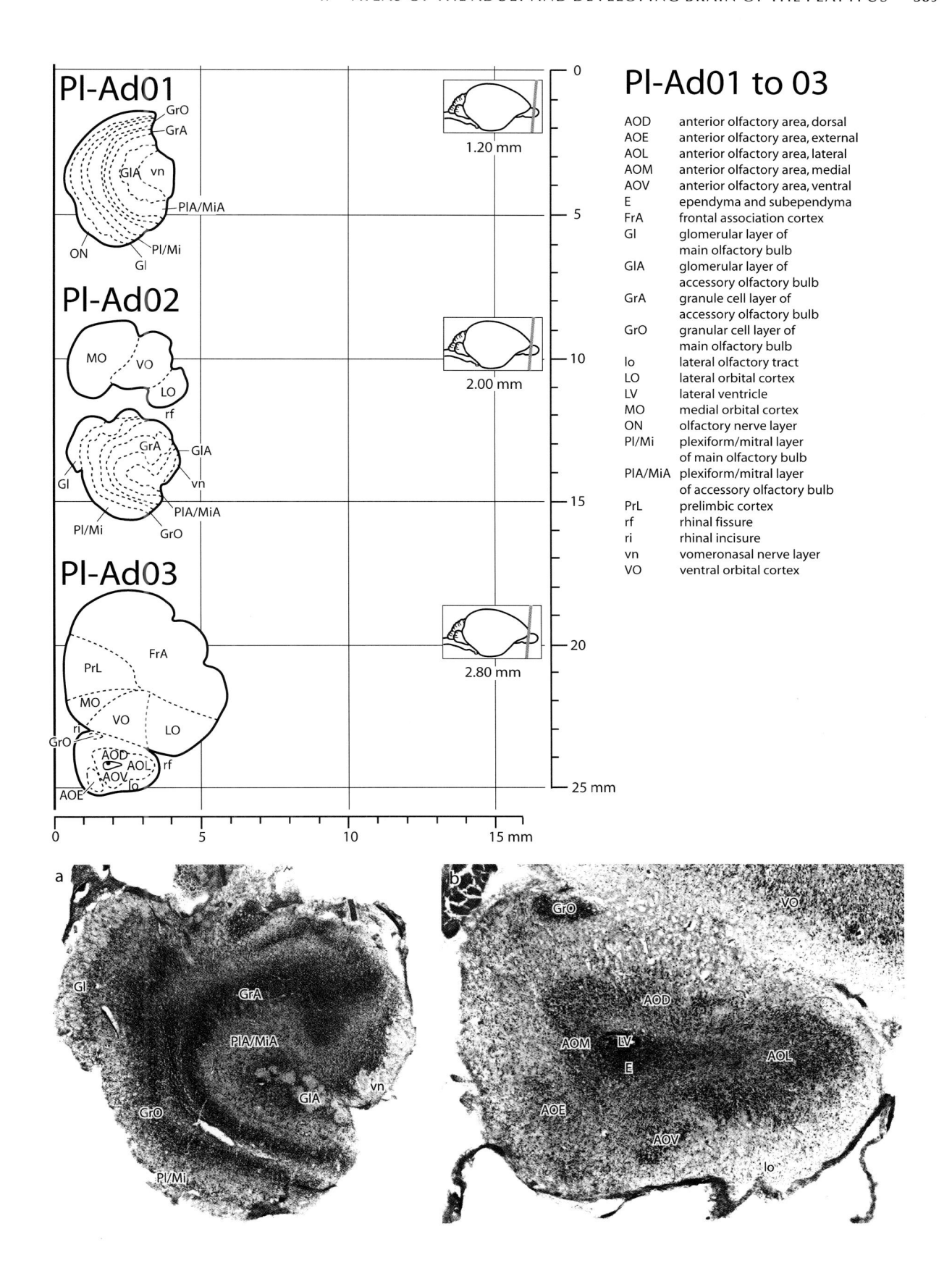

17 – ATLAS OF THE ADULT AND DEVELOPING BRAIN OF THE PLATYPUS 389

Pl-Ad01 to 03

AOD	anterior olfactory area, dorsal
AOE	anterior olfactory area, external
AOL	anterior olfactory area, lateral
AOM	anterior olfactory area, medial
AOV	anterior olfactory area, ventral
E	ependyma and subependyma
FrA	frontal association cortex
Gl	glomerular layer of main olfactory bulb
GlA	glomerular layer of accessory olfactory bulb
GrA	granule cell layer of accessory olfactory bulb
GrO	granular cell layer of main olfactory bulb
lo	lateral olfactory tract
LO	lateral orbital cortex
LV	lateral ventricle
MO	medial orbital cortex
ON	olfactory nerve layer
Pl/Mi	plexiform/mitral layer of main olfactory bulb
PlA/MiA	plexiform/mitral layer of accessory olfactory bulb
PrL	prelimbic cortex
rf	rhinal fissure
ri	rhinal incisure
vn	vomeronasal nerve layer
VO	ventral orbital cortex

390 NEUROBIOLOGY OF MONOTREMES

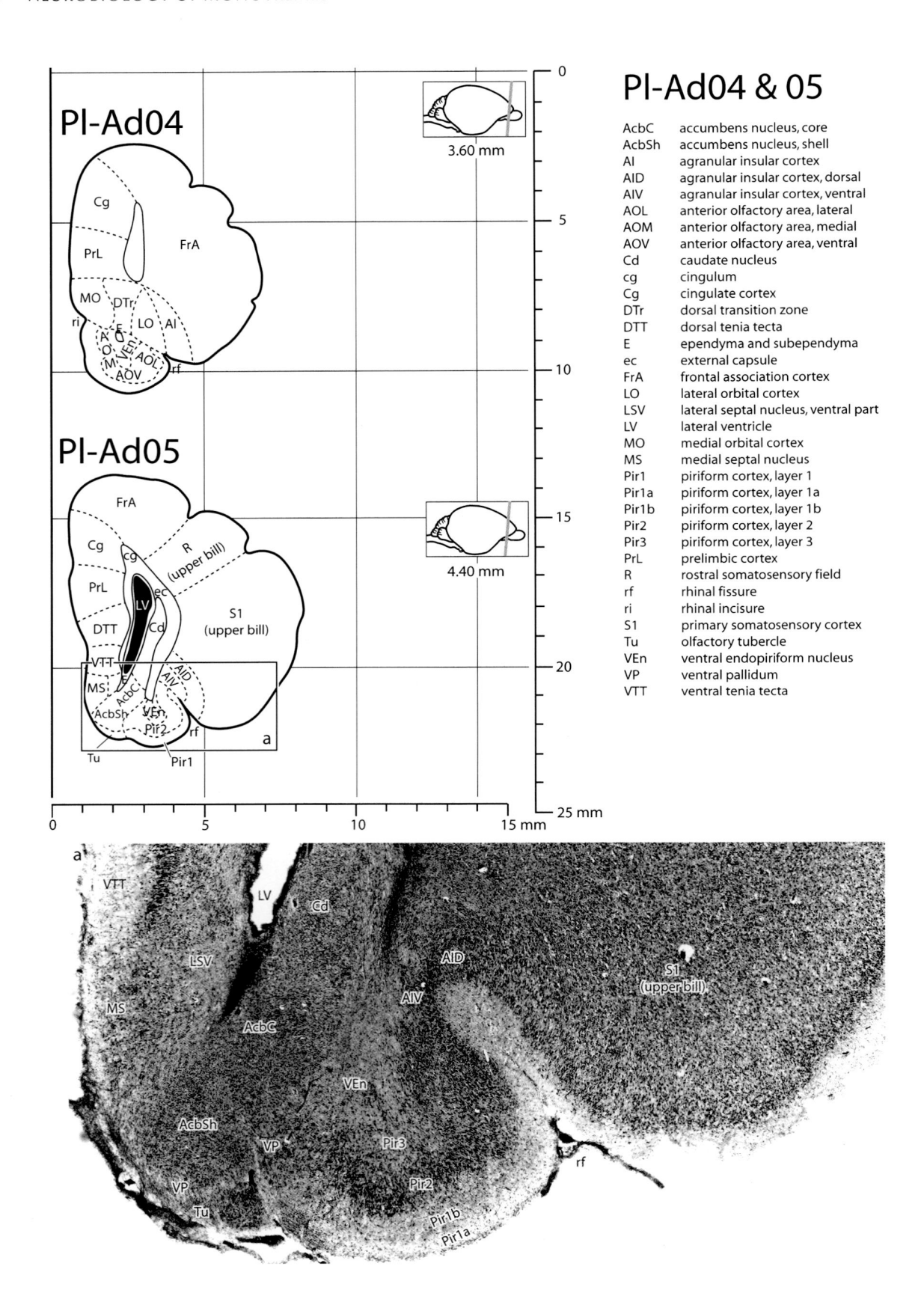

Pl-Ad04 & 05

AcbC	accumbens nucleus, core
AcbSh	accumbens nucleus, shell
AI	agranular insular cortex
AID	agranular insular cortex, dorsal
AIV	agranular insular cortex, ventral
AOL	anterior olfactory area, lateral
AOM	anterior olfactory area, medial
AOV	anterior olfactory area, ventral
Cd	caudate nucleus
cg	cingulum
Cg	cingulate cortex
DTr	dorsal transition zone
DTT	dorsal tenia tecta
E	ependyma and subependyma
ec	external capsule
FrA	frontal association cortex
LO	lateral orbital cortex
LSV	lateral septal nucleus, ventral part
LV	lateral ventricle
MO	medial orbital cortex
MS	medial septal nucleus
Pir1	piriform cortex, layer 1
Pir1a	piriform cortex, layer 1a
Pir1b	piriform cortex, layer 1b
Pir2	piriform cortex, layer 2
Pir3	piriform cortex, layer 3
PrL	prelimbic cortex
R	rostral somatosensory field
rf	rhinal fissure
ri	rhinal incisure
S1	primary somatosensory cortex
Tu	olfactory tubercle
VEn	ventral endopiriform nucleus
VP	ventral pallidum
VTT	ventral tenia tecta

Pl-Ad06 & 07

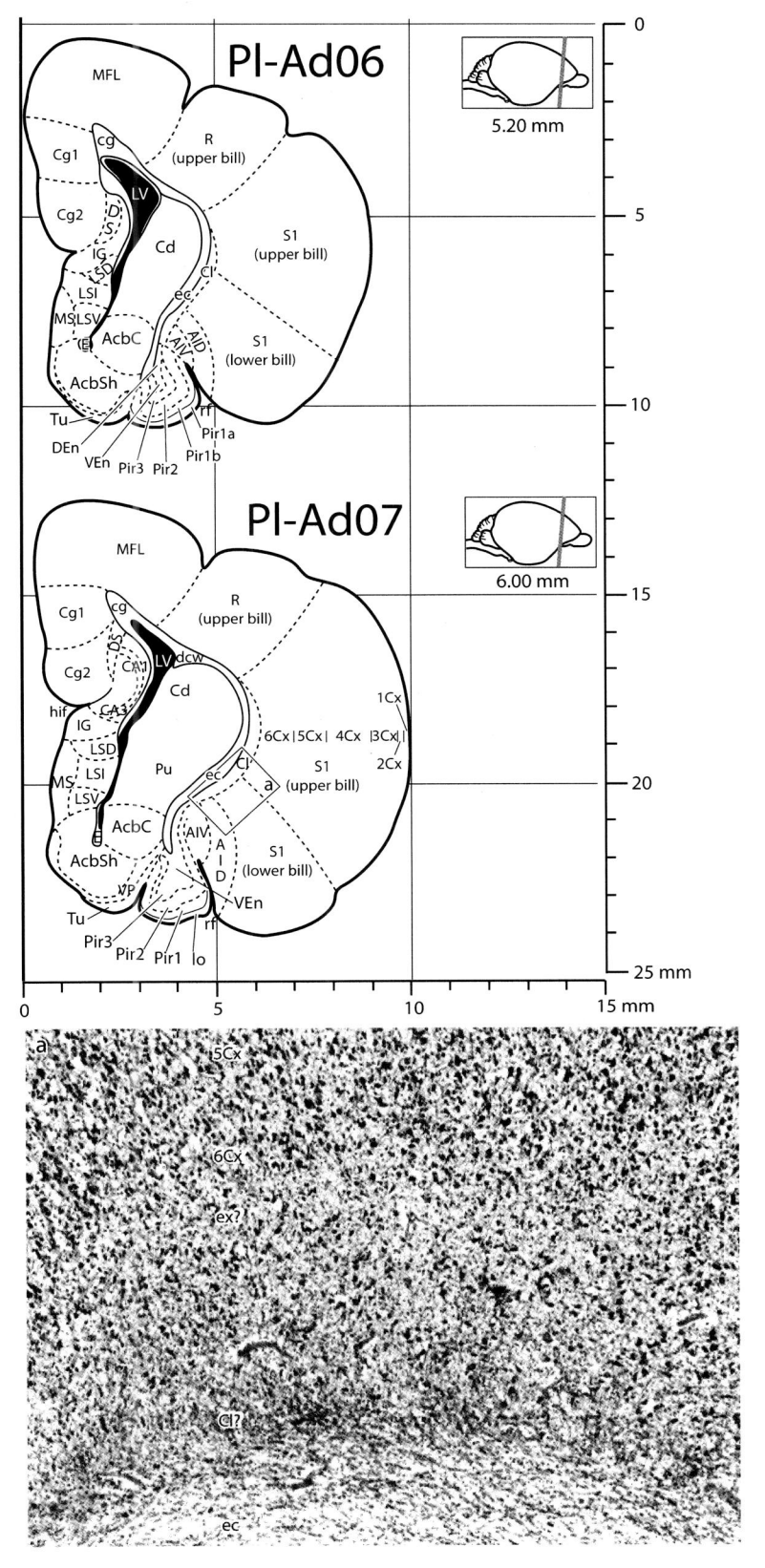

1Cx	layer 1 of cerebral cortex
2Cx	layer 2 of cerebral cortex
3Cx	layer 3 of cerebral cortex
4Cx	layer 4 of cerebral cortex
5Cx	layer 5 of cerebral cortex
6Cx	layer 6 of cerebral cortex
AcbC	accumbens nu., core
AcbSh	accumbens nu., shell
AID	agranular insular cortex, dorsal
AIV	agranular insular cortex, ventral
CA1	field CA1 of hippocampus
CA3	field CA3 of hippocampus
Cd	caudate nucleus
cg	cingulum
Cg1	cingulate cortex, area 1
Cg2	cingulate cortex, area 2
Cl	claustrum
dcw	deep cerebral white matter
DEn	dorsal endopiriform nucleus
DS	dorsal subiculum
E	ependyma and subependyma
ec	external capsule
ex	extreme capsule
hif	hippocampal fissure
IG	indusium griseum
lo	lateral olfactory tract
LSD	lateral septal nu., dorsal part
LSI	lateral septal nu., intermediate
LSV	lateral septal nu., ventral part
LV	lateral ventricle
MFL	motor cortex forelimb
MS	medial septal nucleus
Pir1	piriform cortex, layer 1
Pir1a	piriform cortex, layer 1a
Pir1b	piriform cortex, layer 1b
Pir2	piriform cortex, layer 2
Pir3	piriform cortex, layer 3
Pu	putamen
R	rostral somatosensory field
rf	rhinal fissure
S1	primary somatosensory cortex
Tu	olfactory tubercle
VEn	ventral endopiriform nucleus
VP	ventral pallidum

Pl-Ad08

aca	anterior commissure, anterior
AcbC	accumbens nucleus, core
AcbSh	accumbens nucleus, shell
AID	agranular insular cortex, dorsal
AIV	agranular insular cortex, ventral
CA1	field CA1 of hippocampus
CA3	field CA3 of hippocampus
Cd	caudate nucleus
Cg1	cingulate cortex, area 1
Cg2	cingulate cortex, area 2
Cl	claustrum
DEn	dorsal endopiriform nucleus
DG	dentate gyrus
DS	dorsal subiculum
ec	external capsule
LSD	lateral septal nu., dorsal part
LSI	lateral septal nu., intermediate part
LSV	lateral septal nu., ventral part
LV	lateral ventricle
MFL	motor cortex forelimb
MS	medial septal nucleus
Nv	navicular postolfactory nu.
Pir1a	piriform cortex, layer 1a
Pir1b	piriform cortex, layer 1b
Pir2	piriform cortex, layer 2
Pir3	piriform cortex, layer 3
Pu	putamen
R	rostral somatosensory field
rf	rhinal fissure
S1	primary somatosensory cortex
SHi	septohippocampal nucleus
Tu	olfactory tubercle
VEn	ventral endopiriform nucleus
VP	ventral pallidum

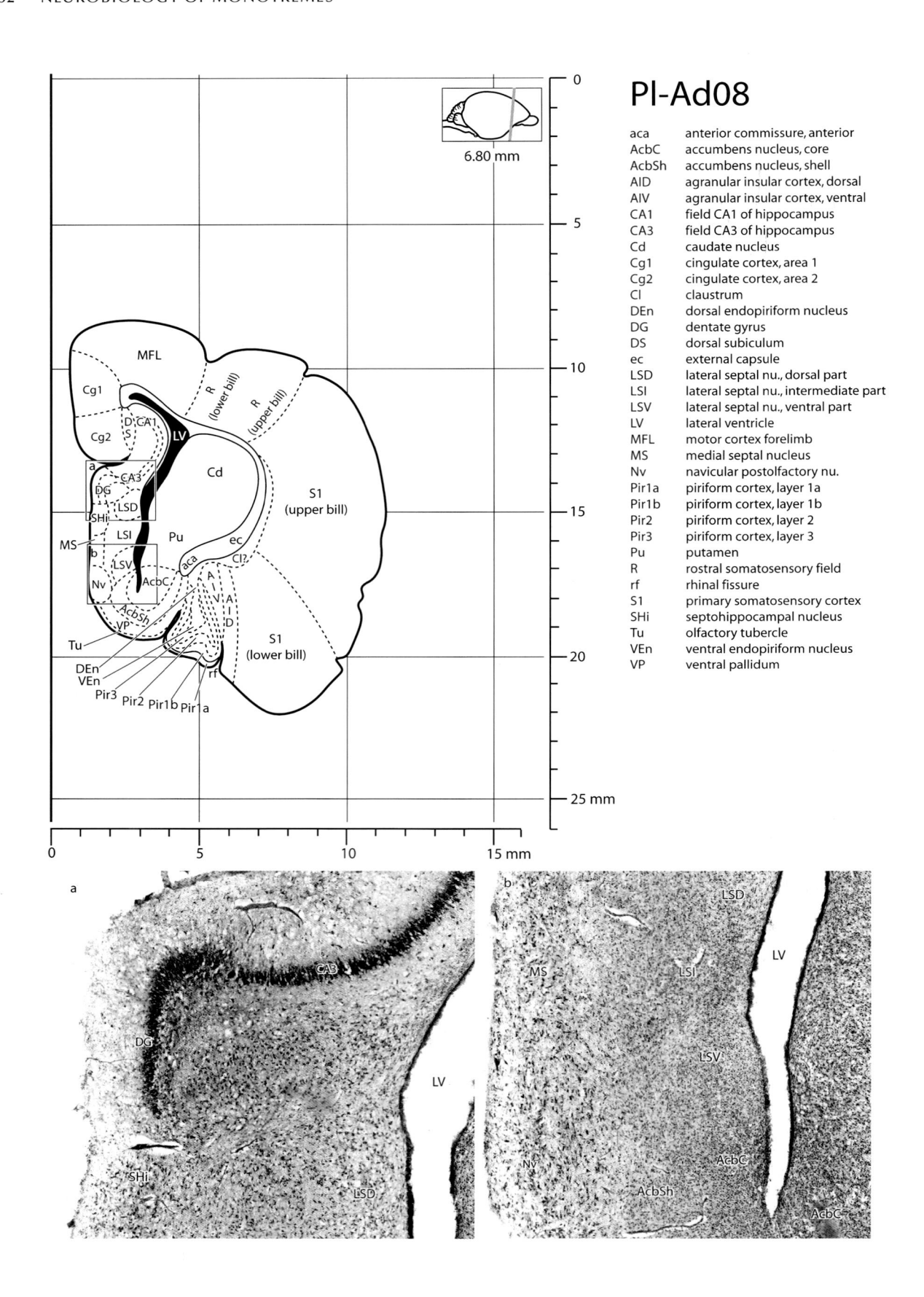

Pl-Ad09

ac	anterior commissure
aca	anterior commissure, anterior
AcbC	accumbens nucleus, core
AIP	agranular insular cortex, posterior
alv	alveus of the hippocampus
CA1	field CA1 of hippocampus
CA2	field CA2 of hippocampus
CA3	field CA3 of hippocampus
CB	cell bridges of the ventral striatum
Cd	caudate nucleus
cg	cingulum
Cg1	cingulate cortex, area 1
Cg2	cingulate cortex, area 2
Cl	claustrum
DEn	dorsal endopiriform nucleus
DG	dentate gyrus
DS	dorsal subiculum
ec	external capsule
HDB	nu. of horizontal limb of diagonal band
LPO	lateral preoptic area
LSD	lateral septal nu., dorsal part
LSI	lateral septal nu., intermediate part
LSV	lateral septal nu., ventral part
LV	lateral ventricle
MFL	motor cortex forelimb
MS	medial septal nucleus
Pir1	piriform cortex, layer 1
Pir1a	piriform cortex, layer 1a
Pir1b	piriform cortex, layer 1b
Pir2	piriform cortex, layer 2
Pir3	piriform cortex, layer 3
Pu	putamen
R	rostral somatosensory field
rf	rhinal fissure
S1	primary somatosensory cortex
SHi	septohippocampal nucleus
ST	bed nu. of stria terminalis
VDB	nu. of vertical limb of diagonal band
VEn	ventral endopiriform nucleus
VP	ventral pallidum

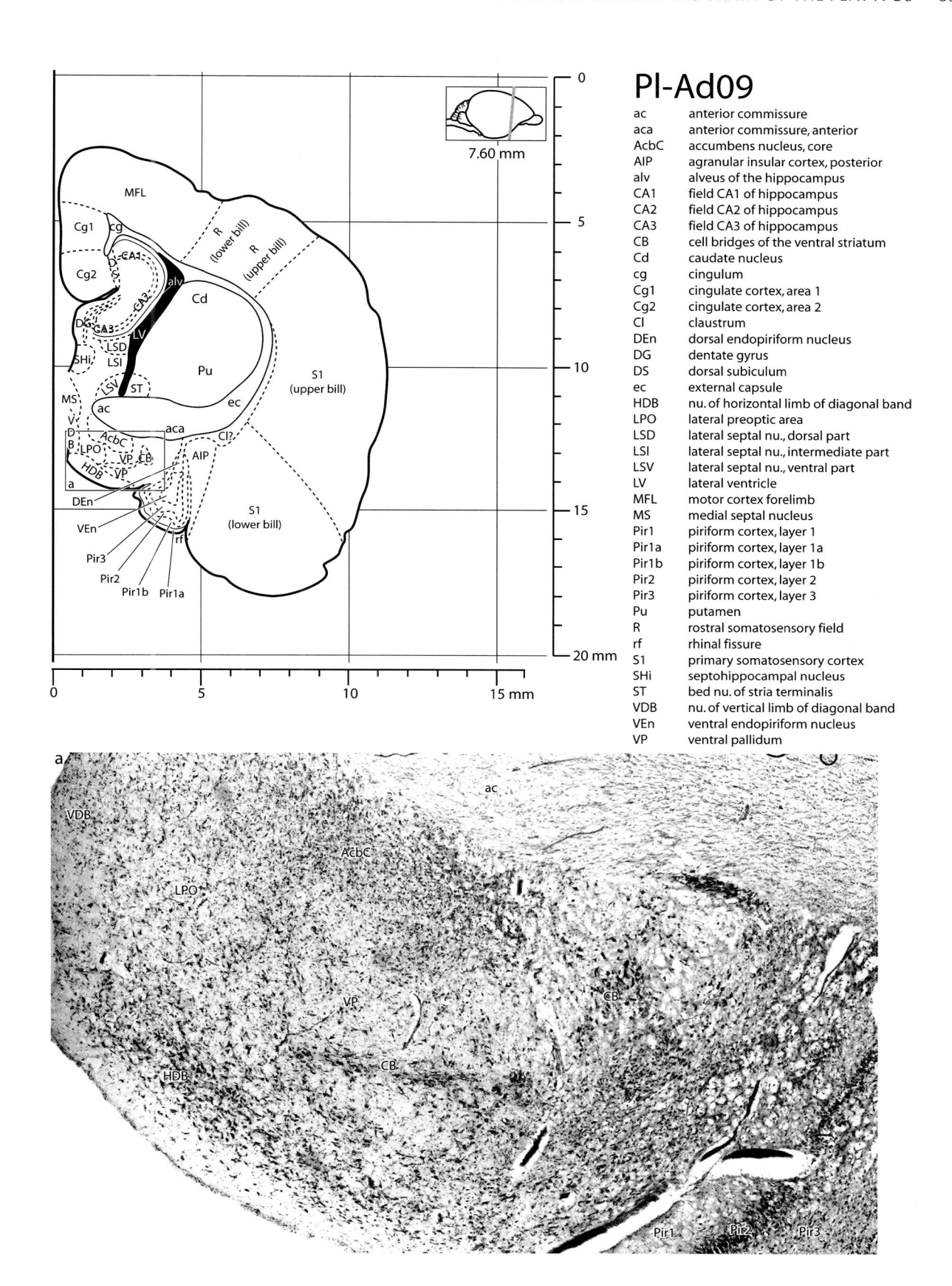

Pl-Ad10

1 to 6Cx	layers of cerebral cortex	Pu	putamen
ac	anterior commissure	R	rostral somato-sensory field
AIP	agranular insular cortex, posterior part	rf	rhinal fissure
alv	alveus of the hippocampus	S1	primary somato-sensory cortex
CA1 to 3	field CA1 to 3 of hippocampus		
CB	cell bridges of ventral striatum	SFi	septofimbrial nu.
Cd	caudate nucleus	ST	bed nucleus of stria terminalis
cg	cingulum		
Cg1, 2	cingulate cortex, areas 1, 2	StHy	striohypothalamic nu.
Cl	claustrum	VEn	ventral endopiriform nucleus
DEn	dorsal endopiriform nucleus		
DG	dentate gyrus	VP	ventral pallidum
DS	dorsal subiculum		
ec	external capsule		
fi	fimbria of the hippocampus		
GI	granular insular cortex		
GP	globus pallidus		
HDB	nu. of horizontal limb of diagonal band		
hif	hippocampal fissure		
ic	internal capsule		
IPAC	interstitial nucleus of the posterior limb ac		
LPO	lateral preoptic area		
LSD	lateral septal nu., dorsal part		
LSI	lateral septal nu., intermediate part		
LV	lateral ventricle		
M	motor cortex		
MCPO	magnocellular preoptic nucleus		
MFL	motor cortex forelimb		
MPA	medial preoptic area		
MPO	medial preoptic nucleus		
Pir1 to 3	layers of piriform cortex		

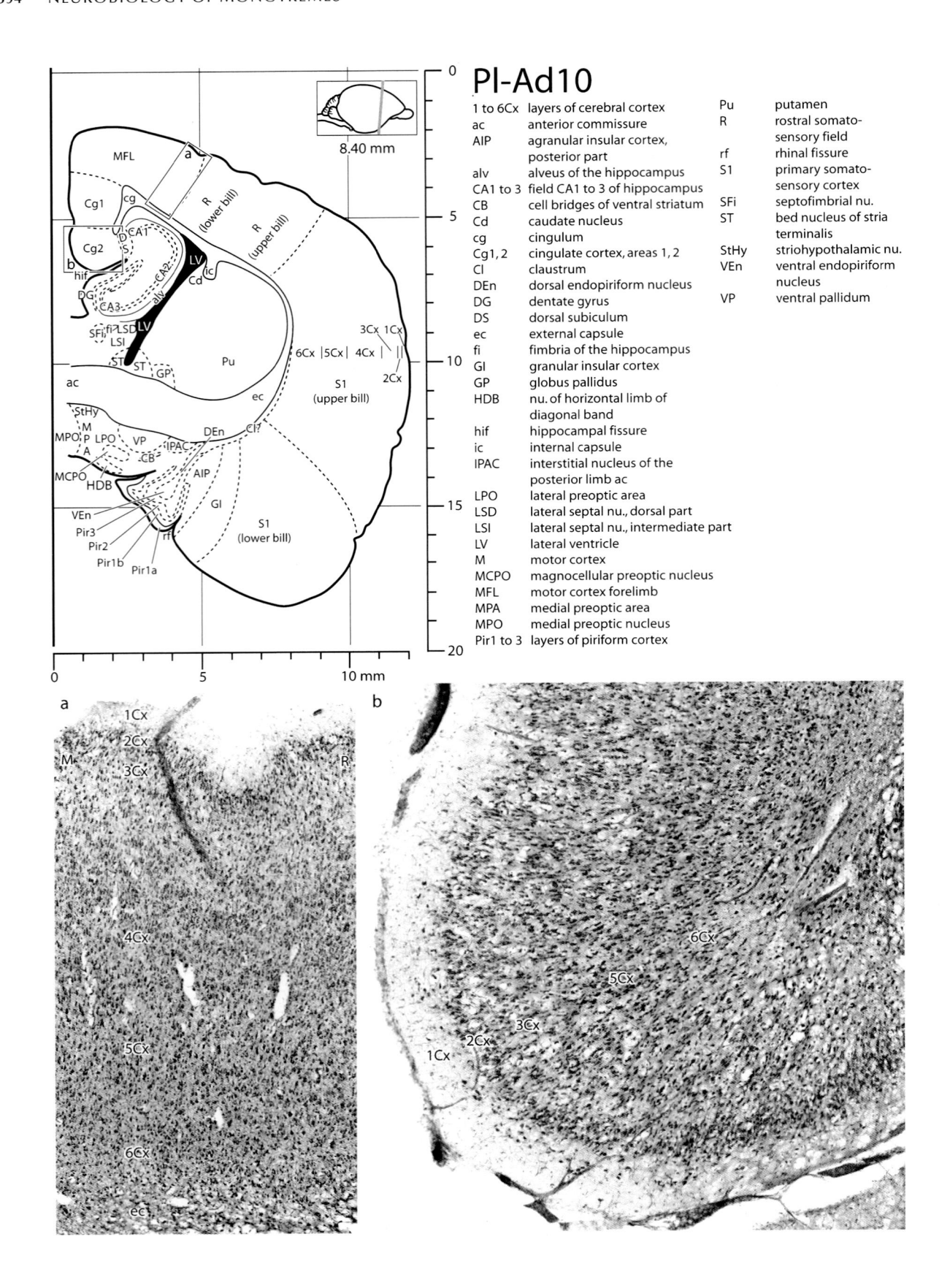

Pl-Ad11

9.20 mm

1Cx	layer 1 of cerebral cortex
2Cx	layer 2 of cerebral cortex
3Cx	layer 3 of cerebral cortex
4Cx	layer 4 of cerebral cortex
5Cx	layer 5 of cerebral cortex
6Cx	layer 6 of cerebral cortex
3V	3rd ventricle
ac	anterior commissure
acp	anterior commissure, posterior
AIP	agranular insular cortex, posterior
alv	alveus of the hippocampus
CA1	field CA1 of hippocampus
CA2	field CA2 of hippocampus
CA3	field CA3 of hippocampus
Cd	caudate nucleus
cg	cingulum
Cg1	cingulate cortex, area 1
Cg2	cingulate cortex, area 2
Cl	claustrum
DS	dorsal subiculum
EA	extended amygdala
ec	external capsule
ex	extreme capsule
f	fornix
GI	granular insular cortex
GP	globus pallidus
GrDG	granular dentate gyrus
ic	internal capsule
IPAC	interstitial nucleus of the posterior limb ac
LPO	lateral preoptic area
LSI	lateral septal nucleus, intermediate part
LV	lateral ventricle
MCPO	magnocellular preoptic nucleus
MFL	motor cortex forelimb
MoDG	molecular dentate gyrus
MPOL	medial preoptic nu., lateral
MPOM	medial preoptic nu., medial
Pir1	piriform cortex, layer 1
Pir2	piriform cortex, layer 2
Pir3	piriform cortex, layer 3
Pu	putamen
R	rostral somatosensory field
rf	rhinal fissure
S1	primary somatosensory cortex
SCh	suprachiasmatic nu.
SFi	septofimbrial nu.
SHy	septohypothalamic nu.
ST	bed nu. of the stria terminalis
st	stria terminalis
StHy	striohypothalamic nu.
VEn	ventral endopiriform nu.
VLPO	ventrolateral preoptic nu.
VP	ventral pallidum

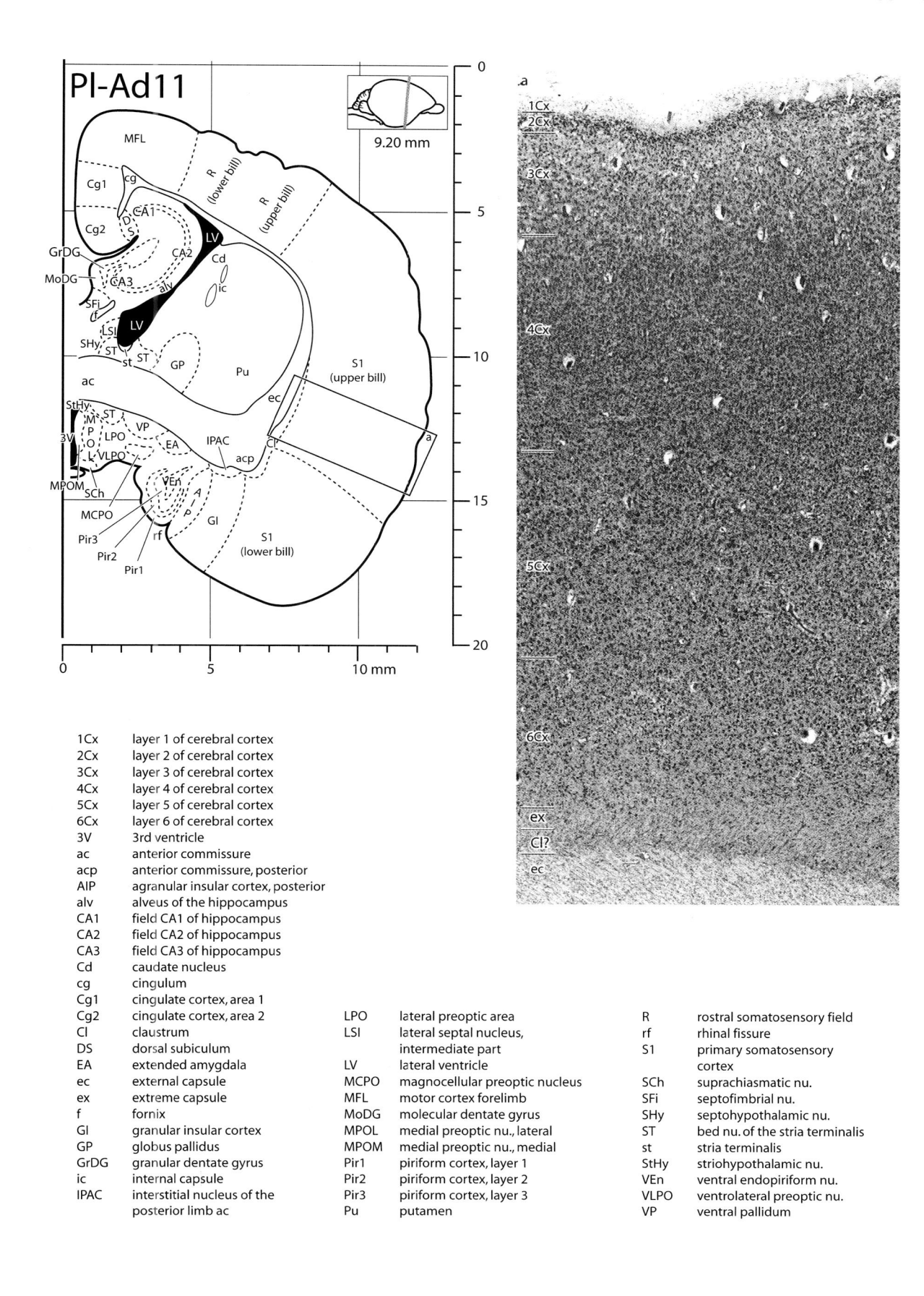

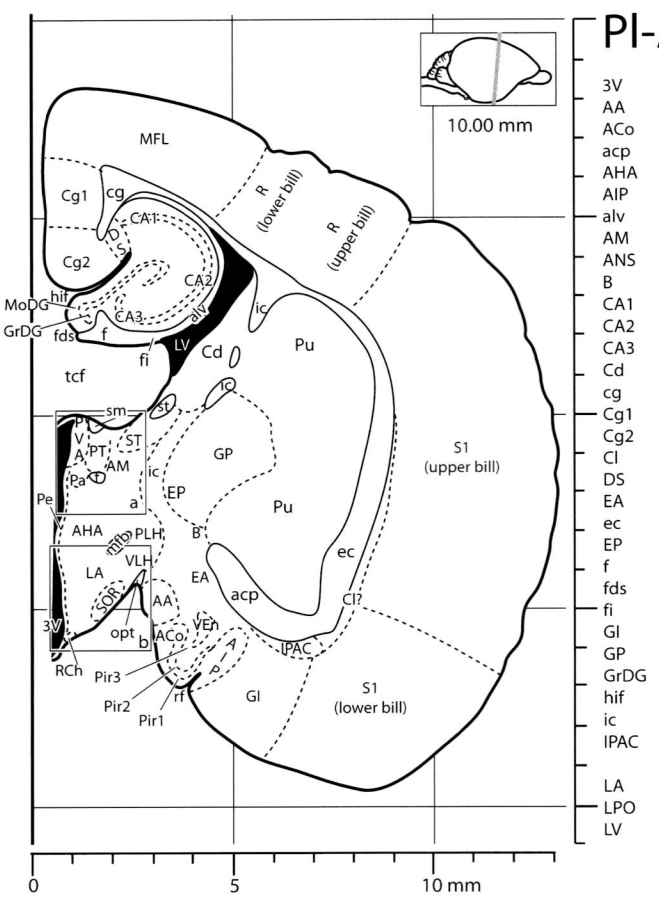

Pl-Ad12

3V	3rd ventricle	ME	median eminence
AA	anterior amygdaloid area	mfb	medial forebrain bundle
ACo	anterior cortical amyg. area	MFL	motor cortex forelimb
acp	anterior commissure, post.	MoDG	molecular dentate gyrus
AHA	anterior hypothal. area, ant.	opt	optic tract
AIP	agranular insular cx, post	Pa	paraventricular hypo. nu.
alv	alveus of the hippocampus	Pe	periventricular hypo. nu.
AM	anteromedial thalamic nu.	Pir1	piriform cortex, layer 1
ANS	access neurosecretory nu.	Pir2	piriform cortex, layer 2
B	basal nucleus (Meynert)	Pir3	piriform cortex, layer 3
CA1	field CA1 of hippocampus	PLH	peduncular part lat. hypo.
CA2	field CA2 of hippocampus	PT	paratenial thalamic nu.
CA3	field CA3 of hippocampus	Pu	putamen
Cd	caudate nucleus	PVA	paraventric. thal. nu., ant.
cg	cingulum	R	rostral somato. field
Cg1	cingulate cortex, area 1	RCh	retrochiasmatic area
Cg2	cingulate cortex, area 2	rf	rhinal fissure
Cl	claustrum	S1	primary somatosensory cortex
DS	dorsal subiculum		
EA	extended amygdala	sm	stria medullaris
ec	external capsule	SOR	supraoptic, retrochiasm.
EP	entopeduncular nucleus	st	stria terminalis
f	fornix	ST	bed nu. of stria terminalis
fds	fimbriodentate sulcus	StHy	striohypothalamic nu.
fi	fimbria of the hippocampus	STMPI	ST, posterior intermed.
GI	granular insular cortex	STMPL	ST, med. divn, posterolat.
GP	globus pallidus	STMPM	ST, med. divn, posteromed.
GrDG	granular dentate gyrus	tcf	transverse cerebral fissure
hif	hippocampal fissure	VEn	ventral endopiriform nu.
ic	internal capsule	VLH	ventrolateral hypothalamic nucleus
IPAC	interstitial nucleus of posterior limb of anterior commissure		
LA	lateroanterior hypothalamic nucleus		
LPO	lateral preoptic area		
LV	lateral ventricle		

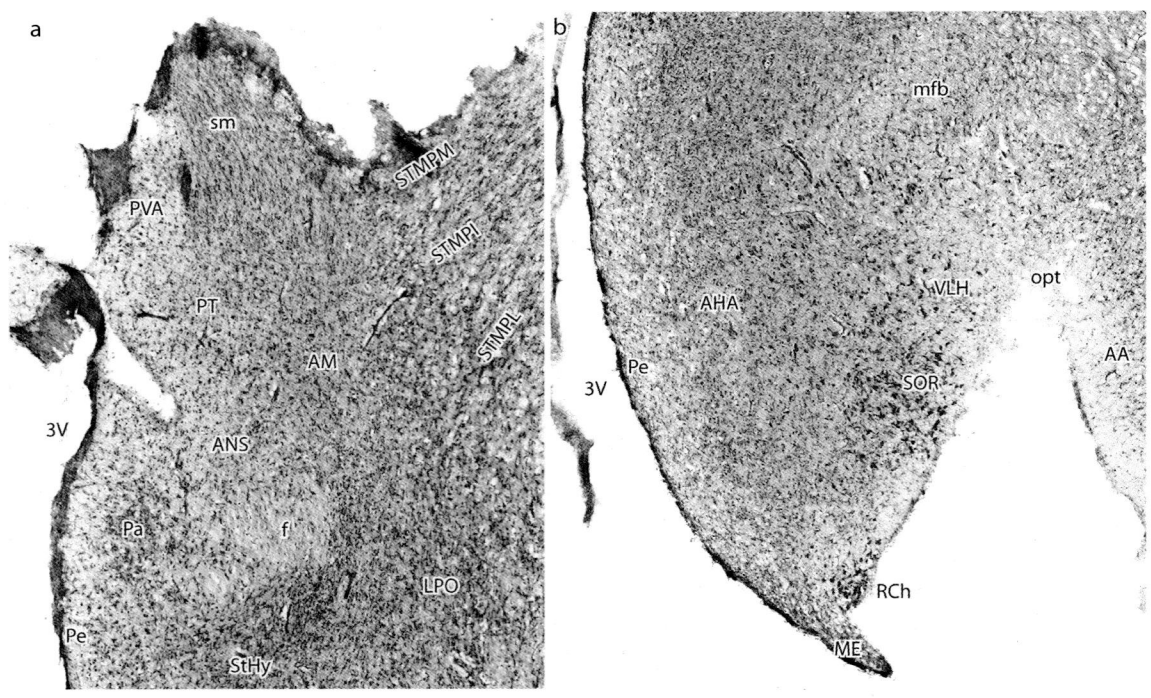

Pl-Ad13

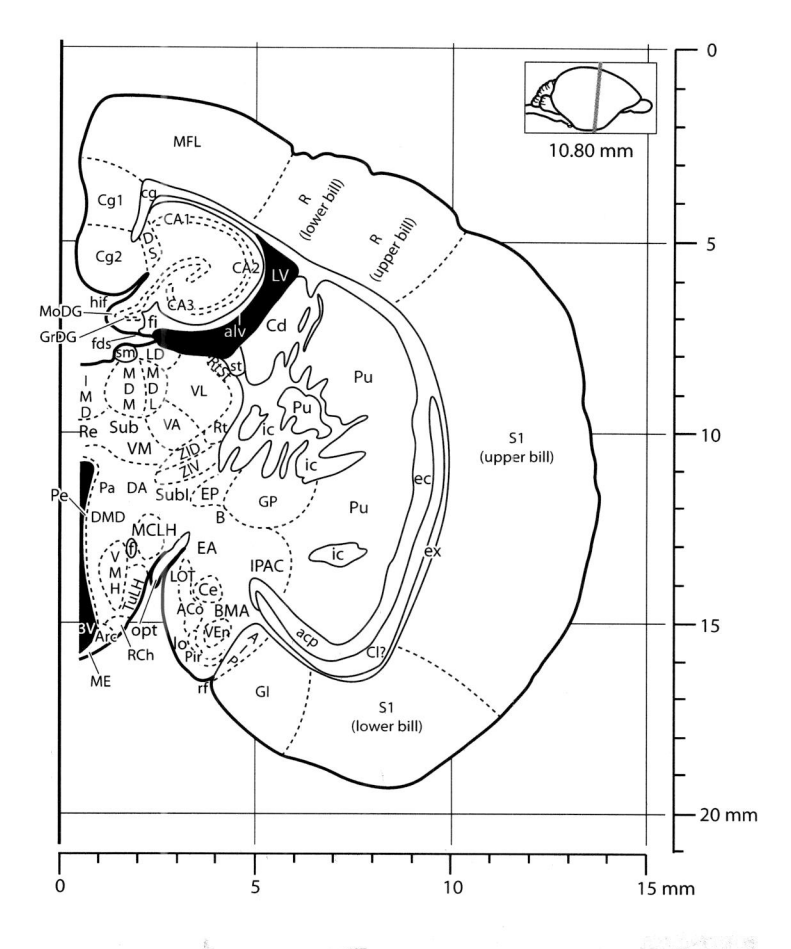

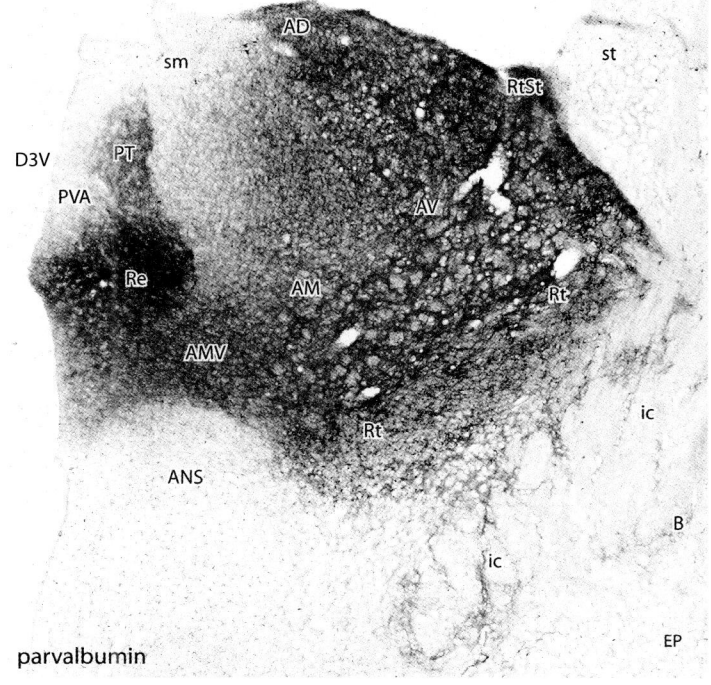

3V	3rd ventricle
ACo	anterior cortical amygdaloid area
acp	anterior commissure, posterior
AD	anterodorsal thalamic nu.
AIP	agranular insular cortex, posterior
alv	alveus of the hippocampus
AM	anteromedial thalamic nu.
AMV	anteromedial thalamic nu., ventral
ANS	access. neurosecretory nu.
Arc	arcuate hypothalamic nu.
AV	anteroventral thalamic nu.
B	basal nucleus (Meynert)
BMA	basomedial amygdaloid nu., anterior
CA1 to 3	field CA1 to 3 of hippocampus
Cd	caudate nucleus
Ce	central amygdaloid nu.
cg	cingulum
Cg1, 2	cingulate cortex, areas 1, 2
Cl	claustrum
D3V	dorsal 3rd ventricle
DA	dorsal hypothalamic area
DMD	dorsomedial hypothalamic nu., dorsal
DS	dorsal subiculum
EA	extended amygdala
ec	external capsule
EP	entopeduncular nu.
ex	extreme capsule
f	fornix
fds	fimbriodentate sulcus
fi	fimbria of the hippocampus
GI	granular insular cortex
GP	globus pallidus
GrDG	granular dentate gyrus
hif	hippocampal fissure
ic	internal capsule
IMD	intermediodorsal thalamic nu.
IPAC	interstitial nucleus of post. limb ac
LD	laterodorsal thalamic nu.
lo	lateral olfactory tract
LOT	nu. of lateral olfactory tract
LV	lateral ventricle
MCLH	magnocell. nu. lat. hypothal.
MDL	mediodorsal thalamic nu., lat.
MDM	mediodorsal thalamic nu., med.
ME	median eminence
MFL	motor cortex forelimb
MoDG	molecular dentate gyrus
opt	optic tract
Pa	paraventricular hypothalamic nu.
Pe	periventricular hypothalamic nu.
Pir	piriform cortex
PT	paratenial thalamic nu.
Pu	putamen
PVA	paraventricular thalamic nu., anterior
R	rostral somatosensory field
RCh	retrochiasmatic area
Re	reuniens thalamic nu.
rf	rhinal fissure
Rt	reticular nucleus (prethalamus)
RtSt	reticulostrial nu.
S1	primary somatosensory cortex
sm	stria medullaris
st	stria terminalis
Sub	submedius thalamic nu.
SubI	subincertal nu.
TuLH	tuberal region of lateral hypothalamus
VA	ventral anterior thalamic nu.
VEn	ventral endopiriform nu.
VL	ventrolateral thalamic nu.
VM	ventromedial thalamic nu.
VMH	ventromedial hypothalamic nu.
ZID	zona incerta, dorsal part
ZIV	zona incerta, ventral part

Pl-Ad14

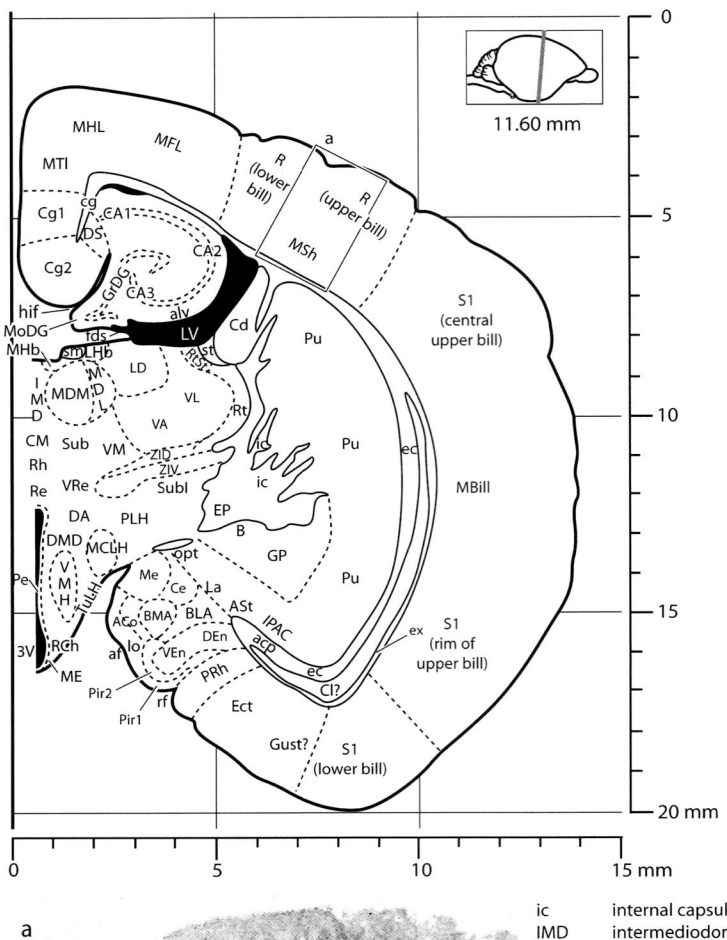

1 to 6Cx	layers 1 to 6 of cerebral cortex
3V	3rd ventricle
ACo	anterior cortical amygdaloid area
acp	anterior commissure, posterior limb
af	amygdaloid fissure
alv	alveus of the hippocampus
ASt	amygdalostriatal transition area
B	basal nucleus (Meynert)
BLA	basolateral amygdaloid nucleus, anterior part
BMA	basomedial amygdaloid nucleus, anterior part
CA1	field CA1 of hippocampus
CA2	field CA2 of hippocampus
CA3	field CA3 of hippocampus
Cd	caudate nucleus
Ce	central amygdaloid nucleus
cg	cingulum
Cg1	cingulate cortex, area 1
Cg2	cingulate cortex, area 2
Cl	claustrum
CM	central medial thalamic nucleus
DA	dorsal hypothalamic area
DEn	dorsal endopiriform nucleus
DMD	dorsomedial hypothalamic nucleus, dorsal part
DS	dorsal subiculum
ec	external capsule
Ect	ectorhinal cortex
EP	entopeduncular nucleus
ex	extreme capsule
fds	fimbriodentate sulcus
GP	globus pallidus
GrDG	granular layer of dentate gyrus
Gust	gustatory cortex
hif	hippocampal fissure
ic	internal capsule
IMD	intermediodorsal thal. nu.
IPAC	interstitial nucleus of posterior limb of ac
La	lateral amygdaloid nu.
LD	laterodorsal thalamic nu.
LHb	lateral habenular nu.
lo	lateral olfactory tract
LV	lateral ventricle
MBill	motor cortex bill
MCLH	magnocell. nu. lat. hypothal.
MDL	mediodorsal thal. nu., lat.
MDM	mediodorsal thal. nu., med.
Me	medial amygdaloid nu.
ME	median eminence
MFL	motor cortex forelimb
MHb	medial habenular nu.
MHL	motor cortex hindlimb
MoDG	molecular layer of dentate gyrus
MSh	motor cortex shoulder
MTl	motor cortex tail
opt	optic tract
Pe	periventricular hypothalamic nu.
Pir1	piriform cortex, layer 1
Pir2	piriform cortex, layer 2
PLH	peduncular part of lateral hypothalamus
PRh	perirhinal cortex
Pu	putamen
R	rostral somatosensory field
RCh	retrochiasmatic area of hypothalamus
Re	reuniens thalamic nu.
rf	rhinal fissure
Rh	rhomboid thalamic nu.
Rt	reticular nucleus (prethalamus)
RtSt	reticulostrial nucleus
S1	primary somatosensory cortex
sm	stria medullaris
st	stria terminalis
Sub	submedius thalamic nu.
SubI	subincertal nu.
TuLH	tuberal region of lateral hypothalamus
VA	ventral anterior thalamic nu.
VEn	ventral endopiriform nu.
VL	ventrolateral thalamic nu.
VM	ventromedial thalamic nu.
VMH	ventromedial hypothal. nu.
VRe	ventral reuniens thalamic nu.
ZID	zona incerta, dorsal part
ZIV	zona incerta, ventral part

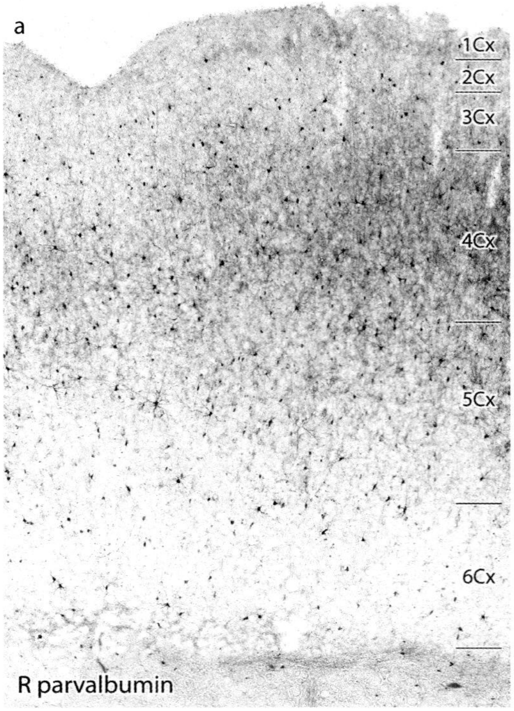

R parvalbumin

Pl-Ad15

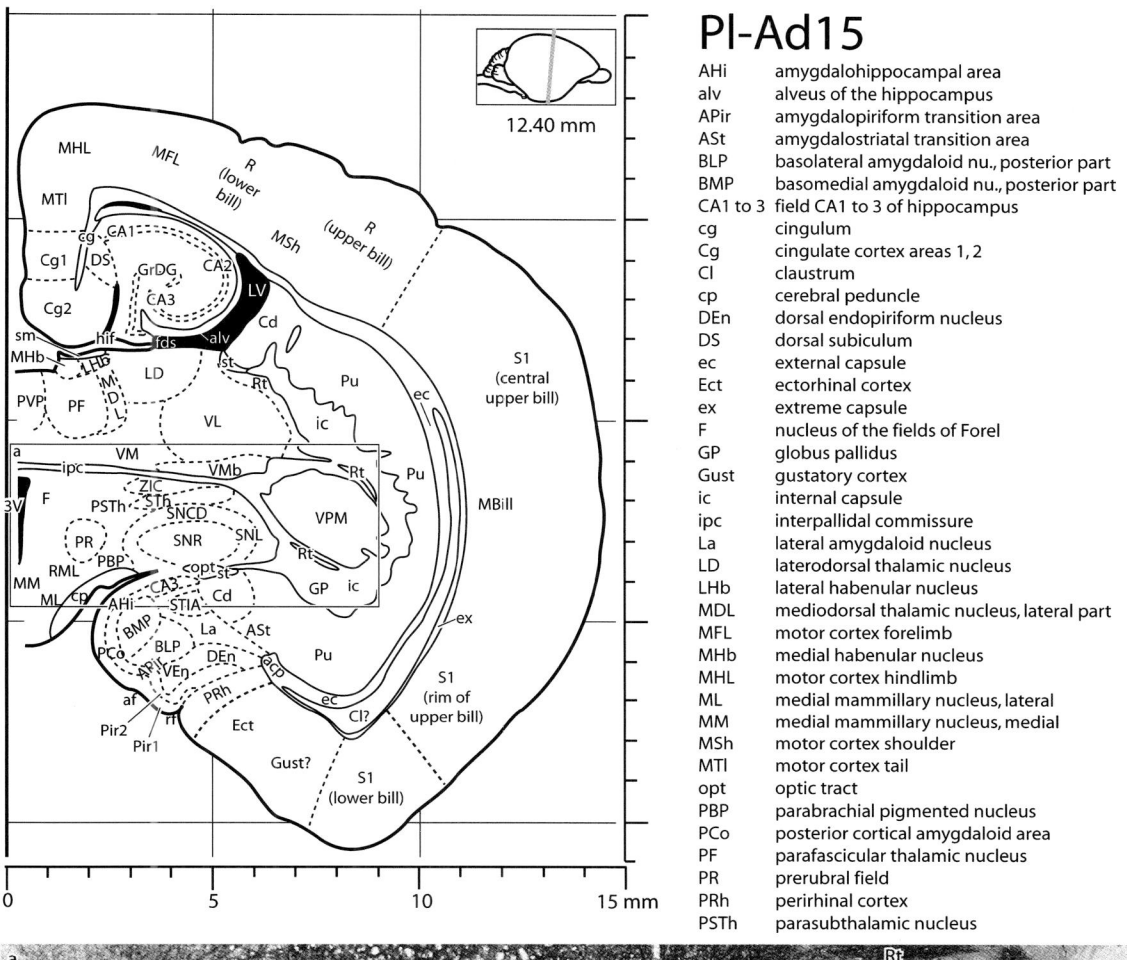

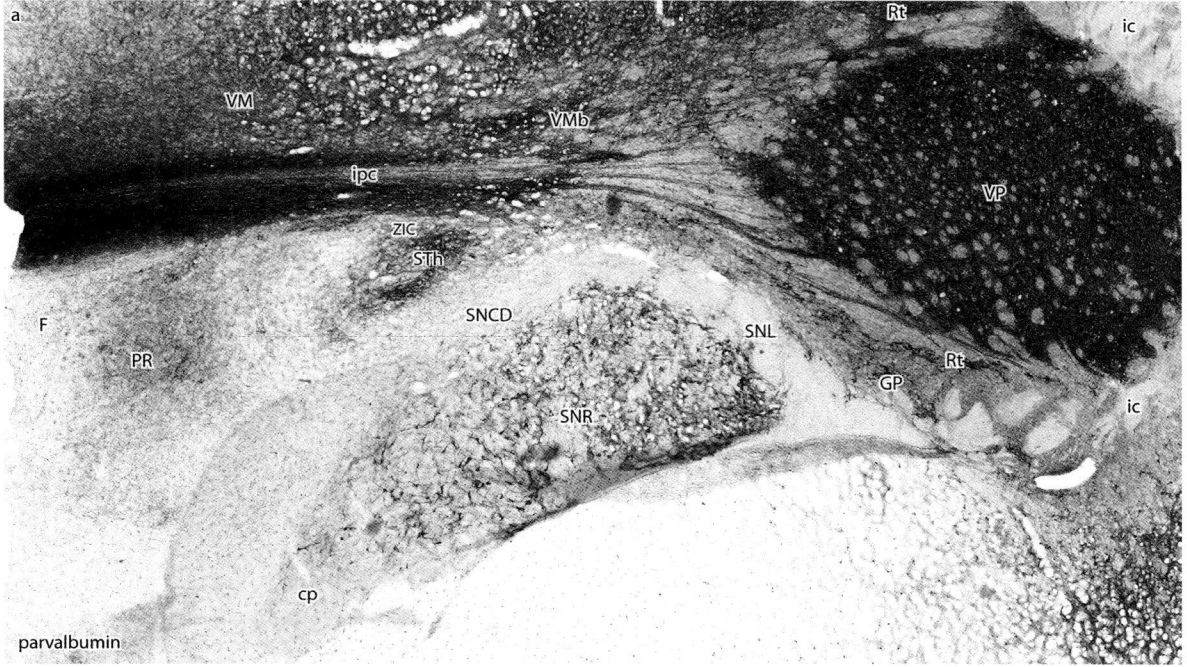

AHi	amygdalohippocampal area
alv	alveus of the hippocampus
APir	amygdalopiriform transition area
ASt	amygdalostriatal transition area
BLP	basolateral amygdaloid nu., posterior part
BMP	basomedial amygdaloid nu., posterior part
CA1 to 3	field CA1 to 3 of hippocampus
cg	cingulum
Cg	cingulate cortex areas 1, 2
Cl	claustrum
cp	cerebral peduncle
DEn	dorsal endopiriform nucleus
DS	dorsal subiculum
ec	external capsule
Ect	ectorhinal cortex
ex	extreme capsule
F	nucleus of the fields of Forel
GP	globus pallidus
Gust	gustatory cortex
ic	internal capsule
ipc	interpallidal commissure
La	lateral amygdaloid nucleus
LD	laterodorsal thalamic nucleus
LHb	lateral habenular nucleus
MDL	mediodorsal thalamic nucleus, lateral part
MFL	motor cortex forelimb
MHb	medial habenular nucleus
MHL	motor cortex hindlimb
ML	medial mammillary nucleus, lateral
MM	medial mammillary nucleus, medial
MSh	motor cortex shoulder
MTl	motor cortex tail
opt	optic tract
PBP	parabrachial pigmented nucleus
PCo	posterior cortical amygdaloid area
PF	parafascicular thalamic nucleus
PR	prerubral field
PRh	perirhinal cortex
PSTh	parasubthalamic nucleus

12.40 mm

parvalbumin

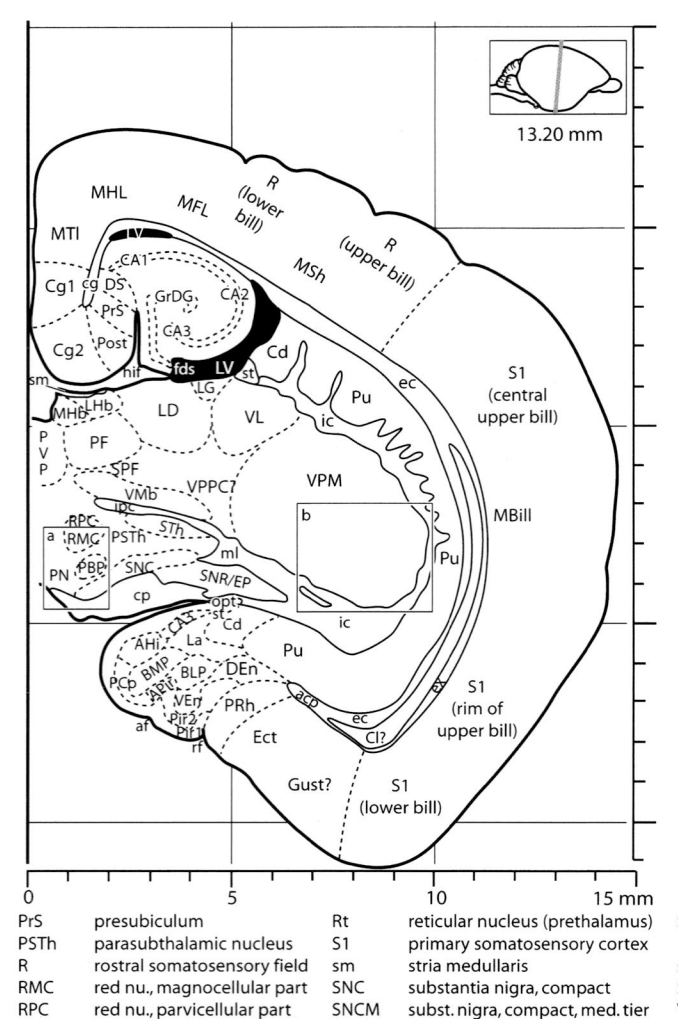

Pl-Ad16

13.20 mm

acp	anterior commissure, posterior limb
af	amygdaloid fissure
AHi	amygdalohippocampal area
APir	amygdalopiriform transition area
BLP	basolateral amygdaloid nu., post.
BMP	basomedial amygdaloid nu., post.
cg	cingulum
Cl	claustrum
cp	cerebral peduncle
DEn	dorsal endopiriform nucleus
DS	dorsal subiculum
ec	external capsule
Ect	ectorhinal cortex
ex	extreme capsule
fds	fimbriodentate sulcus
GrDG	granular dentate gyrus
Gust	gustatory cortex
hif	hippocampal fissure
ic	internal capsule
ipc	interpallidal commissure
La	lateral amygdaloid nucleus
LD	laterodorsal thalamic nucleus
LG	lateral geniculate nucleus
LHb	lateral habenular nucleus
LV	lateral ventricle
MBill	motor cortex bill
MFL	motor cortex forelimb
MHb	medial habenular nucleus
MHL	motor cortex hindlimb
ml	medial lemniscus
MSh	motor cortex shoulder
MTl	motor cortex tail
PBP	parabrachial pigmented nucleus
PCo	posterior cortical amygdaloid area
PF	parafascicular thalamic nucleus
PN	paranigral nucleus of the VTA
PRh	perirhinal cortex

PrS	presubiculum	Rt	reticular nucleus (prethalamus)	SPF	subparafascicular thalamic nucleus	VL	ventrolateral thal. nu.
PSTh	parasubthalamic nucleus	S1	primary somatosensory cortex			VMb	ventromedial basal nu.
R	rostral somatosensory field	sm	stria medullaris	st	stria terminalis	VPM	VPM thalamic nucleus
RMC	red nu., magnocellular part	SNC	substantia nigra, compact	STh	subthalamic nucleus	VPPC	ventral post. nu. thal., parvicellular part
RPC	red nu., parvicellular part	SNCM	subst. nigra, compact, med. tier	VEn	ventral endopiriform nu.		

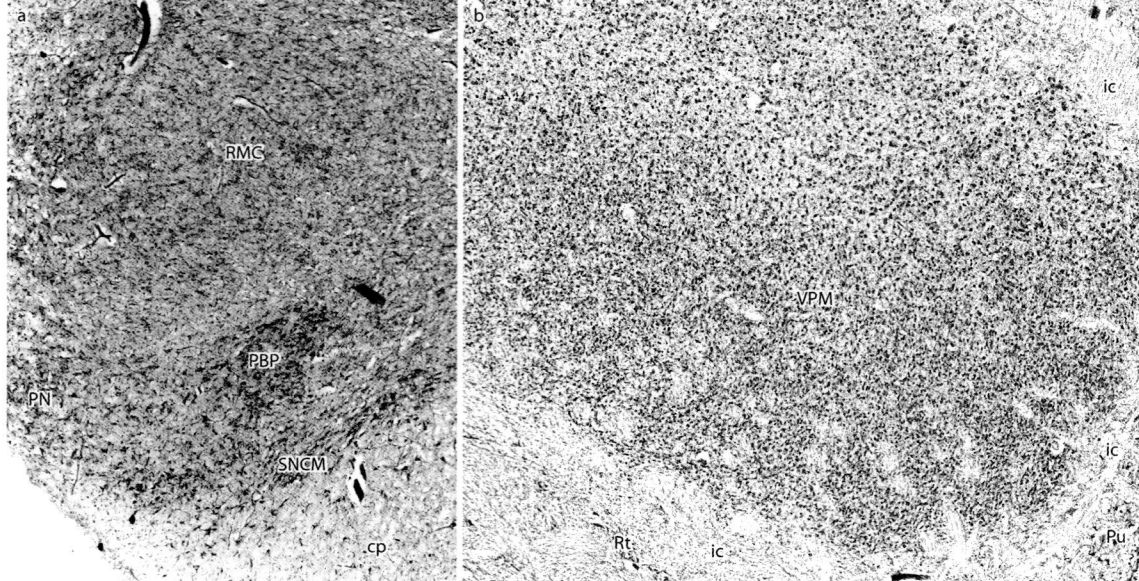

Pl-Ad17

APT	anterior pretectal nucleus
CA	field CA of hippocampus
Cg1,2	cingulate cortex, areas 1, 2
Cl	claustrum
ec	external capsule
Ect	ectorhinal cortex
fds	fimbriodentate sulcus
GrDG	granular dentate gyrus
Gust	gustatory cortex
hif	hippocampal fissure
ic	internal capsule
LEnt	lateral entorhinal cortex
LG	lateral geniculate nucleus
LP	lateral posterior thalamic nucleus
MBill	motor cortex bill
MEnt	medial entorhinal cortex
MFL	motor cortex forelimb
ml	medial lemniscus
MSh	motor cortex shoulder
OT	nucleus of the optic tract
p1PAG	p1 periaqueductal grey
PaR	pararubral nu.
PaS	parasubiculum
PBP	parabrachial pigmented nucleus
PF	parafascicular thalamic nucleus
Po	post. thalamic nuclear group
Pr5pc	principal sensory trigeminal nu., parvicellular
PrC	precommissural nucleus
PRh	perirhinal cortex
R	rostral somatosensory field
RMC	red nucleus, magnocellular part
RPC	red nucleus, parvicellular part
Rt	reticular nucleus (prethalamus)
s5	sensory root of trigeminal nerve
SNCD	substantia nigra, compact part, dorsal tier
SNL	substantia nigra, lateral
SNR	substantia nigra, reticular
SNR/EP	subst. nigra reticular/entopeduncular nu.
STh	subthalamic nucleus
VMb	ventromedial basal nucleus
VPM	ventral posteromed thalamic nucleus

Dorsal thalamus parvalbumin

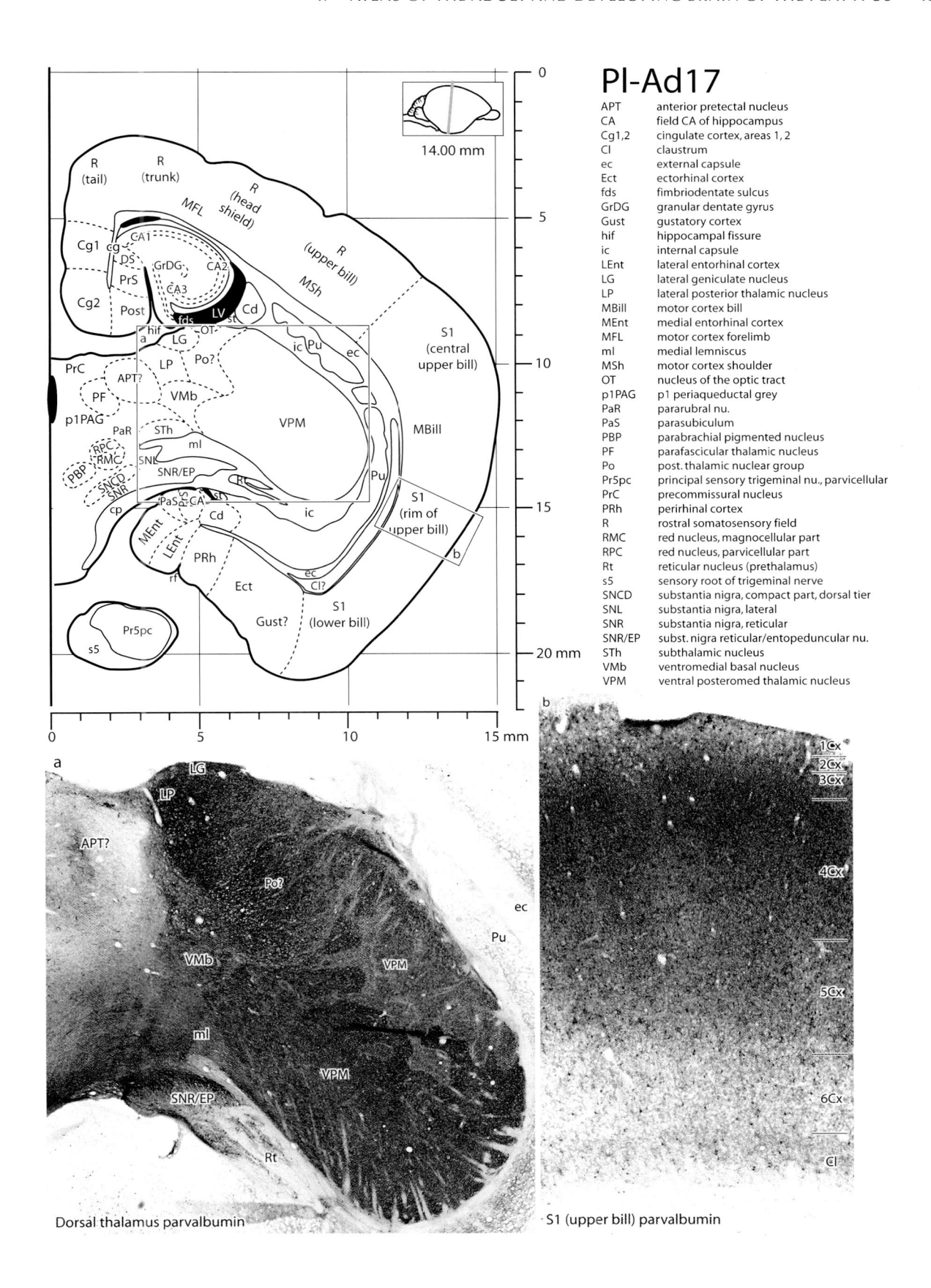

S1 (upper bill) parvalbumin

Pl-Ad18

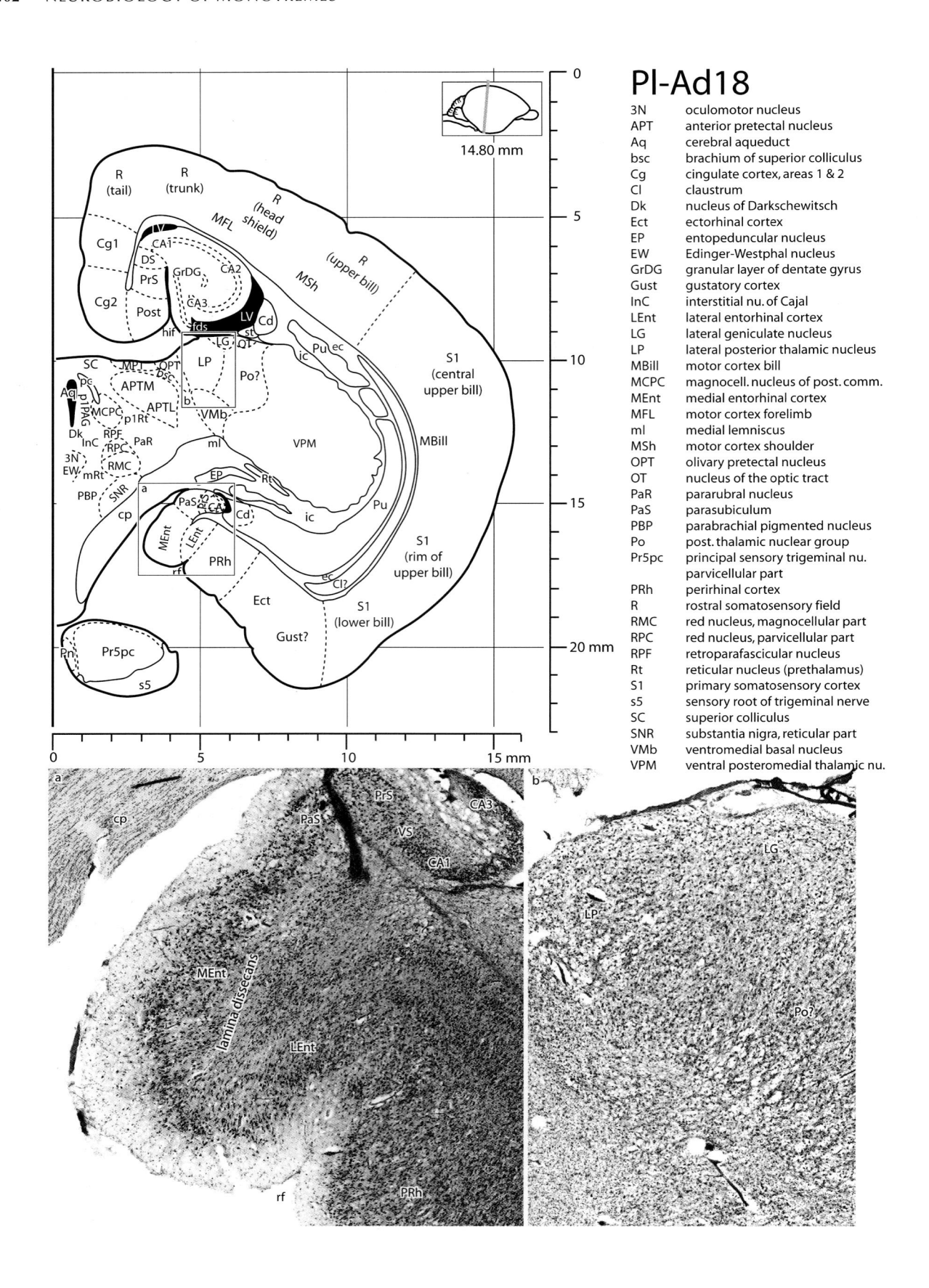

3N	oculomotor nucleus
APT	anterior pretectal nucleus
Aq	cerebral aqueduct
bsc	brachium of superior colliculus
Cg	cingulate cortex, areas 1 & 2
Cl	claustrum
Dk	nucleus of Darkschewitsch
Ect	ectorhinal cortex
EP	entopeduncular nucleus
EW	Edinger-Westphal nucleus
GrDG	granular layer of dentate gyrus
Gust	gustatory cortex
InC	interstitial nu. of Cajal
LEnt	lateral entorhinal cortex
LG	lateral geniculate nucleus
LP	lateral posterior thalamic nucleus
MBill	motor cortex bill
MCPC	magnocell. nucleus of post. comm.
MEnt	medial entorhinal cortex
MFL	motor cortex forelimb
ml	medial lemniscus
MSh	motor cortex shoulder
OPT	olivary pretectal nucleus
OT	nucleus of the optic tract
PaR	pararubral nucleus
PaS	parasubiculum
PBP	parabrachial pigmented nucleus
Po	post. thalamic nuclear group
Pr5pc	principal sensory trigeminal nu. parvicellular part
PRh	perirhinal cortex
R	rostral somatosensory field
RMC	red nucleus, magnocellular part
RPC	red nucleus, parvicellular part
RPF	retroparafascicular nucleus
Rt	reticular nucleus (prethalamus)
S1	primary somatosensory cortex
s5	sensory root of trigeminal nerve
SC	superior colliculus
SNR	substantia nigra, reticular part
VMb	ventromedial basal nucleus
VPM	ventral posteromedial thalamic nu.

Pl-Ad19

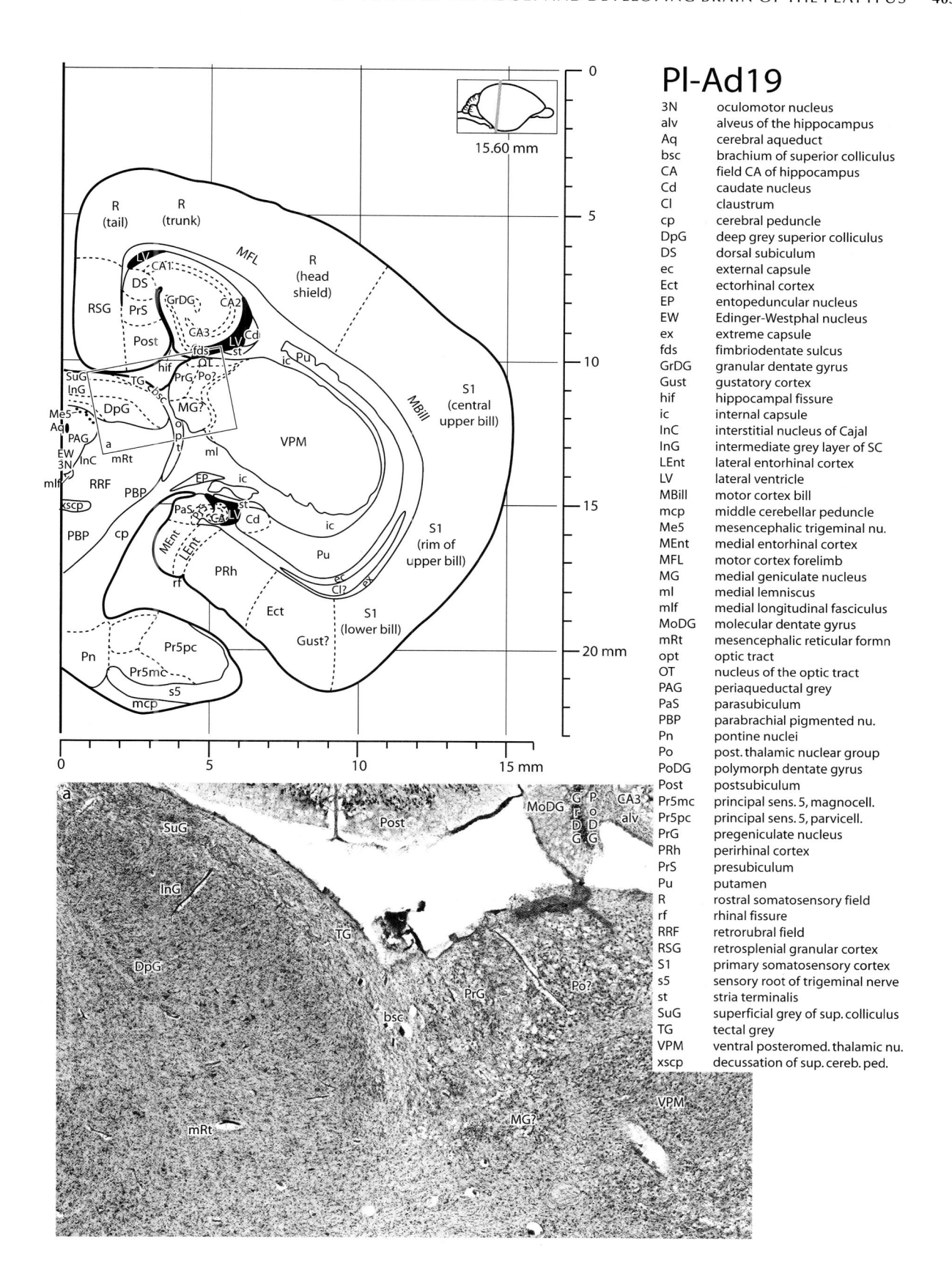

3N	oculomotor nucleus
alv	alveus of the hippocampus
Aq	cerebral aqueduct
bsc	brachium of superior colliculus
CA	field CA of hippocampus
Cd	caudate nucleus
Cl	claustrum
cp	cerebral peduncle
DpG	deep grey superior colliculus
DS	dorsal subiculum
ec	external capsule
Ect	ectorhinal cortex
EP	entopeduncular nucleus
EW	Edinger-Westphal nucleus
ex	extreme capsule
fds	fimbriodentate sulcus
GrDG	granular dentate gyrus
Gust	gustatory cortex
hif	hippocampal fissure
ic	internal capsule
InC	interstitial nucleus of Cajal
InG	intermediate grey layer of SC
LEnt	lateral entorhinal cortex
LV	lateral ventricle
MBill	motor cortex bill
mcp	middle cerebellar peduncle
Me5	mesencephalic trigeminal nu.
MEnt	medial entorhinal cortex
MFL	motor cortex forelimb
MG	medial geniculate nucleus
ml	medial lemniscus
mlf	medial longitudinal fasciculus
MoDG	molecular dentate gyrus
mRt	mesencephalic reticular formn
opt	optic tract
OT	nucleus of the optic tract
PAG	periaqueductal grey
PaS	parasubiculum
PBP	parabrachial pigmented nu.
Pn	pontine nuclei
Po	post. thalamic nuclear group
PoDG	polymorph dentate gyrus
Post	postsubiculum
Pr5mc	principal sens. 5, magnocell.
Pr5pc	principal sens. 5, parvicell.
PrG	pregeniculate nucleus
PRh	perirhinal cortex
PrS	presubiculum
Pu	putamen
R	rostral somatosensory field
rf	rhinal fissure
RRF	retrorubral field
RSG	retrosplenial granular cortex
S1	primary somatosensory cortex
s5	sensory root of trigeminal nerve
st	stria terminalis
SuG	superficial grey of sup. colliculus
TG	tectal grey
VPM	ventral posteromed. thalamic nu.
xscp	decussation of sup. cereb. ped.

Pl-Ad20

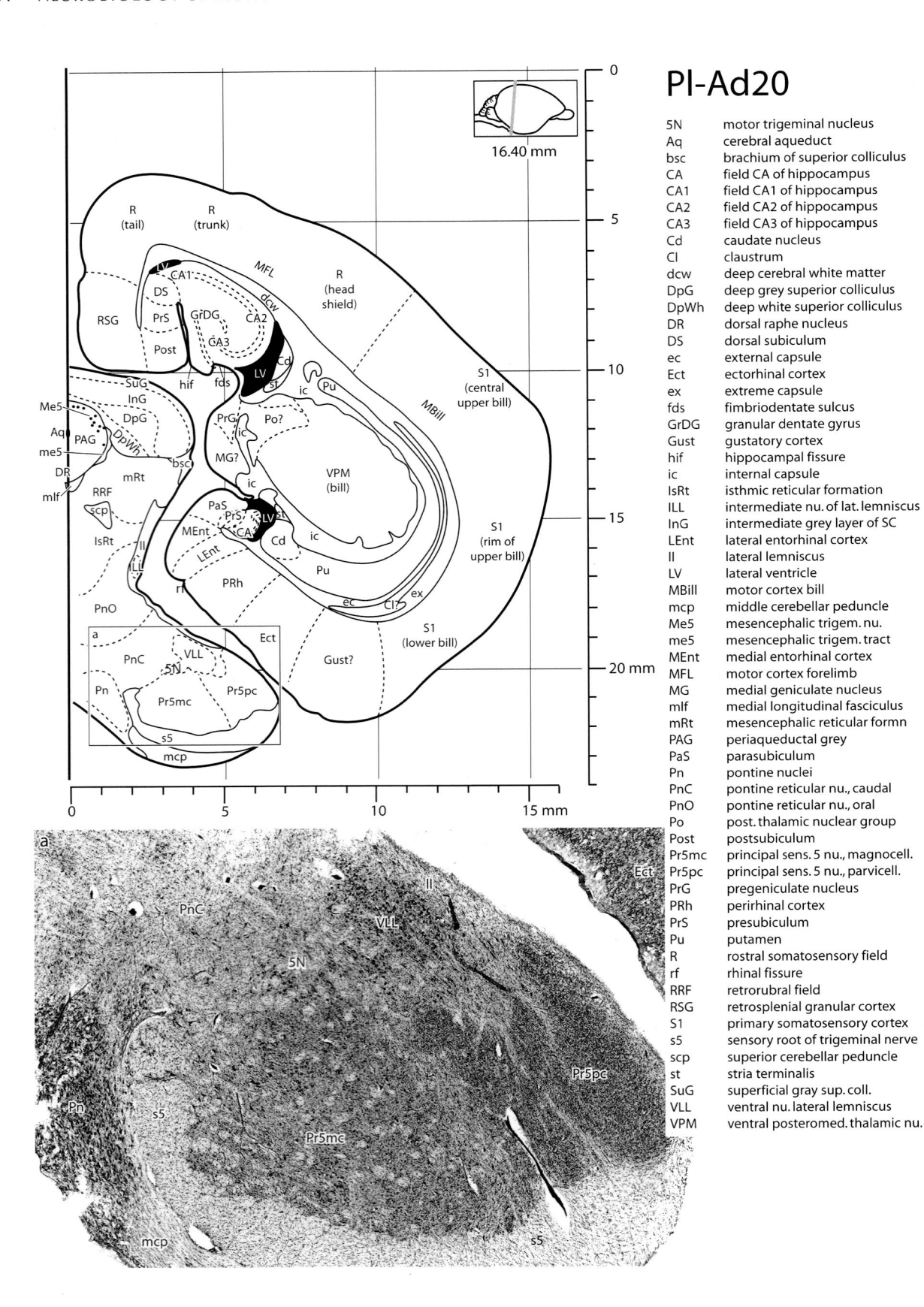

5N	motor trigeminal nucleus
Aq	cerebral aqueduct
bsc	brachium of superior colliculus
CA	field CA of hippocampus
CA1	field CA1 of hippocampus
CA2	field CA2 of hippocampus
CA3	field CA3 of hippocampus
Cd	caudate nucleus
Cl	claustrum
dcw	deep cerebral white matter
DpG	deep grey superior colliculus
DpWh	deep white superior colliculus
DR	dorsal raphe nucleus
DS	dorsal subiculum
ec	external capsule
Ect	ectorhinal cortex
ex	extreme capsule
fds	fimbriodentate sulcus
GrDG	granular dentate gyrus
Gust	gustatory cortex
hif	hippocampal fissure
ic	internal capsule
IsRt	isthmic reticular formation
ILL	intermediate nu. of lat. lemniscus
InG	intermediate grey layer of SC
LEnt	lateral entorhinal cortex
ll	lateral lemniscus
LV	lateral ventricle
MBill	motor cortex bill
mcp	middle cerebellar peduncle
Me5	mesencephalic trigem. nu.
me5	mesencephalic trigem. tract
MEnt	medial entorhinal cortex
MFL	motor cortex forelimb
MG	medial geniculate nucleus
mlf	medial longitudinal fasciculus
mRt	mesencephalic reticular formn
PAG	periaqueductal grey
PaS	parasubiculum
Pn	pontine nuclei
PnC	pontine reticular nu., caudal
PnO	pontine reticular nu., oral
Po	post. thalamic nuclear group
Post	postsubiculum
Pr5mc	principal sens. 5 nu., magnocell.
Pr5pc	principal sens. 5 nu., parvicell.
PrG	pregeniculate nucleus
PRh	perirhinal cortex
PrS	presubiculum
Pu	putamen
R	rostral somatosensory field
rf	rhinal fissure
RRF	retrorubral field
RSG	retrosplenial granular cortex
S1	primary somatosensory cortex
s5	sensory root of trigeminal nerve
scp	superior cerebellar peduncle
st	stria terminalis
SuG	superficial gray sup. coll.
VLL	ventral nu. lateral lemniscus
VPM	ventral posteromed. thalamic nu.

Pl-Ad21

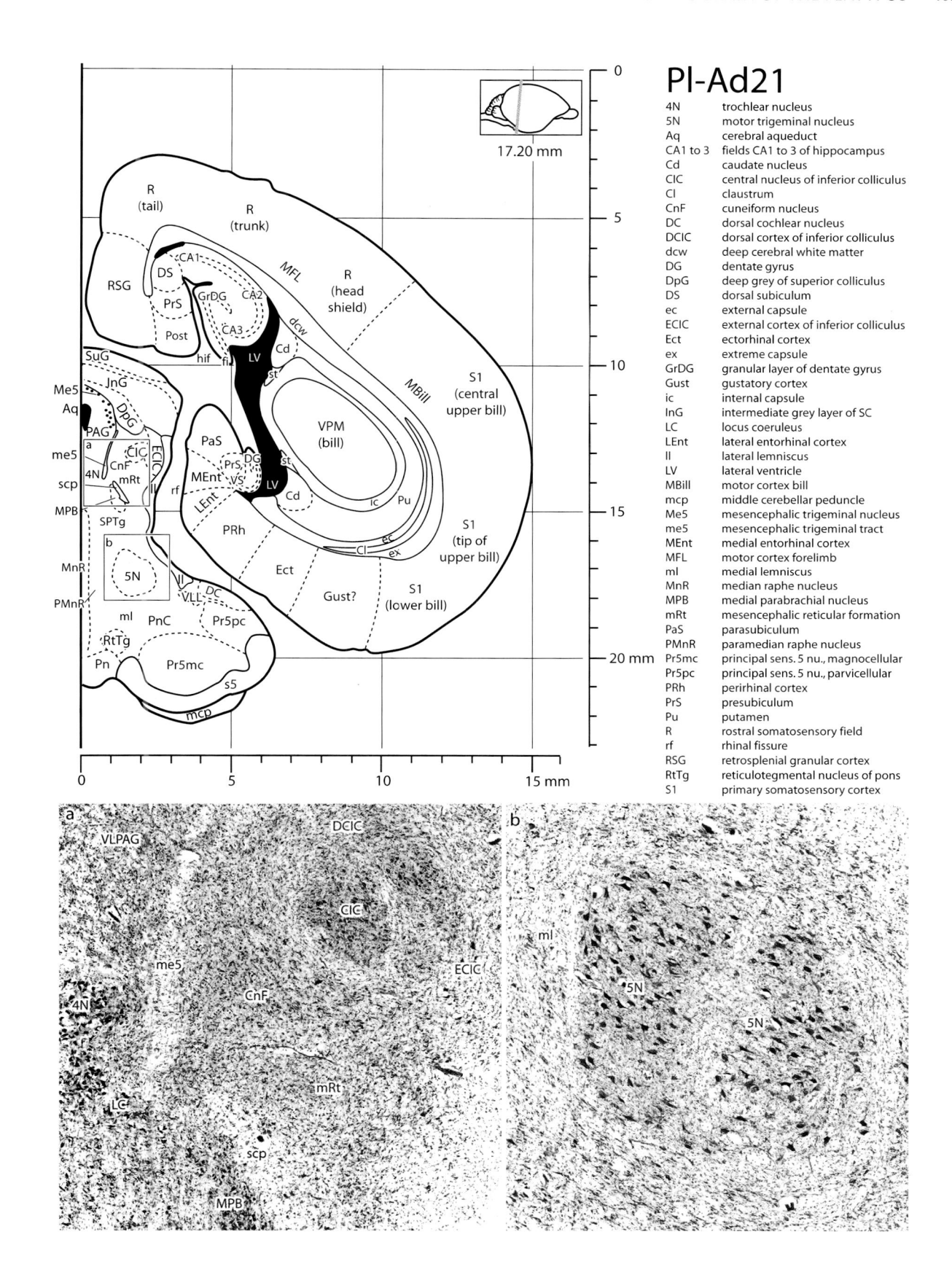

4N	trochlear nucleus
5N	motor trigeminal nucleus
Aq	cerebral aqueduct
CA1 to 3	fields CA1 to 3 of hippocampus
Cd	caudate nucleus
CIC	central nucleus of inferior colliculus
Cl	claustrum
CnF	cuneiform nucleus
DC	dorsal cochlear nucleus
DCIC	dorsal cortex of inferior colliculus
dcw	deep cerebral white matter
DG	dentate gyrus
DpG	deep grey of superior colliculus
DS	dorsal subiculum
ec	external capsule
ECIC	external cortex of inferior colliculus
Ect	ectorhinal cortex
ex	extreme capsule
GrDG	granular layer of dentate gyrus
Gust	gustatory cortex
ic	internal capsule
InG	intermediate grey layer of SC
LC	locus coeruleus
LEnt	lateral entorhinal cortex
ll	lateral lemniscus
LV	lateral ventricle
MBill	motor cortex bill
mcp	middle cerebellar peduncle
Me5	mesencephalic trigeminal nucleus
me5	mesencephalic trigeminal tract
MEnt	medial entorhinal cortex
MFL	motor cortex forelimb
ml	medial lemniscus
MnR	median raphe nucleus
MPB	medial parabrachial nucleus
mRt	mesencephalic reticular formation
PaS	parasubiculum
PMnR	paramedian raphe nucleus
Pr5mc	principal sens. 5 nu., magnocellular
Pr5pc	principal sens. 5 nu., parvicellular
PRh	perirhinal cortex
PrS	presubiculum
Pu	putamen
R	rostral somatosensory field
rf	rhinal fissure
RSG	retrosplenial granular cortex
RtTg	reticulotegmental nucleus of pons
S1	primary somatosensory cortex

Pl-Ad22

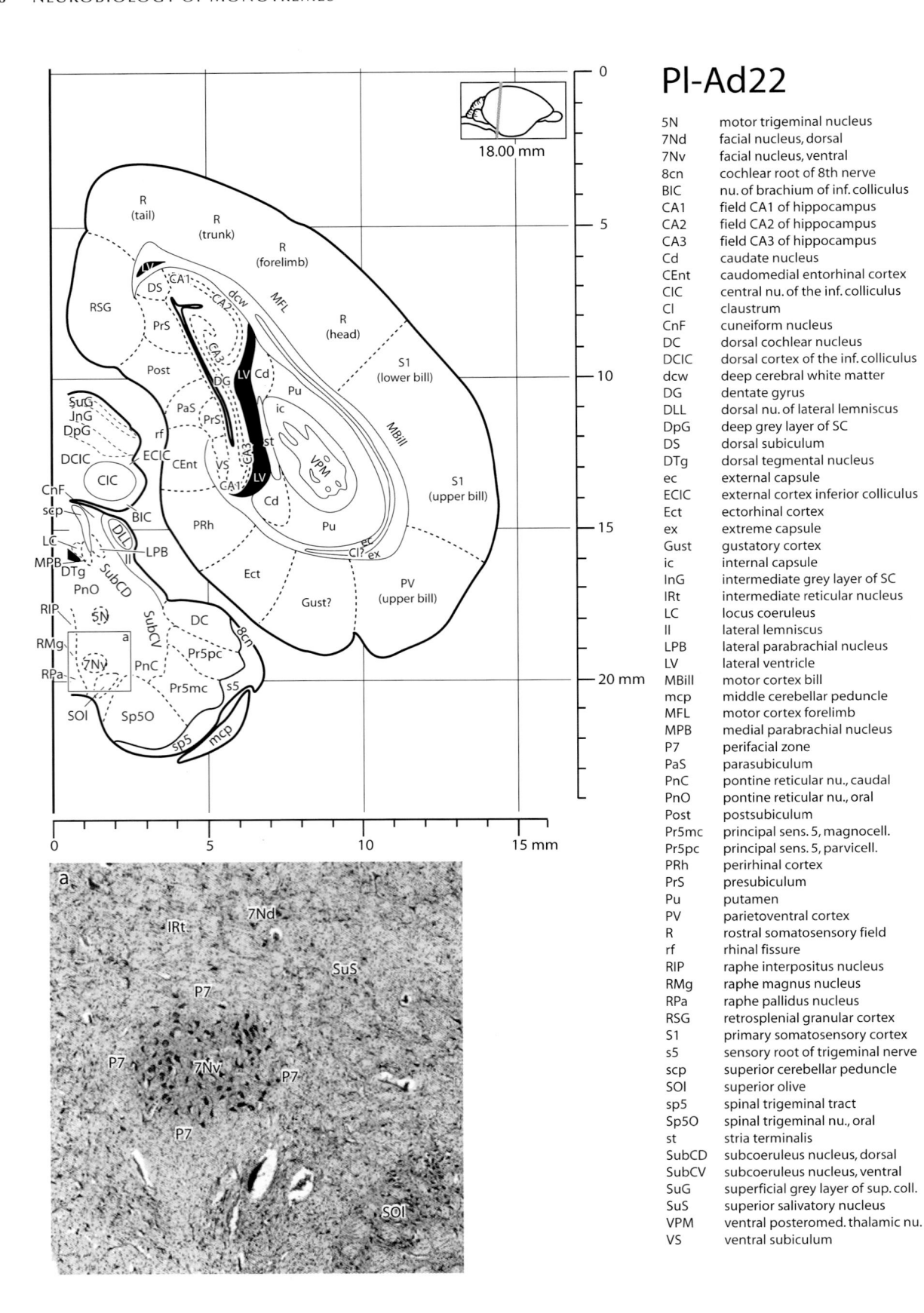

5N	motor trigeminal nucleus
7Nd	facial nucleus, dorsal
7Nv	facial nucleus, ventral
8cn	cochlear root of 8th nerve
BIC	nu. of brachium of inf. colliculus
CA1	field CA1 of hippocampus
CA2	field CA2 of hippocampus
CA3	field CA3 of hippocampus
Cd	caudate nucleus
CEnt	caudomedial entorhinal cortex
CIC	central nu. of the inf. colliculus
Cl	claustrum
CnF	cuneiform nucleus
DC	dorsal cochlear nucleus
DCIC	dorsal cortex of the inf. colliculus
dcw	deep cerebral white matter
DG	dentate gyrus
DLL	dorsal nu. of lateral lemniscus
DpG	deep grey layer of SC
DS	dorsal subiculum
DTg	dorsal tegmental nucleus
ec	external capsule
ECIC	external cortex inferior colliculus
Ect	ectorhinal cortex
ex	extreme capsule
Gust	gustatory cortex
ic	internal capsule
InG	intermediate grey layer of SC
IRt	intermediate reticular nucleus
LC	locus coeruleus
ll	lateral lemniscus
LPB	lateral parabrachial nucleus
LV	lateral ventricle
MBill	motor cortex bill
mcp	middle cerebellar peduncle
MFL	motor cortex forelimb
MPB	medial parabrachial nucleus
P7	perifacial zone
PaS	parasubiculum
PnC	pontine reticular nu., caudal
PnO	pontine reticular nu., oral
Post	postsubiculum
Pr5mc	principal sens. 5, magnocell.
Pr5pc	principal sens. 5, parvicell.
PRh	perirhinal cortex
PrS	presubiculum
Pu	putamen
PV	parietoventral cortex
R	rostral somatosensory field
rf	rhinal fissure
RIP	raphe interpositus nucleus
RMg	raphe magnus nucleus
RPa	raphe pallidus nucleus
RSG	retrosplenial granular cortex
S1	primary somatosensory cortex
s5	sensory root of trigeminal nerve
scp	superior cerebellar peduncle
SOl	superior olive
sp5	spinal trigeminal tract
Sp5O	spinal trigeminal nu., oral
st	stria terminalis
SubCD	subcoeruleus nucleus, dorsal
SubCV	subcoeruleus nucleus, ventral
SuG	superficial grey layer of sup. coll.
SuS	superior salivatory nucleus
VPM	ventral posteromed. thalamic nu.
VS	ventral subiculum

Pl-Ad23

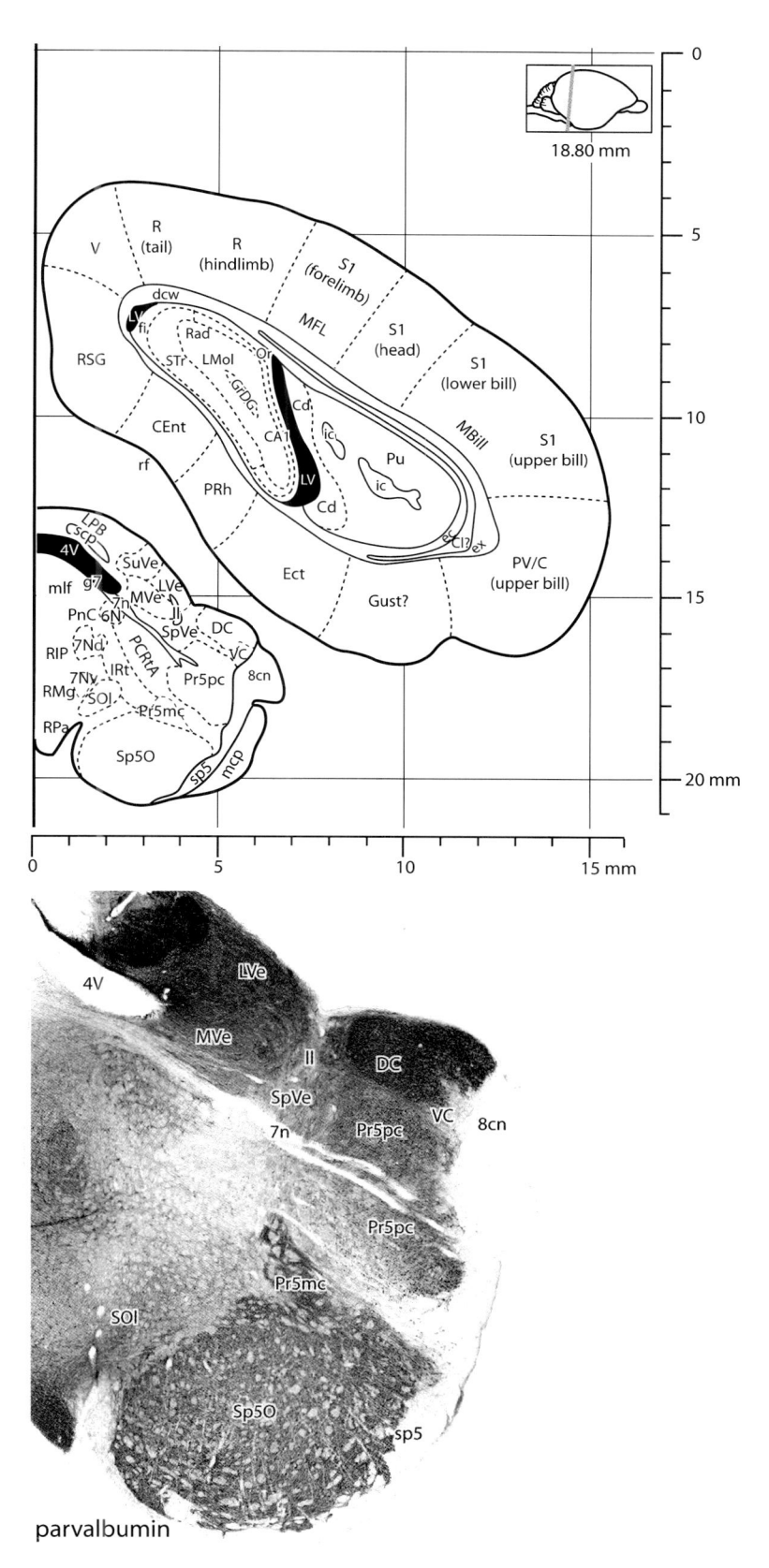

4V	4th ventricle
6N	abducens nucleus
7n	facial nerve
7Nd	facial nucleus, dorsal
7Nv	facial nucleus, ventral
8cn	cochlear root of 8th nerve
CA1	field CA1 of hippocampus
Cd	caudate nucleus
CEnt	caudomedial entorhinal cortex
Cl	claustrum
DC	dorsal cochlear nucleus
dcw	deep cerebral white matter
ec	external capsule
Ect	ectorhinal cortex
ex	extreme capsule
fi	fimbria of the hippocampus
g7	genu of the facial nerve
GrDG	granular dentate gyrus
Gust	gustatory cortex
ic	internal capsule
IRt	intermediate reticular nucleus
ll	lateral lemniscus
LMol	lacunosum moleculare layer
LPB	lateral parabrachial nucleus
LV	lateral ventricle
LVe	lateral vestibular nucleus
MBill	motor cortex bill
mcp	middle cerebellar peduncle
MFL	motor cortex forelimb
mlf	medial longitudinal fasciculus
MVe	medial vestibular nucleus
Or	oriens layer of the hippocampus
PCRtA	parvicellular reticular nu., alpha
PnC	pontine reticular nu., caudal
Pr5mc	principal sens. 5, magnocellular
Pr5pc	principal sens. 5, parvicellular
PRh	perirhinal cortex
Pu	putamen
PV	parietoventral cortex
R	rostral somatosensory field
Rad	radiatum layer of hippocampus
rf	rhinal fissure
RIP	raphe interpositus nucleus
RMg	raphe magnus nucleus
RPa	raphe pallidus nucleus
RSG	retrosplenial granular cortex
S1	primary somatosensory cortex
scp	superior cerebellar peduncle
SOl	superior olive
sp5	spinal trigeminal tract
Sp5O	spinal trigeminal nu., oral part
SpVe	spinal vestibular nucleus
STr	subiculum, transition area
SuVe	superior vestibular nucleus
V	primary visual cortex
VC	ventral cochlear nucleus

parvalbumin

Pl-Ad24

3Cb	lobule 3 of the Cb vermis
4V	4th ventricle
8cn	cochlear root of 8th nerve
8vn	vestibular root of 8th nerve
Au/S1	auditory/somato. bimodal cortex
CA1	field CA1 of hippocampus
Cl	claustrum
dcw	deep cerebral white matter
DPGi	dorsal paragigantocellular nucleus
ec	external capsule
Ect	ectorhinal cortex
ex	extreme capsule
fi	fimbria of the hippocampus
Gi	gigantocellular reticular nucleus
GiA	gigantocellular retic. nu., alpha
IRt	intermediate reticular nucleus
LPB	lateral parabrachial nucleus
LPGiA	lateral paragigantocell. nu., alpha
LV	lateral ventricle
LVe	lateral vestibular nucleus
mcp	middle cerebellar peduncle
mlf	medial longitudinal fasciculus
MVe	medial vestibular nucleus
Or	oriens layer of the hippocampus
PCRt	parvicellular reticular nucleus
Pr	prepositus nucleus
PRh	perirhinal cortex
Pu	putamen
PV	parietoventral cortex
Rad	radiatum layer hippocampus
RMg	raphe magnus nucleus
RPa	raphe pallidus nucleus
RSG	retrosplenial granular cortex
S1	primary somatosensory cortex
scp	superior cerebellar peduncle
Sol	solitary nucleus
sp5	spinal trigeminal tract
Sp5I	spinal trigeminal nu., interpolar
Sp5O	spinal trigeminal nu., oral
SpVe	spinal vestibular nucleus
STr	subiculum, transition area
SuVe	superior vestibular nucleus
V	primary visual cortex
VeCb	vestibulocerebellar nucleus

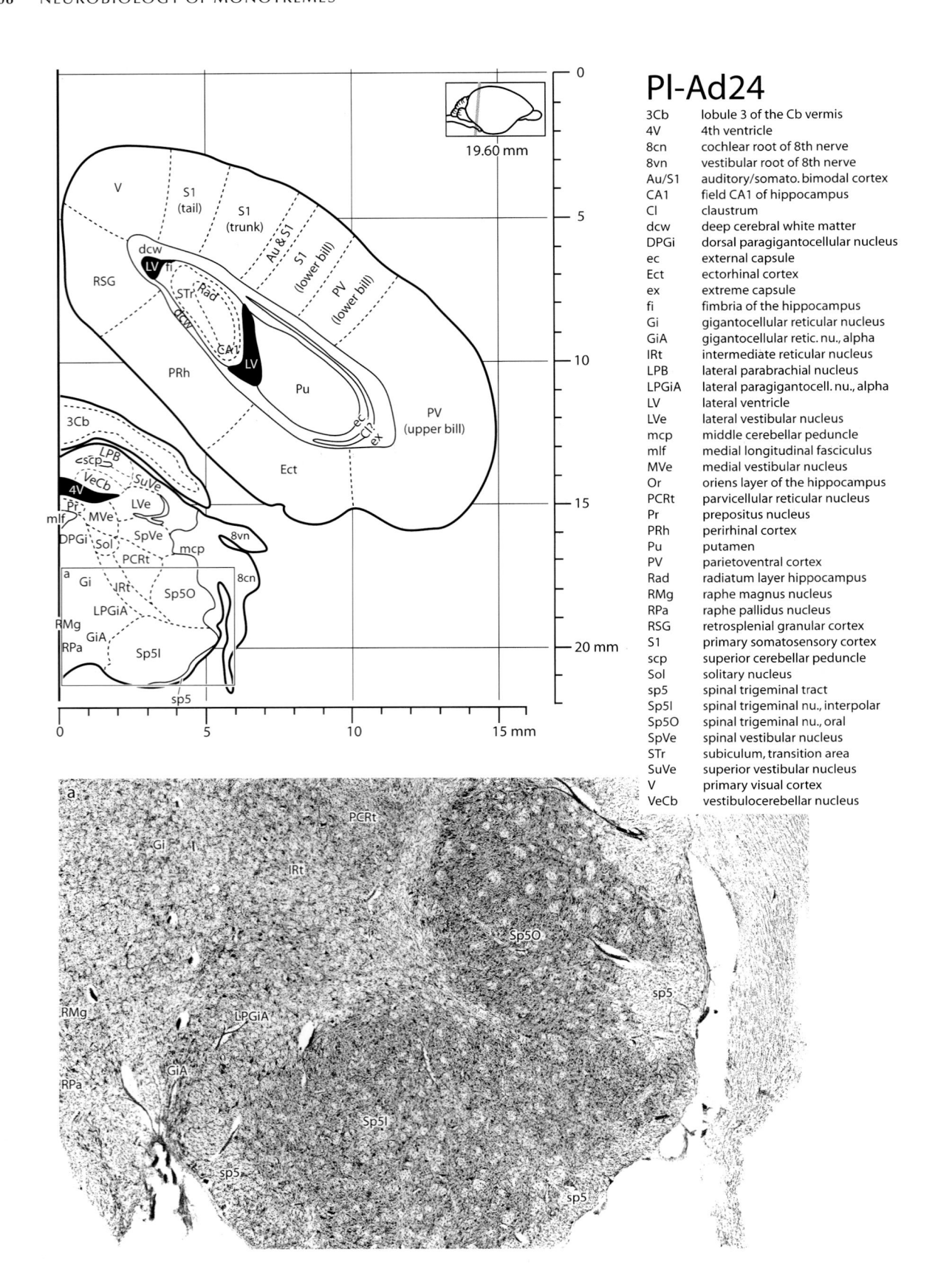

Pl-Ad25

2Cb	lobule 2 of the Cb vermis
3Cb	lobule 3 of the Cb vermis
4Cb	lobule 4 of the Cb vermis
5Cb	lobule 5 of the Cb vermis
6Cb	lobule 6 of the Cb vermis
4V	4th ventricle
8n	vestibulocochlear nerve
Amb	ambiguus nucleus
Au	auditory cortex
Au/S1	auditory/somatosensory bimodal cx
Cl	claustrum
dcw	deep cerebral white matter
ec	external capsule
Ect	ectorhinal cortex
ex	extreme capsule
Gi	gigantocellular reticular nucleus
GiV	gigantocellular reticular nu., ventral
GrCb	granule cell layer of the Cb
I8	interstitial nucleus of 8th nerve
icp	inferior cerebellar peduncle
IntA	interposed cerebellar nu., anterior
IO	inferior olivary nucleus
IRt	intermediate reticular nucleus
Lat	lateral cerebellar nucleus
LPGi	lateral paragigantocellular nucleus
LV	lateral ventricle
LVe	lateral vestibular nucleus
mcp	middle cerebellar peduncle
Med	medial cerebellar nucleus
MoCb	molecular layer of the cerebellum
MVe	medial vestibular nucleus
Occ	occipital bone
PCRt	parvicellular reticular nucleus
PFl	paraflocculus
Pk	Purkinje cell layer of the cerebellum
Pr	prepositus nucleus
Pu	putamen
PV	parietoventral cortex
RMg	raphe magnus nucleus
RPa	raphe pallidus nucleus
RSG	retrosplenial granular cortex
scp	superior cerebellar peduncle
Sol	solitary nucleus
sp5	spinal trigeminal tract
Sp5I	spinal trigeminal nu., interpolar part
Sp5O	spinal trigeminal nu., oral part
SpVe	spinal vestibular nucleus
SuVe	superior vestibular nucleus
V	primary visual cortex
VCP	ventral cochlear nucleus, posterior

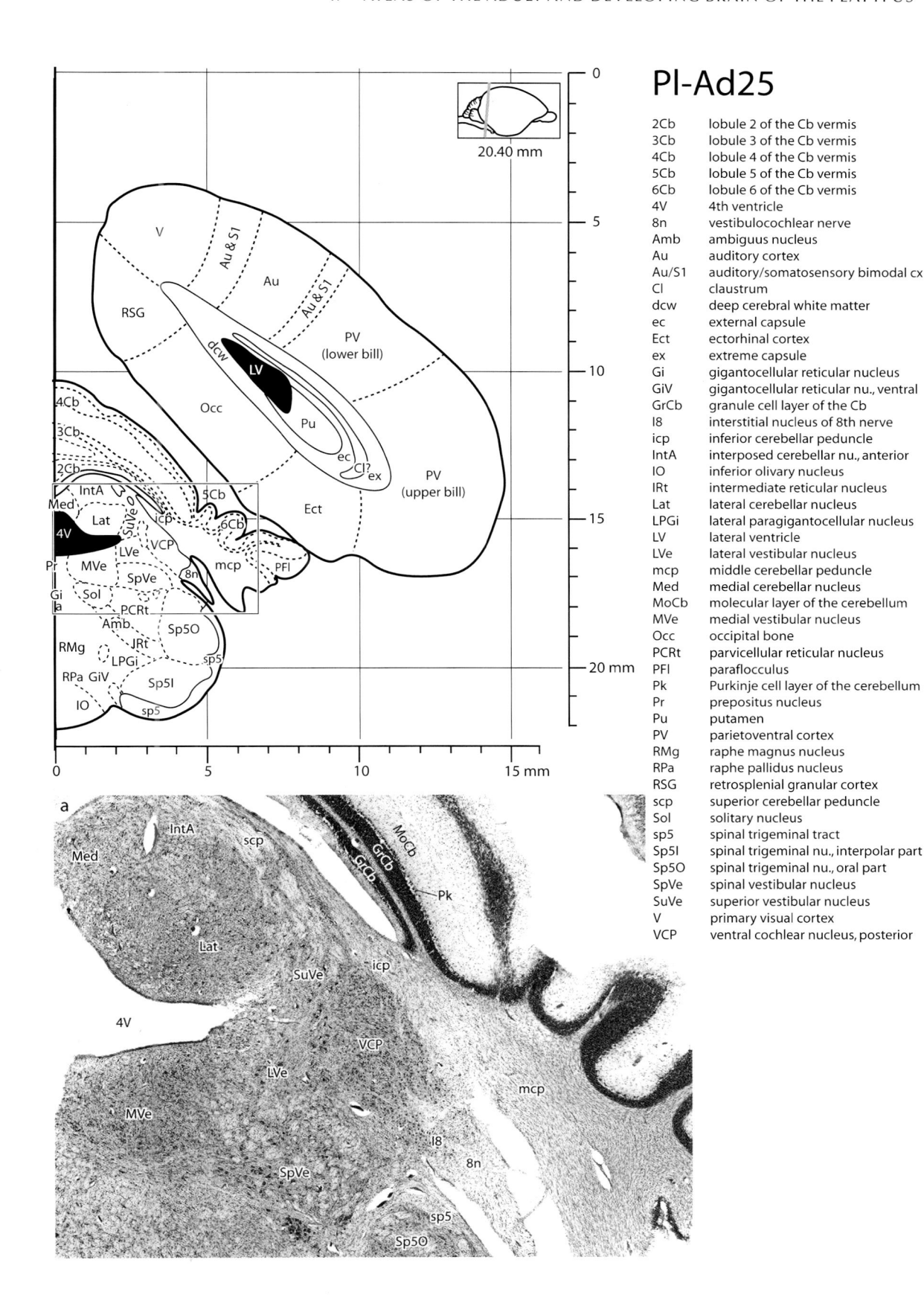

Pl-Ad26

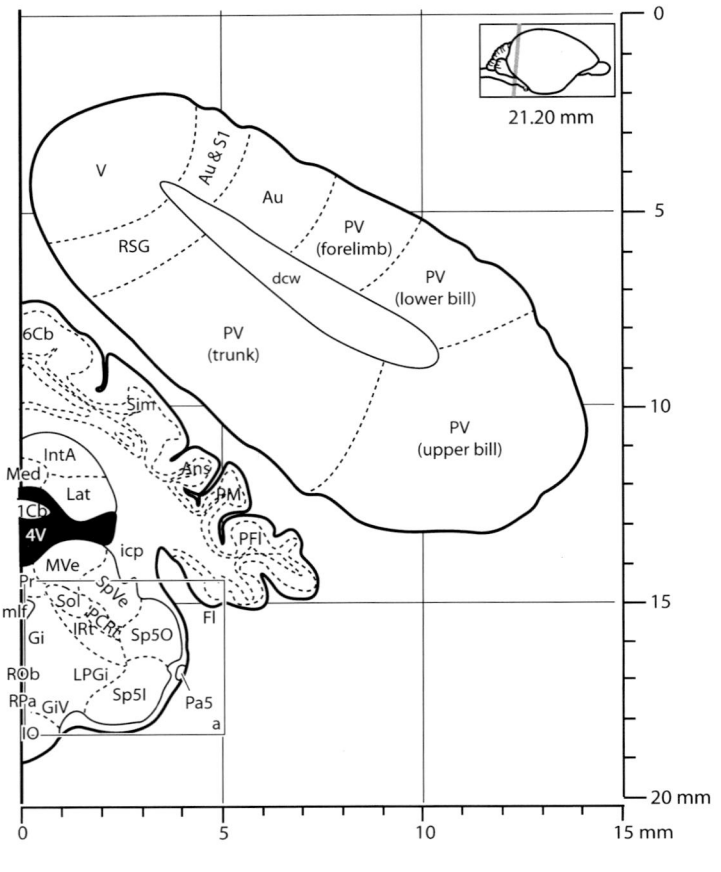

1,6Cb	lobules 1,6 of the Cb vermis
Ans	ansiform lobule of cerebellum
Au	auditory cortex
Au/S1	auditory/somatosensory bimodal cortex
dcw	deep cerebral white matter
DMSp5	dorsomedial spinal trigeminal nucleus
Fl	flocculus of the cerebellum
Gi	gigantocellular reticular nucleus
GiV	gigantocellular reticular nucleus, ventral
GrCb	granule cell layer of the cerebellum
icp	inferior cerebellar peduncle
IntA	interposed cerebellar nucleus, anterior
IO	inferior olivary nucleus
IRt	intermediate reticular nucleus
Lat	lateral cerebellar nucleus
LPGi	lateral paragigantocellular nucleus
mlf	medial longitudinal fasciculus
MoCb	molecular layer of the cerebellum
MVe	medial vestibular nucleus
Pa5	paratrigeminal nucleus
PCRt	parvicellular reticular nucleus
PFl	paraflocculus
Pk	Purkinje cell layer of the cerebellum
PM	paramedian lobule
Pr	prepositus nucleus
PV	parietoventral cortex
ROb	raphe obscurus nucleus
RPa	raphe pallidus nucleus
RSG	retrosplenial granular cortex
Sim	simple lobule of the cerebellum
Sol	solitary nucleus
sol	solitary tract
sp5	spinal trigeminal tract
Sp5I	spinal trigeminal nucleus, interpolar part
Sp5O	spinal trigeminal nucleus, oral part
SpVe	spinal vestibular nucleus
V	primary visual cortex

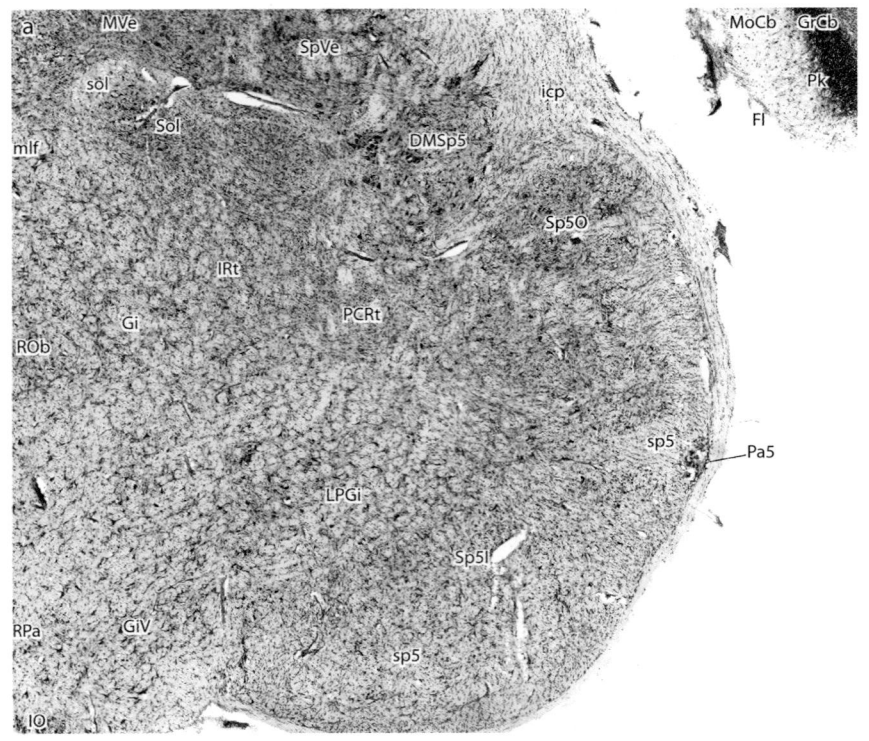

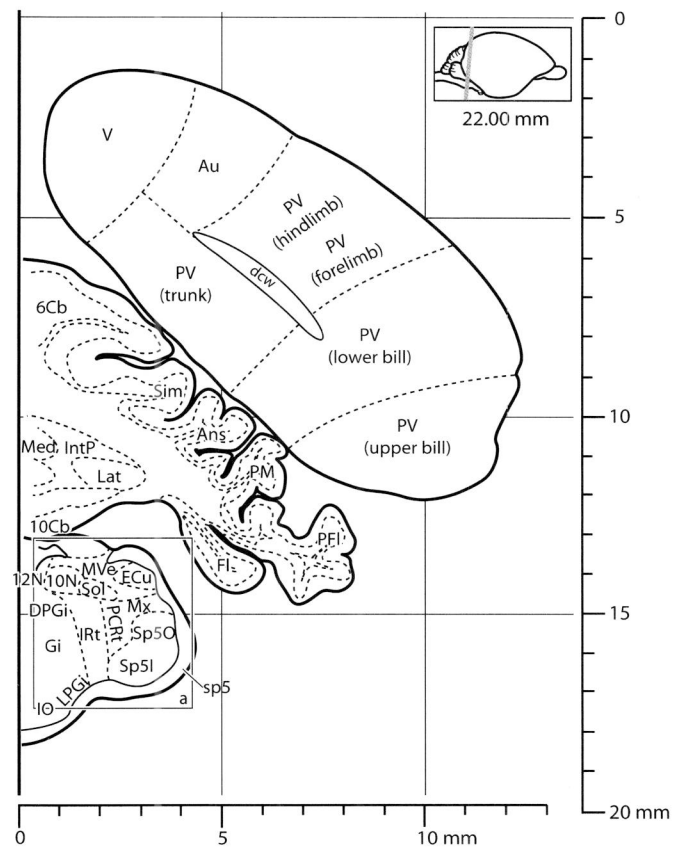

Pl-Ad27

6Cb	lobule 6 of the Cb vermis
10Cb	lobule 10 of the Cb vermis
4V	4th ventricle
10N	vagus nerve nucleus
12N	hypoglossal nucleus
Amb	ambiguus nucleus
Ans	ansiform lobule of the cerebellum
Au	auditory cortex
cu	cuneate fasciculus
dcw	deep cerebral white matter
DPGi	dorsal paragigantocellular nucleus
ECu	external cuneate nucleus
Fl	flocculus of the cerebellum
Gi	gigantocellular reticular nucleus
icp	inferior cerebellar peduncle
IntP	interposed cerebellar nucleus, posterior
IO	inferior olivary nucleus
IRt	intermediate reticular nucleus
Lat	lateral cerebellar nucleus
LPGi	lateral paragigantocellular nucleus
Med	medial cerebellar nucleus
MVe	medial vestibular nucleus
Mx	matrix region of the medulla
PCRt	parvicellular reticular nucleus
PFl	paraflocculus
PM	paramedian lobule of the cerebellum
PV	parietoventral cortex
Sim	simple lobule of the cerebellum
Sol	solitary nucleus
sol	solitary tract
sp5	spinal trigeminal tract
Sp5I	spinal trigeminal nucleus, interpolar part
Sp5O	spinal trigeminal nucleus, oral part
SpVe	spinal vestibular nucleus
V	primary visual cortex

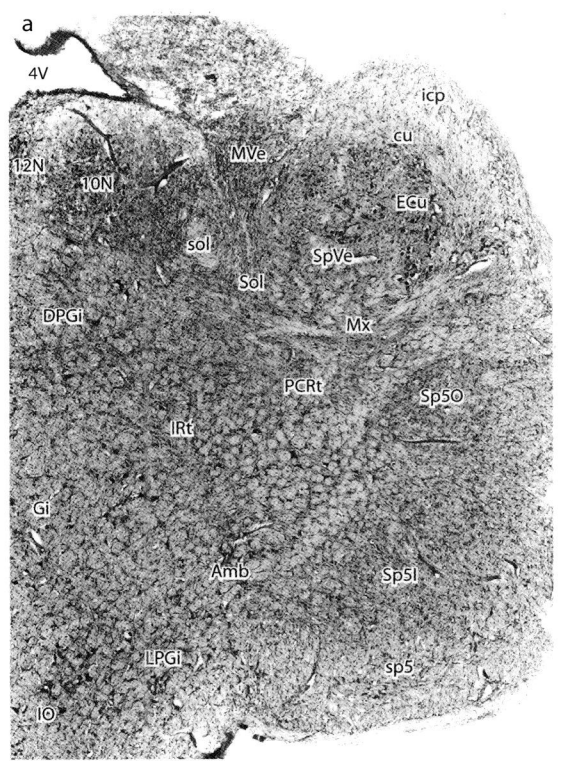

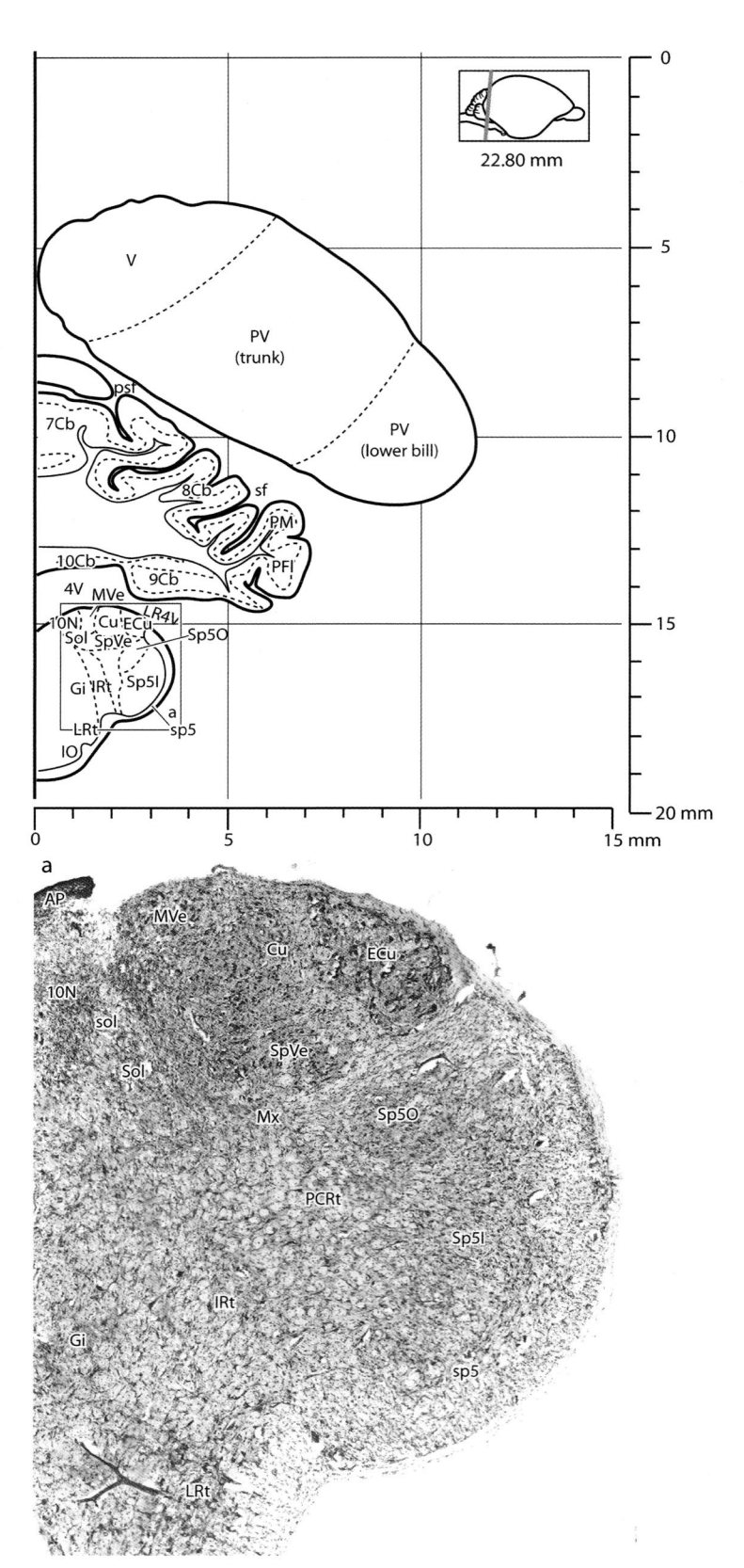

Pl-Ad28

6Cb	lobule 6 of the Cb vermis
7Cb	lobule 7 of the Cb vermis
8Cb	lobule 8 of the Cb vermis
9Cb	lobule 9 of the Cb vermis
10Cb	lobule 10 of the Cb vermis
4V	4th ventricle
10N	vagus nerve nucleus
AP	area postrema
Cu	cuneate nucleus
ECu	external cuneate nucleus
Gi	gigantocellular reticular nucleus
IO	inferior olivary nucleus
IRt	intermediate reticular nucleus
LR4V	lateral recess of 4th ventricle
LRt	lateral reticular nucleus
MVe	medial vestibular nucleus
Mx	matrix region of the medulla
PCRt	parvicellular reticular nucleus
PFl	paraflocculus
PM	paramedian lobule
psf	posterior superior fissure
PV	parietoventral cortex
sf	secondary fissure
sol	solitary tract
Sol	solitary nucleus
sp5	spinal trigeminal tract
Sp5I	spinal trigeminal nucleus, interpolar part
Sp5O	spinal trigeminal nucleus, oral part
SpVe	spinal vestibular nucleus
V	primary visual cortex

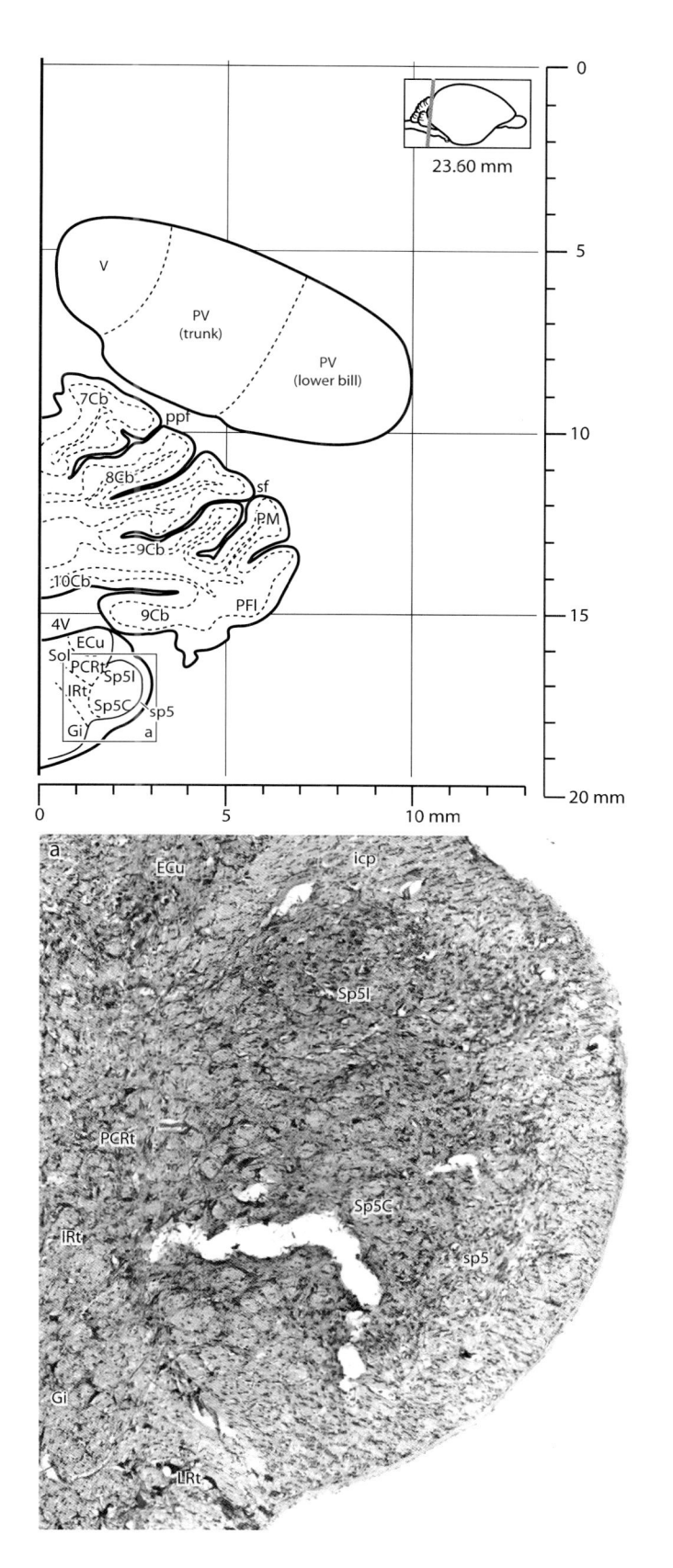

Pl-Ad29

7Cb	lobule 7 of the Cb vermis
8Cb	lobule 8 of the Cb vermis
9Cb	lobule 9 of the Cb vermis
10Cb	lobule 10 of the Cb vermis
4V	4th ventricle
ECu	external cuneate nucleus
Gi	gigantocellular reticular nucleus
icp	inferior cerebellar peduncle
IRt	intermediate reticular nucleus
LRt	lateral reticular nucleus
PCRt	parvicellular reticular nucleus
PFl	paraflocculus of the cerebellum
PM	paramedian lobule of the cerebellum
ppf	prepyramidal fissure of the cerebellum
PV	parietoventral cortex
sf	secondary fissure of the cerebellum
Sol	solitary nucleus
sp5	spinal trigeminal tract
Sp5C	spinal trigeminal nucleus, caudal part
Sp5I	spinal trigeminal nucleus, interpolar part
V	primary visual cortex

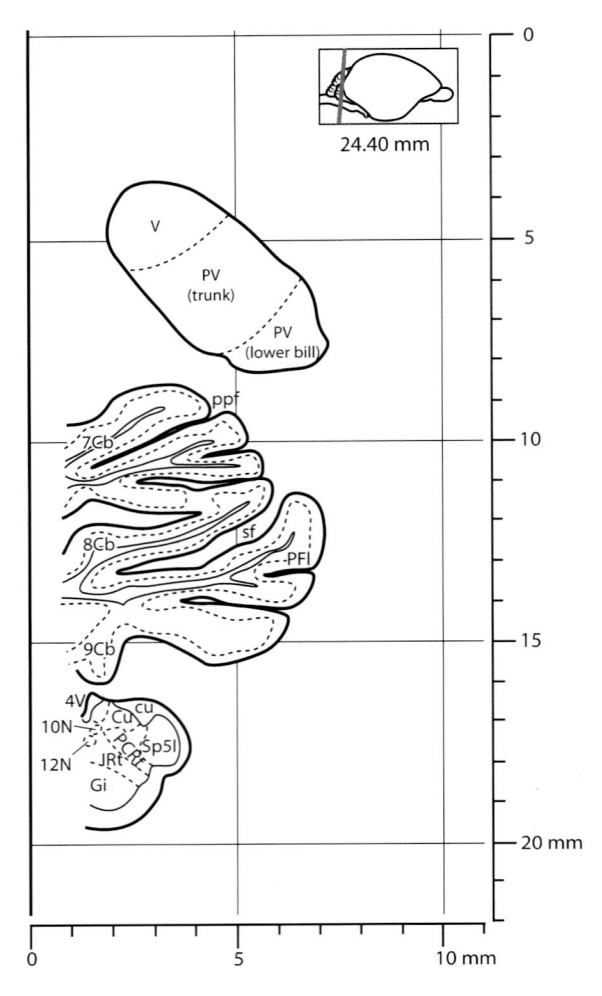

Pl-Ad30

7Cb	lobule 7 of the Cb vermis
8Cb	lobule 8 of the Cb vermis
9Cb	lobule 9 of the Cb vermis
4V	4th ventricle
10N	vagus nerve nucleus
12N	hypoglossal nucleus
cu	cuneate fasciculus
Cu	cuneate nucleus
ECu	external cuneate nucleus
Gi	gigantocellular reticular nucleus
IRt	intermediate reticular nucleus
Mx	matrix region of the medulla
PCRt	parvicellular reticular nucleus
PFl	paraflocculus of cerebellum
ppf	prepyramidal fissure
PV	parietoventral cortex
sf	secondary fissure
Sol	solitary nucleus
sp5	spinal trigeminal tract
Sp5l	spinal trigeminal nucleus, interpolar part
V	primary visual cortex

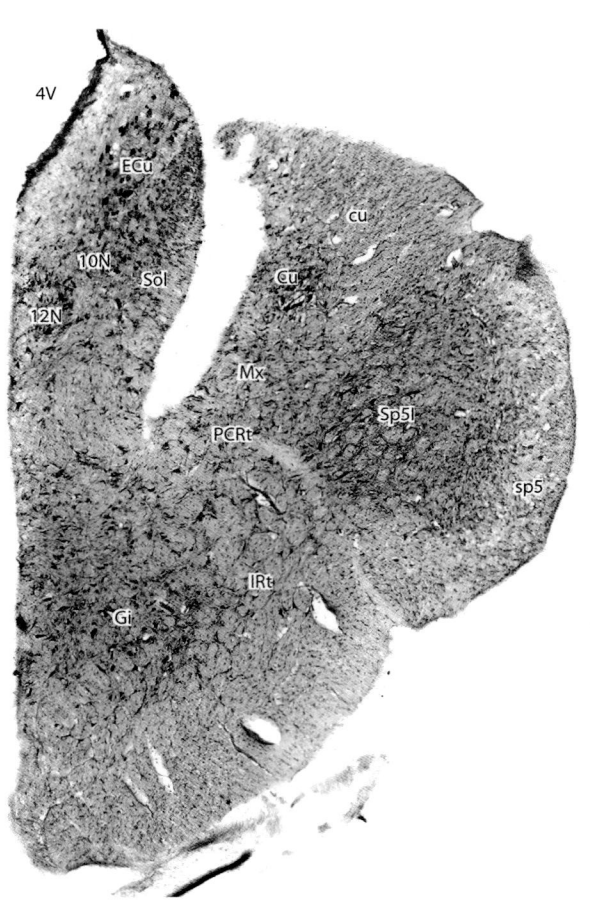

Pl-Ha01

*	denotes precursor
5fr	frontal branch of ophthalmic division of trigeminal nerve
5oph	ophthalmic division of trigeminal nerve
Conjunct	conjunctival sac
Cx	cerebral cortex
EthB	ethmoid bone
LV	lateral ventricle
MaxB	maxillary bone (maxilla)
ne	neuroepithelium
OB	olfactory bulb
olf	olfactory nerve
ON	olfactory nerve layer
OV	olfactory ventricle
Pig	pigment layer of the eye
PrePl	preplate of cortex
PrOrb	preoptic root of orbital bone
Vent	ventricular space of the eye
Vom	vomer

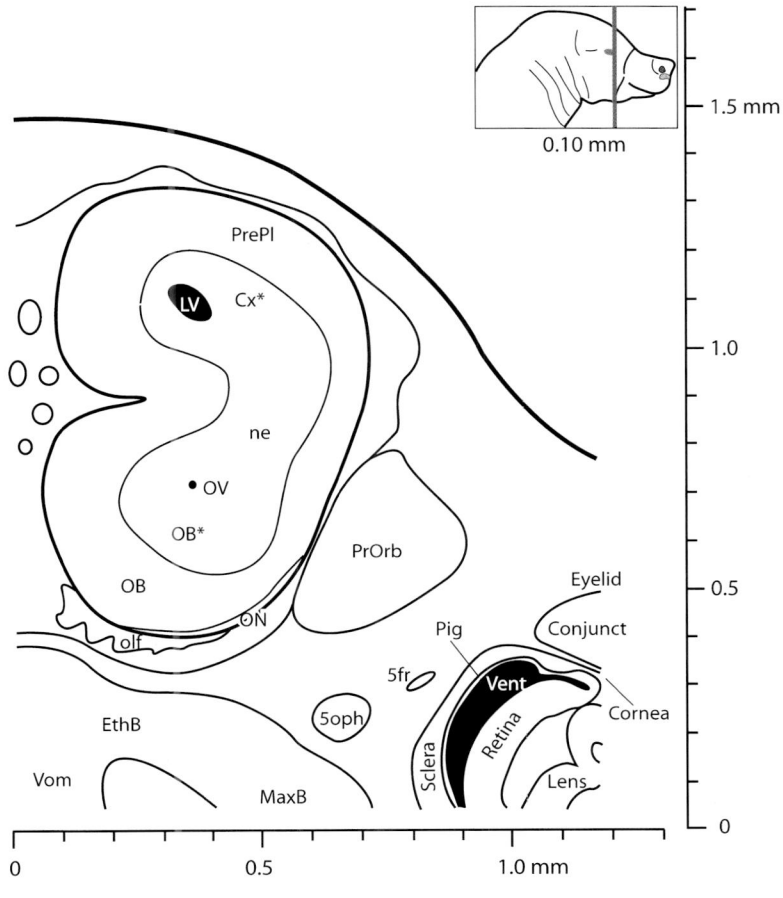

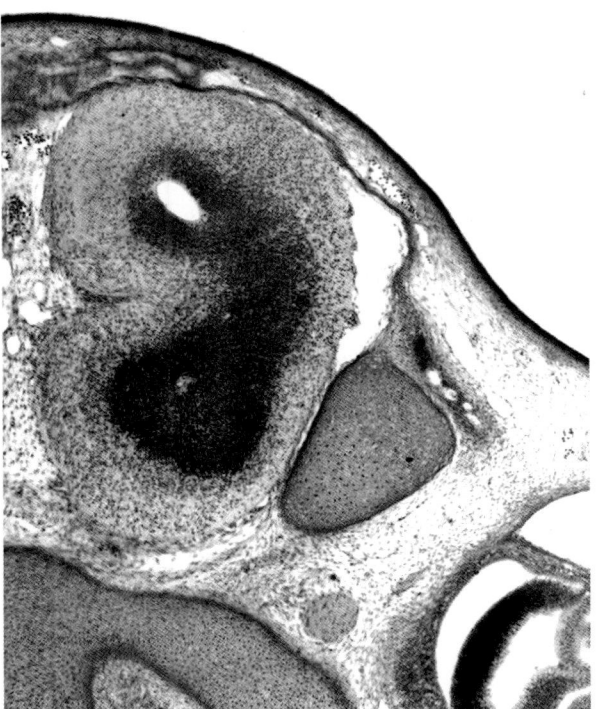

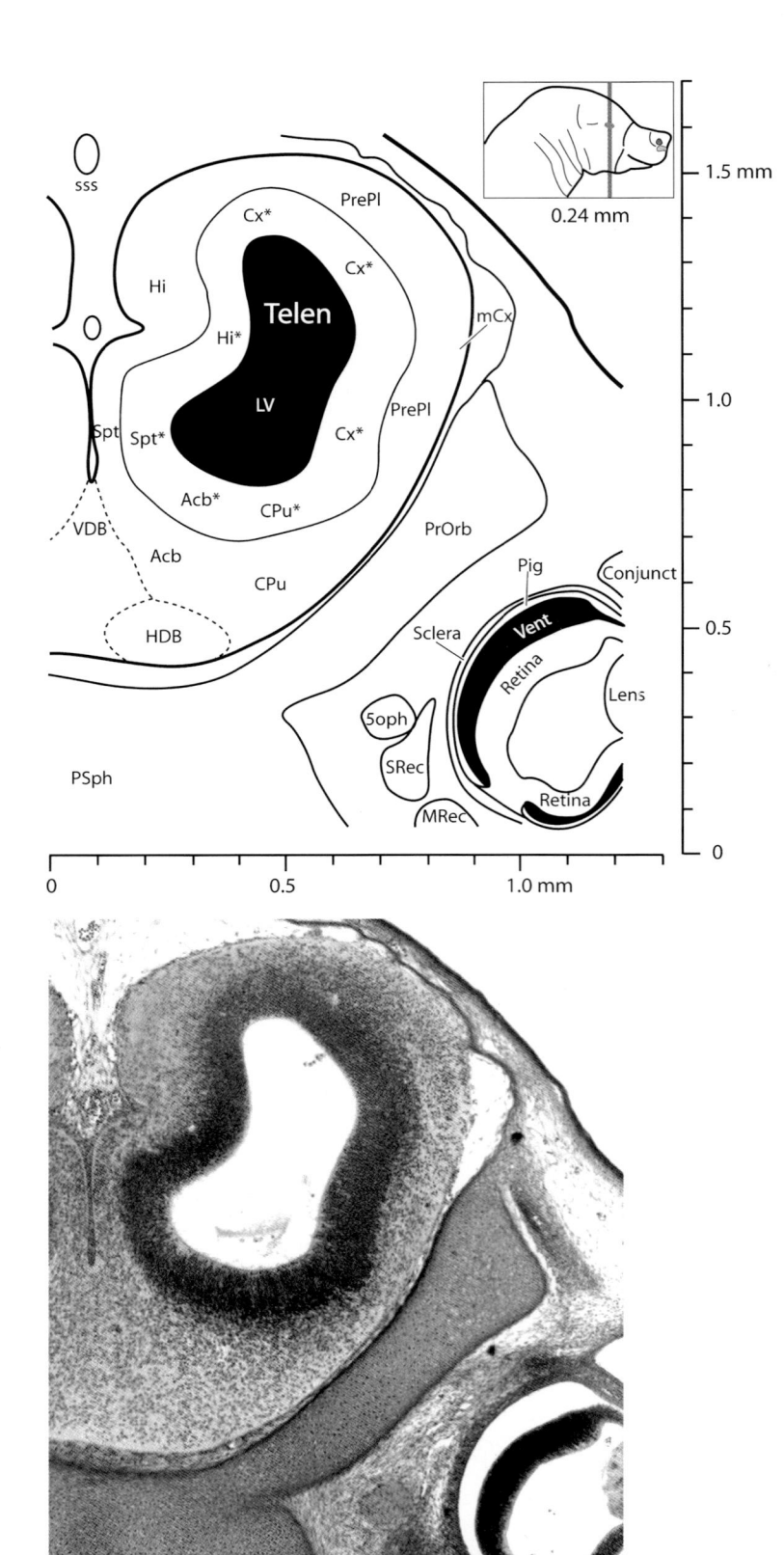

Pl-Ha02

*	denotes precursor
5oph	ophthalmic division of trigeminal nerve
Acb	accumbens nucleus
Conjunct	conjunctival sac
CPu	caudate putamen (striatum)
Cx	cerebral cortex
HDB	nucleus of horizontal limb of diagonal band
Hi	hippocampus
LV	lateral ventricle
mCx	marginal zone of developing cortex
MRec	medial rectus muscle
Pig	pigment layer of the eye
PrePl	preplate of cortex
PrOrb	preoptic root of orbital bone
PSph	presphenoid bone
Spt	septum of brain
SRec	superior rectus muscle
sss	superior sagittal sinus
Telen	telencephalon
VDB	nucleus of vertical limb of diagonal band
Vent	ventricular space of the eye

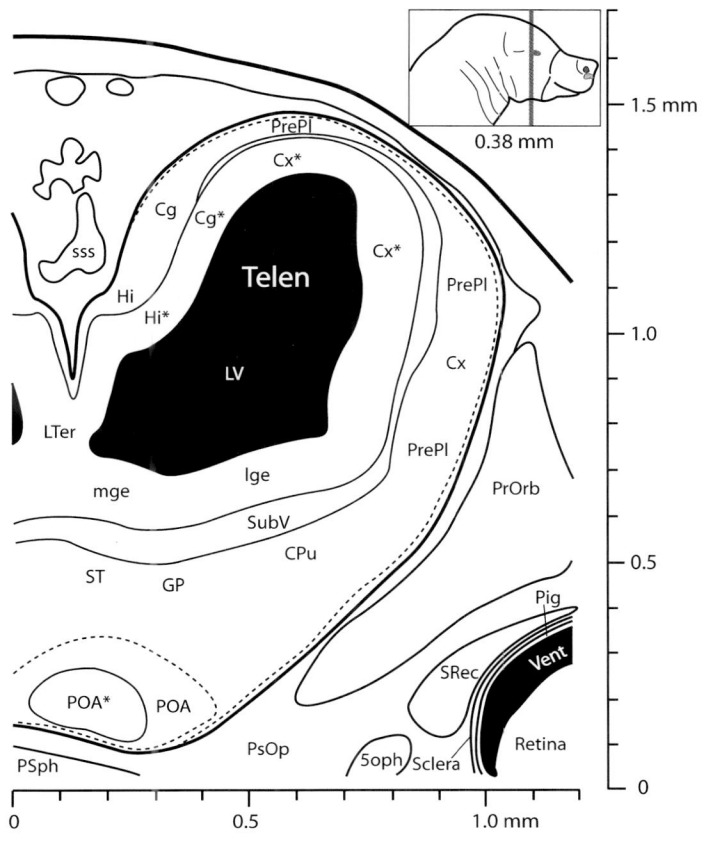

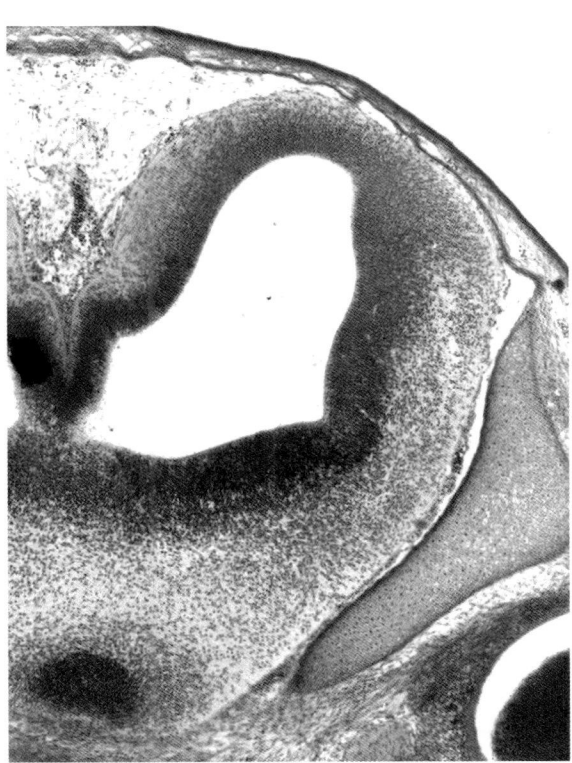

Pl-Ha03

*	denotes precursor
5oph	ophthalmic division of trigeminal nerve
Cg	cingulate cortex
CPu	caudate putamen (striatum)
Cx	cerebral cortex
GP	globus pallidus
Hi	hippocampus
lge	lateral ganglionic eminence of developing subpallium
LTer	lamina terminalis
LV	lateral ventricle
mge	medial ganglionic eminence of developing subpallium
Pig	pigment layer of the eye
POA	preoptic area
PrePl	preplate of cortex
PrOrb	preoptic root orbital bone
PsCp	pseudo-optic foramen
PSph	presphenoid bone
SRec	superior rectus muscle
sss	superior sagittal sinus
ST	bed nucleus of stria terminalis
SubV	subventricular layer of developing cortex
Telen	telencephalon
Vent	ventricular space of the eye

Pl-Ha04

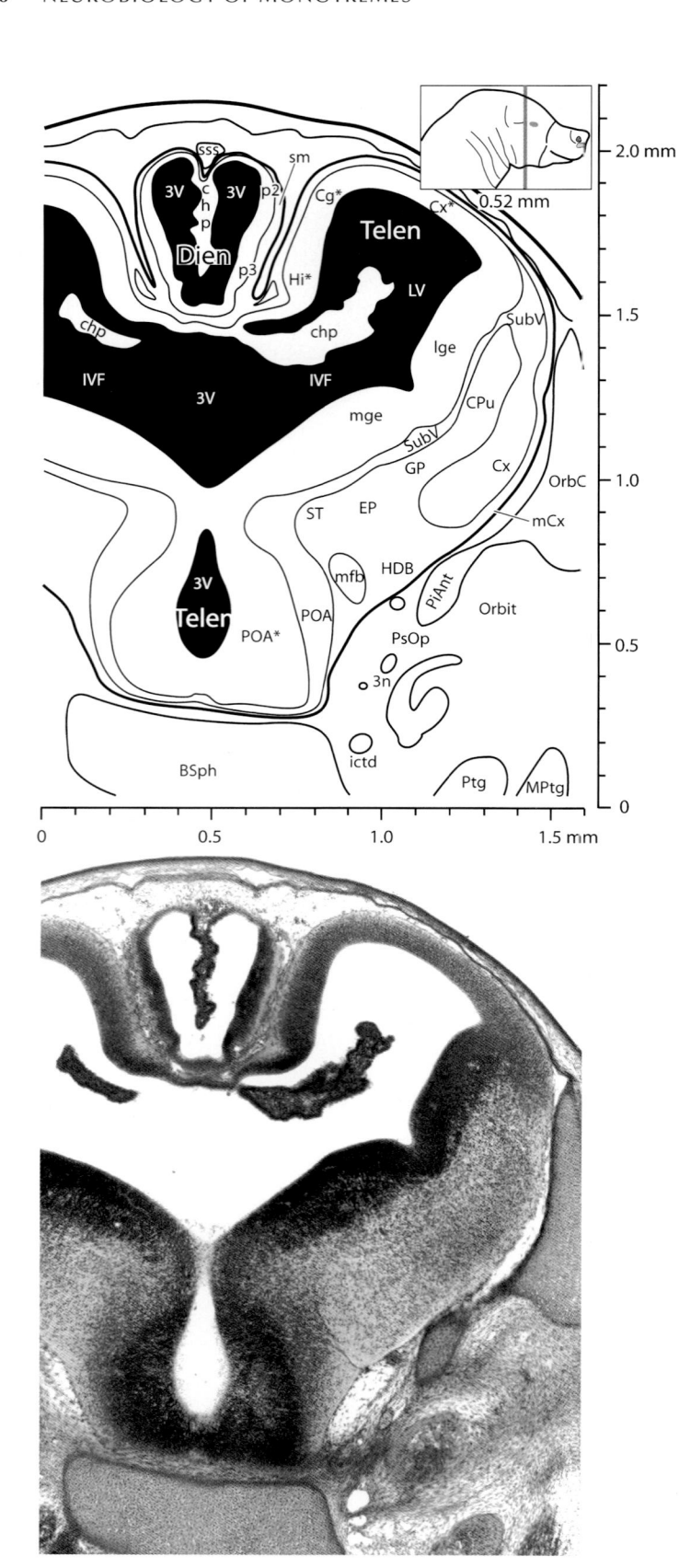

*	denotes precursor
3n	oculomotor nerve
3V	3rd ventricle
BSph	basisphenoid bone
Cg	cingulate cortex
chp	choroid plexus
CPu	caudate putamen (striatum)
Cx	cerebral cortex
Dien	diencephalon
EP	entopeduncular nucleus
GP	globus pallidus
HDB	nucleus of horizontal limb of diagonal band
Hi	hippocampus
ictd	internal carotid artery
IVF	interventricular foramen
lge	lateral ganglionic eminence of developing subpallium
LV	lateral ventricle
mCx	marginal zone of developing cx
mfb	medial forebrain bundle
mge	medial ganglionic eminence of developing subpallium
MPtg	medial pterygoid muscle
OrbC	orbital cartilage
p2	prosomere 2
p3	prosomere 3
PiAnt	pila antotica
POA	preoptic area
PsOp	pseudo-optic foramen
Ptg	pterygoid process of sphenoid
sm	stria medullaris thalami
sss	superior sagittal sinus
ST	bed nucleus of stria terminalis
SubV	subventricular layer of developing cortex
Telen	telencephalon

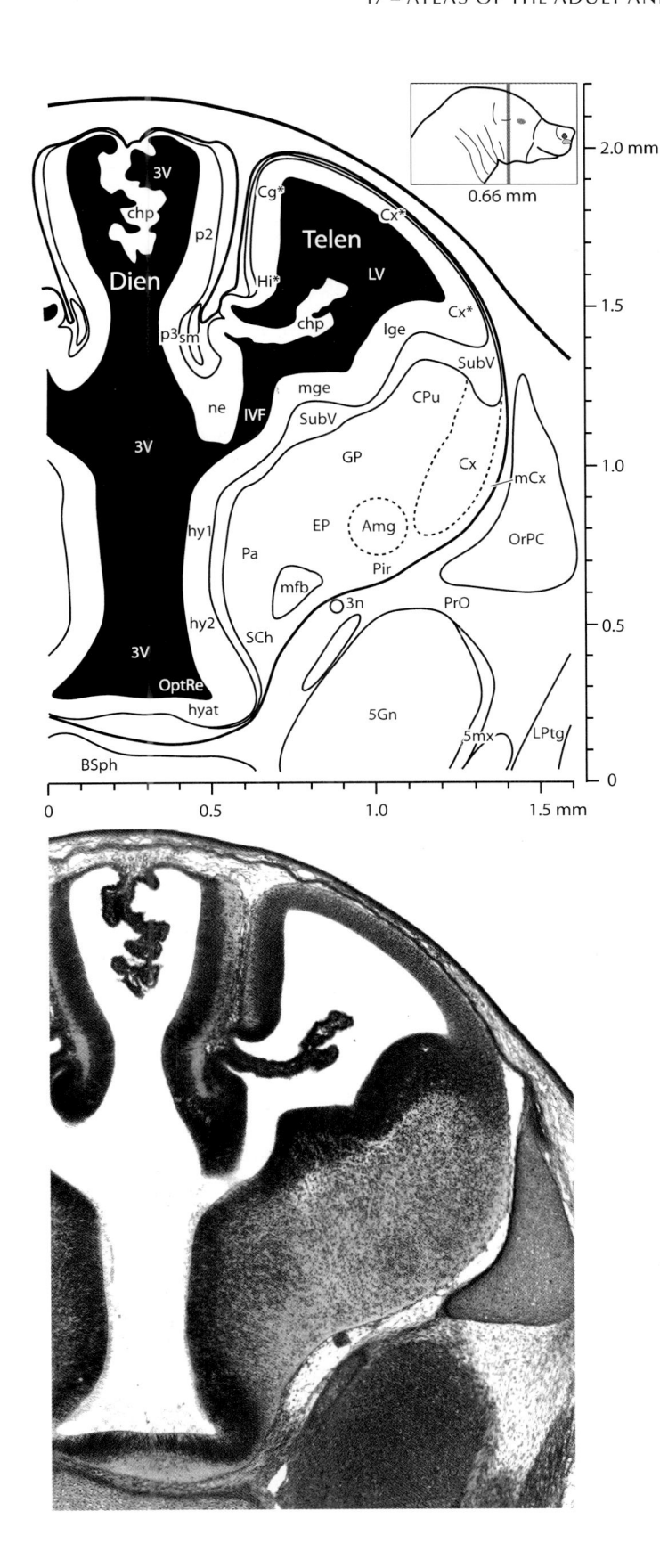

Pl-Ha05

*	denotes precursor
3n	oculomotor nerve
3V	3rd ventricle
5Gn	trigeminal ganglion
5mx	maxillary division of trigeminal nerve
Amg	amygdala
BSph	basisphenoid bone
Cg	cingulate cortex
chp	choroid plexus
CPu	caudate putamen (striatum)
Cx	cerebral cortex
Dien	diencephalon
EP	entopeduncular nucleus
GP	globus pallidus
Hi	hippocampus
hy1	peduncular hypothalamus
hy2	tuberal hypothalamus
hyat	acroterminal hypothalamus
IVF	interventricular foramen
lge	lateral ganglionic eminence of developing subpallium
LPtg	lateral pterygoid muscle
LV	lateral ventricle
mCx	marginal zone of developing cortex
mfb	medial forebrain bundle
mge	medial ganglionic eminence of developing subpallium
ne	neuroepithelium
OptRe	optic recess of third ventricle
OrPC	orbitoparietal commissure
p2	prosomere 2
p3	prosomere 3
Pa	paraventricular hypothalamic nucleus
Pir	piriform cortex
PrO	pro-otic foramen
SCh	suprachiasmatic nucleus of hypothalamus
sm	stria medullaris thalami
SubV	subventricular layer of developing cortex
Telen	telencephalon

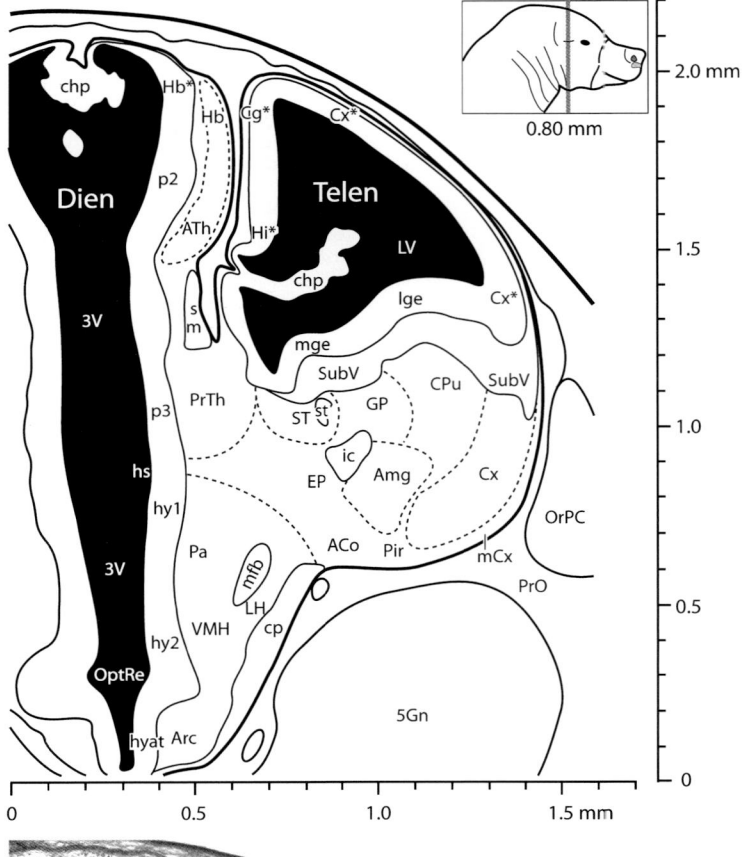

Pl-Ha06

*	denotes precursor
3V	3rd ventricle
5Gn	trigeminal ganglion
ACo	anterior cortical amygdaloid area
Amg	amygdala
Arc	arcuate hypothalamic nucleus
ATh	anterior thalamic region
Cg	cingulate cortex
chp	choroid plexus
cp	cerebral peduncle
CPu	caudate putamen (striatum)
Cx	cerebral cortex
Dien	diencephalon
EP	entopeduncular nucleus
GP	globus pallidus
Hb	habenular nuclei
Hi	hippocampus
hs	hypothalamic sulcus
hy1	peduncular hypothalamus
hy2	tuberal hypothalamus
hyat	acroterminal hypothalamus
ic	internal capsule
lge	lateral ganglionic eminence of developing subpallium
LH	lateral hypothalamic area
LV	lateral ventricle
mCx	marginal zone of developing cortex
mfb	medial forebrain bundle
mge	medial ganglionic eminence of developing subpallium
OptRe	optic recess of third ventricle
OrPC	orbitoparietal commissure
p2	prosomere 2
p3	prosomere 3
Pa	paraventricular hypothalamic nucleus
Pir	piriform cortex
PrO	pro-otic foramen
PrTh	prethalamus
sm	stria medullaris thalami
ST	bed nucleus of stria terminalis
st	stria terminalis
SubV	subventricular layer of developing cortex
Telen	telencephalon
VMH	ventromedial hypothalamic nucleus

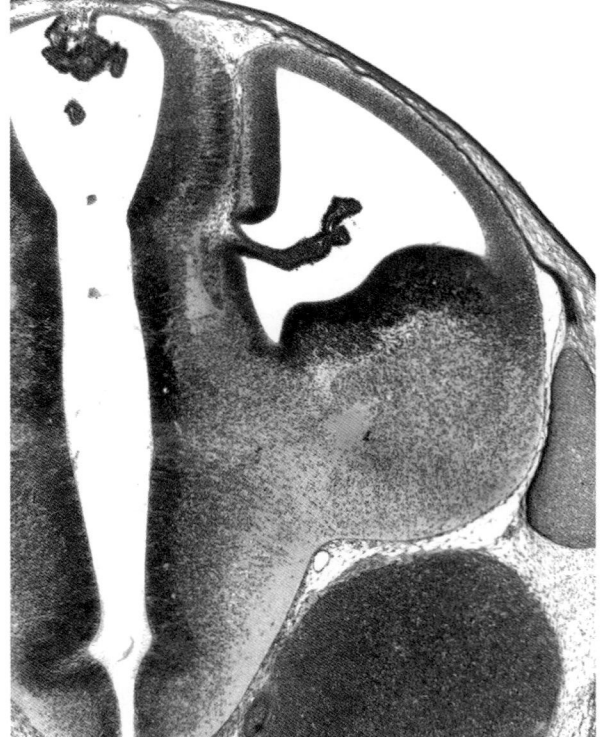

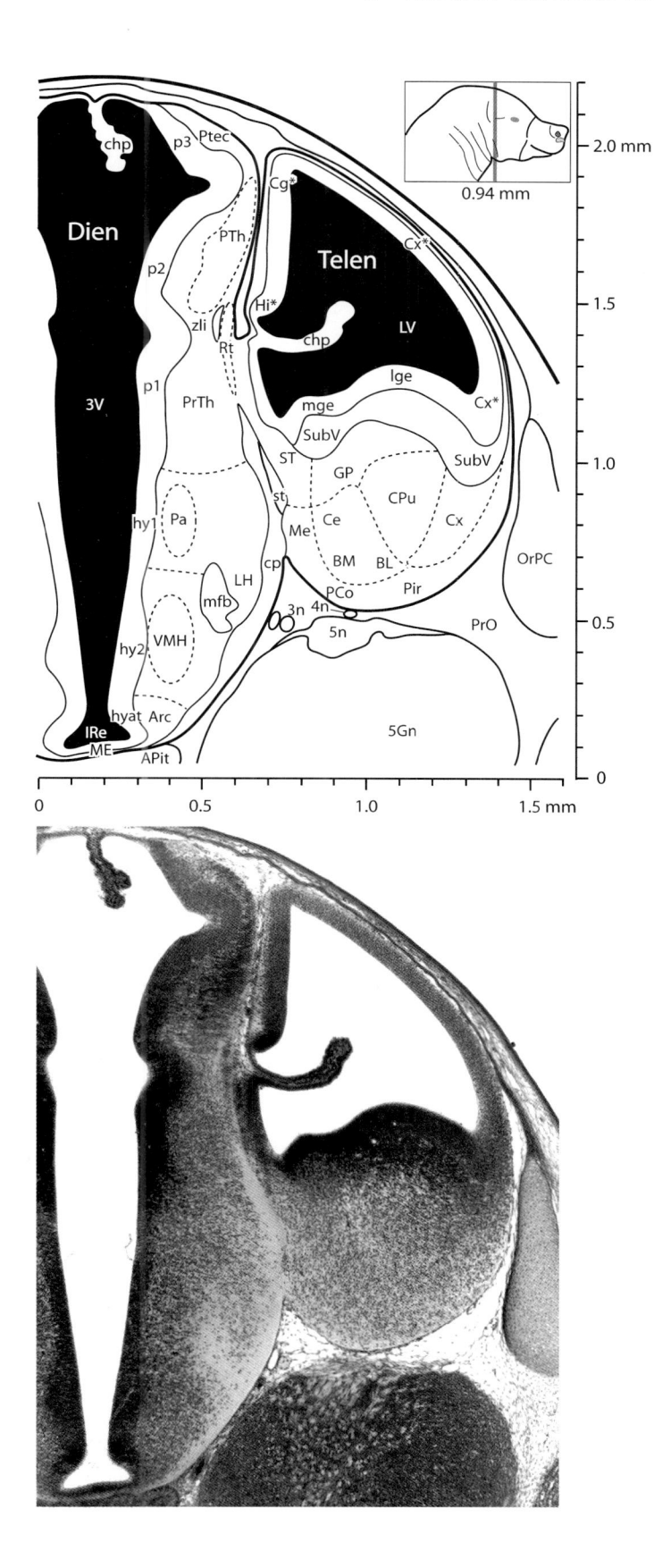

Pl-Ha07

*	denotes precursor
3n	oculomotor nerve
3V	3rd ventricle
4n	trochlear nerve
5Gn	trigeminal ganglion
5n	trigeminal nerve
APit	anterior lobe pituitary
Arc	arcuate hypothalamic nucleus
BL	basolateral amygdaloid nucleus
BM	basomedial amygdaloid nucleus
Ce	central amygdaloid nucleus
Cg	cingulate cortex
chp	choroid plexus
cp	cerebral peduncle
CPu	caudate putamen (striatum)
Cx	cerebral cortex
Dien	diencephalon
GP	globus pallidus
Hi	hippocampus
hy1	peduncular hypothalamus
hy2	tuberal hypothalamus
hyat	acroterminal hypothalamus
IRe	infundibular recess of third ventricle
lge	lateral ganglionic eminence of developing subpallium
LH	lateral hypothalamic area
LV	lateral ventricle
Me	medial amygdaloid nucleus
ME	median eminence
mfb	medial forebrain bundle
mge	medial ganglionic eminence of developing subpallium
OrPC	orbitoparietal commissure
p1	prosomere 1
p2	prosomere 2
p3	prosomere 3
Pa	paraventricular hypothalamic nucleus
PCo	posterior cortical amygdaloid area
Pir	piriform cortex
PrO	pro-otic foramen
PrTh	prethalamus
Ptec	pretectum
PTh	posterior thalamus
Rt	reticular nucleus (prethalamus)
ST	bed nucleus of stria terminalis
st	stria terminalis
SubV	subventricular layer of developing cortex
Telen	telencephalon
VMH	ventromedial hypothalamic nucleus
zli	zona limitans interthalamica

Pl-Ha08

*	denotes precursor
3n	oculomotor nerve
3V	3rd ventricle
4n	trochlear nerve
5Gn	trigeminal ganglion
5n	trigeminal nerve
APit	anterior lobe pituitary
Arc	arcuate hypothalamic nucleus
BL	basolateral amygdaloid nucleus
Ce	central amygdaloid nucleus
Cg	cingulate cortex
chp	choroid plexus
cp	cerebral peduncle
CPu	caudate putamen (striatum)
Cx	cerebral cortex
Dien	diencephalon
DM	dorsomedial hypothalamic nucleus
GP	globus pallidus
Hi	hippocampus
hs	hypothalamic sulcus
hy1	peduncular hypothalamus
hy2	tuberal hypothalamus
hyat	acroterminal hypothalamus
ic	internal capsule
IRe	infundibular recess of third ventricle
lge	lateral ganglionic eminence of developing subpallium
LH	lateral hypothalamic area
LV	lateral ventricle
ME	median eminence
Me	medial amygdaloid nucleus
mfb	medial forebrain bundle
mge	medial ganglionic eminence of developing subpallium
OrPC	orbitoparietal commissure
p1	prosomere 1
p2	prosomere 2
p3	prosomere 3
PCo	posterior cortical amygdaloid area
Pir	piriform cortex
Pr5	principal sensory trigeminal nucleus
PrTh	prethalamus
Ptec	pretectum
PTh	posterior thalamus
Rt	reticular nucleus (prethalamus)
s5	sensory root trigeminal nerve
ST	bed nucleus of stria terminalis
STh	subthalamic nucleus
SubV	subventricular layer of developing cortex
Telen	telencephalon
VMH	ventromedial hypothalamic nucleus
ZI	zona incerta
zli	zona limitans interthalamica

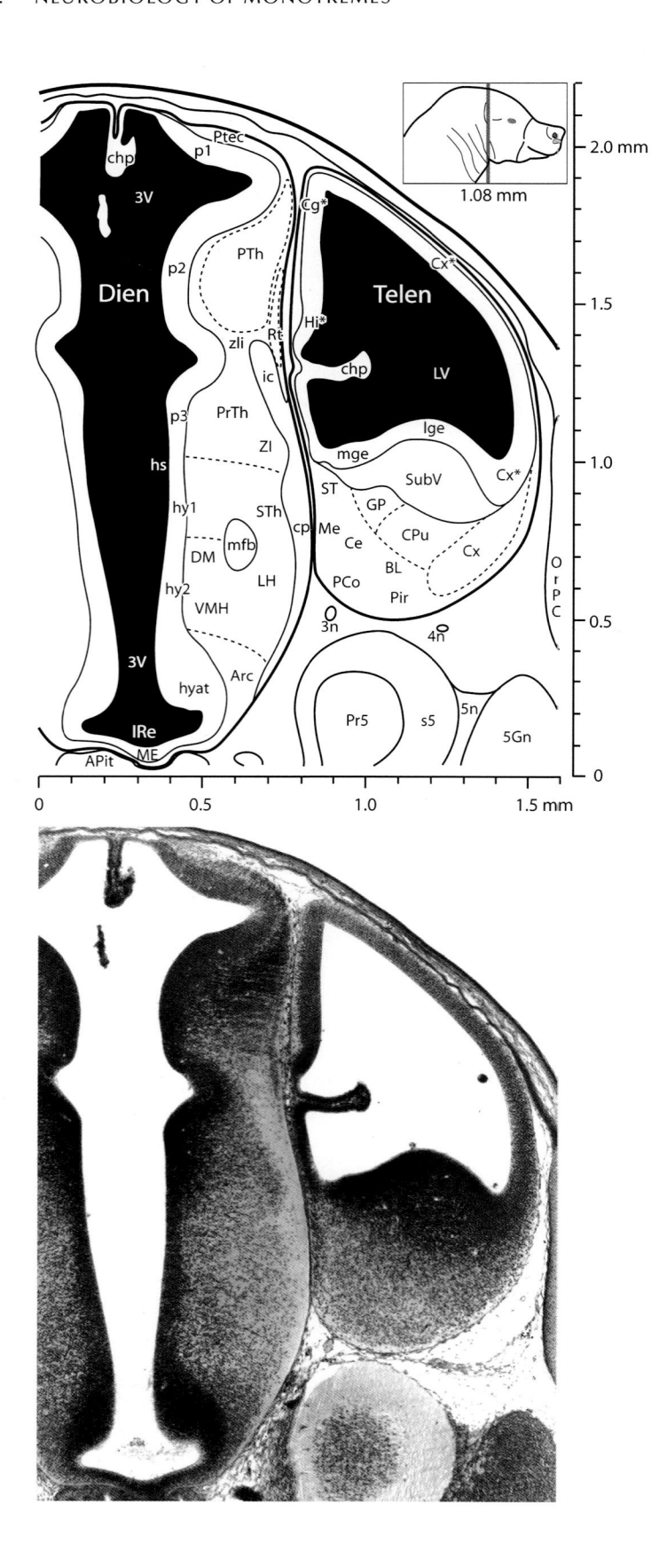

Pl-Ha09

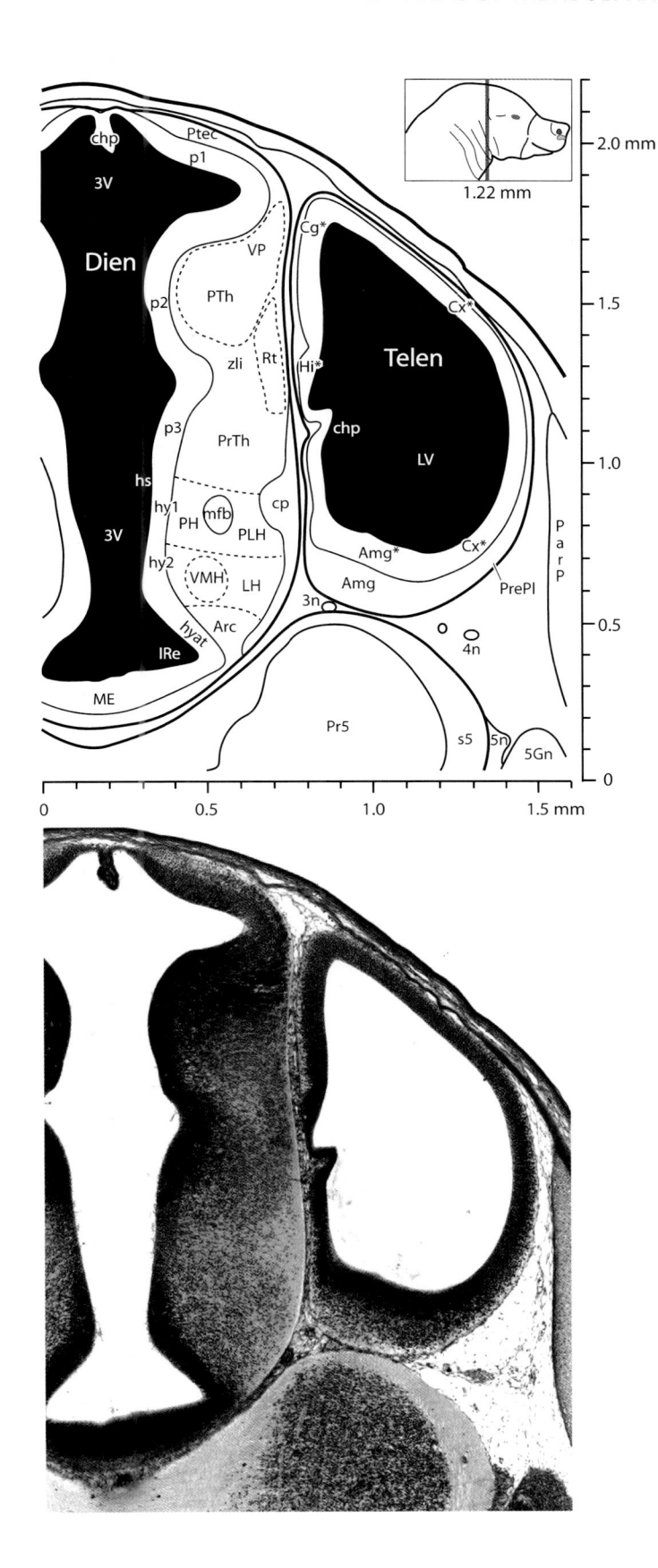

*	denotes precursor
3n	oculomotor nerve
3V	3rd ventricle
4n	trochlear nerve
5Gn	trigeminal ganglion
5n	trigeminal nerve
Amg	amygdala
Arc	arcuate hypothalamic nucleus
Cg	cingulate cortex
chp	choroid plexus
cp	cerebral peduncle
Cx	cerebral cortex
Dien	diencephalon
Hi	hippocampus
hs	hypothalamic sulcus
hy1	peduncular hypothalamus
hy2	tuberal hypothalamus
hyat	acroterminal hypothalamus
IRe	infundibular recess of third ventricle
LH	lateral hypothalamic area
LV	lateral ventricle
ME	median eminence
mfb	medial forebrain bundle
p1	prosomere 1
p2	prosomere 2
p3	prosomere 3
ParP	parietal plate
PH	posterior hypothalamic nucleus
PLH	peduncular part of lateral hypothalamus
Pr5	principal sensory trigeminal nucleus
PrePl	preplate of cortex
PrTh	prethalamus
Ptec	pretectum
PTh	posterior thalamus
Rt	reticular nucleus (prethalamus)
s5	sensory root of trigeminal nerve
Telen	telencephalon
VMH	ventromedial hypothalamic nucleus
VP	ventral pallidum
zli	zona limitans interthalamica

Pl-Ha10

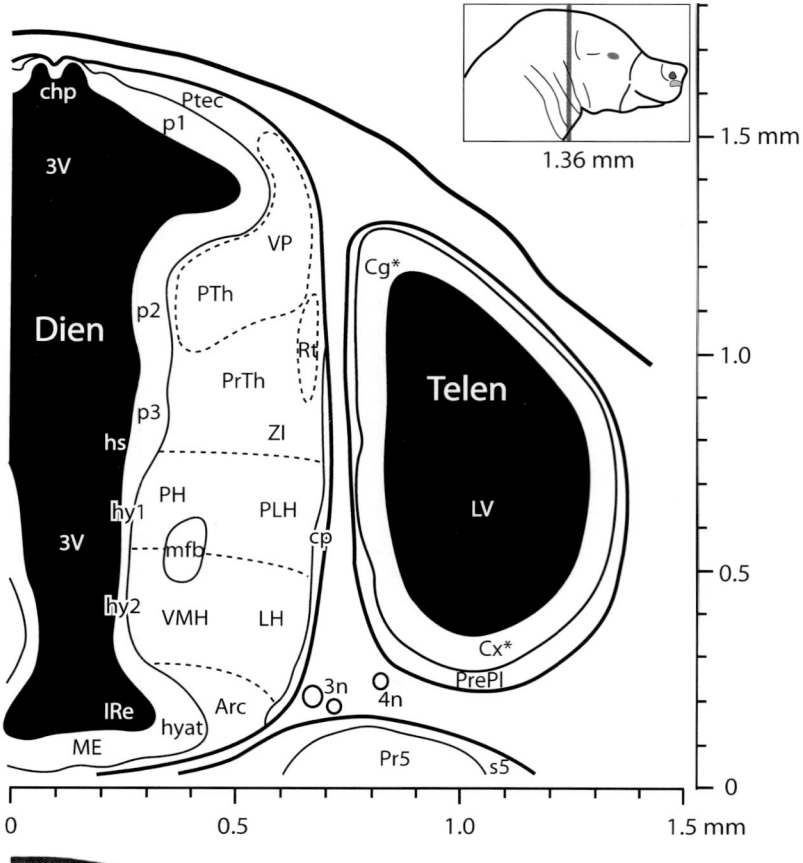

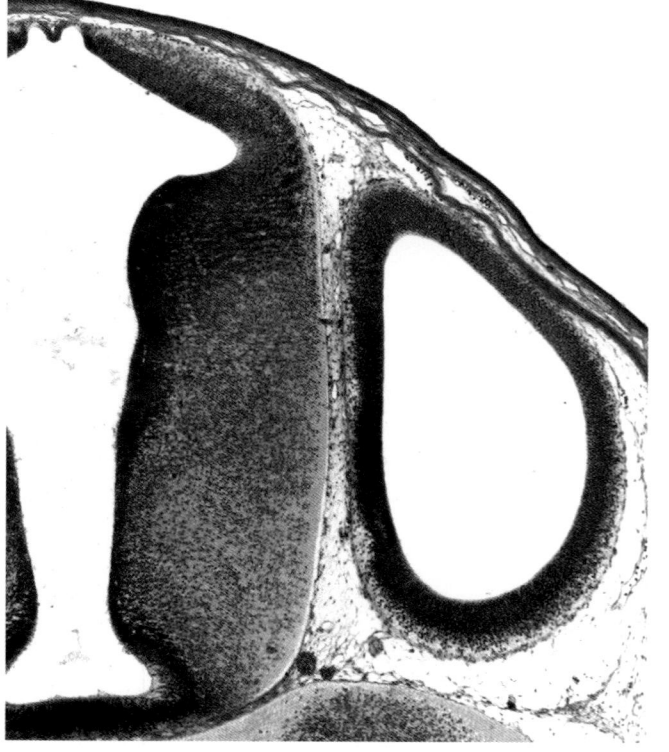

*	denotes precursor
3n	oculomotor nerve
3V	3rd ventricle
4n	trochlear nerve
Arc	arcuate hypothalamic nucleus
Cg	cingulate cortex
chp	choroid plexus
cp	cerebral peduncle
Cx	cerebral cortex
Dien	diencephalon
hs	hypothalamic sulcus
hy1	peduncular hypothalamus
hy2	tuberal hypothalamus
hyat	acroterminal hypothalamus
IRe	infundibular recess of third ventricle
LH	lateral hypothalamic area
LV	lateral ventricle
ME	median eminence
mfb	medial forebrain bundle
p1	prosomere 1
p2	prosomere 2
p3	prosomere 3
PH	posterior hypothalamic nucleus
PLH	peduncular part of lateral hypothalamus
Pr5	principal sensory trigeminal nucleus
PrePl	preplate of cortex
PrTh	prethalamus
Ptec	pretectum
PTh	posterior thalamus
Rt	reticular nucleus (prethalamus)
s5	sensory root of trigeminal nerve
Telen	telencephalon
VMH	ventromedial hypothalamic nucleus
VP	ventral pallidum
ZI	zona incerta

Pl-Ha11

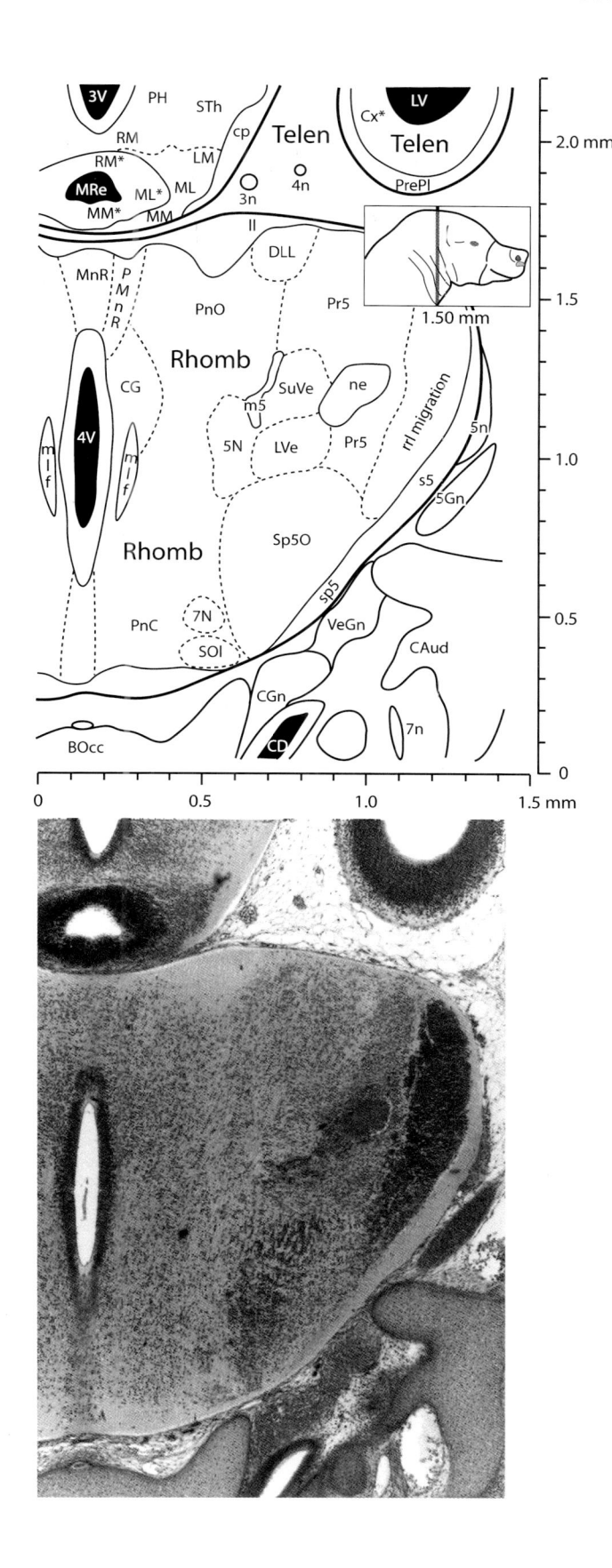

*	denotes precursor
3n	oculomotor nerve
3V	3rd ventricle
4n	trochlear nerve
4V	4th ventricle
5Gn	trigeminal ganglion
5N	motor trigeminal nucleus
5n	trigeminal nerve
7n	facial nerve
7N	facial nucleus
BOcc	basioccipital bone
CAud	canalicular part of auditory capsule
CD	cochlear duct
CG	central grey
CGn	cochlear (spiral) ganglion
cp	cerebral peduncle
Cx	cerebral cortex
DLL	dorsal nucleus of lateral lemniscus
ll	lateral lemniscus
LM	lateral mammillary nucleus
LV	lateral ventricle
LVe	lateral vestibular nucleus
m5	motor root trigeminal nerve
ML	medial mammillary nucleus, lateral part
mlf	medial longitudinal fasciculus
MM	medial mammillary nucleus, medial part
MnR	median raphe nucleus
MRe	mammillary recess of third ventricle
ne	neuroepithelium
PH	posterior hypothalamic nucleus
PMnR	paramedian raphe nucleus
PnC	pontine reticular nucleus, caudal part
PnO	pontine reticular nucleus, oral part
Pr5	principal sensory trigeminal nucleus
PrePl	preplate of cortex
Rhomb	rhombencephalon
RM	retromammillary nucleus
rrl	rostral rhombic lip
s5	sensory root of trigeminal nerve
SOl	superior olive
sp5	spinal trigeminal tract
Sp5O	spinal trigeminal nucleus, oral part
STh	subthalamic nucleus
SuVe	superior vestibular nucleus
Telen	telencephalon
VeGn	vestibular nerve ganglion

Pl-Ha12

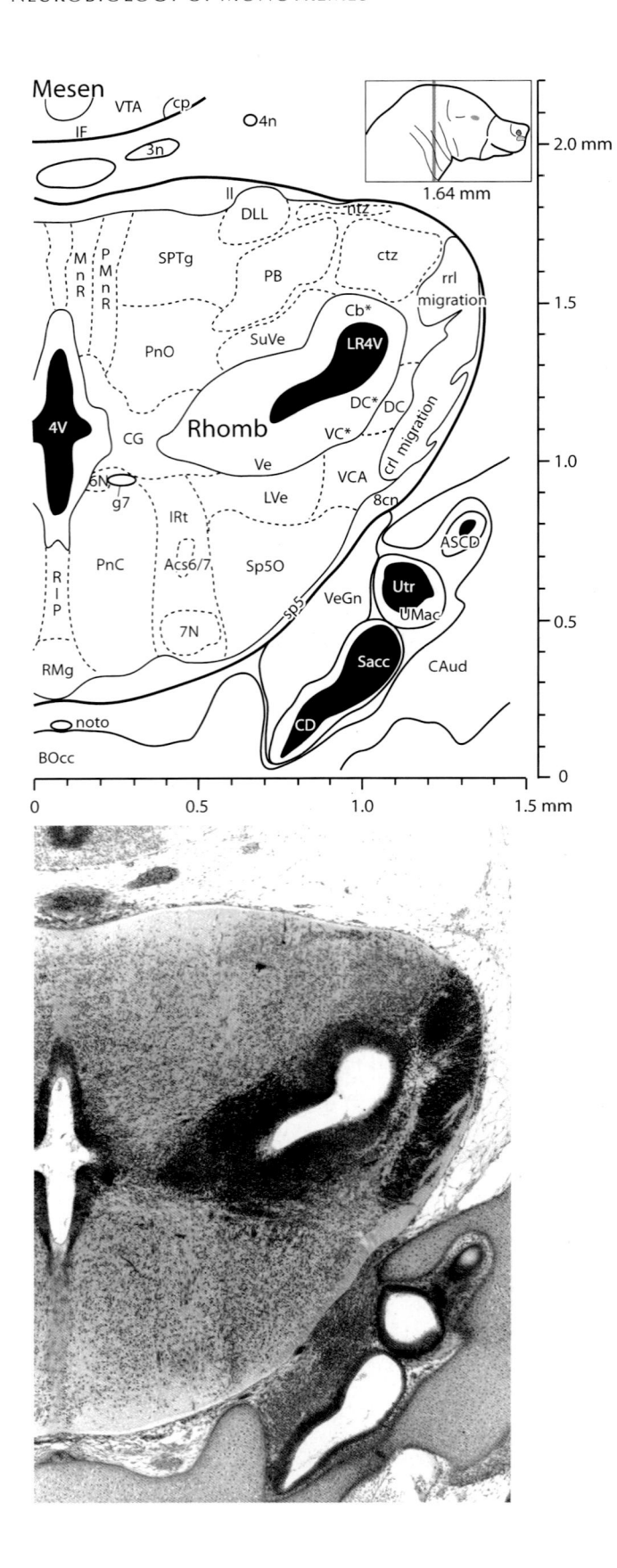

*	denotes precursor
3n	oculomotor nerve
4n	trochlear nerve
4V	4th ventricle
6N	abducens nucleus
7N	facial nucleus
8cn	cochlear root of 8th nerve
Acs6/7	accessory abducens/facial nucleus
ASCD	anterior semicircular duct
BOcc	basioccipital bone
CAud	canalicular part of auditory capsule
Cb	cerebellum
CD	cochlear duct
CG	central grey
cp	cerebral peduncle
crl	caudal rhombic lip
ctz	cortical transitory zone of the cerebellum
DC	dorsal cochlear nucleus
DLL	dorsal nucleus of lateral lemniscus
g7	genu of the facial nerve
IF	interfascicular nucleus
IRt	intermediate reticular nucleus
ll	lateral lemniscus
LR4V	lateral recess of 4th ventricle
LVe	lateral vestibular nucleus
Mesen	mesencephalon
MnR	median raphe nucleus
noto	notochord
ntz	nuclear transitory zone of the cerebellum
PB	parabrachial complex
PMnR	paramedian raphe nucleus
PnC	pontine reticular nucleus, caudal
PnO	pontine reticular nucleus, oral
Rhomb	rhombencephalon
RIP	raphe interpositus nucleus
RMg	raphe magnus nucleus
rrl	rostral rhombic lip
Sacc	saccule
sp5	spinal trigeminal tract
Sp5O	spinal trigeminal nucleus, oral
SPTg	subpeduncular tegmental nucleus
SuVe	superior vestibular nucleus
UMac	macula of the utricle
Utr	utricle
VC	ventral cochlear nucleus
VCA	ventral cochlear nucleus, anterior
Ve	vestibular neuroepithelium
VeGn	vestibular nerve ganglion
VTA	ventral tegmental area

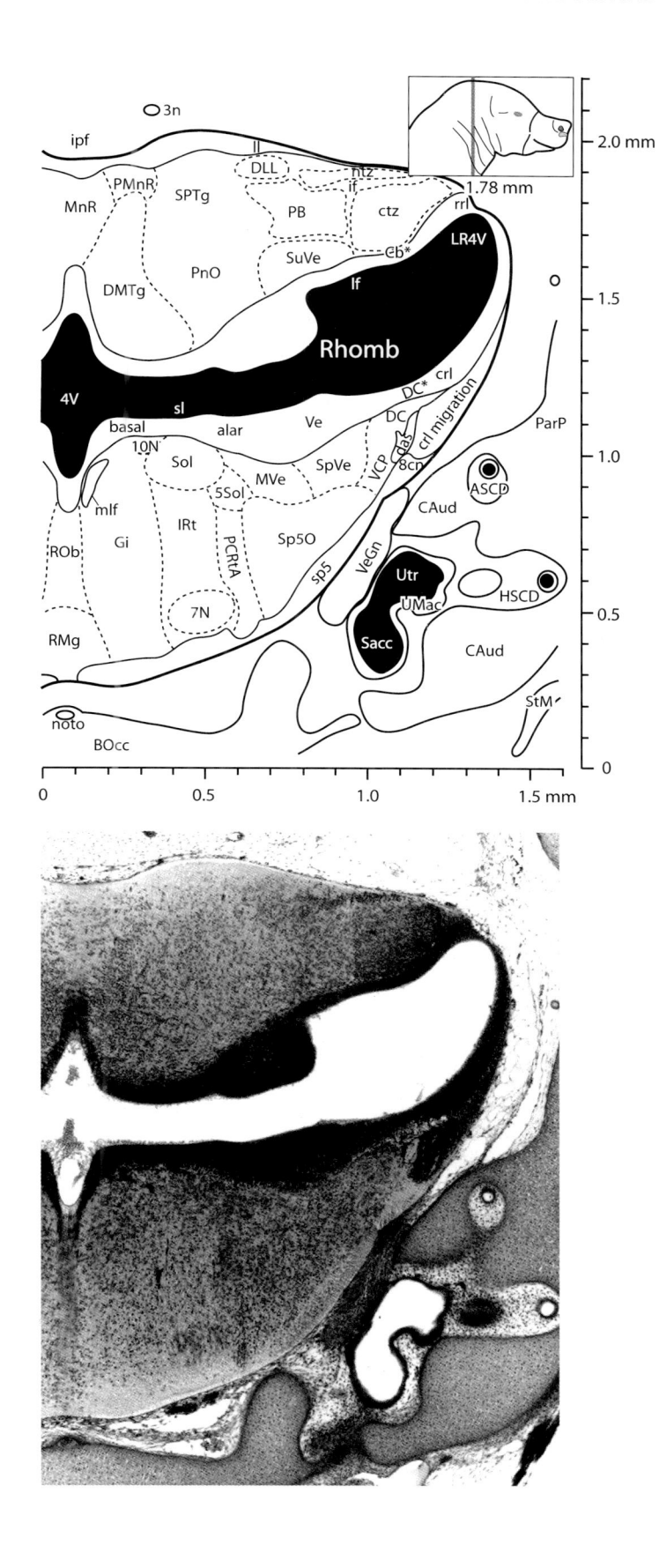

Pl-Ha13

*	denotes precursor
10N	vagus nerve nucleus
3n	oculomotor nerve
4V	4th ventricle
5Sol	trigeminal-solitary transition zone
7N	facial nucleus
8cn	cochlear root of 8th nerve
alar	alar plate
ASCD	anterior semicircular duct
basal	basal plate
BOcc	basioccipital bone
CAud	canalicular part of auditory capsule
Cb	cerebellum
crl	caudal rhombic lip
ctz	cortical transitory zone of cerebellum
das	dorsal acoustic stria
DC	dorsal cochlear nucleus
DLL	dorsal nucleus of lateral lemniscus
DMTg	dorsomedial tegmental area
Gi	gigantocellular reticular nucleus
HSCD	horizontal semicircular duct
if	intermediate fibres of cerebellum
ipf	interpeduncular fossa
IRt	intermediate reticular nucleus
lf	lateral fissure of cerebellum
ll	lateral lemniscus
LR4V	lateral recess of 4th ventricle
mlf	medial longitudinal fasciculus
MnR	median raphe nucleus
MVe	medial vestibular nucleus
noto	notochord
ntz	nuclear transitory zone of cerebellum
ParP	parietal plate
PB	parabrachial complex
PCRtA	parvicell reticular nucleus, alpha
PMnR	paramedian raphe nucleus
PnO	pontine reticular nucleus, oral
Rhomb	rhombencephalon
RMg	raphe magnus nucleus
ROb	raphe obscurus nucleus
rrl	rostral rhombic lip
Sacc	saccule
sl	sulcus limitans
Sol	solitary nucleus
sp5	spinal trigeminal tract
Sp5O	spinal trigeminal nucleus, oral
SPTg	subpeduncular tegmental nucleus
SpVe	spinal vestibular nucleus
StM	sternomastoid muscle
SuVe	superior vestibular nucleus
UMac	macula of the utricle
Utr	utricle
VCP	ventral cochlear nucleus, posterior
Ve	vestibular neuroepithelium
VeGn	vestibular nerve ganglion

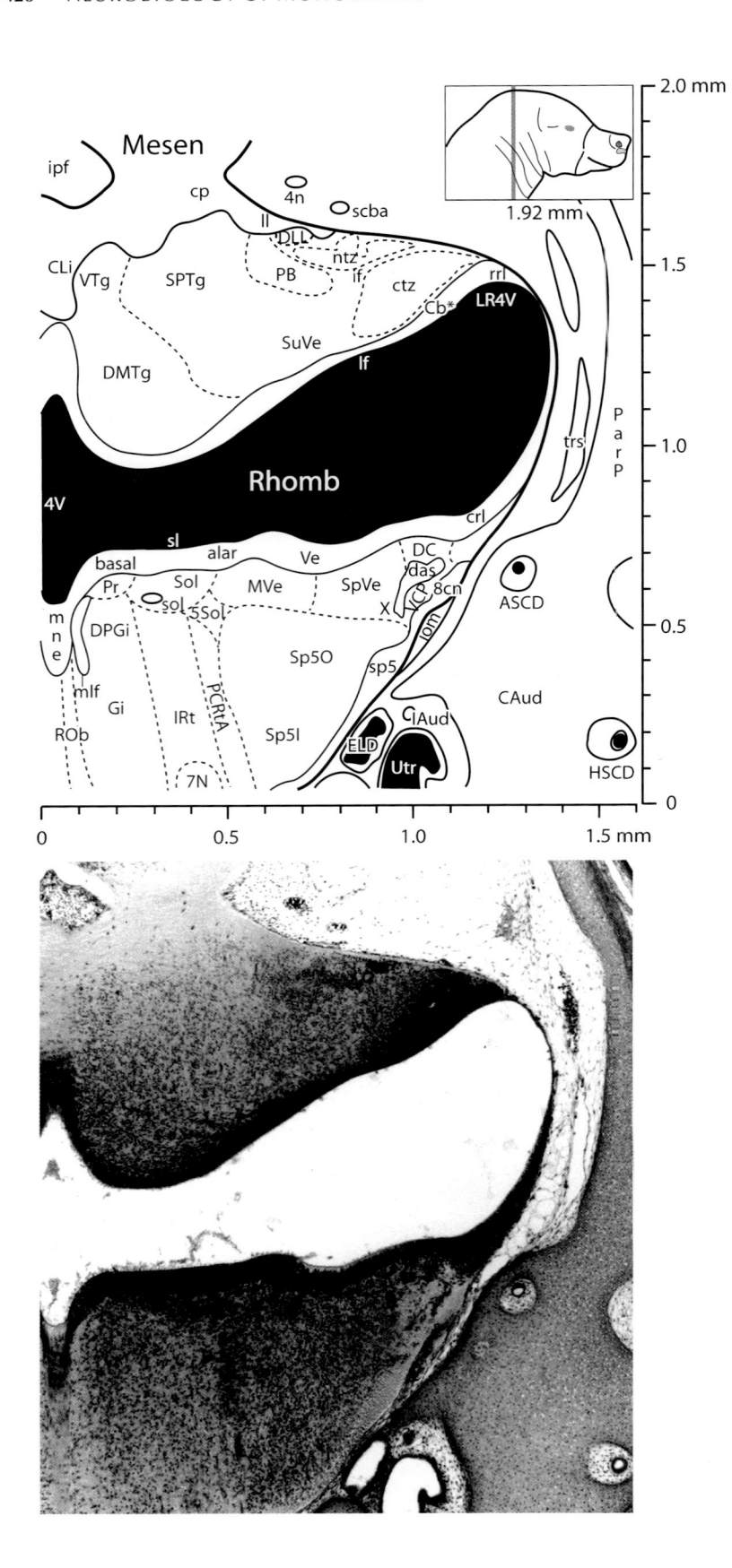

Pl-Ha14

*	denotes precursor
4n	trochlear nerve
4V	4th ventricle
5Sol	trigeminal-solitary transition zone
7N	facial nucleus
8cn	cochlear root of 8th nerve
alar	alar plate
ASCD	anterior semicircular duct
basal	basal plate
CAud	canalicular part of auditory capsule
Cb	cerebellum
CLi	caudal linear nucleus of the raphe
cp	cerebral peduncle
crl	caudal rhombic lip
ctz	cortical transitory zone of cerebellum
das	dorsal acoustic stria
DC	dorsal cochlear nucleus
DLL	dorsal nucleus of lateral lemniscus
DMTg	dorsomedial tegmental area
DPGi	dorsal paragigantocellular nucleus
ELD	endolymphatic duct
Gi	gigantocellular reticular nucleus
HSCD	horizontal semicircular duct
IAud	internal auditory meatus
if	intermediate fibres of cerebellum
iom	inferior olivary migration
ipf	interpeduncular fossa
IRt	intermediate reticular nucleus
lf	lateral fissure of cerebellum
ll	lateral lemniscus
LR4V	lateral recess of 4th ventricle
Mesen	mesencephalon
mlf	medial longitudinal fasciculus
mne	median neuroepithelium
MVe	medial vestibular nucleus
ntz	nuclear transitory zone of cerebellum
ParP	parietal plate
PB	parabrachial complex
PCRtA	parvicellular reticular nucleus, alpha
Pr	prepositus nucleus
Rhomb	rhombencephalon
ROb	raphe obscurus nucleus
rrl	rostral rhombic lip
scba	superior cerebellar artery
sl	sulcus limitans
Sol	solitary nucleus
sol	solitary tract
sp5	spinal trigeminal tract
Sp5I	spinal trigeminal nucleus, interpolar
Sp5O	spinal trigeminal nucleus, oral
SPTg	subpeduncular tegmental nucleus
SpVe	spinal vestibular nucleus
SuVe	superior vestibular nucleus
trs	transverse venous sinus
Utr	utricle
VCP	ventral cochlear nucleus, posterior
Ve	vestibular neuroepithelium
VTg	ventral tegmental nucleus
X	nucleus X

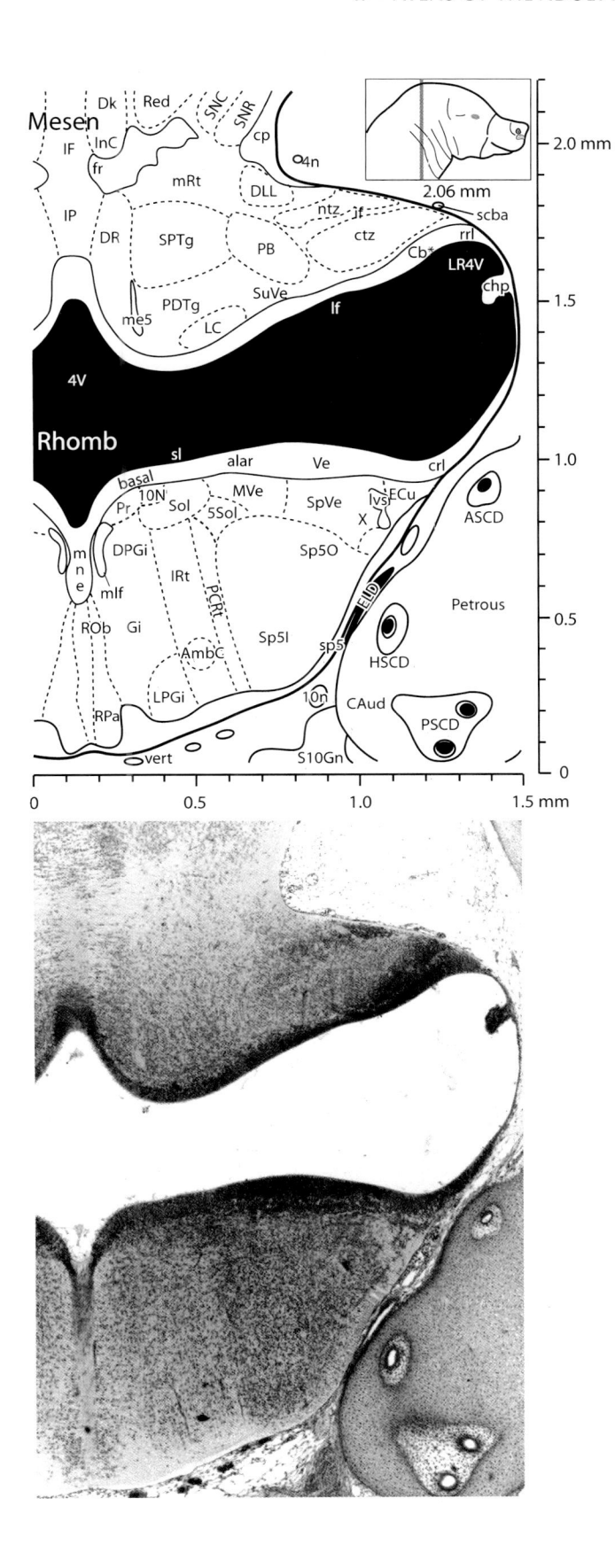

Pl-Ha15

*	denotes precursor
10n	vagus nerve
10N	vagus nerve nucleus
4n	trochlear nerve
4V	4th ventricle
5Sol	trigeminal-solitary transition zone
alar	alar plate
AmbC	ambiguus nucleus, compact part
ASCD	anterior semicircular duct
basal	basal plate
CAud	canalicular part of auditory capsule
Cb	cerebellum
chp	choroid plexus
cp	cerebral peduncle
crl	caudal rhombic lip
ctz	cortical transitory zone of cerebellum
Dk	nu. of Darkschewitsch
DLL	dorsal nucleus of lateral lemniscus
DPGi	dorsal paragigantocellular nucleus
DR	dorsal raphe nucleus
ECu	external cuneate nucleus
ELD	endolymphatic duct
fr	fasciculus retroflexus
Gi	gigantocellular reticular nucleus
HSCD	horizontal semicircular duct
if	intermediate fibres of cerebellum
IF	interfascicular nucleus
InC	interstitial nucleus of Cajal
IP	interpeduncular nucleus
IRt	intermediate reticular nucleus
LC	locus coeruleus
lf	lateral fissure
LPGi	lateral paragigantocellular nucleus
LR4V	lateral recess of 4th ventricle
lvs	lateral vestibulospinal tract
me5	mesencephalic trigeminal tract
Mesen	mesencephalon
mlf	medial longitudinal fasciculus
mne	median neuroepithelium
mRt	mesencephalic reticular formation
MVe	medial vestibular nucleus
ntz	nuclear transitory zone of cerebellum
PB	parabrachial complex
PCRt	parvicellular reticular nucleus
PDTg	posterodorsal tegmental nucleus
Petrous	petrous temporal bone
Pr	prepositus nucleus
PSCD	posterior semicircular duct
Red	red nucleus
Rhomb	rhombencephalon
ROb	raphe obscurus nucleus
RPa	raphe pallidus nucleus
rrl	rostral rhombic lip
S10Gn	superior ganglion of the vagus nerve
scba	superior cerebellar artery
sl	sulcus limitans
SNC	substantia nigra, compact
SNR	substantia nigra, reticular
Sol	solitary nucleus
sp5	spinal trigeminal tract
Sp5I	spinal trigeminal nucleus, interpolar
Sp5O	spinal trigeminal nucleus, oral
SPTg	subpeduncular tegmental nucleus
SpVe	spinal vestibular nucleus
SuVe	superior vestibular nucleus
Ve	vestibular neuroepithelium
vert	vertebral artery
X	nucleus X

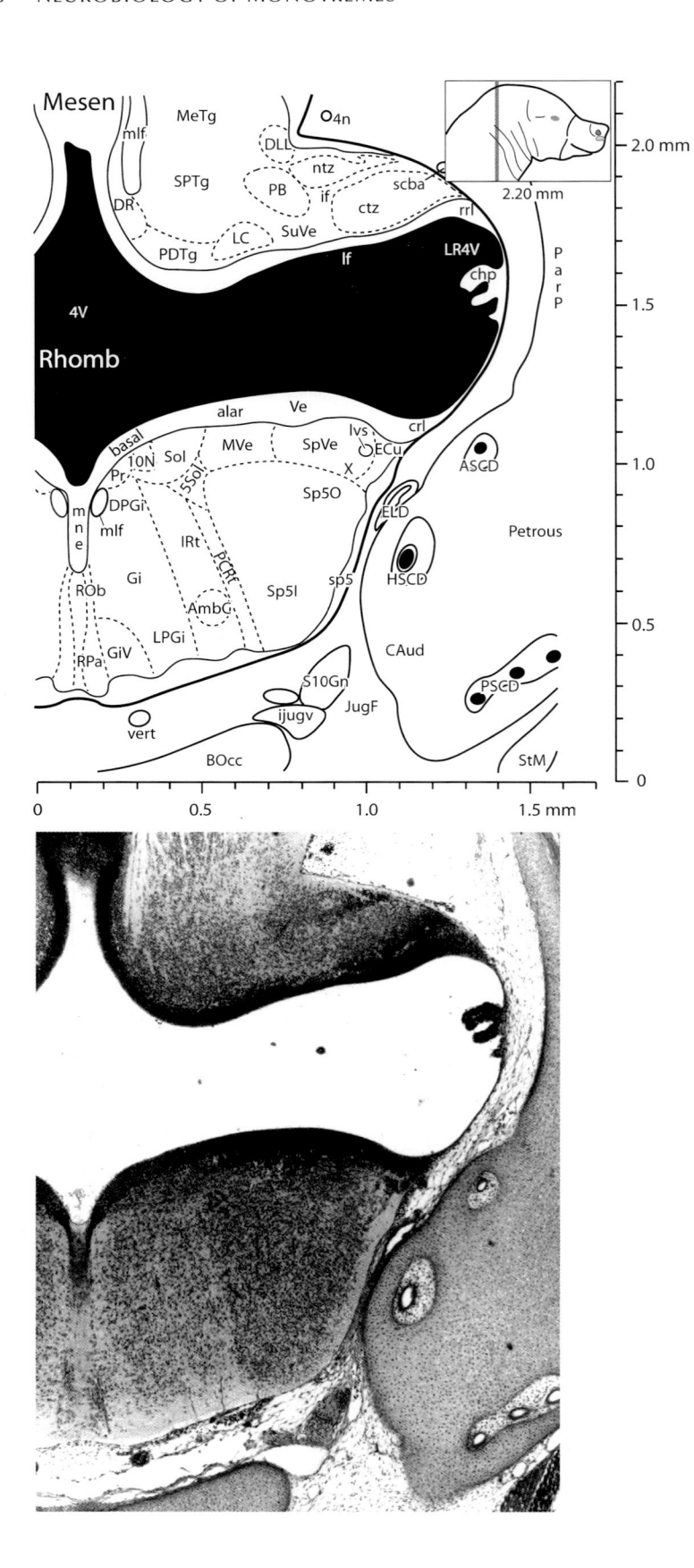

Pl-Ha16

*	denotes precursor
10N	vagus nerve nucleus
4n	trochlear nerve
4V	4th ventricle
5Sol	trigeminal-solitary transition zone
alar	alar plate
AmbC	ambiguus nucleus, compact part
ASCD	anterior semicircular duct
basal	basal plate
BOcc	basioccipital bone
CAud	canalicular part of auditory capsule
chp	choroid plexus
crl	caudal rhombic lip
ctz	cortical transitory zone of cerebellum
DLL	dorsal nucleus of lateral lemniscus
DPGi	dorsal paragigantocellular nucleus
DR	dorsal raphe nucleus
ECu	external cuneate nucleus
ELD	endolymphatic duct
Gi	gigantocellular reticular nucleus
GiV	gigantocellular reticular nucleus, ventral
HSCD	horizontal semicircular duct
if	intermediate fibres of cerebellum
ijugv	internal jugular vein
IRt	intermediate reticular nucleus
JugF	jugular foramen
LC	locus coeruleus
lf	lateral fissure of cerebellum
LPGi	lateral paragigantocellular nucleus
LR4V	lateral recess of 4th ventricle
lvs	lateral vestibulospinal tract
Mesen	mesencephalon
MeTg	mesencephalic tegmentum
mlf	medial longitudinal fasciculus
mne	median neuroepithelium
MVe	medial vestibular nucleus
ntz	nuclear transitory zone of cerebellum
ParP	parietal plate
PB	parabrachial complex
PCRt	parvicellular reticular nucleus
PDTg	posterodorsal tegmental nucleus
Petrous	petrous temporal bone
Pr	prepositus nucleus
PSCD	posterior semicircular duct
Rhomb	rhombencephalon
ROb	raphe obscurus nucleus
RPa	raphe pallidus nucleus
rrl	rostral rhombic lip
S10Gn	superior ganglion of vagus nerve
scba	superior cerebellar artery
Sol	solitary nucleus
sp5	spinal trigeminal tract
Sp5I	spinal trigeminal nucleus, interpolar
Sp5O	spinal trigeminal nucleus, oral
SPTg	subpeduncular tegmental nucleus
SpVe	spinal vestibular nucleus
StM	sternomastoid muscle
SuVe	superior vestibular nucleus
Ve	vestibular neuroepithelium
vert	vertebral artery
X	nucleus X

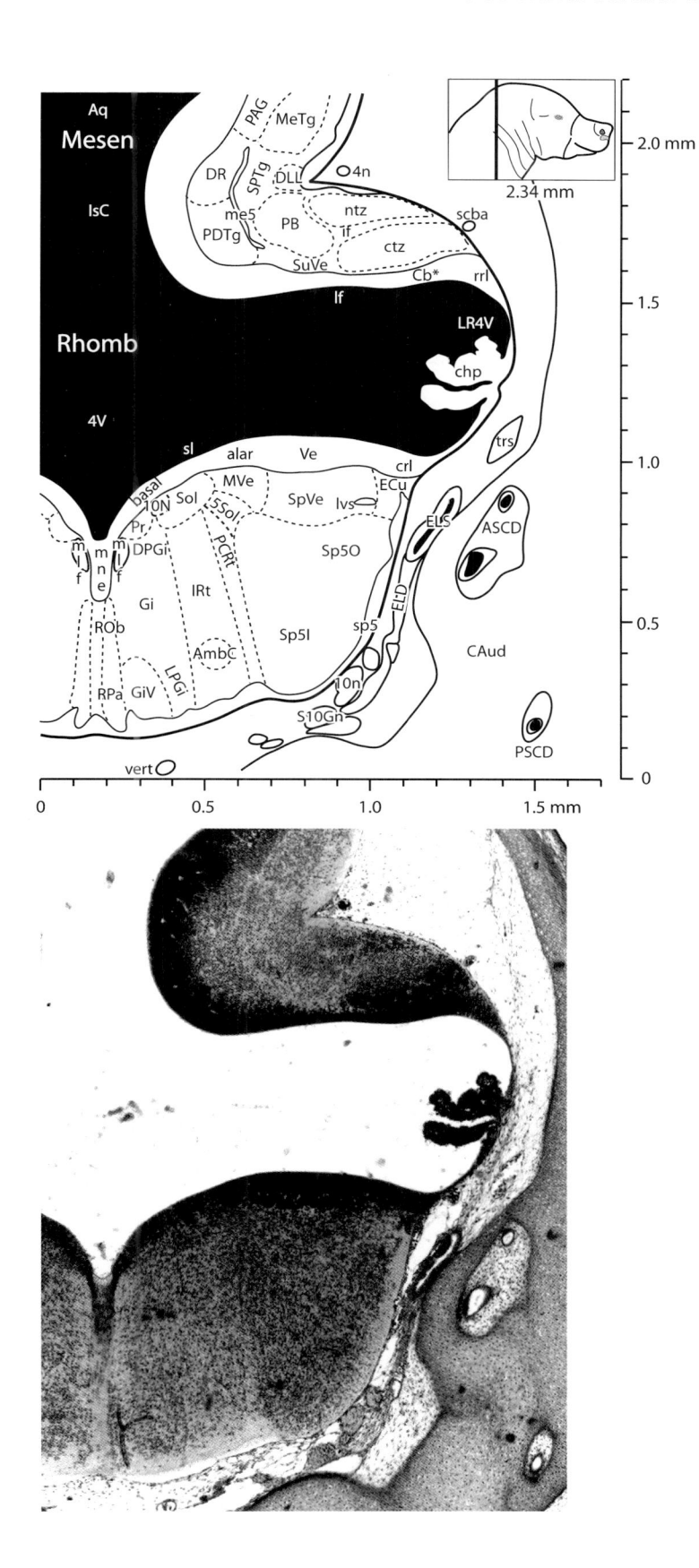

Pl-Ha17

*	denotes precursor
10N	vagus nerve nucleus
10n	vagus nerve
4n	trochlear nerve
4V	4th ventricle
5Sol	trigeminal-solitary transition zone
alar	alar plate
AmbC	ambiguus nucleus, compact part
Aq	cerebral aqueduct
ASCD	anterior semicircular duct
basal	basal plate
CAud	canalicular part of auditory capsule
Cb	cerebellum
chp	choroid plexus
crl	caudal rhombic lip
ctz	cortical transitory zone of cerebellum
DLL	dorsal nu. of lateral lemniscus
DPGi	dorsal paragigantocellular nucleus
DR	dorsal raphe nucleus
ECu	external cuneate nucleus
ELD	endolymphatic duct
ELS	endolymphatic sac
Gi	gigantocellular reticular nucleus
GiV	gigantocellular reticular nucleus, ventral
if	intermediate fibres of cerebellum
IRt	intermediate reticular nucleus
IsC	isthmic canal
lf	lateral fissure
LPGi	lateral paragigantocellular nucleus
LR4V	lateral recess of 4th ventricle
lvs	lateral vestibulospinal tract
me5	mesencephalic trigeminal tract
Mesen	mesencephalon
MeTg	mesencephalic tegmentum
mlf	medial longitudinal fasciculus
mne	median neuroepithelium
MVe	medial vestibular nucleus
ntz	nuclear transitory zone of cerebellum
PAG	periaqueductal grey
PB	parabrachial complex
PCRt	parvicellular reticular nucleus
PDTg	posterodorsal tegmental nucleus
Pr	prepositus nucleus
PSCD	posterior semicircular duct
Rhomb	rhombencephalon
ROb	raphe obscurus nucleus
RPa	raphe pallidus nucleus
rrl	rostral rhombic lip
S10Gn	superior ganglion of vagus nerve
scba	superior cerebellar artery
sl	sulcus limitans
Sol	solitary nucleus
sp5	spinal trigeminal tract
Sp5I	spinal trigeminal nucleus, interpolar
Sp5O	spinal trigeminal nucleus, oral
SPTg	subpeduncular tegmental nucleus
SpVe	spinal vestibular nucleus
SuVe	superior vestibular nucleus
trs	transverse venous sinus
Ve	vestibular neuroepithelium
vert	vertebral artery

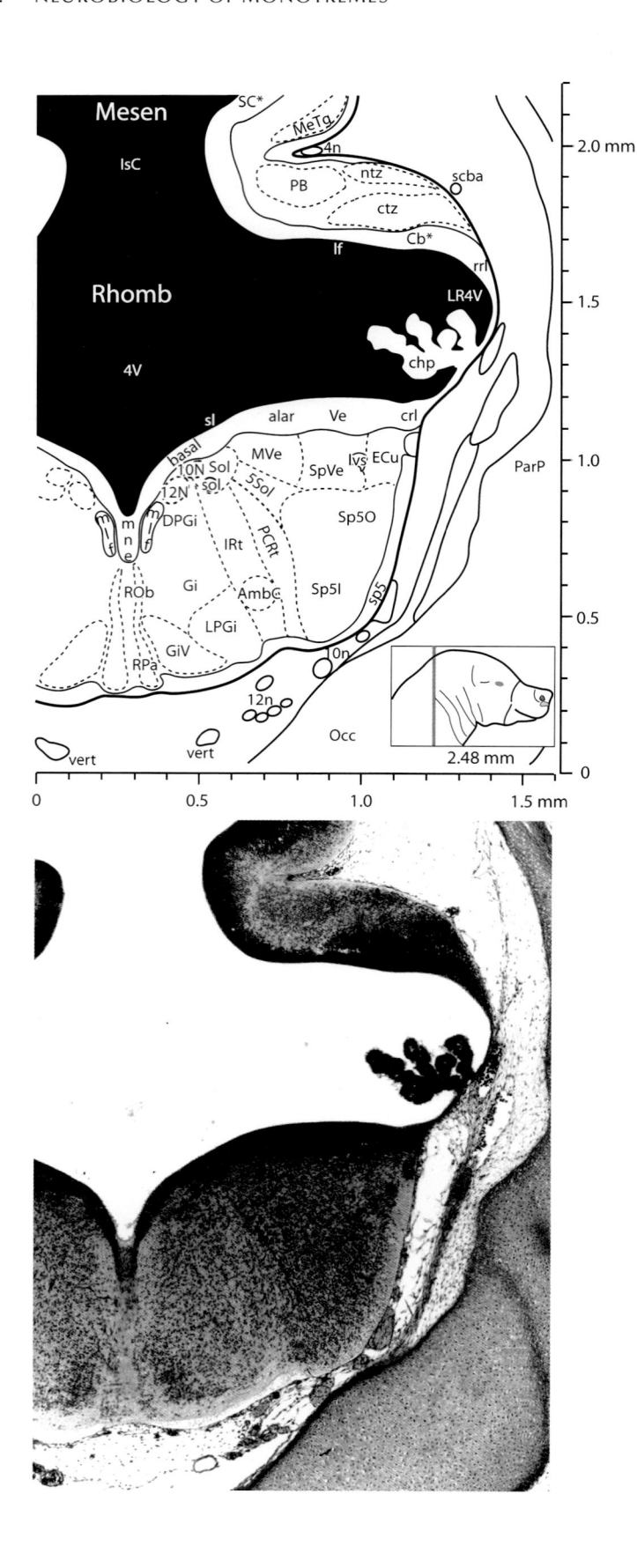

Pl-Ha18

*	denotes precursor
10n	vagus nerve
10N	vagus nerve nucleus
12n	hypoglossal nerve
12N	hypoglossal nucleus
4n	trochlear nerve
4V	4th ventricle
5Sol	trigeminal-solitary transition zone
alar	alar plate
AmbC	ambiguus nucleus, compact part
basal	basal plate
Cb	cerebellum
chp	choroid plexus
crl	caudal rhombic lip
ctz	cortical transitory zone of cerebellum
DPGi	dorsal paragigantocellular nucleus
ECu	external cuneate nucleus
Gi	gigantocellular reticular nucleus
GiV	gigantocellular reticular nucleus, ventral
IRt	intermediate reticular nucleus
IsC	isthmic canal
lf	lateral fissure
LPGi	lateral paragigantocellular nucleus
LR4V	lateral recess of 4th ventricle
lvs	lateral vestibulospinal tract
Mesen	mesencephalon
MeTg	mesencephalic tegmentum
mlf	medial longitudinal fasciculus
mne	median neuroepithelium
MVe	medial vestibular nucleus
ntz	nuclear transitory zone of cerebellum
Occ	occipital bone
ParP	parietal plate
PB	parabrachial complex
PCRt	parvicellular reticular nucleus
Rhomb	rhombencephalon
ROb	raphe obscurus nucleus
RPa	raphe pallidus nucleus
rrl	rostral rhombic lip
SC	superior colliculus
scba	superior cerebellar artery
sl	sulcus limitans
Sol	solitary nucleus
sol	solitary tract
sp5	spinal trigeminal tract
Sp5I	spinal trigeminal nucleus, interpolar
Sp5O	spinal trigeminal nucleus, oral
SpVe	spinal vestibular nucleus
Ve	vestibular neuroepithelium
vert	vertebral artery

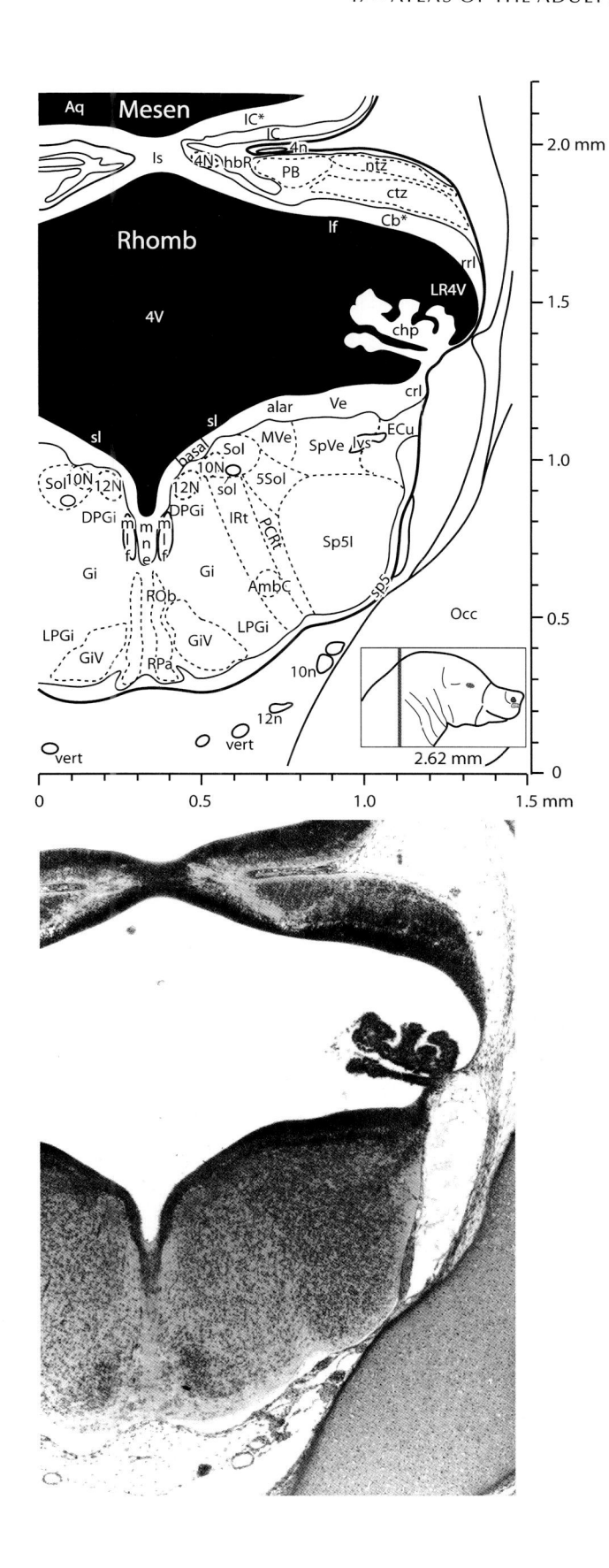

Pl-Ha19

*	denotes precursor
10n	vagus nerve
10N	vagus nerve nucleus
12n	hypoglossal nerve
12N	hypoglossal nucleus
4n	trochlear nerve
4N	trochlear nucleus
4V	4th ventricle
5Sol	trigeminal-solitary transition zone
alar	alar plate
AmbC	ambiguus nucleus, compact part
Aq	cerebral aqueduct
basal	basal plate
Cb	cerebellum
chp	choroid plexus
crl	caudal rhombic lip
ctz	cortical transitory zone of cerebellum
DPGi	dorsal paragigantocellular nucleus
ECu	external cuneate nucleus
Gi	gigantocellular reticular nucleus
GiV	gigantocellular reticular nucleus, ventral
hbR	hook bundle of Russell
IC	inferior colliculus
IRt	intermediate reticular nucleus
Is	isthmus
lf	lateral fissure
LPGi	lateral paragigantocellular nucleus
LR4V	lateral recess of 4th ventricle
lvs	lateral vestibulospinal tract
Mesen	mesencephalon
mlf	medial longitudinal fasciculus
mne	median neuroepithelium
MVe	medial vestibular nucleus
ntz	nuclear transitory zone of cerebellum
Occ	occipital bone
PB	parabrachial complex
PCRt	parvicellular reticular nucleus
Rhomb	rhombencephalon
ROb	raphe obscurus nucleus
RPa	raphe pallidus nucleus
rrl	rostral rhombic lip
sl	sulcus limitans
Sol	solitary nucleus
sol	solitary tract
sp5	spinal trigeminal tract
Sp5I	spinal trigeminal nucleus, interpolar
SpVe	spinal vestibular nucleus
Ve	vestibular neuroepithelium
vert	vertebral artery

Pl-Ha20

*	denotes precursor
10n	vagus nerve
10N	vagus nerve nucleus
12n	hypoglossal nerve
12N	hypoglossal nucleus
4V	4th ventricle
5Sol	trigeminal-solitary transition zone
alar	alar plate
AmbC	ambiguus nucleus, compact part
Aq	cerebral aqueduct
basal	basal plate
Cb	cerebellum
chp	choroid plexus
crl	caudal rhombic lip
DPGi	dorsal paragigantocellular nucleus
ECu	external cuneate nucleus
Gi	gigantocellular reticular nucleus
GiV	gigantocellular reticular nucleus, ventral
IC	inferior colliculus
IRt	intermediate reticular nucleus
LPGi	lateral paragigantocellular nucleus
lvs	lateral vestibulospinal tract
Mesen	mesencephalon
mlf	medial longitudinal fasciculus
mne	median neuroepithelium
MVe	medial vestibular nucleus
ntz	nuclear transitory zone of cerebellum
Occ	occipital bone
PCRt	parvicellular reticular nucleus
Rhomb	rhombencephalon
ROb	raphe obscurus nucleus
RPa	raphe pallidus nucleus
rrl	rostral rhombic lip
sl	sulcus limitans
Sol	solitary nucleus
sol	solitary tract
sp5	spinal trigeminal tract
Sp5C	spinal trigeminal nucleus, caudal part
Sp5I	spinal trigeminal nucleus, interpolar part
SpVe	spinal vestibular nucleus
Ve	vestibular neuroepithelium
vert	vertebral artery

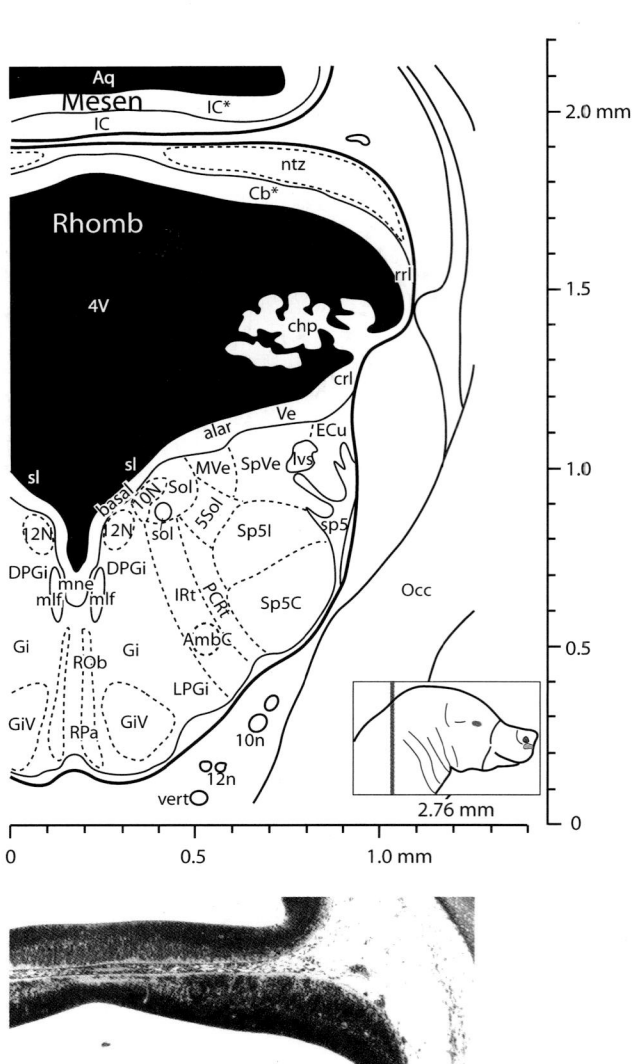

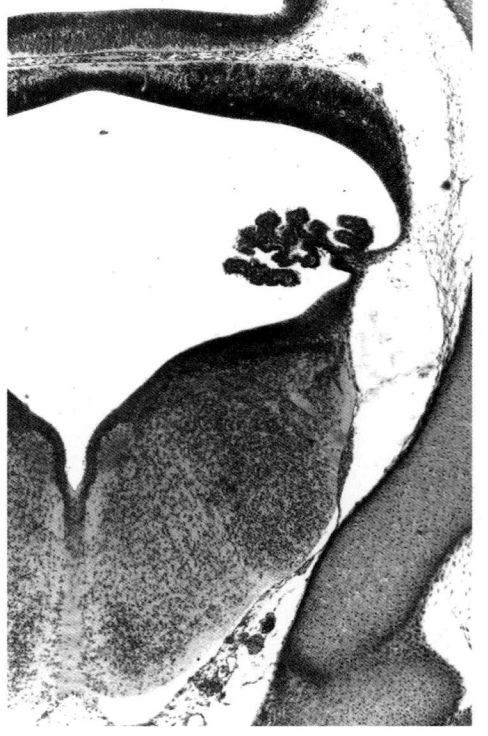

Pl-Ha21

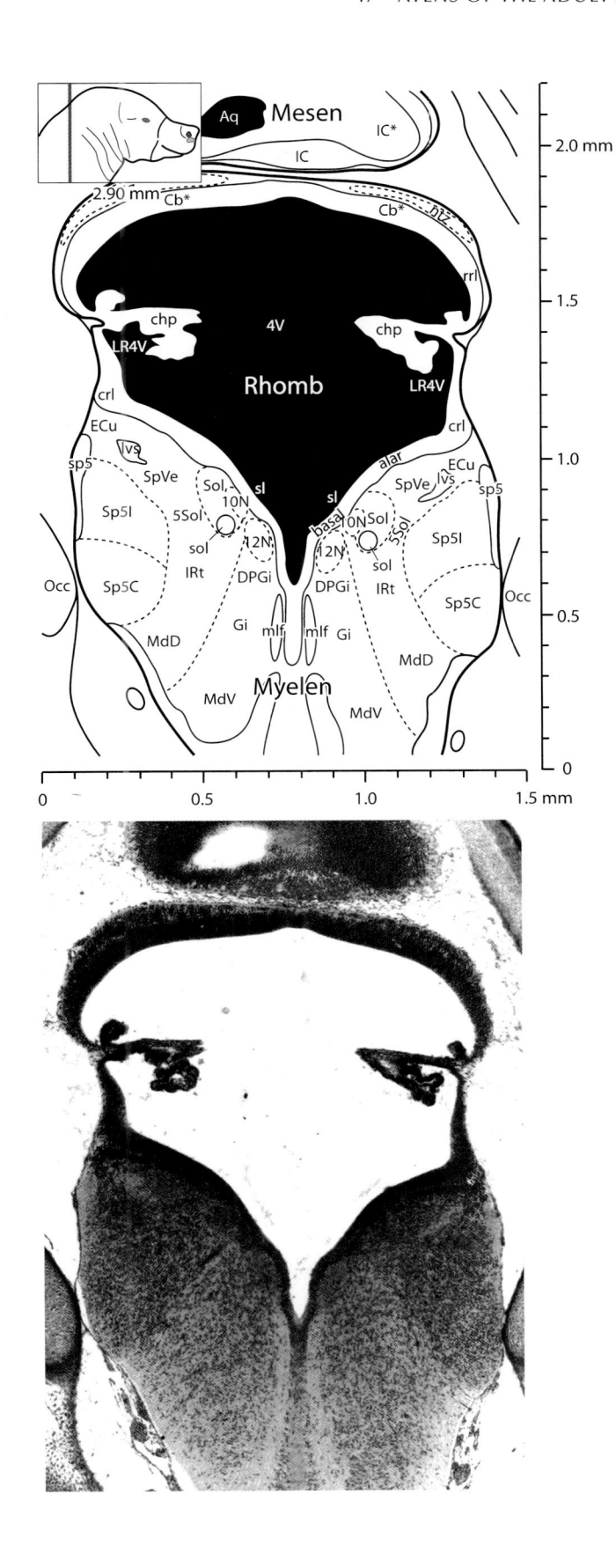

*	denotes precursor
10N	vagus nerve nucleus
12N	hypoglossal nucleus
4V	4th ventricle
5Sol	trigeminal-solitary transition zone
alar	alar plate
Aq	cerebral aqueduct
basal	basal plate
Cb	cerebellum
chp	choroid plexus
crl	caudal rhombic lip
DPGi	dorsal paragigantocellular nucleus
ECu	external cuneate nucleus
Gi	gigantocellular reticular nucleus
IC	inferior colliculus
IRt	intermediate reticular nucleus
LR4V	lateral recess of 4th ventricle
lvs	lateral vestibulospinal tract
MdD	medullary reticular nucleus, dorsal
MdV	medullary reticular nucleus, ventral
Mesen	mesencephalon
mlf	medial longitudinal fasciculus
Myelen	myelencephalon
ntz	nuclear transitory zone of cerebellum
Occ	occipital bone
Rhomb	rhombencephalon
rrl	rostral rhombic lip
sl	sulcus limitans
Sol	solitary nucleus
sol	solitary tract
sp5	spinal trigeminal tract
Sp5C	spinal trigeminal nucleus, caudal part
Sp5I	spinal trigeminal nucleus, interpolar part
SpVe	spinal vestibular nucleus

LIST OF ABBREVIATIONS USED IN BRAIN AND EMBRYO ATLAS PLATES

A
A5 noradrenaline cells **A5**
abducens nerve **6n**
accessory nucleus abducens/facial **Acs6/7**
accessory neurosecretory nuclei **ANS**
accumbens nucleus **Acb**
accumbens nucleus, core **AcbC**
accumbens nucleus, shell **AcbSh**
acroterminal hypothalamus **hyat**
agranular insular cortex **AI**
    dorsal part **AID**
    posterior part **AIP**
    ventral part **AIV**
alar plate of neural tube **alar**
alveus of the hippocampus **alv**
ambiguus nucleus **Amb**
    compact part **AmbC**
    subcompact part **AmbSC**
amygdala **Amg**
amygdalohippocampal area **AHi**
amygdaloid fissure **af**
amygdalopiriform transition area **APir**
amygdalostriatal transition area **ASt**
ansiform lobule of cerebellum **Ans**
anterior amygdaloid area **AA**
anterior cerebral artery **acer**
anterior commissure **ac**
    anterior part **aca**
    posterior part **acp**
anterior cortical amygdaloid area **ACo**
anterior hypothalamic area, anterior part **AHA**
anterior lobe of the pituitary **APit**
anterior nucleus of thalamus **Ant**
anterior olfactory area **AO**
    dorsal part **AOD**
    external part **AOE**
    lateral part **AOL**
    medial part **AOM**
    ventral part **AOV**
anterior pretectal nucleus **APT**
    dorsal part **APTD**
    lateral part **APTL**
    medial part **APTM**
    ventral part **APTV**
anterior semicircular duct **ASCD**
anterior spinal artery **asp**
anterior thalamic region **ATh**
anterobasal nucleus **AB**
anterodorsal thalamic nucleus **AD**
anteromedial thalamic nucleus **AM**
    ventral part **AMV**
anteroventral thalamic nucleus **AV**
arcuate hypothalamic nucleus **Arc**
arcuate medullary nucleus **Ar**
area postrema **AP**
arytenoid cartilage **Ary**
auditory cortex **Au**
auditory/somato. bimodal sensory cortex **Au/S1**

B
B9 serotonin cells **B9**
Barrington's nucleus **Bar**
basal nucleus (of Meynert) **B**
basal plate of neural tube **basal**
basal ventromedial nucleus **VMb**
basilar artery **bas**
basioccipital bone **BOcc**
basisphenoid bone **BSph**
basolateral amygdaloid nucleus **BL**
    anterior part **BLA**
    lateral part **BLL**
    medial part **BLM**
    posterior part **BLP**
basomedial amygdaloid nucleus **BM**
    anterior part **BMA**
    posterior part **BMP**
bed nucleus of accessory olfactory tract **BAOT**
bed nucleus of the stria terminalis **ST**
    intraamygdaloid division **STIA**
    lateral division, dorsal part **STLD**

lateral division, posterior part **STLP**
lateral division, ventral part **STLV**
medial division, anterior part **STMA**
medial division, posterointermed. part **STMPI**
medial division, posterolateral part **STMPL**
medial division, posteromedial part **STMPM**
medial division, ventral part **STMV**
brachium of the inferior colliculus **bic**
brachium of the superior colliculus **bsc**

C
canalicular part of auditory capsule **CAud**
cardinal vein **cdnv**
caudal linear nucleus of the raphe **CLi**
caudal rhombic lip **crl**
caudate nucleus **Cd**
caudate putamen (striatum) **CPu**
caudomedial entorhinal cortex **CEnt**
cavernous sinus **cav**
cell bridges of the ventral striatum **CB**
central amygdaloid nucleus **Ce**
    capsular part **CeC**
    lateral division **CeL**
    medial division **CeM**
central canal **CC**
central grey **CG**
central medial thalamic nucleus **CM**
central nucleus of the inferior colliculus **CIC**
centrolateral thalamic nucleus **CL**
cerebellum **Cb**
cerebral aqueduct **Aq**
cerebral cortex (developing) **Cx**
cerebral peduncle **cp**
choroid plexus **chp**
cingulate cortex **Cg**
    area 1 **Cg1**
    area 2 **Cg2**
cingulum **cg**
Clarke's column/nucleus dorsalis **ClC**
claustrum **Cl**
cochlear (spiral) ganglion **CGn**
cochlear duct **CD**
cochlear root of the vestibulocochlear nerve **8cn**
commissure of the superior colliculus **csc**
common ventricular chamber of heart **cvent**
conjunctival sac **Conjunct**
cornu Ammonis of hippocampus **CA**
cortex-amygdala transition zone **CxA**

cortical transitory zone of cerebellum **ctz**
cribriform plate of the ethmoid **CrP**
cuneate fasciculus **cu**
cuneate nucleus **Cu**
cuneiform nucleus **CnF**

D
decussation of superior cerebellar peduncle **xscp**
decussation of trochlear nerve **d4n**
deep cerebral white matter **dcw**
deep grey layer of the superior colliculus **DpG**
deep white layer of the superior colliculus **DpWh**
dentate gyrus **DG**
diencephalon **Dien**
dorsal third ventricle **D3V**
dorsal acoustic stria **das**
dorsal aorta **daorta**
dorsal cochlear nucleus **DC**
dorsal cortex of the inferior colliculus **DCIC**
dorsal endopiriform nucleus **DEn**
dorsal hippocampal commissure **dhc**
dorsal hypothalamic area **DA**
dorsal nucleus of the lateral lemniscus **DLL**
dorsal paragigantocellular nucleus **DPGi**
dorsal raphe nucleus **DR**
dorsal root ganglion **DRG**
dorsal roots of spinal cord **dr**
dorsal subiculum **DS**
dorsal tegmental nucleus **DTg**
dorsal tenia tecta **DTT**
dorsal transition zone **DTr**
dorsolateral periaqueductal grey **DLPAG**
dorsomedial hypothalamic nucleus **DM**
    dorsal part **DMD**
    ventral part **DMV**
dorsomedial periaqueductal grey **DMPAG**
dorsomedial spinal trigeminal nucleus **DMSp5**
dorsomedial tegmental area **DMTg**

E
ectorhinal cortex **Ect**
Edinger-Westphal nucleus **EW**
endolymphatic duct **ELD**
endolymphatic sac **ELS**
entopeduncular nucleus **EP**
entorhinal cortex **Ent**
ependyma and subependymal layer **E**
ethmoid bone **EthB**

extended amygdala **EA**
external capsule **ec**
external cortex of the inferior colliculus **ECIC**
external cuneate nucleus **ECu**
extreme capsule **ex**

F
facial nerve **7n**
facial nucleus **7N**
    dorsal part **7Nd**
    ventral part **7Nv**
facioacoustic ganglion **fag**
fasciculus retroflexus **fr**
field CA1 of the hippocampus **CA1**
field CA2 of the hippocampus **CA2**
field CA3 of the hippocampus **CA3**
fimbria of the hippocampus **fi**
fimbriodentate sulcus of hippocampus **fds**
flocculus of cerebellum **Fl**
fornix **f**
frontal association cortex **FrA**
frontal bone **Fro**
frontal br. of ophthalmic division of trigeminal nerve **5fr**
frontal cortex **Fr**
    area 1 **Fr1**
    area 2 **Fr2**
    area 3 **Fr3**

G
gelatinous layer of Sp5C **Ge5**
geniculate ganglion **Gen**
genu of the facial nerve **g7**
gigantocellular reticular nucleus **Gi**
    α part **GiA**
    ventral part **GiV**
globus pallidus **GP**
glomerular layer of accessory olfactory bulb **GlA**
glomerular layer of olfactory bulb **Gl**
glossopharyngeal nerve **9n**
gracile fasciculus **gr**
gracile nucleus **Gr**
granular insular cortex **GI**
granule cell cluster **GCC**
granule cell layer of accessory olfactory bulb **GrA**
granule cell layer of cerebellum **GrCb**
granule cell layer of dentate gyrus **GrDG**

granule cell layer of olfactory bulb **GrO**
gustatory cortex **Gust**

H
habenular nuclei **Hb**
hamulus of the pterygoid bone **HPtg**
Harderian gland **HardG**
hippocampal fissure **hif**
hippocampus **Hi**
hook bundle of Russell **hbR**
horizontal semicircular duct **HSCD**
hypoglossal nerve **12n**
hypoglossal nucleus **12N**
hypothalamic sulcus **hs**

I
indusium griseum **IG**
inferior cerebellar peduncle **icp**
inferior colliculus **IC**
inferior olivary migration **iom**
inferior olivary nucleus **IO**
    β subnucleus **IOBe**
    dorsal nucleus **IOD**
    medial nucleus **IOM**
    subnucleus A of IOM **IOA**
    ventral nucleus **IOV**
inferior rectus muscle **IRec**
infralimbic cortex **IL**
infundibular recess of third ventricle **IRe**
inner blade of dentate gyrus **IBl**
insular cortex **Ins**
interanterodorsal thalamic nucleus **IAD**
interfascicular nucleus **IF**
intermediate fibres of cerebellum **if**
intermediate grey layer of superior colliculus **InG**
intermediate nucleus of the lateral lemniscus **ILL**
intermediate reticular nucleus **IRt**
intermediodorsal thalamic nucleus **IMD**
intermediolateral cell column of SpC **IML**
intermediomedial nucleus of SpC **IMSpC**
internal arcuate fibres **ia**
internal auditory meatus **IAud**
internal basilar nucleus of the spinal cord **IB**
internal capsule **ic**
internal carotid artery **ictd**
internal jugular vein **ijugv**
interpallidal commissure **ipc**
interpeduncular fossa **ipf**

interpeduncular nucleus **IP**
   caudal subnucleus **IPC**
   rostral subnucleus **IPR**
interposed cerebellar nucleus, anterior part **IntA**
interposed cerebellar nucleus, posterior part **IntP**
interstitial nucleus of Cajal **InC**
interstitial nucleus of posterior limb of ac **IPAC**
interstitial nucleus of vestibulocochlear nerve **I8**
intertrigeminal nucleus **I5**
interventricular foramen **IVF**
isthmic canal **IsC**
isthmic reticular formation **IsRt**
isthmus **Is**
isthmic neuromere **is**

J
jugular foramen **JugF**

K
Kölliker-Fuse nucleus **KF**

L
lacunosum moleculare layer of hippocampus **LMol**
lamina terminalis **LTer**
lateral (dentate) cerebellar nucleus **Lat**
lateral amygdaloid nucleus **La**
lateral corticospinal tract of spinal cord **lcs**
lateral entorhinal cortex **LEnt**
lateral fissure of developing Cb **lf**
lateral ganglionic eminence, embryonic brain **lge**
lateral geniculate nucleus **LG**
lateral habenular nucleus **LHb**
lateral hypothalamic area **LH**
lateral lemniscus **ll**
lateral mammillary nucleus **LM**
lateral olfactory tract **lo**
lateral orbital cortex **LO**
lateral parabrachial nucleus **LPB**
lateral paragigantocellular nucleus **LPGi**
   α part **LPGiA**
lateral periaqueductal grey **LPAG**
lateral posterior thalamic nucleus **LP**
lateral preoptic area **LPO**
lateral pterygoid muscle **LPtg**
lateral recess of the fourth ventricle **LR4V**
lateral rectus muscle **LRec**
lateral reticular nucleus **LRt**
lateral septal nucleus **LS**
   dorsal part **LSD**
   intermediate part **LSI**
   ventral part **LSV**
lateral spinal nucleus **LSp**
lateral stripe of the striatum **LSS**
lateral terminal nucleus (pretectum) **LT**
lateral ventricle **LV**
lateral vestibular nucleus **LVe**
lateral vestibulospinal tract **lvs**
lateroanterior hypothalamic nucleus **LA**
laterodorsal thalamic nucleus **LD**
layer 1 of cerebral cortex **1Cx**
layer 2 of cerebral cortex **2Cx**
layer 3 of cerebral cortex **3Cx**
layer 4 of cerebral cortex **4Cx**
layer 5 of cerebral cortex **5Cx**
layer 6 of cerebral cortex **6Cx**
left atrium **LAtr**
Lissauer's zone **Liss**
lobule 1 of the cerebellar vermis **1Cb**
lobule 2 of the cerebellar vermis **2Cb**
lobule 3 of the cerebellar vermis **3Cb**
lobule 4 of the cerebellar vermis **4Cb**
lobule 5 of the cerebellar vermis **5Cb**
lobule 6 of the cerebellar vermis **6Cb**
lobule 7 of the cerebellar vermis **7Cb**
lobule 8 of the cerebellar vermis **8Cb**
lobule 9 of the cerebellar vermis **9Cb**
lobule 10 of the cerebellar vermis **10Cb**
locus coeruleus **LC**
longitudinal fasciculus of the pons **lfp**

M
macula of the utricle **UMac**
magnocellular nucleus of lateral hypothalamus **MCLH**
magnocellular nucleus of post. commissure **MCPC**
magnocellular preoptic nucleus **MCPO**
mammillary body **MB**
mammillary recess of the third ventricle **MRe**
mammillothalamic tract **mt**
mandibular process of arch 1 **1mand**
manipul'n/rostral somatosens. area **M/R transition**
marginal zone of developing cortex **mCx**
matrix region of the medulla **Mx**
maxillary bone (maxilla) **MaxB**
maxillary process of arch 1 **1max**
medial amygdaloid nucleus **Me**

medial arch of inferior olive **MA**
medial cerebellar nucleus **Med**
medial entorhinal cortex **MEnt**
medial forebrain bundle **mfb**
medial ganglionic eminence, embryonic brain **mge**
medial geniculate nucleus **MG**
medial habenular nucleus **MHb**
medial lemniscus **ml**
medial longitudinal fasciculus **mlf**
medial mammillary nucleus, lateral part **ML**
medial mammillary nucleus, medial part **MM**
medial orbital cortex **MO**
medial parabrachial nucleus **MPB**
medial preoptic area **MPA**
medial preoptic nucleus **MPO**
   lateral part **MPOL**
   medial part **MPOM**
medial pretectal nucleus **MPT**
medial pterygoid muscle **MPtg**
medial rectus muscle **MRec**
medial septal nucleus **MS**
medial tuberal nucleus **MTu**
medial vestibular nucleus **MVe**
median eminence **ME**
median neuroepithelium of brainstem **mne**
median nucleus of spinal cord **Mn**
median preoptic nucleus **MnPO**
median raphe nucleus **MnR**
mediodorsal thalamic nucleus **MD**
   lateral part **MDL**
   medial part **MDM**
medullary reticular nucleus **MD**
   dorsal part **MdD**
   ventral part **MdV**
mesencephalic neuromere **mes**
mesencephalic reticular formation **mRt**
mesencephalic tegmentum **MeTg**
mesencephalic trigeminal nucleus **Me5**
mesencephalic trigeminal tract **me5**
mesencephalon **Mesen**
middle cerebellar peduncle **mcp**
molecular layer of the cerebellum **MoCb**
molecular layer of the dentate gyrus **MoDG**
motor cortex unspecified **M**
   bill area **MBill**
   forelimb area **MFL**
   head area **MHd**
   hindlimb area **MHL**
   shoulder area **MSh**
   tail area **MTl**
   tongue area **MTo**
motor root of the trigeminal nerve **m5**
motor trigeminal nucleus **5N**
myelencephalon **Myelen**

N
nasal cavity **NasC**
nasal placode **nplac**
nasal septum **NSpt**
navicular postolfactory nucleus **Nv**
neuroepithelium **ne**
notochord **noto**
nuclear transitory zone of cerebellum **ntz**
nucleus of Darkschewitsch **Dk**
nucleus of the brachium of inferior colliculus **BIC**
nucleus of the fields of Forel **F**
nucleus of horizontal limb of diagonal band **HDB**
nucleus of the lateral olfactory tract **LOT**
nucleus of the optic tract **OT**
nucleus of the trapezoid body **Tz**
nucleus of vertical limb of diagonal band **VDB**
nucleus X **X**

O
occipital association cortex **OccA**
occipital bone **Occ**
occipital condyle **OccC**
oculomotor nerve **3n**
oculomotor nucleus **3N**
   parvicellular part **3PC**
olfactory bulb **OB**
olfactory epithelium **olfepith**
olfactory nerve fibres **olf**
olfactory nerve layer **ON**
olfactory tubercle **Tu**
olfactory ventricle (olfact. part of LV) **OV**
olivary pretectal nucleus **OPT**
olivocerebellar tract **oc**
optic chiasm **och**
optic nerve **2n**
optic recess of third ventricle **OptRe**
optic stalk **os**
optic tract **opt**
oral cavity **Oral**
orbital cartilage **OrbC**
orbitoparietal commissure **OrPC**

oriens layer of the hippocampus **Or**
outer blade of dentate gyrus **OBl**

P
p1 periaqueductal grey **p1PAG**
p1 reticular formation **p1Rt**
parabrachial complex **PB**
parabrachial pigmented nucleus of the VTA **PBP**
parafascicular thalamic nucleus **PF**
paraflocculus of cerebellum **PFl**
paramedian lobule **PM**
paramedian raphe nucleus **PMnR**
paramedian reticular nucleus **PMn**
paranigral nucleus of the VTA **PN**
parapyramidal nucleus **PPy**
pararubral nucleus **PaR**
parasubiculum **PaS**
parasubthalamic nucleus **PSTh**
paratenial thalamic nucleus **PT**
paratrigeminal nucleus **Pa5**
paraventricular hypothalamic nucleus **Pa**
paraventricular thalamic nucleus, anterior part **PVA**
paraventricular thalamic nucleus, posterior part **PVP**
parietal bone **ParB**
parietal plate (chondrocranium) **ParP**
parietoventral area of sensory cortex **PV**
parvicellular reticular nucleus **PCRt**
    α part **PCRtA**
peduncular embryonic hypothalamus **hy1**
peduncular part of lateral hypothalamus **PLH**
pedunculotegmental nucleus **PTg**
periaqueductal grey **PAG**
perifacial zone **P7**
perifornical nucleus **PeF**
peripeduncular nucleus **PP**
perirhinal cortex **PRh**
periventricular hypothalamic nucleus **Pe**
petrous temporal bone **Petrous**
pharyngeal pouch 1 of embryo **P1**
pharyngeal pouch 2 of embryo **P2**
pharyngeal pouch 3 of embryo **P3**
pigment layer of the eye **Pig**
pila antotica **PiAnt**
pineal gland **Pi**
pineal recess of third ventricle **PiRe**
piriform area unspecified **PirA**
piriform area 1 **PirA1**
piriform area 2 **PirA2**

piriform area 3 **PirA3**
piriform area 4 **PirA4**
piriform cortex **Pir**
    layer 1 **Pir1**
    layer 1a **Pir1a**
    layer 1b **Pir1b**
    layer 2 **Pir2**
    layer 3 **Pir3**
plexiform/mitral layer of olfactory bulb **Pl/Mi**
plexiform/mitral layer of access. olf. bulb **PlA/MiA**
polymorph layer of the dentate gyrus **PoDG**
pontine nuclei **Pn**
pontine reticular nucleus, caudal part **PnC**
pontine reticular nucleus, oral part **PnO**
posterior cerebral artery **pcer**
posterior commissure **pc**
posterior cortical amygdaloid area **PCo**
posterior hypothalamic nucleus **PH**
posterior intralaminar thalamic nucleus **PIL**
posterior limitans thalamic nucleus **PLi**
posterior lobe of pituitary **PPit**
posterior semicircular duct **PSCD**
posterior spinal artery **psp**
posterior superior fissure of Cb **psf**
posterior thalamic nuclear group **Po**
    triangular part **PoT**
posterior thalamus **PTh**
posterodorsal tegmental nucleus **PDTg**
posterolateral cortical amygdaloid area **PLCo**
posteromedial cortical amygdaloid area **PMCo**
postsubiculum **Post**
precommissural nucleus **PrC**
pregeniculate nucleus **PrG**
prelimbic cortex **PrL**
preoptic area **POA**
preoptic root of orbital bone **PrOrb**
preplate of developing cortex **PrePl**
prepositus nucleus **Pr**
prepyramidal fissure **ppf**
prerubral field **PR**
presphenoid bone **PSph**
presubiculum **PrS**
pretectum **Ptec**
prethalamus (prosomere 3) **PrTh**
primary somatosensory cortex **S1**
    forelimb region **S1FL**
    head region **S1Hd**
    hindlimb region **S1HL**

snout **S1Sn**
snout base **S1SnB**
snout tip **S1SnT**
tongue region **S1To**
tail region **S1Tl**
trunk region **S1Tr**
primary visual cortex **V**
principal sensory trigeminal nucleus **Pr5**
    dorsolateral part **Pr5DL**
    magnocellular part **Pr5mc**
    parvicellular part **Pr5pc**
    ventromedial part **Pr5VM**
pro-otic foramen **PrO**
prosencephalon **Pros**
prosomere 1 **p1**
prosomere 2 **p2**
prosomere 3 **p3**
pseudo-optic foramen **PsOp**
pterygoid process of the sphenoid bone **Ptg**
Purkinje cell layer of the cerebellum **Pk**
putamen **Pu**
pyramidal cell layer of the hippocampus **Py**

R

radiatum layer of the hippocampus **Rad**
raphe interpositus nucleus **RIP**
raphe magnus nucleus **RMg**
raphe obscurus nucleus **ROb**
raphe pallidus nucleus **RPa**
red nucleus **Red**
    magnocellular part **RMC**
    parvicellular part **RPC**
respiratory epithelium **respepith**
reticular nucleus (prethalamus) **Rt**
reticulostrial nucleus **RtSt**
reticulotegmental nucleus of the pons **RtTg**
retrochiasmatic area **RCh**
retromammillary nucleus **RM**
    lateral part **RML**
    medial part **RMM**
retroparafascicular nucleus **RPF**
retrorubral field **RRF**
retrosplenial granular cortex **RSG**
reuniens thalamic nucleus **Re**
Rexed's lamina 1 **R1**
Rexed's lamina 2 **R2**
    external **R2e**
    internal **R2i**

Rexed's lamina 3 **R3**
Rexed's lamina 4 **R4**
Rexed's lamina 5 **R5**
Rexed's lamina 6 **R6**
Rexed's lamina 7 **R7**
Rexed's lamina 8 **R8**
Rexed's lamina 9 **R9**
Rexed's lamina 10 **R10**
rhabdoid nucleus **Rbd**
rhinal fissure **rf**
rhinal incisure **ri**
rhombencephalon **Rhomb**
rhombic lip **rl**
rhomboid thalamic nucleus **Rh**
rhombomere 1 **r1**
rhombomere 2 **r2**
rhombomere 4 **r4**
rhombomere 5 **r5**
rhombomere 6 **r6**
rhombomere 7 **r7**
right atrium **RAtr**
rostral linear nucleus (midbrain) **RLi**
rostral rhombic lip **rrl**
rostral somatosensory field of cortex **R**

S

saccule **Sacc**
second pharyngeal arch **2arch**
secondary fissure **sf**
sensory root of the trigeminal nerve **s5**
septofimbrial nucleus **SFi**
septohippocampal nucleus **SHi**
septohypothalamic nucleus **SHy**
septum **Spt**
simple lobule of Cb **Sim**
solitary nucleus **Sol**
solitary tract **sol**
sphenopalatine ganglion **SphPal**
spinal cord **SpC**
spinal trigeminal nucleus **Sp5**
    caudal part **Sp5C**
    interpolar part **Sp5I**
    oral part **Sp5O**
spinal trigeminal tract **sp5**
spinal vestibular nucleus **SpVe**
sternomastoid muscle **StM**
stria medullaris (thalami) **sm**
stria terminalis **st**

striohypothalamic nucleus **StHy**
subcoeruleus nucleus, dorsal part **SubCD**
subcoeruleus nucleus, ventral part **SubCV**
subiculum, transition area **STr**
subincertal nucleus **SubI**
submedius thalamic nucleus **Sub**
subparafascicular thalamic nucleus **SPF**
    parvicellular part **SPFPC**
subpeduncular tegmental nucleus **SPTg**
substantia nigra **SN**
    compact part **SNC**
    compact part, dorsal tier **SNCD**
    compact part, medial tier **SNCM**
    lateral part **SNL**
    reticular part **SNR**
    reticular part/entopeduncular nucleus **SNR/EP**
subthalamic nucleus **STh**
subventricular layer of developing cortex **SubV**
sulcus α of cerebral cortex **α**
sulcus β of cerebral cortex **β**
sulcus γ of cerebral cortex **γ**
sulcus δ of cerebral cortex **δ**
sulcus ε of cerebral cortex **ε**
sulcus ζ of cerebral cortex **ζ**
sulcus η of cerebral cortex **η**
sulcus μ of cerebral cortex **μ**
sulcus o of cerebral cortex **o**
sulcus limitans **sl**
superficial grey layer of SC **SuG**
superior cerebellar artery **scba**
superior cerebellar peduncle **scp**
superior colliculus **SC**
superior ganglion of glossopharyngeal nerve **S9Gn**
superior ganglion of the vagus nerve **S10Gn**
superior oblique muscle **SOb**
superior olive **SOl**
superior rectus muscle **SRec**
superior sagittal sinus **sss**
superior salivatory nucleus **SuS**
superior vestibular nucleus **SuVe**
suprachiasmatic nucleus **SCh**
suprageniculate thalamic nucleus **SG**
supraoculomotor periaqueductal grey **Su3**
supraoptic nucleus, retrochiasmatic part **SOR**
supratrigeminal nucleus **Su5**

T
tectal grey formation (midbrain) **TG**
tectospinal tract **ts**

telencephalon **Telen**
temporal association cortex **TeA**
temporalis muscle **TempM**
tenia tecta **TT**
terminal nerve **term**
third pharyngeal arch **3arch**
third ventricle **3V**
transverse cerebral fissure **tcf**
transverse fibres of the pons **tfp**
transverse venous sinus **trs**
trigeminal ganglion **5Gn**
trigeminal nerve **5n**
    maxillary division **5mx**
    ophthalmic division **5oph**
trigeminal transition zone **5Tr**
trigeminal-solitary transition zone **5Sol**
trochlear nerve **4n**
trochlear nucleus **4N**
tuberal developing hypothalamus **hy2**
tuberal region of lateral hypothalamus **TuLH**

U
utricle **Utr**

V
vagus nerve **10n**
vagus nerve nucleus **10N**
ventral anterior thalamic nucleus **VA**
ventral cochlear nucleus **VC**
    anterior part **VCA**
    posterior part **VCP**
ventral endopiriform nucleus **VEn**
ventral horn **VH**
ventral nucleus of the lateral lemniscus **VLL**
ventral orbital cortex **VO**
ventral pallidum **VP**
ventral posterior thalamic nucleus, parvi. **VPPC**
ventral posterolateral thalamic nucleus **VPL**
ventral posteromedial thalamic nucleus **VPM**
ventral reuniens thalamic nucleus **VRe**
ventral subiculum **VS**
ventral tegmental area **VTA**
ventral tegmental nucleus **VTg**
ventral tenia tecta **VTT**
ventral third ventricle **V3V**
ventricular space of the eye **Vent**
ventrolateral hypothalamic nucleus **VLH**
ventrolateral periaqueductal grey **VLPAG**
ventrolateral preoptic nucleus **VLPO**

LIST OF ABBREVIATIONS USED IN BRAIN AND EMBRYO ATLAS PLATES

ventrolateral thalamic nucleus **VL**
ventromedial hypothalamic nucleus **VMH**
ventromedial preoptic nucleus **VMPO**
ventromedial thalamic nucleus **VM**
vertebral artery **vert**
vestibular nerve ganglion **VeGn**
vestibular nuclei or neuroepithelium **Ve**
vestibular root of vestibulocochlear nerve **8vn**
vestibulocerebellar nucleus **VeCb**
vestibulocochlear nerve **8n**

vomer **Vom**
vomeronasal nerve **vn**

Z
zona incerta **ZI**
   caudal part **ZIC**
   dorsal part **ZID**
   ventral part **ZIV**
zona limitans interthalamica **zli**

INDEX OF BRAIN AND EMBRYO ATLAS PLATES

α	sulcus α of cerebral cortex	Ec-Ad08 to 24
β	sulcus β of cerebral cortex	Ec-Ad08 to 19
γ	sulcus γ of cerebral cortex	Ec-Ad04 to 12
δ	sulcus δ of cerebral cortex	Ec-Ad04 to 19
ε	sulcus ε of cerebral cortex	Ec-Ad02 to 09
ζ	sulcus ζ of cerebral cortex	Ec-Ad22 to 26
η	sulcus η of cerebral cortex	Ec-Ad11 to 22
μ	sulcus μ of cerebral cortex	Ec-Ad17 to 25
o	sulcus o of cerebral cortex	Ec-Ad26
1Cb	lobule 1 of the cerebellar vermis	Ec-Ad25, Pl-Ad26
1Cx	layer 1 of cerebral cortex	Ec-Ad05, Pl-Ad06–07, Pl-Ad10, 11, 14 & 17
1mand	mandibular process of arch 1	Ec-InA03 & 04
1max	maxillary process of arch 1	Ec-InA02 & 03
2arch	second pharyngeal arch	Ec-InA03 & 04
2Cb	lobule 2 of the cerebellar vermis	Ec-Ad24 & 25, Pl-Ad25
2Cx	layer 2 of cerebral cortex	Ec-Ad05, Pl-Ad06–07, Pl-Ad10, 11 & 17
2n	optic nerve	Ec-Ad10
3arch	third pharyngeal arch	Ec-InA04
3Cb	lobule 3 of the cerebellar vermis	Ec-Ad23 to 25, Pl-Ad24 & 25
3Cx	layer 3 of cerebral cortex	Ec-Ad05, Pl-Ad06–07, Pl-Ad10, 11, 14 & 17
3N	oculomotor nucleus	Ec-Ad20 & 21, Ec-InB18 & 19, Pl-Ad18 & 19
3n	oculomotor nerve	Ec-Ad17 & 18, Ec-InB04 to 19, Pl-Ha04 to 13
3PC	oculomotor nucleus, parvicell. part	Ec-Ad21
3V	third ventricle	Ec-Ad11 to 18, Ec-InA01 & 02, Ec-InB15 to 17, Pl-Ad11 to 15, Pl-Ha04 to 11
4Cb	lobule 4 of the cerebellar vermis	Ec-Ad23 to 25, Pl-Ad25
4Cx	layer 4 of cerebral cortex	Pl-Ad06–07, 10, 11, 14 & 17
4N	trochlear nucleus	Pl-Ad21, Pl-Ha19
4n	trochlear nerve	Ec-Ad22, Ec-InB07 to 24, Pl-Ha07 to 19
4V	fourth ventricle	Ec-Ad23 to 30, Ec-InA01 to 04, Ec-InB16 to 31, Pl-Ad23 to 30, Pl-Ha11 to 21
5Cb	lobule 5 of the cerebellar vermis	Ec-Ad26, Pl-Ad25
5Cx	layer 5 of cerebral cortex	Ec-Ad05, Pl-Ad06–07, Pl-Ad10, Pl-Ad11, Pl-Ad14, Pl-Ad17
5fr	front. br. of ophthalmic trigeminal nerve	Pl-Ha01
5Gn	trigeminal ganglion Ec-InA02,	Ec-InB10 to 15, Pl-Ha05 to 09 & 11
5mx	trigeminal nerve, maxillary divn	Ec-InB08 & 09, Pl-Ha05
5N	motor trigeminal nucleus	Ec-Ad23 & 24, Ec-InB14 to 16, Pl-Ad20 to 22, Pl-Ha11
5n	trigeminal nerve	Ec-Ad22, Ec-InB13 to 15, Pl-Ha07 to 09 &11
5oph	trigeminal nerve, ophthalmic divn	Ec-InB01 to 09, Pl-Ha01 to 03

5Sol	trigeminal-solitary transition zone	Ec-InB19 to 31, Pl-Ha13 to 21
5Tr	trigeminal transition zone	Ec-InB18
6Cb	lobule 6 of the cerebellar vermis	Ec-Ad26, Pl-Ad25 to 28
6Cx	layer 6 of cerebral cortex	Ec-Ad05, Pl-Ad06–07, 10, 11, 14 & 17
6N	abducens nucleus	Ec-Ad25, Ec-InB18, Pl-Ad23, Pl-Ha12
6n	abducens nerve	Ec-InB08 to 12 & 16
7Cb	lobule 7 of the cerebellar vermis	Ec-Ad27 & 28, Pl-Ad28 to 30
7N	facial nucleus	Ec-InB19 to 23, Pl-Ha11 to 14
7n	facial nerve	Ec-Ad25 & 26, Ec-InB17, Pl-Ad23, Pl-Ha11
7Nd	facial nucleus, dorsal part	Ec-Ad25, Pl-Ad22 & 23
7Nv	facial nucleus, ventral part	Ec-Ad25 & 26, Pl-Ad22 & 23
8Cb	lobule 8 of the cerebellar vermis	Ec-Ad27 & 28, Pl-Ad28 to 30
8cn	cochlear root of vestibulococh. nerve	Pl-Ad22 to 24, Pl-Ha12 to 14
8n	vestibulocochlear nerve	Ec-Ad25, Ec-InB18 to 20, Pl-Ad25
8vn	vestibular root of vestibulococh. nerve	Ec-Ad24 to 26, Pl-Ad24
9Cb	lobule 9 of the cerebellar vermis	Ec-Ad27, Ec-Ad28 to 30, Pl-Ad28 to 30
9n	glossopharyngeal nerve	Ec-InB22 & 23
10Cb	lobule 10 of the cerebellar vermis	Ec-Ad25 to 29, Pl-Ad27 to 29
10N	vagus nerve nucleus	Ec-Ad27 to 30, Ec-InB22, Ec-InB23 to 31, Pl-Ad27 to 30, Pl-Ha13 to 21
10n	vagus nerve	Ec-Ad28, Ec-InB24 to 30, Pl-Ha15, 17 to 20
12N	hypoglossal nucleus	Ec-Ad27 to 30, Ec-InB25 to 31, Pl-Ad27 & 30, Pl-Ha18 to 21
12n	hypoglossal nerve	Ec-Ad27 to 29, Ec-InB28 to 30, Pl-Ha to 20

A

A5	A5 noradrenaline cells	Ec-Ad24
AA	anterior amygdaloid area	Pl-Ad12
AB	anterobasal nucleus	Ec-InB10 & 11
ac	anterior commissure	Ec-Ad08 to 11, Pl-Ad09 to 11
aca	anterior commissure, anterior part	Ec-Ad06 & 07, Pl-Ad08 & 09
Acb	accumbens nucleus	Ec-Ad05, Ec-InB04, Pl-Ha02
AcbC	accumbens nucleus, core	Ec-Ad06 & 07, Pl-Ad04 to 09
AcbSh	accumbens nucleus, shell	Ec-Ad06 & 07, Pl-Ad04 to 08
acer	anterior cerebral artery	Ec-Ad04
ACo	anterior cortical amygdaloid area	Ec-Ad12 & 13, Ec-InB08 to 10, Pl-Ad12 to 14, Pl-Ha06
acp	anterior commissure, posterior part	Ec-Ad11, Pl-Ad11 to 16
Acs6/7	accessory nucleus abducens/facial	Pl-Ha12
AD	anterodorsal thalamic nucleus	Pl-Ad13
af	amygdaloid fissure	Pl-Ad14 to 16
AHA	anterior hypothalamic area, ant. part	Ec-InB09, Pl-Ad12
AHi	amygdalohippocampal area	Ec-Ad14 & 15, Pl-Ad15 & 16
AI	agranular insular cortex	Pl-Ad04–05
AID	agranular insular cortex, dorsal part	Pl-Ad04 to 08
AIP	agranular insular cortex, posterior part	Pl-Ad09 to 13
AIV	agranular insular cortex, ventral part	Pl-Ad04 to 08
alar	alar plate of neural tube	Ec-InA02 to 05, Ec-InB16 to 31, Pl-Ha13 to 21
alv	alveus of the hippocampus	Ec-Ad09, Ec-Ad11 to 18, Pl-Ad09 to 15, Pl-Ad19
AM	anteromedial thalamic nucleus	Pl-Ad12 & 13

Amb	ambiguus nucleus	Ec-Ad27, Ec-Ad28, Ec-Ad30, Pl-Ad25, Pl-Ad27
AmbC	ambiguus nucleus, compact part	Ec-InB26 to 31, Pl-Ha15 to 20
AmbSC	ambiguus nucleus, subcompact part	Ec-InB24 & 25
Amg	amygdala	Ec-InB12 & 13, Pl-Ha05, Pl-Ha06, Pl-Ha09
AMV	anteromedial thalamic nucleus, vent. part	Pl-Ad13
Ans	ansiform lobule of cerebellum	Pl-Ad26 & 27
ANS	accessory neurosecretory nuclei	Ec-Ad12, Pl-Ad12 & 13
Ant	anterior nucleus of thalamus	Ec-Ad13 & 14
AOD	anterior olfactory area, dorsal part	Ec-Ad01 & 02, Pl-Ad01–03
AOE	anterior olfactory area, ext. part	Pl-Ad01–03
AOL	anterior olfactory area, lateral part	Pl-Ad01 to 05
AOM	anterior olfactory area, medial part	Pl-Ad01 to 05
AOV	anterior olfactory area, ventral part	Ec-Ad01 & 02, Pl-Ad01 to 05
AP	area postrema	Pl-Ad28
APir	amygdalopiriform transition area	Ec-Ad14, Pl-Ad15, Pl-Ad16
APit	anterior lobe of the pituitary	Ec-Ad14, Ec-InB12 & B13, Pl-Ha07 & 08
APT	anterior pretectal nucleus	Ec-Ad18, Pl-Ad17
APTD	anterior pretectal nucleus, dorsal part	Ec-Ad19
APTL	anterior pretectal nucleus, lateral part	Pl-Ad18
APTM	anterior pretectal nucleus, medial part	Pl-Ad18
APTV	anterior pretectal nucleus, ventral part	Ec-Ad19
Aq	cerebral aqueduct	Ec-Ad19 to 22, Ec-InA01, Ec-InB13 to 15, Pl-Ad18 to 21, Pl-Ha17 to 21
Ar	arcuate medullary nucleus	Ec-Ad27 to 29
Arc	arcuate hypothalamic nucleus	Ec-Ad12 to 14, Ec-InB11 & 12, Pl-Ad13, Pl-Ha06 to 10
Ary	arytenoid cartilage	Ec-InA05
ASCD	anterior semicircular duct	Ec-InB17 to 21, Pl-Ha12 to 17
asp	anterior spinal artery	Ec-Ad32
ASt	amygdalostriatal transition area	Ec-Ad12 to 14, Pl-Ad14 & 15
ATh	anterior thalamic region	Ec-InB07 to 09, Pl-Ha06
Au	auditory cortex	Ec-Ad21 to 25, Pl-Ad25 to 27
Au/S1	auditory/somato bimodal sens. cortex	Pl-Ad24 to 26
AV	anteroventral thalamic nucleus	Pl-Ad13
B		
B	basal nucleus (Meynert)	Ec-Ad13, Pl-Ad12 to 14
B9	B9 serotonin cells	Ec-Ad21
BAOT	bed nucleus of access. olfact. tract	Ec-Ad11 to 13
Bar	Barrington's nucleus	Ec-Ad23
bas	basilar artery	Ec-InB14
basal	basal plate of neural tube	Ec-InA02 to 05, Ec-InB19 to 31, Pl-Ha13 to 21
BIC	nucleus of brachium of inferior colliculus	Ec-Ad22, Pl-Ad22
bic	brachium of the inferior colliculus	Ec-Ad22
BL	basolateral amygdaloid nucleus	Pl-Ha07 & 08
BLA	basolateral amygdaloid nucleus, ant.	Pl-Ad14
BLL	basolateral amygdaloid nucleus, lateral	Ec-Ad11 to 13
BLM	basolateral amygdaloid nucleus, med.	Ec-Ad12 & 13
BLP	basolateral amygdaloid nucleus, post.	Ec-Ad14, Pl-Ad15 & 16

BM	basomedial amygdaloid nucleus	Pl-Ha07
BMA	basomedial amygdaloid nucleus, ant.	Ec-Ad11 to 13, Pl-Ad13 & 14
BMP	basomedial amygdaloid nucleus, post.	Ec-Ad14 & 15, Pl-Ad15 & 16
BOcc	basioccipital bone	Ec-InB15 to 31, Pl-Ha11 to 13, Pl-Ha16
bsc	brachium of the superior colliculus	Ec-Ad20, Pl-Ad18, to 20
BSph	basisphenoid bone	Ec-InB08 to 13, Pl-Ha04 & 05

C

CA	cornu Ammonis of hippocampus	Ec-Ad16, Pl-Ad17 to 20
CA1	field CA1 of the hippocampus	Ec-Ad09 to 20, Pl-Ad06 to 24
CA2	field CA2 of the hippocampus	Ec-Ad13 to 19, Pl-Ad09 to 22
CA3	field CA3 of the hippocampus	Ec-Ad09 to 20, Pl-Ad06 to 22
CAud	canalicular part of auditory capsule	Ec-InB16 to 24, Pl-Ha11 to 17
cav	cavernous sinus	Ec-InB10 to 12
CB	cell bridges of the ventral striatum	Pl-Ad09 & 10
Cb	cerebellum	Ec-InB16 to 29, Pl-Ha12 to 21
CC	central canal	Ec-Ad31 to 33
CD	cochlear duct	Ec-InB16 & 17, Pl-Ha11 & 12
Cd	caudate nucleus	Ec-Ad04 to 18, Pl-Ad04 to 23
cdnv	cardinal vein	Ec-InA05
Ce	central amygdaloid nucleus	Ec-Ad11 to 13, Ec-InB12, Pl-Ad13 &14, Pl-Ha07 & 08
CeC	central amygdaloid nucleus, capsular	Ec-Ad14
CeL	central amygdaloid nucleus, lateral divn	Ec-Ad12 & 14
CeM	central amygdaloid nucleus, med. divn	Ec-Ad12
CEnt	caudomedial entorhinal cortex	Ec-Ad21 to 23, Pl-Ad22 & 23
CG	central grey	Ec-Ad24, Ec-InB17 to 21, Pl-Ha11 & 12
Cg	cingulate cortex	Ec-Ad02 to 07, Ec-InB06 to 10, Pl-Ad04–05, Pl-Ha03 to 10
cg	cingulum	Ec-Ad06 to 17, Pl-Ad04 to 17
Cg1	cingulate cortex, area 1	Ec-Ad08 to 21, Pl-Ad06 to 18
Cg2	cingulate cortex, area 2	Ec-Ad08 to18, Pl-Ad06 to 18
CGn	cochlear (spiral) ganglion	Ec-InB18 & 19, Pl-Ha11
chp	choroid plexus	Ec-Ad27 to 30, Ec-InB05 to 30, Pl-Ha04 to 21
CIC	central nucleus of inf. colliculus	Ec-Ad22, Pl-Ad21 & 22
CL	centrolateral thalamic nucleus	Ec-Ad15 & 16
Cl	claustrum	Ec-Ad05 to 10, Pl-Ad06 to 25
ClC	Clarke's column/nucleus dorsalis	Ec-Ad32
CLi	caudal linear nucleus of the raphe	Ec-Ad19 to 21, Pl-Ha14
CM	central medial thalamic nucleus	Ec-Ad14 to 16, Pl-Ad14
CnF	cuneiform nucleus	Ec-Ad22, Pl-Ad21 & 22
Conjunct	conjunctival sac	Ec-InB02 to 06, Pl-Ha01 & 02
cp	cerebral peduncle	Ec-Ad13 to 20, Ec-InB11 to 21, Pl-Ad15 to 19, Pl-Ha06 to 15
CPu	caudate putamen (striatum)	Ec-InB04 to 11, Pl-Ha02 to 08
crl	caudal rhombic lip	Ec-InB18 to 31, Pl-Ha12 to 21
CrP	cribriform plate of the ethmoid	Ec-InB01, Ec-InB04
csc	commissure of superior colliculus	Ec-Ad20
ctz	cortical transit. zone of cerebellum	Ec-InB16 to 25, Pl-Ha12 to 19
Cu	cuneate nucleus	Ec-Ad27 to 30, Pl-Ad28 & 30
cu	cuneate fasciculus	Ec-Ad27 to 31, Pl-Ad27 & 30

cvent	common ventric. chamber of heart	Ec-InA05
Cx	cerebral cortex (developing)	Ec-InB01 to 15, Pl-Ha01 to 11
CxA	cortex-amygdala transition zone	Ec-Ad11 to 13

D

D3V	dorsal third ventricle	Ec-Ad13 to 17, Ec-InB05 to 09, Ec-InB11 & 12, Pl-Ad13
d4n	decussation of trochlear nerve	Ec-Ad23
DA	dorsal hypothalamic area	Ec-Ad12, Pl-Ad13 & 14
daorta	dorsal aorta	Ec-InA05
das	dorsal acoustic stria	Ec-InB16 & 17, Pl-Ha13 & 14
DC	dorsal cochlear nucleus	Ec-Ad26, Ec-InB16 to 21, Pl-Ad21 to 23, Pl-Ha12 to 14
DCIC	dorsal cortex of inferior colliculus	Ec-Ad22, Pl-Ad21 & 22
dcw	deep cerebral white matter	Ec-Ad21 to 25, Pl-Ad06–07, Pl-Ad20 to 27
DEn	dorsal endopiriform nucleus	Ec-Ad05 to 10, Pl-Ad06 to 10, Pl-Ad14 to 16
DG	dentate gyrus	Ec-Ad09 to 20, Pl-Ad08 to 10, Pl-Ad21 & 22
dhc	dorsal hippocampal commissure	Ec-Ad10 & 11
Dien	diencephalon	Ec-InB06 to 13, Ec-InB17, Pl-Ha04 to 10
Dk	nucleus of Darkschewitsch	Ec-Ad19, Pl-Ad18, Pl-Ha15
DLL	dorsal nucleus of lateral lemniscus	Ec-Ad21, Ec-InB17, Ec-InB18 to 23, Pl-Ad22, Pl-Ha11 to 17
DLPAG	dorsolateral periaqueductal grey	Ec-Ad20 & 21
DM	dorsomedial hypothalamic nucleus	Pl-Ha08
DMD	dorsomedial hypothalamic nucleus, dorsal	Ec-Ad12 to 14, Pl-Ad13 & 14
DMPAG	dorsomedial periaqueductal grey	Ec-Ad20 & 21
DMSp5	dorsomedial spinal trigeminal nucleus	Ec-Ad25 & 26, Ec-InB19 to 31, Pl-Ad26
DMTg	dorsomedial tegmental area	Ec-Ad23 & 24, Ec-InB19 to 21, Pl-Ha13 & 14
DMV	dorsomedial hypothalamic nucleus, ventral	Ec-Ad14
DpG	deep grey layer of sup. colliculus	Ec-Ad20 & 21, Pl-Ad19 to 22
DPGi	dorsal paragigantocellular nucleus	Ec-Ad25 & 26, Ec-InB21 to 28, Pl-Ad24 & 27, Pl-Ha14 to 21
DpWh	deep white layer of sup. colliculus	Pl-Ad20
DR	dorsal raphe nucleus	Ec-Ad20 to 22, Ec-InB19 to 24, Pl-Ad20, Pl-Ha15 to 17
dr	dorsal roots of spinal cord	Ec-Ad31 & 32
DRG	dorsal root ganglion	Ec-InA05
DS	dorsal subiculum	Ec-Ad09 to 21, Pl-Ad06 to 22
DTg	dorsal tegmental nucleus	Ec-Ad23, Ec-InB24, Pl-Ad22
DTr	dorsal transition zone	Ec-Ad03, Pl-Ad04–05
DTT	dorsal tenia tecta	Pl-Ad04–05

E

E	ependyma and subependymal layer	Ec-Ad02 to 06, Pl-Ad01 to 07
EA	extended amygdala	Ec-Ad11 to 14, Pl-Ad11 to 13
ec	external capsule	Ec-Ad05 to 11, Pl-Ad04 to 25
ECIC	external cortex of inf. colliculus	Ec-Ad22, Pl-Ad21 & 22
Ect	ectorhinal cortex	Pl-Ad14 to 25
ECu	external cuneate nucleus	Ec-Ad27 to 29, Ec-InB25 to 31, Pl-Ad27 to 30, Pl-Ha15 to 21
ELD	endolymphatic duct	Ec-InB21 & 22, Ec-InB26, Pl-Ha14 to 17
ELS	endolymphatic sac	Ec-InB23 to 27, Pl-Ha17
Ent	entorhinal cortex	Ec-Ad14 & 15
EP	entopeduncular nucleus	Ec-Ad13 to 15, Ec-InB06 to 10, Pl-Ad12 to 19, Pl-Ha04 to 06

EthB	ethmoid bone	Ec-InB01 to 06, Pl-Ha01
EW	Edinger-Westphal nucleus	Ec-InB18 & 19, Pl-Ad18 & 19
ex	extreme capsule	Ec-Ad05 to 10, Pl-Ad06 to 25
Eyelid	eyelid	Ec-InB02 & 03

F

F	nucleus of the fields of Forel	Ec-Ad16, Pl-Ad15
f	fornix	Ec-Ad10 to 14, Pl-Ad11 to 13
fag	facioacoustic ganglion	Ec-InA03
fds	fimbriodentate sulcus of Hi	Ec-Ad17 & 18, Pl-Ad12 to 20
fi	fimbria of the hippocampus	Ec-Ad10 to 17, Pl-Ad10 to 24
Fl	flocculus	Ec-Ad25, 26, & 29, Pl-Ad26 & 27
Fr	frontal cortex, unspecified	Ec-Ad01
fr	fasciculus retroflexus	Ec-Ad15, Ec-Ad17 & 18, Pl-Ha15
Fr1	frontal cortex, area 1	Ec-Ad02 to 19
Fr2	frontal cortex, area 2	Ec-Ad02 to 09
Fr3	frontal cortex, area 3	Ec-Ad02 to 10
FrA	frontal association cortex	Pl-Ad01 to 05
Fro	frontal bone	Ec-InB01 to 03

G

g7	genu of the facial nerve	Ec-Ad26, Pl-Ad23, Pl-Ha12
GCC	granule cell cluster	Ec-Ad07
Ge5	gelatinous layer of the Sp5C	Ec-Ad30
Gen	geniculate ganglion	Ec-InB16
GI	granular insular cortex	Pl-Ad10 to 13
Gi	gigantocellular reticular nucleus	Ec-Ad25 to 29, Ec-InB19 to 31, Pl-Ad24 to 30, Pl-Ha13 to 21
GiA	gigantocellular reticular nucleus, α	Ec-InB18 to 20, Ec-InB21 to 25, Pl-Ad24
GiV	gigantocellular reticular nucleus, vent.	Ec-InB26 to 30, Pl-Ad25 & 26, Pl-Ha16 to 20
Gl	glomerular layer of olfactory bulb	Ec-Ad01 to 04, Pl-Ad01–03
GlA	glomerular layer access. olf. bulb	Ec-Ad01, Pl-Ad01–03
GP	globus pallidus	Ec-Ad09 to 15, Ec-InB05 to 11, Pl-Ad10 to 15, Pl-Ha03 to 08
Gr	gracile nucleus	Ec-Ad28 to 30
gr	gracile fasciculus	Ec-Ad29 to 33
GrA	granule cell layer access. olf. bulb	Pl-Ad01–03
GrCb	granule cell layer of cerebellum	Pl-Ad25 & 26
GrDG	granule cell layer of dentate gyrus	Ec-Ad17, Pl-Ad11 to 23
GrO	granule cell layer of olfactory bulb	Ec-Ad01 to 04, Pl-Ad01–03
Gust	gustatory cortex	Ec-Ad15 to 17, Pl-Ad14 to 23

H

HardG	Harderian gland	Ec-InB02, Ec-InB05
Hb	habenular nuclei	Ec-InB05, Pl-Ha06
hbR	hook bundle of Russell	Pl-Ha19
HDB	nucleus of horizontal limb diag. band	Ec-Ad08 to 10, Ec-InB05 to 07, Pl-Ad09 & 10, Pl-Ha02 & 04
Hi	hippocampus	Ec-InB04 to 12, Ec-InB13, Pl-Ha02 to 09
hif	hippocampal fissure	Ec-Ad09 to 20, Pl-Ad10 to 21
HPtg	hamulus of the pterygoid bone	Ec-InB10 & B11

INDEX OF BRAIN AND EMBRYO ATLAS PLATES

hs	hypothalamic sulcus	Ec-Ad14, Ec-InB10 to 13, Pl-Ha06 to 10
HSCD	horizontal semicircular duct	Ec-InB19 & 20, Pl-Ha13 to 16
hy1	peduncular hypothalamus	Ec-InB09 to 13, Pl-Ha05 to 10
hy2	tuberal hypothalamus	Ec-InB09 to 13, Pl-Ha05 to 10
hyat	acroterminal hypothalamus	Ec-InA02, Ec-InB09 to 13, Pl-Ha05 to 10

I

I5	intertrigeminal nucleus	Ec-Ad24
I8	interstitial nucleus of 8n	Ec-Ad25, Pl-Ad25
ia	internal arcuate fibres	Ec-Ad27 & 28
IAD	interanterodorsal thalamic nucleus	Ec-Ad13
IAud	internal auditory meatus	Ec-InB19, Ec-InB20, Pl-Ha14
IB	internal basilar nucleus of SpC	Ec-Ad31
IBl	inner blade of dentate gyrus	Ec-Ad17
IC	inferior colliculus	Ec-Ad23, 24 & 28, Pl-Ha19 to 21
ic	internal capsule	Ec-Ad12 to 18, Ec-InB09 to 11, Pl-Ad10 to 23, Pl-Ha06 & 08
icp	inferior cerebellar peduncle	Ec-Ad26 & 27, Pl-Ad25 to 29
ictd	internal carotid artery	Ec-InB12 & 13, Pl-Ha04
IF	interfascicular nucleus	Pl-Ha12 & 15
if	intermediate fibres of cerebellum	Ec-InB16 to 25, Pl-Ha13 to 17
IG	indusium griseum	Ec-Ad07 & 08, Pl-Ad06–07
ijugv	internal jugular vein	Pl-Ha16
IL	infralimbic cortex	Ec-Ad06 & 07
ILL	intermediate nucleus of lateral lemniscus	Ec-Ad22, Ec-InB14 to 16, Pl-Ad20
IMD	intermediodorsal thalamic nucleus	Ec-Ad15 & 16, Pl-Ad13 & 14
IML	intermediolateral nucleus of SpC	Ec-Ad32 & 33
IMSpC	intermediomedial nucleus of SpC	Ec-Ad31 to 33
InC	interstitial nucleus of Cajal	Ec-Ad19, Pl-Ad18 & 19, Pl-Ha15
InG	intermediate grey layer of SC	Ec-Ad20 & 21, Pl-Ad19 to 22
Ins	insular cortex	Ec-Ad03 to 11
IntA	interposed cerebellar nucleus, anterior	Ec-Ad24 & 25, Pl-Ad25 & 26
IntP	interposed cerebellar nucleus, posterior	Ec-Ad26, Pl-Ad27
IO	inferior olivary nucleus	Ec-Ad27 to 30, Pl-Ad25 to 28
IOA	inferior olive, subnucleus A of med. nucleus	Ec-Ad29
IOBe	inferior olive, β subnucleus	Ec-Ad29
IOD	inferior olive, dorsal nucleus	Ec-Ad27 to 29
IOM	inferior olive, medial nucleus	Ec-Ad28
iom	inferior olivary migration	Pl-Ha14
IOV	inferior olive, ventral nucleus	Ec-Ad27 & 28
IP	interpeduncular nucleus	Ec-InB19 to 21, Pl-Ha15
IPAC	interstitial nucleus of post. limb of ac	Ec-Ad09 to 11, Pl-Ad10 to 14
IPC	interpeduncular nucleus, caudal subnucleus	Ec-Ad19
ipc	interpallidal commissure	Pl-Ad15 & 16
ipf	interpeduncular fossa	Ec-Ad17 & 18, Ec-InB21 & 22, Pl-Ha13 & 14
IPR	interpeduncular nucleus, rostral subnucleus	Ec-Ad17 & 18
IRe	infundibular recess of third ventricle	Ec-Ad12 & 13, Ec-InB12 & 13, Pl-Ha07 to 10
IRec	inferior rectus muscle	Ec-InB04 to 08
Iris	iris	Ec-InB05

IRt	intermediate reticular nucleus	Ec-Ad24 to 30, Ec-InB17 to 31, Pl-Ad22 to 30, Pl-Ha12 to 21
Is	isthmus	Pl-Ha19
is	isthmic neuromere	Ec-InA01
IsC	isthmic canal	Ec-InB25, Pl-Ha17 & 18
IsRt	isthmic reticular formation	Pl-Ad20
IVF	interventricular foramen	Ec-InB05 to 08, Pl-Ha04 & 05

J

JugF	jugular foramen	Ec-InB23 to 25, Pl-Ha16

K

KF	Kölliker-Fuse nucleus	Ec-Ad22 & 23, Ec-InB15 & 16

L

LA	lateroanterior hypothalamic nucleus	Pl-Ad12
La	lateral amygdaloid nucleus	Ec-Ad11 to 14, Ec-InB12, Pl-Ad14 to 16
Lat	lateral (dentate) cerebellar nucleus	Ec-Ad24 to 26, Pl-Ad25 to 27
LAtr	left atrium	Ec-InA05
LC	locus coeruleus	Ec-Ad23 & 24, Ec-InB23 to 25, Pl-Ad21 & 22, Pl-Ha15 & 16
lcs	lateral corticospinal tract of SpC	Ec-Ad24 to 33
LD	laterodorsal thalamic nucleus	Ec-Ad15 to 17, Pl-Ad13 to 16
Lens	lens	Ec-InB03 & 04
LEnt	lateral entorhinal cortex	Ec-Ad16 to 20, Pl-Ad17 to 21
lf	lateral fissure of cerebellum	Ec-InB16 to 25, Pl-Ha13 to 19
lfp	longitudinal fasciculus of the pons	Ec-Ad21 & 22
LG	lateral geniculate nucleus	Ec-Ad16 to 20, Pl-Ad16 to 18
lge	lateral ganglionic eminence	Ec-InB05 to 12, Pl-Ha03 to 08
LH	lateral hypothalamic area	Ec-InB10 to 13, Pl-Ha06 to 10
LHb	lateral habenular nucleus	Ec-Ad16 to 18, Pl-Ad14 to 16
Liss	Lissauer's zone	Ec-Ad31 to 33
ll	lateral lemniscus	Ec-Ad20 to 24, Ec-InB14 to 18, Pl-Ad20 to 23, Pl-Ha11 to 14
LM	lateral mammillary nucleus	Ec-Ad15 & 16, Ec-InB14 & 16, Pl-Ha11
LMol	lacunosum moleculare layer of Hi	Ec-Ad17, Pl-Ad23
LO	lateral orbital cortex	Pl-Ad03 to 05
lo	lateral olfactory tract	Ec-Ad06, Pl-Ad03 to 07, Pl-Ad13 & 14
LOT	nucleus of the lateral olfactory tract	Pl-Ad13
LP	lateral posterior thalamic nucleus	Ec-Ad17 to 20, Pl-Ad17 & 18
LPAG	lateral periaqueductal grey	Ec-Ad20 to 22
LPB	lateral parabrachial nucleus	Ec-Ad22 & 23, Pl-Ad22 to 24
LPGi	lateral paragigantocellular nucleus	Ec-InB19 to 31, Pl-Ad25 to 27, Pl-Ha15 to 20
LPGiA	lateral paragigantocell. nucleus, α	Ec-Ad26, Pl-Ad24
LPO	lateral preoptic area	Ec-Ad08 to 11, Pl-Ad09 to 12
LPtg	lateral pterygoid muscle	Ec-InB09 to 11, Pl-Ha05
LR4V	lateral recess of the fourth ventricle	Ec-InB17 to 26, Pl-Ad28, Pl-Ha12 to 21
LRec	lateral rectus muscle	Ec-InB05 to 08
LRt	lateral reticular nucleus	Ec-Ad28 to 30, Pl-Ad28 & 29
LS	lateral septal nucleus	Ec-Ad09
LSD	lateral septal nucleus, dorsal part	Ec-Ad06 to 09, Pl-Ad07 to 10

LSI	lateral septal nucleus, intermediate part	Ec-Ad07 & 08, Pl-Ad07 to 11
LSp	lateral spinal nucleus	Ec-Ad31 to 33
LSS	lateral stripe of the striatum	Ec-Ad07 & 08
LSV	lateral septal nucleus, ventral part	Ec-Ad06 & 08, Pl-Ad05 to 09
LT	lateral terminal nucleus	Ec-Ad20
LTer	lamina terminalis	Ec-InA02, Pl-Ha03
LV	lateral ventricle	Ec-Ad04 to 21, Ec-InB01 to 14, Pl-Ad03 to 25, Pl-Ha01 to 11
LVe	lateral vestibular nucleus	Ec-Ad24 & 25, Ec-InB16 to 23, Pl-Ad23 to 25, Pl-Ha11 & 12
lvs	lateral vestibulospinal tract	Ec-InB22 to 31, Pl-Ha15 to 21

M

M	motor cortex unspecified	Pl-Ad10
m5	motor root of the trigeminal nerve	Ec-Ad22 to 24, Pl-Ha11
MaxB	maxillary bone (maxilla)	Pl-Ha01
MB	mammillary body	Ec-InB15, Ec-InB18
MBill	motor cortex bill	Pl-Ad14 to 23
MCLH	magnocellular nucleus of lateral hypothalamus	Ec-Ad13, Pl-Ad13 & 14
mcp	middle cerebellar peduncle	Ec-Ad20 to 24, Pl-Ad19 to 25
MCPC	magnocellular nucleus of post. comm.	Ec-Ad19, Pl-Ad18
MCPO	magnocellular preoptic nucleus	Ec-Ad11, Pl-Ad10 & 11
mCx	marginal zone of developing cortex	Ec-InB04 & 05, Pl-Ha02 to 06
MD	mediodorsal thalamic nucleus	Ec-Ad14 to 16
MdD	medullary reticular nucleus, dorsal part	Ec-Ad30, Pl-Ha21
MDL	mediodorsal thalamic nucleus, lateral	Pl-Ad13 to 15
MDM	mediodorsal thalamic nucleus, medial	Pl-Ad13 & 14
MdV	medullary reticular nucleus, ventral	Ec-Ad30, Pl-Ha21
ME	median eminence	Ec-Ad12 to 14, Pl-Ad12 to 14, Pl-Ha07 to 10
Me	medial amygdaloid nucleus	Ec-Ad12 to 14, Ec-InB09 to 12, Pl-Ad14, Pl-Ha07 & 08
Me5	mesencephalic trigeminal nucleus	Ec-Ad20 & 21, Ec-InB18 to 21, Pl-Ad19 to 21
me5	mesencephalic trigeminal tract	Ec-Ad22, Ec-InB18 & 19, Pl-Ad20 & 21, Pl-Ha15 & 17
Med	medial cerebellar nucleus	Ec-Ad24 to 26, Pl-Ad25 to 27
MEnt	medial entorhinal cortex	Ec-Ad16 to 20, Pl-Ad17 to 21
mes	mesencephalic neuromere	Ec-InA01, Ec-InB13 to 15
Mesen	mesencephalon	Ec-InA01, Ec-InB18 to 28, Pl-Ha12 to 21
MeTg	mesencephalic tegmentum	Pl-Ha16 to 18
mfb	medial forebrain bundle	Ec-InB08 to 17, Pl-Ad12, Pl-Ha04 to 10
MFL	motor cortex forelimb	Ec-Ad17 to 20, Pl-Ad06 to 23
MG	medial geniculate nucleus	Ec-Ad19 & 20, Pl-Ad19 & 20
mge	medial ganglionic eminence	Ec-InB05 to 10, Pl-Ha03 to 08
MHb	medial habenular nucleus	Ec-Ad16 to 18, Pl-Ad14 to 16
MHd	motor cortex head	Ec-Ad11 to 20
MHL	motor cortex hindlimb	Ec-Ad18 to 20, Pl-Ad14 to 16
ML	medial mammillary nucleus, lateral	Ec-Ad15 & 16, Ec-InB14 to 17, Pl-Ad15, Pl-Ha11
ml	medial lemniscus	Ec-Ad14 to 26, Pl-Ad16 to 21
mlf	medial longitudinal fasciculus	Ec-Ad19 to 30, Ec-InB18 to 31, Pl-Ad19 to 26, Pl-Ha11 to 21
MM	medial mammillary nucleus, medial	Ec-Ad15 & 16, Ec-InB14 to 17, Pl-Ad15, Pl-Ha11
Mn	median nucleus of spinal cord	Ec-Ad31 & 32

mne	median neuroepithelium brainstem	Ec-InB20 to 28, Pl-Ha14 to 20
MnPO	median preoptic nucleus	Ec-Ad08 & 10
MnR	median raphe nucleus	Ec-Ad20 to 22, Ec-InB15 to 18, Pl-Ad21, Pl-Ha11 to 13
MO	medial orbital cortex	Ec-Ad02 to 05, Pl-Ad03 to 05
MoCb	molecular layer of cerebellum	Pl-Ad25 & 26
MoDG	molecular layer of dentate gyrus	Ec-Ad17 & 19, Pl-Ad11 to 14, Pl-Ad19
MPA	medial preoptic area	Ec-Ad09 to 11, Pl-Ad10
MPB	medial parabrachial nucleus	Ec-Ad22 & 23, Pl-Ad21 & 22
MPO	medial preoptic nucleus	Ec-Ad11, Pl-Ad10
MPOL	medial preoptic nucleus, lateral part	Pl-Ad11
MPOM	medial preoptic nucleus, medial part	Pl-Ad11
MPT	medial pretectal nucleus	Ec-Ad19, Pl-Ad18
MPtg	medial pterygoid muscle	Pl-Ha04
M/R	M/R transition somatosens. cortex	Ec-Ad09 to 20
MRe	mammillary recess of third ventricle	Ec-InB14 to 17, Pl-Ha11
MRec	medial rectus muscle	Ec-InB07 & 08, Pl-Ha02
mRt	mesencephalic reticular formation	Ec-Ad20 & 21, Ec-InB18 & 19, Pl-Ad18 to 21, Pl-Ha15
MS	medial septal nucleus	Ec-Ad08, Pl-Ad04–05, Pl-Ad06–07, Pl-Ad08, Pl-Ad09
MSh	motor cortex shoulder	Pl-Ad14 to 18
mt	mammillothalamic tract	Ec-Ad15
MTl	motor cortex tail	Ec-Ad18, Ec-Ad21, Pl-Ad14 to 16
MTo	motor cortex tongue	Ec-Ad08 to 17
MTu	medial tuberal nucleus	Ec-Ad14
MVe	medial vestibular nucleus	Ec-Ad26, Ec-InB17 to 27, Pl-Ad23 to 28, Pl-Ha13 to 20
Mx	matrix region of the medulla	Ec-Ad27 to 29, Pl-Ad27 to 30
Myelen	myelencephalon	Pl-Ha21

N

NasC	nasal cavity	Ec-InB01 to 06
ne	neuroepithelium	Ec-InB15 to 20, Ec-InB21, Ec-InB25, Pl-Ha01, Pl-Ha05, Pl-Ha11
noto	notochord	Ec-InA02 to 05, Ec-InB16 to 26, Pl-Ha12 & 13
nplac	nasal placode	Ec-InA01 & 02
NSpt	nasal septum	Ec-InB05 & 06
ntz	nuclear transit. zone of cerebellum	Ec-InB16 to 25, Ec-InB28, Pl-Ha12 to 21
Nv	navicular postolfactory nucleus	Ec-Ad05 to 07, Pl-Ad08

O

OB	olfactory bulb	Ec-Ad05, Ec-InB01 to 03, Pl-Ha01
oc	olivocerebellar tract	Ec-Ad27 & 28
Occ	occipital bone	Ec-InB26 to 30, Pl-Ad25, Pl-Ha18 to 21
OccA	occipital association cortex	Ec-Ad22 & 26
OccC	occipital condyle	Ec-InB31
och	optic chiasm	Ec-Ad10 & 11
olf	olfactory nerve fibres	Ec-InB01 & 02, Pl-Ha01
olfepith	olfactory epithelium	Ec-InB01 to 06
ON	olfactory nerve layer	Ec-Ad01 to 04, Pl-Ad01–03, Pl-Ha01
OPT	olivary pretectal nucleus	Ec-Ad19, Pl-Ad18

opt	optic tract	Ec-Ad12 to 20, Pl-Ad12 to 16, Pl-Ad19
OptRe	optic recess of third ventricle	Ec-Ad10, Ec-InB08 to 10, Pl-Ha05 & 06
Or	oriens layer of the hippocampus	Ec-Ad17, Pl-Ad23 & 24
Oral	oral cavity	Ec-InA03, Ec-InB07
OrbC	orbital cartilage	Ec-InB09, Pl-Ha04
OrPC	orbitoparietal commissure	Ec-InB10 to 13, Pl-Ha05 to 08
os	optic stalk	Ec-InA02, Ec-InB09 & 10
OT	nucleus of the optic tract	Ec-Ad19 & 20, Pl-Ad17 to 19
OV	olfactory ventricle (olfactory LV)	Ec-InB02 & 03, Pl-Ha01

P

P1	pharyngeal pouch 1 of embryo	Ec-InA03
p1	prosomere 1	Ec-InA01, Ec-InB10 to 16, Pl-Ha07 to 10
p1PAG	p1 periaqueductal grey	Ec-Ad18, Ec-InB14 & 15, Pl-Ad17 & 18
p1Rt	p1 reticular formation	Ec-Ad19, Pl-Ad18
P2	pharyngeal pouch 2 of embryo	Ec-InA04
p2	prosomere 2	Ec-InA01, Ec-InB06 to 13, Pl-Ha04 to 10
P3	pharyngeal pouch 3 of embryo	Ec-InA04
p3	prosomere 3	Ec-InA01, Ec-InB06 to 13, Pl-Ha04 to 10
P7	perifacial zone	Pl-Ad22
Pa	paraventricular hypothalamic nucleus	Ec-Ad12, Pl-Ad12 & 13, Pl-Ha05 to 07
Pa5	paratrigeminal nucleus	Ec-Ad27, Pl-Ad26
PAG	periaqueductal grey	Ec-Ad19 to 22, Pl-Ad19 to 21, Pl-Ha17
PaR	pararubral nucleus	Pl-Ad17 & 18
ParB	parietal bone	Ec-InB12 to 20
ParP	parietal plate (chondrocranium)	Ec-InB13 to 23, Pl-Ha09, Pl-Ha13, Pl-Ha14, Pl-Ha16, Pl-Ha18
PaS	parasubiculum	Ec-Ad16 to 21, Pl-Ad17 to 22
PB	parabrachial complex	Ec-InB17 to 25, Pl-Ha12 to 19
PBP	parabrachial pigmented nucleus VTA	Ec-Ad15 to 19, Ec-InB20 & 21, Pl-Ad15 to 19
pc	posterior commissure	Ec-Ad19, Ec-InB13 to 15, Pl-Ad18
pcer	posterior cerebral artery	Ec-InB21
PCo	posterior cortical amygdaloid area	Ec-InB11 & 13, Pl-Ad15 & 16, Pl-Ha07 & 08
PCRt	parvicellular reticular nucleus	Ec-Ad27 to 29, Ec-InB24 to B31, Pl-Ad24 to 30, Pl-Ha15 to 20
PCRtA	parvicellular reticular nucleus, α	Ec-Ad25 & 26, Ec-InB15 to 23, Pl-Ad23, Pl-Ha13 & 14
PDTg	posterodorsal tegmental nucleus	Ec-Ad24, Ec-InB22 & 23, Pl-Ha15 to 17
Pe	periventricular hypothalamic nucleus	Ec-Ad10 to 14, Pl-Ad12 to 14
PeF	perifornical nucleus	Ec-Ad14
Petrous	petrous temporal bone	Ec-InB17, Ec-InB20 to 23, Pl-Ha15 & 16
PF	parafascicular thalamic nucleus	Ec-Ad17 & 18, Pl-Ad15 to 17
PFl	paraflocculus of cerebellum	Ec-Ad24 to 28, Pl-Ad25 to 30
PH	posterior hypothalamic nucleus	Ec-Ad14 to 17, Pl-Ha09 to 11
Pi	pineal gland	Ec-Ad19
PiAnt	pila antotica	Pl-Ha04
Pig	pigment layer of the eye	Ec-InB04 to 06, Pl-Ha01 to 03
PIL	posterior intralaminar thalamic nucleus	Ec-Ad19 & 20
Pir	piriform cortex	Ec-Ad01, Ec-InB05 to 11, Pl-Ad13, Pl-Ha05 to 08

Pir1	piriform cortex, layer 1	Pl-Ad04 to 16
Pir1a	piriform cortex, layer 1a	Pl-Ad04 to 10
Pir1b	piriform cortex, layer 1b	Pl-Ad04 to 10
Pir2	piriform cortex, layer 2	Pl-Ad04 to 16
Pir3	piriform cortex, layer 3	Pl-Ad04 to 12
PirA	piriform area unspecified	Ec-Ad14
PirA1	piriform area 1	Ec-Ad01 to 13
PirA2	piriform area 2	Ec-Ad01 to 13
PirA3	piriform area 3	Ec-Ad01 to 10
PirA4	piriform area 4	Ec-Ad01 to 10
PiRe	pineal recess of third ventricle	Ec-Ad19
Pk	Purkinje cell layer of cerebellum	Pl-Ad25 & 26
Pl/Mi	plexiform/mitral layer of olf. bulb	Ec-Ad01 to 04, Pl-Ad01 to 03
PlA/MiA	plexiform/mitral layer of acc. OB	Ec-Ad01, Pl-Ad01 to 03
PLCo	posterolateral cortical amyg. area	Ec-Ad14 & 15, Ec-InB12
PLH	peduncular part of lateral hypothalamus	Ec-Ad12 to 15, Ec-InB11 to 13, Pl-Ad12 & 14, Pl-Ha09 & 10
PLi	posterior limitans thalamic nucleus	Ec-Ad19 & 20
PM	paramedian lobule	Ec-Ad26 to 28, Pl-Ad26 to 29
PMCo	posteromedial cortical amyg. area	Ec-Ad14 & 15
PMn	paramedian reticular nucleus	Ec-Ad27 to 29
PMnR	paramedian raphe nucleus	Ec-Ad20 & 21, Ec-InB15 to 18, Pl-Ad21, Pl-Ha11 to 13
PN	paranigral nucleus of the VTA	Ec-InB18 to 21, Pl-Ad16
Pn	pontine nuclei	Ec-Ad20 to 24, Pl-Ad18 to 21
PnC	pontine reticular nucleus, caudal	Ec-Ad23 & 24, Ec-InB15 to18, Pl-Ad20 to 23, Pl-Ha11 & 12
PnO	pontine reticular nucleus, oral part	Ec-Ad20 to 22, Ec-InB14 to 18, Pl-Ad20 & 22, Pl-Ha11 to 13
Po	posterior thalamic nuclear group	Ec-Ad17 to 20, Pl-Ad17 to 20
POA	preoptic area	Ec-InB06 to 08, Pl-Ha03 & 04
PoDG	polymorph layer of dentate gyrus	Ec-Ad17 to 20, Pl-Ad19
Post	postsubiculum	Ec-Ad19 to 23, Pl-Ad16 to 22
PoT	posterior thalamic nucleus, triangular	Ec-Ad19
PP	peripeduncular nucleus	Ec-Ad20
ppf	prepyramidal fissure	Ec-Ad27 & 28, Pl-Ad29 & 30
PPit	posterior lobe of pituitary	Ec-InB13
PPy	parapyramidal nucleus	Ec-InB19 to 23
PR	prerubral field	Ec-Ad16 to 18, Pl-Ad15
Pr	prepositus nucleus	Ec-Ad26, Ec-InB19 to 24, Pl-Ad24 to 26, Pl-Ha14 to 17
Pr5	principal sensory trigeminal nucleus	Ec-Ad22 to 24, Ec-InB13 to 18, Pl-Ha08 to 11
Pr5DL	Pr5, dorsolateral part	Ec-Ad23 & 24
Pr5mc	Pr5, magnocellular part	Pl-Ad19 to 23
Pr5pc	Pr5, parvicellular part	Pl-Ad17 to 23
Pr5VM	Pr5, ventromedial part	Ec-Ad23 & 24
PrC	precommissural nucleus	Pl-Ad17
PrePl	preplate of developing cortex	Ec-InB01 & 15, Pl-Ha01 to 03, Pl-Ha09 to 11
PrG	pregeniculate nucleus	Ec-Ad19 & 20, Pl-Ad19 & 20
PRh	perirhinal cortex	Pl-Ad14 to 24
PrL	prelimbic cortex	Ec-Ad02 to 07, Pl-Ad03 to 05
PrO	pro-otic foramen	Ec-InB10 to 15, Pl-Ha05 to 07
PrOrb	preoptic root of orbital bone	Ec-InB01 to 08, Pl-Ha01 to 03

Pros	prosencephalon	Ec-InA01 & 02
PrS	presubiculum	Ec-Ad16 to 21, Pl-Ad16 to 22
PrTh	prethalamus (prosomere 3)	Ec-InB08 to 13, Pl-Ha06 to 10
PSCD	posterior semicircular duct	Ec-InB21 to 24, Pl-Ha15 to 17
psf	posterior superior fissure of Cb	Pl-Ad28
PsOp	pseudo-optic foramen	Ec-InB08 & 09, Pl-Ha03 & 04
psp	posterior spinal artery	Ec-Ad31 & 32
PSph	presphenoid bone	Ec-InB07, Pl-Ha02 & 03
PSTh	parasubthalamic nucleus	Ec-Ad14, Pl-Ad15 & 16
PT	paratenial thalamic nucleus	Ec-Ad12, Pl-Ad12 & 13
Ptec	pretectum	Ec-InB10 to 17, Pl-Ha07 to 10
PTg	pedunculotegmental nucleus	Ec-Ad22, Ec-InB17 to 21
Ptg	pterygoid process of sphenoid bone	Pl-Ha04
PTh	posterior thalamus	Ec-InB07 to 13, Pl-Ha07 to 10
Pu	putamen	Ec-Ad05 to 18, Pl-Ad07 to 25
PV	parietoventral area of sensory cortex	Ec-Ad05 to 21, Pl-Ad22 to 30
PVA	paraventricular thalamic nucleus, anterior	Ec-Ad12 to 14, Pl-Ad12 to 13
PVP	paraventricular thalamic nucleus, posterior	Ec-Ad15 to 18, Pl-Ad15 & 16
Py	pyramidal cell layer of Hi	Ec-Ad17

R

R	rostral somatosensory field of cortex	Ec-Ad05 to 21, Pl-Ad04 to 23
R1	Rexed's lamina 1	Ec-Ad31 to 33
r1	rhombomere 1	Ec-InA01 & 02
r2	rhombomere 2	Ec-InA02
R2e	Rexed's lamina 2, external	Ec-Ad31 to 33
R2i	Rexed's lamina 2, internal	Ec-Ad31 to 33
R3	Rexed's lamina 3	Ec-Ad31 to 33
R4	Rexed's lamina 4	Ec-Ad31 to 33
r4	rhombomere 4	Ec-InA03
R5	Rexed's lamina 5	Ec-Ad31 to 33
r5	rhombomere 5	Ec-InA03
R6	Rexed's lamina 6	Ec-Ad31 & 33
r6	rhombomere 6	Ec-InA03 & 04
R7	Rexed's lamina 7	Ec-Ad31 to 33
r7	rhombomere 7	Ec-InA04
R8	Rexed's lamina 8	Ec-Ad31 to 33
R9	Rexed's lamina 9	Ec-Ad31 to 33
R10	Rexed's lamina 10	Ec-Ad31 to 33
Rad	radiatum layer of the hippocampus	Ec-Ad17, Pl-Ad23 & 24
RAtr	right atrium	Ec-InA05
Rbd	rhabdoid nucleus	Ec-Ad20 & 22
RCh	retrochiasmatic area	Pl-Ad12 to 14
Re	reuniens thalamic nucleus	Ec-Ad13 & 14, Pl-Ad13 & 14
Red	red nucleus	Ec-InB18 & 19, Pl-Ha15
respepith	respiratory epithelium	Ec-InB02 to 05
rf	rhinal fissure	Ec-Ad01 to 17, Ec-InB02 to 09, Pl-Ad03 to 23
Rh	rhomboid thalamic nucleus	Pl-Ad14

Rhomb	rhombencephalon	Ec-InA01 to 04, Ec-InB17 to 31, Pl-Ha11 to 21
ri	rhinal incisure	Ec-InB01 to 03, Pl-Ad01 to 05
RIP	raphe interpositus nucleus	Pl-Ad22 & 23, Pl-Ha12
rl	rhombic lip	Ec-InA02 to 04
RLi	rostral linear nucleus (midbrain)	Ec-Ad18 & 19
RM	retromammillary nucleus	Ec-Ad16, Ec-InB14 to 17, Pl-Ha11
RMC	red nucleus, magnocellular part	Ec-Ad19, Pl-Ad16 to 18
RMg	raphe magnus nucleus	Ec-Ad24 to 26, Ec-InB16 to 26, Pl-Ad22 to 25, Pl-Ha12 & 13
RML	retromammillary nucleus, lateral	Ec-Ad15, Pl-Ad15
RMM	retromammillary nucleus, medial	Ec-Ad15
ROb	raphe obscurus nucleus	Ec-Ad27 to 29, Ec-InB21 to 29, Pl-Ad26, Pl-Ha13 to 20
RPa	raphe pallidus nucleus	Ec-Ad25 to 30, Ec-InB26 to 29, Pl-Ad22, Pl-Ad23 to 26, Pl-Ha15 to 20
RPC	red nucleus, parvicellular part	Ec-Ad19, Pl-Ad16 to 18
RPF	retroparafascicular nucleus	Pl-Ad18
RRF	retrorubral field	Ec-Ad20 & 21, Pl-Ad19 & 20
rrl	rostral rhombic lip	Ec-InB16 to 25, Pl-Ha11 to 21
RSG	retrosplenial granular cortex	Ec-Ad19 to 26, Pl-Ad19 to 26
Rt	reticular nucleus (prethalamus)	Ec-Ad12 to 16, Pl-Ad13, Pl-Ad14 to 18, Pl-Ha07 to 10
RtSt	reticulostrial nucleus	Ec-Ad13 to 15, Pl-Ad13 & 14
RtTg	reticulotegmental nucleus of pons	Ec-Ad22 & 23, Pl-Ad21
S		
S1	primary somatosensory cortex	Pl-Ad05 to 24
S1FL	primary somato. cortex, forelimb region	Ec-Ad20 & 21
S1Hd	primary somato. cortex, head region	Ec-Ad17 to 20
S1HL	primary somato. cortex, hindlimb region	Ec-Ad21 to 25
S1Sn	primary somato. cortex, snout	Ec-Ad17 & 18
S1SnB	primary somato. cortex, snout base	Ec-Ad16
S1SnT	primary somato. cortex, snout tip	Ec-Ad09 to 16
S1Tl	primary somato. cortex, tail region	Ec-Ad22 to 25
S1To	primary somato. cortex, tongue region	Ec-Ad05 to 18
S1Tr	primary somato. cortex, trunk region	Ec-Ad21 to 25
s5	sensory root of trigeminal nerve	Ec-Ad22, Ec-InB13 to 15, Pl-Ad17 to 22, Pl-Ha08 to 11
S9Gn	sup. gang. of glossopharyngeal nerve	Ec-InB23
S10Gn	superior ganglion of vagal nerve	Ec-InB24, Ec-InB26, Ec-InB31, Pl-Ha15 to 17
Sacc	saccule	Ec-InB18 to 20, Pl-Ha12 & 13
SC	superior colliculus	Ec-Ad20 & 21, Ec-InB13 to 15, Pl-Ad18, Pl-Ha18
scba	superior cerebellar artery	Pl-Ha14 to 18
SCh	suprachiasmatic nucleus	Ec-Ad11, Pl-Ad11, Pl-Ha05
scp	superior cerebellar peduncle	Ec-Ad21 to 25, Pl-Ad20 to 25
sf	secondary fissure	Ec-Ad27 & 28, Pl-Ad28 to 30
SFi	septofimbrial nucleus	Pl-Ad10 & 11
SG	suprageniculate thalamic nucleus	Ec-Ad20
SHi	septohippocampal nucleus	Pl-Ad08 & 09
SHy	septohypothalamic nucleus	Ec-Ad08, Pl-Ad11
Sim	simple lobule of cerebellum	Pl-Ad26 & 27
sl	sulcus limitans	Ec-InA02 to 05, Ec-InB19 to 31, Pl-Ha13 to 21

sm	stria medullaris (thalami)	Ec-Ad12 to 18, Ec-InB05 & 06, Pl-Ad12 to 16, Pl-Ha04 to 06
SNC	substantia nigra, compact part	Ec-InB18 & 19, Pl-Ad16, Pl-Ha15
SNCD	SNC, dorsal tier	Ec-Ad15 to 19, Pl-Ad15 & 17
SNCM	SNC, medial tier	Pl-Ad16
SNL	substantia nigra, lateral part	Ec-Ad16 to 19, Pl-Ad15 & 17
SNR	substantia nigra, reticular part	Ec-Ad14 to 19, Ec-InB18 to 21, Pl-Ad15 to 18, Pl-Ha15
SNR/EP	SN, reticular part/entopedunc. nucleus	Pl-Ad16 &17
SOb	superior oblique muscle	Ec-InB03 & 05
SOl	superior olive	Ec-Ad24 & 25, Ec-InB15 to 17, Pl-Ad22 & 23, Pl-Ha11
Sol	solitary nucleus	Ec-Ad26 to 30, Ec-InB18 to 31, Pl-Ad24 to 30, Pl-Ha13 to 21
sol	solitary tract	Ec-Ad26 to 30, Ec-InB22 to 31, Pl-Ad26 to 28, Pl-Ha14 to 21
SOR	supraoptic nucleus, retrochiasmatic	Ec-Ad12 & 13, Pl-Ad12
sp5	spinal trigeminal tract	Ec-Ad23 to 30, Ec-InB15 to 31, Pl-Ad22 to 30, Pl-Ha11 to 21
Sp5C	spinal trigeminal nucleus, caudal	Ec-Ad30, Pl-Ad29, Pl-Ha20 & 21
Sp5I	spinal trigeminal nucleus, interpolar	Ec-Ad27 to 29, Ec-InB23 to 31, Pl-Ad24 to 30, Pl-Ha14 to 21
Sp5O	spinal trigeminal nucleus, oral	Ec-Ad25 & 26, Ec-InB19 to 23, Pl-Ad22 to 28, Pl-Ha11 to 18
SpC	spinal cord	Ec-InA05
SPF	subparafascicular thalamic nucleus	Ec-Ad17 & 18, Pl-Ad16
SPFPC	subparafascicular thalamic nucleus, parvi.	Ec-Ad19
SphPal	sphenopalatine ganglion	Ec-InB07, Ec-InB08
Spt	septum	Ec-InB04, Pl-Ha02
SPTg	subpeduncular tegmental nucleus	Ec-InB22 & 23, Pl-Ad21, Pl-Ha12 to 17
SpVe	spinal vestibular nucleus	Ec-Ad26, Ec-InB22 to 31, Pl-Ad23 to 28, Pl-Ha13 to 21
SRec	superior rectus muscle	Ec-InB03 to 07, Pl-Ha02 & 03
sss	superior sagittal sinus	Ec-InB04, Pl-Ha02 to 04
ST	bed nucleus of the stria terminalis	Ec-InB05 to 11, Pl-Ad09, Pl-Ad10 to 12, Pl-Ha03 to 08
st	stria terminalis	Ec-Ad11 to 18, Ec-InB11 & 12, Pl-Ad11 to 22, Pl-Ha06 & 07
STh	subthalamic nucleus	Ec-Ad14, Pl-Ad15 to 17, Pl-Ha08, Pl-Ha11
StHy	striohypothalamic nucleus	Ec-Ad12, Pl-Ad10 to 12
STIA	bed nucleus st, intraamygdaloid divn	Pl-Ad15
STLD	bed nucleus st, lateral divn, dorsal	Ec-Ad08 to 11
STLP	bed nucleus st, lateral divn, posterior	Ec-Ad08 to 12
STLV	bed nucleus st, lateral divn, ventral	Ec-Ad08 to 11
StM	sternomastoid muscle	Pl-Ha13, Pl-Ha16
STMA	bed nucleus st, medial divn, anterior	Ec-Ad08 to 10
STMPI	bed nucleus st, medial divn, posteroint.	Pl-Ad12
STMPL	bed nucleus st, medial divn, posterolateral	Ec-Ad12, Pl-Ad12
STMPM	bed nucleus st, medial divn, postmed.	Ec-Ad12, Pl-Ad12
STMV	bed nucleus st, medial divn, ventral	Ec-Ad10 & 11
STr	subiculum, transition area	Pl-Ad23 & 24
Su3	supraoculomotor periaqued. grey	Ec-Ad20 & 21
Su5	supratrigeminal nucleus	Ec-InB15 & 16
Sub	submedius thalamic nucleus	Pl-Ad13 & 14
SubCD	subcoeruleus nucleus, dorsal part	Ec-Ad23, Pl-Ad22
SubCV	subcoeruleus nucleus, ventral part	Ec-Ad23, Pl-Ad22
SubI	subincertal nucleus	Ec-Ad13, Pl-Ad13 & 14
SubV	subventricular layer developing cortex	Ec-InB11 & 12, Pl-Ha03 to 08
SuG	superficial grey layer of SC	Ec-Ad20 & 21, Pl-Ad19 to 22

SuS	superior salivatory nucleus	Pl-Ad22
SuVe	superior vestibular nucleus	Ec-Ad24 to 26, Ec-InB16 to 24, Pl-Ad23 to 25, Pl-Ha11 to 17

T

tcf	transverse cerebral fissure	Ec-Ad11 to 19
TeA	temporal association cortex	Ec-Ad17 to 26
Telen	telencephalon Ec-InB01 to 13,	Pl-Ha02 to 11
TempM	temporalis muscle	Ec-InB10
term	terminal nerve	Ec-InB02
tfp	transverse fibres of the pons	Ec-Ad20 to 24
TG	tectal grey formation (midbrain)	Ec-Ad20, Pl-Ad19
trs	transverse venous sinus	Ec-InB16 to 19, Pl-Ha14, Pl-Ha17
ts	tectospinal tract	Ec-Ad28 to 30
TT	tenia tecta	Ec-Ad04 to 06
Tu	olfactory tubercle	Ec-Ad05 to 10, Ec-InB04, Pl-Ad04 to 08
TuLH	tuberal region of lateral hypothalamus	Ec-Ad12 & 13, Pl-Ad13 & 14
Tz	nucleus of the trapezoid body	Ec-Ad23

U

UMac	macula of the utricle	Ec-InB18, Pl-Ha12 & 13
Utr	utricle	Ec-InB18 to 22, Pl-Ha12 to 14

V

V	primary visual cortex	Ec-Ad21 to 25, Pl-Ad23 to 30
V3V	ventral third ventricle	Ec-Ad10 to 17, Ec-InB08 to 13
VA	ventral anterior thalamic nucleus	Pl-Ad13 & 14
VC	ventral cochlear nucleus	Ec-InB19 & 20, Pl-Ad23, Pl-Ha12
VCA	ventral cochlear nucleus, anterior part	Ec-InB17, Pl-Ha12
VCP	ventral cochlear nucleus, posterior part	Ec-InB18 to 20, Pl-Ad25, Pl-Ha13 & 14
VDB	nucleus of vertical limb diag. band	Ec-Ad07 & 08, Ec-InB04 & 05, Pl-Ad09, Pl-Ha02
Ve	vestibular nucleus or neuroepithelium	Ec-InB17 to 23, Ec-InB30, Pl-Ha12 to 20
VeCb	vestibulocerebellar nucleus	Ec-Ad24 & 25, Pl-Ad24
VeGn	vestibular nerve ganglion	Ec-InB18 to 20, Pl-Ha11 to 13
VEn	ventral endopiriform nucleus	Ec-Ad05 to 10, Pl-Ad05 to 16
Vent	ventricular space of the eye	Pl-Ha01 to 03
vert	vertebral artery	Ec-InB24 to 31, Pl-Ha15 to 20
VH	ventral horn	Ec-InB31
VL	ventrolateral thalamic nucleus	Ec-Ad15 to 17, Pl-Ad13 to 16
VLH	ventrolateral hypothalamic nucleus	Pl-Ad12
VLL	ventral nucleus of lateral lemniscus	Ec-InB13 & 14, Pl-Ad20 & 21
VLPAG	ventrolateral periaqueductal grey	Ec-Ad20 to 22, Pl-Ad21
VLPO	ventrolateral preoptic nucleus	Pl-Ad11
VM	ventromedial thalamic nucleus	Ec-Ad14 to 16, Pl-Ad13 to 15
VMb	basal ventromedial nucleus	Pl-Ad15 to 18
VMH	ventromedial hypothalamic nucleus	Ec-Ad12 to 14, Pl-Ad13, Pl-Ad14, Pl-Ha06 to 10
VMPO	ventromedial preoptic nucleus	Ec-Ad10
vn	vomeronasal nerve	Ec-Ad01, Pl-Ad01–03
VO	ventral orbital cortex	Ec-Ad02 to 04, Pl-Ad01–03

Vom	vomer	Ec-InB01 to 06, Pl-Ha01
VP	ventral pallidum	Ec-Ad05 to 11, Pl-Ad04–05, Pl-Ad07 to 11, Pl-Ad15, Pl-Ha09 & 10
VPL	ventral posterolateral thalamic nucleus	Ec-Ad14 to 18
VPM	ventral posteromedial thalamic nucleus	Ec-Ad14 to 18, Pl-Ad15 to 22
VPPC	ventral posterior thalamic nucleus, parvi.	Ec-Ad17 & 18, Pl-Ad16
VRe	ventral reuniens thalamic nucleus	Ec-Ad14, Pl-Ad14
VS	ventral subiculum	Ec-Ad16 to 21, Pl-Ad18, Pl-Ad21, Pl-Ad22
VTA	ventral tegmental area	Ec-InB18 & 19, Pl-Ha12
VTg	ventral tegmental nucleus	Ec-InB17 to 21, Pl-Ha14
VTT	ventral tenia tecta	Pl-Ad04–05

X

X	nucleus X	Ec-InB22 to 24, Pl-Ha14 to 16
xscp	decussation of scp	Ec-Ad19, Pl-Ad19

Z

ZI	zona incerta	Ec-Ad13 to 18, Ec-InB13 to 17, Pl-Ha08, Pl-Ha10
ZIC	zona incerta, caudal part	Pl-Ad15
ZID	zona incerta, dorsal part	Pl-Ad13 & 14
ZIV	zona incerta, ventral part	Pl-Ad13 & 14
zli	zona limitans interthalamica	Ec-InB11 & 12, Pl-Ha07 to 09

REFERENCES

Abbie AA (1934) The brainstem and cerebellum of *Echidna aculeata*. *Philosophical Transactions of the Royal Society of London. Series B, Biological Sciences* **224**, 1–74.

Abbie AA (1938) The excitable cortex in the Monotremata. *The Australian Journal of Experimental Biology and Medical Science* **16**, 143–152.

Abbie AA (1940) Cortical lamination in the monotremata. *The Journal of Comparative Neurology* **72**, 429–467.

Abdel-Mannan O, Cheung AFP, Molnár Z (2008) Evolution of cortical neurogenesis. *Brain Research Bulletin* **75**, 398–404.

Abensperg-Traun M (1989) Some observations on the duration of lactation and movements of a *Tachyglossus aculeatus acanthion* (Monotremata: Tachyglossidae) from Western Australia. *Australian Mammalogy* **12**, 33–34.

Aboitiz F, Montiel J (2007) Origin and evolution of the vertebrate telencephalon, with special reference to the mammalian neocortex. *Advances in Anatomy, Embryology, and Cell Biology* **193**, 1–112.

Addens JL, Kurotsu T (1936) The pyramidal tract of echidna. *Proceedings of the Koninklijke Akademie van Wetenschappen Te Amsterdam* **39**, 1142–1151.

Aiello LC, Wheeler P (1995) The expensive-tissue hypothesis: the brain and the digestive system in human and primate evolution. *Current Anthropology* **36**, 199–221.

Aitkin LM, Johnstone BM (1972) Middle ear function in a monotreme: the echidna (*Tachyglossus aculeatus*). *The Journal of Experimental Zoology* **180**, 245–250.

Akiyama S (1998) Molecular ecology of the platypus (*Ornithorhynchus anatinus*). PhD thesis. La Trobe University, Australia.

Akoev GN (1995) Electroreceptors: involvement of excitatory amino acids in synaptic transmission. *Comparative Biochemistry and Physiology. Part A, Molecular & Integrative Physiology* **110**, 217–222.

Allin EF (1975) Evolution of the mammalian middle ear. *Journal of Morphology* **147**, 403–438.

Allison T, Goff WR (1972) Electrophysiological studies of echidna, *Tachyglossus aculeatus*. 3. Sensory and interhemispheric evoked responses. *Archives Italiennes de Biologie* **110**, 195–216.

Allison T, Van Twyver H, Goff WR (1972) Electrophysiological studies of echidna, *Tachyglossus aculeatus*. 1. Waking and sleep. *Archives Italiennes de Biologie* **110**, 145–184.

Alonso JR, Brinon JG, Crespo C, Bravo IG, Arevalo R, Aijon J (2001) Chemical organization of the macaque monkey olfactory bulb: II Calretinin, calbindin D-28k, parvalbumin, and neurocalcin immunoreactivity. *The Journal of Comparative Neurology* **432**, 389–407.

Altman J, Bayer SA (1978a) Prenatal development of the cerebellar system in the rat. I. Cytogenesis and histogenesis of the deep nuclei and the cortex of the cerebellum. *The Journal of Comparative Neurology* **179**, 23–48.

Altman J, Bayer SA (1978b) Prenatal development of the cerebellar system in the rat. II. Cytogenesis and histogenesis of the inferior olive, pontine gray, and the precerebellar reticular nuclei. *The Journal of Comparative Neurology* **179**, 49–76.

Altman J, Bayer SA (1979) Development of the diencephalon in the rat. IV. Quantitative study of the time of origin of neurons and internuclear chronological gradients in the thalamus. *The Journal of Comparative Neurology* **188**, 455–472.

Altman J, Bayer SA (1985a) Embryonic development of the rat cerebellum. I. Delineation of the cerebellar primordium and early cell movements. *The Journal of Comparative Neurology* **231**, 1–26.

Altman J, Bayer SA (1985b) Embryonic development of the rat cerebellum. II. Translocation and regional distribution of the deep neurons. *The Journal of Comparative Neurology* **231**, 27–41.

Altman J, Bayer SA (1986) The development of the rat hypothalamus. *Advances in Anatomy, Embryology, and Cell Biology* **100**, 1–178.

Altman J, Bayer SA (1987a) Development of the precerebellar nuclei in the rat. I. The precerebellar neuroepithelium of the rhombencephalon. *The Journal of Comparative Neurology* **257**, 477–489.

Altman J, Bayer SA (1987b) Development of the precerebellar nuclei in the rat. II. The intramural olivary migratory stream and the neurogenetic organization of the inferior olive. *The Journal of Comparative Neurology* **257**, 490–512.

Altman J, Bayer SA (1987c) Development of the precerebellar nuclei in the rat. III. The posterior precerebellar extramural migratory stream and the lateral reticular and external cuneate nuclei. *The Journal of Comparative Neurology* **257**, 513–528.

Altman J, Bayer SA (1987d) Development of the precerebellar nuclei in the rat. IV. The anterior precerebellar extramural migratory stream and the nucleus reticularis tegmenti pontis and the basal pontine gray. *The Journal of Comparative Neurology* **257**, 529–552.

Ambrosiani J, Armengol JA, Martinez S, Puelles L (1996) The avian inferior olive derives from the alar neuroepithelium of the rhombomeres 7 and 8: an analysis by using chick-quail chimeric embryos. *Neuroreport* **17**, 1285–1288.

Andres KH, von Düring M (1984) The platypus bill. A structural and functional model of pattern-like arrangement of different cutaneous sensory receptors. In *Sensory Receptor Mechanisms*. (Eds W Hamann and A Iggo) pp. 81–89. World Scientific Publishing Company, Singapore.

Andres KH, von Düring M (1988) Comparative anatomy of vertebrate electroreceptors. *Progress in Brain Research* **74**, 113–131.

Andres KH, von Düring M, Iggo A, Proske U (1991) The anatomy and fine structure of the echidna *Tachyglossus aculeatus* snout with respect to its different trigeminal sensory receptors including the electroreceptors. *Anatomy and Embryology* **184**, 371–393.

Archer M, Plane MD, Pledge NS (1978) Additional evidence for interpreting the Miocene *Obdurodon insignis* Woodburne and Tedford, 1975, to be a fossil platypus (Ornithorhynchidae: Monotremata) and a reconsideration of the status of *Ornithorhynchus agilis* De Vis, 1885. *Australian Zoologist* **20**, 9–27.

Archer M, Flannery TF, Ritchie A, Molnar RE (1985) First Mesozoic mammal from Australia – an Early Cretaceous monotreme. *Nature* **318**, 363–366.

Archer M, Jenkins FA, Jr, Hand SJ, Murray P, Godthelp H (1992) Description of the skull and non-vestigial dentition of a Miocene platypus (*Obdurodon dicksoni* n. sp.) from Riversleigh, Australia, and the problem of monotreme origins. In *Platypus and Echidnas*. (Ed. ML Augee) pp. 15–27. Royal Zoological Society of New South Wales, Sydney.

Archibald JD (2011) *Extinction and Radiation. How the Fall of the Dinosaurs Led to the Rise of Mammals.* Johns Hopkins University Press, Baltimore.

Ariëns-Kappers CUA, Huber GC, Crosby EC (1960). *The Comparative Anatomy of the Nervous System, Including Man.* Hafner, New York.

Armstrong WE (2004) Hypothalamic supraoptic and paraventricular nuclei. In *The Rat Nervous System*. 3rd edn. (Ed. G Paxinos) pp. 369–388. Elsevier, San Diego.

Armstrong CL, Hopkins DA (1998) Neurochemical organization of paratrigeminal nucleus projections to the dorsal vagal complex in the rat. *Brain Research* **785**, 49–57.

Ashwell KWS (2006a) Chemoarchitecture of the monotreme olfactory bulb. *Brain, Behavior and Evolution* **67**, 69–84.

Ashwell KWS (2006b) Cyto- and chemoarchitecture of the monotreme olfactory tubercle. *Brain, Behavior and Evolution* **67**, 85–102.

Ashwell KWS (2008a) Topography and chemoarchitecture of the striatum and pallidum in a monotreme, the short–beaked echidna (*Tachyglossus aculeatus*). *Somatosensory & Motor Research* **25**, 171–187.

Ashwell KWS (2008b) Encephalization of Australian and New Guinean Marsupials. *Brain, Behavior and Evolution* **71**, 181–199.

Ashwell KWS (2010a) Motor system and spinal cord. In *Neurobiology of Australian Marsupials*. (Ed. K Ashwell) pp. 202–215. Cambridge University Press, Cambridge.

Ashwell KWS (2010b) Diencephalon and associated structures: prethalamus; thalamus; hypothalamus; pituitary gland; epithalamus; and pretectal area. In *Neurobiology of Australian Marsupials*. (Ed.

K Ashwell) pp. 95–117. Cambridge University Press, Cambridge.

Ashwell KWS (2012a) The olfactory system. In *The Mouse Nervous System*. (Eds C Watson, G Paxinos, L Puelles) pp. 653–660. Elsevier, Amsterdam.

Ashwell KWS (2012b) Development of the olfactory pathways in platypus and echidna. *Brain, Behavior and Evolution* **79**, 45–56.

Ashwell KWS (2012c) Development of the dorsal and ventral thalamus in platypus (*Ornithorhychus anatinus*) and short-beaked echidna (*Tachyglossus aculeatus*). *Brain Structure & Function* **217**, 577–589.

Ashwell KWS (2012d) Development of the spinal cord and peripheral nervous system in platypus (*Ornithorhynchus anatinus*) and short-beaked echidna (*Tachyglossus aculeatus*). *Somatosensory & Motor Research* **29**, 13–27.

Ashwell KWS (2012e) Development of the hypothalamus and pituitary in platypus (*Ornithorhynchus anatinus*) and short-beaked echidna (*Tachyglossus aculeatus*). *Journal of Anatomy* **221**, 9–20.

Ashwell KWS (2012f) Development of the cerebellum in the platypus (*Ornithorhynchus anatinus*) and short-beaked echidna (*Tachyglossus aculeatus*). *Brain, Behavior and Evolution* **79**, 237–251.

Ashwell KWS, Hardman CD (2012a) Distinct development of the cerebral cortex in platypus and echidna. *Brain, Behavior and Evolution* **79**, 57–72.

Ashwell KWS, Hardman CD (2012b) Distinct development of the trigeminal sensory nuclei in platypus and echidna. *Brain, Behavior and Evolution* **79**, 261–274.

Ashwell KWS, Paxinos G (2005) Cyto- and chemoarchitecture of the dorsal thalamus of the monotreme *Tachyglossus aculeatus*, the short beaked echidna. *Journal of Chemical Neuroanatomy* **30**, 161–183.

Ashwell KWS, Paxinos G (2007) The pretectal nuclei in two monotremes: the short-beaked echidna (*Tachyglossus aculeatus*) and the platypus (*Ornithorhynchus anatinus*). *Brain Structure & Function* **212**, 359–369.

Ashwell KWS, Paxinos G (2008) *Atlas of the Developing Rat Nervous System*. 3rd edn. Elsevier, London.

Ashwell KWS, Phillips JM (2006) The anterior olfactory nucleus and piriform cortex of the echidna and platypus. *Brain, Behavior and Evolution* **67**, 203–227.

Ashwell KWS, Zhang L-L (1997) Cyto- and myeloarchitectonic organisation of the spinal cord of an echidna (*Tachyglossus aculeatus*). *Brain, Behavior and Evolution* **49**, 276–294.

Ashwell KWS, Waite PME, Marotte LR (1996a) Ontogeny of the projection tracts and commissural fibres in the forebrain of the tammar wallaby (*Macropus eugenii*): timing in comparison with other mammals. *Brain, Behavior and Evolution* **47**, 8–22.

Ashwell KWS, Marotte LR, Lixin L, Waite PME (1996b) Anterior commissure of the wallaby (*Macropus eugenii*): adult morphology and development. *The Journal of Comparative Neurology* **366**, 478–494.

Ashwell KWS, Hardman CD, Paxinos G (2004) The claustrum is not missing from all monotreme brains. *Brain, Behavior and Evolution* **64**, 223–241.

Ashwell KWS, Hardman CD, Paxinos G (2005) Cyto- and chemoarchitecture of the amygdala of a monotreme, *Tachyglossus aculeatus* (the short beaked echidna). *Journal of Chemical Neuroanatomy* **30**, 82–104.

Ashwell KW, Lajevardi SE, Cheng G, Paxinos G (2006a) The hypothalamic supraoptic and paraventricular nuclei of the echidna and platypus. *Brain, Behavior and Evolution* **68**, 197–217.

Ashwell KWS, Hardman CD, Paxinos G (2006b) Cyto- and chemoarchitecture of the sensory trigeminal nuclei of the echidna, platypus and rat. *Journal of Chemical Neuroanatomy* **31**, 81–107.

Ashwell KWS, Paxinos G, Watson CRR (2007a) Cyto- and chemoarchitecture of the cerebellum of the short–beaked echidna (*Tachyglossus aculeatus*). *Brain, Behavior and Evolution* **70**, 71–89.

Ashwell KWS, Paxinos G, Watson CRR (2007b) Precerebellar and vestibular nuclei of the short-beaked echidna (*Tachyglossus aculeatus*). *Brain Structure & Function* **212**, 209–221.

Ashwell KWS, Marotte LR, Cheng G (2008) Development of the olfactory system in a wallaby (*Macropus eugenii*). *Brain, Behavior and Evolution* **71**, 216–230.

Ashwell KWS, Hardman CD, Giere P (2012) Distinct development of peripheral trigeminal pathways in the platypus (*Ornithorhynchus anatinus*) and short-beaked echidna (*Tachyglossus aculeatus*). *Brain, Behavior and Evolution* **79**, 113–127.

Aston-Jones G (2004) Locus coeruleus, A5 and A7 noradrenergic cell groups. In *The Rat Nervous*

System. 3rd edn. (Ed. G Paxinos) pp. 259–294. Elsevier, San Diego.

Atkins AM, Krause WJ (1971) An unusual basement membrane underlying intestinal epithelium of the platypus *Ornithorhynchus anatinus*. *Experientia* **27**, 686–688.

Augee ML (1976) Heat tolerance of monotremes. *Journal of Thermal Biology* **1**, 181–184.

Augee ML (2008) Short-beaked echidna. In *The Mammals of Australia*. (Eds S van Dyck and R Strahan) pp. 37–39. Reed New Holland, Sydney.

Augee ML, Gooden BA (1992) Evidence for electroreception from field studies of the echidna *Tachyglossus aculeatus*. In *Platypus and Echidnas*. (Ed. ML Augee) pp. 211–215. The Royal Zoological Society of New South Wales, Sydney.

Augee ML, Elsner RW, Gooden BA, Wilson PR (1971a) Respiratory and cardiac responses of a burrowing animal, the echidna. *Respiration Physiology* **11**, 327–334.

Augee ML, Fink G, Smith GC (1971b) Morphological and experimental studies on pars distalis of echidna (*Tachyglossus aculeatus*). *Journal of Anatomy* **108**, 208.

Augee ML, Bergin TL, Morris C (1978) Observations on behaviour of echidnas at Taronga Zoo. In *Monotreme Biology: The Australian Zoologist Special Symposium*. (Ed. ML Augee). Published in *The Australian Zoologist* **20**, 121–129.

Augee ML (1976) Heat tolerance of monotremes. *Journal of Thermal Biology* **1**, 181–184.

Augee M, Gooden B, Musser AM (2006) *Echidna: Extraordinary Egg-laying Mammal*. CSIRO Publishing, Melbourne.

Augee ML, Carrick FN, Grant TR, Temple-Smith PD (2008) Order Monotremata: platypus and echidnas. In *The Mammals of Australia*. (Eds S Van Dyck and R Strahan). 3rd edn. pp. 30–31. Reed New Holland, Sydney.

Augee ML (2008) Short-beaked echidna. In *The Mammals of Australia*. (Eds S van Dyck and R Strahan) pp. 37–39. Reed New Holland, Sydney.

Baillie JEM, Turvey ST, Waterman C (2009) Survival of Attenborough's long-beaked echidna *Zaglossus attenboroughi* in New Guinea. *Oryx* **43**, 146–148.

Baird JA, Hales JRS, Lang WJ (1974) Thermoregulatory responses to injection of monoamines, acetylcholine and prostaglandins into a lateral cerebral ventricle of echidna. *The Journal of Physiology* **236**, 539–548.

Bakker AJ, Parkinson AL, Head SI (2005) Contractile properties of single-skinned skeletal muscle fibres of the extensor digitorum longus muscle of the Australian short-nosed echidna. *Australian Journal of Zoology* **53**, 237–240.

Bangma GC, ten Donkelaar HJ (1982) Afferent projections of the cerebellum in various types of reptiles. *The Journal of Comparative Neurology* **207**, 255–273.

Barton RA, Dean P (1993) Comparative evidence indicating neural specialization for predatory behaviour in mammals. *Proceedings of The Royal Society. Series B, Biological Sciences* **254**, 63–68.

Baudinette RV, Runciman SIC, Frappell PF, Gannon BJ (1988) Development of the marsupial cardiorespiratory system. In *The Developing Marsupial: Models for Biomedical Research*. (Eds CH Tyndale-Biscoe, PA Janssens) pp. 132–147. Springer Verlag, Berlin.

Beard LA, Grigg GC (2000) Reproduction in the short–beaked echidna, *Tachyglossus aculeatus*: field observations at an elevated site in south-east Queensland. *Proceedings of the Linnean Society of New South Wales* **122**, 89–99.

Beard LA, Grigg GC, Augee ML (1992) Reproduction by echidnas in a cold climate. In *Platypus and Echidnas*. (Ed. ML Augee) pp. 93–100. Royal Zoological Society of New South Wales, Sydney.

Beazley LD, Arrese C, Hunt DM (2010) Visual system. In *Neurobiology of Australian Marsupials*. (Ed. KWS Ashwell) pp. 155–166. Cambridge University Press, Cambridge.

Berger RJ, Nicol SC, Andersen NA, Phillips NH (1995) Paradoxical sleep in the echidna. *Sleep Research* **24A**, 199.

Bethge P, Munks S, Nicol S (2001) Energetics of foraging and locomotion in the platypus *Ornithorhynchus anatinus*. *Journal of Comparative Physiology. Part B. Biochemical, Systemic, and Environmental Physiology*. **V171**, 497–506.

Bethge P, Munks S, Otley H, Nicol S (2003) Diving behaviour, dive cycles and aerobic dive limit in the platypus *Ornithorhynchus anatinus*. *Comparative Biochemistry and Physiology. Part A, Molecular & Integrative Physiology* **136**, 799–809.

Bethge P, Munks S, Otley H, Nicol S (2004) Platypus burrow temperatures at a subalpine Tasmanian lake. *Proceedings of the Linnean Society of New South Wales* **125**, 273–276.

Bethge P, Munks S, Otley H, Nicol SC (2009) Activity patterns and sharing of time and space of platypuses, Ornithorhynchus anatinus, in a subalpine Tasmanian lake *Journal of Mammalogy* **90**, 1350–1356.

Bjarkam CR, Sørensen JC, Geneser FA (1997) Distribution and morphology of serotonin-immunoreactive neurons in the brainstem of the New Zealand white rabbit. *The Journal of Comparative Neurology* **380**, 507–519.

Bohringer RC (1976) Bill receptors in the platypus Ornithorhynchus anatinus. *Journal of Anatomy* **121**, 417.

Bohringer RC (1977) The trigeminal nerve of the platypus, Ornithorhynchus anatinus. *Journal of Anatomy* **124**, 532.

Bohringer RC (1981) Cutaneous receptors in the bill of the platypus (Ornithorhynchus anatinus). *Australian Mammalogy* **4**, 93–105.

Bohringer RC, Rowe MJ (1977) The organization of the sensory and motor areas of cerebral cortex in the platypus (Ornithorhynchus anatinus). *The Journal of Comparative Neurology* **174**, 1–14.

Bojnik E, Babos F, Magyar A, Borsodi A, Benyhe S (2010) Bioinformatic and biochemical studies on the phylogenetic variability of proenkephalin-derived octapeptides. *Neuroscience* **165**, 542–552.

Bonaparte JF (1990) New Late Cretaceous mammals from the Los Alamitos Formation, northern Patagonia. *National Geographic Research* **6**, 63–93.

Bowker RM, Abbott LC (1990) Quantitative re-evaluation of descending serotonergic and non-serotonergic projections from the medulla of the rabbit: evidence for extensive co-existence of serotonin and peptides in the same spinally projecting neurons, but not from the nucleus raphe magnus. *Brain Research* **512**, 15–25.

Braekevelt CR, Hollenberg MJ (1970) The development of the retina of the albino rat. *The American Journal of Anatomy* **127**, 281–302.

Brannian J, Cloak C (1985) Observations of daily activity patterns in two captive short-nosed echidnas Tachyglossus aculeatus. *Zoo Biology* **4**, 75–81.

Brattstrom BH (1973) Social and maintenance behavior of the echidna, Tachyglossus aculeatus. *Journal of Mammalogy* **54**, 50–70.

Brice PH (2009) Thermoregulation in monotremes: riddles in a mosaic. *Australian Journal of Zoology* **57**, 255–263.

Briñón JG, Weruaga E, Crespo C, Porteros A, Arévalo R, Aijón J, Alonso JR (2001) Calretinin-, neurocalcin-, and parvalbumin-immunoreactive elements in the olfactory bulb of the hedgehog (*Erinaceus europaeus*). *The Journal of Comparative Neurology* **429**, 554–570.

Brooks E, Waters E, Farrington L, Canal MM (2011) Differential hypothalamic tyrosine hydroxylase distribution and activation by light in adult mice reared under different light conditions during the suckling period. *Brain Structure & Function* **216**, 357–370.

Bruesch SR, Arey LB (1942) The number of myelinated and unmyelinated fibres in the optic nerve of vertebrates. *The Journal of Comparative Neurology* **77**, 631–665.

Buchmann OLK, Rhodes J (1979) Instrumental discrimination: reversal learning in the monotreme Tachyglossus aculeatus setosus. *Animal Behaviour* **27**, 1048–1053.

Burke D, Cieplucha C, Cass J, Russell F, Fry G (2002) Win-shift and win-stay learning in the short-beaked echidna (Tachyglossus aculeatus). *Animal Cognition* **5**, 79–84.

Burrell H (1927) *The Platypus*. Angus and Robertson, Sydney.

Butcher LL, Woolf NJ (2004) Cholinergic neurons and networks revisited. In *The Rat Nervous System*. 3rd edn. (Ed. G Paxinos) pp. 1257–1268. Elsevier, San Diego.

Butler AB, Hodos W (2005) *Comparative Vertebrate Neuroanatomy. Evolution and Adaptation*. 2nd edn. Wiley, Hoboken, New Jersey.

Butler AB, Molnar Z, Manger PR (2002) Apparent absence of claustrum in monotremes: implications for forebrain evolution in amniotes. *Brain, Behavior and Evolution* **60**, 230–240.

Byrne RW, Bates LA (2011) Cognition in the wild: exploring animal minds with observational evidence. *Biology Letters* **7**, 619–622.

Caffé AR, van Ryen PC, van der Woude TP, van Leeuwen FW (1989) Vasopressin and oxytocin systems in the brain and upper spinal cord of Macaca fascicularis. *The Journal of Comparative Neurology* **287**, 302–325.

Caldwell WH (1887) The embryology of the Monotremata and Marsupialia. Part 1. *Philosophical Transactions of the Royal Society of London. Series B, Biological Sciences* **178**, 463–486.

Camens AB (2010) Were early Tertiary monotremes really all aquatic? Inferring paleobiology and phylogeny from a depauperate fossil record. *Proceedings of the National Academy of Sciences of the United States of America* **107**, E12

Campbell CBG, Hayhow WR (1971) Primary optic pathways in echidna, *Tachyglossus aculeatus* – experimental degeneration study. *The Journal of Comparative Neurology* **143**, 119–136.

Campbell CBG, Hayhow WR (1972) Primary optic pathways in the duckbill platypus, *Ornithorhynchus anatinus*: an experimental degeneration study. *The Journal of Comparative Neurology* **145**, 195–208.

Carletti B, Rossi F (2008) Neurogenesis in the cerebellum. *The Neuroscientist* **14**, 91–100.

Carmichael ST, Price JL (1996) Connectional networks within the orbital and medial prefrontal cortex of macaque monkeys. *The Journal of Comparative Neurology* **371**, 179–207.

Carney RSE, Bystron I, López-Bendito G, Molnár Z (2007) Comparative analysis of extra-ventricular mitoses at early stages of cortical development in rat and human. *Brain Structure & Function* **212**, 37–54.

Carrick FN, Hughes RL (1978) Reproduction in male monotremes. In *Monotreme Biology: The Australian Zoologist Special Symposium.* (Ed. ML Augee). Published in *The Australian Zoologist* **20**, 211–231.

Carter AM (2008) Sources for comparative studies of placentation. I. Embryological collections. *Placenta* **29**, 95–98.

Caverson MM, Ciriello J, Calaresu FR, Krukoff TL (1987) Distribution and morphology of vaspressin-, neurophysin II-, and oxytocin- immunoreactive cell bodies in the forebrain of the cat. *The Journal of Comparative Neurology* **259**, 211–236.

Celio MR (1990) Calbindin D-28k and parvalbumin in the rat nervous system. *Neuroscience* **35**, 375–475.

Charvet CJ, Striedter GF (2011) Causes and consequences of expanded subventricular zones. *The European Journal of Neuroscience* **34**, 988–993.

Charvet CJ, Owerkowicz T, Striedter GF (2009) Phylogeny of the telencephalic subventricular zone in sauropsids: evidence for the sequential evolution of pallial and subpallial subventricular zones. *Brain, Behavior and Evolution* **73**, 285–294.

Chauvet MT, Hurpet D, Chauvet J, Acher R (1980) Evolution of vasopressin in marsupial – a new hormone phenypressin (Phe-2-Arg-8-vasopressin) found in macropodidae. *Comptes Rendus Hebdomadaires des Seances de L'Academie des Sciences Series D* **291**, 541–543.

Chauvet MT, Hurpet D, Michel G, Chauvet MT, Carrick FN, Acher R (1985) The neurohypophyseal hormones of the egg-laying mammals. Identification of arginine vasopressin in the platypus (*Ornithorhynchus anatinus*). *Biochemical and Biophysical Research Communications* **127**, 277–282.

Chen CS, Anderson LM (1985) The inner ear structures of the echidna. An SEM study. *Experientia* **41**, 1324–1326.

Cheng G, Marotte LR, Mai JK, Ashwell KW (2002) Early development of the hypothalamus of a wallaby (*Macropus eugenii*). *The Journal of Comparative Neurology* **453**, 199–215.

Cheng G, Marotte LR, Ashwell KW (2003) Cyto- and chemoarchitecture of the hypothalamus of a wallaby (*Macropus eugenii*) with special emphasis on oxytocin and vasopressinergic neurons. *Anatomy and Embryology* **207**, 233–253.

Cheung AFP, Pollen AA, Tavare A, DeProto J, Molnár Z (2007) Comparative aspects of cortical neurogenesis in vertebrates. *Journal of Anatomy* **211**, 164–176.

Cheung AFP, Kondo S, Abdel-Mannan O, Chodroff RA, Sirey TM, Bluy LE, Webber N, DeProto J, Karlen SJ, Krubitzer L, Stolp HB, Saunders NR, Molnár Z (2010) The subventricular zone is the developmental milestone of a 6-layered neocortex: comparisons in metatherian and eutherian mammals. *Cerebral Cortex* **20**, 1071–1081.

Choi CQ (2009) Extreme monotremes. How egg-laying mammals survived their live-birthing competitors. *Scientific American* **301**, 21–22.

Choy VJ, Watkins WB (1977) Immunocytochemical study of the hypothalamo-neurohypophysial system. II Distribution of neurophysin, vasopressin and oxytocin in the normal and osmotically stimulated rat. *Cell and Tissue Research* **180**, 467–490.

Ciriello J, Calaresu FR (1980) Role of paraventricular and supraoptic nuclei in central cardiovascular regulation in the cat. *The American Journal of Physiology* **239**, R137–R142.

Clutton-Brock TH (1989) Mammalian mating systems. *Proceedings of the Royal Society of London. Series B, Biological Sciences* **236**, 339–372.

Comans PE, McLennan IS, Mark RF, Hendry IA (1988) Mammalian motoneuronal development: effect of peripheral deprivation on motoneuron numbers in a marsupial. *The Journal of Comparative Neurology* **270**, 111–120.

Connolly JH, Obendorf DL (1998) Distribution, captures and physical characteristics of the platypus (*Ornithorhynchus anatinus*) in Tasmania. *Australian Mammalogy* **20**, 231–237.

Crespo C, Blasco-Ibáñez JM, Marqués-Mari AI, Martinez-Guijarro FJ (2001) Parvalbumin-containing interneurons do not innervate granule cells in the olfactory bulb. *Neuroreport* **12**, 2553–2556.

Dahlström A, Fuxe K (1964a) Localization of monoamines in the lower brainstem. *Experientia* **20**, 398–399.

Dahlström A, Fuxe K (1964b) Evidence for the existence of monoamine-containing neurons in the central nervous system. I. Demonstration of monoamines in the cell bodies of brainstem neurons. *Acta Physiologica Scandinavica* **62** (Suppl 232), 1–155.

Dann JF, Buhl EH (1995) Patterns of connectivity in the neocortex of the echidna (*Tachyglossus aculeatus*). *Cerebral Cortex* **5**, 363–373.

Darwin C (1859) *On the Origin of Species by Means of Natural Selection, or the Preservation of Favoured Races in the Struggle for Life*. 1st edn. John Murray, London.

Davies WL, Carvalho LS, Cowing JA, Beazley LD, Hunt DM, Arrese CA (2007) Visual pigments of the platypus: a novel route to mammalian colour vision. *Current Biology* **17**, R161–R163.

Dawson TJ, Fanning D, Bergin TJ (1978) Metabolism and temperature regulation in the New Guinea monotreme *Zaglossus bruijnii*. In *Monotreme Biology: The Australian Zoologist Special Symposium*. (Ed. ML Augee). Published in *The Australian Zoologist* **20**, 99–103.

Dawson TJ, Grant T, Fanning D (1979) Standard metabolism of monotremes and the evolution of homeothermy. *Australian Journal of Zoology* **27**, 511–515.

Dawson CA, Jhamandas JH, Krukoff TL (1998) Activation by systemic angiotensin II of neurochemically identified neurons in rat hypothalamic paraventricular nucleus. *Journal of Neuroendocrinology* **10**, 453–459.

de Beer GR, Fell WA (1936) The development of the monotremata. Part III. The development of the skull of *Ornithorhynchus*. *Transactions of the Zoological Society of London* **23**, 1–42.

De-La-Warr M, Serena M (1999) Observations of platypus (*Ornithorhynchus anatinus*) mating behaviour. *Victorian Naturalist* **116**, 172–174.

de Olmos JS, Beltramino CA, Alheid G (2004) Amygdala and extended amygdala of the rat: a cytoarchitectonical, fibroarchitectonical and chemoarchitectonical survey. In *The Rat Nervous System*. 3rd edn. (Ed. G Paxinos) pp. 509–603. Elsevier, San Diego.

Deaner R, Isler K, Burkart J, van Schaik C (2007) Overall brain size, and not encephalization quotient, best predicts cognitive ability across nonhuman primates. *Brain, Behavior and Evolution* **70**, 115–124.

Dillon LS (1962) Comparative notes on the cerebellum of the monotremes. I. Contribution toward a phylogeny of the mammalian brain. *The Journal of Comparative Neurology* **118**, 343–353.

Diogo R, Abdala V, Lonergan N, Wood BA (2008) From fish to modern human – comparative anatomy, homologies and evolution of the head and neck musculature. *Journal of Anatomy* **213**, 391–424.

Divac I (1995) Monotremunculi and brain evolution. *Trends in Neurosciences* **18**, 2–4.

Divac I, Holst M-C, Nelson J, McKenzie JS (1987a) Afferents of the frontal cortex in the echidna (*Tachyglossus aculeatus*). Indication of an outstandingly large prefrontal cortex. *Brain, Behavior and Evolution* **30**, 303–320.

Divac I, Pettigrew JD, Holst MC, McKenzie JS (1987b) Efferent connections of the prefrontal cortex of echidna (*Tachyglossus aculeatus*). *Brain, Behavior and Evolution* **30**, 321–327.

Dixson AF, Anderson MJ (2004) Sexual behavior, reproductive physiology and sperm competition in male mammals. *Physiology & Behavior* **83**, 361–371.

Dobroruka LJ (1960) Einige Beobachtungen an Ameisenigeln, *Echidna aculeata* Shaw (1792). *Zeitschrift für Tierpsychologie* **17**, 178–181.

Doran GA (1973) The lingual musculature of the echidna, *Tachyglossus aculeatus*. *Anatomischer Anzeiger* **133**, 468–476.

Doran GA, Baggett H (1970) The vascular stiffening mechanism in the tongue of the echidna (*Tachyglossus aculeatus*). *Journal of Anatomy* **106**, 203.

Doré FY, Dumas C (1987) Psychology of animal cognition: Piagetian studies. *Psychological Bulletin* **102**, 219–233.

Drager UC, Olsen J (1981) Ganglion cell distribution in the retina of the mouse. *Investigative Ophthalmology & Visual Science* **20**, 285–293.

Dreher Z, Robinson SR, Distler C (1992) Müller cells in vascular and avascular retinae. A survey of 7 mammals. *The Journal of Comparative Neurology* **323**, 59–80.

Dun WS (1895) Notes on the occurrence of monotreme remains in the Pliocene of New South Wales. *Records of the Geological Survey of New South Wales* **4**, 118–126.

Dunbar RIM, Shultz S (2007) Evolution in the social brain. *Science* **317**, 1344–1347.

Edmeades R, Baudinette RV (1975) Energetics of locomotion in a monotreme, echidna *Tachyglossus aculeatus*. *Experientia* **31**, 935–936.

Ekström P (1987) Distribution of choline acetyltransferase-immunoreactive neurons in the brain of a cyprinid teleost (*Phoxinus phoxinus* L.) *The Journal of Comparative Neurology* **256**, 494–515.

Elliot Smith G (1896a) Jacobson's organ and the olfactory bulb in *Ornithorhynchus*. *Anatomischer Anzeiger* **11**, 161–166.

Elliot Smith G (1896b) The structure of the cerebral hemisphere of *Ornithorhynchus*. *Journal of Anatomy and Physiology* **30**, 465–487.

Elliot Smith G (1899) Further observations on the anatomy of the brain in the Monotremata. *Journal of Anatomy and Physiology* **33**, 309–342.

Elliot Smith G (1903) On the morphology of the cerebral commissures in the Vertebrata, with special reference to an aberrant commissure in the brain of certain reptiles. *Transactions of the Linnean Society (London) 2nd Series Zoology* **8**, 455–500.

Elliot Smith G (1910) The Arris and Gale lectures. On some problems relating to the evolution of the brain. *Lancet* **175**, 147–153.

Elliot Smith G (1919) A prelimary note on the morphology of the corpus striatum and the origin of the neopallium. *Journal of Anatomy* **53**, 271–291.

Ellsworth AF (1974) *Reassessment of Muscle Homologies and Nomeclature in Conservative Amniotes, the Echidna,* Tachyglossus aculeatus, *the Opossum,* Didelphis, *and the Tuatara,* Sphenodon. Robert E Krieger, New York.

Else PL, Hulbert AJ (1985) An allometric comparison of the mitochondria of mammalian and reptilian tissues – the implications for the evolution of endothermy. *Journal of Comparative Physiology. Part B. Biochemical, Systemic and Environmental Physiology* **156**, 3–11.

Elston GN, Manger PR, Pettigrew JD (1999) Morphology of pyramidal neurons in cytochrome oxidase modules of the S-I bill representation of the platypus. *Brain, Behavior and Evolution* **53**, 87–101.

Fallon JH (1983) The islands of Calleja complex of rat basal forebrain II: connections of medium and large sized cells. *Brain Research Bulletin* **10**, 775–793.

Fallon JH, Loughlin SE, Ribak CE (1983) The islands of Calleja complex of rat basal forebrain. III. Histochemical evidence for a striatopallidal system. *The Journal of Comparative Neurology* **218**, 91–120.

Faragher RA, Grant TR, Carrick FN (1979) Food of the platypus (*Ornithorhynchus anatinus*) with notes on the food of brown trout (*Salmo trutta*) in the Shoalhaven River, NSW. *Australian Journal of Ecology* **4**, 171–179.

Feakes MJ, Hodgkin EP, Strahan R, Waring H (1950) The effect of posterior lobe pituitary extracts on the blood pressure of *Ornithorhynchus* (duck-billed platypus). *The Journal of Experimental Biology* **27**, 50–58.

Feldman ML (1984) Morphology of the neocortical pyramidal neuron. In *Cerebral Cortex Vol 1. Cellular Components of the Cerebral Cortex*. (Eds A Peters and EG Jones) pp. 123–200. Plenum, New York.

Ferguson A, Turner B (2012) Reproductive parameters and behaviour of captive short-beaked echidna (*Tachyglossus aculeatus acanthion*) at Perth Zoo. *Australian Mammalogy* Published online: 26 October 2012. doi:10.1071/AM12022

Ferguson IA, Hardman CD, Marotte LR, Salardini A, Halasz P, Vu D, Waite PME (1999) Serotonergic neurons in the brainstem of the wallaby, *Macropus eugenii*. *The Journal of Comparative Neurology* **411**, 535–549.

Ferner K, Zeller U, Renfree MB (2009) Lung development of monotremes: evidence for the mammalian morphotype. *The Anatomical Record* **292**, 190–201.

Fink G, Smith GC, Augee ML (1975) Histochemical, ultrastructural and hormonal studies on pars distalis of echidna (*Tachyglossus aculeatus*). *Cell and Tissue Research* **159**, 531–540.

Finlay BL, Hinz F, Darlington RB (2011) Mapping behavioural evolution onto brain evolution: the strategic roles of conserved organization in individuals and species. *Philosophical Transactions of the Royal Society of London. Series B, Biological Sciences* **366**, 2111–2123.

Fish FE, Frappell PB, Baudinette RV, MacFarlane PM (2001) Energetics of terrestrial locomotion of the platypus *Ornithorhynchus anatinus*. *The Journal of Experimental Biology* **204**, 797–803.

Fjällbrant TT, Manger PR, Pettigrew JD (1998) Some related aspects of platypus electroreception: temporal integration behaviour, electroreceptive thresholds and directionality of the bill acting as an antenna. *Philosophical Transactions of the Royal Society of London. Series B, Biological Sciences* **353**, 1211–1219.

Flannery TF, Groves CP (1998) A revision of the genus *Zaglossus* (Monotremata, Tachyglossidae), with description of new species and subspecies. *Mammalia* **62**, 367–396.

Flannery TF, Archer M, Rich TH, Jones R (1995) A new family of monotremes from the Cretaceous of Australia. *Nature* **377**, 418–420.

Fleay D (1944) *We Breed the Platypus*. Robertson & Mullins, Melbourne.

Fleay D (1980) *Paradoxical Platypus: Hobnobbing with Duckbills*. 2nd edn. Friends of Fleays Association, Brisbane.

Flower WH (1865) On the commissures of the cerebral hemispheres of the marsupialia and monotremata as compared with those of placental mammals. *Philosophical Transactions of the Royal Society of London. Series B, Biological Sciences* **155**, 633–652.

Flynn TT, Hill JP (1942) The later stages of cleavage and the formation of the primary germ layers in the Monotremata. *Proceedings of the Zoological Society of London. Series A. General and Experimental* **111**, 233–253.

Flynn JJ, Parrish JM, Rakotosamimanana B, Simpson WF, Wyss AR (1999) A Middle Jurassic mammal from Madagascar. *Nature* **401**, 57–60.

Folley SJ, Knaggs GS (1966) Milk ejection activity (oxytocin) in the external jugular vein blood of the cow, goat and sow, in relation to the stimulus of milking and suckling. *The Journal of Endocrinology* **34**, 197–214.

Forasiepi AM, Martinelli AG (2003) Femur of a monotreme (Mammalia, Monotremata) from the Early Paleocene Salamanca Formation of Patagonia, Argentina. *Ameghiniana* **40**, 625–630.

Fortin M, Marchand R, Parent A (1998) Calcium-binding proteins in primate cerebellum. *Neuroscience Research* **30**, 155–168.

Fox RC, Meng J (1997) An x-radiographic and SEM study of the osseous inner ear of multituberculates and monotremes (Mammalia): implications for mammalian phylogeny and evolution of hearing. *Zoological Journal of the Linnean Society* **121**, 249–291.

Francis AJP, De Alwis C, Peach L, Redman JR (1999) Circadian activity rhythms in the Australian platypus, *Ornithorhynchus anatinus* (Monotremata). *Biological Rhythm Research* **30**, 91–103.

Frankenberg S, Schneider NY, Fletcher TP, Shaw G, Renfree M (2011) Identification of two distinct genes at the vertebrate *TRPC2* locus and their characterization in a marsupial and a monotreme. *BMC Molecular Biology* **12**, 39.

Frappell PB (2003) Ventilation and metabolic rate in the platypus: insights into the evolution of the mammalian breathing pattern. *Comparative Biochemistry and Physiology. Part A, Molecular & Integrative Physiology* **136**, 943–955.

Frappell PB, Franklin CE, Grigg GC (1994) Ventilatory and metabolic responses to hypoxia in the echidna, *Tachyglossus aculeatus*. *The American Journal of Physiology* **267**, R1510–R1515.

Freitag J, Ludwig G, Andreini I, Rossler P, Breer H (1998) Olfactory receptors in aquatic and terrestrial vertebrates. *Journal of Comparative Physiology. Part A. Neuroethology, Sensory, Neural, and Behavioral Physiology* **183**, 635–650.

Fuxe K, Ungerstedt U (1968) Histochemical studies on the distribution of catecholamines and 5-hydroxytryptamine after intraventricular injections. *Histochemie. Histochemistry. Histochimie* **13**, 16–28.

Gardner J, Serena M (1995) Spatial organization and movement patterns of adult male platypus, *Ornithorhynchus anatinus* (Monotremata, Ornithorhynchidae). *Australian Journal of Zoology* **43**, 91–103.

Garey LJ, Winkelmann E, Brauer K (1985) Golgi and Nissl studies of the visual cortex of the bottlenosed dolphin. *The Journal of Comparative Neurology* **240**, 305–321.

Gasse H, Meyer W (1995) Neuron-specific enolase as a marker of hypothalamo-neurohypophyseal development in postnatal *Monodelphis domestica* (Marsupialia). *Neuroscience Letters* **189**, 54–56.

Gates GR (1973) Vision in the monotreme echidna. MSc thesis. Monash University, Australia.

Gates GR (1978) Vision in the monotreme echidna. In *Monotreme Biology: The Australian Zoologist Special Symposium.* (Ed. ML Augee). Published in *The Australian Zoologist* **20**, 147–169.

Gates GR, Saunders JC, Bock GR, Aitkin LM, Elliott MA (1974) Peripheral auditory function in the platypus. *The Journal of the Acoustical Society of America* **56**, 152–156.

Gerfen CR (2004) Basal ganglia. In *The Rat Nervous System.* 3rd edn. (Ed. G Paxinos) pp. 455–508. Elsevier, San Diego.

Giere P, Zeller U (2005) Transfer of the Hubrecht laboratory collection from Utrecht to the Museum für Naturkunde, Berlin. *Placenta* **26**, A15.

Gittleman JL (1986) Carnivore brain size, behavioral ecology, and phylogeny. *Journal of Mammalogy* **67**, 23–36.

Glezer II, Morgane PJ (1990) Ultrastructure of synapses and Golgi analysis of neurons in neocortex of the lateral gyrus (visual cortex) of the dolphin and pilot whale. *Brain Research Bulletin* **24**, 401–427.

Glusman G, Bahar A, Sharon D, Pilpel Y, White J, Lancet D (2000) The olfactory receptor gene superfamily: data mining, classification, and nomenclature. *Mammalian Genome* **11**, 1016–1023.

Gogebakan A, Talas ZS, Ozdemir I, Sahna E (2012) Role of propolis on tyrosine hydroxylase activity and blood pressure in nitric oxide synthase-inhibited hypertensive rats. *Clinical and Experimental Hypertension* **34**, 424–428.

Goldby F (1939) An experimental investigation of the motor cortex and pyramidal tract of *Echidna aculeata*. *Journal of Anatomy* **73**, 509–524.

Gongora J, Swan AB, Chong AY, Ho SYW, Damayanti CS, Kolomyjec S, Grant T, Miller E, Blair D, Furlan E, Gust N (2012) Genetic structure and phylogeography of platypuses revealed by mitochondrial DNA. *Journal of Zoology* **286**, 110–119.

González-Lagos C, Sol D, Reader SM (2010) Large-brained mammals live longer. *Journal of Evolutionary Biology* **23**, 1064–1074.

Gould SJ (1991) To be a platypus. In *Bully for Brontosaurus*. pp. 269–280. Hutchinson Radius, London.

Gould SJ (1992) Bligh's Bounty. In *Bully for Brontosaurus*. pp. 281–293. Hutchinson Radius, London.

Grant T (2004) Captures, capture mortality, age and sex ratios of platypuses, *Ornithorhynchus anatinus*, during studies over 30 years in the upper Shoalhaven River in New South Wales. *Proceedings of the Linnean Society of New South Wales* **125**, 217–226.

Grant TR (1983) Body temperatures of free-ranging platypuses, *Ornithorhynchus anatinus* (Monotremata), with observations on their use of burrows. *Australian Journal of Zoology* **31**, 117–122.

Grant TR (1984) *The Platypus*. New South Wales University Press, Kensington.

Grant TR (1989) Ornithorhynchidae. In *Fauna of Australia. Vol 1B Mammalia.* (Eds DW Walton and BJ Richardson) pp. 436–450. Australian Government Publishing Service, Canberra.

Grant TR (2007) *The Platypus*. 4th edn. CSIRO Publishing, Collingwood.

Grant TR, Carrick FN (1978) Some aspects of the ecology of the platypus, *Ornithorhynchus anatinus*, in the upper Shoalhaven River, New South Wales. *Australian Zoologist* **20**, 181–199.

Grant TR, Dawson TJ (1978) Temperature regulation in the platypus, *Ornithorhynchus anatinus*; production and loss of metabolic heat in air and water. *Physiological Zoology* **51**, 315–332.

Grant TR, Temple-Smith PD (1983) Size, seasonal weight change and growth in platypuses, *Ornithorhynchus anatinus*, (Monotremata: Ornithorhynchidae), from rivers and lakes of New South Wales. *Australian Mammalogy* **6**, 51–60.

Grant TR, Temple-Smith PD (1998) Field biology of the platypus (*Ornithorhynchus anatinus*): historical and current perspectives. *Philosophical Transactions of the Royal Society of London. Series B, Biological Sciences* **353**, 1081–1091.

Grant TR, Griffiths M, Leckie RMC (1983) Aspects of lactation in the platypus, *Ornithorhynchus anatinus* (Monotremata), in waters of eastern New South Wales. *Australian Journal of Zoology* **31**, 881–889.

Grant TR, Griffiths M, Temple-Smith PD (2004) Breeding in a free-ranging population of platypuses, *Ornithorhynchus anatinus*, in the upper Shoalhaven

River, New South Wales – a 27 Year Study. *Proceedings of the Linnean Society of New South Wales* **125**, 227–234.

Gray AA (1908) An investigation on the anatomical structure and relationships of the labyrinth in the reptile, the bird and the mammal. *Proceedings of the Royal Society of London. Series B, Biological Sciences* **80**, 507–528.

Green DM, Swets JA (1974) *Signal Detection Theory and Psychophysics*. Robert E Kreiger, New York.

Green B, Griffiths M, Newgrain K (1985) Intake of milk by suckling echidnas. (*Tachyglossus aculeatus*). *Comparative Biochemistry and Physiology. Part A, Physiology* **81**, 441–444.

Green B, Griffths M, Newgrain K (1992) Seasonal patterns in water, sodium and energy turnover in free-living echidnas, *Tachyglossus aculeatus* (Mammalia: Monotremata). *Journal of Zoology* **227**, 351–365.

Gregory JE, Iggo A, McIntyre AK, Proske U (1987a) Sensory receptors in the bill of the platypus (*Ornithorhynchus anatinus*). *The Journal of Physiology* **382**, 120.

Gregory JE, Iggo A, McIntyre AK, Proske U (1987b) Electroreceptors in the platypus. *Nature* **326**, 386–387.

Gregory JE, Iggo A, McIntyre AK, Proske U (1988) Receptors in the bill of the platypus. *The Journal of Physiology* **400**, 349–366.

Gregory JE, Iggo A, McIntyre AK, Proske U (1989a) Responses of electroreceptors in the platypus bill to steady and alternating potentials. *The Journal of Physiology* **408**, 391–404.

Gregory JE, Iggo A, McIntyre AK, Proske U (1989b) Responses of electroreceptors in the snout of the echidna. *The Journal of Physiology* **414**, 521–538.

Gresser EB, Noback CV (1935) The eye of the monotreme, *Echidna hystrix*. *Journal of Morphology* **58**, 279–284.

Griffiths M (1965) Rate of growth and intake of milk in a suckling echidna. *Comparative Biochemistry and Physiology* **16**, 383–392.

Griffiths M (1968) *Echidnas*. Pergamon, London.

Griffiths M (1978) *The Biology of the Monotremes*. Academic Press, New York.

Griffiths M (1989) Tachyglossidae. In *Fauna of Australia. Vol 1B Mammalia*. (Eds DW Walton and BJ Richardson) pp. 407–435. Australian Government Publishing Service, Canberra.

Griffiths M (1999) Monotremes. In *Encyclopedia of Reproduction. Vol 3*. (Eds E Nobil and JD Neill) pp. 295–302. Academic Press, San Diego.

Griffiths M, McIntosh DL, Coles REA (1969) The mammary gland of the echidna, *Tachyglossus aculeatus*, with observations on the incubation of the egg and on the newly hatched young. *Journal of Zoology* **158**, 371–386.

Griffiths M, Elliott MA, Leckie RMC, Schoefl GI (1973) Observations of comparative anatomy and ultrastructure of mammary-glands and on fatty-acids of triglycerides in platypus and echidna milk fats. *Journal of Zoology* **169**, 255–279.

Griffiths M, Greenslade PJM, Miller M, Kerle JA (1990) The diet of the spiny ant-eater *Tachyglossus aculeatus acanthion* in tropical habitats in the Northern Territory. *The Beagle, Records of the Northern Territory Museum of Arts and Sciences* **7**, 79–90.

Griffiths M, Wells RT, Barrie DJ (1991) Observations on the skulls of fossil and extant echidnas (Monotremata: Tachyglossidae). *Australian Mammalogy* **14**, 87–101.

Grigg GC, Beard L (1996) Heart rates and respiratory rates of free-ranging echidnas – evidence for metabolic inhibition during hibernation? In *Adaptations to the Cold: Tenth International Hibernation Symposium*. (Eds F Geiser F, AJ Hulbert, SC Nicol) pp. 13–21. University of New England Press, Armidale.

Grigg GC, Beard LA, Augee ML (1989) Hibernation in a monotreme, the echidna *Tachyglossus aculeatus*. *Comparative Biochemistry and Physiology. Part A, Molecular & Integrative Physiology* **92**, 609–612.

Griffiths M, Wells RT, Barrie DJ (1991) Observations on the skulls of fossil and extant echidnas (Monotremata: Tachyglossidae). *Australian Mammalogy* **14**, 87–101.

Grigg GC, Beard LA, Barnes JA, Perry LI, Fry GJ, Hawkins M (2003) Body temperature in captive long–beaked echidnas (*Zaglossus bartoni*). *Comparative Biochemistry and Physiology. Part A, Molecular & Integrative Physiology* **136**, 911–916.

Groenewegen HJ (1988) Organization of the afferent connections of the mediodorsal thalamic nucleus in the rat, related to mediodorsal-prefrontal topography. *Neuroscience* **24**, 379–431.

Grothe B, Carr CE, Casseday JH, Fritzsch B, Köppl C (2004) The evolution of central pathways and their neural processing patterns. In *Evolution of the Ver-*

tebrate Auditory System. (Eds GA Manley, AN Popper and RR Fay) pp. 289–359. Springer Verlag, New York.

Grove S, Richards K, Spencer C, Yaxley B (2006) What lives under large logs in Tasmanian eucalypt forest? *The Tasmanian Naturalist* **128**, 86–93.

Grus WE, Shi P, Zhang J (2007) Largest vertebrate vomeronasal type 1 receptor gene repertoire in the semiaquatic platypus. *Molecular Biology and Evolution* **24**, 2153–2157.

Grützner F, Rens W, Tsend-Ayush E, El-Mogharbel N, O'Brien PCM, Jones RC, Ferguson-Smith MA, Marshall-Graves JA (2004) In the platypus a meiotic chain of ten sex chromosomes shares genes with the bird Z and mammal X chromosomes. *Nature* **432**, 913–917.

Gunn RM (1884) On the eye of *Ornithorhynchus paradoxus*. *Journal of Anatomy and Physiology* **17**, 400–405.

Gurdjian FS (1927) The diencephalon of the albino rat. *The Journal of Comparative Neurology* **43**, 1–114.

Gust N, Handasyde K (1995) Seasonal variation in the ranging behavior of the platypus (*Ornithorhynchus anatinus*) on the Goulburn River, Victoria. *Australian Journal of Zoology* **43**, 193–208.

Haacke W (1885) On the marsupial ovum, the mammary pouch and the male milk gland in *Echidna hystrix*. *Proceedings of the Royal Society of London. Series B, Biological Sciences* **38**, 72–74.

Haber SN, Adler A, Bergman H (2012) The basal ganglia. In *The Human Nervous System.* 3rd edn. (Eds JK Mai, G Paxinos) pp. 678–738. Academic Press, San Diego.

Haight JR, Murray PF (1981) The cranial endocast of the early Miocene marsupial, *Wynyardia bassiana*: an assessment of taxonomic relationships based upon comparisons with recent forms. *Brain, Behavior and Evolution* **19**, 17–36.

Hall LS, Hughes RL (1985) The embryological development and cytodifferentiation of the anterior pituitary in the marsupial, *Isoodon macrourus*. *Anatomy and Embryology* **172**, 353–363.

Hamilton-Smith E (1968) Platypuses in caves. *Victorian Naturalist* **85**, 292.

Handasyde KA, McDonald IR, Evans BK (1992) Seasonal changes in plasma concentration of progesterone in free-ranging platypus (*Ornithorhynchus anatinus*). In *Platypus and Echidnas.* (Ed. ML Augee) pp. 75–79. Royal Zoological Society of New South Wales, Sydney.

Handasyde KA, McDonald IR, Evans BK (2003) Plasma glucocorticoid concentrations in free-ranging platypuses (*Ornithorhynchus anatinus*): response to capture and patterns in relation to reproduction. *Comparative Biochemistry and Physiology. Part A, Molecular & Integrative Physiology* **136**, 895–902.

Hanström B (1954) Further studies on the hypophysis and on the hypothalamic neurosecretory tracts in the Monotremata. *Kungl. Fysiografiska Sällskapets Handlingar* **65**, 1–13.

Harkmark W (1954) Cell migrations from the rhombic lip to the inferior olive, the nucleus raphe and the pons; a morphological and experimental investigation on chick embryos. *The Journal of Comparative Neurology* **100**, 115–209.

Harris RL, Davies NW, Nicol SC (2012) Chemical composition of odorous secretions in the Tasmanian short-beaked echidna (*Tachyglossus aculeatus setosus*). *Chemical Senses* **37**, 819–836.

Harrison S (1997) The feeding ecology of the echidna, *Tachyglossus aculeatus*, in the Strathbogie Ranges, north-eastern Victoria. BSc Hons thesis. University of Melbourne, Australia.

Harrison PH, Porter M (1992) Development of the brachial spinal cord in the marsupial *Macropus eugenii* (tammar wallaby). *Brain Research. Developmental Brain Research* **70**, 139–144.

Hassiotis M, Ashwell KWS (2003) Neuronal classes in the isocortex of a monotreme, the Australian echidna (*Tachyglossus aculeatus*). *Brain, Behavior and Evolution* **61**, 6–27.

Hassiotis M, Paxinos G, Ashwell KWS (2003) The anatomy of the cerebral cortex of the echidna (*Tachyglossus aculeatus*). *Comparative Biochemistry and Physiology. Part A, Molecular & Integrative Physiology* **136**, 827–850.

Hassiotis M, Paxinos G, Ashwell KWS (2004a) Anatomy of the central nervous system of the Australian echidna. *Proceedings of the Linnean Society of New South Wales* **125**, 287–300.

Hassiotis M, Paxinos G, Ashwell KWS (2004b) Cyto- and chemoarchitecture of the cerebral cortex of the Australian echidna (*Tachyglossus aculeatus*). I. Areal organisation. *The Journal of Comparative Neurology* **475**, 493–517.

Hassiotis M, Paxinos G, Ashwell KWS (2005) Cyto- and chemoarchitecture of the cerebral cortex of the Australian echidna (*Tachyglossus aculeatus*). II. Laminar organisation and synaptic density. *The Journal of Comparative Neurology* **482**, 94–122.

Hawkins M, Battaglia A (2009) Breeding behaviour of the platypus (*Ornithorhynchus anatinus*) in captivity. *Australian Journal of Zoology* **57**, 283–293.

Hayden S, Bekaert M, Crider TA, Mariani S, Murphy WJ, Teeling EC (2010) Ecological adaptation determines functional mammalian olfactory subgenomes. *Genome Research* **20**, 1–9.

Head SI, Bakker AJ, Parkinson AL (2000) The contractile properties of single skinned skeletal muscle fibres of the echidna. *The Journal of Physiology* **523**, 280P–280P.

Hediger H, Kummer H (1961) Das Verhalten der Schnabeligel (Tachyglossidae). *Handbuch der Zoologie* **8**, 1–8.

Helgen KM, Portela Miguez R, Kohen J, Helgen L (2012) Twentieth century occurrence of the long-beaked echidna *Zaglossus bruijnii* in the Kimberley region of Australia. *ZooKeys* **255**, 103–132.

Hendry SH, Jones EG, De Felipe J, Schmechel D, Brandon C, Emson PC (1984) Neuropeptide-Y containing neurons of the cerebral cortex are also GABAergic. *Proceedings of the National Academy of Sciences of the United States of America* **81**, 6526–6530.

Herculano-Houzel S, Lent R (2005) Isotropic fractionator: a simple, rapid method for the quantification of total cell and neurons numbers in the brain. *The Journal of Neuroscience* **25**, 2518–2521.

Hines M (1929) The brain of *Ornithorhynchus anatinus*. *Philosophical Transactions of the Royal Society. Series B, Biological Sciences* **217**, 155–288.

Hiroi N (1995) Compartmental organization of calretinin in the rat striatum. *Neuroscience Letters* **197**, 223–226.

Ho SM (1997) Rhythmic motor activity and interlimb co-ordination in the developing pouch young of a wallaby (*Macropus eugenii*). *The Journal of Physiology* **501**, 623–636.

Ho SM (1998) Strychnine- and bicuculline-induced changes in the firing pattern of motoneurones during *in vitro* fictive locomotion reveal a possible N-methyl-D-aspartic acid (NMDA)-mediated suppression of motor discharge in wallaby (*Macropus eugenii*) pouch young. *Somatosensory & Motor Research* **15**, 325–332.

Ho SM, Stirling RV (1998) Development of muscle afferents in the spinal cord of the tammar wallaby. *Brain Research. Developmental Brain Research* **106**, 79–91.

Hobson JA, Pace-Schott EF (2003) Sleep, dreaming and wakefulness. In *Fundamental Neuroscience*. 2nd edn. (Eds LR Squire, FE Bloom, SK McConnell, JL Roberts, NC Spitzer and MJ Zigmond) pp. 1085–1108. Elsevier, London.

Hof PR, Glezer II, Condé F, Flagg RA, Rubin MB, Nimchinsky EA, Vogt Weisenhorn DM (1999) Cellular distribution of the calcium-binding proteins parvalbumin, calbindin, and calretinin in the neocortex of mammals: phylogenetic and developmental patterns. *Journal of Chemical Neuroanatomy* **16**, 77–116.

Hoffman PN, Cleveland DW, Griffin JW, Landes PW, Cowan NJ, Price DL (1987) Neurofilament gene expression: a major determinant of axonal caliber. *Proceedings of the National Academy of Sciences of the United States of America* **84**, 3472–3476.

Hofman MA (1982) Encephalization in mammals in relation to the size of the cerebral cortex. *Brain, Behavior and Evolution* **20**, 84–96.

Hofman MA (1983) Energy metabolism, brain size and longevity in mammals. *The Quarterly Review of Biology* **58**, 495–512.

Holland N, Jackson SM (2002) Reproductive behaviour and food consumption associated with the captive breeding of platypus (*Ornithorhynchus anatinus*). *Journal of Zoology* **256**, 279–288.

Holst M-C (1986) *The Olivocerebellar Projection in a Marsupial and a Monotreme*. PhD thesis. The University of New South Wales, Sydney.

Holst M-C, Watson CRR (1973) Some ultrastructural features of inferior olivary nucleus of echidna. *Journal of Anatomy* **114**, 153.

Hope R, Cooper S, Wainwright B (1990) Globin macromolecular sequences in marsupials and monotremes. *Australian Journal of Zoology* **37**, 289–313.

Horner CH (1993) Plasticity of the dendritic spine. *Progress in Neurobiology* **41**, 281–321.

Howell AB (1936) The musculature of antebrachium and manus in the platypus. *The American Journal of Anatomy* **59**, 425–431.

Hu Y, Meng J, Wang Y, Li C (2005) Large Mesozoic mammals fed on young dinosaurs. *Nature* **433**, 149–152.

Hughes RL, Hall LS (1998) Early development and embryology of the platypus. *Philosophical Transactions of the Royal Society of London. Series B, Biological Sciences* **353**, 1101–1114.

Hughes RL, Carrick FN, Shorey CD (1975) Reproduction in the platypus *Ornithorhynchus anatinus* with particular reference to the evolution of viviparity. *Journal of Reproduction and Fertility* **43**, 374–375.

Hulbert AJ, Grant TR (1983) A seasonal study of body condition and water turnover in a free-living population of platypuses, *Ornithorhynchus anatinus* (Monotremata). *Australian Journal of Zoology* **31**, 109–116.

Hulbert AJ, Beard LA, Grigg GC (2008) The exceptional longevity of an egg-laying mammal, the short-beaked echidna (*Tachyglossus aculeatus*) is associated with peroxidation-resistant membrane composition. *Experimental Gerontology* **43**, 729–733.

Hulse SH (2006) Postscript: an essay on the study of cognition in animals. In *Comparative Cognition: Experimental Explorations of Animal Intelligence*. (Eds EO Wasserman and TR Zentall) pp. 668–678. Oxford University Press, Oxford.

Hunt DM, Carvalho LS, Cowing JA, Parry JW, Wilkie SE, Davies WL, Bowmaker JK (2007) Spectral tuning of shortwave-sensitive visual pigments in vertebrates. *Photochemistry and Photobiology* **83**, 303–310.

Hurum JH, Luo Z-X, Kielan-Jaworowska Z (2006) Were mammals originally venomous? *Acta Palaeontologica Polonica* **51**, 1–11.

Hwang YC, Hinsman EJ, Roesel OF (1975) Calibre spectra of fibres in the fasciculus gracilis of the cat cervical spinal cord: a quantitative electron microscopic study. *The Journal of Comparative Neurology* **162**, 195–204.

Iggo A, McIntyre AK, Proske U (1983) Sensory receptors in the snout of the echidna. *The Journal of Physiology* **345**, P60.

Iggo A, McIntyre AK, Proske U (1985) Responses of mechanoreceptors and thermoreceptors in skin of the snout of the echidna *Tachyglossus aculeatus*. *Proceedings of the Royal Society of London. Series B, Biological Sciences* **223**, 261–277.

Iggo A, Proske U, McIntyre AK, Gregory JE (1988) Cutaneous electroreceptors in the platypus: a new mammalian receptor. *Progress in Brain Research* **74**, 133–138.

Iggo A, Gregory JE, Proske U (1992) The central projection of electrosensory information in the platypus. *The Journal of Physiology* **447**, 449–465.

Iggo A, Gregory JE, Proske U (1996) Studies of mechanoreceptors in skin of the snout of the echidna *Tachyglossus aculeatus*. *Somatosensory & Motor Research* **13**, 129–138.

Iqbal J, Jacobson CD (1995) Ontogeny of oxytocin-like immunoreactivity in the Brazilian opossum brain. *Brain Research. Developmental Brain Research* **90**, 1–16.

Isler K, van Schaik CP (2006) Metabolic costs of brain size evolution. *Biology Letters* **2**, 557–560.

Isler K, van Schaik CP (2009a) The expensive brain: a framework for explaining evolutionary changes in brain size. *Journal of Human Evolution* **57**, 392–400.

Isler K, van Schaik CP (2009b) Why are there so few smart mammals (but so many smart birds)? *Biology Letters* **5**, 125–129.

Jenkins FA (1970) Limb movements in a monotreme (*Tachyglossus aculeatus*). A cineradiographic analysis. *Science* **168**, 1473–1475.

Ji Q, Luo Z, Yuan C-X, Wible JR, Zhang J-P, Georgi JA (2002) The earliest known placental mammal. *Nature* **416**, 816–822.

Ji Q, Luo Z-X, Yuan C-X, Tabrum AR (2006) A swimming mammaliaform from the Middle Jurassic and ecomorphological diversification of early mammals. *Science* **311**, 1123–1127.

Ji Q, Luo Z-X, Zhang X, Yuan C-X, Xu L (2009) Evolutionary development of the middle ear in Mesozoic therian mammals. *Science* **326**, 278–281.

Jia C, Halpern M (2004) Calbindin D-28k, parvalbumin, and calretinin immunoreactivity in the main and accessory olfactory bulbs of the gray short-tailed opossum, *Monodelphis domestica*. *Journal of Morphology* **259**, 271–280.

Johnson D (2004) *The Geology of Australia*. Cambridge University Press, Cambridge.

Johnson C (2006) *Australia's Mammal Extinctions: A 50 000 Year History*. Cambridge University Press, Cambridge.

Johnson JI, Kirsch JAW, Switzer RC, III (1982a) Phylogeny through brain traits: fifteen characters which adumbrate mammalian genealogy. *Brain, Behavior and Evolution* **20**, 72–83.

Johnson JI, Switzer RC, III Kirsch JAW (1982b) Phylogeny through brain traits: the distribution of categorizing characters in contemporary mammals. *Brain, Behavior and Evolution* **20**, 97–117.

Johnson JI, Kirsch JAW, Reep RL, Switzer RC, III (1994) Phylogeny through brain traits: more characters for the analysis of mammalian evolution. *Brain, Behavior and Evolution* **43**, 319–347.

Johnson MS, Thomson SC, Speakman JR (2001) Limits to sustained energy intake. II. Inter-relationships between resting metabolic rate, life-history traits and morphology in *Mus musculus*. *The Journal of Experimental Biology* **204**, 1937–1946.

Jones EG (2007) *The Thalamus*. 2nd edn. Cambridge University Press, Cambridge.

Jørgensen JM, Locket NA (1995) The inner ear of the echidna *Tachyglossus aculeatus*: the vestibular sensory organs. *Proceedings of the Royal Society of London. Series B, Biological Sciences* **260**, 183–189.

Kaas JH (2012) Somatosensory system. In *The Human Nervous System*. (Eds JK Mai and G Paxinos) pp. 1074–1109. Elsevier, San Diego.

Kakuta S, Oda S, Takayanagi M, Kishi K (1998) Parvalbumin immunoreactive neurons in the main olfactory bulb of the house musk shrew, *Suncus murinus*. *Brain, Behavior and Evolution* **52**, 285–291.

Kalinichenko SG, Pushchin II (2008) Calcium-binding proteins in the cerebellar cortex of the bottlenose dolphin and harbour porpoise. *Journal of Chemical Neuroanatomy* **35**, 364–370.

Kaneko T, Caria MA, Asanuma H (1994) Information processing within the motor cortex. II. Intracortical connections between neurons receiving somatosensory cortical input and motor output neurons of the cortex. *The Journal of Comparative Neurology* **345**, 172–184.

Kawata M, Sano Y (1982) Immunohistochemical identification of the oxytocin and vasopressin neurons in the hypothalamus of the monkey (*Macaca fuscata*). *Anatomy and Embryology* **165**, 151–167.

Keast JR (1993) Innervation of the monotreme gastrointestinal tract. A study of peptide and catecholamine distribution. *The Journal of Comparative Neurology* **334**, 228–240.

Keast JR, Furness JB, Costa M (1985) Distribution of certain peptide-containing nerve fibres and endocrine cells in the gastrointestinal mucosa in five mammalian species. *The Journal of Comparative Neurology* **236**, 403–422.

Kenny GC, Scheelings FT (1979) Observations of the pineal region of non-eutherian mammals. *Cell and Tissue Research* **198**, 309–324.

Kielan-Jaworowska Z, Crompton AW, Jenkins FA, Jr (1987) The origin of egg-laying mammals. *Nature* **326**, 871–873.

Kielan-Jaworowska Z, Cifelli R, Luo Z-X (2004) *Mammals from the Age of Dinosaurs: Origins, Evolution and Structure*. Columbia University Press, New York.

Kirkcaldie MTK (2012) Neocortex. In *The Mouse Nervous System*. (Eds C Watson, G Paxinos and L Puelles) pp. 52–111. Elsevier, San Diego.

Kishida T (2008) Pattern of the divergence of olfactory receptor genes during tetrapod evolution. *PLoS ONE* **3**, e2385.

Kiss JZ (1988) Dynamism of the chemoarchitecture in the hypothalamic paraventricular nucleus. *Brain Research Bulletin* **20**, 699–708.

Kiss JZ, Martos J, Palkovits M (1991) Hypothalamic paraventricular nucleus: a quantitative analysis of cytoarchitectonic subdivisions in the rat. *The Journal of Comparative Neurology* **313**, 563–573.

Klamt M, Thompson R, Davis J (2011) Early response of the platypus to climate warming. *Global Change Biology* **17**, 3011–3018.

Koh JMS, Bansal PS, Torres AM, Kuchel PW (2009) Platypus venom: source of novel compounds. *Australian Journal of Zoology* **57**, 203–210.

Koh JMS, Haynes L, Belov K, Kuchel PW (2011) l-to-d-peptide isomerase in male echidna venom. *Australian Journal of Zoology* **58**, 284–288.

Koizumi M, Sakai T (1997) On the morphology of the brachial plexus of the platypus (*Ornithorhynchus anatinus*) and the echidna (*Tachyglossus aculeatus*). *Journal of Anatomy* **190**, 447–455.

Kolmer W (1925) Zur Organologie and mikroskopischen Anatomie von Proechidna (*Zaglossus*) bruijnii 1. *Mitteilung Zeistchrift für Wissenschaft Zoologie*. **125**, 448–482.

Kolomyjec SH (2010) The history and relationships of northern platypus (*Ornithorhynchus anatinus*) populations: a molecular approach. PhD thesis. James Cook University, Australia.

Kooy FH (1916) The inferior olive in vertebrates. *Folia Neurobiologica* **10**, 205–369.

Kosaka K, Heizmann CW, Kosaka T (1994) Calcium binding protein parvalbumin-immunoreactive neurons in the rat olfactory bulb. 1. Distribution and structural features in adult rat. *Experimental Brain Research* **99**, 191–204.

Koutcherov I, Mai JK, Ashwell KWS, Paxinos G (2000) The organization of the human paraventricular hypothalamic nucleus. *The Journal of Comparative Neurology* **423**, 299–318.

Kozlowski J, Konarzewski M (2005) West, Brown and Enquist's model of allometric scaling again: the same questions remain. *Functional Ecology* **19**, 739–743.

Krause WJ (1971) Brunner's glands of the duckbilled platypus (*Ornithorhynchus anatinus*). *The American Journal of Anatomy* **132**, 147–166.

Krause WJ (2009) Morphological and histochemical observations on the crural gland-spur apparatus of the echidna (*Tachyglossus aculeatus*) together with comparative observations on the femoral gland-spur apparatus of the duckbilled platypus (*Ornithorhyncus anatinus*). *Cells, Tissues, Organs* **191**, 336–354.

Krause WJ, Leeson CR (1974) The gastric mucosa of two monotremes: the duck-billed platypus and echidna. *Journal of Morphology* **142**, 285–300.

Krefft G (1868) On the discovery of a new and gigantic fossil species of *Echidna* in Australia. *Annals & Magazine of Natural History* **1**, 113–114.

Krettek JE, Price JL (1977) The cortical projections of the mediodorsal nucleus and adjacent thalamic nuclei in the rat. *The Journal of Comparative Neurology* **171**, 157–191.

Krieg WJS (1932) The hypothalamus of the albino rat. *The Journal of Comparative Neurology* **55**, 19–89.

Kriegstein A, Noctor S, Martinez-Cerdeno V (2006) Patterns of neural stem and progenitor cell division may underlie evolutionary cortical expansion. *Nature Reviews. Neuroscience* **7**, 883–890.

Krubitzer L (1998) What can monotremes tell us about brain evolution? *Philosophical Transactions of the Royal Society. Series B, Biological Sciences* **353**, 1127–1146.

Krubitzer L, Manger P, Pettigrew J, Calford M (1995) Organization of somatosensory cortex in monotremes. In search of the prototypical plan. *The Journal of Comparative Neurology* **351**, 261–306.

Kuchel LJ (2003) The energetics and patterns of torpor in free-ranging *Tachyglossus aculeatus* from a warm-temperate climate. PhD thesis. University of Queensland, Brisbane.

Kuhn H-J (1971) Die Entwicklung und Morphologie des Schädels von *Tachyglossus aculeatus*. *Abhandlungen der Senckenbergischen Naturforschenden Gesellschaft* **528**, 1–192.

Kullander K, Carlson B, Hallbook F (1997) Molecular phylogeny and evolution of the neurotrophins from monotremes and marsupials. *Journal of Molecular Evolution* **45**, 311–321.

Kullberg M, Hallstrom BM, Arnason U, Janke A (2008) Phylogenetic analysis of 1.5Mbp and platypus EST data refute the Marsupionta hypothesis and unequivocally support Monotremata as sister group to Marsupialia/Placentalia. *Zoologica Scripta* **37**, 115–127.

Künzle H (2005) The striatum in the hedgehog tenrec: histochemical organization and cortical afferents. *Brain Research* **1034**, 90–113.

Ladevèze S, de Muizon C, Colbert M, Smith T (2010) 3D computational imaging of the petrosal of a new multituberculate mammal from the Late Cretaceous of China and its paleobiologic inferences. *Comptes Rendus. Palévol* **9**, 319–330.

Ladhams A, Pickles JO (1996) Morphology of the monotreme organ of Corti and macula lagena. *The Journal of Comparative Neurology* **366**, 335–347.

Lambeth L, Blunt MJ (1975) Electron-microscopic study of monotreme neuroglia. *Acta Anatomica* **93**, 115–125.

Langner G, Scheich, H ((1986)) Electroreceptive cortex of platypus marked by 2-deoxyglucose. *First Int. Cong. Neuroethol.*, 63.

Larsell O (1970) *The Comparative Anatomy and Histology of the Cerebellum from Monotremes through Apes.* (Ed. J Jansen) University of Minnesota Press, Minneapolis.

Leamey CA, Flett DL, Ho SM, Marotte LR (2007) Development of structural and functional connectivity in the thalamocortical somatosensory pathway in the wallaby. *The European Journal of Neuroscience* **25**, 3058–3070.

Leary T, Seri L, Flannery T, Wright D, Hamilton S, Helgen KM, Singadan R, Menzies J, Allison A, James R, Aplin K, Salas L, Dickman C (2008)

Zaglossus bartoni. In IUCN 2010. IUCN Red List of Threatened Species. Version 2010.4.

Lechan RM, Nestler JL, Jacobson S, Reichlin S (1980) The hypothalamic 'tuberoinfundibular' system of the rat as demonstrated by horseradish peroxidase (HRP) microiontophoresis. *Brain Research* **195**, 13–27.

Lefèvre CM, Sharp JA, Nicholas KR (2009) Characterisation of monotreme caseins reveals lineage-specific expansion of an ancestral casein locus in mammals. *Reproduction, Fertility and Development* **21**, 1015–1027.

Lefèvre CM, Sharp JA, Nicholas KR (2010) Evolution of lactation: ancient origin and extreme adaptations of the lactation system. *Annual Review of Genomics and Human Genetics* **11**, 219–238.

Lende RA (1964) Representation in cerebral cortex of primitive mammal. Sensorimotor, visual and auditory fields in the echidna (*Tachyglossus aculeatus*). *Journal of Neurophysiology* **27**, 37–48.

Leonard CM (1969) The connections of the dorsomedial nuclei. *Brain, Behavior and Evolution* **6**, 524–541.

Lesku JA, Meyer LCR, Fuller A, Maloney SK, Dell'Omo G, Vyssotski AL, Rattenborg NC (2011) Ostriches sleep like platypuses. *PLoS ONE* **6**, e23203.

Li G, Luo Z-X (2006) A Cretaceous symmetrodont therian with some monotreme-like postcranial features. *Nature* **439**, 195–200.

Locket NA (1985) Echidna retina contains cones and Landolt's clubs. *Journal of Anatomy* **143**, 217.

Lu Qui IJ, Fox CA (1976) The supraoptic nucleus and the supraopticohypophysial tract in the monkey (*Macaca mulatta*). *The Journal of Comparative Neurology* **168**, 7–40.

Luo Z-X (2007) Transformation and diversification in early mammal evolution. *Nature* **450**, 1011–1019.

Luo Z, Ketten DR (1991) CT scanning and computerized reconstruction of the inner ear of multituberculate mammals. *Journal of Vertebrate Paleontology* **11**, 220–228.

Luo Z-X, Wible JR (2005) A Late Jurassic digging mammal and early mammalian diversification. *Science* **308**, 103–107.

Luo Z-X, Cifelli RL, Kielan-Jaworowska Z (2001) Dual origin of tribosphenic mammals. *Nature* **409**, 53–57.

Luo Z-X, Kielan-Jaworowska Z, Cifelli R (2002) In quest for a phylogeny of Mesozoic mammals. *Acta Palaeontologica Polonica* **47**, 1–78.

Luo Z-X, Ji Q, Wible JR, Yuan C-X (2003) An Early Cretaceous tribosphenic mammal and metatherian evolution. *Science* **302**, 1934–1940.

Macrini TE, Rowe T, Archer M (2006) Description of a cranial endocast from a fossil platypus, *Obdurodon dicksoni* (Monotremata, Ornithorhynchidae), and the relevance of endocranial characters to monotreme monophyly. *Journal of Morphology* **267**, 1000–1015.

Mahns DA, Coleman GT, Ashwell KWS, Rowe MJ (2003) Tactile sensory function in the forearm of the monotreme *Tachyglossus aculeatus*. *The Journal of Comparative Neurology* **459**, 173–185.

Mai JK, Forutan F (2012) Thalamus. In *The Human Nervous System*. (Eds JK Mai and G Paxinos) pp. 618–677. Elsevier, San Diego.

Mai JK, Lensing-Höhn S, Ende AA, Sofroniew MV (1997) Developmental organization of neurophysin neurons in the human brain. *The Journal of Comparative Neurology* **385**, 477–489.

Malmierca MS, Merchán MA (2004) Auditory system. In *The Rat Nervous System*. 3rd edn. (Ed. G Paxinos) pp. 997–1082. Elsevier, San Diego.

Manger PR (1994) Platypus electroreception: Neuroethology of a novel mammalian sensory system. PhD thesis. The University of Queensland, Australia.

Manger PR, Hughes RL (1992) Ultrastructure and distribution of epidermal sensory receptors in the beak of the echidna, *Tachyglossus aculeatus*. *Brain, Behavior and Evolution* **40**, 287–296.

Manger PR, Pettigrew JD (1995) Electroreception and the feeding behaviour of platypus (*Ornithorhynchus anatinus*, Monotremata, Mammalia). *Proceedings of the Royal Society of London. Series B, Biological Sciences* **347**, 359–381.

Manger PR, Pettigrew JD (1996) Ultrastructure, number, distribution and innervation of electroreceptors and mechanoreceptors in the bill skin of the platypus, *Ornithorhynchus anatinus*. *Brain, Behavior and Evolution* **48**, 27–54.

Manger PR, Calford MB, Pettigrew JD (1996) Properties of electrosensory neurons in the cortex of the platypus (*Ornithorhynchus anatinus*): implications for processing of electrosensory stimuli. *Proceedings of the Royal Society of London. Series B, Biological Sciences* **263**, 611–617.

Manger PR, Collins R, Pettigrew JD (1997) Histological observations on presumed electroreceptors and mechanoreceptors in the beak skin of the long-beaked echidna. *Zaglossus bruijnii. Proceedings of the Royal Society of London. Series B, Biological Sciences* **264**, 165–172.

Manger PR, Hall LS, Pettigrew JD (1998a) The development of the external features of the platypus (*Ornithorhynchus anatinus*). *Philosophical Transactions of the Royal Society. Series B, Biological Sciences* **353**, 1115–1125.

Manger PR, Keast JR, Pettigrew JD, Troutt L (1998b) Distribution and putative function of autonomic nerve fibres in the bill skin of the platypus (*Ornithorhynchus anatinus*). *Philosophical Transactions of the Royal Society. Series B, Biological Sciences* **353**, 1159–1170.

Manger PR, Collins R, Pettigrew JD (1998c) The development of the electroreceptors of the platypus (*Ornithorhynchus anatinus*). *Philosophical Transactions of the Royal Society. Series B, Biological Sciences* **353**, 1171–1186.

Manger PR, Fahringer HM, Pettigrew JD, Siegel JM (2002a) The distribution and morphological characteristics of cholinergic cells in the brain of monotremes as revealed by ChAT immunohistochemistry. *Brain, Behavior and Evolution* **60**, 275–297.

Manger PR, Fahringer HM, Pettigrew JD, Siegel JM (2002b) The distribution and morphological characteristics of catecholaminergic cells in the brain of monotremes as revealed by tyrosine hydroxylase immunohistochemistry. *Brain, Behavior and Evolution* **60**, 298–314.

Manger PR, Fahringer HM, Pettigrew JD, Siegel JM (2002c) The distribution and morphological characteristics of serotonergic cells in the brain of monotremes. *Brain, Behavior and Evolution* **60**, 315–332.

Marín O, Smeets WJAJ, González A (1997) Distribution of choline acetyltransferase immunoreactivity in the brain of anuran (*Rana perezi, Xenopus laevis*) and urodele (*Pleurodeles waltl*) amphibians. *The Journal of Comparative Neurology* **382**, 499–534.

Marin-Padilla M (1978) Dual origin of the mammalian neocortex and evolution of the cortical plate. *Anatomy and Embryology* **152**, 109–126.

Marotte LR, Sheng X-M (2000) Neurogenesis and the identification of developing layers in the visual cortex of the wallaby (*Macropus eugenii*). *The Journal of Comparative Neurology* **416**, 131–142.

Marotte LR, Leamey CA, Waite PME (1997) Time-course of development of the wallaby trigeminal pathway: III. Thalamocortical and corticothalamic projections. *The Journal of Comparative Neurology* **387**, 194–214.

Martin CJ (1898) Cortical physiology in *Ornithorhynchus*. *The Journal of Physiology* **23**, 383–385. [Supplementary]

Martin CJ (1903) Thermal adjustment and respiratory exchange in monotremes and marsupials. A study in the development of homoeothermism. *Philosophical Transactions of the Royal Society of London. Series B, Biological Sciences* **195**, 1–37.

Martin T, Rauhut OWM (2005) Mandible and dentition of *Asfaltomylos patagonicus* (Australosphenida, Mammalia) and evolution of tribosphenic teeth. *Journal of Vertebrate Paleontology* **25**, 414–425.

Martínez-García F, Novejarque A, Gutiérrez-Castellanos N, Lanuza E (2012) Piriform cortex and amygdala. In *The Mouse Nervous System*. (Ed. C Watson, G Paxinos and L Puelles) pp. 140–172. Elsevier, San Diego.

Maseko BC, Jacobs B, Spocter MA, Sherwood CC, Hof PR, Manger PR (2013) Qualitative and quantitative aspects of the microanatomy of the African elephant cerebellar cortex. *Brain, Behavior and Evolution* **81**, 40–55.

Mason WT, Ho YW, Eckenstein F, Hatton GI (1983) Mapping of cholinergic neurons associated with rat supraoptic neurons: combined immunocytochemical and histochemical identification. *Brain Research Bulletin* **11**, 617–626.

McAllan BM, Joss JMP, Firth BT (1991) Phase delay of the natural photoperiod alters reproductive timing in the marsupial *Antechinus stuartii*. *Journal of Zoology* **225**, 633–646.

McDonald IR, Handasyde KA, Evans BK (1992) Adrenal function in the platypus. In *Platypus and Echidnas*. (Ed. ML Augee) pp. 127–133. Royal Zoological Society of New South Wales, Sydney.

McFarlane JR, Carrick FN (1992) Androgen concentrations in male platypuses. In *Platypus and Echidnas*. (Ed. ML Augee) pp. 69–74. Royal Zoological Society of New South Wales, Sydney.

McKay WJS (1894) The morphology of the muscles of the shoulder girdle in monotremes. *Proceedings of the Linnean Society of New South Wales* **19**, 262–360.

McKenna MC, Bell SK (1997) *Classification of Mammals Above the Species Level*. Columbia University Press, New York.

McLachlan-Troup TA, Dickman CR, Grant TR (2010) Diet and dietary selectivity of the platypus in relation to season, sex and macroinvertebrate assemblages. *Journal of Zoology* **280**, 237–246.

McLeod AL (1993) Movement, home range and burrow usage, diel activity and juvenile dispersal of platypuses, *Ornithorhynchus anatinus*, on the Duckmaloi Weir, NSW. BSc (Hons) thesis. Charles Sturt University, Bathurst, Australia.

McMenamin PG (2007) The unique paired retinal vascular pattern in marsupials: structural, functional and evolutionary perspectives based on observations in a range of species. *The British Journal of Ophthalmology* **91**, 1399–1405.

McNab BK (1984) Physiological convergence amongst ant-eating and termite-eating mammals. *Journal of Zoology* **203**, 485–510.

McNab BK (2002) *The Physiological Ecology of Vertebrates: A View from Energetics*. Cornell University Press, New York.

McNab BK, Eisenberg JF (1989) Brain size and its relation to the rate of metabolism in mammals. *American Naturalist* **133**, 157–167.

McNamara P, Nunn CL, Barton RA (2010) Introduction. In *Evolution of Sleep*. (Eds P McNamara, RA Barton and CL Nunn) pp. 1–11. Cambridge University Press, Cambridge UK.

Medina L, Abellán A (2012) Subpallial structures. In *The Mouse Nervous System*. (Eds C Watson, G Paxinos and L Puelles) pp. 173–220. Elsevier, San Diego.

Medina L, Reiner A (1994) Distribution of choline acetyltransferase activity in the pigeon brain. *The Journal of Comparative Neurology* **342**, 497–537.

Medina L, Smeets WJAJ, Hoogland PV, Puelles L (1993) Distribution of choline acetyltransferase immunoreactivity in the brain of the lizard *Gallotia galloti*. *The Journal of Comparative Neurology* **331**, 261–285.

Meisami E, Bhatnagar KP (1998) Structure and diversity in mammalian accessory olfactory bulb. *Microscopy Research and Technique* **43**, 476–499.

Menétrey D, Basbaum AI (1987) Spinal and trigeminal projections to the nucleus of the solitary tract: a possible substrate for somatovisceral and viscerovisceral reflex activation. *The Journal of Comparative Neurology* **255**, 439–450.

Meng J, Wyss AR (1995) Monotreme affinities and low-frequency hearing suggested by multituberculate ear. *Nature* **377**, 141–144.

Meyer J (1981) A quantitative comparison of the parts of the brains of two Australian marsupials and some eutherian mammals. *Brain, Behavior and Evolution* **18**, 60–71.

Mikula S, Manger PR, Jones EG (2008) The thalamus of the monotremes: cyto- and myeloarchitecture and chemical neuroanatomy. *Philosophical Transactions of the Royal Society. Series B. Biological Sciences* **363**, 2415–2440.

Millhouse OE (1987) Granule cells of the olfactory tubercle and the question of the islands of Calleja. *The Journal of Comparative Neurology* **265**, 1–24.

Mills DM, Shepherd RK (2001) Distortion product otoacoustic emission and auditory brainstem responses in the echidna (*Tachyglossus aculeatus*). *Journal of the Association for Research in Otolaryngology* **2**, 130–146.

Mink JW, Blumenschine RJ, Adams DB (1981) Ratio of central nervous system to body metabolism in vertebrates: its constancy and functional basis. *American Journal of Physiology. Regulatory, Integrative and Comparative Physiology* **241**, R203–R212.

Molnár Z (2011) Evolution of cerebral cortical development. *Brain, Behavior and Evolution* **78**, 94–107.

Molnár Z, Métin C, Stoykova A, Tarabykin V, Price DJ, Francis F, Meyer G, Dehay C, Kennedy H (2006) Comparative aspects of cerebral cortical development. *The European Journal of Neuroscience* **23**, 921–934.

Montiel JF, Wang WZ, Oeschger FM, Hoerder-Suabedissen A, Tung WL, Garcia-Moreno F, Holm IE, Villalón A, Molnár Z (2011) Hypothesis on the dual origin of the mammalian subplate. *Frontiers in Neuroanatomy* **5**, article 25, 1–10–

Morest DK (1970) The pattern of neurogenesis in the retina of the rat. *Zeitschrift fur Anatomie und Entwicklungsgeschichte* **131**, 45–67.

Morris JR, Lasek RJ (1982) Stable polymers of the axonal cytoskeleton: the axoplasmic ghost. *The Journal of Cell Biology* **92**, 192–198.

Morrow G, Nicol SC (2009) Cool sex? Hibernation and reproduction overlap in the echidna. *PLoS ONE* **4**, e6070.

Morrow GE, Nicol SC (2013) Maternal care in the Tasmanian echidna *Tachyglossus aculeatus setosus*. *Australian Journal of Zoology* **60**, 289–298. Published online: 22 January 2013. doi:10.1071/ZO12066

Morrow G, Andersen NA, Nicol SC (2009) Reproductive strategies of the short-beaked echidna – a review with new data from a long-term study on the Tasmanian subspecies (*Tachyglossus aculeatus setosus*). *Australian Journal of Zoology* **57**, 275–282.

Moutin E, Raynaud F, Roger J, Pellegrino E, Homburger V, Bertaso F, Ollendorff V, Bockaert J, Fagni L, Perroy J (2012) Dynamic remodeling of scaffold interactions in dendritic spines controls synaptic excitability. *The Journal of Cell Biology* **198**, 251–263.

Mueller P, Diamond J (2001) Metabolic rate and environmental productivity: well provisioned animals evolved to run and idle fast. *Proceedings of the National Academy of Sciences of the United States of America* **98**, 12550–12555.

Munks S, Eberhard R, Duhig N (2004) Nests of the platypus *Ornithorhynchus anatinus* in a Tasmanian cave. *The Tasmanian Naturalist* **126**, 55–58.

Murray PF (1978a) A Pleistocene spiny anteater from Tasmania (Monotremata: Tachyglossidae). *Papers and Proceedings of the Royal Society of Tasmania* **112**, 39–27.

Murray PF (1978b) Late Cenozoic monotreme anteaters. *Australian Zoologist* **20**, 29–55.

Murray PF (1981) A unique jaw mechanism in the echidna, *Tachyglossus aculeatus* (Monotremata). *Australian Journal of Zoology* **29**, 1–5.

Musser AM (1998) Evolution, biogeography and paleoecology of the Ornithorhynchidae. *Australian Mammalogy* **20**, 147–162.

Musser AM (2003) Review of the monotreme fossil record and comparison of paleontological and molecular data. *Comparative Biochemistry and Physiology. Part A, Molecular & Integrative Physiology* **136**, 927–942.

Musser AM (2006a) *Investigations into the Evolution of Australian Mammals with a Focus on Monotremata*. PhD thesis. University of New South Wales, Australia.

Musser AM (2006b) Furry egg-layers: monotreme relationships and radiations. In *Evolution and Biogeography in Australasia*. (Eds J Merrick and G Hickey) pp. 523–550. Australian Scientific Publishing, Sydney.

Musser AM, Archer A (1998) New information about the skull and dentary of the Miocene platypus *Obdurodon dicksoni*, and a discussion of ornithorhynchid relationships. *Philosophical Transactions of the Royal Society of London. Series B, Biological Sciences* **353**, 1063–1079.

Musser AM, Temple-Smith P (2008) Trouble in paradise: challenges facing monotremes. *Australian Institute of Biology Newsletter* **8**, 25–31.

Nadarajah B, Alifragis P, Wong ROL, Parnavelas JG (2003) Neuronal migration in the developing cerebral cortex: observations based on real-time imaging. *Cerebral Cortex* **13**, 607–611.

Nelson TE, King JS, Bishop GA (1997) Distribution of tyrosine hydroxylase-immunoreactive afferents to the cerebellum differs between species. *The Journal of Comparative Neurology* **379**, 443–454.

New JG (1997) The evolution of vertebrate electrosensory systems. *Brain, Behavior and Evolution* **50**, 244–252.

Nichols DH, Bruce LL (2006) Migratory routes and fates of cells transcribing the Wnt-1 gene in the murine hindbrain. *Developmental Dynamics* **235**, 285–300.

Nicol S (1992) Blood viscosity of the echidna, *Tachyglossus aculeatus*. In *Platypus and Echidnas*. (Ed. ML Augee) pp. 140–144. Royal Zoological Society of New South Wales, Sydney.

Nicol S, Andersen NA (1996) Hibernation in the echidna: not an adaptation to the cold? In *Adaptations to the Cold. Tenth International Hibernation Symposium*. (Eds F Gleiser, AJ Hubert and SC Nicol) pp. 7–12. University of New England Press, Armidale.

Nicol SC, Andersen NA (2000) Patterns of hibernation of echidnas in Tasmania. In *Life in the Cold: Eleventh International Hibernation Symposium*. (Eds G Heldmaier, M Klingenspor) pp. 21–28, Springer Verlag, Berlin.

Nicol SC, Andersen NA (2002) The timing of hibernation in Tasmanian echidnas: why do they do it when they do it? *Comparative Biochemistry and Physiology. Part B, Biochemistry & Molecular Biology* **131**, 603–611.

Nicol SC, Andersen NA (2003) Control of breathing in the echidna (*Tachyglossus aculeatus*) during hibernation. *Comparative Biochemistry and Physiology. Part A, Molecular & Integrative Physiology* **136**, 917–925.

Nicol SC, Andersen NA (2007) The life history of an egg-laying mammal, the echidna (*Tachyglossus aculeatus*). *Ecoscience* **14**, 275–285.

Nicol SC, Morrow GE (2012) Sex and seasonality: reproduction in the echidna (*Tachyglossus aculeatus*). In *Living in a Seasonal World*. (Eds T Ruf, C Bieber, W Arnold and E Millesi) pp. 143–153. Springer Verlag, Berlin Heidelberg.

Nicol S, Andersen NA, Mesch U (1992) Metabolic rate and ventilatory pattern in the echidna during hibernation and arousal. In *Platypus and Echidnas*. (Ed. ML Augee) pp. 150–159. Royal Zoological Society of New South Wales, Sydney.

Nicol SC, Andersen NA, Phillips NH, Berger RJ (2000) The echidna manifests typical characteristics of rapid eye movement sleep. *Neuroscience Letters* **283**, 49–52.

Nicol S, Andersen NA, Jones SM (2005) Seasonal variations in reproductive hormones in free-ranging echidnas (*Tachyglossus aculeatus*): interaction between reproduction and hibernation. *General and Comparative Endocrinology* **144**, 204–210.

Nicol SC, Morrow G, Andersen NA (2008) Hibernation in monotremes – a review. In *Hypometabolism in Animals: Torpor, Hibernation and Cryobiology*. (Eds BG Lovegrove and AE McKechnie) pp. 251–262. University of KwaZulu-Natal, Pietermaritzburg.

Nicol SC, Vanpé C, Sprent J, Morrow G, Andersen NA (2011) Spatial ecology of a ubiquitous Australian anteater, the short-beaked echidna (*Tachyglossus aculeatus*). *Journal of Mammalogy* **92**, 101–110.

Nieuwenhuys R (1994) The neocortex. An overview of its evolutionary development, structural organization and synaptology. *Anatomy and Embryology* **190**, 307–337.

Niimura Y, Nei M (2007) Extensive gains and losses of olfactory receptor genes in mammalian evolution. *PLoS ONE* **2**, e708.

O'Day KJ (1938) The visual cells of the platypus (*Ornithorhynchus*). *The British Journal of Ophthalmology* **22**, 321–328.

O'Day KJ (1952) Observations on the eye of the monotreme. *Transactions of the Ophthalmological Society of Australia* **12**, 95–104.

Oftedal OT (2002) The mammary gland and its origin during synapsid evolution. *Journal of Mammary Gland Biology and Neoplasia* **7**, 225–252.

Olsson Herrin R (2009) Distribution and individual characteristics of the platypus (*Ornithorhynchus anatinus*) in the Plenty River, southeast Tasmania. MSc thesis. University of Tasmania, Hobart.

Opiang MD (2009) Home ranges, movement and den use in long-beaked echidnas, *Zaglossus bartoni*, from Papua New Guinea. *Journal of Mammalogy* **90**, 340–346.

Othmar L, Buchmann K, Rhodes J (1978) Instrumental learning in the echidna *Tachyglossus aculeatus setosus*. In *Monotreme Biology: The Australian Zoologist Special Symposium*. (Ed. ML Augee). Published in *The Australian Zoologist* **20**, 131–145.

Otley HM, Munks SA, Hindell MA (2000) Activity patterns, movements and burrows of platypuses (*Ornithorhynchus anatinus*) in a sub-alpine Tasmanian lake. *Australian Journal of Zoology* **48**, 701–713.

Palomero-Gallagher N, Zilles K (2004) Isocortex. In *The Rat Nervous System*. 3rd edn. (Ed. G Paxinos) pp. 729–757. Elsevier, San Diego.

Pascual R, Archer M, Ortiz Jaureguizar E, Prado JL, Godthelp H, Hand SJ (1992a) First discovery of monotremes in South America. *Nature* **356**, 704–706.

Pascual R, Archer M, Ortiz Jaureguizar E, Prado JL, Godthelp H, Hand SJ (1992b) The first non-Australian monotreme: an early Paleocene South American platypus (Monotremata: Ornithorhynchidae). In *Platypus and Echidnas*. (Ed. ML Augee) pp. 1–14. Royal Zoological Society of New South Wales, Sydney.

Pascual R, Goin FJ, Balarino L, Udrizar Sauthier DE (2002) New data on the Paleocene monotreme *Monotrematum sudamericanum*, and the convergent evolution of triangulate molars. *Acta Palaeontologica Polonica* **47**, 487–492.

Patullo BW, Macmillan DL (2004) The relationship between body size and the field potentials generated by swimming crayfish. *Comparative Physiology and Biochemistry. Part A. Molecular and Integrative Physiology* **139**, 77–81.

Paulli S (1900) Über die Pneumaticität des Schadels bei den Säugethieren. I. Über den Bau des Siebbeins. Über die Morphologie des Siebbeins und die der Pneumaticität bei den Monotremen and die Marsupialern. *Gegenbaurs Morphologisches Jahrbuch* **28**, 147–178.

Paxinos G, Watson CRR (2007) *The Rat Brain in Stereotaxic Co-ordinates.* 6th edn. Elsevier Academic, London.

Paxinos G, Huang X-F, Toga AW (2000) *The Rhesus Monkey Brain In Stereotaxic Co-ordinates.* Academic Press, San Diego.

Paxinos G, Watson C, Carrive P, Kirkcaldie M, Ashwel KWS (2009) *Chemoarchitectonic Atlas of the Rat Brain.* 2nd edn. Elsevier Academic, London.

Paxinos G, Xu-Feng H, Sengul G, Watson C (2012) Organization of brainstem nuclei. In *The Human Nervous System.* (Eds JK Mai and G Paxinos) pp. 260–327. Elsevier, San Diego.

Pearson HS (1926) Pelvic and thigh muscles of *Ornithorhynchus. Journal of Anatomy* **60**, 152–163.

Pettigrew JD (1999) Electroreception in monotremes. *The Journal of Experimental Biology* **202**, 1447–1454.

Pettigrew JD, Manger PR, Fine SLB (1998) The sensory world of the platypus. *Philosophical Transactions of the Royal Society. Series B, Biological Sciences* **353**, 1199–1210.

Phillips MJ, Bennett TH, Lee MSY (2009) Molecules, morphology, and ecology indicate a recent, amphibious ancestry for echidnas. *Proceedings of the National Academy of Sciences of the United States of America* **106**, 17089–17094.

Phillips MJ, Bennett TH, Lee MSY (2010) Reply to Camens: How recently did modern monotremes diversify? *Proceedings of the National Academy of Sciences of the United States of America* **107**, E13.

Pickles JO (1992) Scanning electron-microscopy of the echidna – morphology of a primitive mammalian cochlea. *Auditory Physiology and Perception. Advances in the Biosciences* **83**, 101–107.

Pierce F, Mattiske J, Menkhorst P (2007) A probable case of twins in the short-beaked echidna *Tachyglossus aculeatus* (Tachyglossidae: Monotremata), with observations on suckling of young after their emergence from the nursery burrow. *Victorian Naturalist* **124**, 332–340.

Pihlström H (2008) Comparative anatomy and physiology of chemical senses in aquatic mammals. In *Sensory Evolution on the Threshold: Adaptation in Secondarily Aquatic Vertebrates.* (Eds JGM Thewissen and S Nummela) pp. 95–111. University of California Press, Berkeley.

Pihlström H, Fortelius M, Hemilä S, Forsman R, Reuter T (2005) Scaling of mammalian ethmoid bones can predict olfactory organ size and performance. *Proceedings of the Royal Society. Series B, Biological Sciences* **272**, 957–962.

Pirlot P, Nelson J (1978) Volumetric analysis of monotreme brains. In *Monotreme Biology: The Australian Zoologist Special Symposium.* (Ed. ML Augee). Published in *The Australian Zoologist* **20**, 171–179.

Pledge NS (1980) Giant echidnas in South Australia. *South Australian Naturalist* **55**, 27–30.

Plowright CMS, Reid S, Kilian T (1998) Finding hidden food: behavior on visible displacement tasks by mynahs (*Gracula religiosa*) and pigeons (*Columba livia*). *Journal of Comparative Psychology* **112**, 13–25.

Poulton EB (1885) On the tactile terminal organs and other structures in the bill of *Ornithorhynchus. The Journal of Physiology* **5**, 15–16.

Powers AS, Reiner A (1993) The distribution of cholinergic neurons in the central nervous system of turtles. *Brain, Behavior and Evolution* **41**, 326–345.

Preston BT, Stevenson IR, Pemberton JM, Coltman DW, Wilson K (2003) Overt and covert competition in a promiscuous mammal: the importance of weaponry and testes size to male reproductive success. *Proceedings of the Royal Society. Series B, Biological Sciences* **270**, 633–640.

Price JL (2003) Comparative aspects of amygdala connectivity. *Annals of the New York Academy of Sciences* **985**, 50–58.

Pridmore PA, Rich TH, Vickers-Rich P, Gambaryan PP (2005) A tachyglossid-like humerus from the Early Cretaceous of south-eastern Australia. *Journal of Mammalian Evolution* **12**, 359–378.

Proske U, Gregory JE (2003) Electrolocation in the platypus – some speculations. *Comparative Biochemistry and Physiology. Part A, Molecular & Integrative Physiology* **136**, 821–825.

Proske U, Gregory JE (2004) The role of push rods in platypus and echidna – some speculations. *Proceedings of the Linnean Society of New South Wales* **125**, 319–326.

Proske U, Gregory JE, Iggo A (1992) Activity in the platypus brain evoked by weak electrical stimulation of the bill. In *Platypus and Echidnas*. (Ed. ML Augee) pp. 204–210. Royal Zoological Society of New South Wales, Sydney.

Proske U, Gregory JE, Iggo A (1998) Sensory receptors in monotremes. *Philosophical Transactions of the Royal Society. Series B, Biological Sciences* **353**, 1187–1198.

Puelles L, Kuwana E, Puelles E, Bulfone A, Shimamura K, Keleher J, Smiga S, Rubenstein JLR (2000) Pallial and subpallial derivatives in the embryonic chick and mouse telencephalon, traced by the expression of the genes Dlx-2, Emx-1, Nkx-2.1, Pax-6, and Tbr-1. *The Journal of Comparative Neurology* **424**, 409–438.

Puelles L, Martínez S, Martínez-de-la-Torre M, Rubenstein JLR (2004) Gene maps and related histogenetic domains in the forebrain and midbrain. In *The Rat Nervous System*. 3rd edn. (Ed. G Paxinos) pp. 3–125. Elsevier, San Diego.

Puelles L, Martinez-de-la-Torre M, Bardet S, Rubenstein JLR (2012a) Hypothalamus. In *The Mouse Nervous System*. (Eds C Watson, G Paxinos and L Puelles) pp. 221–312. Elsevier, San Diego.

Puelles L, Martinez-de-la-Torre M, Ferran J-L, Watson CRR (2012b) Diencephalon. In *The Mouse Nervous System*. (Eds C Watson, G Paxinos and L Puelles) pp. 313–336. Elsevier, San Diego.

Puelles E, Martinez-de-la-Torre M, Watson CRR, Puelles L (2012c) Midbrain. In *The Mouse Nervous System*. (Eds C Watson, G Paxinos and L Puelles) pp. 337–359. Elsevier, San Diego.

Puzzolo E, Mallamaci A (2010) Cortico-cerebral histogenesis in the opossum *Monodelphis domestica*: generation of a hexalaminar neocortex in the absence of a basal proliferative compartment. *Neural Development* **5**, 8.

Quilliam TA (1979) Tactile and teletactile organs in the muzzle of the duck-billed platypus (*Ornithorhynchus anatinus*). *The Anatomical Record* **193**, 754.

Quiroga JC (1979) The brains of two mammal-like reptiles (Cynodontia – Therapsida). *Journal für Hirnforschung* **20**, 341–350.

Quiroga JC (1980a) The brain of the mammal-like reptile *Probainognathus jenseni* (Therapsida, Cynodontia). A correlative paleo-neurological approach to the neocortex at the reptile-mammal transition. *Journal für Hirnforschung* **21**, 299–336.

Quiroga JC (1980b) Further studies on cynodont endocasts (Reptilia – Therapsida). *Zeitschrift für Mikroskopische und Anatomische Forschung* **94**, 580–592.

Quiroga JC (1984) The endocranial cast of the advanced mammal-like reptile *Therioherpeton cargnini* (Therapsida – Cynodontia) from the middle Triassic of Brazil. *Journal für Hirnforschung* **25**, 285–290.

Rahmann H, Hilbig R, Probst W, Muhleisen M (1984) Brain gangliosides and thermal adaptation in vertebrates. *Advances in Experimental Biology and Medicine* **174**, 395–404.

Rahmann H, Hilbig R, Geiser F (1986) Brain gangliosides in monotremes, marsupials and placentals: phylogenetic and thermoregulatory aspects. *Comparative Biochemistry and Physiology. Part B, Biochemistry & Molecular Biology* **83**, 151–157.

Rapaport DH, Robinson SR, Stone J (1985) Cytogenesis in the developing retina of the cat. *Australian and New Zealand Journal of Ophthalmology* **13**, 113–124.

Ray JP, Price JL (1992) The organization of the thalamocortical connections of the mediodorsal thalamic nucleus in the rat, related to the ventral forebrain–prefrontal cortex topography. *The Journal of Comparative Neurology* **323**, 167–197.

Rechtschaffen A, Bergmann BM (2002) Sleep deprivation in the rat: an update of the 1989 paper. *Sleep* **25**, 18–24.

Rechtschaffen A, Bergmann BM, Everson CA, Kushida CA, Gilliland MA (2002) Sleep deprivation in the rat: X. Integration and discussion of the findings. *Sleep* **25**, 68–87.

Redford KH (1987) Ants and termites as food: patterns of mammalian myrmecophagy. In *Current Mammalogy*. (Ed. HH Genoways) pp. 349–399. Plenum Press, New York.

Regidor J, Divac I (1987) Architectonics of the thalamus in the echidna (*Tachyglossus aculeatus*). Search for the mediodorsal nucleus. *Brain, Behavior and Evolution* **30**, 328–341.

Renfree MB, Papenfuss AT, Shaw G, Pask AJ (2009) Eggs, embryos and the evolution of imprinting: insights from the platypus genome. *Reproduction, Fertility and Development* **21**, 935–942.

Rens W, O'Brien PCM, Grützner F, Clarke O, Graphodatskaya D, Tsend-Ayush E, Trifonov VA, Skelton H, Wallis MC, Johnston S, Veyrunes F, Graves

JAM, Ferguson-Smith MA (2007) The multiple sex chromosomes of platypus and echidna are not completely identical and several share homology with the avian Z. *Genome Biology* **8**, R243.

Ribak CE, Fallon JH (1982) The islands of Calleja complex of rat basal forebrain. I. Light and electron microscopic observations. *The Journal of Comparative Neurology* **205**, 207–218.

Rich TH, Vickers-Rich P, Constantine A, Flannery TF, Kool L, van Klaveren N (1997) A tribosphenic mammal from the Mesozoic of Australia. *Science* **278**, 1438–1442.

Rich TH, Vickers-Rich P, Constantine A, Flannery TF, Kool L, van Klaveren N (1999) Early Cretaceous mammals from Flat Rocks, Victoria, Australia. *Records of the Queen Victoria Museum* **106**, 1–35.

Rich TH, Flannery TF, Trusler P, Kool L, van Klaveren N, Vickers-Rich P (2001a) A second tribosphenic mammal from the Early Cretaceous Flat Rocks site, Victoria, Australia. *Records of the Queen Victoria Museum* **110**, 1–9.

Rich TH, Vickers-Rich P, Trusler P, Flannery TF, Cifelli RL, Constantine A (2001b) Monotreme nature of the Australian early Cretaceous mammal *Teinolophos trusleri*. *Acta Palaeontologica Polonica* **46**, 113–118.

Rich TH, Flannery TF, Trusler P, Kool L, van Klaveren NA, Vickers-Rich P (2002) Evidence that monotremes and ausktribosphenids are not sister-groups. *Journal of Vertebrate Paleontology* **22**, 466–469.

Rich TH, Hopson JA, Musser AM, Flannery TF, Vickers-Rich P (2005) Independent origins of middle ear bones in monotremes and therians. *Science* **307**, 910–914.

Rich TH, Vickers-Rich P, Flannery TF, Kear BP, Cantrill DJ, Komarower P, Kool L, Pickering D, Trusler P, Morton S, van Klaveren N, Fitzgerald EMG (2009) An Australian multituberculate and its palaeobiogeographic implications. *Acta Palaeontologica Polonica* **54**, 1–6.

Richardson MK, Narraway J (1999) A treasure house of comparative embryology. *The International Journal of Developmental Biology* **43**, 591–602.

Rismiller PD (1992) Field observations on Kangaroo island echidnas (*Tachyglossus aculeatus multiaculeatus*) during the breeding season. In *Platypus and Echidnas*. (Ed. ML Augee) pp. 101–105. Royal Zoological Society of New South Wales, Sydney.

Rismiller PD, McKelvey MW (1996) Sex, torpor and activity in temperate climate echidnas. In *Adaptations to the Cold: Tenth International Hibernation Symposium*. (Eds F Geiser F, AJ Hulbert and SC Nicol) pp. 23–30. University of New England Press, Armidale.

Rismiller PD, McKelvey MW (2000) Frequency of breeding and recruitment in the short-beaked echidna, *Tachyglossus aculeatus*. *Journal of Mammalogy* **81**, 1–17.

Rismiller PD, McKelvey MW (2003) Body mass, age and sexual maturity in short-beaked echidnas, *Tachyglossus aculeatus*. *Comparative Biochemistry and Physiology. Part A, Molecular & Integrative Physiology* **136**, 851–865.

Rismiller PD, McKelvey MW (2009) Activity and behaviour of lactating echidnas (*Tachyglossus aculeatus multiaculeatus*) from hatching of egg to weaning of young. *Australian Journal of Zoology* **57**, 265–273.

Robinson SR (1987) Ontogeny of the area centralis in the cat. *The Journal of Comparative Neurology* **255**, 50–67.

Robinson SR, Rapaport DH, Stone J (1985) Cell division in the developing cat retina occurs in two zones. *Brain Research* **351**, 101–109.

Rodriguez CI, Dymecki SM (2000) Origin of the precerebellar system. *Neuron* **27**, 475–486.

Rogers JH (1989) Immunoreactivity for calretinin and other calcium-binding proteins in cerebellum. *Neuroscience* **31**, 711–721.

Rose RW, Nevison CM, Dixson AF (1997) Testes weight, body weight and mating systems in marsupials and monotremes. *Journal of Zoology* **243**, 523–531.

Roth G, Dicke U (2005) Evolution of the brain and intelligence. *Trends in Cognitive Sciences* **9**, 250–257.

Rowe MJ (1990) Organization of the cerebral cortex in monotremes and marsupials. In *Cerebral Cortex*. Vol. 8B. (Eds EG Jones and A Peters) pp. 263–334. Plenum, New York.

Rowe MJ, Mahns DA, Bohringer RC, Ashwell KWS, Sahai V (2003) Tactile neural mechanisms in monotremes. *Comparative Biochemistry and Physiology. Part A, Molecular & Integrative Physiology* **136**, 883–893.

Rowe MJ, Mahns DA, Sahai V (2004) Monotreme tactile mechanisms: from sensory nerves to cerebral cortex. *Proceedings of the Linnean Society of New South Wales* **125**, 301–317.

Rowe T, Rich TH, Vickers-Rich P, Springer M, Woodburne MO (2008) The oldest platypus and its bearing on divergence timing of the platypus and echidna clades. *Proceedings of the National Academy of Sciences of the United States of America* **105**, 1238–1242.

Saban R (1969) Extrinsic ocular musculature in echidna (*Tachyglossus aculeatus* Shaw). *Comptes Rendus Hebdomadaires des Séances de l'Académie des Sciences. Série D, Sciences Naturelles* **268**, 2351–2354.

Saper C (2012) Hypothalamus. In *The Human Nervous System*. 3rd edn. (Eds JK Mai and G Paxinos) pp. 548–583. Elsevier, San Diego.

Saunders JC, Chen CS, Pridmore PA (1971a) Successive habit-reversal learning in monotreme *Tachyglossus aculeatus* (echidna). *Animal Behaviour* **19**, 552–555.

Saunders JC, Teague J, Slonim D, Pridmore PA (1971b) A position habit in the monotreme *Tachyglossus aculeatus* (the spiny anteater). *Australian Journal of Psychology* **23**, 47–51.

Scheich H, Langner G, Tidemann C, Coles RB, Guppy A (1986) Electroreception and electrolocation in platypus. *Nature* **319**, 401–402.

Schimchowitsch S, Stoeckel ME, Vigny A, Porte A (1983) Oxytocinergic neurons with tyrosine hydroxylase-like immunoreactivity in the paraventricular nucleus of the rabbit hypothalamus. *Neuroscience Letters* **43**, 55–59.

Schimchowitsch S, Moreau C, Laurent F, Stoeckel M-E (1989) Distribution and morphometric characteristics of oxytocin- and vasopressin- immunoreactive neurons in the rabbit hypothalamus. *The Journal of Comparative Neurology* **285**, 304–324.

Schmidt-Nielsen K, Dawson TJ, Crawford EC (1966) Temperature regulation in the echidna (*Tachyglossus aculeatus*). *Journal of Cellular Physiology* **67**, 63–71.

Schneider NY (2011) The development of the olfactory organs in newly hatched monotremes and neonate marsupials. *Journal of Anatomy* **219**, 229–242.

Schuster E (1910) Preliminary note upon the cell lamination of the cerebral cortex of *Echidna*, with an enumeration of the fibres in the cranial nerves. *Proceedings of the Royal Society of London. Series B, Containing Papers of a Biological Character* **82**, 113–123.

Schwab IR, McMenamin P (2005) How do I fit it? *The British Journal of Ophthalmology* **89**, 129.

Semon R (1894) Die Embryonalhüllen der Monotremen und Marsupialier. *Denkschrift Medizin Naturwissenschaften Geschichte Jena* **5**, 19–74.

Serena M (1994) Use of time and space by platypus (*Ornithorhynchus anatinus:* Monotremata) along a Victorian stream. *Journal of Zoology* **232**, 117–131.

Serena M, Williams GA (2013) Movement and range size of the platypus (*Ornithorhynchus anatinus*) inferred from mark-recapture studies. *Australian Journal of Zoology*, (In press).

Serena M, Thomas JL, Williams GA, Officer RCE (1998) Use of stream and river habitats by the platypus, *Ornithorhynchus anatinus*, in an urban fringe environment. *Australian Journal of Zoology* **46**, 267–282.

Serena M, Worley M, Swinnerton M, Williams GA (2001) Effect of food availability and habitat on the distribution of platypus (*Ornithorhynchus anatinus*) foraging activity. *Australian Journal of Zoology* **49**, 263–277.

Sherlock DA, Field PM, Raisman G (1975) Retrograde transport of horseradish peroxidase in the magnocellular neurosecretory system of the rat. *Brain Research* **88**, 403–414.

Shi P, Zhang J (2006) Contrasting modes of evolution between vertebrate sweet/umami receptor genes and bitter receptor genes. *Molecular Biology and Evolution* **23**, 292–300.

Shi P, Zhang J (2007) Comparative genomic analysis identifies an evolutionary shift of vomeronasal receptor gene repertoires in the vertebrate transition from water to land. *Genome Research* **17**, 166–174.

Shi P, Bielawski JP, Yang H, Zhang YP (2005) Adaptive diversification of vomeronasal receptor 1 genes in rodents. *Journal of Molecular Evolution* **60**, 566–576.

Shipley MT, Ennis M, Puche A (2004) Olfactory system. In *The Rat Nervous System*. 3rd edn. (Ed. G Paxinos) pp. 923–964. Elsevier, San Diego.

Short RV (2009) The miraculous platypus. *Australian Journal of Zoology* **57**, v–vi.

Sieg AE, O'Connor MP, McNair JN, Grant BW, Agosta SJ, Dunham AE (2009) Mammalian metabolic allometry: do intraspecific variation, phylogeny, and regression models matter? *American Naturalist* **174**, 720–733.

Siegel JM (1995) Phylogeny and the function of REM sleep. *Behavioural Brain Research* **69**, 29–34.

Siegel JM, Manger PR, Nienhuis R, Fahringer HM, Pettigrew JD (1996) The echidna *Tachyglossus aculeatus* combines REM and non-REM aspects in a single sleep state: Implications for the evolution of sleep. *The Journal of Neuroscience* **16**, 3500–3506.

Siegel JM, Manger PR, Nienhuis R, Fahringer HM, Pettigrew JD (1998) Monotremes and the evolution of rapid eye movement sleep. *Philosophical Transactions of the Royal Society. Series B, Biological Sciences* **353**, 1147–1157.

Siegel JM, Manger PR, Nienhuis R, Fahringer HM, Shalita T, Pettigrew JD (1999) Sleep in the platypus. *Neuroscience* **91**, 391–400.

Simerly RB (2004) Anatomical substrates of hypothalamic integration. In *The Rat Nervous System*. 3rd edn. (Ed. G Paxinos) pp. 335–368. Elsevier, San Diego.

Smeets WJAJ, Steinbusch HWM (1988) Distribution of serotonin immunoreactivity in the forebrain and midbrain of the lizard *Gecko gecko*. *The Journal of Comparative Neurology* **271**, 419–434.

Smiley JF, McGinnis JP, Javitt DC (2000) Nitric oxide synthase interneurons in the monkey cerebral cortex are subsets of the somatostatin, neuropeptide Y, and calbindin cells. *Brain Research* **863**, 205–212.

Smith GC (1971) Median eminence of echidna: fluorescence, histochemical, neurosecretory and ultrastructural correlations. *Journal of Anatomy* **108**, 207–208.

Smith KK (1994) Development of craniofacial musculature in *Monodelphis* domestica (Marsupialia, Didelphidae). *Journal of Morphology* **222**, 149–173.

Smith KK (2001) The evolution of mammalian development. *Bulletin of the Museum of Comparative Zoology* **156**, 119–135.

Smith GC, Osborne LW, Simpson RW (1970) A comparative study of distribution of amines in perivascular zone of median eminence of echidna, rabbit and rat. *Journal of Anatomy* **106**, 207.

Smith AP, Welham GS, Green SW (1989) Seasonal foraging activity and microhabitat selection by echidnas (*Tachyglossus aculeatus*) on the New England tableland. *Australian Journal of Ecology* **14**, 457–466.

Sol D (2009) Revisiting the cognitive buffer hypothesis for the evolution of large brains. *Biology Letters* **5**, 130–133.

Spoor F, Thewissen JGM (2008) Comparative and functional anatomy of balance in aquatic mammals. In *Sensory Evolution on the Threshold. Adaptations in Secondarily Aquatic Vertebrates*. (Ed. JGM Thewissen and S Nummela) pp. 257–284. University of California Press, Berkeley.

Sprent JA (2011) Diet, spatial ecology and energetics of echidnas: the significance of habitat and seasonal variation. PhD thesis. University of Tasmania, Hobart.

Sprent JA, Nicol SC (2012) The influence of habitat on home range size of the short beaked echidna. *Australian Journal of Zoology* **60**, 46–53.

Sprent JA, Andersen NA, Nicol SC (2006) Latrine use by the short-beaked echidna, *Tachyglossus aculeatus*. *Australian Mammalogy* **28**, 131–133.

Stephan H, Andy OJ (1977) Quantitative comparison of the amygdala in insectivores and primates. *Acta Anatomica* **98**, 130–153.

Stone J (1983) Topographical organization of the retina in a monotreme, Australian spiny anteater *Tachyglossus aculeatus*. *Brain, Behavior and Evolution* **22**, 175–184.

Striedter G (2005) *Principles of Brain Evolution*. Sinauer, Sunderland.

Sunderland S (1941) The vascular pattern in the central nervous system of the monotremes and Australian marsupials. *The Journal of Comparative Neurology* **75**, 123–129.

Swanson LW, Kuypers HG (1980) The paraventricular nucleus of the hypothalamus: cytoarchitectonic subdivisions and organization of projections to the pituitary, dorsal vagal complex, and spinal cord as demonstrated by retrograde fluorescence double-labeling methods. *The Journal of Comparative Neurology* **194**, 555–570.

Swanson LW, Sawchenko PE, Lind RW (1986) Regulation of multiple peptides in CRF parvocellular neurosecretory neurons: implications for the stress response. *Progress in Brain Research* **68**, 169–190.

Switzer RC, III, Johnson JI (1977) Absence of mitral cells in monolayer in monotremes. Variations in olfactory bulbs. *Acta Anatomica* **99**, 36–42.

Taggart DA, Breed WG, Temple-Smith PD, Purvis A, Shimmin G (1998) Reproduction, mating strategies and sperm competetion in marsupials and monotremes. In *Sperm Competition and Sexual Selection*. (Eds TR Birkhead and AP Møller) pp. 623–666. Academic Press, London.

Tago H, McGeer PL, Bruce G, Hersh LB (1987) Distribution of choline acetyltransferase-containing neurons of the hypothalamus. *Brain Research* **415**, 49–62.

Taylor NG, Manger PR, Pettigrew JD, Hall LS (1992) Electromyographic potentials of a variety of platypus prey items: an amplitude and frequency analysis. In *Platypus and Echidnas*. (Ed. ML Augee) pp. 216–224. Royal Zoological Society of New South Wales, Sydney.

Temple-Smith PD (1973) *Seasonal Breeding Biology of the Platypus, Ornithorhynchus anatinus (Shaw, 1799) with Special Reference to the Male*. PhD thesis. Australian National University, Canberra.

Temple-Smith PD, Grant T (2001) Uncertain breeding: a short history of reproduction in monotremes. *Reproduction, Fertility and Development* **13**, 487–497.

Thakkar MM, Datta S (2010) The evolution of REM sleep. In *Evolution of Sleep*. (Eds P McNamara, RA Barton and CL Nunn) pp. 197–217. Cambridge, Cambridge.

Thulborn RA, Turner S (2003) The last dicynodont: an Australian Cretaceous relict. *Proceedings of the Royal Society. B. Biological Sciences* **270**, 985–993.

Tindal JS, Beyer C, Sawyer CH (1963) Milk ejection reflex and maintenance of lactation in the rabbit. *Endocrinology* **72**, 720–724.

Toida K, Kosaka K, Heizmann CW, Kosaka T (1996) Electron microscopic serial-sectioning/reconstruction study of parvalbumin-containing neurons in the external plexiform layer of the rat olfactory bulb. *Neuroscience* **72**, 449–466.

Tononi G, Cirelli C (2006) Sleep function and synaptic homeostasis. *Sleep Medicine Reviews* **10**, 49–62.

Tracey DJ (2004) Ascending and descending pathways in the spinal cord. In *The Rat Nervous System*. 3rd edn. (Ed. G Paxinos) pp. 149–164. Elsevier, San Diego.

Tsang YM, Chiong F, Kuznetsov D, Kasarskis E, Geula C (2000) Motor neurons are rich in non-phosphorylated neurofilaments: cross-species comparison and alterations in ALS. *Brain Research* **861**, 45–58.

Ulinski PS (1984) Thalamic projections to the somatosensory cortex of the echidna, *Tachyglossus aculeatus*. *The Journal of Comparative Neurology* **229**, 153–170.

Unger JW, Lange W (1991) Immunohistochemical mapping of neurophysins and calcitonin gene-related peptide in the human brainstem and spinal cord. *Journal of Chemical Neuroanatomy* **4**, 299–309.

van Hartevelt TJ, Kringelbach ML (2012) The olfactory system. In *The Human Nervous System*. 3rd edn. (Eds JK Mai and G Paxinos) pp. 1219–1238. Elsevier, San Diego.

van Rheede T, Bastiaans T, Boone DN, Hedges SB, de Jong WW, Madsen O (2006) The platypus is in its place: nuclear genes and indels confirm the sister group relation of monotremes and therians. *Molecular Biology and Evolution* **23**, 587–597.

Vater M, Kössl M (2011) Comparative aspects of cochlear functional organization in mammals. *Hearing Research* **273**, 89–99.

Vater M, Meng J, Fox RC (2004) Hearing organ evolution and specialization: early and later mammals. In *Evolution of the Vertebrate Auditory System*. (Eds GA Manley, AN Popper and RR Fay) pp. 256–288. Springer Verlag, New York.

Vidal P-P, Sans A (2004) Vestibular system. In *The Rat Nervous System*. 3rd edn. (Ed. G Paxinos) pp. 965–996. Elsevier, San Diego.

Voogd J (2004) Cerebellum. In *The Rat Nervous System*. 3rd edn. (Ed. G Paxinos) pp. 205–242. Elsevier, San Diego.

Voss H (1963) Besitzen die Monotremen (*Echidna* und *Ornithorhynchus*) ein – oder mehrfaserige Muskelspindlen? *Anatomischer Anzeiger* **113**, 255–258.

Vyazovskiy VV, Olcese U, Lazimy YM, Faraguna U, Esser SK, Williams JC, Cirelli C, Tononi G (2009) Cortical firing and sleep homeostasis. *Neuron* **63**, 865–878.

Wakefield MJ, Anderson M, Chang E, Wei KJ, Kaul R, Graves JAM, Grützner F, Deeb SS (2008) Cone visual pigments of monotremes: filling the phylogenetic gap. *Visual Neuroscience* **25**, 257–264.

Waldvogel HJ, Faull RLM (1993) Compartmentalization of parvalbumin immunoreactivity in the human striatum. *Brain Research* **610**, 311–316.

Wallis M (2012) Molecular evolution of the neurohypophysial hormone precursors in mammals: comparative genomics reveals novel mammalian oxytocin and vasopressin analogues. *General and Comparative Endocrinology* **179**, 313–318.

Walls GL (1942) *The Vertebrate Eye and its Adaptive Radiation*. Cranbrook Press, Hills, Michigan.

Walter LR (1988) Appendicular musculature in the echidna *Tachyglossus aculeatus* (Monotremata, Tachyglossidae). *Australian Journal of Zoology* **36**, 65–81.

Wang WZ, Hoerder-Suabedissen A, Oeschger FM, Bayatti N, Ip BK, Lindsay S, Supramaniam V, Srinivasan L, Rutherford M, Møllgard K, Clowry GJ, Molnár Z (2010) Subplate in the developing cortex of mouse and human. *Journal of Anatomy* **217**, 368–380.

Wang WZ, Oeschger FM, Montiel JF, García-Moreno F, Hoerder-Suabedissen A, Krubitzer L, Ek CJ, Saunders NR, Reim K, Villalón A, Molnár Z (2011) Comparative aspects of subplate zone studied with gene expression in sauropsids and mammals. *Cerebral Cortex* **21**, 2187–2203.

Warren AA, Kool L, Cleeland M, Rich TH, Vickers-Rich P (1991) An early Cretaceous labyrinthodont. *Alcheringa* **15**, 327–332.

Watkins WB, Choy VJ (1977) Immunocytochemical study of the hypothalamo-neurohypophyseal system. III. Localization of oxytocin- and vasopressin-containing neurons in the pig hypothalamus. *Cell and Tissue Research* **180**, 491–503.

Watson CRR (2012a) Hindbrain. In *The Mouse Nervous System*. (Eds C Watson, G Paxinos and L Puelles) pp. 398–423. Elsevier, San Diego.

Watson CRR (2012b) Visual system. In *The Mouse Nervous System*. (Eds C Watson, G Paxinos and L Puelles) pp. 646–652. Elsevier, San Diego.

Watson CRR, Qi Y (2012) Neurosecretory nuclei of the hypothalamus and preoptic area. In *The Mouse Nervous System*. (Eds C Watson, G Paxinos and L Puelles) pp. 520–527. Elsevier, San Diego.

Watson CRR, Provis JM, Bohringer R (1977) The subdivisions of the trigeminal nucleus of the platypus, *Ornithorhynchus anatinus*: a comparative study. *Journal of Anatomy* **124**, 533.

Weckerly FW (1998) Sexual-size dimorphism: influence of mass and mating systems in the most dimorphic mammals. *Journal of Mammalogy* **79**, 33–52.

Weidman TA, Kuwabara T (1969) Development of the rat retina. *Investigative Ophthalmology* **8**, 60–69.

Weisbecker V, Goswami A (2010) Brain size, life history, and metabolism at the marsupial/placental dichotomy. *Proceedings of the National Academy of Sciences of the United States of America* **107**, 16216–16221.

Weisbecker V, Goswami A (2011) Marsupials indeed confirm an ancestral mammalian pattern: A reply to Isler. *BioEssays* **33**, 358–361.

Welker W, Lende RA (1980) Thalamocortical relationships in echidna (*Tachyglossus aculeatus*). In *Comparative Neurology of the Telencephalon*. (Ed. SOE Ebbesson) pp. 449–481. Plenum, New York.

Werneburg I, Sánchez-Villagra MR (2011) The early development of the echidna, *Tachyglossus aculeatus* (Mammalia: Monotremata), and patterns of mammalian development. *Acta Zoologica (Stockholm)* **92**, 75–88.

Westerman M, Edwards D (1992) DNA hybridization and the phylogeny of monotremes. In *Platypus and Echidnas*. (Ed. ML Augee) pp. 28–34. The Royal Zoological Society of New South Wales, Sydney.

White CR, Seymour RS (2004) Does basal metabolic rate contain a useful signal? Mammalian BMR allometry and correlations with a selection of physiological, ecological and life-history variables. *Physiological and Biochemical Zoology* **77**, 929–941.

White CR, Seymour RS (2005) Allometric scaling of mammalian metabolism. *The Journal of Experimental Biology* **208**, 1611–1619.

Whittington C, Belov K (2007) Platypus venom: a review. *Australian Mammalogy* **29**, 57–62.

Wible JR, Rougier GW (2000) Cranial anatomy of *Kryptobaatar dashzevegi* (Mammalia, Multituberculata), and its bearing on the evolution of mammalian characters. *Bulletin of the American Museum of Natural History* **247**, 1–124.

Williams GA, Serena M, Grant TR (2012) Age-related change in spurs and spur sheaths of the platypus (*Ornithorhynchus anatinus*). *Australian Mammalogy*. Published online: 26 October 2012. doi:10.1071/AM12011

Wilson DE, Reeder DM (Eds) (2005) *Mammal Species of the World. A Taxonomic and Geographic Reference*. 3rd edn. Johns Hopkins University Press, Baltimore.

Wislocki GB, Campbell ACP (1937) The unusual manner of vascularisation of the brain of the opossum (*Didelphys virginiana*). *The Anatomical Record* **67**, 177–191.

Witter MP, Amaral DG (2004) Hippocampal formation. In *The Rat Nervous System*. 3rd edn. (Ed. G Paxinos) pp. 635–704. Elsevier, San Diego.

Wong E, Whittington C, Papenfuss T, Nicol S, Warren WC, Belov K (2012) The evolutionary origins of monotreme crural glands. *Toxicon* **60**, 122–123.

Wong-Riley MT (1989) Cytochrome oxidase: an endogenous metabolic marker for neuronal activity. *Trends in Neurosciences* **12**, 94–101.

Woodburne MO, Tedford RH (1975) The first Tertiary monotreme from Australia. *American Museum Novitates* **2588**, 1–11.

Woodburne MO, Tedford RH, Archer M, Turnbull WD, Plane MD, Lundelius EL (1985) Biochronology of the continental mammal record of Australia and New Guinea. Special Publication. *South Australia Department of Mines and Energy* **5**, 347–363.

Woodburne MO, Rich TH, Springer MS (2003) The evolution of tribospheny and the antiquity of mammalian clades. *Molecular Phylogenetics and Evolution* **28**, 360–385.

Yamada J, Krause WJ (1983) An immunohistochemical survey of endocrine cells and nerves in the proximal small intestine of the platypus, *Ornithorhynchus anatinus*. *Cell and Tissue Research* **234**, 153–164.

Yeh KY, Wu CH, Tai MY, Tsai YF (2011) *Ginkgo biloba* extract enhances noncontact erection in rats: the role of dopamine in the paraventricular nucleus and the mesolimbic system. *Neuroscience* **189**, 199–206.

Yew DT, Luo CB, Heizmann CW, Chan WY (1997) Differential expression of calretinin, calbindin D-28k and parvalbumin in the developing human cerebellum. *Brain Research. Developmental Brain Research* **103**, 37–45.

Young HM, Pettigrew JD (1991) Cone photoreceptors lacking oil droplets in the retina of the echidna, *Tachyglossus aculeatus* (Monotremata). *Visual Neuroscience* **6**, 409–420.

Young HM, Vaney DI (1990) The retinae of prototherian mammals possess neuronal types that are characteristic of nonmammalian retinae. *Visual Neuroscience* **5**, 61–66.

Young JM, Massa HF, Hsu L, Trask BJ (2010) Extreme variability among mammalian V1R gene families. *Genome Research* **20**, 10–18.

Zeiss CJ, Schwab IR, Murphy CJ, Dubietzig RW (2011) Comparative retinal morphology of the platypus. *Journal of Morphology* **272**, 949–957.

Zeller U (1988) The lamina cribrosa of *Ornithorhynchus* (Monotremata, Mammalia). *Anatomy and Embryology* **178**, 513–519.

Zeller U (1989) Die Entwicklung und Morphologie des Schadels von *Ornithorhynchus anatinus*: (Mammalia, Prototheria, Monotremata). *Abhandlungen der Senckenbergischen Naturforschenden Gesellschaft* **545**, 1–188 . Verlag Waldemar Kramer, Frankfurt.

Zhao DQ, Lu CL, Ai HB (2011) The role of catecholaminergic neurons in the hypothalamus and medullary visceral zone in response to restraint water-immersion stress in rats. *The Journal of Physiological Sciences* **61**, 37–45.

Ziehen T (1897) Das Centralnervensystem der Monotremen und Marsupialier. Teil I. Makroskopische Anatomie. *Jena Denkschriften* **6**, 168–187.

Ziehen T (1901) To the furrows and lobes of the cerebellum in echidna. *Monatsschrift für Psychiatrie und Neurologie* **10**, 143–149.

Zilles K, Armstrong E, Moser KH, Schleicher A, Stephan H (1989) Gyrification in the cerebral cortex of primates. *Brain, Behavior and Evolution* **34**, 143–150.

GLOSSARY

Accommodation The change in the shape of the lens of the eye that allows focusing on near objects. Accommodation depends on changes in the activation of a smooth muscle, known as the ciliary muscle, within the anterior eye.

Acetylcholine A neurotransmitter used by cholinergic pathways in the brain and at the neuromuscular junction. It is often abbreviated as ACh and is formed from the esterification of choline and acetyl coenzyme A.

Acetylcholinesterase An enzyme that cleaves the neurotransmitter acetylcholine into acetate and choline. It is often used as a chemoarchitectural marker, to distinguish functionally significant brain regions.

Adrenaline Also called epinephrine. A catecholamine neurotransmitter and hormone produced by the modified neurons of the adrenal medulla and by some central nervous system neurons.

Afferent A type of axon or pathway that carries information into the central nervous system, implying a sensory or input function.

Allocortex The type of cerebral cortex that has between three and five layers. Allocortex includes both the hippocampus (archicortex) and olfactory cortex (paleocortex).

Altricial Literally meaning 'requiring nourishment' –describes young requiring extensive care after birth or hatching.

Amygdala An almond-shaped group of neurons (named from the Latin for 'almond') located mainly in the temporal region of the brain. The amygdala is part of the limbic system and is concerned with assigning emotional significance during learning.

Anterior commissure A large fibre bundle in the monotreme forebrain that connects the two cerebral hemispheres. It develops in the rostral end of the third ventricle.

Antidiuretic hormone (ADH) Also called vasopressin. A hormone released into the circulation from axon terminals of the posterior pituitary (neurohypophysis). It controls reabsorption of water by the kidneys and can increase peripheral vascular resistance to maintain arterial blood pressure.

Apomorphy A derived characteristic or trait of a clade.

Areola Also called milk patches. The region of pigmented belly skin of a female monotreme where milk is secreted to the skin surface.

Australasian ecozone One of the eight large biogeographic divisions of the Earth's land surface. It comprises Australia, Tasmania, New Guinea, New Zealand and neighbouring islands.

Autapomorphy A derived characteristic or trait that is unique to a particular clade.

Axon The long output process of a neuron that carries the nerve impulse (action potential) away from the cell body. It may be coated with myelin to increase the velocity of transmission of the action potentials.

BMR Basal metabolic rate. The amount of energy expended by an animal at rest in a thermally neutral environment and while in a post-absorptive state. Usually measured as oxygen consumption, may be expressed in watts or kJoules.

Branchiomeric Referring to either: (1) skeletal muscle derived from the mesoderm of the embryonic pharyngeal arches 1 to 4; or (2) the groups of motor neurons in the brainstem that supply those muscles.

Calbindin A calcium-binding protein found within some neurons in the central nervous system. Calbindin acts as a calcium-stabilising protein and intracellular messenger in many neurons (e.g. Purkinje cells of the cerebellum and dopaminergic neurons of the substantia nigra).

Calretinin A calcium-binding protein found within some neurons in the central nervous system. In therians it is found in local circuit neurons of the cerebral cortex and thalamus.

Cauda equina The collected dorsal and ventral roots that extend from the end of the spinal cord to the vertebral foramina. In the short-beaked echidna, these are very long and collectively as thick as the spinal cord itself.

Caudal Literally meaning 'towards the tail'. The term refers to a direction towards the end of the spinal cord.

Caudatoputamen The combined components of the dorsal striatum. They may be distinct structures (i.e. caudate and putamen) in some placentals (e.g. primates).

Cerebellar peduncles Three large bundles of axons that connect the cerebellum with the brainstem. The peduncles (superior, middle and inferior) contain cerebellar inputs and outputs and connect to the midbrain, pons and medulla oblongata, respectively.

Cerebellum A structure with a folded surface (cerebellar cortex) and deep cerebellar nuclei (dentate, interposed and fastigial nuclei) that is attached to the brainstem. It uses information from the vestibular sensory apparatus, proprioreceptors of the muscles and joints, and other senses (visual and auditory) to coordinate motor activity.

Cerebrocerebellum Part of the cerebellum (anatomically the lateral part of the cerebellar hemisphere and the lateral deep cerebellar nucleus) that is engaged in looped circuits involving the cortex, pontine nuclei and motor thalamic nuclei. It coordinates fine movements of the distal limbs.

Cerebrospinal fluid (CSF) The clear, colourless fluid that fills the subarachnoid space around the brain, and the ventricular system within the brain. The CSF is mainly produced by the choroid plexus of the lateral, third and fourth ventricles and is reabsorbed into the systemic veins by arachnoid granulations associated with the major dural venous sinuses.

C fibre A type of sensory axon that is unmyelinated, slowly conducting (less than 2 m/s) and concerned with pain, temperature, itch and muscle burn or cramp.

Chemoarchitect/ure, -onics The study of brain regions based on the distribution of functionally significant brain chemicals (enzymes, calcium-binding proteins, neurotransmitters).

Choroid plexus One of several highly vascular structures within the lateral, third and fourth ventricles that produce most of the CSF. The choroid plexus consists of a cuboidal choroidal epithelium surrounding connective tissue and fenestrated choroidal capillaries.

Claustrum Named from the Latin for 'shutter' or 'lock'. The claustrum is a thin sheet of grey matter lying deep to the insular cortex and separated from the striatum by the external capsule. The claustrum and associated endopiriform nucleus are nuclear derivatives of the embryonic pallium.

Cloaca A common opening for the rectal and urogenital orifices, so that faeces, urine and reproductive products all emerge through this orifice. Although a characteristic of monotremes, some marsupials and the placental beaver also have a cloaca.

Colliculus From the Latin for 'little hill'. Superior colliculi are visual in function, whereas inferior colliculi are part of the ascending auditory pathway.

Commissural Pertaining to axonal connections between the two sides of the brain. For the forebrain of monotremes, these pass through the anterior, hippocampal or posterior commissures.

Conspecific Belonging to the same species.

Contralateral Referring to the opposite side (of the body). This term is used with respect to representations of the body surface or musculature in, or on, the brain; or the course of sensory or motor pathways in the central nervous system.

Cortex The layered surfaces of the cerebral or cerebellar hemispheres. The term 'cortex' is derived from the Latin for the rind (of a fruit). Cortical surfaces may be folded to maximise surface area and hence processing capacity, by increasing the number of the constituent neuronal columns in the cortical sheet. See also pallium.

Corticofugal Literally meaning 'fleeing the cortex'. It refers to axons leaving the cerebral cortex.

Corticopetal Literally meaning 'seeking the cortex'. It refers to axons approaching and entering the cortex.

Corticospinal tract A bundle of axons running from neuronal cell bodies in the cerebral cortex (isocortex) to the spinal cord.

Crepuscular Being active around dawn or dusk. Echidnas in hot climates tend to be crepuscular to avoid the intense heat of the middle of the day.

Cytoarchitect/ure, -onics The process of distinguishing between different brain regions on the basis of differences in the distribution, staining intensity and size of neurons and glia.

Cytochrome oxidase An important enzyme in the oxidative metabolism of neurons, usually abbreviated as CO. It is a useful marker in differentiating sensory and non-sensory cortical regions, because it reveals areas of high oxidative metabolic rate (e.g. clustered axon terminals).

Dendrite The process of a neuron that receives incoming information.

Diencephalon The part of the brain derived from the three embryonic prosomeres. It includes the pretectum, thalamus and prethalamus, ventrally; as well as the epithalamus dorsally.

Dopamine A catecholamine neurotransmitter used by neurons of the substantia nigra and ventral tegmental area of the midbrain in pathways to the striatum, frontal cortex and limbic regions.

Dorsal Literally meaning 'towards the back'. A direction towards the back of the animal.

Dorsal columns Longitudinal bundles of axons in the dorsal white matter of the spinal cord. These ascending somatosensory pathways (fasciculus gracilis and cuneatus) are concerned with conscious proprioception, vibration and discriminative (i.e. high spatial resolution) aspects of touch.

Ectoderm One of the three layers of the embryonic germ disc. It gives rise to the epithelium of the skin and the cells of the peripheral and central nervous system.

Efferents Axonal pathways either projecting out of the central nervous system to muscles or glands, or projecting out of one brain region to another. Those projecting out of the CNS serve a motor function in the broadest sense of the word.

Egg-tooth A structure that develops during embryonic life in monotremes to allow the hatchling to break through the embryonic membranes. It has a core of mesoderm that pushes out the buccal epithelium of the upper jaw to form a toothlet. An enamel organ develops within the epithelial covering and the toothlet becomes attached to the premaxilla.

Encephalisation A measure of relative brain size, after the effect of body size on brain size has been excluded. The encephalisation quotient is the ratio of actual brain volume or mass over the expected brain volume or mass for a reference animal of that body size.

Endoderm One of the three layers of the embryonic germ disc. It forms the lining of the yolk sac, gut epithelium, respiratory epithelium and associated glands.

Entopeduncular nucleus A part of the dorsal pallidum that is homologous to the medial globus pallidus of primates. The entopeduncular nucleus consists of large neurons embedded in the lateral parts of the internal capsule. The entopeduncular nucleus receives input from the lateral globus pallidus and the compact part of the substantia nigra and projects to the motor and intralaminar nuclei of the thalamus.

Entorhinal cortex Literally meaning 'the cortex within the rhinal fissure', but in many mammals the entorhinal cortex extends onto the dorsal bank of the rhinal fissure. The entorhinal cortex is characterised by an acellular layer 4Cx (the lamina dissecans), so that neurons are grouped into bands above and below this. The entorhinal cortex is connected with the hippocampal formation.

Eutherian A type of mammal where the placenta is a major organ of exchange for sustaining the developing young before birth. Often used interchangeably with the term 'placental mammal', but the latter term is less satisfactory because marsupial young are also supported by a placenta during prenatal development.

External capsule A shell of white matter external (outside) the putamen. In mammals with an obvious claustrum, the external capsule intervenes between the claustrum and putamen.

In monotremes it contains commissural axons destined for the anterior commissure.

External granular layer　A secondary proliferative zone that develops over the cerebellum during the lactational phase of monotreme development. It is derived from the rhombic lip and gives rise to many smaller neurons of the cerebellum.

Fasciculus　A term derived from the Latin for 'little bundle'. It refers to a collection of axons that serve a common purpose, e.g. the fasciculus gracilis.

Foliation　Transversely running folds of the surface of the cerebellar cortex.

Fornix　An arching fibre bundle that joins the hippocampal formation with the septal area and hypothalamus. It is a key component of the Papez circuit of the limbic system.

Ganglion　In the strictest sense this means a collection of neuronal cell bodies outside the central nervous system, but the term is occasionally used with reference to special clusters of neurons inside the brain (e.g. basal ganglia). Ganglia may be sensory or motor (autonomic effector) in function.

Ganglionic eminences　Protrusions of the proliferative zone of the ventral surface of the lateral ventricle that produce neurons of the cortex, striatum and pallidum. The medial ganglionic eminence produces neurons for the globus pallidus and bed nuclei of the stria terminalis. The lateral ganglionic eminence produces neurons for the striatum, amygdala, olfactory tubercle and cerebral cortex.

Germinal trigone　A region of undifferentiated proliferative cells at the lateral rim of the fourth ventricle. It includes the rostral rhombic lip.

GL (Greatest length)　The longest dimension of an embryo or hatchling.

Glabrous　A type of skin that is normally without hair (e.g. palm, sole and digit pads).

Globus pallidus　Part of the dorsal pallidum of the basal ganglia. Its name comes from the Latin for 'pale globe'. The lateral globus pallidus receives input from the dorsal striatum and projects to the subthalamic nucleus.

Glutamate　An amino acid that serves as the most prevalent excitatory neurotransmitter in the vertebrate nervous system.

***Griffonia simplicifolia* isolectin B4**　A naturally occurring molecule derived from the plant *Griffonia* (or *Bandeiraea*) *simplicifolia*. It has high affinity for the α-D-galactose sugar on the cell surface of some neurons and glia. When the molecule is tagged with a marker enzyme like horseradish peroxidase, it can be used to label small calibre nociceptive axons (C fibres) in the dorsal horn of the spinal cord and the trigeminal sensory nuclei.

Hippocampus　More correctly known as the hippocampal formation. It is a part of the allocortex that has a three-layered structure and can be divided into the dentate gyrus, hippocampus proper and subiculum. It is critically important in consolidation of new declarative memories.

Homeothermy　The maintenance of a constant internal body temperature despite changes in the environmental temperature.

Homology　The fundamental similarity of structures (e.g. brain nuclei) in different organisms that is thought to be the result of their evolution from a precursor structure in a common ancestor.

Hypothalamus　The part of the forebrain that develops rostral to the prethalamus. It is divided into peduncular, terminal and acroterminal zones on the basis of gene expression patterns and has intimate connections with the endocrine system through the pituitary gland. It plays a central role in the maintenance of a constant internal body environment (homeostasis) and reproduction.

Immunoreactivity　The use of specific antibodies to label naturally occurring brain chemicals (e.g. calcium-bonding proteins, neurotransmitters, neurofilament proteins, enzymes) and identify neuronal bodies and axons where those chemicals occur. It allows the identification of functional pathways and the distinction between brain regions using chemoarchitecture.

Inferior olivary nuclear complex　A set of neuronal clusters in the ventral medulla oblongata that project axons to the cerebellum (olivocerebellar tract). They are derived from the rhombic lip during post-hatching development and play a critical role in motor learning. Dorsal, principal and ventral groups are identified in vertebrates.

Insula Part of the cerebral cortex that lies immediately dorsal to the rhinal fissure and on the outside of the putamen. It is transitional in structure between the six-layered isocortex and three-layered olfactory allocortex.

Internal capsule The sheet or bundle of axons that connects the cerebral cortex and deeper parts of the neuraxis (e.g. thalamus, striatum). It passes through the dorsal striatum and contains both ascending and descending pathways.

Ipsilateral Referring to the same side of the body. It is the opposite term to contralateral and is used to describe representations of body parts in, or on, the brain and/or the course of pathways.

Isocortex The part of the laminated pallium of the forebrain that has six layers at some stage of development. The region is broadly equivalent to the cortex described by the older, phylogenetically invalid term, 'neocortex'.

Lagomorphs Hares and rabbits.

Latency The period of delay between a stimulus and a neural response.

Lateral To the side, i.e. away from the midline.

Lentic ecosystems Relatively still terrestrial waters such as lakes and ponds.

Limbic system A group of structures lying around the rim or edge of the telencephalon. The name comes from the Latin for 'fringe' or 'hem'. The limbic system can be divided into two subsystems and their component pathways: one centred around the amygdala, the other around the hippocampus.

Lipotyphlids A group of insectivores including shrews, moles and hedgehogs.

Locus coeruleus A column of blue–black pigmented noradrenergic neurons located in the brainstem on the lateral floor of the most rostral fourth ventricle. Its name comes from the Latin for 'sky blue place'.

Lotic ecosystems Flowing terrestrial water, such as rivers and streams.

Macrosmatic Referring to animals with large and specialised olfactory systems, e.g. the short- and long-beaked echidnas.

Mammillary body Small, paired elevations of the ventral surface of the hypothalamus. The mammillary bodies are part of the Papez circuit involved in consolidating declarative memory.

Mammillothalamic tract A bundle of axons that arises from neurons in the mammillary bodies of the hypothalamus. These axons reach the anterior nucleus of the thalamus and are part of the Papez circuit.

Mantle layer The region outside the proliferative zone of the developing nervous system, where young neurons settle and begin to differentiate.

Manus The distal end of the forelimb, equivalent to the human hand.

Marsupial Literally meaning a type of mammal in which the young are born in an immature state and are protected and nourished during lactation in a pouch (marsupium) or a depression on the maternal belly, but the term is not entirely satisfactory because young echidnas, which are monotremes, also develop in a maternal pouch. However, the most distinctive anatomical features of marsupials pertain to the presence of lateral vaginae and the relative positions of the genitourinary passages in both sexes. The term 'metatherian' may also be used.

Medial forebrain bundle A diffuse bundle of axons that runs throughout the length of the forebrain. It is identifiable from the late incubation subphase of life in monotremes and reciprocally connects ventral telencephalic structures with the caudal hypothalamus.

Megalibgwilia A genus of extinct, large-bodied, long-beaked echidnas that lived in Greater Australia during the Pleistocene.

Meninges The three layers of membranes that cover the brain. From outside to inside these are the dura mater, arachnoid and pia mater.

Meroblastic ovum An egg that divides only in part. The term refers to a type of egg where cleavage division is confined to only the animal pole, due to the presence of a large amount of yolk, as seen in monotremes.

Mesencephalon Equivalent to the midbrain, a region of the neuraxis that lies between the diencephalon and the isthmus. The mesencephalon is divided into the tectum, periaqueductal grey and midbrain tegmentum.

Mesoderm The middle of the three layers of the embryonic germ disc. It gives rise to components of the circulatory, musculoskeletal and genitourinary systems as well as connective tissue structures throughout the body. Microglia of the brain and spinal cord are also probably derived from this layer.

Metatherian Marsupial.

Microsmatic Referring to animals with very limited olfactory capabilities, e.g. cetaceans.

Monotreme A type of mammal characterised by the development of the young in a leathery-skinned egg, followed by hatching in an immature state and dependence on maternal milk for nutritional support. The term 'monotreme' is a reference to the presence of a single opening for genitourinary and digestive tracts, but that can also be seen in some metatherians.

Mucous sensory gland A type of sensory receptor in the bill or beak of monotremes. It consists of a coiled secretory gland duct in the dermis and an innervated papillary portion at the dermal/epidermal junction. The gland duct continues upwards to open at the epidermal surface. Mucous sensory glands are part of the electrosensory system.

Müller cells Elongated glial cells of the retina, with processes extending to both the inner vitreal surface and the outer surface, where contact is made with pigment epithelium. Müller cells may dedifferentiate into multipotent progenitor cells to subsequently produce a variety of retinal cells.

Musculotopic organisation The arrangement of a group of motor neurons in a motor nucleus, column or cortical sheet, such that motor neurons controlling particular muscles are grouped together in a discrete part of the nucleus, column or cortical sheet, and separated from those supplying other muscles. Musculotopy is often a continuous mapping process, such that the body part representations retain a similar spatial relationship to each other as they do on the body itself.

Myrmecophage An animal whose diet is principally ants and termites.

NADPH diaphorase An abbreviation for reduced nicotinamide adenine dinucleotide phosphate diaphorase, an enzyme found in neurons that use nitric oxide as a neurotransmitter.

Neural crest Ectodermal cells that migrate from the rim or edge of the folding neural plate during embryonic life. They give rise to sensory and autonomic ganglion cells, Schwann cells and adrenal medullary cells, among others.

Neural plate A flattened tadpole-shaped region of the embryonic ectoderm that gives rise to components of the central and peripheral nervous systems.

Neural tube A tubular structure formed during embryogenesis by folding of the neural plate. Neurons, astrocytes and oligodendrocytes are derived from the dividing cells of the wall of the neural tube.

Neuraxis The sequence of the parts of the central nervous system laid out from the forebrain to spinal cord. The path of the neuraxis is curved by flexures at the spinomedullary junction, rhombencephalon, midbrain and rostral hypothalamus.

Neuroepithelium When used in neuroembryology, this term refers to the wall of the early neural tube. It is a region of rapidly dividing cells that will give rise to the neurons and most glia of the adult brain. The term may also be applied to specialised sensory epithelia of the adult, e.g. hair cells of the inner ear.

Neurofilament An assembly of structural proteins in the axon and cell body that help maintain the configuration and structural strength of neurons.

Neurogenesis The process whereby immature neurons are produced by cell division in the proliferative zones of the developing brain and spinal cord (e.g. ventricular and subventricular zones, and external granular layer of the cerebellum).

Neuromeres Developmental segments (prosomeres, mesomeres, rhombomeres) of the neural tube that give rise to regions of the adult nervous system (diencephalon, midbrain and hindbrain, respectively). Boundaries between neuromeres are identified and defined on the basis of gene expression patterns.

Nissl stain A type of acidic histological stain that binds the rough endoplasmic reticulum (protein manufacturing organelle) of the neuron cell body. It is useful in analysing the cytoarchitecture of the brain.

Nociception The perception of painful stimuli, usually the result of the action of damaging agents on the skin or internal body structures.

Noradrenaline   Also called norepinephrine. A catecholamine neurotransmitter used by neurons of the locus coeruleus and postganglionic neurons of the sympathetic nervous system.

Notochord   A rod-like structure of mesodermal origin that underlies the developing nervous system and induces formation of the embryonic neural tube.

Nucleus   May refer to: (1) the cellular organelle that contains the DNA and chromosomes; or (2) a group of neurons in the central nervous system that serve a similar function.

Nucleus accumbens   Part of the ventral striatum located around the ventral angle of the lateral ventricle. It plays a central role in the pathways that reward and reinforce behaviour.

Olfactory bulb   The rostral parts of the laminated forebrain that are concerned with the processing of sensory input from the olfactory epithelium of the nasal cavity. They project to more caudal parts of the olfactory system (e.g. the anterior olfactory area, olfactory tubercle and piriform cortex).

Olfactory tubercle   A partially layered region of the ventral telencephalon. It receives olfactory input and is considered part of the ventral striatum.

Os caruncle   The bony core of the prominence on the dorsal tip of the snout that forms during incubation phase development in monotremes and is used to break through the embryonic membranes at hatching.

Oviparous   Giving birth to young in an egg phase.

Oxytocin   A hormone used in mammals to drive milk ejection and contraction of the smooth muscle of the reproductive tract. It is also important in central pathways involved in mating and maternal behaviour.

Pacinian corpuscle   A rapidly adapting, vibration-sensitive mechanoreceptor found in the dermis of the skin, mesenteries of the gut, membranes around bones and tissue around muscle fibres. Each is approximately 1 mm in length and has many concentric layers like the skins of an onion. A single, large, myelinated nerve fibre supplies each corpuscle.

Pallidum   Part of the basal forebrain. The dorsal pallidum is the globus pallidus/entopeduncular nucleus and the ventral pallidum is part of the substantia innominata of the ventral forebrain.

Pallium   The most dorsal and lateral part of the forebrain. It is mainly formed into a sheet-like structure known as the cerebral cortex, but may include some nuclear elements, e.g. claustrum, endopiriform nucleus and parts of the amygdala.

Panniculus carnosus   A sheet of skeletal muscle attached to the skin in the platypus and echidnas. It is thicker over the dorsum than the belly. It allows the monotremes to achieve striking changes in body shape, including rolling into a ball in the case of the short-beaked echidna.

Paraphyletic   A group of organisms that contains all the descendants of the last common ancestor of a group's members minus a small number of monophyletic groups of descendants.

Parvalbumin   A calcium-binding protein that is often concentrated in the neurons of the visual, somatosensory and auditory pathways. Parvalbumin buffers calcium in highly active neurons that use GABA as a neurotransmitter.

Periallocortex   A region of transitional cortex situated between the six-layered isocortex and the three- to five-layered allocortex. It is found around the rhinal fissure (insula) and the medial isocortical border.

Photopic vision   Vision under strong light conditions. In many primates it allows colour and high acuity vision and is dependent on cone photoreceptors.

Photoreceptor   A light-sensitive cell in the external part of the retina. The two kinds of photoreceptors are the rods and cones.

Pineal gland   A gland located dorsal to the caudal third ventricle. It may be included in the epithalamus of the diencephalon and plays a central role in the control of circadian rhythms and the seasonal regulation of reproduction.

Pituitary gland   The so-called master gland of the endocrine system. It has anterior (adenohypophysis) and posterior (neurohypophysis) components and is under direct control of the hypothalamus.

Placental   A type of mammal where the placenta is a major organ of exchange for sustaining the developing young before birth. It is used for convenience in this book, but is not an entirely satisfactory term because some marsupials also make extensive use of a placenta during development. See also eutherian.

Plesiomorphic Sharing a character state with an ancestral clade; primitive.

Pons A region of the hindbrain defined by the presence of the pontine nuclei that project to the cerebrocerebellum. The pons has no defined counterpart within the segments of the embryonic brain and its adult extent varies between mammals depending on the size of the cerebrocerebellar system.

Pouch Or marsupium. An enclosed space with mammary glands on the belly of a female mammal that is used to protect and nourish immature young. Most familiar with respect to metatherians (marsupials), but the echidnas also have a pouch-like abdominal depression that they use to protect and carry their immature young.

Prosencephalon The part of the embryonic brain that gives rise to the forebrain. It is divided into three prosomeres that form the diencephalon, plus unevaginated and evaginated parts of the telencephalon that give rise to the hypothalamus, septum, striatum, pallidum and pallium (cerebral cortex).

Prosomeres Three segments of the neural tube (p1, p2, p3 from caudal to rostral) that give rise respectively to the pretectum, (dorsal) thalamus and prethalamus (ventral thalamus).

Puggle An early hatchling or juvenile platypus or echidna.

Push-rod mechanoreceptor A type of sensory apparatus in the beak or bill of monotremes. Each consists of a rod of tightly flattened spinous cells separated from the surrounding epidermis by a zone of loosely packed cells. The mobile push rod is associated with four types of nerve endings: two kinds of vesicle chain receptors, Merkel cell complexes and lamellated corpuscles.

Rapidly adapting unit Sensory receptors that respond best to high frequency oscillation of the stimulus (e.g. Paciniform corpuscles). They respond only briefly to the onset and offset of a sustained stimulus.

Receptive field The region of skin or visual field in which a stimulus elicits a response from a given sensory neuron.

Remak bundle Clusters of C fibre axons surrounded by a Schwann cell without myelination.

Retinal ganglion cell The output neurons from the retina. Their axons reach the hypothalamus, tectum, pretectum and dorsal thalamus.

Retinofugal Literally 'fleeing the retina'. Refers to retinal ganglion cell axons that leave the retina.

Reversal learning tasks An operant technique where an animal is first trained to make one type of discrimination (e.g. choosing a white object), and is then required to learn the reverse discrimination (e.g. choosing a black object). It is a test of behavioural flexibility.

Rhombencephalon The component of the neural tube that gives rise to the hindbrain. It is divided into segments (rhombomeres) during embryonic development.

Rhombic lip The rim of the embryonic rhombencephalon. It gives rise to four migratory streams that provide small neurons of the cerebellum and many nuclei of the hindbrain (e.g. cochlear, trigeminal and precerebellar nuclei).

Rhombomeres Rostrocaudal segments of the embryonic neural tube that give rise to the regions and nuclei of the hindbrain.

Rostral Literally meaning 'towards the beak'. A direction towards the front end of the central nervous system.

Sauropsida A group of amniotes that includes all reptiles and birds and their fossil ancestors.

Scotopic vision Vision under very low light levels, usually dependent on rod photoreceptors.

Septum The region of forebrain medial to the rostral end of the lateral ventricle. The two sides of the septum (also called the septal area) are fused across the midline to a variable extent in mammals. The septum is most closely linked functionally with the limbic system and hypothalamus.

Serotonin 5-hydroxytryptamine. A neurotransmitter used by the raphe nuclei of the midbrain and hindbrain to modulate function throughout the forebrain, brainstem and spinal cord.

Serous sensory gland A type of sensory gland in the bill or beak of monotremes. It is similar in structure to the mucous sensory gland, but is smaller and has fewer axon terminals. It may be part of the electrosensory system.

Slowly adapting unit   A sensory receptor that gives a sustained response to a stimulus that is maintained for a prolonged period, e.g. a second.

Somite   A segment of the paraxial mesoderm of the body of the embryo. It is divided into dermatome, myotome and sclerotome components, giving rise to dermal, muscular and skeletal elements.

Spinocerebellum   Part of the cerebellum (anatomically the midline and adjacent cerebellar cortex and deep interposed nucleus) that uses proprioceptive information from the limbs (joint position and muscle tension) to coordinate proximal limb movements and postural adjustments.

Striatum   Part of the basal forebrain. It is usually divided into the dorsal striatum (caudate/putamen) and the ventral striatum (nucleus accumbens and parts of the olfactory tubercle).

Synapomorphy   A shared derived trait found among two or more species and their most recent common ancestor.

Tectum   The roof or dorsal part of the midbrain. It includes the superior and inferior colliculi.

Tegmentum   The central part of the midbrain ventral to the tectum and periaqueductal grey (midbrain tegmentum), or the central part of the pons and medulla (pontine and medullary tegmentum).

Thalamus   The part of the diencephalon that provides a relay centre between the caudal neuraxis or special sense organs and the cerebral cortex. It is derived from prosomere 2 of the embryonic brain. It is also known as the dorsal thalamus.

Therian   A collective term for marsupial and placental mammals. Monotremes are regarded as the sister group of therians (van Rheede *et al.* 2006).

Threshold   In a physiological context, this is the smallest sensory stimulus that will elicit a response from a sensory receptor.

Trigeminal nerve   The major somatosensory nerve of the vertebrate head. It carries electrosensory information from the bill or beak of monotremes into the brainstem. It is named because of its three divisions (ophthalmic, maxillary and mandibular).

Ventral   Literally meaning 'towards the belly'. A direction towards the underside of the body.

Vestibulocerebellum   Part of the cerebellum (anatomically the flocculus, nodule and fastigial nucleus) involved in processing information on the orientation and angular acceleration of the head to coordinate activity of the extraocular and axial muscles.

Zaglossus   A genus of long-beaked echidna found in the central cordillera of the island of New Guinea (the Vogelkop peninsula of Irian Jaya to Mt Simpson in eastern Papua). Several species and/or subspecies are recognised: *bartoni, diamondi, bruijnii*.

Zygote   The fertilised egg (ovum) before cleavage has begun.

APPENDIX TABLES AND FIGURES

Introduction

In this section, supplementary information on the development of monotremes and the morphology of the central nervous system is collected in tables. Data in the tables are derived from measurements of archived specimens (eggs, embryos, hatchlings, and adult brains and spinal cords) held at the Museum für Naturkunde in Berlin (MfN), the National Museum of Australia in Canberra (NMA) and the Australian Museum in Sydney (AustMus). The National Museum of Australia does not permit publication of photographs of their specimens, so it is not possible to show images of their eggs, hatchlings, brains and spinal cords. I am very grateful to both the Museum für Naturkunde and the Australian Museum for the opportunity to use photographs of the specimens in their collection elsewhere in this book.

The developmental tables summarise the key events in the main neural systems of evolutionary interest for monotreme neuroscience. These events have been plotted against changes in greatest length of embryos and hatchlings and the presumed age based on a system that counts the days before hatching (H-x days) and after hatching (PH for post-hatching, x days) (see Chapter 3). The ages of embryos and hatchlings are, of necessity, approximate, because of the uncertainty of the timetable of monotreme development, particularly during the intrauterine period, but also with respect to the tempo of development during the incubation phase. The events in neural development have been grouped according to three zones of the neuraxis: the telencephalon (Appendix Fig. 1); the eye and diencephalon (Appendix Fig. 2); and the brainstem, spinal cord and peripheral nervous system (Appendix Fig. 3).

Appendix Table 1. Diameters of platypus eggs

Specimen number	Museum[1]	Species	Long axis diameter (mm)	Short axis diameter (mm)
1984.0010.0474	NMA	O. anatinus	12.6	11.9
1984.0010.0475a	NMA	O. anatinus	13.3	11.9
1984.0010.0475b	NMA	O. anatinus	13.2	12.1
1984.0010.0842a	NMA	O. anatinus	17.6	16.1
1984.0010.0842b	NMA	O. anatinus	17.8	15.4
1984.0010.2280a	NMA	O. anatinus	18.1	16.7
1984.0010.2280b	NMA	O. anatinus	17.6	17.1
Mean ± s.d.			15.74 ± 2.55	14.46 ± 2.39

1 NMA – National Museum of Australia.

Appendix Table 2. Dimensions of platypus hatchlings

Specimen number	Museum[1]	Age (days PH)[2]	Mass (g)	GL (mm)[3]	Head length (mm)	DCL (mm)[4]	Bill length (mm)	Max bill width (mm)	Tail length (mm)
1984.0010.2204b	NMA	0	na	11.7	3.8	17.6	1.2	na	na
1984.0010.2204c	NMA	0	na	11.9	4.6	19.0	1.2	na	na
1984.0010.2204a	NMA	0	na	12.8	4.1	19.1	na	na	na
1984.0010.0842	NMA	1	na	16.1	7.2	34.5	1.2	na	4.7
1984.0010.0667b	NMA	2	na	18.5	7.3	36.9	1.4	2.1	5.0
1984.0010.0667a	NMA	2	na	19.3	7.3	37.6	1.4	na	4.1
1984.0010.2349a	NMA	5	na	24.7	11.4	54.2	1.8	na	5.5
1984.0010.0478a	NMA	4	na	25.0	8.8	46.1	1.5	na	5.2
M5017	AustMus	4?	3	25.7	10.6	49.5	2.9	4.3	4.6
1984.0010.0478b	NMA	4	na	26.7	9.7	45.7	1.8	na	4.4
1984.0010.2349b	NMA	5	na	26.9	11.9	56.1	2.2	na	6.7
1984.0010.2349c	NMA	5	na	28.5	11.0	56.3	2.2	na	5.7
1984.0010.0479	NMA	5	na	30.0	12.4	64.6	1.9	na	5.8
1984.0010.0480a	NMA	6	na	32.9	14.5	73.7	2.8	na	7.6
1984.0010.0480b	NMA	6	na	35.3	14.4	71.1	2.8	na	8.5
1984.0010.0668b	NMA	7	na	40.2	19.9	104.4	5.4	6.4	12.0
1984.0010.0668a	NMA	7	na	42.2	21.3	106.9	5.0	6.5	14.7
1984.0010.0481a	NMA	10	na	44.8	21.5	117.2	4.4	na	11.5
1984.0010.0481c	NMA	10	na	47.6	19.5	108.9	3.4	na	11.9
1984.0010.0481b	NMA	10	na	47.9	21.5	114.5	4.0	na	10.9
1984.0010.0669b	NMA	12	na	63.6	31.7	157.9	10.1	10.8	21.5
1984.0010.2277	NMA	14	na	67.6	28.3	149.9	8.6	9.4	17.1
1984.0010.0669a	NMA	12	na	69.7	28.1	152.9	7.7	10.2	19.7
1984.0010.0482	NMA	24	na	69.8	37.8	191.1	11.4	11.8	22.6
1984.0010.0483	NMA	28	na	73.4	35.9	212.9	11.4	na	23.6
1984.0010.0831	NMA	35?	na	82.5	38.7	228.9	13.5	16.8	26.8
1984.0010.0331	NMA	42	na	87.2	38.9	219.0	14.3	13.1	30.6
1984.0010.2348	NMA	42	na	90.3	40.6	231.5	14.8	14.9	29.5
M2783	AustMus	44?	170	91.7	45.3	244	16.0	17.2	37.6
M2781	AustMus	45?	199	97.2	47.9	261	20.7	19.6	45.4

(Continued)

Appendix Table 2. (Continued)

Specimen number	Museum[1]	Age (days PH)[2]	Mass (g)	GL (mm)[3]	Head length (mm)	DCL (mm)[4]	Bill length (mm)	Max bill width (mm)	Tail length (mm)
M2780	AustMus	46?	250	100.6	49.1	286	21.3	20.1	49.6
M2782	AustMus	48?	210	105	52.5	278	21.4	20.4	47.2
1984.0010.0670	NMA	49	na	107.9	34.0	185.7	13.7	17.6	45.5
1984.0010.0330	NMA	42	na	110.3	37.4	220.5	12.4	13.7	26.9
M2779	AustMus	70?	375	130.3	61.1	337	24.9	23.2	56.7
1984.0010.2276	NMA	98	na	152.2	56.2	287.7	23.7	20.3	52.5
1984.0010.0333	NMA	98	na	na	57.9	235.9	25.9	29.5	59.3
1984.0010.1280	NMA	120+?	na	184.7	96.8	416.7	34.9	40.8	na

1 NMA – National Museum of Australia. AustMus – Australian Museum.
2 No precise ages are available for those specimens marked with '?'.
3 GL – greatest length.
4 DCL – dorsal contour length.
na – not available.

Appendix Table 3. Dimensions of short-beaked echidna hatchlings

Specimen number	Museum[1]	Age (days PH)[2]	Mass (g)	GL[3] (mm)	Head length (mm)	DCL[4] (mm)	Beak length (mm)	Max beak width (mm)	Tail length (mm)
M5014	AustMus	4–6?	2.0	24.8	12.0	47.5	3.4	3.5	2.9
1984.0010.0768	NMA	4	na	28.2	8.8	na	1.4	na	3.6
MO78	MfN Berlin	6–8?	11.6	38.6	19.9	92.0	6.7	5.5	8.9
MO191	MfN Berlin	6–8?	11.0	39.9	na	108.0	6.4	5.5	6.6
M2788	AustMus	6–8?	11.0	41.1	23.6	93.5	6.3	5.8	8.6
MO81	MfN Berlin	8–10?	25.3	47.6	26.2	122.0	7.6	6.4	11.8
MO88	MfN Berlin	8–10?	28.7	48.4	27.1	132.0	7.8	6.4	11.7
1984.0010.0716	NMA	9	na	55.3	19.6	na	7.0	6.7	10.4
MO190	MfN Berlin	15?	43.1	55.8	31.9	140.0	8.1	6.7	9.9
M4206	AustMus	25?	61.0	69.2	na	136.0	na	na	12.3
M32751	AustMus	25?	62.0	74.1	33.9	166.0	10.1	7.2	13.5
M1338	AustMus	30?	96.0	76.7	39.6	166.0	11.6	8.8	17.3
MO85	MfN Berlin	30?	116.9	80.6	41.5	195.0	10.3	8.0	18.1
M22581	AustMus	30?	109.0	85.2	41.1	184.0	10.8	8.7	12.5
MO84	MfN Berlin	40?	143.3	90.5	41.7	222.0	10.4	8.4	19.6
M2165	AustMus	50?	255.0	103.0	54.6	252.0	14.9	9.5	24.0
M25916	AustMus	50?	132.0	103.8	52.2	210.0	15.8	8.9	21.9
M22580	AustMus	60?	489.0	114.0	62.7	301.0	18.8	10.5	25.8
M14096	AustMus	60?	251.0	140.6	57.3	231.0	15.8	10.2	23.1
M6814	AustMus	60?	442.0	144.0	64.2	264.0	15.1	9.9	29.7

1 NMA – National Museum of Australia; MfN Berlin – Museum für Naturkunde, Berlin; AustMus – Australian Museum.
2 No precise ages are available for those specimens marked with '?'. Those ages are estimates based on known body weights and with reference to Rismiller and McKelvey (2003).
3 GL – greatest length of embryo or hatchling.
4 DCL – dorsal contour length of embryo or hatchling.
na – not available.

Appendix Table 4. Dimensions of platypus brain and spinal cord

Specimen number	Museum[1]	Telen length[2] (mm)	Telen width[2] (mm)	AC area[3] (mm²)	OB ventral surface area[4] (mm²)	Cb width[5] (mm)	Cb length[5] (mm)	Cb height[5] (mm)	Cb vermal area[5] (mm²)	Brainstem width (mm)	Brainstem length (mm)	BrachSpC width[6] (mm)
1984.0010.0211	NMA	22.6	29.4	na	23.4	19.6	8.0	10.3	na	15.8	16.2	4.2
1984.0010.0212	NMA	23.3	31.8	3.8	na	18.1	10.9	10.6	66.5	na	15.3	na
1984.0010.0214	NMA	na	na	na	23.8	20.2	14.0	na	na	15.7	15.4	na
1984.0010.0215	NMA	na	na	na	30.4	28.1	na	na	na	21.9	20.5	na
1984.0010.0459	NMA	28.5	na	7.1	na	na	13.4	11.1	106.5	na	18.9	na
1984.0010.0468	NMA	28.4	na	5.4	na	na	9.6	10.9	86.9	na	15.6	na
Mean		25.7	30.6	5.4	25.9	21.5	11.2	10.7	86.6	17.8	17.0	4.2
s.d.		3.2	1.7	1.7	3.9	4.5	2.5	0.4	20.0	3.6	2.2	na

1 NMA – National Museum of Australia.
2 Telen – Telencephalon length and width.
3 AC – area of anterior commissure in mm² in a mid sagittal section.
4 OB – ventral surface area of both olfactory bulbs in mm² (bulk of the contact area for olfactory nerve fibres).
5 Cb – cerebellum width, length, height and vermal area.
6 BrachSpC – brachial spinal cord.
na – not available.

Appendix Table 5. Dimensions of short-beaked echidna brain and spinal cord

Specimen number	Museum[1]	Telen length[2] (mm)	Telen width[2] (mm)	AC area[3] (mm²)	OB ventral surface area[4] (mm²)	Cb width[5] (mm)	Cb length[5] (mm)	Cb height[5] (mm)	Cb vermal area[5] (mm²)	Brainstem width (mm)	Brainstem length (mm)	BrachSpC width[6] (mm)
1984.0010.0455	NMA	31.7	39.7	6.3	161.1	20.7	13.6	na	104.4	12.1	21.9	na
1984.0010.0728	NMA	31.6	na	na	172.2	na	13.5	na	na	15.4	20.7	7.2
1984.0010.1224	NMA	29.9	36.0	6.5	na	18.2	12.1	14.3	70.8	na	17.7	5.8
1984.0010.1484	NMA	32.4	na	na	197.1	na	na	18.8	na	14.6	20.8	7.3
1984.0010.1488	NMA	33.3	na	na	190.0	22.9	na	na	na	14.9	21.3	na
1984.0010.1490	NMA	37.8	47.1	na	227.7	27.4	na	na	na	14.9	22.1	7.8
1984.0010.1491	NMA	31.9	43.2	na	235.9	25.2	14.2	16.2	na	15.6	21.2	7.8
1984.0010.1492	NMA	36.7	39.1	na	233.2	25.9	13.4	17.3	na	13.1	19.8	7.1
1984.0010.1493	NMA	40.1	44.2	na	303.3	29.1	na	17.6	na	14.8	24.0	na
1984.0010.1498a	NMA	35.0	na	na	na	na	12.7	na	94.1	na	19.8	na
1984.0010.1498b	NMA	34.5	na	12.6	na	na	13.6	na	100.7	na	na	na
1984.0010.1515	NMA	33.9	na	8.9	na	na	13.4	10.3	97.6	na	17.0	5.4
1984.0010.1591	NMA	35.5	na	na	na	na	na	15.4	na	na	na	na
1984.0010.1594	NMA	31.9	37.0	na	na	18.8	12.2	na	na	na	na	na
Mean		34.0	40.9	8.6	215.1	23.5	13.2	15.7	93.5	14.4	20.6	6.9
s.d.		2.8	4.1	2.9	45.4	4.0	0.7	2.8	13.3	1.2	2.0	0.9

1 NMA – National Museum of Australia.
2 Telen – Telencephalon length and width.
3 AC – area of anterior commissure in mm² in a mid sagittal section.
4 OB – olfactory bulb ventral surface area in mm² (contact area for olfactory nerve fibres).
5 Cb – cerebellum width, length, height and vermal area.
6 BrachSpC – brachial spinal cord.

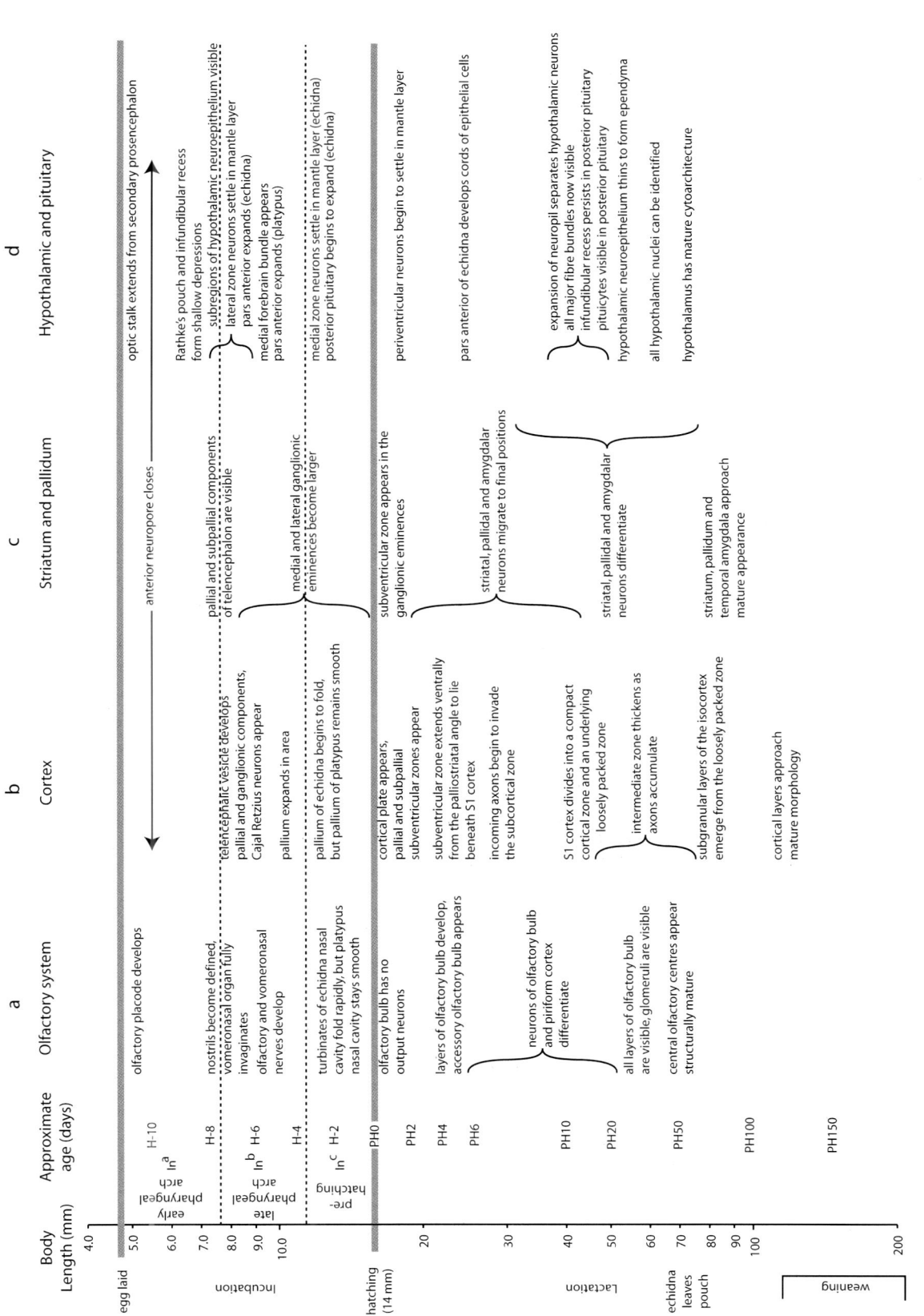

APPENDIX TABLES AND FIGURES 511

Body Length (mm)	Approximate age (days)	a Olfactory system	b Cortex	c Striatum and pallidum	d Hypothalamic and pituitary
4.0	egg laid				
5.0	H-10 (early pharyngeal arch) [In a]	olfactory placode develops		← anterior neuropore closes →	optic stalk extends from secondary prosencephalon
6.0					Rathke's pouch and infundibular recess form shallow depressions
7.0	H-8	nostrils become defined, vomeronasal organ fully invaginates	telencephalic vesicle develops	pallial and subpallial components of telencephalon are visible	subregions of hypothalamic neuroepithelium visible
8.0	H-6 (late pharyngeal arch) [In b]	olfactory and vomeronasal nerves develop	pallial and ganglionic components, Cajal Retzius neurons appear		lateral zone neurons settle in mantle layer; pars anterior expands (echidna); medial forebrain bundle appears; pars anterior expands (platypus)
9.0			pallium expands in area		
10.0	H-4			medial and lateral ganglionic eminences become larger	
	H-2 (pre-hatching) [In c]	turbinates of echidna nasal cavity fold rapidly, but platypus nasal cavity stays smooth	pallium of echidna begins to fold, but pallium of platypus remains smooth		medial zone neurons settle in mantle layer (echidna); posterior pituitary begins to expand (echidna)
	PH0 hatching (14 mm)	olfactory bulb has no output neurons	cortical plate appears, pallial and subpallial subventricular zones appear		
20	PH2			subventricular zone appears in the ganglionic eminences	
	PH4	layers of olfactory bulb develop, accessory olfactory bulb appears	subventricular zone extends ventrally from the palliostriatal angle to lie beneath S1 cortex		periventricular neurons begin to settle in mantle layer
	PH6		incoming axons begin to invade the subcortical zone		pars anterior of echidna develops cords of epithelial cells
30		neurons of olfactory bulb and piriform cortex differentiate		striatal, pallidal and amygdalar neurons migrate to final positions	
40	PH10		S1 cortex divides into a compact cortical zone and an underlying loosely packed zone		expansion of neuropil separates hypothalamic neurons; all major fibre bundles now visible; infundibular recess persists in posterior pituitary; pituicytes visible in posterior pituitary; hypothalamic neuroepithelium thins to form ependyma
50	PH20 lactation	all layers of olfactory bulb are visible, glomeruli are visible	intermediate zone thickens as axons accumulate	striatal, pallidal and amygdalar neurons differentiate	
60					
70	PH50 echidna leaves pouch	central olfactory centres appear structurally mature	subgranular layers of the isocortex emerge from the loosely packed zone	striatum, pallidum and temporal amygdala approach mature appearance	all hypothalamic nuclei can be identified
80					
90	PH100		cortical layers approach mature morphology		hypothalamus has mature cytoarchitecture
100					
200	PH150 weaning				

Appendix Fig. 1: Time-line diagram summarising the sequence of developmental events in the monotreme olfactory system (a), cerebral cortex (b), striatum and pallidum (c) and hypothalamus and pituitary (d). Details in (a) to (d) are based on a figure from Ashwell (2012b) (with permission from John Wiley and Sons), a figure from Ashwell and Hardman (2012a) (with permission from S. Karger AG, Basel), observations in Chapter 6 (not previously published), and observations in Ashwell (2012e), respectively.

512 NEUROBIOLOGY OF MONOTREMES

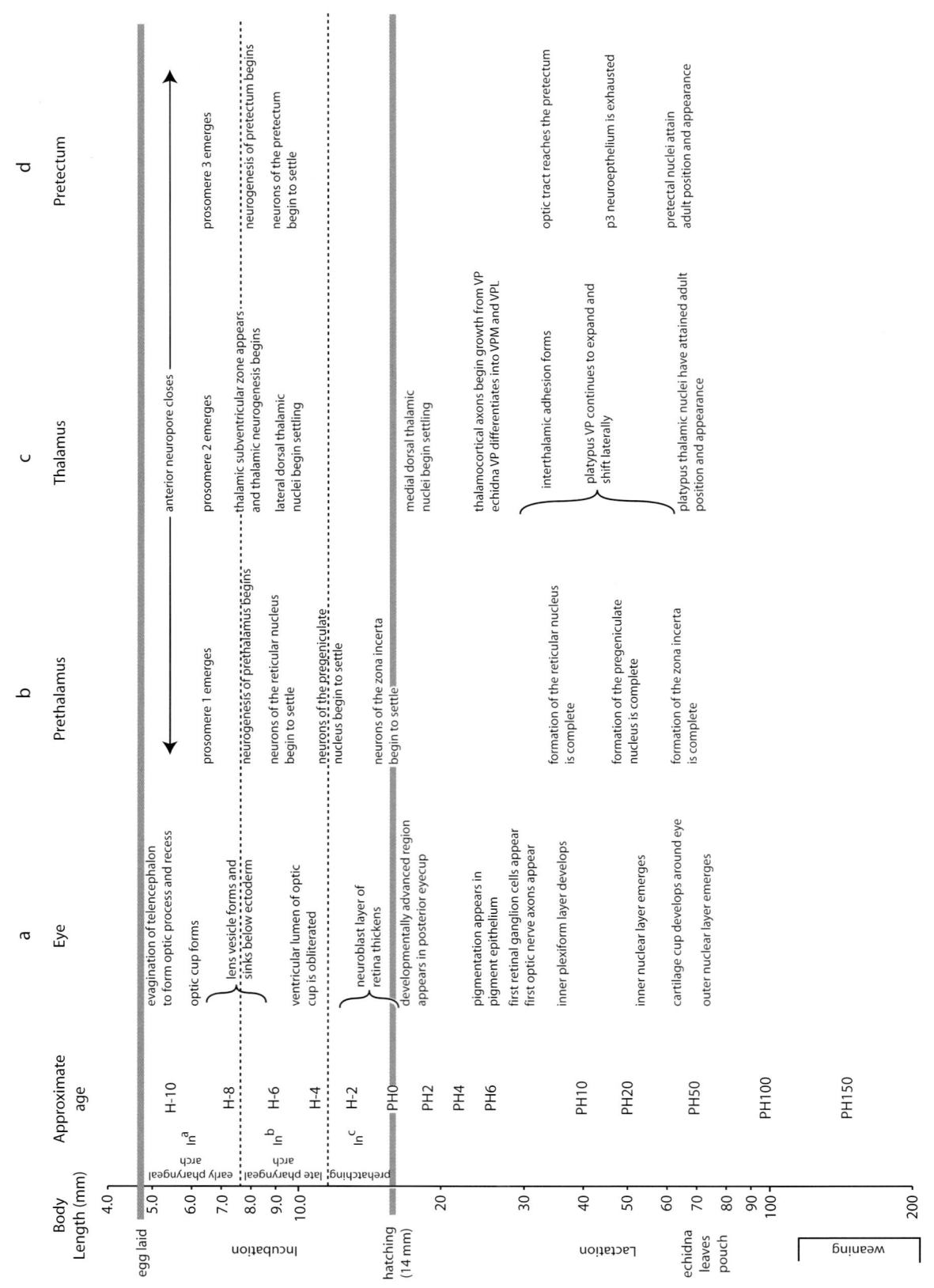

Body Length (mm)	Approximate age	a Eye	b Prethalamus	c Thalamus	d Pretectum
4.0			← anterior neuropore closes →		
5.0 (egg laid)	H-10	evaluation of telencephalon to form optic process and recess			
6.0		optic cup forms	prosomere 1 emerges	prosomere 2 emerges	prosomere 3 emerges
7.0 (Incubation, early pharyngeal arch)	H-8	lens vesicle forms and sinks below ectoderm	neurogenesis of prethalamus begins	thalamic subventricular zone appears and thalamic neurogenesis begins	neurogenesis of pretectum begins
8.0					
9.0 (Incubation, late pharyngeal arch)	H-6	ventricular lumen of optic cup is obliterated	neurons of the reticular nucleus begin to settle	lateral dorsal thalamic nuclei begin settling	neurons of the pretectum begin to settle
10.0	H-4		neurons of the pregeniculate nucleus begin to settle		
	H-2	neuroblast layer of retina thickens	neurons of the zona incerta begin to settle		
hatching (14 mm)	PH0	developmentally advanced region appears in posterior eyecup		medial dorsal thalamic nuclei begin settling	
	PH2				
	PH4	pigmentation appears in pigment epithelium			
20	PH6	first retinal ganglion cells appear; first optic nerve axons appear		thalamocortical axons begin growth from VP echidna VP differentiates into VPM and VPL	
30					
40	PH10	inner plexiform layer develops	formation of the reticular nucleus is complete	interthalamic adhesion forms	optic tract reaches the pretectum
50 (Lactation)	PH20	inner nuclear layer emerges	formation of the pregeniculate nucleus is complete	platypus VP continues to expand and shift laterally	p3 neuroepithelium is exhausted
60	PH50	cartilage cup develops around eye	formation of the zona incerta is complete	platypus thalamic nuclei have attained adult position and appearance	pretectal nuclei attain adult position and appearance
70 (echidna leaves pouch)		outer nuclear layer emerges			
80					
90	PH100				
100					
200 (weaning)	PH150				

Appendix Fig. 2: Time-line diagram summarising the sequence of developmental events in the monotreme eye (a), prethalamus (b), thalamus (c) and pretectum (d). Details are based on observations in Chapter 8 (not previously published) for the eye (a), and on findings in Ashwell (2012c) for (b) to (d).

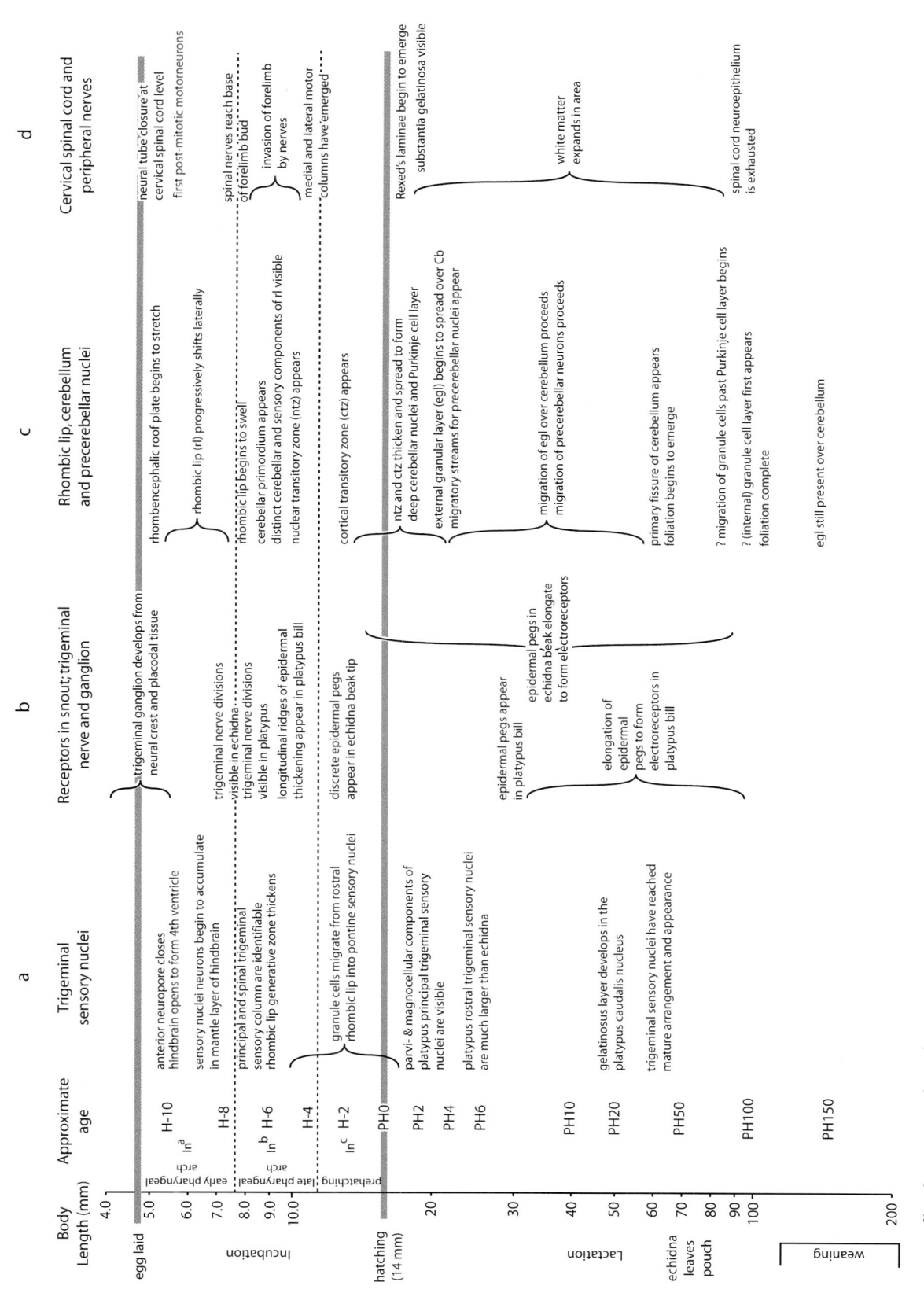

Appendix Fig. 3: Time-line diagram summarising the sequence of developmental events in the monotreme trigeminal sensory nuclei (a), snout receptors and trigeminal ganglion (b), cerebellar and precerebellar structures (c) and cervical spinal cord and peripheral nervous system (d). Details are based on findings from Ashwell and Hardman (2012b) for the trigeminal sensory nuclei (a), Ashwell et al. (2012) for trigeminal receptors and trigeminal ganglion (b), Ashwell (2012f) for the cerebellum (c) and Ashwell (2012d) for the spinal cord and peripheral nervous system (d).

INDEX

accommodation 491
acetylcholine 58, 87, 254, 261, 491
    and thermoregulation 254
acetylcholinesterase 491
acoustico-lateralis system 181
adenohypophysis 252
adrenaline 87, 253, 491
afferents
    general somatic 49, 85
    general visceral 49, 85
    special somatic 85
    special visceral 85
    visceral 268
alar plate 42, 49, 51, 53, 57, 81, 90, 231, 448
allocortex 132, 141, 154, 280, 491
    development 39, 57
α-casein 258
α-lactalbumin 258
altricial 2, 104, 257, 491
amygdala 491
    and hypothalamus 123
    connections 89, 111, 142
    cortical 123, 245
    development 129
    extended 123
    function 123
    in olfaction 235
    in sleep 278
    structure 107, 123, 126
angular acceleration 221, 499
anterior commissure 257, 260, 281, 491
anterior neuropore 172
    closure 38, 90, 231
anterior olfactory area 61, 237, 243, 247, 291, 497
antidiuretic hormone 252, 259, 491
apomorphy 491
aquatic adaptation 3, 16, 177
archicortex 57, 132, 491
area
    preoptic 43, 51, 57, 85, 95, 108, 120, 129, 259, 266, 277, 281
        ventral tegmental 43, 53, 88, 120, 142, 493
areola 31, 39, 46, 81, 94, 205, 257, 264, 491
arginine vasopressin 259
astrocytes
    fibrous 59
    protoplasmic 59
auditory brainstem responses 227
Ausktribosphenids 13
Australasian ecozone 20, 491

Australosphenida 1, 13
autapomorphy, definition 491
autonomic control of platypus bill 267
autonomic function 58
autonomic nervous system 267
axon 491

basal metabolic rate 20, 28, 315, 491
basal plate 79, 82, 90, 231
behavioural reinforcement 87, 130
behavioural thermoregulation 251, 254
Bergmann glia 98
β-casein 258
β-lactoglobulin 258
bill development 201
blastomeres 37
blood-brain barrier 59
blood-CSF barrier 59
BMR 20, 491
body posture 74, 76, 77, 276
body temperature 2, 21, 153, 208, 276, 280, 286, 315, 494
Bowman's gland development 246
brachial plexus 305, 308, 309, 311
brachium of the superior colliculus 53, 115, 169
brain-derived neurotrophic factor 58
brain size 19, 28, 47, 62, 67, 493
    and longevity 29
brainstem
    at hatching 94
    development 90
branchiomeric 491
breeding 17, 23, 27, 31, 35, 194, 253, 255
burrows 4, 12, 17, 21, 25, 27, 32, 35, 40, 162, 187, 206, 226

C fibre 74, 196, 204, 208, 492
calbindin, definition 491
calcitonin gene-related peptide 188, 269
Caldwell WH 32, 316
calretinin, definition 492
canal, external auditory 219, 222
casein gene cluster 257
casein variants 257
Castorocauda 12
cauda equina 48, 74, 76, 492
caudatoputamen 58, 97, 125
cell
    amacrine 163, 166, 167, 177
    basket 101, 133, 150
    bipolar, of retina 163, 166

    bitufted 152
    Cajal-Retzius 151, 157
    deep short axon 241
    ependymal 59, 258
    Golgi 98
    horizontal 245
    Lugaro 98
    mitral 63, 237, 241, 292
    mitral cell development 247
    Müller 166, 496
    photosensitive retinal ganglion 255
    Purkinje 46, 98, 100, 104, 491
    radial glia 166
    retinal ganglion 166, 167
    satellite 241
    tufted 237, 247
    tufted cell development 247
    unipolar brush 98, 104
    van Gehuchten 242
cell column
    intermediolateral 49, 75, 82
    intermediomedial 49
cerebellar peduncles 49, 51, 71, 88, 101, 120, 183, 228, 492
cerebellum 100, 492
    and electroreception in fish 180
    ansiform lobule 97
    cortex 97, 100
    corticonuclear zones 100, 105
    development 70, 104
    external granular layer 104, 494, 496
    fissures 97
    paraflocculus 97–101
    paramedian lobule 97, 102
    structure 96
    vermis 98
    vestibular input 97
cerebral peduncle 71, 76, 85, 248
cerebrocerebellum *see* pontocerebellum
cerebrospinal fluid 59, 260, 492
cervical plexus 299, 304, 307
cervical sinus 38
chemical communication 18, 26
chemoarchitecture 98, 126, 127, 242, 494
Cherax destructor 190
choline-O-acetyl-transferase 89
cholinergic neurons 87, 89, 130, 261
    in sleep 282
    of forebrain 107
choroid plexus 45, 59, 492
ciliary muscle 70, 86, 162, 491
circadian rhythm 120, 255

claustrum
    in monotremes 143
    in the echidna 144
    in the platypus 146, 159
cloaca 18, 27, 32
cloacal secretions 18, 32
    gender differences 238
cochlea 220
    inner hair cells 223
    microphonic potentials 227
    outer hair cells 223
cochlear amplifier 223, 227
cochlear duct 220–3, 231–2
cognitive function 132, 277
colliculus
    inferior 228, 281
    superior 51–4, 75, 169–71
colour vision 167, 177
commissure
    habenular 120
    posterior 55, 108, 120
conspecifics 226, 235, 238, 250, 278
cornea
    development 173
    platypus 162
    sclerified in echidna 163
cornu Ammonis 154, 155, 158, 245
cortex
    amygdaloid 237
    anterior cingulate 143
    association 58, 112, 125, 140, 278
    association, connections 141
    auditory 111, 137, 219, 229
    bundling of dendrites 152, 153, 159
    capillary volume fraction 153
    cerebellar 102
    columnar connections 142, 173, 229
    columnar organisation 229
    development 175
    dorsal insular 249
    dorsolateral prefrontal 141, 278
    entorhinal 61, 117, 135, 141, 245
    evolution 133, 136, 159, 293, 296
    frontal 111, 118, 125, 141–2
    frontal, chemical sense integration centre 245–6
    frontal, connections 145
    GABA 150, 155
    gyrencephalic 131, 214
    gyrification 67, 133
    infralimbic 146
    insular 132, 143, 237
    layers 152
    limbic 115, 146
    lissencephalic 61, 65, 131, 289, 297
    mitochondrial density 131
    mitochondrial volume 153
    motor 66, 111, 139, 151
    motor, and cerebellum 111
    multimodal association 141
    NADPH-diaphorase 151, 155
    neuron types 147, 155
    neuropeptide Y 151

nitric oxide 151
nitric oxide synthase 151
olfactory 57, 237, 242, 247
orbital 142, 237
orbitofrontal 111, 245, 252
parahippocampal 133
parapiriform 133
piriform 60, 111, 243–4, 250
platypus and ranging hypothesis 212
posterior cingulate 278
preplate 155, 157
problems with multiple somatosensory fields 213
somatosensory 46, 66, 113, 118, 131, 136, 150, 173, 212
somatostatin 123, 151
subdivisions 136
subiculum 154
subplate 131, 155–8
synaptic density 154
synaptic morphology 131, 154
types of neurons 147, 211–12
visual 111, 137, 147, 162, 171
cortical plate 45, 155–8, 232
counter-current heat exchange 253
courtship behaviour 18, 24, 27, 238
cribriform plate 66, 73, 225, 239, 240, 289
crural spur 12
cuneate fasciculus development 81
Cynodontia 12, 13, 15, 64, 65, 150
cytochrome oxidase 110, 118, 137, 147, 200, 203, 229

decision making 132, 142, 192
declarative memory 154
    and sleep 278
delta waves 277
dendritic spines 147, 149–51, 244
dental pattern
    and monotreme evolution 7, 9, 14, 16
dentary 4, 7, 65, 288, 296
dentate gyrus 154, 155, 158
development
    and evolution 216
    forelimb nerves 72
    peripheral nerve 69
diencephalon 54, 107, 113
    development 55, 57
    subdivisions 107
discriminative touch 48, 78, 181, 184, 204, 207
dopamine 43, 54, 87, 123, 125, 272
dorsal columns 49, 69, 78, 207
dorsal root ganglion 74, 189
    development 79
dorsal roots 43, 49, 72, 74, 79
    axon calibre 207
dorsal thalamus 54, 62, 109, 142

ear
    bone conduction 219, 221, 226, 228, 232
    cribriform plate 225
    development 230

evolution 233
external 219, 221
internal structure 220
membranous labyrinth 221
ossicle 16, 220, 222, 227
place coding of frequencies 223
semicircular canals 220, 221, 226
vestibular apparatus 221, 223, 226
echidna
    home ranges 23–5
    long-beaked 5, 23, 46, 67
    maternal care 18, 27
    mating behaviour 27, 46
    venom 26
    weaning 28
echidna beak
    electroreceptors 28, 193
    mechanoreceptor 194–6
    receptor development 201
    receptor distribution 196
    sensory gland apparatus 197
    thermoreceptors 194
ectoderm 37, 59, 173
efferents
    branchiomotor 85, 87, 89
    general somatic 49, 84, 92
    general visceral 49
    somatic 84
    visceromotor 75, 85, 87
egg-laying 2, 35, 264
egg tooth 39
Eimer organ 185
electroencephalogram 277
electroreception 3, 18, 180
electroreceptor
    ampullary 180, 185
    physiology 185, 190
    tuberous 180
electrosensation
    and crustacean prey 189–90
    and platypus prey 192
    evolution 62, 70, 216
    in long-beaked echidnas 197
    in short-beaked echidnas 193
    role in the wild 199
    spatial summation 191
    temporal summation 191, 199
encephalisation 18, 62, 65, 133, 277
endoderm 37
endolymph 221, 231–2
endoturbinal 65, 239
energy costs of brain 77
enkephalin 123, 269, 270, 272
enteric nervous system 79, 267–72
ependyma 59, 258
epinephrine *see* adrenaline
epithalamus 43, 54, 102
    components 117
    definition 120
epitrichial claws 40
ethmoidal turbinal 238
ethmoid bone 238, 240
Eucalyptus 294

eutherian 239, 297
eutriconodonts 3, 12
expensive tissue hypothesis 28
external capsule 146
external ear
    curvature 221
    in echidna 222
    in platypus 222
    reinforcement 221
external granular layer 104
eye
    accommodation 83, 161–3, 177
    aquatic adaptation 177
    cartilaginous cup 162, 175
    choroid 162
    ciliary body 162
    cornea 162–3, 172–3
    development 173
    echidna 162
    platypus 163
    tarsal plate 163

fasciculus
    cuneatus 78
    gracilis 78, 181, 207
    medial longitudinal 97
    retroflexus 55, 108, 118, 130
fenestra vestibuli 220–2
fields of Forel 43
5-hydroxytryptamine *see* serotonin
flocculus 97–101
forelimb bud 72, 79
forelimb development 94
fornix 126, 259, 262
    development 232
fossil record 4, 6, 11, 288
Fruitafossor 12–13

GABA 55, 58, 118, 125
GABAergic neurons
    in sleep 278
galanin 188, 269
ganglion
    autonomic 58, 272
    cochlear 225, 228
    facioacoustic 231
    myenteric 269
    submucosal 269
    superior cervical 189, 255
    trigeminal 38, 65, 189
    vestibular 101
ganglionic eminence 45, 57, 129, 155, 157
gastrin releasing peptide in gut 269
Gbx-2 42
genioglossus 40, 73
germinal trigone *see* rhombic lip
glabrous skin 205–6, 211
gland
    pineal 55, 120, 255
    pituitary 67, 252, 258
glia 61
glial limiting membrane 59
globus pallidus 58, 113, 123, 125

glucocorticoids 252
gluconeogenesis 253
glutamate 58, 282
glutamatergic neurons 104, 108, 118
    in sleep 282
gonadotrophin-releasing-hormone 70
Gondwana, separation of 13
granule cell cluster 58, 130, 242
Greater Australia 293
greatest length of embryo 34
Griffonia simplicifolia isolectin B4 74, 203, 207
growth of echidna 35, 41, 72
gut epithelium 269
gut, submucous plexus 269
gymnotiform fish 181
gyrencephaly 131, 155

Haacke W 32
habenulointerpeduncular tract *see* fasciculus retroflexus
hair 2, 22, 41
hair cells of inner ear 221–3
heterothermy in echidna 253
hibernation 18, 23, 254
    and mating 27, 35
    and seasons 35
Hill JP 105
hindbrain 45
    at hatching 87, 89, 104, 204
    development 38, 49, 53
    function 87, 89
hindlimb, development 38, 41, 72
hippocampal formation 132, 154
    development 158
histamine and sleep 252
homeostasis 58, 123, 252
homeothermy of platypus 253, 277
homology 97, 118, 119, 205, 212
hook bundle of Russell 104
hyaloid vessels 173
hyoglossus 40
hypothalamic output 252
hypothalamo-hypophyseal portal system 252, 259, 262
hypothalamo-pituitary-adrenocortical axis 252
hypothalamo-pituitary axis 251
hypothalamus 43, 45, 54, 252, 255
    acroterminal segment 51, 55, 57
    and endocrine system 252
    catecholaminergic neurons 260
    cholinergic neurons 130, 261
    development 259, 265, 267
    function 161, 252
    input 252
    magnocellular neurosecretory system 252, 259, 263
    oxytocinergic neurons 259, 261, 263
    parvicellular neurosecretory system 252, 262, 267
    peduncular segment 57
    preoptic area 58, 260

    serotonergic neurons 260
    structure 251
    terminal 57

incubation phase 33, 45
incus 3, 220
    development 227
inferior colliculus 51, 54, 228–9
infundibular process 265
infundibular recess of third ventricle 257, 265
infundibular stalk 45, 87, 258, 265
inner ear 219
insemination 32, 35
interhemispheric transfer 162
internal acoustic meatus 73, 225
internal capsule 71, 119
    development 122, 209
interocular transfer 19
intrauterine phase 31, 35, 37, 42
island of Calleja 130, 242
isocortex 57, 132
    development 155, 157
isthmic organiser 54
isthmus 45, 53, 89, 104

Jacobson's organ *see* vomeronasal organ
jaw mechanism 304
juvenile period 40

Kangaroo Island 18, 24, 27
κ-casein 258
Kollikodon ritchiei 3, 4, 13, 16
Kryoryctes cadburyi 6, 13, 15

lactation 17, 24, 27, 238, 251, 261
    asynchronous concurrent 258
    evolution 261, 264
lactational phase 90, 130
lagena 220–7
lagomorphs 146, 171, 244
lamina cribrosa 238, 249
lamina dissecans 245
latency of electroreceptors 185, 197
lateral lemniscus 228
    development 232
lateral striatal stripe 125
LCR gene 168
learning 18, 19, 87
    and cholinergic neurons 129
lens fibres 173
lens placode 172
lens, platypus 177
lens vesicle 38, 173
lentic ecosystems 20
limbic system 57, 154, 252
linear acceleration 221
lipotyphlids 65, 244
lobe, flocculonodular 97
locomotion
    mechanism 77
    metabolic cost 78

locus coeruleus 51, 87
   in sleep 278, 282
long-beaked echidna
   evolution 11
   receptors in beak 199
lotic ecosystems 20, 25
lungs 31, 41, 79
LWS opsins 168
LWS pigment gene 168

macroglia 59, 61
macrosmatic 238, 243
macula lagena 223, 226
   development 232
   innervation 223
malleus 3, 221
   development 227
   handle 221
   processus gracilus 221
mammary areolae 31, 39, 94
mammary gland 32
   evolution 257
   structure 39, 257
mammary hair 264
mammary lobule 264
mammillary body 257
mammo-pilo-sebaceous units 264
mandibular canal 3–7, 288
mantle layer 90, 121, 175, 232
manus 38, 77, 213, 307
marsupial 2
marsupium *see* pouch
massa intermedia 55, 112, 117
mating system 24
mating trains 24, 255
maxilloturbinal 238–40
mechanoreceptors in echidna beak 194
Meckel's cartilage 7, 232
medial forebrain bundle 123
medial lemniscus 181
median eminence 259, 263
medulla oblongata 48
Megalibgwilia 11, 294
Megalibgwilia owenii see Megalibgwilia ramsayi
Megalibgwilia ramsayi 11
melatonin 55, 255, 260
meroblastic ovum 32
mesencephalon *see* midbrain
mesoderm 37, 59, 299
mesomere 1 52, 54
mesomere 2 52, 54
mesonephros 41
Mesozoic mammals 1–3, 12–13
metatherian 167, 293
metencephalon development 48, 52
microglia 59
microsmatic 235, 237, 243
midbrain 42
middle ear 3, 6, 221, 227
milk proteins 258
mitochondria 153, 183
monoaminergic neurons 87, 112, 117

Monotremata, definition 1, 13
Monotrematum sudamericanum 9, 286
monotreme
   body temperature 2, 20, 154, 251, 253
   fossil record 1–13
   origins 1, 6, 62, 285
   plesiomorphic features 1–2
   species 1–3
monotreme egg 1, 18, 28
Morganucodon 288
mormyriform fish 181
morphology-based diagnoses 16
mucous sensory gland 183, 187
multituberculates 3, 219, 227
muscle
   constrictor pupillae 162
   extraocular 85, 162
   palpebral 162
   retractor bulbi 163
   retractor membranae nictitantis 162
   stapedius 85
   styloglossus 40, 73, 300
   styloideus 85
   superficial facial 70, 85, 259, 302
   superior oblique 73, 84, 93, 162
muscles
   facial 70, 299, 300, 302
   of forelimb 304
   of hindlimb 313
   of larynx 85, 92, 181, 303
   of mastication 73, 85, 93, 279
   of pectoral girdle 304
   of tongue 299
   palatine 85, 304
muscularis mucosa 269
myelencephalon development 48
myotubules 41
myrmecophage 12, 16, 29
Myrtaceae 294

NADPH diaphorase 123, 209, 241, 243
nasal cavity
   in echidna 239
   in platypus 238
nasal turbinal 289
nerve
   abducens 67
   axillary 305, 309
   cochlear 225
   cranial 70, 299
   facial 51, 85, 300
   glossopharyngeal 51, 73, 181, 300
   hypoglossal 70, 97, 102
   infraorbital 181, 190, 197
   mesencephalic root of the trigeminal 202
   musculospiral *see* nerve, radial
   oculomotor 67, 299
   olfactory 238, 246
   optic 70, 167
   optic, development of 45
   phrenic 307, 309
   radial 309

   spinal accessory 51, 70, 300
   supracoracoid 309
   terminal 70
   trigeminal 3, 51, 60, 62, 70, 76, 181, 188, 201, 289
   trochlear 84, 300
   ulnar 205, 307, 309
   vagus 49, 51, 79, 89, 271, 300
   vestibular 231
   vestibulocochlear 183, 221, 226, 229, 300
nerve growth factor 58
nest 18, 20, 23, 27, 94
neural crest 42, 72, 79, 81, 202, 231, 298
neural plasticity 87
neural plate 38, 42, 72
neural tube 38, 41, 43, 79, 81, 90, 121, 247, 272
neuraxis 55, 71, 122, 170, 247, 249
neuroendocrine function 126
neuroendocrine system 237
neuroepithelium 45, 121, 173, 175, 246, 265
neurofilament 100, 110–13, 137, 139, 141, 150, 241
neurogenesis 79, 90, 121, 155, 158, 173, 176, 205, 215, 232, 290, 292, 297–8
neurohypophysis 42, 57, 252, 258, 263
neuromere 46, 51, 57, 175
neuropeptide Y 123, 151, 188
   in gut 269, 270, 272
neurophysin 259, 263
neuropore
   anterior (rostral) 38, 90, 172, 231
   closure 38, 42, 90
   posterior (caudal) 42, 79
neurotrophins 58–9
neurulation 42
newborn
   marsupials 39, 40, 41, 104, 122, 205
   monotremes 31, 40
New Guinea, in monotreme evolution 2, 6, 293, 295
nictitating membrane 162
Nkx-2.1 42
Nkx-6.1 42
nociception 75
nodule, of cerebellum 97, 100
noradrenaline 87, 254
norepinephrine *see* noradrenaline
Nothofagus 11, 294
notochord 43
nucleus
   abducens 43, 51, 84, 89, 93, 97
   accumbens 58, 97, 112, 123, 126, 143, 281
   ambiguus 51, 85, 91, 93
   anterior hypothalamic 267
   anterior, of thalamus 110, 115
   anterior pretectal 108, 115, 117, 170
   arcuate 88, 100, 252
   basal, of Meynert 58, 107, 127, 130, 142
   bed, of the accessory olfactory tract 126, 237
   bed, of the stria terminalis 123, 127, 129
   caudal linear, and serotonin 89, 91

circularis, of hypothalamus   263
cochlear   42, 92, 183, 228, 231
cuneatus   78, 181, 184
deep cerebellar   46, 90, 96–8
dorsal column   75, 208
dorsal motor, of the vagus   84, 85, 92
dorsal raphe   51, 89, 91
dorsal terminal   109, 169
dorsomedial, of hypothalamus   267
Edinger-Westphal   54, 85, 93, 163
endopiriform   145, 244
entopeduncular   115, 120, 123, 126, 129
facial   51, 85, 91, 93, 203, 228
fastigial   97, 101
gracilis   181
habenular   43, 45, 54, 89, 95, 97, 117, 120
hypoglossal   84, 87, 89, 92, 93, 97
inferior olivary   71, 87, 91, 100, 102, 104
intermediomedial   49, 74
internal basilar   74, 75
interpeduncular   51, 55, 88, 90, 97, 108, 118, 120
interposed deep cerebellar   97
interstitial, of Cajal   54, 108
interstitial, of the posterior limb of the anterior commissure   125, 145
intralaminar   54, 110, 117
lacrimal   84, 89, 93
lateral deep cerebellar   97, 98
lateral geniculate   57, 62, 113, 119, 169, 173
lateral habenular   117, 120
lateral posterior thalamic   113, 115, 117
lateral reticular   42, 100, 103, 104, 105
lateral spinal   74, 81
laterodorsal, of thalamus   209
laterodorsal tegmental   89, 93, 97, 281, 282
locus coeruleus   51, 87, 88, 89, 93, 278, 282
magnocellular, of the lateral hypothalamus   267
magnocellular reticular   83, 85, 267
medial deep cerebellar   98
medial geniculate   110, 113, 117, 219, 228, 229
medial habenular   89, 117
medial pretectal   108, 115, 170
medial terminal   169
median raphe   88, 91, 282
mediodorsal thalamic   142, 237
mesencephalic, of the trigeminal nerve   181, 183, 203
midline, of thalamus   117, 122
motor thalamic, and cerebellum   113
motor trigeminal   51, 84, 88, 93, 97, 183, 203
oculomotor   54, 85, 87, 93, 171, 176
of Bechterew   231
of Darkschewitsch   108
of the diagonal brand of Broca   130
of the H field of Forel   118, 119, 120
of the lateral lemniscus   183, 228

of the optic tract   108, 115, 169, 170
of the posterior commissure   115
of the solitary tract   43, 84, 92, 184
of the trigeminal spinal tract   71, 92, 188, 202, 267
olivary pretectal   108, 115, 117, 170
parabigeminal   51, 89, 93
parabrachial   100, 103, 142, 231, 249
parafascicular, of the thalamus   108, 110, 112, 117, 118
paraventricular, of hypothalamus   262, 267
paraventricular, of thalamus   120
parvicellular extension of VPM   246, 249
pedunculopontine tegmental   89, 93, 142
pontine   42, 49, 55, 90, 97
posterior, of thalamus   117, 229
precerebellar   42, 49, 103
pregeniculate   55, 57, 62, 113, 119, 169, 177
preoptic area   57, 58, 108, 260
pretectal   108, 115, 161, 170
principal trigeminal sensory   51, 100, 183, 200, 202, 291
raphe magnus   51, 89, 91
raphe obscurus   51, 87, 89, 91
raphe pallidus   51, 89, 91
reticular   55, 115, 118, 119, 121
reticulotegmental   100, 103, 105
retinorecipient   113, 119, 169, 177
retroambiguus   257
sensory trigeminal   77, 203, 208
septal   58, 107, 123, 130, 252, 257
subgeniculate   118, 119
submedius   113
superior and inferior salivatory   84, 89, 93
superior olivary complex   51, 100, 219, 228
suprachiasmatic   175, 255
suprachiasmatic, and photoperiod detection   257
supraoptic   237, 257, 262, 264
trapezoid   51, 228
trigeminal development   202, 216
trigeminal motor   85, 91
trigeminal sensory   42, 61, 85, 92, 101, 183, 188, 202, 216
trigeminal sensory development   92, 204, 216
trochlear   43, 51, 84, 93, 97
ventral anterior thalamic   113, 117, 124, 126
ventral lateral thalamic   113, 115, 183
ventral posterior thalamic   46, 54, 110, 112, 117, 122, 183, 208, 210, 229
ventromedial, of hypothalamus   237, 267
ventromedial, of thalamus   113
vestibular   75, 90, 101, 103, 231
viscerosensory   43, 74, 84, 92, 109
visual   107, 113, 169

Obdurodon dicksoni   4, 7, 65
and olfaction   247
odorants   237, 239, 242, 246, 293
olfaction
    and cloacal secretions   238
    at time of hatching   238
    development   246
    in wild   19
    receptor gene repertoire   239
    role of   238
olfactory bulb   43, 51, 61, 63, 142, 241
    accessory   241, 242
    at hatching   247
    development   247
    folding in echidnas   241
olfactory epithelium
    accessory   237
    development   248
    main   39, 237, 239, 247
olfactory glomeruli   241
    development   235
olfactory placode   246
olfactory region, anterior   243, 244
olfactory tubercle   51, 123, 129, 130, 237, 242, 243
olfactory vesicle   241
oligodendrocytes   59, 61
operant techniques   19, 132
opsin   168, 177
optic chiasm   51, 87, 97, 169, 176, 257
optic cup   33, 172, 173
optic recess   45, 173
optic stalk   45, 57, 121, 173, 176
optic tract   62, 109, 113, 162, 169, 175
optic vesicle   172
oral plate of marsupials   39, 41
organ of Corti   221
Ornithorhynchidae   2, 7, 9, 14, 286
    composition   9, 14
    definition   16
ornithorhynchids, in Patagonia   2, 6, 9, 11, 14
os caruncle   36, 38, 39, 41, 79, 94
otic pit   38
otic placode   231
otic vesicle   33
otoacoustic emissions   227
otoconia   223
otocyst   231, 232
oval window   221
ovary   32
oviparous   32
oxytocin   252, 258, 261, 262, 264

Pacinian corpuscle   186, 194, 206, 207
paleocortex   57, 132
pallidum   55, 58
    dorsal   123, 124
    ventral   58, 123, 129
pallium   43, 57, 108, 129, 132, 144, 157, 241, 245
panniculus carnosus   78, 208, 304, 309
    sensory input   208

parahippocampal gyrus in sleep 278
pars distalis of pituitary 258, 265
pars intermedia of pituitary 258, 266
pars nervosa of pituitary 258
pars tuberalis of pituitary 257, 266
pathway
    accessory olfactory 126, 238, 247, 250
    corticospinal 48, 71, 76
    corticostriatal 125, 142, 151
    corticothalamic 46, 110, 119
    cuneocerebellar 101, 184
    dorsal column 49, 69, 75, 78, 181, 207, 208
    dorsal spinocerebellar 74, 75, 81, 101, 184
    electrosensory 181
    geniculocortical 171, 175
    hypothalamo-neurohypophyseal 252, 261, 263, 265
    lateral olfactory 120, 126, 237, 243, 248
    lateral vestibulospinal 75, 103, 231
    main olfactory 237
    mammillothalamic 110, 115
    medial forebrain 123, 265
    olivocerebellar 101, 102, 103
    orexinergic 252
    pontocerebellar 97
    reticulocerebellar 101
    reticulospinal 39, 77, 188, 282
    reticulospinal development 94
    retinohypothalamic 255, 257
    rubrospinal 48, 75, 77
    somatosensory 179, 181, 203, 207
    spinoreticulothalamic 181, 207
    spinothalamic 81, 181, 207
    tectocerebellar 101, 171
    tectospinal 48, 75, 171, 177
    thalamocortical 46, 118, 142, 157, 215
    thalamocortical development 119, 122, 297
    trigeminocerebellar 101
    ventral amygdalofugal 126
    ventral spinocerebellar 101
pattern generator 77, 83, 94, 105
Pax genes 42
periaqueductal grey 53, 71, 88, 91, 108, 115, 117, 202
perilymph 221, 228
peripheral nerve development 72
periphery-driven sensory evolution 202
pharyngeal arch 33, 38, 42, 84, 231, 299
pharyngeal cleft 38
pheromones 237, 257
photopic vision 166
photoreceptor 163, 164
    cone 161, 291
    rod 163
    spatial density 161, 165
pineal gland 55, 120
    and photoperiod detection 120
    function 255, 257
    structure 120
piriform lobe 61, 146, 244

pituitary gland 67, 252, 258
    development 265
    position 67, 258
Platypoda 7, 16
platypus
    aquatic adaptations 3
    burrows 4, 267, 279
    diet 22
    distribution 20
    feeding 22
    home range 17, 25
    maternal care 27
    mating behaviour 26
    mating system 25
    nesting burrow 21, 27
    spurs 26
    temperature tolerance 21
    venom 3, 26
    weaning 28
platypus bill
    as directional antenna 191
    erectile tissue 189
    integration of electrosensory and tactile information 191
    Merkel cells 185, 186, 187, 194
    mucous non-sensory glands 183, 185, 187, 196, 197
    mucous sensory glands 183, 185, 187, 189, 267
    Paciniform corpuscles 185, 195, 196, 197
    parasympathetic innervation 188
    push-rod mechanoreceptor 185, 187, 194, 195, 196, 286
    receptor development 201
    reflexive closing of sensory pores 188
    serous glands 185
    substance P 188
    sympathetic vasoconstrictor axons 188
    vesicle chains 185
platypus-echidna divergence 14
platysma 70, 73, 302
Pleistocene 2, 6, 9, 11, 12, 48, 66, 67, 294, 295
plesiomorphic features 1, 2, 3
pontine tegmentum in sleep 278, 279
pontocerebellum 97, 492, 498
posterior commissure 45, 55, 108, 115, 120, 121
pouch 18, 22, 27, 31, 32, 39, 41, 83
predatory behaviour 76
pre-embryonic stage 35, 37, 41
preisthmus 52, 54
preoptic area *see* area, preoptic
preplate 45, 155, 157
prerubral field 119
pretectum
    boundaries 107, 108
    development 43, 45
    of monotremes 109
    size in different mammals 109
    size in monotremes 109, 170
    structure 108
    visual input 170

prethalamus 43, 55, 107, 113, 169
    embryonic origin 118
primary brain vesicles 38, 45
primitive streak 37, 42
procedural memory, and sleep 278
progesterone, and reproductive cycles 255
proneuromere 45, 47, 54
proprioception 69, 73, 78, 181, 184, 207
prosencephalon 42, 121, 172, 265
prosomere 1 43, 52, 54, 108, 170
prosomere 2 43, 54, 55, 120, 290, 292
prosomere 3 43, 54, 55, 108, 118
prostaglandin E series, and thermoregulation 254
puggle 31, 39
Purkinje cell *see* cell, Purkinje
push rod organs 185, 186
pyramidal neuron 123, 131, 147
    atypical 131
    evolution 150, 151
    in platypus 147
    neurofilament protein 150

range detection and platypus cortex 212
rapidly adapting unit 186, 194, 206
Rathke's pouch 45, 258, 265
receptor
    odorant 239
    taste 84, 249
    T1R, T2R 249
    vomeronasal 239, 250
receptor-like proteins, formyl peptide 239
receptors, postcranial 205
receptor, trace amine-associated 239
Remak bundle 195
REM sleep 275–7
reproductive cycles, and hypothalamus 237, 252, 255
reproductive timing and resources 24
restiform body 101
rete mirabile 253
reticular formation 42
    and sleep 278
    development 94, 267
    gigantocellular 83
    medullary 75
    pontine 94
retina
    area centralis 163, 166, 167
    development 173
    developmentally advanced region 172–4
    echidna 165, 166, 167
    fovea 163, 166, 173
    ganglion cell 161, 164
    inner nuclear layer 163, 166, 175
    inner plexiform layer 166, 175
    input to hypothalamus 255, 257
    outer nuclear layer 166, 175
    platypus 166, 167
    serotonin 167
    visual streak 163, 166, 167, 172
retinal 168

retinofugal axons   109, 113, 170
reversal learning tasks   132
Rexed's laminae   69, 74, 75, 81, 207
rhombencephalon   88
    development   42, 88, 90, 92
rhombic lip   38, 42, 46. 104, 105, 204
    and cerebellar development   53
    and trigeminal nuclei
        development   91–2
    derivatives   104, 205
    development   90
    migration   90
rhombomeres   42, 45, 51, 67, 204
Riversleigh   9–11, 65

saccule   221, 226, 231
    development   232
Sahul   293
sauropsida   2
scala media   221, 222, 223
scala tympani   221, 228
scala vestibuli   221
scarab beetle   294
Schiff base   168
scotopic vision   167
seasonal breeding   17, 23, 24, 255
semicircular ducts   221, 226
    development   232
septum see nucleus, septal
serotonergic neurons   83, 89, 91, 260
    in gut   269
    in sleep   282
    lateralisation   89
serous sensory glands   187, 287
sexual dimorphism   22, 25, 27, 58
sleep
    and catecholaminergic neurons   282
    circuitry   281
    definition   276
    evolution   283
    features   275–6
    function   278
    in ostrich   283
    in short-beaked echidna   279, 280
    in the platypus   278
    non-REM   277, 279
    paradoxical   275, 277, 281
    physiology   119, 129, 275, 281, 282
    slow-wave   277
sleep deprivation   278
slowly adapting units   186
SMI-32 antibody   100, 151, 209, 231, 263
snout receptor development   201
social behaviour   18, 58, 130
somatosensation   48, 57, 136, 205, 207
somite   38, 299
somitogenesis   41
specialisation   1
    of echidnas   9, 48, 62, 201
    of platypus   3, 67, 166, 183, 203
sperm competition   24
spinal cord
    cervical   74, 83, 91, 94

    development   69, 79, 83
    echidna   72, 74, 208
    gross anatomy   72
    length   48, 71
    opioids   208
    platypus   48, 81
    structure   69, 72
spinal lemniscus   111, 181
spinocerebellum   97
spur secretions   26, 238
stapes   220, 227
    footplate   221
stereovilli   221, 223, 226
Steropodon galmani   4, 7, 8, 285
Steropodon, relationships   6–7
Steropodontidae   7, 14, 16
stratum zonale   170, 171
stressful situations   252, 253
stria medullaris thalami   54, 117, 120
stria terminalis   123, 126, 129, 237
striatopallidal complex   107–9, 123, 125, 129
striatum   54, 58
    dorsal   58, 123, 125
    ventral   58, 107, 124, 242, 243
strioscme   125
subcommissural organ   43, 108, 120
subpallium components   129
subplate
    evolution   293
    gene markers   298
substance P in gut   272
substantia nigra   43, 53, 87, 88, 120, 125
subventricular zone see zone, subventricular
suckling   39, 40
sulcus limitans   42, 57, 79, 90
superior colliculus   51, 53, 75
    development   175
    structure   170
superior medullary velum   84, 101
sweat glands   20
swimming   139
SWS1 opsin gene   168, 177
SWS2 opsin gene   161, 168, 177
symmetrodonts   3, 12, 13
synapomorphy   3, 14, 16, 228

Tachyglossa, definition   16
Tachyglossidae   2, 7, 15, 16, 158
Tachyglossus aculeatus
    acanthion   5, 22
    aculeatus   5
    lawesii   5
    multiaculeatus   5, 35
    setosus   5, 22, 35
    subspecies   27, 48
tail bud   38
tandem gene duplication   239
tectal grey   52, 54, 108, 170
tectum   43, 170
tegmentum   43, 45, 52, 85, 93, 94, 278
Teinolophos trusleri   3, 6, 7, 13, 14, 286, 288
telencephalic vesicle   38, 54, 57, 157

telencephalon
    derivatives   108
    development   215
    subdivisions   48, 57, 108
tenia tecta   140, 146, 237
testosterone and reproductive cycles   255
thalamus   43, 45, 55
    and sleep   277
    development   55, 210
    dorsal   43, 107
    evolution   118, 143
    functional components   55, 111, 130
    neurogenesis   209–10
    structure at hatching   210
    ventral see prethalamus
therapsids and lactation   264
thermoregulation   253, 254
    and hypothalamus   252
    development   31, 39
theta rhythm   123, 277
three-layered embryo   38, 242
T-maze test   19, 132
tongue
    circumvallate papillae   249
    muscles   299
    taste   249
tract, solitary   43, 84, 87, 92, 184, 231
tract, trigeminal spinal   71, 92, 183, 200, 202
Tritylodon   3
tunnel of Corti   221–3
tympanic bone   220–2, 231
tympanic membrane   220
tyrosine hydroxylase   87

utricle   221, 225
    development   232
    macula of   225
    recess of   225

vascularisation   61, 131
vasoactive intestinal peptide   188, 269
vasopressin see antidiuretic hormone
ventral tegmental area   43, 53, 88, 120, 142
vertebral column, echidna   72, 74
vesicular acetylcholine transporter   89
vestibular apparatus   221, 223, 226
    crista neglecta   226
    kinocilium   225
    sensory cells   226
    stereovilli   226
vestibular system development   231–2
vestibulocerebellum   97, 229
vibration   48, 181, 187, 194, 206, 221
visceral nociception   75, 78
visual acuity   161, 166
visual discrimination   19, 132, 162
visual magnification   171
visual processing   171
visual system
    aquatic adaptation   177
    evolution   161
vocalisation, by young   226
vomeronasal organ   187, 237, 246, 248

wakefulness 275, 279
   definition 277
weaning 21, 24, 28, 40
webbing 40, 77
   development 40
whey acidic protein 258

yolk sac 40

Zaglossus bartoni 254
Zaglossus bruijnii 2, 5, 11, 197, 253, 287
Zaglossus genus 2, 4. 5
Zaglossus hacketti 11, 67
Zaglossus robustus 12, 67
zona incerta 55, 57, 108, 117–20, 260
zone
   cortical transitory 104
   Lissauer's 74
   nuclear transitory 104
   subventricular 45, 121, 129, 131, 155, 157, 158, 175, 290, 293, 297, 298
   ventricular germinal 45, 90, 104, 121, 157, 297
zygote, cleavage of 37